Soil Biology

Volume 52

Series Editor

Ajit Varma, Amity Institute of Microbial Technology,
Amity University Uttar Pradesh, Noida, UP, India

More information about this series at http://www.springer.com/series/5138

Bhoopander Giri • Ram Prasad • Ajit Varma
Editors

Root Biology

Springer

Editors
Bhoopander Giri
Department of Botany
Swami Shraddhanand College
University of Delhi
Delhi, India

Ram Prasad
Amity Institute of Microbial Technology
Amity University
Noida, Uttar Pradesh, India

Ajit Varma
Amity Institute of Microbial Technology
Amity University
Noida, Uttar Pradesh, India

ISSN 1613-3382 ISSN 2196-4831 (electronic)
Soil Biology
ISBN 978-3-030-09354-9 ISBN 978-3-319-75910-4 (eBook)
https://doi.org/10.1007/978-3-319-75910-4

Printed on acid-free paper

This Springer imprint is published by the registered company Springer International Publishing AG part of Springer Nature.
The registered company address is: Gewerbestrasse 11, 6330 Cham, Switzerland

Preface

Roots are typically underground, complex, and essential parts of the plants, which perform an array of functions, viz., anchor the plant body to the soil for physical and mechanical support, absorb water and mineral nutrients from the soil, and translocate them to other parts of the plant for conducting a myriad of metabolic activities. In certain cases, roots serve as storage organs. In the rhizosphere, a dynamic zone dominated by a variety of organic compounds, root interacts with a sort of organisms. Plant roots have an extraordinary capacity to respond to both pathogenic and plant growth-promoting microorganisms. Certain soil microorganisms in the rhizosphere communicate with plant root and establish intimate associations. Among these, mycorrhizal fungi and nitrogen-fixing rhizobacteria are identical. These microbes improve nutrient status of their hosts, protect against potentially harmful microbes and abiotic environmental stresses, thereby improving plant fitness and yield. At the onset of symbiosis, mycorrhizal fungi modulate plant root morphology, physiology, and biochemistry and alter plant's functioning. Indeed, certain phytohormones largely impact growth and functioning of plant roots by modulating root morphology or other physiological processes. Under current scenarios of climate change, which is expected to aggravate in future too, the functioning of such potential microorganisms is promising as water and soil nutrient availability is predicted to be steadily declining largely due to exertion of physiological drought conditions; consequently, plant roots could rigorously face scarcity of water and mineral nutrients leading to decreased plant growth and productivity.

During recent decades, enormous advances have been made in the area of root research that tempt us to provide a comprehensive and updated overview of recent progress made in this field; thus, we conceived the book *Root Biology*. We approached leading scholars working in the area of plant root biology and invited them to develop articles in the areas of root system architecture, anchorage in the soil, biology of root formation, physiology and molecular developments of root, mutualistic symbioses, rhizosphere root interactions, biological control of root pathogens, and stress biology.

This volume contains 21 thought-provoking chapters written by 75 leading researchers in the area of their specialization highlighting latest research and future prospects. Chapter 1 highlights the role of phytohormones in tailoring underground plant root system architecture; Chap. 2 sheds light on the role of strigolactones in root development and interaction of strigolactones with other phytohormones in determining root architecture; Chap. 3 discusses effects of mineral nutrients and phytohormones on root hair growth and development; Chap. 4 deals with morphological and symbiotic root modifications for mineral acquisition from nutrient-poor soils; Chap. 5 discusses root exudates and microbial communities that drive mineral dissolution and the formation of nanosize minerals in soils in relation to soil C storage; Chap. 6 describes influence of the root exudates on plant growth-promoting rhizobacteria and pathogens; Chap. 7 provides an overview of different aspects of control of soil-borne pathogens that induce root diseases; Chap. 8 deals with the role of nematophagous fungi in the control of root-knot and cyst nematodes; Chap. 9 describes the role of fungi in improving date palm tolerance to drought, salinity, and vascular *Fusarium*-induced wilt; Chap. 10 uncovers the role of mycorrhizal symbiosis in amelioration of salt stress; Chap. 11 describes the role of bacterial and fungal endophytes in protecting plants from drought stress; Chap. 12 focuses on the role of ACC deaminase-producing bacteria in amelioration of abiotic stress; Chap. 13 highlights importance of root–microbe interactions for plant growth and development; Chap. 14 deals with the importance of rhizosphere communities in relation to plant growth promotion and sustainable agriculture; Chap. 15 covers morphology and physiological aspects of symbiotic plant–microbe interactions and their ecological significance; Chap. 16 sheds light on the impact of climate change on root–pathogen interaction; Chap. 17 deals with mycorrhizal fungi and their responses to nutrient enrichment in the terrestrial ecosystems; Chap. 18 provides most recent information on the improved or depressed effect of arbuscular mycorrhiza fungi on plant growth; Chap. 19 discusses the role of arbuscular mycorrhiza in conservation of endangered tropical legume trees; Chap. 20 sheds light on the paradigm shift from mycorrhizosphere to rhizosphere microbiome and also deals with diversity, interaction, and management of mycorrhizal microbiome for better plant health and crop productivity; and Chap. 21 talks about the influence of arbuscular mycorrhizal fungi on growth of different species and provenances of jujube plants.

We are highly thankful to Dr Bhawna Saxena, Swami Shraddhanand College, University of Delhi, Delhi, for her valuable help in incorporating editorial changes during the preparation of this volume. We wish to thank Springer officials, particularly William F Curtis, Eric Schmitt, Hanna Hensler-Fritton, Man-Thi Tran, Isabel Ullmann, and Bibhuti Bhusan Sharma for their generous support and efforts in accomplishing this volume. We are highly delighted and thankful to all our contributing authors for their vigorous support and outstanding cooperation to write altruistically these authoritative and valuable chapters. We specially thank our families for consistent support and encouragement.

With a bouquet of information on different aspects of root biology, we hope this book is a valuable resource for the students of different strata; researchers and academicians, working in the field of plant sciences, agriculture, microbiology, and fungal biology; and the scholars interested in strengthening their knowledge in the area of plant–soil or plant–microbe interactions.

New Delhi, India — Bhoopander Giri
Noida, India — Ram Prasad
Noida, India — Ajit Varma

Contents

Chapter 1
Insights into Pivotal Role of Phytohormonal Cross Talk in Tailoring Underground Plant Root System Architecture

Priyanka Singla and Surinder Kaur

1.1 Introduction

A plant's root system—hidden below the soil surface—is a structural anchor, a sensor, and a site for providing anchorage to the shoot, for responding to various environmental stresses, and for uptaking water and nutrient from the soil, respectively (Nibau et al. 2008; Petricka et al. 2012). Root system architecture (RSA), i.e., root elongation and branching, determines flexibility of the entire root system toward changing extrinsic as well as intrinsic cues (Singh et al. 2014; Gupta et al. 2015). Development of underground roots is thus one of the most dynamic processes in the plant's lifecycle whose metabolic complexities are associated with the perception and integration of these cues into signaling pathways directed by "hormonal cross talk" (Liu et al. 2014). Cellular concentration of hormones is an outcome of numerous aspects such as hormone biosynthesis, activation, inactivation, and degradation and long- and short-range transport by differential rate of influx- and efflux-related carrier proteins (Del Bianco et al. 2013). Hormones coupled with regulatory and target genes (activating downstream signal transductions) form a network, in which pertinent genes regulate hormonal activities and hormones standardize gene expression, thus forming "hormonal cross talk" (Depuydt and Hardtke 2011). Root development in plants culminates via auxin biosynthesis, transport, and signaling, where auxin synthesis generates a source, transport creates a gradient or local accretion, and finally the perception or reaction controls root development (Benfey et al. 2010). Besides auxin, phytohormones like cytokinin (CK), gibberellin (GA),

Both authors have equally contributed.

P. Singla · S. Kaur (✉)
Department of Botany, SGTB Khalsa College, University of Delhi, Delhi, India

B. Giri et al. (eds.), *Root Biology*, Soil Biology 52,
https://doi.org/10.1007/978-3-319-75910-4_1

brassinosteroid (BR), ethylene, abscisic acid (ABA), jasmonic acid (JA), polyamines (PA), and strigolactones (SL) also integrate into these three significant processes to activate cascades of events involved in root development (Jung and McCouch 2013; Saini et al. 2013).

The entire process of root organogenesis commences with embryogenesis generating a primitive blueprint of root apical meristem (RAM) followed by positioning and configuration of stem-cell niche (SCN), continuance of mitotic activity in proximal meristem (PM), and regulated elongation and differentiation of cells departuring the meristematic area (Ruzicka et al. 2009; Giehl et al. 2014; Street et al. 2015). Sustained activity of these meristems along with the occurrence of committed cells that dedifferentiate and generate new meristems allows the sessile plants to retain developmental plasticity in response to abiotic as well as biotic factors, subsequently marking the overall fitness of plant (Peret et al. 2014). The root system is comprised of the primary root (PR), established during embryogenesis, and postembryonically initiated lateral roots (LRs) branching of the PR and root hair (RH) system (Celenza et al. 1995; Verstraeten et al. 2014) which eventually increases the mechanical strength of roots and allows them to explore wider rhizosphere (Nibau et al. 2008).

The increasing awareness in root biology has radiated from the accelerating requirement of crop production, diminishing land resources, unfavorable soil, and environmental conditions (De Smet et al. 2012). Therefore, it is pertinent to attain detailed insights into the underlying complexities involved in the physiological and genetic factors controlling the vital target trait, i.e., competent root architecture formation to enable a much-needed new green revolution in regions characterized by low-input agriculture and where food insecurity is a persistent threat. Key approach to address this objective is to gather information from various researches analyzing transgenic plants with modified expression of hormone biosynthetic enzymes and the characterization of mutants with reduced/altered hormone biosynthesis and/or sensitivity. Thus, this chapter aims to comprehend the cross-regulatory networks and intricacies of phytohormones regulating the self-organization and dynamism of overall root development, i.e., primary root (PR) growth, lateral root (LR) growth, and root hair (RH) development.

1.2 Role of Hormones in Primary Root (PR) Development: Longitudinal and Radial Patterning

1.2.1 Longitudinal Patterning

In dicotyledonous plants upon germination, root development originates in the basal cells of the heart-stage embryo with primary root meristem (Giehl et al. 2014), which begins with the specification of a single cell—hypophysis (the uppermost cell of the suspensor)—dividing asymmetrically to form stem-cell niche (SCN) that contains an

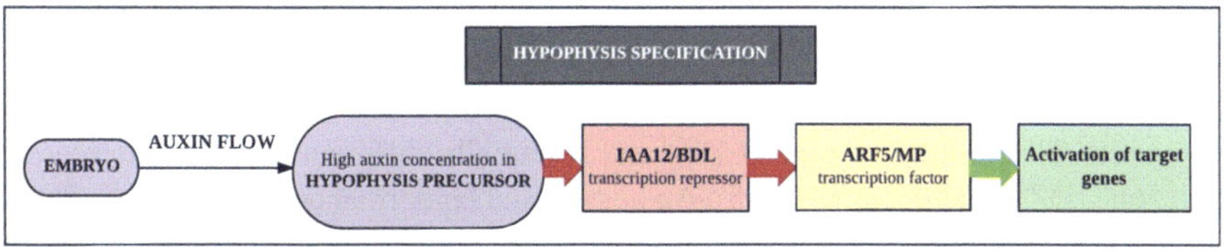

Fig. 1.1 Auxin-mediated specification of hypophysis at embryonic stage (Diagrammatic representation of inference drawn from various researches). Black and green arrows indicate stimulation, and red arrows indicate inhibition

upper lens-shaped cell giving rise to the quiescent center (QC) and a cluster of four larger basal cell forming the columella stem cells [CSCs] (Scheres et al. 1994; Bianco et al. 2013). Hypophysis specification is linked to auxin flux (from the embryo to the hypophysis precursor)-mediated degradation of the transcription repressor, IAA12/BDL, thus releasing auxin response factor (ARF) transcription factor, ARF5/MP, from inhibition and allowing the activation of target genes (Hamann et al. 2002; Friml et al. 2003) (Fig. 1.1). Mutations which affect auxin allocation and signaling hinder the specification of the hypophysis, resulting in no root meristem formation and thus leading to a rootless phenotype (Hamann et al. 1999). In order to ensure proper organization of the root SCN, auxin-mediated suppression of cytokinin (CK) signaling in the embryonic basal cell lineage of the hypophysis is induced by the expression of two type-A genes of *Arabidopsis* response regulator family, *ARR7* and *ARR15* (Muller and Sheen 2008). Further during early embryogenesis, activity of the cytokinin-sensitive synthetic promoter two-component output sensor (TCS) in lens-shaped cell prominently marks tissue-specific phosphorelay outputs entailed for the successful development of hypophysis-derived daughter cells into an operational root stem-cell system (Bianco et al. 2013). Thus, the core of root apical meristem (RAM) is SCN, comprising of the organizing center—a small group of slowly dividing cells known as the QC, the surrounding CSCs, and a mitotically active population of derived cells from CSCs (Heidstra and Sabatini 2014; Street et al. 2015). Stem cells divide asymmetrically to give rise to two daughter cells: one that preserves stem cell identity and another one undertakes supplementary divisions and progresses toward differentiation (Sena et al. 2009). QC cells obstruct differentiation of the columella initials directly in contact with them, thus allowing division of these initials, indicating that QC specifically maintains stemness of initial cells (i.e., an undifferentiated state) thus underlining the indeterminate growth of the root throughout the lifespan (van den Berg et al. 1997). RAM is the crucial reservoir of all cells constituting a PR; any disturbance in the spatial arrangement and functional continuance of RAM will eventually impinge on PR growth (Giehl et al. 2014). Till this phase of root development, the division rate of stem cell daughters prevails over their differentiation, allowing meristem building (Moubayidin et al. 2010). However, as cells progress further away from the SCN, the rate of derived cell division slows and produces transit-amplifying cells, which divide in the proximal meristem [PM] (Nieuwland et al. 2009; Petricka et al. 2012). This position where cell elongation begins in relation to the SCN position determines the size of RAM which is in a straight line with the rate of root growth (Beemster and Baskin 1998). Cell differentiation is activated at the transition zone (TZ), from where cells go into the elongation/differentiation zone (EDZ) and terminally differentiate into differentiation zone [DZ] (Pacifici et al. 2015). An augmented epidermal cell length demarcates the switch between the meristematic and elongation zones, and the root hair (RH) appearance marks the beginning of DZ (Overvoorde et al. 2010). For establishment of continuous root growth, the meristem sets its final size at 5 days post-germination (dpg), creating a dynamic stability between cell division and differentiation (Dello Ioio et al. 2007). However, an unmanaged meristem would either overconsume valuable energy for growth or differentiate completely and turn inactive for the entire lifespan of the plant (Pacifici et al. 2015). Co-ordination flanked by TZ and SCN is regulated by the *SCARECROW (SCR)* gene (Sabatini et al. 2003; Moubayidin et al. 2013). In the QC, SCR directly binds to and negatively regulates *ARR1*, which, sequentially, controls production of auxin by modulating the expression of auxin biosynthesis gene, i.e., *ASB1*. Auxin produced by ASB1 in the QC is further distributed along the meristem via PINs and is adequate to stimulate *ARR1* expression in the TZ, thus sustaining cell differentiation via *SHY2* (Dello Ioio et al. 2008; Moubayidin et al. 2013). This intricate network of hormonal interactions allocates equilibrium between cell division and differentiation, consequently setting the final RAM size (Pacifici et al. 2015) (Fig. 1.2).

1.2.2 Radial Patterning

PR consists of various defined cell types systematized in concentric layers where epidermis, cortex, endodermis, and pericycle surround the central vascular cylinder (Dolan et al. 1993). Radial patterning in PR resembles the array of sets of initials present in the early torpedo stage. The root radial organization is affirmed by cell division patterns established in the embryo and comprises three fundamental tissues: the dermal, ground, and vascular tissues. These tissues are arranged in concentric layers, i.e., proximal side of QC has a tier of cells comprising pericycle and vascular tissue initials, at the radial flanks are initials forming endodermis and cortex, and at the distal side of QC are the initials forming root cap and epidermis. RAM keeps on adding new cells to the preexisting files thus extending back into the mature root (Casson and Lindsey 2003). After embryogenesis, *TORNADO1* and *TORNADO2* define epidermal and lateral root cap specification (Cnops et al. 2000), and *trn1* and *trn2* mutants show rigorous root twisting due to reduced root elongation. Defective division of the initial or the daughter cells, often on only one side of the root, leads to such phenotypes in these mutants; defective distribution of auxin is the chief contributor (Casson and Lindsey 2003). These mutants have lateral root cap-like cells in the epidermis, demonstrating programmed cell death in elongation zone (EZ). Thus, *TRN* genes act either as negative regulators of lateral root cap fate in the epidermis or as positive regulators of epidermal cell fate (Casson and Lindsey 2003).

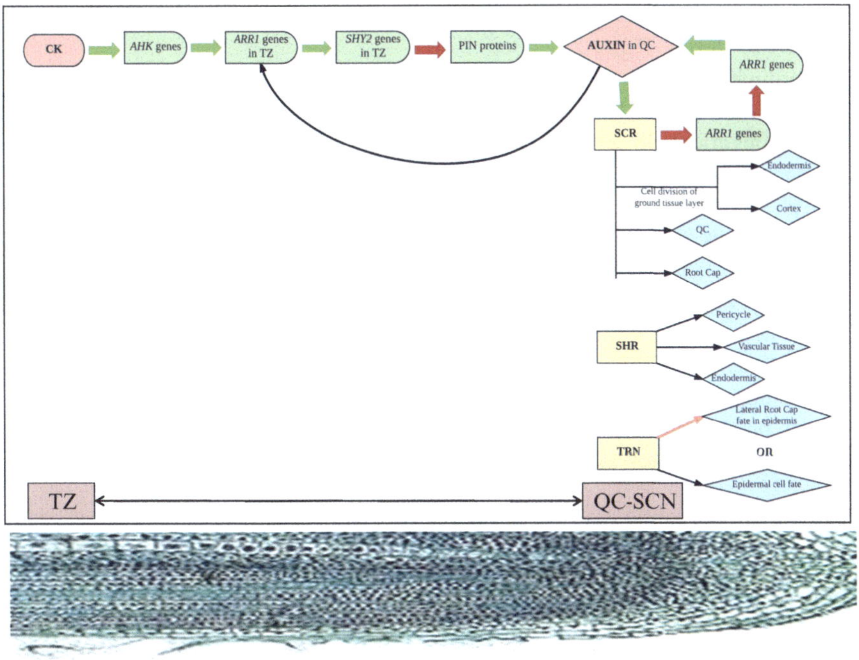

Fig. 1.2 Antagonistic roles of auxin and cytokinins at stem-cell niche (SCN) and transition zone (TZ) for determining appropriate development of longitudinal and radial patterning in root system architecture (Diagrammatic representation of inference drawn from various researches). Black and green arrows indicate stimulation, and red arrows indicate inhibition

Further, mutants of *scarecrow* (*scr*) and *short root* (*shr*) display defective root growth with a missing ground tissue layer due to the absence of a periclinal division during the early embryonic heart stage. *shr* roots miss the endodermis, while *scr* roots shared both endodermis and cortex attributes. *SHR* is required for specification of the endodermis, while *SCR* is required for the cell division that generates the separate endodermis and cortex cell layers (Scheres et al. 1995). *SCARECROW* (*SCR*) gene encoding a transcription factor (Di Laurenzio et al. 1996) is initially localized to the single ground tissue layer of the embryo, but following division is quickly restricted to the endodermis, indicating its requirement for the cell division of both the ground meristem layer in the embryo and the cortical/endodermal initial cell. SCR function is required not only for ground tissue patterning but also for QC and root cap formation after hypophyseal division (Wysocka-Diller et al. 2000). However, *SHORT-ROOT* (*SHR*) gene is expressed exclusively in the pericycle and vascular tissue of the root, including the vascular initial cells, but not in ground tissue, and is required for maintenance of *SCR* expression in the meristem and differentiated cells. Plants overexpressing *SHR* have supernumerary cell layers in the root due to extra cell divisions in the meristem (Helariutta et al. 2000), and these extra layers showed attributes of the endodermis as well as express *SCR*, further affirming that *SHR* is required for endodermis cell specification activation (Casson and Lindsey 2003) (Fig. 1.2).

Coordinated divisions of initial cells forming radial pattern of concentric layers (Garay-Arroyo et al. 2012) and the generation of new cells at the root tip, followed by cell elongation and differentiation, lead to spatiotemporal transcript maps of PR (Birnbaum et al. 2003) which finally emerge and grow gravitropically into their rhizosphere (Beemster and Baskin 1998). Indeterminate growth and development of roots results in all developmental stages being present in distinguishable regions along the root and all stages of root development are apparent at all times (Esau 1977; Benfey et al. 1993). The plant root is thus a highly dynamic structure which is orchestrated by fine tuning of pathways involving hormones and developmental genes thus ensuring root development and functioning. Hormones are among the major endogenous regulators of root growth, where an accurate root development depends on the intricate cross talk among each and every hormone's biosynthetic and signal transduction pathway (Pacifici et al. 2015).

1.3 Role of Auxins

Key to the establishment and function of the polar RAM is auxin biosynthesis (Ikeda et al. 2009) and "polar auxin transport (PAT) machinery," i.e., both auxin influx carriers like AUX1 and AUX1-LIKE (LAX) (Swarup et al. 2005, 2008) and the PIN auxin efflux carriers (Grieneisen et al. 2007; Krupinski and Jonsson 2010; Moore et al. 2015) which play main roles in the configuration of auxin gradients (Band et al. 2012). Genetic or pharmacological manipulation of PAT that disturbs the preservation of the auxin maximum dramatically modifies the patterning of the root tip (Friml et al. 2002; Benjamins and Scheres 2008). In *Arabidopsis* roots, cellular patterning is coordinated when shoot-derived auxin accumulates at its upper limit centered on the QC and columella cells (Friml et al. 2002) and creates gradient at the root apex providing positional information vital for continuation of accurate cell division (i.e., inhibited cell differentiation), polarity, and fate (Sabatini et al. 1999; Overvoorde et al. 2010; Del Bianco and Kepinski 2011). Auxins are transported actively to roots via two pathways: long-distance and short-distance pathways (Teale et al. 2006). In long-distance pathway, auxins are rapidly bulk flown through mature membraneless phloem channels (Tsurumi and Wada 1980); however in short-distance pathway, cell-to-cell auxin transport is mediated by specific auxin influx and efflux carriers (Vanneste and Friml 2009). The track of auxin flux within tissues is driven by the coordinated asymmetrical membrane allocation of efflux carriers like pin-formed (PIN) proteins and p-glycoprotein (PGP) ABC transporter family engaged in both influx and efflux of auxins (Zazımalova et al. 2010). Localization and retention of PIN proteins on plasma membrane are mediated by vesicle-cycling machinery through *GNOM* gene, which encodes ADP-ribosylation factor GTPase guanine exchange factor (ARF-GEF) and also through the reversible protein phosphorylation mechanism involving pinoid kinase (PID) and protein phosphatase 2A (PP2A)

(Michniewicz et al. 2007). Auxin efflux carrier permeability may be sufficient to generate the gradient in the absence of auxin biosynthesis in the root (Wabnik et al. 2010; Clark et al. 2014). Multiple combinations of *pin* mutants display shorter root meristems and delayed root growth (Blilou et al. 2005) thus emphasizing that active auxin transport especially efflux maintains optimum cellular auxin concentration required for root growth and development (Mravec et al. 2011). Before auxin can be transported out of the cell, it first needs to get in via the auxin import carrier AUX1, and mutation of these carriers also interferes with auxin uptake and impairs auxin-mediated root development (Vanneste and Friml 2009). In addition, the auxin transport inhibitors (ATI), such as 1-naphthylphthalamic acid (NPA), tri-iodobenzoic acid (TIBA), and 2-(1-pyrenoyl) benzoic acid (PBA), restrain auxin efflux and thus block polar auxin progress between cells (Dhonukshe et al. 2008). Moreover, root-generated auxin considerably contributes in the protection of this auxin gradient and maxima required for standard root development (Ljung et al. 2005; Ikeda et al. 2009; Petersson et al. 2009). Genes encoding constituents of auxin biosynthetic pathway are expressed in the root, e.g., *YUCCA* genes encoding a flavin monooxygenase involved in tryptophan-dependent auxin biosynthesis. Cheng et al. (2006) substantiated that plants with quadruple *yuc1yuc4yuc10yuc11* mutation in the *YUCCA* genes could not develop a RAM, indicating the importance of root auxin biosynthesis (Pacifici et al. 2015). Independent mutants, *wei8* (Stepanova et al. 2008), *sav3* (Tao et al. 2008), and *tir2* (Yamada et al. 2009), with auxin-related phenotypes are mutated for gene *taa* encoding an aminotransferase (TRYPTOPHAN AMINOTRANSFERASE of *Arabidopsis*, TAA) that catalyzes the tryptophan transamination into indole pyruvic acid, which can be subsequently converted to IAA. Low levels of IAA in these mutants correlate with their defects in gravitropism and vascular tissue differentiation (Yamada et al. 2009).

This auxin gradient regulates the expression of specific genes such as the *PLETHORA* family (Aida et al. 2004) and *WOX5* (Sarkar et al. 2007). Two apetala2 (AP2)-domain transcription factors, PLETHORA (PLT) 1 and 2, accumulate in auxin dose-dependent manner, where their highest expression overlaps with the auxin maximum in the RAM and specifies the positioning of the SCN. On the other hand, the lowest PLT abundance determines where cells start to differentiate (Galinha et al. 2007). Parallel to PLT1 and PLT2, the transcription factors SHORT ROOT (SHR) and SCARECROW (SCR) are also involved in the maintenance of the SCN (Helariutta et al. 2000) and in the specification of the endodermis (Cui et al. 2007). In addition to their role in RAM patterning, SHR and SCR have also been found to regulate the expression of the cell cycle regulator *CYCLIN D6;1* (*CYCD6;1*) directly, thus interconnecting patterning and growth in the root tip (Sozzani et al. 2010). Furthermore, in roots, other well-characterized auxin dose-associated phenotypes are (1) increase in PR length, (2) increase in number of lateral root (LR) primordia, (3) increase in length of epidermal-derived root hairs, and (4) the response to gravity (Ishida et al. 2008; Peret et al. 2009). Auxin is then redistributed back up through the lateral root cap and into the proximal meristem (Vieten et al. 2005).

On receiving the auxin, the receptor like transport inhibitor response 1 (TIR1/AFB—TRANSPORT INHIBITOR 1/AUXIN SIGNALING F-BOX PROTEIN) binds to the auxins to reduce signaling cascade (Kepinski and Leyser 2005). There are two related families of protein: the transcription repressors like auxin/indole-3-acetic acid (AUX/IAA) and auxin response DNA-binding transcription factors (ARF) which act either as activators (ARF5, ARF7, ARF8, and ARF19) or repressors (ARF1 and ARF2) of transcription (Liscum and Reed 2002). At low auxin levels, via a classical negative feedback loop, AUX/IAA binds to and represses ARF (Tiwari et al. 2001). Moreover at low auxin level, *topless* (*TPL*) corepresses auxin-regulated transcription by mediating binding of AUX/IAA proteins to ARF (Szemenyei et al. 2008). However at higher concentrations, auxin binds directly to the TIR1/AFB receptors and negatively regulates Aux/IAA repressor by promoting their ubiquitin-dependent proteolysis [by the 26S proteasome] (Kepinski and Leyser 2005; Santner and Mark Estelle 2009). This results in the ARF activation that subsequently binds to the promoter of auxin-responsive genes involved in root growth and development process (Tian et al. 2003; Guilfoyle and Hagen 2007; Lau et al. 2009). Loss-of-function mutations of *ARF5/MONOPTEROS (ARF5/MP)* and gain-of-function mutations of *IAA12/BODENLOS (IAA12/BDL)* interfere with the proper orientation of embryonic cell division planes and thus disrupt formation of the root meristem (Hamann et al. 2002). In *A. thaliana*, mitogen-activated protein kinases, i.e., MAPK kinase-7 (MAPK7) and MAPK12, negatively regulate PAT and signaling (Dai et al. 2006; Zhang et al. 2008; Lee et al. 2009); however, indole-3-butyric acid-response 5 (IBR5) phosphatase dephosphorylates and inactivates MPK12 leading to the upregulation of auxin-responsive genes involved in root development (Lee et al. 2009).

Cell elongation in higher root zones and indeed development of PR largely depend on other plant hormone signaling pathways and their interaction with auxin biosynthesis and transport (Ubeda-Tomás et al. 2012; Liu et al. 2014) [detailed in Fig. 1.3]. After seed germination, growth of RAM is regulated by amplification in the rate of cell differentiation relative to division, and modulation of cytokinin signaling plays a fundamental role in this event (Moubayidin et al. 2010).

1.4 Role of Cytokinins

A genetic framework of antagonistic interactions between proliferation-supporting auxin signaling and differentiation-promoting cytokinin (CK) input is accountable for regulating equilibrium between division and differentiation of cells leaving RAM (Dello Ioio et al. 2008; Bielach et al. 2012). In addition, suppression of the ARR7 and ARR15 components of the CK signaling pathway in the basal cell at the early globular stage plays a vital role in functional SCN establishment (Moubayidin et al. 2010). Mimicking the CK response in these cells disrupts the established embryo pattern, while manipulation of CK signaling later than the embryonic heart stage does not interfere with QC specification and SCN organization (Muller and Sheen

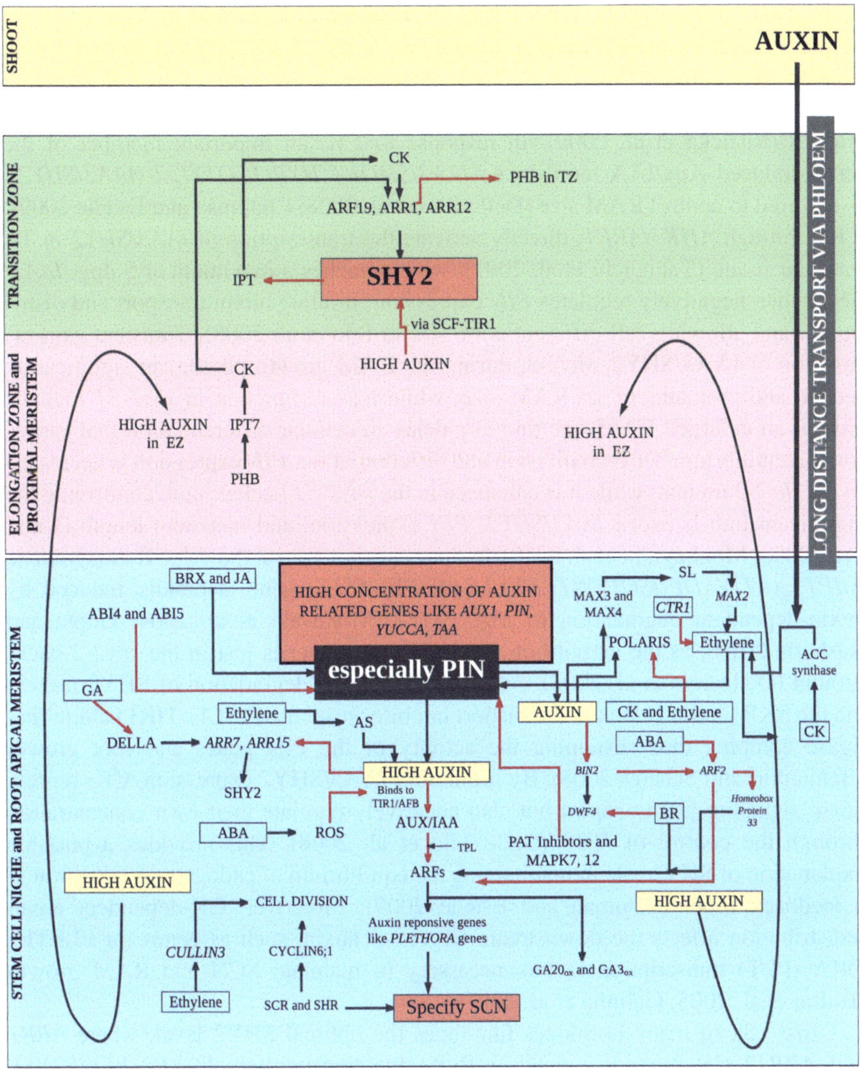

Fig. 1.3 Cross talk of various phytohormones for determining appropriate longitudinal patterning of primary root (PR) in root system architecture (Diagrammatic representation of inference drawn from various researches). Black arrows indicate stimulation, and red arrows indicate inhibition

2008). CK modifies the expression of several *PIN* genes and regulates cell-to-cell auxin transport and thus the actual level of auxin in the cells (Della Rovere et al. 2013; Zhang et al. 2013). In the root, CK application inhibits root growth and induces shoot meristem enlargement (Lindsay et al. 2006). Mutants in CK biosynthesis genes, in the CK receptor *AHK3* gene, and in the type-B *ARR* genes *ARR1* and *ARR12* display strongly enhanced root growth and enlarged meristems (Dello Ioio et al. 2007, 2008). On the other hand, CK transcription network is regulated by a negative feedback loop to control its response to root development via auxin signal transduction pathway (Saini et al. 2013). Auxin controls output of the CK signaling pathway by modulating transcription of its negative regulators, i.e., *ARR7* and *ARR15* (Ruzicka et al. 2009). In response to CK, an important member of the auxin-induced Aux/IAA family, i.e., *IAA3/SHORT HYPOCOTYL 2* (*IAA3/SHY2*), is required to control RAM size (Dello Ioio et al. 2008; Chapman and Estelle 2009). CKs, through *AHK3/ARR1*, directly activate the transcription of *IAA3/SHY2* in TZ vascular tissue (Taniguchi et al. 2007), which reaches a maximum at 5 dpg. *IAA3/SHY2* then negatively regulates *PIN* expression, limiting auxin transport and distribution and allowing cell differentiation (Dello Ioio et al. 2008). Transient gain-of-function of IAA3/SHY2, *shy2-2*, during the RAM growth phase, can significantly reduce and prematurely set RAM size, while loss-of-function in *shy2-31* mutants exhibit an enlarged RAM size due to a delay in cellular differentiation that causes loss of equilibrium between division and differentiation. *PIN* expression is decreased in the *shy2-2* mutant, while it is enhanced in the *shy2-31* background, confirming the significant link between *IAA3/SHY2*, *PIN* expression, and meristem length (Knox et al. 2003; Moubayidin et al. 2010). In the vascular tissue at the TZ, CK biosynthetic *AtIPT5* (*ATP/ADP-ISOPENTENYLTRANSFERASES*) gene is rapidly induced by auxin-dependent degradation of *IAA3/SHY2* (Miyawaki et al. 2004; Gupta and Rashotte 2012), as the activity of the *AtIPT5* promoter is lost in the *shy2-2* background (Dello Ioio et al. 2008). Auxin facilitates the degradation of SHY2 protein via the SKP-Cullin-F-Box and transport inhibitor response1 (SCF–TIR1) ubiquitin–ligase complex thus sustaining the activity of the PIN genes and root growth (Benjamins and Scheres 2008). By promoting IAA3/SHY2 expression, CKs repress auxin signaling and transport but also negatively regulate their own concentration through the control of *IPT5* (Dello Ioio et al. 2008). This provides a possible explanation of SHY2 role in maintaining the equilibrium of endogenous CK through a feedback loop (Chapman and Estelle 2009). Moreover, CK-dependent auxin redistribution affects the downstream targets of auxin, such as genes for PLETHORA (PLT) transcription factors necessary to maintain SCN and RAM growth (Blilou et al. 2005; Galinha et al. 2007).

Cross talk of many hormones fine-tunes the optimal SHY2 level, where *ARR1* and *ARR12* CK-dependent genes activate the transcription *AUXIN RESPONSE FACTOR19* (*ARF19*) and *SHY2* transcription and thus promote cell differentiation of meristematic cells at the TZ (Moubayidin et al. 2010; Perilli et al. 2013). PHABULOSA (PHB), a member of CLASS III HOMEODOMAIN-LEUCINE ZIPPER (HD-ZIP III) transcription factors, regulates RAM length by inducing CK biosynthesis via isopentenyl transferase7 (IPT7) in the PM of the root, thus activating ARR1 at the TZ. ARR1, in turn, represses the expression of PHB at the TZ vasculature, thus restricting PHB expression to the distal part of the PM (Bianco et al. 2013). At the same time, BREVIS RADIX (BRX), a protein implicated in the brassinosteroids (BR) pathway and a transcription coactivator in the vasculature, controls PAT by inducing *PIN3* expression (Scacchi et al. 2010). Through IAA3/SHY2, CKs are able to control BRX expression and, additionally, polar auxin

2006). Further, GA helps PAT by modulating the turnover of PIN1 proteins required for root gravitropism; as in GA-deficient *A. thaliana*, auxin transport was hindered due to the reduced level of PIN proteins resulting in impaired root gravitropic responses (Willige et al. 2011).

1.6 Role of Brassinosteroids

Signaling of brassinosteroids (BR) includes its binding to BRI1 receptor which stimulates the interaction of BRI1 with BAK1 receptor forming BRI1/BAK1 complex. This complex inhibits glycogen synthase kinase 3/SHAGGY (GSK3/SHAGGY) serine/threonine kinase encoded by *BIN2* gene (a negative regulator of BR signaling) thus activating BR responsive genes such as transcription factors, brassinazole-resistant1 (BZR1) and BRI-EMS suppressor1 (BES1) (Wang et al. 2005b; Yang et al. 2011), which further upregulate BR biosynthetic genes, such as *DWF4* and *CPD*, respectively (Chung et al. 2011). Genetic analysis of *brl1 brl3 bak1-3* triple mutants revealed that BAK1, BRL1, and BRL3 signaling modulates root growth and development by contributing to the cellular activities of provascular and QC cells (Fabregas et al. 2013). Although auxin utilizes *BIN2* for auxin-dependent upregulation of DWF4, there is antagonistic role of BR and auxins; as when optimum amount of BR is synthesized, *DWF4* gene is feedback inhibited by BR (Saini et al. 2013). Further, BIN2-mediated phosphorylation of ARF2 results in loss of DNA-binding repression activities of ARF2 leading to root development. Auxin-insensitive mutants of *AUX/IAA* genes, such as *iaa7/axr2* and *iaa17/axr3*, also showed altered BR sensitivity in roots and inhibition of root growth in these mutants upon treatment with exogenous brassinolide, evidenced cross talk between BR and auxin signaling pathway (Nakamura et al. 2006). Further, exogenous application of BR induces the expression of various auxin-responsive genes involved in root development (AXR3/IAA17, AXR2/IAA7, SLR/IAA14), while BR signaling mutant and biosynthetic mutant *det2* significantly reduce gene expression during root development (Kim et al. 2006). BRs are also known to play role in PAT by affecting the cellular localization of auxin efflux carriers, such as PIN3 and PIN4, and influx carriers, AUX1/LAXs (Hacham et al. 2011).

1.7 Role of Abscisic Acid

Abscisic acid (ABA) plays important role in the maintenance of PR growth at low water potential by preventing excess ethylene production (Sharp 2002). The pyrabactin resistance 1 (PYR1) and PYR1-like proteins (PYLs), also known as the regulatory component of ABA receptor (RCAR) family proteins (Nibau et al. 2008), mediate the ability of ABA to inhibit PR growth through the PP2C-SnRK2 pathway (Antoni et al. 2013; Zhao et al. 2014). Auxin has both antagonistic and synergistic effects on responses to ABA. High concentrations of ABA antagonize auxin by promoting reactive oxygen species production and reduce the auxin reporter *ProDR5:GUS* expression in roots leading to reduction in auxin concentration or response. However, ABA upregulates the expression of auxin response factor *ARF2*, which suppresses the expression of *HOMEOBOX PROTEIN 33*, resulting in the inhibition of root growth (Wang et al. 2011). Contrastingly, ABA can also augment auxin signaling by activating auxin-responsive promoters to repress embryonic axis elongation (Belin et al. 2009). Plants with mutations in the *auxin-resistant* (*AXR*) genes, *axr2-1* and *axr3-1*, and auxin transport mutants *aux1* and *pin2* are insensitive to the effects of both ABA and auxin on embryonic axis elongation and root growth (Wilson et al. 1990; Liu et al. 2013). *ABSCISIC ACID INSENSITIVE5* (*ABI5*) encodes a transcription factor belonging to the basic leucine zipper (bZIP) family and represses polar auxin flux through *PIN1* inhibition, thereby controlling root growth (Yuan et al. 2014). However, at the same time, auxin counteracts this repression, directing SHY2 degradation and thus allowing the maintenance of PIN activity and the consequent induction of cell division (Dello Ioio et al. 2008). Auxin transport to the root via PIN1 is limited under osmotic stress together with enhanced PIN2 levels, leading to reduced auxin concentrations in RAM (Rowe et al. 2016). In this aspect, ABA plays a synergistic role with CK as CK-induced ABSCISIC ACID INSENSITIVE4 (ABI4), a protein encoding an ABA-regulated AP2 domain transcription factor, represses *PIN1* expression (Shkolnik-Inbar and Bar-Zvi 2010). Lower auxin concentrations lead to a reduction in meristem size and reduced root growth (Thole et al. 2014). Roots of IAA-resistant mutants, i.e., *axr2-1/iaa7* gain-of-function mutants, have been found to be resistant to exogenous ABA, whereas *slr-1/iaa14* mutant was found to be hypersensitive to ABA in PR growth inhibition assays (Fukaki et al. 2002). ROP-interactive CRIB motif-containing protein 1 (RIC1) is an important component of the intricate signaling network between auxin and ABA, where RIC1 positively regulates auxin responses but negatively regulates ABA responses (Choi et al. 2012).

1.8 Role of Ethylene

Auxin–ethylene cross talk explains ethylene-dependent root growth stimulated by auxin biosynthesis as they synergistically interact in the regulation of root gravitropism (Buer et al. 2006) and root growth (Rahman et al. 2001), thereby inducing ectopic accretion of auxin in outer layers of the root (Ruzicka et al. 2007). Auxin upregulates ethylene biosynthesis by activating transcription of *1-aminocyclopropane-1-carboxylate synthase* (*ACS*) gene which encodes a key enzyme catalyzing a rate-limiting step of ethylene production (Abel et al. 1995; Muday et al. 2012). A weak *ethylene insensitive 1* (*wei1*) mutant harbors a recessive mutation in an auxin receptor gene *TRANSPORT INHIBITOR RESPONSE1 [tir7]* (Alonso et al. 2003), and ethylene also upregulates the expression of *weak ethylene insensitive 2* (*wei2*) and *wei7* genes encoding subunits (a and b) of anthranilate synthase (AS), an enzyme that catalyzes the rate-limiting formation of anthranilate from chorismate during tryptophan synthesis (Stepanova et al. 2005). *WEAK ETHYLENE INSENSITIVE* (*wei2 and wei7*) and *TRANSPORT INHIBITOR RESPONSE 7* (*tir7-1*) mutants have also been reported to suppress the high auxin phenotypes of *SUPERROOT* (*sur1* or *sur2*) by disrupting AS synthesis and thus lowering their capacity to produce indole-3-acetic acid, which impairs root growth (Ljung et al. 2005). Chen and Xiong (2009) substantiated that only shoot-derived auxin could not maintain the planar polarity of root epidermal cells and that root-derived auxin is required for proper postembryonic root growth, as mutants of *pdx1* genes encoding pyridoxal phosphate had impaired capacity to convert tryptophan to IAA, thus forming short PR with a reduced RAM size which could not be rescued by shoot-derived auxin. Moreover, local auxin biosynthesis and ethylene signaling are inhibited by ethylene signaling protein Constitutive Triple Response1 (CTR1), acting as concentration-dependent repressor of auxin biosynthesis (Muday et al. 2012). In the study of Le et al. (2001), ethylene-insensitive mutants *ctr1-3* and *eln2-1* exhibited increased cell elongation, and the constitutive ethylene-response mutant *ctr1-1* reduced cell elongation in the root compared with the WT. Cross talk of auxin and ethylene is a critical component for ethylene-induced inhibition of root cell elongation as ethylene stimulates auxin biosynthesis in the root as well as upregulates the major PAT components, such as PIN2/AUX1, leading to the stimulation of basipetal auxin transport toward the EZ, and the increased auxin activity in EZ serves to inhibit cell elongation (Swarup et al. 2007; Strader and Bartel 2008). The ability of auxin to induce ethylene and ethylene to modulate PAT suggests a mechanism by which root elongation can be controlled by a feedback loop (Casson and Lindsey 2003).

CK induces ethylene biosynthesis by stabilizing ACC synthase enzymes, and this facilitates the ability of CK to repress cell expansion (Ruzicka et al. 2009). Although ethylene facilitates but is not the absolute requirement of CK control of meristem cell proliferation, as increased CK concentration could overcome the role of ethylene. Ethylene, like CK, induces divisions at the QC (Zhang et al. 2013), and though CK response at QC is not dependent on ethylene signaling, but it remains determined if ethylene also facilitates this response at lower CK concentrations. ARR1 and SHY2 are the points of intersection between the ethylene and CK signaling pathways in their control of RAM cell proliferation (Dello Ioio et al. 2008). CK-induced ethylene production could account for the inhibitory effects of CK on root and hypocotyl elongation (Cary et al. 1995). *CULLIN3* genes regulate cell proliferation at RAM through an ethylene-dependent mechanism, as *ein2-1* and *ein3-1* mutations could revert the reduced RAM size of a *cullen3* mutation (Thomann et al. 2009). The *PLS* gene of *Arabidopsis* transcribes a short mRNA encoding a 36-amino-acid POLARIS peptide required for correct root growth and vascular development (Casson et al. 2002). Short roots of *pls* mutant with reduced cell elongation are due to reduced auxin and increased hyperresponsive to exogenous cytokinins. The cross talk among auxin, ethylene, and cytokinin could be established here, as increasing the concentration of either ethylene or cytokinin inhibits *PLS* gene expression, while increasing auxin concentrations promotes *PLS* gene expression (Chilley et al. 2006; Liu et al. 2014).

1.9 Role of Polyamines

Polyamines (PAs) are low-molecular-mass polycations, where free PAs (agmatine, putrescine, spermidine, and spermine) (Smith and Davies 1985), soluble-conjugated PAs (caffeoylputrescine and feruloylputrescine), and insoluble-bound PA have been observed in root tissues suggesting their possible role in root development (Hummel et al. 2002). In the studies of Hummel et al. (2002) and Jang et al. (2002), exogenous root application of spermidine and spermine showed enhanced PR growth with neutral or negative effects of putrescine. Palavan-Unsal (1987) reported that decreased arginine decarboxylase (ADC) activity depleted putrescine, spermidine, and spermine levels, thereby decreasing root length in *Phaseolus vulgaris*. However, addition of PA in root induction medium containing Murashige and Skoog (MS) supplemented with NAA and IBA significantly improved root formation and growth of *Citrus sinensis* L. Osb, thereby indicating the possible interaction of auxins and polyamines in regulating root development. However, these processes were inhibited in the presence of PA biosynthesis inhibitor, a-difluoromethylornithine (DFMO) (Mendes et al. 2011). PA biosynthesis shares a common precursor, i.e., S-adenosyl methionine (SAM) with ethylene biosynthesis (Couee et al. 2004), and ethylene regulates auxin biosynthesis, thereby directing possible interactions of PA with ethylene and auxin (Saini et al. 2013).

1.10 Role of Jasmonic Acid

Jasmonic acid (JA) induced the expression of auxin efflux carrier genes, such as *OsPIN1c*, *OsPIN5a*, *OsPIN10a*, and *OsPIN10b* in the roots of *Oryza sativa*, signifying possible role of JA in regulation of auxin efflux carriers and auxin transport (Wang et al. 2009; Monzon et al. 2012). However, studies of Vellosillo et al. (2007) suggested that JA, such as oxylipins [initiated by the action of 9-lipoxygenases (9-LOX) and 13-lipoxygenases (13-LOX)], results in loss of root apical dominance and decreases root elongation in *A. thaliana*. In a study by Monzon et al. (2012) on growth of PR in *Helianthus annuus* seedlings, it was demonstrated that JA inhibits PR and LR growth through auxin-independent pathway, as auxin produced its phenotype even when ibuprofen (JA inhibitor) was applied (Saini et al. 2013).

1.11 Role of Strigolactones

In rice, a major quantitative trait locus on chromosome 1 (qSLB1.1) was identified for the exudation of strigolactones (SL) (Cardoso et al. 2014), and many root architectural traits were mapped in the same region by Topp et al. (2013), suggesting the involvement of this locus in both SL synthesis and RSA (Kapulnik and Koltai 2014). Study of Kapulnik et al. (2011a, b) and Ruyter-spira et al. (2011) demonstrated that SL positively regulates PR development as PR length of SL-deficient and SL-insensitive *A. thaliana* plants was shorter than the WT plants. MAX2, an SL signaling component, leads to the induction of ethylene biosynthesis, signaling, and transport, thus positively and indirectly affecting auxin transport, and maintains auxin optima for SL biosynthesis to regulate root development. Auxin induces SL synthesis in the root, through induction of *MAX3* and *MAX4* expression (Beveridge and Kyozuka 2010; Koltai 2011). SLs mediate auxin flux in the roots as reported in the study of Koltai et al. (2010) where SL regulated PR growth, cell elongation, and RH elongation in the presence of exogenously applied auxin in tomato. Also, GR24 (synthetic SL) application increased PR length in a MAX2-dependent manner under favorable growth conditions, while the PR lengths of the SL-deficient and SL-response mutants were shorter than WT. Reduction in PR length was accompanied by a reduction in cell number in the PR meristem that could be rescued by application of GR24 to SL-deficient, but not SL-response mutants (Ruyter-Spira et al. 2011). Strigolactones regulate PIN protein activity as evidenced by studies of tomato (*Solanum lycopersicum*) roots, in which exogenous supplementation of 2,4-dichlorophenoxyacetic acid (a synthetic auxin that is not secreted by auxin efflux carriers) led to reversion of the GR24-related root effect, suggesting functional involvement of GR24 with auxin export (Kapulnik and Koltai 2014). On analysis of strigolactone and auxin signaling mutants, it can be predicted that during root development auxin signaling acts downstream of strigolactones (Brewer et al. 2013) (Fig. 1.3).

1.12 Role of Hormones in Lateral Roots Development

Besides PR length, the number and length of lateral roots (LRs) represent the other dominant feature of RSA. LRs are the most dynamic and physiologically active part of the root system which allows the plant to explore the highly heterogeneous soil environment and to adapt to changing nutrient and water availability (Gou et al. 2010). Although fully developed LRs are structurally very similar to primary roots, the formation of a LR is a completely postembryonic event (Peret et al. 2009; De Smet 2012; Petricka et al. 2012). It is noteworthy that LRs exhibit some unique features, such as an altered gravitropic response (Rosquete et al. 2013). The formation of a new LR involves a series of tightly coordinated events which start with the pre-initiation by cell cycle reentry of pericycle founder cells and are followed by LR primodium (LRP) establishment, emergence via cell expansion, meristem activation, and maintenance of meristem function (detailed in Fig. 1.4). Priming of pericycle cells for LR initiation takes place very early, in the EDZ adjacent to the TZ, where local auxin responsiveness oscillates with peaks of expression at regular time intervals (Moreno-Risueno et al. 2010). In *Arabidopsis* and most other dicots, LRs are formed only from the xylem pole pericycle cells which are smaller than other pericycle cells, indicating differential cell cycle regulation between pericycle cell types (Rowe et al. 2016). In other species, particularly cereals such as maize, rice, and wheat, LRs arise specifically from the phloem pole pericycle, with additional contributions from the endodermis (Hochholdinger and Zimmermann 2008). Unlike normal pericycle cells are arrested in the G1 phase of the cell cycle; however, those pericycle cells that will give rise to a LR proceed through S phase and are arrested at G2. LR-inducing signals stimulate these cells to undergo proliferative cell divisions (Beeckman et al. 2001). Auxin plays a stimulatory role at four of the checkpoints, i.e., the first (Laskowski et al. 1995), second (Laskowski et al. 1995), third (Bhalerao et al. 2002), and fifth (Celenza et al. 1995). ALF4 and ALF3 mediate the regulatory roles of auxin in the first (initiation) and the fifth (maintenance of meristem function) checkpoint, respectively (Celenza et al. 1995). However, ABA regulates the fourth checkpoint and affects LR development by suppression of auxin response at the other checkpoints. The removal of apical tissues prior to the formation of the first true leaves has very little effect on LR initiation but inhibits LR emergence (Bhalerao et al. 2002), indicating that the emergence of LRs requires leaf-derived IAA (De Smet et al. 2003). AUX1 promotes IAA accumulation in the root apex, thereby influencing the rate of initiation of LRP (phase 1), and later facilitates LR emergence (phase 2) through IAA export from newly formed leaves and/or uptake in LRP (Marchant et al. 2002). However, exogenous application of auxin can activate the whole pericycle to form LRPs, whereas the application of auxin transport inhibitors blocks LR formation without loss of pericycle identity, indicating that all the cells within the pericycle retain the ability to form LRs but only some of them do so (Himanen et al. 2002). Interestingly, the regular spacing of LR in *Arabidopsis* is due to a transient oscillatory increase in AFB-Aux/IAA-ARF activity in pericycle cells that are transiting the basal root meristem, several millimeters distal to the eventual site of LR emergence. This early auxin response just primes the subsets of xylem-pole pericycle cells to respond differently to later auxin activation of primordium initiation in the older root (De Smet et al. 2007). The fact that all xylem-pole pericycle cells can form LR founder cells in response to high levels of auxin suggests that initiating a LR is merely a case of exceeding a threshold auxin concentration (Benkova et al. 2003; Bianco and Kepinski 2011). Secondly, this endogenous system is sensitive to external cues such as gravity. Gravistimulation concentrates auxin (threshold necessary for LR formation to be reached) at a certain point in the root, where LR initiation would, in turn, consume the auxin pool, preventing new LR initiation until the pool had been refilled [requiring acceleration by a new gravistimulation] (Lucas et al. 2007).

LR development involves cell cycle regulation, especially at the initiation, LRP establishment, and meristem activation steps (De Smet et al. 2003). Auxin signaling during LR initiation is closely coupled with auxin binding to receptors—TIR1 and AFB1–3, SINAT5, XBAT32, etc.—which allows them to target AUX/IAA proteins, e.g., SOLITARY ROOT (SLR1)/IAA14/IAA28 for degradation by SKP1-CULLIN1-F-box (SCF) E3 ubiquitin ligases. *IAA14/SLR* controls the entry of these cells into the cell cycle, as dominant negative mutations in this gene inhibit the first anticlinal divisions (De Rybel et al. 2010). This prevents dimerization and inactivation of ARF transcription factors (ARF7 and ARF19), allowing them to bind with promoter elements in *LATERAL ORGAN BOUNDARIES DOMAIN/ASYMMETRIC LEAVES LIKE* (*LBD/ASL*) genes and, in turn, activate the transcription of cell cycle-associated genes or cell division-associated genes and genes involved in auxin signaling, transport, or metabolism (Okushima et al. 2007). Further, VIER F-BOX PROTEINE (VFB) F-box proteins regulate auxin-induced gene expression, and consequently LR formation, by a pathway independent of the auxin receptor TIR1 (Schwager et al. 2007). During LRP development, in response to auxin signaling, a hotspot of auxin is created by relocalization of PIN1 efflux carrier protein via VPS29—a membrane trafficking component to the cell walls nearest to the forming apex—thus accumulating auxin at the apex (Jaillais et al. 2007). The action of the auxin efflux carriers PINs generates a dynamic auxin distribution that is responsible for the establishment of a new self-sustaining apical meristem. PIN1 is initially detected exclusively on the anticlinal sides of the short initial cells, but later it can also be found at the periclinal sides. As the LR develops, PIN1 polarity points increasingly toward the primordium tip to sustain auxin accumulation in the founder cells (Benkova et al. 2003; Marhavŷ et al. 2013). However, only certain pericycle cells usually give rise to LRs; it is crucial that auxin signals are tightly regulated. Interestingly, in a feedback mechanism, auxin stimulates the transcription of ubiquitin ligases [F-box protein CEGENDUO (CEG)] that repress auxin signals, to maintain auxin sensitivity in the pericycle (Dong et al. 2006). ARF proteins can also function as transcriptional repressors, where *arf10/arf16* knockout line produces more LR (Wang et al. 2005a). Conversely, the inhibition of PAT by TIBA and NPA also inhibits LR production (Topping and Lindsey 1997). Basipetal auxin transport is required for LR initiation, where NPA treatment inhibits this process resulting in increased levels of auxin in the root tip and prevents it from reaching the tissues basal to the tip that are competent for LR development. Once a LR primordium has formed, acropetal auxin transport is required to stimulate elongation of the emerging LR (Casimiro et al. 2001) where long-distance transport of IAA may occur primarily via the phloem (Marchant et al. 2002). Interference with indole-3-butyric acid (IBA) efflux from roots by mutation of the *PLEIOTROPIC DRUG RESISTANCE8/ABCG36 ATP Binding Cassette Transporter* (*PDR8*) gene leads to higher IBA content in the root tip that, in turn, increases LR production, suggesting that PDR8 normally inhibits LR formation by transporting away part of the IBA pool in root tips (Strader and Bartel 2009). Conversely, beyond control by auxin signaling, LR formation is regulated independently, as shown by tomato *diageotropica* (*dgt*) mutants (grown in the presence of exogenous auxin) lacking LRs, but pericycle cells maintain their full proliferative capacity (Ivanchenko et al. 2006). A similar mechanism operates in *wol* mutant of *Arabidopsis*, which forms very few LRs even in the presence of auxin, thereby indicating that cell cycle activation is not sufficient for LR initiation to occur (Parizot et al. 2008). In the absence of auxin, only primordial that had at least 3–5 cell layers develops into LRs, and no further development occurs in primordia of less than three cell layers. These observations suggested that LR development prior to the 3–5 cell layer stage was auxin dependent and that development beyond this stage was either auxin independent or auxin self-sufficient (i.e., the primordium is capable of synthesizing the required amount of auxin de novo). However, the phenotype of the *aberrant LR formation 3* (*alf3-1*) mutant indicated that auxin is also required for LR development beyond the 3–5 cell layer stage, i.e., at the postemergence stage (Celenza et al. 1995). Exogenous brassinolide application can increase LR initiation, but this increase can be suppressed in the presence of the PAT inhibitor *N*-1-naphthylphtalamic acid (NPA) (Bao et al. 2004), suggesting that brassinolide induction requires auxin (Malamy 2005). Inhibition of LR development by exogenous ABA occurs immediately after the emergence of LRP from the parent root and prior to the activation of the LR meristem, and inhibition is mediated by an auxin-independent pathway as the inhibition could not be rescued by either exogenous auxin application or elevated auxin synthesis. Moreover, mutation in the *ALF3* gene (encoding a component in the auxin-dependent regulatory pathway for the postemergence LR development) does not affect the sensitivity of LRs to ABA (De Smet et al. 2003).

After cell cycle activation, asymmetrical cell division gives rise to the LR primordium—LRP having a shorter and a longer daughter cell (Malamy and Benfey 1997). The shorter cells are highly sensitive to auxin and express ACR4, a leucine-rich repeat receptor-like kinase that represses cell division in the adjacent pericycle cells (De Smet et al. 2008). The coordinated pattern of cell division is dependent on auxin signaling and on the activity of the *PUCHI* gene in pericycle cells that will form the LRP. *PUCHI* encodes an APETALA2 (AP2) transcription factor that is upregulated by auxin and acts downstream of auxin to restrict the area of cell proliferation within the LRP (Hirota et al. 2007). In rice, the EL5 RING finger ubiquitin E3 ligase maintains cell viability in the developing primordium and acts downstream of auxin, CK, and JA to prevent meristematic cell death (Koiwai et al. 2007). As a result of rounds of cell division, the LRP increases in size, forming a dome-shaped structure that penetrates the external cell layers of the PR. Again, auxin is involved in reprogramming cells adjacent to the new LR primordium to facilitate its emergence as it breaks through endodermis, cortex, and epidermis for moving outside (Peret et al. 2009). To aid in this process, LRP-originated auxin activates cell wall-remodeling enzymes like polygalacturonase (PG) (Gonzalez-Carranza et al. 2007), expansin, and a beta-xylosidase that loosen the adjacent cells (Swarup et al. 2008). Interestingly, during LR emergence, the pectin in the emerging LR remains methylated, whereas the pectin in the overlying parent root tissues becomes demethylated by pectin methylesterases (PMEs) (Laskowski et al. 2006). Mutations in *CELLULASE3/GLYCOSYLHYDROLASE9B3* and *LEUCINE RICH EXTENSIN2* genes revealed defects in the early and late stages of LR development, respectively, and suggested auxin-mediated cell wall remodeling as an essential feature of LR development (Lewis et al. 2013).

On the other hand, downregulation of CK sensitivity in pericycle cells is required for LR priming at early stages in the EDZ of the PM (De Smet et al. 2007). Initiation of LRs is associated with a localized repression of a CK-responsive reporter gene, indicating spatial and temporal regulation of the CK status during LR formation (Lohar et al. 2004). CK suppresses induction of cell division in the root pericycle via regulation of cytokine receptors AHK2, AHK3, and CRE1/AHK4, and the enhanced root system of *ahk2 ahk3* mutants, where PR grows faster with more LRs than in wild-type plants, is due to the reduced endogenous CK content (Werner et al. 2003; Riefler et al. 2006). Induction of *IPT* expression in the zone of LR initiation greatly affects the formation of LR at later stages; however auxin application to *ipt3 ipt5 ipt7* triple mutant roots induces a massive LR initiation, not seen in wild-type roots (Bielach et al. 2012). CK negatively regulates LR formation (Laplaze et al. 2007) by disrupting the patterning in de novo formed LRP and indirectly via an effect on PAT (Bielach et al. 2012). Indeed, exogenous CK application inhibits LR initiation as reported in both *A. thaliana* (Kuderová et al. 2008) and *O. sativa* (Rani Debi et al. 2005) by preventing pericycle cell cycle reentry and disrupting the organization of cell divisions within developing LRPs. Moreover, plants carrying loss-of-function mutations in several type-B *ARR* and *AHK* genes, as well as overexpressing *CKX*, showed enhanced LR formation (Riefler et al. 2006). However, CK activity is required very early in the LR formation process as CK signaling disruption in xylem pole pericycle cells leads to perturbation of the auxin maximum in developing LRPs as a result of the reduced expression of *PIN* auxin transporter genes and mislocalization of PIN proteins (Laplaze et al. 2007). CKs, through the CK RESPONSE 1 (CRE1) receptor, maintain procambial cell identity by regulating the localization of the auxin efflux carriers PIN, thus forcing auxin flow towards the protoxylem. Accumulation of auxin in the protoxylem induces expression of the pseudo-ARABIDOPSIS HIS PHOSPHOTRANSFER (AHP) protein AHP6, which suppresses CK signaling (Kuderová et al. 2008). The ability of CK to affect LR development is strongly stage-dependent as young LR primordia are more sensitive to perturbations in CK activity than are developmentally more advanced primordia (Bielach et al. 2012). 9-cis-epoxycarotenoid dioxygenase genes (involved in ABA biosynthesis) are expressed in pericycle cells surrounding LR initiation sites (Tan et al. 2003), suggesting that ABA may restrain cell proliferation outside the LR initiation site (De Smet et al. 2006). Some auxin-induced LR-initiation genes like AUXIN-INDUCED IN ROOT CULTURES 12 (*AIR12*) and *IAA19* (Vanneste et al. 2005) and KNAT1 homeobox transcription factor (Truernit et al. 2006) are ABA-repressed in LR primordial, suggesting an antagonistic effect of auxin and ABA on LR initiation (Nibau et al. 2008). Curiously in rice, upregulation of a casein kinase 1 gene, *OsCKI*, by both brassinosteroid and abscisic acid (ABA) promoted LR formation by regulating endogenous auxin levels (Liu et al. 2003). Salicylic acid promotes LR initiation, emergence, and growth, possibly via cross talk with CK or auxin (Echevarria-Machado et al. 2007).

The last step of LR formation is the emergence of LRs along with the activation and maintenance of their meristems which is related to the ability of LRs to increase their own auxin synthesis allowing them to escape the apical dominance of the PR and requires PAT to create an auxin maximum at the tip of the LRP (Giehl et al. 2014). During LR development there is an AUX/IAA-dependent repositioning of auxin efflux carriers toward the tip of the newly formed LR, facilitating change in the direction of auxin flow and resulting in LR growth perpendicular to the PR (Sauer et al. 2006). However, sustained development and elongation of LRs still depend on shoot and/or PR-originated auxin, as suggested by slower elongation of LRs upon mutation of the rootward auxin transporter *MDR1* (Spalding et al. 2007). At later developmental stages, LRs resemble primary roots, and it is thought that most of the developmental processes that regulate primary root growth are also involved in LR growth (Giehl et al. 2014). The auxin-resistant mutants *axr1–3*, *axr2*, and *aux1–7*, whose roots continue to elongate on inhibitory concentrations of auxin, all showed induction of LR primordia formation on 1 μM IAA, but further growth was arrested at the emergence stage (Celenza et al. 1995), and mutants were defective in the spatial distribution of LRs as new LRP auxin maximum is required to regulate the activity of several transcription factors (Blilou et al. 2005). LR elongation by cell division from the meristem is controlled by auxin transport, as mutations in the auxin efflux transporter *MDR1* cause nascent LRs to arrest their growth (Wu et al. 2007). Despite the fact that CKs inhibit LR initiation, they have a positive effect on LR elongation in *Arabidopsis* and rice, possibly via stimulation of cell cycle gene expression in an auxin-independent process (Li et al. 2006). Abscisic acid can reversibly block meristem activation postemergence by inhibiting the cell cycle gene expression necessary for meristem activity, leading to LR growth arrest, defining a new auxin-independent checkpoint between LR emergence and meristem activation (Razem et al. 2006). Interestingly, ABA appears to have the opposite effect on LR emergence in legumes, stimulating LR formation in *Medicago* (Liang et al. 2007). LR elongation by cell division from the meristem is controlled by auxin transport, as mutations in the auxin efflux transporter *MDR1* cause nascent LRs to arrest their growth (Wu et al. 2007). Under stress conditions, ABA signaling activates SnRK2 protein kinases to inhibit LR growth after emergence from the primary root. However, even in the case of persistent stress, LR growth eventually recovers from inhibition via PYL8 interaction with the transcription factors MYB77 which target MBSI motif in the promoters of multiple auxin-responsive genes, suggesting a synergistic action of ABA and IAA through their interactions with PYL8 (Zhao et al. 2014). Besides the direct role in LR initiation, ABA mediates the balance between CK and auxin and thus also indirectly inhibits LR formation (Shkolnik-Inbar and Bar-Zvi 2010). Transcription factor ABSCISIC ACID INSENSITIVE 4 (ABI4) mediates ABA and CK inhibition of LR formation via reduction of polar auxin transport (Shkolnik-Inbar and Bar-Zvi 2010; Mu et al. 2017). In the presence of exogenous ABA, LR development is inhibited immediately after emergence of LR primordium indicating the role of ABA in suppression of auxin-responsive LR formation (De Smet et al. 2003). Further, ABA Insensitive3 (ABI3) interacts with ARF or Aux/IAA proteins through conserved B3 binding domain of ABI3 type of transcription factors, and mutation of *ABI3* gene inhibits LR initiation by attenuating auxin response (Brady et al. 2003). GA mutants deficient in either synthesis (Berova and Zlatev 2000) or signaling (Busov et al. 2006) have shown enhanced LR formation, and inhibitors of GA biosynthesis, such as paclobutrazol, can stimulate LR formation (Watson 2004; Grossi et al. 2005). Blockage of GA signaling via heterologous expression of DELLA-less versions of *GAI* and *RGL1* in *Populus* has also elicited an increase in root biomass, likely via LR proliferation (Busov et al. 2006). In contrast, GA-overproducing mutations and exogenous GA applications in aspen (*Populus tremula*) led to suppression of LR formation (Eriksson et al. 2000). Auxin level, including *PIN9* gene, is upregulated both in GA-deficient and GA-insensitive mutants with significantly more LR primordia in these mutants indicating GA–auxin interactions in LR development (Saini et al. 2013). GAs negatively affected LR formation by inhibiting LR primordium initiation by PAT modification (Gou et al. 2010). Auxin-mediated expression of *DWF4* is elevated in the *bin2/dwf12-1D* with concomitant increase in the number of LR generated after auxin treatment in comparison to wild-type or *bri1-5* (BR signaling mutant) background, which suggests that brassinosteroid-insensitive 2 (BIN2) kinase is an important component of auxin signaling, particularly involving signaling for LR development (Maharjan et al. 2011). Brassinosteroids (BRs) regulate LR initiation positively by increasing acropetal auxin transport, where BR perception-defective mutant *brassinosteroid insensitive1* (*bin1*) shows dramatically decreased numbers of LRs (Bao et al. 2004). Recently, the BR signaling negative regulator *BRASSINOSTEROID-INSENSITIVE2* has been reported to regulate LR organogenesis by phosphorylating ARF7 and ARF19 (Cho et al. 2014; Gupta et al. 2015). Enhanced ethylene synthesis or signaling, through *eto1-1* and *ctr1-1* mutations, or through the application of 1-aminocyclopropane-1-carboxylic acid (ACC), negatively impacts LR formation, and this is reversed by the treatment with ethylene antagonist, silver nitrate (Ivanchenko et al. 2008). In contrast, mutants, such as *etr1-3* and *ein2-5*, block ethylene responses that lead to the enhancement of LR formation and render it insensitive to ACC effect (Saini et al. 2013). Further, ACC treatment enhances the expression of auxin efflux carriers, such as *PIN3* and *PIN7*, resulting in elevated auxin transport and preventing the localized accumulation of auxin needed to drive LR formation (Lewis et al. 2011). Further, PA and auxin cross talk promotes LR formation by mitogenic activation and induction of G2/M transitions (Bouchereau et al. 1999). In *A. thaliana*, methyl jasmonate (MeJA) has been reported to repress LR formation in the *jdl/asa1-1* mutant (jasmonate-induced defective LR1) by blocking anthranilate synthase a1 (ASA1)-dependent auxin biosynthesis in the root tissues. JA treatment also reduces PIN1 and PIN2 protein levels in the plasma membrane leading to suboptimal auxin accumulation in the root basal meristem (Sun et al. 2009), while mutants with defective 9-LOX activity showed increased number of LR formation (Vellosillo et al. 2007) indicating the negative effect of JA on LR development (Saini et al. 2013). However, Raya-Gonzalez et al. (2012) suggested that JA promotes LR formation in *Arabidopsis* and rice; as in JA receptor mutant *coi1*, LR-positioning is disturbed, thus indicating that JA acts both through an auxin-dependent and an auxin-independent pathway (Verstraeten et al. 2014). SL has been suggested to negatively regulate LR formation as SL biosynthesis mutants in *A. thaliana*, such as *max3* and *max4*, and SL signaling mutant, *max2*, were shown to have more LR than the WT (Kapulnik et al. 2011a, b;

Ruyter-spira et al. 2011). Treatment of seedlings with GR24 repressed LR formation in the wild type and the strigolactone-synthesis mutants (*max3* and *max4*) but not in the strigolactone-response mutant (*max2*), suggesting that the negative effect of strigolactones on LR formation is MAX2 dependent (Kapulnik et al. 2011a, b; Ruyter-Spira et al. 2011). GR24-mediated reduction of LR formation is through reduction in auxin efflux activity by PIN1, PIN3, and PIN7 proteins and, as a result, alters the auxin optima necessary for LR formation (Ruyter-Spira et al. 2011). Root tip acts as a negative regulator of LR formation, where root tip removal removes a major site of auxin metabolism, and hence an increased auxin concentration in the remainder root is recorded. *PINOID* encodes a protein kinase required for correct auxin response (Benjamins et al. 2001). Since basipetal auxin transport appears to be required for root elongation and LR initiation, an increase in auxin efflux would result in reduced levels in the root tip and LR initiation zone. Low-level inhibition of auxin efflux with NPA may rescue the *35S::PID* phenotype by restoring auxin levels in these regions to more normal levels (Casson and Lindsey 2003) (Fig. 1.4).

1.13 Role of Hormones in Root Hair Development

In natural and agricultural soils, in line to resourcefully acquire nutrients, plants trigger foraging responses comprising morphological changes, such as root hair (RH) formation (Gruber et al. 2013; Giehl and von Wirén 2014). As studied in *Arabidopsis* root epidermis, epidermal cell layer root consists of root hair producing cells (trichoblasts) and hairless cells (atrichoblasts), and the cells implement these distinct fates in a position-dependent manner, where epidermal cells that overlay the junction/anticlinal (radial) wall separating two cortical cell files adopt a RH cell fate, whereas the epidermal cells that lie outside periclinal cortical cell walls differentiate into mature hairless cells (Berger et al. 1998; Rahman et al. 2002). When the cells are still located within the meristematic region of the root, the differentiating RH cells possess relatively dense, intensely staining cytoplasm as compared to the cell destined to become hairless. At later stages of epidermis development, the two differentiating cell types differ in their rate of vacuolation, in the extent of elongation, and ultimately in the presence or absence of root hairs. Thus, the cells of the root epidermis begin to acquire cell-type-specific characteristics in a position-dependent manner at an early stage of epidermis development, and the emergence of RHs is a later event in cell differentiation (Masucci and Schiefelbein 1996). Important stages involved in the formation of a RH are (1) specification of trichoblast cell fate, (2) root hair initiation, (3) bulge formation, and (4) tip growth (Schiefelbein 2000). Site of hair initiation is dependent on functional auxin influx activity (Grebe et al. 2002), as RH initiation was shifted apically in *aux1* (*auxin influx*) mutants, and there was also a much higher proportion of double hair formation than the WT. In contrast, roots of mutants of *EIR1/AtPIN2* carriers (*eir1*) had no defects in the polarity of initiation (detailed in Fig. 1.5). Treatment of Brefeldin A (BFA) [known to inhibit auxin efflux carrier activity (Delbarre et al. 1998) and the

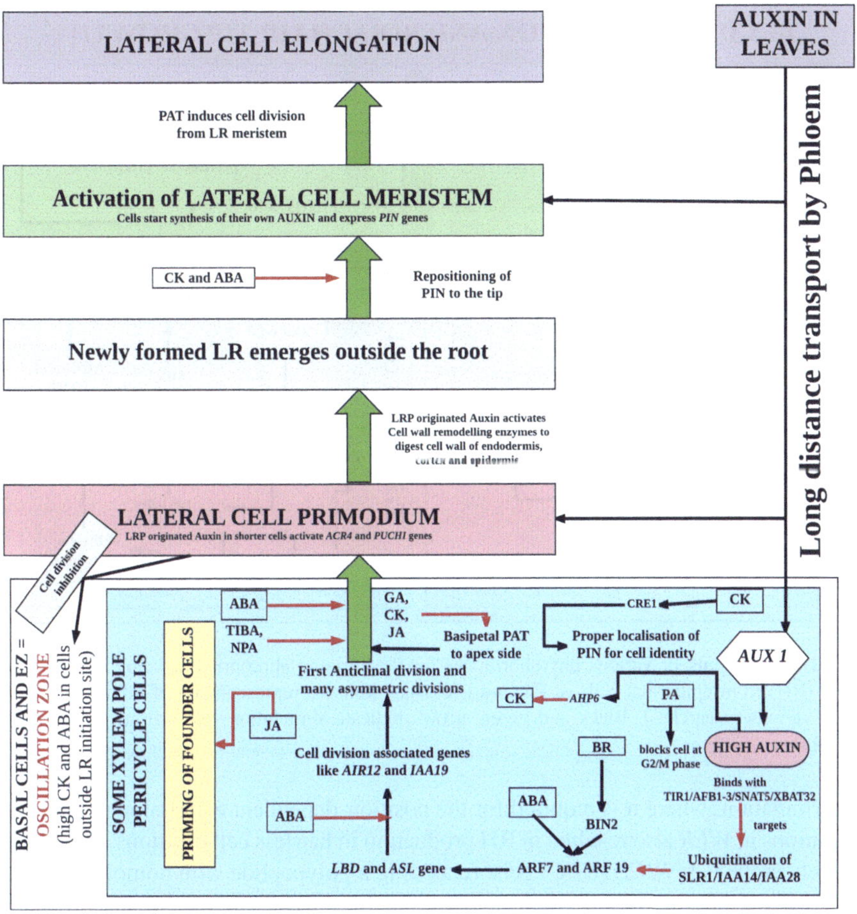

Fig. 1.4 Cross talk of various phytohormones for determining appropriate development of lateral root (LR) from pericycle cells of PR (Diagrammatic representation of inference drawn from various researches). Black and green arrows indicate stimulation, and red arrows indicate inhibition

polar localization of PIN1 (Steinmann et al. 1999)] results in a concentration-dependent apical shift in hair initiation, as well as double hair formation, indicating BFA disrupts the establishment of hair cell polarity by disrupting localization of AUX1 to protophloem and epidermal cells from otherwise usual localization in plasma membrane (Swarup et al. 2001). *Transparent Testa Glabra1* (*ttg1*) gene encoding a putative Wd40 repeat protein and *GLABRA2* (*GL2*) encoding a homeodomain-containing protein act as a negative regulator of RH development in atrichoblast cell files. Interestingly, expression of *GL2* is reduced in a *ttg1* background indicating that *TTG1* is a positive regulator of *GL2* expression (Hung et al. 1998). Also *WEREWOLF* (*WER*) gene, encoding a polypeptide with a MYB-like DNA-binding domain, plays a role in an early stage of epidermal cell fate

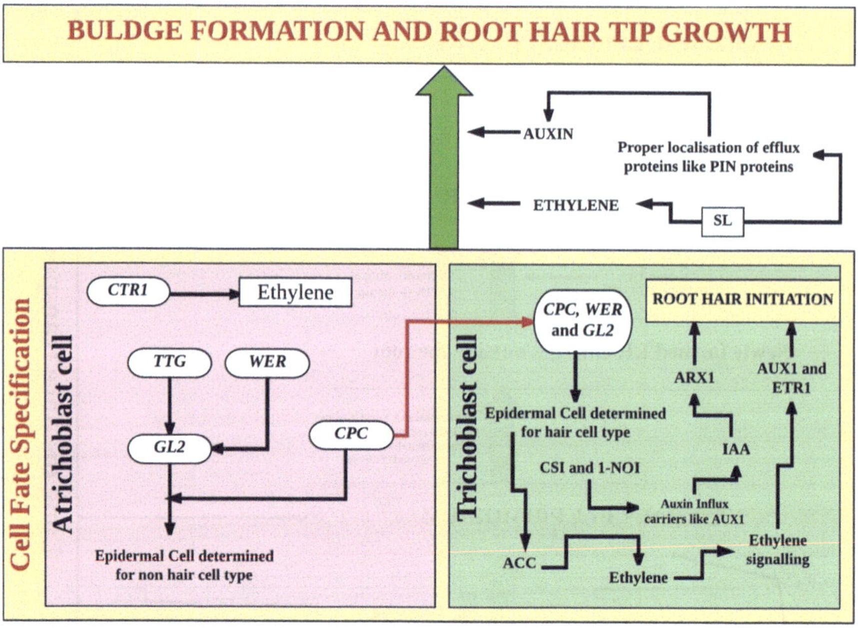

Fig. 1.5 Cross talk of various phytohormones for determining appropriate development of root hairs (RH) from epidermal cells of PR and LR (Diagrammatic representation of inference drawn from various researches). Black and green arrows indicate stimulation, and red arrows indicate inhibition

determination, where it is required for the position-dependent expression of *GL2*, as mutations in *WER* also resulted in RH production in hairless cell positions. However, defects in the *CAPRICE* (*CPC*) gene (encoding a polypeptide with homology to the MYB-like DNA-binding domain but lacking the typical transcriptional activation domain) result in roots with fewer than usual root hairs, indicating that it is a positive regulator of hair formation. Double mutant analysis revealed that the *gl2* mutation was epistatic to *cpc*, while the roots of *ttg1 cpc* and *cpc wer* double mutant plants had an intermediate phenotype indicating that the genes possibly function in independent but opposing pathways (Lee and Schiefelbein 1999). Therefore, according to the model presented by Lee and Schiefelbe, higher levels of *WER* in non-hair cells induce expression of *GL2*, therefore promoting non-hair fate, and also *CPC*, which acts by inhibiting expression of *CPC*, *WER*, and *GL2* in neighboring cells such that they adopt a hair cell fate (Casson and Lindsey 2003).

As a result of the position-dependent action of the TTG/GL2 pathway, the genes that promote root hair formation (including *RHD6* and the ethylene and auxin pathway genes) are proposed to act only in the developing epidermal cells located outside the anticlinal cortical cell walls. The *RHD6* gene plays a central role in root hair initiation, as *rhd6* mutants exhibit a dramatic defect in RH initiation site selection and reduction in RH formation (Masucci and Schiefelbein 1994). However, *rhd6* defect can be suppressed by inclusion of an auxin (indole-3-acetic acid

(IAA)) or the ethylene precursor l-aminocyclopropane-1-carboxylic acid (ACC) in the growth media; thus *RHD6* gene is proposed to promote RH formation by acting through either the ethylene and auxin pathways or a separate pathway (Masucci and Schiefelbein 1996). Mutations in *ethylene overproducer* (*eto*) gene encoding regulators of ethylene biosynthesis like *eto*, *eto2*, *eto3*, and *eto4* led to significantly more RH than the WT, indicating that epidermal cells in the trichoblast and atrichoblast positions have different sensitivities to ethylene (Cao et al. 1999). Plants grown in the presence of ethylene biosynthetic inhibitors, i.e., either aminoethoxyvinylglycine (AVG, an ethylene synthesis inhibitor) or Ag^+ (an inhibitor of ethylene action), display a reduction in the frequency of root hairs which could be countered by including ACC in the media containing AVG (Masucci and Schiefelbein 1994). At very high concentrations of ACC, all epidermal cells, regardless of position, develop root hairs, suggesting that ethylene is also a positive regulator of root hair development (Tanimoto et al. 1995). Thus, at least two pathways are likely to exist to influence root epidermis cell-type differentiation in *Arabidopsis*. One pathway involves the *TTG* and *GL2* gene products (putative negative regulators of root hair cell differentiation), and another pathway involves ethylene and/or auxin and their associated signal transduction cascades (putative positive regulators of root hair cell differentiation). Alternatively, the ethylene/auxin pathway could negatively regulate the *TTG/GL2* pathway in differentiating root hair cells, or *TTG/GL2* pathway controls the patterning of epidermal cell types by negatively regulating the root hair-promoting activities of the ethylene/auxin pathway in a cell position-dependent manner (Masucci and Schiefelbein 1996). The *CTR7* gene encodes a Raf-like protein kinase to negatively regulate the ethylene signal transduction pathway with recessive *ctr7* mutations leading to RH formation in even hairless epidermal cells (Dolan et al. 1994). The dominant mutations of the *axr2* gene confer insensitivity to high concentrations of auxin, ethylene, and ABA, thereby reducing RH formation, and neither ACC nor IAA treatments could induce root hairs in the *axr2* single mutant or the *axr2 rhd6* double mutants (Wilson et al. 1990). Thus, auxin and ethylene hormone pathways converge at, or upstream of, the AXR2 product (Masucci and Schiefelbein 1996). *AUX1* and *ETRl* gene products are each able to contribute in RH formation with each efficient in bypassing defects of each other, but ACC being unable to suppress the *auxl etrl* RH phenotype indicate that these gene products are acting in ethylene signal transduction. However, ability of IAA via ARX1, to suppress the *aux1 etr1* phenotype, indicates that IAA acts through an AUX1-independent pathway (Masucci and Schiefelbein 1996).

After cell specification and initiation, the root hair starts to grow through the process of tip growth. Mutants with altered responses to ethylene and auxin also show defects in RH length, suggesting that these two hormones play indispensable roles regulating root hair morphogenesis (Pitts et al. 1998). Chromosaponin I (CSI) inhibits auxin influx carrier *AUX1* activity and makes the WT *Arabidopsis* roots resistant to ethylene and slows down the root gravitropic response. However, in the auxin influx mutant *aux1-7*, CSI conversely stimulated the uptake of auxin, restored ethylene response, and partially restored the gravitropic response. Application of low concentrations of 1-naphthaleneacetic acid (NAA) restored the ethylene response in *aux1-7*, suggesting that the intracellular level of auxin plays an important role in regulating the ethylene response in *Arabidopsis* root growth (Rahman et al. 2001). Aryloxyalkylcarboxylic acids including 1-naphthoxyacetic acid (1-NOA) in *Arabidopsis* seedlings are useful auxin influx inhibitors as 1-NOA phenocopied the agravitropic *aux1* root phenotype and effect of CSI in disrupting the root gravitropism (Parry et al. 2001; Rahman et al. 2002). Exogenous supplementation of various synthetic SL analogs induced RH elongation in WT *Arabidopsis*, the SL-deficient mutants (*max3* and *max4*) but not in the SL-response mutant *max2*, suggesting that the effect of SLs on RH elongation is mediated by MAX2 (Cohen et al. 2013). Furthermore, response to auxin and ethylene signaling is required, at least in part, for the positive effect of SL on RH elongation; however MAX2-dependent SL signaling is not necessary for the RH elongation induced by auxin (Kapulnik et al. 2011b). The sufficiency of *MAX2* expression under SCR (expressed mainly in the root endodermis and quiescence center) for GR24 sensitivity suggests that SLs act noncell autonomously at short range (Kapulnik and Koltai 2014). Furthermore, in *Arabidopsis* where GR24 treatment induced RH elongation, PIN2 polarization was increased in the plasma membrane of the root epidermis in the wild type by regulating the architecture and dynamics of actin filaments and PIN endocytosis for PIN2 polarization (Pandya-Kumar et al. 2014; Kapulnik and Koltai 2014). Though, SL signaling is not necessary for the RH elongation induced by auxin, but auxin signaling enhances the RH elongation response to SLs: the SL-insensitive mutant *max2* was responsive to auxin, whereas under low GR24 concentrations, the auxin receptor mutant *tir1-1* (Dharmasiri et al. 2005) was less responsive to SLs than the WT (Kapulnik et al. 2011b; Koltai 2011). Ethylene, rather than auxin, is directly involved in the RH response to SLs, as blockage of ethylene biosynthesis by AVG even abolishes the SL effect on RH elongation, whereas root treatment with GR24 elevates transcription of At-ACS2 (Kapulnik et al. 2011b; Koltai 2011) (Fig. 1.5).

1.14 Conclusion and Perspective

Thus, in this chapter, we attempted to construct the delicately regulated spatiotemporal cross talk between phytohormones deciphering the apt mechanisms engaged in the development of primary and lateral root architecture along with the root hair branching pattern. The spatiotemporal variations in the synergistic and/or antagonistic interactions among the same hormones reflect that a universal regulatory mechanism intersects with diverse downstream signaling components to achieve the appropriate developmental responses at different stages. Although in recent years, significant understanding about the physiological and molecular basis of phytohormonal interactions has been gained, yet knowledge attained from only genetically tractable crop plants is certainly not sufficient to fully understand the mechanisms of hormonal cross talk during root organogenesis. In the future the major challenge will be to elucidate how the integration of phytohormonal networks regulates root morphogenesis in other agronomically significant species and how it can be utilized to engineer crop plants that can exist in a range of potentially challenged ecological niches.

References

Abel S, Nguyen MD, Chow W, Theologis A (1995) *ACS4*, a primary indoleacetic acid-responsive gene encoding 1-aminocyclopropane- 1-carboxylate synthase in *Arabidopsis thaliana*. J Biol Chem 270:19093–19099

Achard P, Gusti A, Cheminant S, Alioua M, Dhondt S, Coppens F, Beemster GT, Genschik P (2009) Gibberellin signaling controls cell proliferation rate in *Arabidopsis*. Curr Biol 19:1188–1193

Aida M, Beis D, Heidstra R, Willemsen V, Blilou I, Galinha C, Nussaume L, Noh YS, Amasino R, Scheres B (2004) The *PLETHORA* genes mediate patterning of the *Arabidopsis* root stem cell niche. Cell 119:109–120

Alonso JM, Stepanova AN, Solano R, Wisman E, Ferrari S, Ausubel FM, Ecker JR (2003) Five components of the ethylene-response pathway identified in a screen for weak ethylene-insensitive mutants in *Arabidopsis*. Proc Natl Acad Sci 100:2992–2997

Antoni R, Gonzalez-Guzman M, Rodriguez L, Peirats-Llobet M, Pizzio GA et al (2013) *PYRABACTIN RESISTANCE1-LIKE8* plays an important role for the regulation of abscisic acid signaling in root. Plant Physiol 161:931–941

Band LR, Wells DM, Larrieu A, Sun J, Middleton AM, French AP et al (2012) Root gravitropism is regulated by a transient lateral auxin gradient controlled by a tipping-point mechanism. PNAS 109(12):4668–4673

Bao F, Shen J, Brady SR, Muday GK, Asami T, Yang Z (2004) Brassinosteroids interact with auxin to promote lateral root development in *Arabidopsis*. Plant Physiol 134:1624–1631

Beeckman T, Burssens S, Inze D (2001) The peri-cell-cycle in *Arabidopsis*. J Exp Bot 52:403–411

Beemster GT, Baskin TI (1998) Analysis of cell division and elongation underlying the developmental acceleration of root growth in *Arabidopsis thaliana*. Plant Physiol 116:1515–1526

Belin C, Megies C, Hauserova E, Lopez-Molina L (2009) Abscisic acid represses growth of the *Arabidopsis* embryonic axis after germination by enhancing auxin signaling. Plant Cell 21:2253–2268

Benfey PN, Linstead PJ, Roberts K, Schiefelbein JW, Hauser M-T, Aeschbacher RA (1993) Root development in *Arabidopsis*: four mutants with dramatically altered root morphogenesis. Development 119:57–70

Benfey PN, Bennett M, Schiefelbein J (2010) Getting to the root of plant biology: impact of the *Arabidopsis*. Plant J 61:992–1000

Benjamins R, Scheres B (2008) Auxin: the looping star in plant development. Annu Rev Plant Biol 59:443–465

Benjamins R, Quint AB, Weijers D, Hooykaas P, Offringa R (2001) The PINOID protein kinase regulates organ development in *Arabidopsis* by enhancing polar auxin transport. Development 128:4057–4067

Benkova E, Michniewicz M, Sauer M, Teichmann T, Seifertova D, Jurgens G, Friml J (2003) Local, efflux-dependent auxin gradients as a common module for plant organ formation. Cell 115:591–602

Berger F, Hung C-Y, Dolan L, Schiefelbein J (1998) Control of cell division in the root epidermis of *Arabidopsis thaliana*. Dev Biol 194:235–245

Berova M, Zlatev Z (2000) Physiological response and yield of paclobutrazol treated tomato plants (*Lycopersicon esculentum* Mill.). Plant Growth Regul 30:117–123

Beveridge CA, Kyozuka J (2010) New genes in the strigolactone-related shoot branching pathway. Curr Opin Plant Biol 13:34–39

Bhalerao RP, Eklof J, Ljung K, Marchant A, Bennett M, Sandberg G (2002) Shoot-derived auxin is essential for early lateral root emergence in *Arabidopsis* seedlings. Plant J 29:325–332

Bianco MD, Kepinski S (2011) Context, specificity, and self-organization in auxin response. Cold Spring Harb Perspect Biol 3:a001578

Bianco MD, Giustini L, Sabatini S (2013) Spatiotemporal changes in the role of cytokinin during root development. New Phytol 199:324–338

Bielach A, Duclercq J, Marhavy P, Benkova E (2012) Genetic approach towards the identification of auxin–cytokinin crosstalk components involved in root development. Philos Trans R Soc B 367:1469–1478

Birnbaum K, Shasha DE, Wang JY, Jung JW, Lambert GM, Galbraith DW, Philip NB (2003) A gene expression map of the *Arabidopsis* root. Science 302:1956–1960

Bishopp A, Help H, El-Showk S, Weijers D, Scheres B, Friml J, Benkova E, Mahonen AP, Helariutta Y (2011a) A mutually inhibitory interaction between auxin and cytokinin specifies vascular pattern in roots. Curr Biol 21:917–926

Bishopp A, Lehesranta S, Vaten V, Help H, El-Showk E, Scheres B, Helariutta K, Mahonen AP, Sakakibara H, Helariutta Y (2011b) Phloem-transported cytokinin regulates polar auxin transport and maintains vascular pattern in the root meristem. Curr Biol 21:927–932

Blilou I, Xu J, Wildwater M, Willemsen V, Paponov I, Friml J, Heidstra R, Aida M, Palme K, Scheres B (2005) The *PIN* auxin efflux facilitator network controls growth and patterning in *Arabidopsis* roots. Nature 433:39–44

Bouchereau A, Aziz A, Larher F, Martin-Tanguy J (1999) Polyamines and environmental challenges: recent development. Plant Sci 140:103–125

Brady SM, Sarkar SF, Bonetta D, McCourt P (2003) The *ABSCISIC ACID INSENSITIVE 3* (*ABI3*) gene is modulated by farnesylation and is involved in auxin signalling and lateral root development in *Arabidopsis*. Plant J 34:67–75

Brewer PB, Koltai H, Beveridge CA (2013) Diverse roles of strigolactones in plant development. Mol Plant 6(1):18–28

Brugiere N, Jiao S, Hantke S, Zinselmeier C, Roessler JA, Niu X, Jones RJ, Habben JE (2003) Cytokinin oxidase gene expression in maize is localized to the vasculature, and is induced by cytokinins, abscisic acid, and abiotic stress. Plant Physiol 132:1228–1240

Buer CS, Sukumar P, Muday GK (2006) Ethylene modulates flavonoid accumulation and gravitropic responses in roots of *Arabidopsis thaliana*. Plant Physiol 140:1384–1396

Busov V, Meilan R, Pearce DW, Rood SB, Ma C, Tschaplinski TJ, Strauss SH (2006) Transgenic modification of *gai* or *rgl1* causes dwarfing and alters gibberellins, root growth, and metabolite profiles in *Populus*. Planta 224:288–299

Cao XF, Linstead P, Berger F, Kieber J, Dolan L (1999) Differential sensitivity of epidermal cells is involved in the establishment of cell pattern in the *Arabidopsis* root. Physiol Plant 106:311–317

Cardoso C, Zhang Y, Jamil M, Hepworth J, Charnikhova T, Dimkpa SO et al (2014) Natural variation of rice strigolactone biosynthesis is associated with the deletion of two *MAX1* orthologs. Proc Natl Acad Sci 111:2379–2384

Cary AJ, Liu W, Howell SH (1995) Cytokinin action is coupled to ethylene in its effects on the inhibition of root and hypocotyl elongation in *Arabidopsis thaliana* seedlings. Plant Physiol 107:1075–1082

Casimiro I, Marchant A, Bhalerao RP, Beeckman T, Dhooge S, Swarup R, Graham N, Inzé D, Sandberg G, Casero PJ, Bennett M (2001) Auxin transport promotes *Arabidopsis* lateral root initiation. Plant Cell 13:843–852

Casson SA, Lindsey K (2003) Genes and signalling in root development. New Phytol 158:11–38

Casson SA, Chilley PM, Topping JF, Evans IM, Souter MA, Lindsey K (2002) The *POLARIS* gene of *Arabidopsis* encodes a predicted peptide required for correct root growth and leaf vascular patterning. Plant Cell 14:1705–1721

Celenza JL, Grisafi PL, Fink GR (1995) A pathway for lateral root formation in *Arabidopsis thaliana*. Genes Dev 9:2131–2142

Chapman EJ, Estelle M (2009) Cytokinin and auxin intersection in root meristems. Genome Biol 10:210

Chen H, Xiong L (2009) The short-rooted vitamin B6-deficient mutant *pdx1* has impaired local auxin biosynthesis. Planta 229:1303–1310

Cheng Y, Dai X, Zhao Y (2006) Auxin biosynthesis by the *YUCCA* flavin monooxygenases controls the formation of floral organs and vascular tissues in *Arabidopsis*. Genes Dev 20:1790–1799

Chilley PM, Casson SA, Tarkowski P, Hawkins N, Wang KL, Hussey PJ (2006) The POLARIS peptide of *Arabidopsis* regulates auxin transport and root growth via effects on ethylene signaling. Plant Cell 18:3058–3072

Cho H, Ryu H, Rho S, Hill K, Smith S, Audenaert D, Park J, Han S, Beeckman T, Bennett MJ, Hwang D, De Smet I, Hwang I (2014) A secreted peptide acts on *BIN2*-mediated phosphorylation of ARFs to potentiate auxin response during lateral root development. Nat Cell Biol 16:66–76

Choi Y, Lee Y, Kim SY, Lee Y, Hwang J (2012) *Arabidopsis* ROP interactive CRIB motif-containing protein 1 (RIC1) positively regulates auxin signalling and negatively regulates abscisic acid (ABA) signalling during root development. Plant Cell Environ 36(5):945–955

Chung Y, Maharjan PM, Lee O, Fujioka S, Jang S, Kim B, Takatsuto S, Tsujimoto M, Kim H, Cho S, Park T, Cho H, Hwang I, Choe S (2011) Auxin stimulates *DWARF4* expression and brassinosteroid biosynthesis in *Arabidopsis*. Plant J 66:564–578

Clark NM, de Luis Balaguer MA, Sozzani R (2014) Experimental data and computational modeling link auxin gradient and development in the *Arabidopsis* root. Front Plant Sci 5:328

Cnops G, Wang X, Linstead P, Van Montagu M, Van Lijsebettens M, Dolan L (2000) *TORNADO1* and *TORNADO2* are required for the specification of radial and circumferential pattern in the *Arabidopsis* root. Development 127:3385–3394

Cohen M, Prandi C, Occhiato EG, Tabasso S, Wininger S, Resnick N, Steinberger Y, Koltai H, Kapulnik Y (2013) Structure-function relations of strigolactone analogs: activity as plant hormones and plant interactions. Mol Plant 6:141–152

Couee I, Hummel I, Sulmon C, Gouesbet G, Amrani AE (2004) Involvement of polyamines in root development. Plant Cell Tiss Org Cult 76:1–10

Cui H, Levesque MP, Vernoux T, Jung JW, Paquette AJ, Gallagher KL, Wang JY, Blilou I, Scheres B, Benfey PN (2007) An evolutionarily conserved mechanism delimiting SHR movement defines a single layer of endodermis in plants. Science 316:421–425

Dai Y, Wang H, Li B, Huang J, Liu X, Zhou Y, Mou Z, Li J (2006) Increased expression of *MAP KINASE KINASE7* causes deficiency in polar auxin transport and leads to plant architectural abnormality in *Arabidopsis*. Plant Cell 18:308–320

De Rybel B, Vassileva V, Parizot B, Demeulenaere M, Grunewald W, Audenaert D et al (2010) A novel *aux/IAA28* signaling cascade activates GATA23-dependent specification of lateral root founder cell identity. Curr Biol 20:1697–1706

De Smet I (2012) Lateral root initiation: one step at a time. New Phytol 193:867–873

De Smet I, Signora L, Beeckman T, Inze D, Foyer CH, Zhang H (2003) An abscisic acid sensitive checkpoint in lateral root development of *Arabidopsis*. Plant J 33:543–555

De Smet I, Vanneste S, Inze D, Beeckman T (2006) Lateral root initiation or the birth of a new meristem. Plant Mol Biol 60:871–887

De Smet I, Tetsumura T, De Rybel B, Frey N, Laplaze L, Casimiro I et al (2007) Auxin-dependent regulation of lateral root positioning in the basal meristem of *Arabidopsis*. Development 134:681–690

De Smet I, Vassileva V, De Rybel B, Levesque MP, Grunewald W, Van Damme D et al (2008) Receptor-like kinase *ACR4* restricts formative cell divisions in the *Arabidopsis* root. Science 322:594–597

De Smet I, White PJ, Bengough AG, Dupuy L, Parizot B, Casimiro I et al (2012) Analyzing lateral root development: how to move forward. Plant Cell 24:15–20

Del Bianco M, Kepinski S (2011) Context, specificity and self-organization in auxin response. Cold Spring Harb Perspect Biol 3:a001578

Del Bianco M, Giustini L, Sabatini S (2013) Spatiotemporal changes in the role of cytokinin during root development. New Phytol 199:324–338

Delbarre A, Muller P, Guern J (1998) Short-lived and phosphorylated proteins contribute to carrier-mediated efflux, but not to influx, of auxin in suspension-cultured tobacco cells. Plant Physiol 116:833–844

Della Rovere F, Fattorini L, D'Angeli D, Veloccia A, Falasca G, Altamura MM (2013) Auxin and cytokinin control formation of the quiescent centre in the adventitious root apex of *Arabidopsis*. Ann Bot 112:1395–1407

Dello Ioio R, Linhares FS, Scacchi E, Casamitjana-Martinez E, Heidstra R, Costantino P, Sabatini S (2007) Cytokinins determine *Arabidopsis* root meristem size by controlling cell differentiation. Curr Biol 17:678–682

Dello Ioio R, Nakamura K, Moubayidin L, Perilli S, Taniguchi M, Morita MT, Aoyama T, Costantino P, Sabatini S (2008) A genetic framework for the control of cell division and differentiation in the root meristem. *Science* 322:1380–1384

Depuydt S, Hardtke CS (2011) Hormone signalling crosstalk in plant growth regulation. Curr Biol 21:365–373

Dharmasiri N, Dharmasiri S, Estelle M (2005) The F-box protein *TIR1* is an auxin receptor. Nature 435:441–445

Dhonukshe P, Grigoriev I, Fischer R, Tominaga M, Robinson DG, Hasek J et al (2008) Auxin transport inhibitors impair vesicle motility and actin cytoskeleton dynamics in diverse eukaryotes. PNAS 105:4489–4494

Di Laurenzio L, Wysocka-Diller J, Malamy J, Pysh L, Helariutta Y, Freshour G, Hahn MG, Feldmann KA, Benfey PN (1996) The *SCARECROW* gene regulates an asymmetric cell division that is essential for generating the radial organization of the *Arabidopsis* root. Cell 86:423–433

Doerner P, Jorgensen J-E, You R, Steppuhn J, Lamb C (1996) Control of root growth and development by cyclin expression. Nature 380:520–523

Dolan L, Janmaat K, Willemsen V, Linstead P, Poethig S, Roberts K, Scheres B (1993) Cellular organization of the *Arabidopsis* thaliana root. Development 119:71–84

Dolan L, Duckett C, Grienon C, Linstead P, Schneider K, Lawson E, Dean C, Poethig S, Roberts K (1994) Clonal relations and patterning in the root epidermis of *Arabidopsis*. Development 120:2465–2474

Dong L, Wang L, Zhang Y, Zhang Y, Deng X, Xue Y (2006) An auxin-inducible F-box protein CEGENDUO negatively regulates auxin-mediated lateral root formation in *Arabidopsis*. Plant Mol Biol 60:599–615

Echevarria-Machado I, MRM E-G, Larque-Saavedra A (2007) Responses of transformed *Catharanthus roseus* roots to femtomolar concentrations of salicylic acid. Plant Physiol Biochem 45:501–507

Eriksson ME, Israelsson M, Olsson O, Moritz T (2000) Increased gibberellin biosynthesis in transgenic trees promotes growth, biomass production and xylem fiber length. Nat Biotechnol 18:784–788

Esau K (1977) Anatomy of seed plants. John Wiley & Sons, New York

Fabregas N, Li N, Boeren S, Nash TE, Goshe MB, Clouse SD, de Vries S, Cano-Delgado AI (2013) The *BRASSINOSTEROID INSENSITIVE1-LIKE3* signalosome complex regulates *Arabidopsis* root development. Plant Cell 25:3377–3388

Frigerio M, Alabadi D, Perez-Gomez J, Gracia-Cacel L, Phillips AL, Hedden P, Blazquez MA (2006) Transcriptional regulation of gibberellins metabolism genes by auxin signaling in *Arabidopsis*. Plant Physiol 142:553–563

Friml J, Benkova E, Blilou I, Wisniewska J, Hamann T, Ljung K, Woody S, Sandberg G, Scheres B, Jurgens G, Palme K (2002) *AtPIN4* mediates sink-driven auxin gradients and root patterning in *Arabidopsis*. Cell 108:661–673

Friml J, Vieten A, Sauer M, Weijers D, Schwarz H, Hamann T, Offringa R, Jurgens G (2003) Efflux-dependent auxin gradients establish the apical-basal axis of *Arabidopsis*. Nature 426:147–153

Fu X, Harberd NP (2003) Auxin promotes *Arabidopsis* root growth by modulating gibberellins response. Nature 421:740–743

Fukaki H, Tameda S, Masuda H, Tasaka M (2002) Lateral root formation is blocked by a gain-of-function mutation in the *SOLITARY-ROOT/IAA14* gene in *Arabidopsis*. Plant J 29:153–168

Galinha C, Hofhuis H, Luijten M, Willemsen V, Blilou I, Heidstra R, Scheres B (2007) *PLETHORA* proteins as dose dependent master regulators of *Arabidopsis* root development. Nature 449:1053–1057

Garay-Arroyo A, Sanchez MD, Garcia-PonceB AE, Alvarezuylla ER (2012) Hormone symphony during root growth and development. Dev Dyn 241:1867–1885

Giehl RFH, von Wirén N (2014) Root nutrient foraging. Plant Physiol 166:509–517

Giehl RF, Gruber BD, von Wiren N (2014) It's time to make changes: modulation of root system architecture by nutrient signals. J Exp Bot 65:769–778

Gonzalez-Carranza ZH, Elliott KA, Roberts JA (2007) Expression of polygalacturonases and evidence to support their role during cell separation processes in *Arabidopsis thaliana*. J Exp Bot 58:3719–3730

Gou J, Strauss SH, Tsai CJ, Fang K, Chen Y, Jiang X, Busova VB (2010) Gibberellins regulate lateral root formation in Populus through interactions with auxin and other hormones. Plant Cell 22:623–639

Grebe M, Friml J, Swarup R, Ljung K, Sandberg G, Terlou M, Palme K, Bennett MJ, Scheres B (2002) Cell polarity signalling in *Arabidopsis* involves a BFA-sensitive auxin influx pathway. Curr Biol 12:329–334

Grieneisen VA, Xu J, Maree AFM, Hogeweg P, Scheres B (2007) Auxin transport is sufficient to generate a maximum and gradient guiding root growth. Nature 449:1008–1013

Grossi JA, Moraes PJ, Tinoco SA, Barbosa JG, Finger FL, Cecon PR (2005) Effects of paclobutrazol on growth and fruiting characteristics of *Pitanga* ornamental pepper. Acta Hortic (683):333–336

Gruber BD, Giehl RFH, Friedel S, von Wirén N (2013) Plasticity of the *Arabidopsis* root system under nutrient deficiencies. Plant Physiol 163:161–179

Guilfoyle TJ, Hagen G (2007) Auxin response factors. Curr Opin Plant Biol 10:453–460

Gupta S, Rashotte AM (2012) Down-stream components of cytokinin signaling and the role of cytokinin throughout the plant. Plant Cell Rep 31:801–812

Gupta A, Singh M, Laxmi A (2015) Interaction between glucose and brassinosteroid during the regulation of lateral root development in *Arabidopsis*. Plant Physiol 168:307–320

Hacham Y, Holland N, Butterfield C, Ubeda-Tomas S, Bennett MJ, Chory J, Savaldi-Goldstein S (2011) Brassinosteroid perception in the epidermis controls root meristem size. Development 138:839–848

Hamann T, Mayer U, Jurgens G (1999) The auxin-insensitive bodenlos mutation affects primary root formation and apical-basal patterning in the *Arabidopsis* embryo. Development 126:1387–1395

Hamann T, Benkova E, Baurle I, Kientz M, Jurgens G (2002) The Arabidopsis *BODENLOS* gene encodes an auxin response protein inhibiting *MONOPTEROS*-mediated embryo patterning. Genes Dev 16:1610–1615

Heidstra R, Sabatini S (2014) Plant and animal stem cells: similar yet different. Nat Rev Mol Cell Biol 15:301–312

Helariutta Y, Fukaki H, Wysocka-Diller J, Nakajima K, Jung J, Sena G, Hauser MT, Benfey PN (2000) The *SHORT-ROOT* gene controls radial patterning of the *Arabidopsis* root through radial signaling. Cell 101:555–567

Himanen K, Boucheron E, Vanneste S, de Almeida Engler J, Inze D, Beeckman T (2002) Auxin-mediated cell cycle activation during early lateral root initiation. Plant Cell 14:2339–2351
Hirota A, Kato T, Fukaki H, Aida M, Tasaka M (2007) The auxin-regulated *AP2/EREBP* gene *PUCHI* is required for morphogenesis in the early lateral root primordium of *Arabidopsis*. Plant Cell 19:2156–2168
Hochholdinger F, Zimmermann R (2008) Conserved and diverse mechanisms in root development. Curr Opin Plant Biol 11:70–74
Hummel I, Couee I, El Amrani A, Martin-Tanguy J, Hennion F (2002) Involvement of polyamines in root development at low temperature in the subantarctic cruciferous species *Pringlea antiscorbutica*. J Exp Bot 53:1463–1473
Hung C-Y, Lin Y, Zhang M, Pollock S, Marks MD, Schiefelbein J (1998) A common position-dependent mechanism controls cell-type patterning and *GLABRA2* regulation in the root and hypocotyl epidermis of *Arabidopsis*. Plant Physiol 117:73–84
Ikeda Y, Men S, Fischer U, Stepanova AN, Alonso JM, Ljung K, Grebe M (2009) Local auxin biosynthesis modulates gradient directed planar polarity in *Arabidopsis*. Nat Cell Biol 11:731–738
Ishida T, Kurata T, Okada K, Wada T (2008) A genetic regulatory network in the development of trichomes and root hairs. Annu Rev Plant Biol 59:365–386
Ivanchenko MG, Coffeen WC, Lomax TL, Dubrovsky JG (2006) Mutations in the Diageotropica (*Dgt*) gene uncouple patterned cell division during lateral root initiation from proliferative cell division in the pericycle. Plant J 46:436–447
Ivanchenko MG, Muday GK, Dubrovsky JG (2008) Ethylene–auxin interactions regulate lateral root initiation and emergence in *Arabidopsis thaliana*. Plant J 55:335–347
Jaillais Y, Santambrogio M, Rozier F, Fobis-Loisy I, Miege C, Gaude T (2007) The retromer protein VPS29 links cell polarity and organ initiation in plants. Cell 130:1057–1070
Jang SJ, Choi YJ, Park KY (2002) Effects of polyamines on shoot and root development in *Arabidopsis* seedlings and carnation cultures. Plant Biol J 45:230–236
Jung JKH, McCouch S (2013) Getting to the roots of it: genetic and hormonal control of root architecture. Front Plant Sci 4:186
Kapulnik Y, Koltai H (2014) Strigolactone involvement in root development, response to abiotic stress, and interactions with the biotic soil environment. Plant Physiol 166:560–569
Kapulnik Y, Delaux PM, Resnick N, Mayzlish-Gati E, Wininger S, Bhattacharya C et al (2011a) Strigolactones affect lateral root formation and root-hair elongation in *Arabidopsis*. Planta 233:209–216
Kapulnik Y, Resnick N, Mayzlish-Gati E, Kaplan Y, Wininger S, Hershenhorn J, Koltai H (2011b) Strigolactones interact with ethylene and auxin in regulating root-hair elongation in *Arabidopsis*. J Exp Bot 62:2915–2924
Kepinski S, Leyser O (2005) The Arabidopsis F-box protein TIR1 is an auxin receptor. Nature 435:446–451
Kim H, Park PJ, Hwang HJ, Lee SY, Oh MH, Kim SG (2006) Brassinosteroid signals control expression of the *AXR3/IAA17* gene in the cross-talk point with auxin in root development. Biosci Biotechnol Biochem 70:768–773
Knox K, Grierson CS, Leyser O (2003) *AXR3* and *SHY2* interact to regulate root hair development. Development 130:5769–5777
Koiwai H, Tagiri A, Katoh S, Katoh E, Ichikawa H, Minami E, Nishizawa Y (2007) *RING-H2* type ubiquitin ligase EL5 is involved in root development through the maintenance of cell viability in rice. Plant J 51:92–104
Koltai H (2011) Strigolactones are regulators of root development. New Phytol 190:545–549
Koltai H, Dor E, Hershenhorn J, Joel DM, Weininger S, Lekalla S et al (2010) Strigolactones effect on root growth and root-hair elongation may be mediated by auxin-efflux carriers. J Plant Growth Regul 29:129–136
Krupinski P, Jonsson H (2010) Modeling auxin-regulated development. Cold Spring Harb Perspect Biol 2:a001560

Kuderová A, Urbánková I, Válková M, Malbeck J, Brzobohaty B, Némethová D, Hejátko J (2008) Effects of conditional *IPT*-dependent cytokinin overproduction on root architecture of *Arabidopsis* seedlings. Plant Cell Physiol 49:570–582

Kurakawa T, Ueda N, Maekawa M, Kobayashi K, Kojima M, Nagato Y, Sakakibara H, Kyozuka J (2007) Direct control of shoot meristem activity by a cytokinin-activating enzyme. Nature 445:652–655

Kuroha T, Tokunaga H, Kojima M, Ueda N, Ishida T, Nagawa S, Fukuda H, Sugimoto K, Sakakibara H (2009) Functional analyses of *LONELY GUY* cytokinin-activating enzymes reveal the importance of the direct activation pathway in *Arabidopsis*. Plant Cell 21:3152–3169

Laplaze L, Benkova E, Casimiro I, Maes L, Vanneste S, Swarup R et al (2007) Cytokinins act directly on lateral root founder cells to inhibit root initiation. Plant Cell 19:3889–3900

Laskowski MJ, Williams ME, Nusbaum HC, Sussex IM (1995) Formation of lateral root meristems is a two-stage process. Development 121:3303–3310

Laskowski M, Biller S, Stanley K, Kajstura T, Prusty R (2006) Expression profiling of auxin-treated Arabidopsis roots: toward a molecular analysis of lateral root emergence. Plant Cell Physiol 47:788–792

Lau S, Shao N, Bock R, Jurgens G, De Smet I (2009) Auxin signaling in algal lineages: fact or myth? Trends Plant Sci 14:182–188

Le J, Vandenbussche F, Van Der Straeten D, Verbelen JP (2001) In the early response of *Arabidopsis* roots to ethylene, cell elongation is up- and down-regulated and uncoupled from differentiation. Plant Physiol 125(2):519–522

Lee MM, Schiefelbein J (1999) WEREWOLF, a MYB-related protein in *Arabidopsis*, is a position-dependent regulator of epidermal cell patterning. Cell 99:473–483

Lee JS, Wang S, Sritubtim S, Chen JG, Ellis BE (2009) *Arabidopsis* mitogen-activated protein kinase MPK12 interacts with the MAPK phosphatase IBR5 and regulates auxin signaling. Plant J 57:975–985

Lewis DR, Negi S, Sukumar P, Muday GK (2011) Ethylene inhibits lateral root development, increases IAA transport and expression of *PIN3* and *PIN7* auxin efflux carriers. Development 138:3485–3495

Lewis DR, Olex AL, Lundy SR, Turkett WH, Fetrow JS, Muday GK (2013) A kinetic analysis of the auxin transcriptome reveals cell wall remodeling proteins that modulate lateral root development in *Arabidopsis*. Plant Cell 25:3329–3346

Li X, Mo X, Shou H, Wu P (2006) Cytokinin-mediated cell cycling arrest of pericycle founder cells in lateral root initiation of *Arabidopsis*. Plant Cell Physiol 47:1112–1123

Liang Y, Mitchell DM, Harris JM (2007) Abscisic acid rescues the root meristem defects of the *Medicago truncatula latd* mutant. Dev Biol 304:297–307

Lindsay D, Sawhney V, Bonham-Smith P (2006) Cytokinin-induced changes in *CLAVATA1* and *WUSCHEL* expression temporally coincide with altered floral development in *Arabidopsis*. Plant Sci 170:1111–1117

Liscum E, Reed JW (2002) Genetics of *Aux/IAA* and *ARF* action in plant growth and development. Plant Mol Biol 49:387–400

Liu W, Xu ZH, Luo D, Xue HW (2003) Roles of *OsCKI1*, a rice casein kinase I, in root development and plant hormone sensitivity. Plant J 36:189–202

Liu X, Zhang H, Zhao Y, Feng Z, Li Q, Yang HQ, Luan S, Li J, He ZH (2013) Auxin controls seed dormancy through stimulation of abscisic acid signaling by inducing *ARF*-mediated *ABI3* activation in *Arabidopsis*. Proc Natl Acad Sci 110:15485–15490

Liu J, Rowe J, Lindsey K (2014) Hormonal cross talk for root development: a combined experimental and modeling perspective. Front Plant Sci 5:116

Ljung K, Hull AK, Celenza J, Yamada M, Estelle M, Normanly J, Sandberga G (2005) Sites and regulation of auxin biosynthesis in *Arabidopsis* roots. Plant Cell 17:1090–1104

Lohar DP, Schaff JE, Laskey JG, Kieber JJ, Bilyeu KD, Bird DM (2004) Cytokinins play opposite roles in lateral root formation, and nematode and rhizobial symbioses. Plant J 38:203–214

Lucas M, Godin C, Jay-Allemand C, Laplaze L (2007) Auxin fluxes in the root apex co-regulate gravitropism and lateral root initiation. J Exp Bot 59:56–66
Maharjan PM, Schulz B, Choe S (2011) *BIN2/DWF12* antagonistically transduces brassinosteroid and auxin signals in the roots of *Arabidopsis*. J Plant Biol 54:126–134
Mähönen H, Törmäkangas M, Pischke S, Helariutta K (2006) Cytokinins regulate a bidirectional phosphorelay network in *Arabidopsis*. Curr Biol 16:1116–1122
Malamy JE (2005) Intrinsic and environmental response pathways that regulate root system architecture. Plant Cell Environ 28:67–77
Malamy JE, Benfey PN (1997) Organization and cell differentiation in lateral roots of *Arabidopsis thaliana*. Development 124:33–44
Marchant A, Bhalerao R, Casimiro I, Eklöf J, Casero PJ, Bennett M, Sandberg G (2002) *AUX1* promotes lateral root formation by facilitating Indole-3-Acetic Acid distribution between sink and source tissues in the *Arabidopsis* seedling. Plant Cell 14(3):589–597
Marhavý P, Vanstraelen M, De Rybel B, Zhaojun D, Bennett MJ, Beeckman T, Benková E (2013) Auxin reflux between the endodermis and pericycle promotes lateral root initiation. EMBO J 32:149–158
Masucci JD, Schiefelbein JW (1994) The *rdh6* mutation of *Arabidopsis thaliana* alters root-hair initiation through an auxin-and-ethylene associated process. Plant Physiol 106:1335–1346
Masucci JD, Schiefelbein JW (1996) Hormones act downstream of *TTG* and *GL2* to promote root hair outgrowth during epidermis development in the Arabidopsis root. Plant Cell 8:1505–1517
Matsumoto-Kitano M, Kusumoto T, Tarkowski P, Kinoshita-Tsujimura K, Vaclavikova K, Miyawaki K, Kakimoto T (2008) Cytokinins are central regulators of cambial activity. Proc Natl Acad Sci U S A 105:20027–20031
Mendes AFS, Cidade LC, Otoni WC, Soares-Filho WS, Costa WGC (2011) Role of auxins, polyamines and ethylene in root formation and growth in sweet orange. Biol Plant 55:375–378
Michniewicz M, Zago MK, Abas L, Weijers D, Schweighofer A, Meskiene I et al (2007) Antagonistic regulation of PIN phosphorylation by *PP2A* and *PINOID* directs auxin flux. Cell 130:1044–1056
Miyawaki K, Matsumoto-Kitano M, Kakimoto T (2004) Expression of cytokinin biosynthetic *isopentenyl transferase* genes in *Arabidopsis*: tissue specificity and regulation by auxin, cytokinin, and nitrate. Plant J 37:128–138
Miyawaki K, Tarkowski P, Matsumoto-Kitano M, Kato T, Sato S, Tarkowska D, Tabata S, Sandberg G, Kakimoto T (2006) Roles of *Arabidopsis* ATP/ADP isopentenyl transferases and tRNA isopentenyl transferases in cytokinin biosynthesis. Proc Natl Acad Sci USA 103:16598–16603
Monzon GC, Pinedo M, Lamattina L, Canal L (2012) Sunflower root growth regulation: the role of jasmonic acid and its relation with auxins. Plant Growth Regul 66:129–136
Moore S, Zhang X, Mudge A, Rowe JH, Topping JF, Liu J, Lindsey K (2015) Spatiotemporal modelling of hormonal crosstalk explains the level and patterning of hormones and gene expression in *Arabidopsis thaliana* wild-type and mutant roots. New Phytol 207:1110–1122
Moreno-Risueno MA, Van Norman JM, Moreno A, Zhang J, Ahnert SE, Benfey PN (2010) Oscillating gene expression determines competence for periodic *Arabidopsis* root branching. Science 329:1306–1311
Moubayidin L, Perilli S, Dello Ioio R et al (2010) The rate of cell differentiation controls the *Arabidopsis* root meristem growth phase. Curr Biol 20:1138–1143
Moubayidin L, Di Mambro R, Sozzani R, Pacifici E, Salvi E, Terpstra I et al (2013) Spatial coordination between stem cell activity and cell differentiation in the root meristem. Dev Cell 26:405–415
Mravec J, Petrasek J, Li N, Boeren S, Karlova R, Kitakura S et al (2011) Cell plate restricted association of DRP1A and PIN proteins is required for cell polarity establishment in *Arabidopsis*. Curr Biol 21:1055–1060
Mu Y, Zou M, Sun X, HeB, Xu X, Liu Y, Zhang L, Chi1 W (2017) BASIC PENTACYSTEINE proteins repress *ABSCISIC ACID INSENSITIVE4* expression via direct recruitment of the

Polycomb-Repressive Complex 2 in Arabidopsis root development. Plant Cell Physiol 58 (3):607–621

Muday GK, Rahman A, Binder BM (2012) Auxin and ethylene: collaborators or competitors? Trends Plant Sci 17:181–195

Muller B, Sheen J (2008) Cytokinin and auxin interaction in root stem-cell specification during early embryogenesis. Nature 453:1094–1097

Murase K, Hirano Y, Sun TP, Hakoshima T (2008) Gibberellin induced DELLA recognition by the gibberellin receptor *GID1*. Nature 456:459–463

Nakamura A, Nakajima N, Goda H, Shimada Y, Hayashi K, Nozaki H, Asami T, Yoshida S, Fujioka S (2006) Arabidopsis *Aux/IAA* genes are involved in brassinosteroid-mediated growth responses in a manner dependent on organ type. Plant J 45:193–205

Nibau C, Gibbs DJ, Coates JC (2008) Branching out in new directions: the control of root architecture by lateral root formation. New Phytol 179:595–614

Nieuwland J, Maughan S, Dewitte W, Scofield S, Sanz L, Murray JA (2009) The D-type cyclin *CYCD4;1* modulates lateral root density in *Arabidopsis* by affecting the basal meristem region. Proc Natl Acad Sci 106:22528–22533

Okushima Y, Fukaki H, Onoda M, Theologis A, Tasaka M (2007) *ARF7* and *ARF19* regulate lateral root formation via direct activation of *LBD/ASL* genes in *Arabidopsis*. Plant Cell 19:118–130

Overvoorde P, Fukaki H, Beeckman T (2010) Auxin control of root development. Cold Spring Harb Perspect Biol 2:a001537

Pacifici E, Polverari L, Sabatini S (2015) Plant hormone cross-talk: the pivot of root growth. J Exp Bot 66:1113–1121

Palavan-Unsal N (1987) Polyamine metabolism in the roots of *Phaseolus vulgaris*. Interaction of inhibitors of polyamine biosynthesis with putrescine in growth and polyamine biosynthesis. Plant Cell Physiol 28:565–572

Pandya-Kumar N, Shema R, Kumar M, Mayzlish-Gati E, Levy D, Zemach H et al (2014) Strigolactone analog *GR24* triggers changes in PIN2 polarity, vesicle trafficking and actin filament architecture. New Phytol 202:1184–1196

Parizot B, Laplaze L, Ricaud L, Boucheron-Dubuisson E, Bayle V, Bonke M et al (2008) Diarch symmetry of the vascular bundle in *Arabidopsis* root encompasses the pericycle and is reflected in distich lateral root initiation. Plant Physiol 146:140–148

Parry G, Delbarre A, Marchant A, Swarup R, Napier R, Perrot-Rechenmann C, Bennett MJ (2001) Novel auxin transport inhibitors phenocopy the auxin influx carrier mutation *aux1*. Plant J 25:399–406

Peret B, De Rybel B, Casimiro I, Benkova E, Swarup R, Laplaze L, Beeckman T, Bennett MJ (2009) *Arabidopsis* lateral root development: an emerging story. Trends Plant Sci 14:399–408

Peret B, Desnos T, Jost R, Kanno S, Berkowitz O, Nussaume L (2014) Root architecture responses: In search of phosphate. Plant Physiol 166:1713–1723

Perilli S, Moubayidin L, Sabatini S (2013) Molecular basis of Cytokinin action during Root development. In: Beeckman T (ed) Plant Roots-The Hidden Half, CRC Press, pp 1–12

Petersson SV, Johansson AI, Kowalczyk M, Makoveychuk A, Wang JY, Moritz T et al (2009) An auxin gradient and maximum in the *Arabidopsis* root apex shown by high-resolution cell-specific analysis of IAA distribution and synthesis. Plant Cell 21(6):1659–1668

Petricka JJ, Winter CM, Benfey PN (2012) Control of *Arabidopsis* root development. In: Merchant SS (ed) Annual Review of Plant Biology. Palo Alto, CA: Annual Reviews, pp 563–590

Pitts RJ, Cernac A, Estelle M (1998) Auxin and ethylene promote root hair elongation in *Arabidopsis*. Plant J 16:553–560

Rahman A, Amakawa T, Goto N, Tsurumi S (2001) Auxin is a positive regulator for ethylene-mediated response in the growth of *Arabidopsis* roots. Plant Cell Physiol 42:301–307

Rahman A, Hosokawa S, Oono Y, Amakawa T, Goto N, Tsurumi S (2002) Auxin and ethylene response interactions during *Arabidopsis* root hair development dissected by auxin influx modulators. Plant Physiol 130:1908–1917

Rani Debi B, Taketa S, Ichii M (2005) Cytokinin inhibits lateral root initiation but stimulates lateral root elongation in rice (*Oryza sativa*). J Plant Physiol 162:507–515

Raya-Gonzalez J, Pelagio-Flores R, Lopez-Bucio J (2012) The jasmonate receptor COI1 plays a role in jasmonate-induced lateral root formation and lateral root positioning in *Arabidopsis thaliana*. J Plant Physiol 169:1348–1358

Razem FA, El-Kereamy A, Abrams SR, Hill RD (2006) The RNA-binding protein FCA is an abscisic acid receptor. Nature 439:290–294

Riefler M, Novak O, Strnad M, Schmülling T (2006) *Arabidopsis* cytokinin receptor mutants reveal functions in shoot growth, leaf senescence, seed size, germination, root development, and cytokinin metabolism. Plant Cell 18:40–54

Rosquete MR, von Wangenheim D, Marhavy P, Barbez E, Stelzer EH, Benkova E, Maizel A, Kleine-Vehn J (2013) An auxin transport mechanism restricts positive orthogravitropism in lateral roots. Curr Biol 23:817–822

Rowe JH, Topping JF, Liu J, Lindsey K (2016) Abscisic acid regulates root growth under osmotic stress conditions via an interacting hormonal network with cytokinin, ethylene and auxin. New Phytol 211:225–239

Ruyter-Spira C, Kohlen W, Charnikhova T, van Zeijl A, van Bezouwen L, de Ruijter N et al (2011) Physiological effects of the synthetic strigolactone analog GR24 on root system architecture in *Arabidopsis*: another belowground role for strigolactones? Plant Physiol 155:721–734

Ruzicka K, Ljung K, Vanneste S, Podhorska R, Beechman T, Friml J, Benkova E (2007) Ethylene regulates root growth through effects on auxin biosynthesis and transport-dependent auxin distribution. Plant Cell 19:2197–2212

Ruzicka K, Simaskova M, Duclercq J, Petrasek J, Zazimalova E, Simon S, Friml J, Van Montagu MCE, Benkova E (2009) Cytokinin regulates root meristem activity via modulation of the polar auxin transport. Proc Natl Acad Sci U S A 106:4284–4289

Sabatini S, Beis D, Wolkenfelt H, Murfett J, Guilfoyle T, Malamy J, Benfey P, Leyser O, Bechtold N, Weisbeek P, Scheres B (1999) An auxin-dependent distal organizer of pattern and polarity in the *Arabidopsis* root. Cell 99:463–472

Sabatini S, Heidstra R, Wildwater M, Scheres B (2003) *SCARECROW* is involved in positioning the stem cell niche in the *Arabidopsis* root meristem. Genes Dev 17:354–358

Saini S, Sharma I, Kaur N, Pati PK (2013) Auxin: A master regulator in plant root development. Plant Cell Rep 32:741–757

Santner A, Mark Estelle M (2009) Recent advances and emerging trends in plant hormone signalling. Nature 459:1071–1078

Sarkar AK, Luijten M, Miyashima S, Lenhard M, Hashimoto T, Nakkajima K, Scheres B, Heidstra R, Laux T (2007) Conserved factors regulate signalling in *Arabidopsis thaliana* shoot and root stem cell organizers. Nature 446:811–814

Sauer M, Balla J, Luschnig C, Wisniewska J, Reinohl V, Friml J, Benkova E (2006) Canalization of auxin flow by *Aux/IAA-ARF*-dependent feedback regulation of PIN polarity. Genes Dev 20:2902–2911

Scacchi E, Salinas P, Gujas B, Santuari L, Krogan N, Ragni L, Berleth T, Hardtke CS (2010) Spatio-temporal sequence of cross regulatory events in root meristem growth. Proc Natl Acad Sci U S A 107:22734–22739

Scheres B, Wolkenfelt H, Willemsen V, Terlouw M, Lawson E, Dean C, Weisbeek P (1994) Embryonic origin of the *Arabidopsis* primary root and root meristem initials. Development 120:2475–2487

Scheres B, Di Laurenzio L, Willemsen V, Hauser M-T, Janmaat K, Weisbeek P, Benfey PN (1995) Mutations affecting the radial organisation of the *Arabidopsis* root display specific defects throughout the embryonic axis. Development 121:53–62

Schiefelbein JW (2000) Constructing a plant cell. The genetic control of root hair development. Plant Physiol 124:1525–1531

Schwager KM, Calderon-Villalobos LI, Dohmann EM, Willige BC, Knierer S, Nill C, Schwechheimer C (2007) Characterization of the *VIER F-BOX PROTEINE* genes from *Arabidopsis* reveals their importance for plant growth and development. Plant Cell 19:1163–1178

Sena G, Wang X, Liu HY, Hofhuis H, Birnbaum KD (2009) Organ regeneration does not require a functional stem cell niche in plants. Nature 457:1150–1153

Shani E, Weinstain R, Zhang Y, Castillejo C, Kaiserli E, Chory J, Tsien RY, Estelle M (2013) Gibberellins accumulate in the elongating endodermal cells of *Arabidopsis* root. Proc Natl Acad Sci USA 110:4834–4839

Sharp RE (2002) Interaction with ethylene: changing views on the role of abscisic acid in root and shoot growth responses to water stress. Plant Cell Environ 25:211–222

Shimada A, Ueguchi-Tanaka M, Nakatsu T, Nakajima M, Naoe Y, Ohmiya H, Kato H, Matsuoka M (2008) Structural basis for gibberellins recognition by its receptor *GID1*. Nature 456:520–523

Shkolnik-Inbar D, Bar-Zvi D (2010) *ABI4* mediates abscisic acid and cytokinin inhibition of lateral root formation by reducing polar auxin transport in *Arabidopsis*. Plant Cell 22:3560–3573

Silverstone AL, Chang C, Krol E, Sun TP (1997) Developmental regulation of the gibberellin biosynthetic gene GA1 in *Arabidopsis thaliana*. The Plant J 12:9–19

Singh M, Gupta A, Laxmi A (2014) Glucose control of root growth direction in *Arabidopsis thaliana*. J Exp Bot 65:2981–2993

Smith MA, Davies PJ (1985) Separation and quantification of polyamines in plant tissue by high performance liquid chromatography of their dansyl derivatives. Plant Physiol 78:89–91

Sozzani R, Cui H, Moreno-Risueno MA, Busch W, Van Norman JM, Vernoux T, Brady SM, Dewitte W, Murray JA, Benfey PN (2010) Spatiotemporal regulation of cell-cycle genes by *SHORTROOT* links patterning and growth. Nature 466:128–132

Spalding EP, Wu G, Lewis DR (2007) Mutations in *Arabidopsis* multidrug resistance-like ABC transporters separate the roles of acropetal and basipetal auxin transport in lateral root development. Plant Cell 19:1826–1837

Steinmann T, Geldner N, Grebe M, Mangold S, Jackson CL, Paris S, Galweiler L, Palme K, Jurgens G (1999) Coordinated polar localisation of auxin efflux carrier PIN1 by *GNOM ARF GEF*. Science 286:316–318

Stepanova AN, Hoyt JM, Hamilton AA, Alonso JM (2005) A link between ethylene and auxin uncovered by the characterization of two root-specific ethylene-insensitive mutants in *Arabidopsis*. Plant Cell 17:2230–2242

Stepanova AN, Robertson-Hoyt J, Yun J, Benavente LM, Xie DY, Dolezal K, Schlereth A, Jurgens G, Alonso JM (2008) *TAA1*-mediated auxin biosynthesis is essential for hormone crosstalk and plant development. Cell 133:177–191

Strader LC, Bartel B (2008) A new path to auxin. Nat Chem Biol 4:337–339

Strader LC, Bartel B (2009) The Arabidopsis *PLEIOTROPIC DRUG RESISTANCE8/ABCG36* ATP binding cassette transporter modulates sensitivity to the auxin precursor indole-3-butyric acid. Plant Cell 21:1992–2007

Street IH, Aman S, Zubo Y, Ramzan A, Wang X, Shakeel SN, Kieber JJ, Schaller GE (2015) Ethylene inhibits cell proliferation of the *Arabidopsis* Root Meristem. Plant Physiol 169:338–350

Sun J, Xu Y, Ye S, Jiang H, Chen Q, Liu F, Zhou W, Chen R, Li X, Tietz O, Wu X, Cohen JD, Palme K, Li C (2009) Arabidopsis *ASA1* is important for jasmonate-mediated regulation of auxin biosynthesis and transport during lateral root formation. Plant Cell 21:1495–1511

Swarup R, Friml J, Marchant A, Ljung K, Sandberg G, Palme K, Bennett M (2001) Localization of the auxin permease *AUX1* suggests two functionally distinct hormone transport pathways operate in the *Arabidopsis* root apex. Genes Dev 15:2648–2653

Swarup R, Kramer EM, Perry P, Knox K, Leyser HM, Haseloff J, Beemster GT, Bhalerao R, Bennett MJ (2005) Root gravitropism requires lateral root cap and epidermal cells for transport and response to a mobile auxin signal. Nat Cell Biol 7:1057–1065

Swarup R, Perry P, Hagenbeek D, Van Der Straeten D, Beemster GT, Sandberg G, Bhalerao R, Ljung K, Bennet MJ (2007) Ethylene upregulates auxin biosynthesis in *Arabidopsis* seedlings to enhance inhibition of root cell elongation. Plant Cell 19:2186–2196

Swarup K, Benkova E, Swarup R, Casimiro I, Peret B, Yang Y et al (2008) The auxin influx carrier *LAX3* promotes lateral root emergence. Nat Cell Biol 10:946–954

Szemenyei H, Hannon M, Long JA (2008) *TOPLESS* mediates auxin dependent transcriptional repression during *Arabidopsis* embryogenesis. Science 319:1384–1386

Tan BC, Joseph LM, Deng WT, Liu L, Li QB, Cline K, McCarty DR (2003) Molecular characterization of the *Arabidopsis 9-cis epoxycarotenoid dioxygenase* gene family. Plant J 35:44–56

Taniguchi M, Sasaki N, Tsuge T, Aoyama T, Oka A (2007) *ARR1* directly activates cytokinin response genes that encode proteins with diverse regulatory functions. Plant Cell Physiol 48:263–277

Tanimoto M, Roberts K, Dolan L (1995) Ethylene is a positive regulator of root hair development in *Arabidopsis thaliana*. Plant J 8:943–948

Tao Y, Ferrer JL, Ljung K, Pojer F, Hong F, Long JA et al (2008) Rapid synthesis of auxin via a new tryptophan-dependent pathway is required for shade avoidance in plants. Cell 133:164–176

Teale WD, Paponov IA, Palme K (2006) Auxin in action: signalling, transport and the control of plant growth and development. Nat Rev Mol Cell Biol 7:847–859

Thole JM, Beisner ER, Liu J, Venkova SV, Strader LC (2014) Abscisic acid regulates root elongation through the activities of auxin and ethylene in *Arabidopsis thaliana G3*-Genes Genomes. Genetics 4:1259–1274

Thomann A, Lechner E, Hansen M, Dumbliauskas E, Parmentier Y, Kieber J, Scheres B, Genschik P (2009) *Arabidopsis CULLIN3* genes regulate primary root growth and patterning by ethylene-dependent and -independent mechanisms. PLoS Genet 5:e1000328

Tian Q, Nagpal P, Reed JW (2003) Regulation of Arabidopsis *SHY2/IAA3* protein turnover. Plant J 36:643–651

Tiwari SB, Wang XJ, Hagen G, Guilfoyle TJ (2001) Aux/IAA proteins are active repressors, and their stability and activity are modulated by auxin. Plant Cell 13:2809–2822

Topp CN, Iyer-Pascuzzi AS, Anderson JT, Lee CR, Zurek PR, Symonova O et al (2013) 3D phenotyping and quantitative trait locus mapping identify core regions of the rice genome controlling root architecture. Proc Natl Acad Sci USA 110:E1695–E1704

Topping JF, Lindsey K (1997) Promoter trap markers differentiate structural and positional components of polar development in *Arabidopsis*. Plant Cell 9:1713–1725

Truernit E, Siemering KR, Hodge S, Grbic V, Haseloff J (2006) A map of *KNAT* gene expression in the *Arabidopsis* root. Plant Mol Biol 60:1–20

Tsurumi S, Wada S (1980) Transport of shoot- and cotyledon-applied indole-3-acetic acid to *Vicia faba* root. Plant Cell Physiol 21:803–816

Ubeda-Tomas S, Beemster GTS, Bennett MJ (2012) Hormonal regulation of root growth: integrating local activities into global behaviour. Trends Plant Sci 17:326–331

van den Berg C, Willemsen V, Hendriks G, Weisbeek P, Scheres B (1997) Short-range control of cell differentiation in the *Arabidopsis* root meristem. Nature 390:287–289

Vanneste S, Friml J (2009) Auxin: a trigger for change in plant development. Cell 136:1005–1016

Vanneste S, De Rybel B, Beemster GT, Ljung K, De Smet I, Van Isterdael G, Naudts M, Iida R, Gruissem W, Tasaka M, Inzé D, Fukaki H, Beeckman T (2005) Cell cycle progression in the pericycle is not sufficient for *SOLITARY ROOT/IAA14*-mediated lateral root initiation in *Arabidopsis thaliana*. Plant Cell 17:3035–3050

Vellosillo T, Martinez M, Lopez MA, Vicente J, Cascon T, Dolan L, Hamberg M, Castresana C (2007) Oxylipins produced by the 9-lipoxygenase pathway in *Arabidopsis* regulate lateral root development and defence responses through a specific signalling cascade. Plant Cell 19:831–846

Verstraeten I, Schotte S, Geelen D (2014) Hypocotyl adventitious root organogenesis differs from lateral root development. Front Plant Sci 5:495

Vieten A, Vanneste S, Wisniewska J, Benkova E, Benjamins R, Beeckman T, Luschnig C, Friml J (2005) Functional redundancy of *PIN* proteins is accompanied by auxin-dependent cross-regulation of *PIN* expression. Development 132:4521–4531

Voegele A, Linkies A, Muller K, Leubner-Metzger G (2011) Members of the gibberellin receptor gene family *GID1* (*GIBBERELLININSENSITIVE DWARF1*) play distinct roles during *Lepidium sativum* and *Arabidopsis thaliana* seed germination. J Exp Bot 62:5131–5147

Wabnik K, Kleine-Vehn J, Balla J, Sauer M, Naramoto S, Reinohl V, Merks RMH, Govaerts W, Friml J (2010) Emergence of tissue polarization from synergy of intracellular and extracellular auxin signaling. Mol Syst Biol 6:447

Wang JW, Wang LJ, Mao YB, Cai WJ, Xue HW, Chen XY (2005a) Control of root cap formation by Micro RNA targeted auxin response factors in *Arabidopsis*. Plant Cell 17:2204–2216

Wang X, Goshe MB, Soderblom EJ, Phinney BS, Kuchar JA, Li J, Asami T, Yoshida S, Huber SC, Clouse SD (2005b) Identification and functional analysis of in vivo phosphorylation sites of the *Arabidopsis BRASSINOSTEROID INSENSITIVE1* receptor kinase. Plant Cell 17:1685–1703

Wang JR, Hu H, Wang GH, Li J, Chen JY, Wu P (2009) Expression of *PIN* Genes in Rice (*Oryza sativa* L.): tissue specificity and regulation by hormones. Mol Plant 2:823–831

Wang L, Hua D, He J, Duan Y, Chen Z, Hong X, Gong Z (2011) *Auxin Response Factor2* (*ARF2*) and its regulated homeodomain gene *HB33* mediate abscisic acid response in *Arabidopsis*. PLoS Genet 7:e1002172

Watson G (2004) Effect of transplanting and paclobutrazol on root growth of 'Green Column' black maple and 'Summit' green ash. J Environ Hortic 22:209–212

Werner T, Motyka V, Laucou V, Smets R, Onckelen HV, Schmuelling T (2003) Cytokinin-deficient transgenic *Arabidopsis* plants show functions of cytokinins in the regulation of shoot and root meristem activity. Plant Cell 15:2532–2550

Willige BC, Isono E, Richter R, Zourelidou M, Schwechheimer C (2011) Gibberellin regulates *PIN-FORMED* abundance and is required for auxin transport-dependent growth and development in *Arabidopsis thaliana*. Plant Cell 23:2184–2195

Wilson AK, Pickett FB, Turner JC, Estelle M (1990) A dominant mutation in *Arabidopsis* confers resistance to auxin, ethylene and abscisic acid. Mol Gen Genet 222:377–383

Wu G, Lewis DR, Spalding EP (2007) Mutations in Arabidopsis multidrug resistance-like ABC transporters separate the roles of acropetal and basipetal auxin transport in lateral root development. Plant Cell 19:1826–1837

Wysocka-Diller JW, Helariutta Y, Fukaki H, Malamy JE, Benfey PN (2000) Molecular analysis of *SCARECROW* function reveals a radial patterning mechanism common to root and shoot. Development 127:595–603

Yamada M, Greenham K, Prigge MJ, Jensen PJ, Estelle M (2009) The *TRANSPORT INHIBITOR RESPONSE2* gene is required for auxin synthesis and diverse aspects of plant development. Plant Physiol 151:168–179

Yang CJ, Zhang C, Lu YN, Jin JQ, Wang XL (2011) The mechanisms of brassinosteroids' action: from signal transduction to plant development. Mol Plant 4:588–600

Yuan H, Hao Q, Li W, Yan C, Yan N, Yin P (2014) Identification and characterization of ABA receptors in *Oryza sativa*. PLoS One 9(4):e95246

Zazımalova E, Murphy AS, Yang H, Hoyerova K, Hosek P (2010) Auxin transporters-why so many? Cold Spring Harb Perspect Biol 2:a001552

Zhang X, Xiong Y, DeFraia C, Schmelz E, Mou Z (2008) The *Arabidopsis* MAP kinase kinase 7. A crosstalk point between auxin signaling and defense responses? Plant Signal Behav 3:272–274

Zhang W, Swarup R, Bennett M, Schaller GE, Kieber JJ (2013) Cytokinin induces cell division in the quiescent center of the *Arabidopsis* root apical meristem. Curr Biol 23:1979–1989

Zhao Y, Xing L, Wang X, Hou Y-J, Gao J, Wang P et al (2014) The ABA receptor PYL8 promotes lateral root growth by enhancing MYB77-dependent transcription of auxin-responsive genes. Sci Signal 7:ra53. https://doi.org/10.1126/scisignal.2005051

Chapter 2
Effects of Strigolactones on Plant Roots

Adrianus P. Claassens and Paul N. Hills

2.1 Introduction

Strigolactones (SLs) are a relatively recently discovered class of phytohormones that are derived from the cleavage of a carotenoid precursor. SLs play numerous roles in plant growth and development, both above- and belowground, such as seed germination, control of root and shoot architecture and structure, modulation of hormone fluxes, and mycorrhizal symbiosis (Xie et al. 2010; Waters et al. 2017). SLs are found in a wide range of plants and model organisms such as *Arabidopsis thaliana*, *Oryza sativa*, *Pisum sativum*, *Solanum lycopersicum*, *Petunia hybrida*, *Dendranthema grandiflorum*, and *Nicotiana tabacum* that are used to study SLs and their effects on growth regulation (Xie et al. 2010; Ruyter-Spira et al. 2013; Matthys et al. 2016; Waters et al. 2017).

Strigolactones were first identified as germination stimulants for the parasitic weeds *Striga* spp., *Phelipanche* spp., and *Orobanche* (Cook et al. 1966; Xie et al. 2009). Later, it was observed that strigolactones were exuded by plant roots and increased hyphal branching of arbuscular mycorrhizal fungi (AMF) (Akiyama et al. 2005). The abovementioned fungi form a symbiotic relationship with plants; as many as 80% of land-based plants are presumed to have some version of such a relationship (Akiyama et al. 2005; Bouwmeester et al. 2007). The general structure of natural strigolactones consists of four connected rings, labeled A, B, C, and D (Fig. 2.1). The C–D rings are the most critical for SL activity and its associated effects in the plant. These two rings are attached to one another via an enol-ether bridge (Ćavar et al. 2015). The A–B rings are known to be more varied between natural SLs. These variations occur through the attachment of different side groups

A. P. Claassens · P. N. Hills (✉)
Department of Genetics, Institute for Plant Biotechnology, Stellenbosch University, Matieland, South Africa
e-mail: phills@sun.ac.za

B. Giri et al. (eds.), *Root Biology*, Soil Biology 52,
https://doi.org/10.1007/978-3-319-75910-4_2

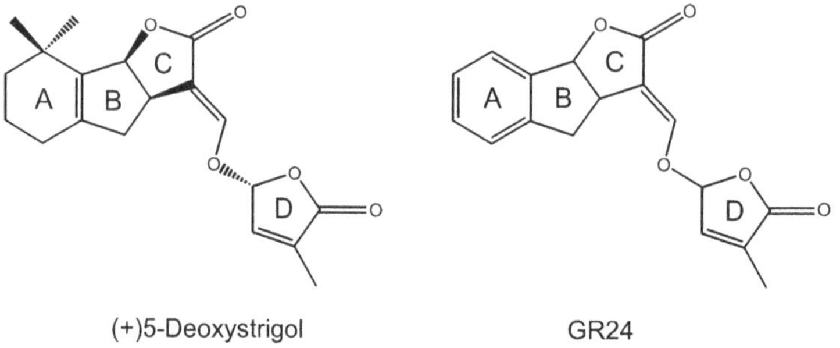

Fig. 2.1 One of the last known precursors in the biosynthetic pathway for natural SLs, 5-deoxystrigol, and the synthetic SL GR24

to the A–B rings (Mangnus et al. 1992; Matusova 2005; Xie et al. 2010). Orobanchol, orobanchyl acetate, methoxyorobanchol, 7-hydroxy-orobanchol, sorgomol, sorgolactone, strigol, strigyl acetate, and methoxystrigol are all believed to be manufactured from the same precursor molecule, 5-deoxystrigol (Fig. 2.1). Although not all the steps in the SL biosynthetic pathway have been fully elucidated (Ruyter-Spira et al. 2013), major progress has been made in gaining an understanding of the main steps.

2.2 Strigolactone Biosynthesis

Most of our current understanding of the SL biosynthetic pathway has been derived from the use of *Arabidopsis*, rice, pea, and petunia mutants. Generally, these mutants are referred to as *max* (*more axillary branching*) in *Arabidopsis*, *d* (*dwarf*) in rice, and *rms* (*ramosus*) in pea. For the sake of simplicity, we will mainly use the *Arabidopsis* gene names for general discussions, although reference to other orthologues will be made when referring to specific research. A full list of orthologues of genes involved in SL biosynthesis, and signaling for these model species may be found in Table 2.1. Knowledge gained from these mutants elucidated that D27, MAX3, MAX4, and MAX1 play specific roles in the biosynthetic pathway, whereas D14, MAX2, and the SMXL family play prominent roles in SL perception and signal transduction (Fig. 2.2) (Ruyter-Spira et al. 2013; Zhang et al. 2013). The synthesis of SLs begins with the isomerase activity of DWARF27 (D27) on all-*trans*-β-carotene, producing 9-*cis*-β-carotene. Two carotenoid cleavage dioxygenases, MAX3 and MAX4, act consecutively on 9-*cis*-β-carotene to liberate carlactone, which is believed to be the last shared precursor of SLs (Schwartz et al. 2001; Bouvier et al. 2003). Carlactone was confirmed to be the precursor molecule for SL biosynthesis and has been shown to have some degree of strigolactone-like activity (Seto et al. 2014). In *Arabidopsis*, the cytochrome P450 enzyme MAX1 acts on carlactone to produce carlactonoic acid (CLA). Further downstream of this reaction lie considerable gaps in our knowledge

Table 2.1 Orthologues of genes for known strigolactone biosynthetic and signal transduction proteins in *Arabidopsis thaliana*, *Oryza sativa* (rice), and *Pisum sativum* (pea)

	Arabidopsis	*Oryza sativa*	*Pisum sativum*	
Protein function	Genes			Reference(s)
Strigolactone biosynthesis				
Carotenoid isomerase	*AtD27*	*OsD27*		Ishikawa et al. (2005); Lin et al. (2009); Waters et al. (2012)
Carotenoid cleavage dioxygenase	*AtMAX3*	*OsD17*	*PsRMS5*	Booker et al. (2004); Ishikawa et al. (2005); Zou et al. (2006)
Carotenoid cleavage dioxygenase	*AtMAX4*	*OsD10*	*PsRMS1*	Sorefan (2003); Ishikawa et al. (2005); Arite et al. (2007)
Cytochrome P450	*AtMAX1*	*OsMAX1*		Booker et al. (2005); Zhang et al. (2014); Abe et al. (2014)
2-Oxoglutarate- and Fe(II)-dependent dioxygenase	*AtLBO*	*Os01g0935400*		Brewer et al. (2016)
Strigolactone signaling				
F-box protein: SCF complex	*AtMAX2*	OsD3	*PsRMS4*	Beveridge et al. (1996); Ishikawa et al. (2005); Johnson et al. (2006); Stirnberg et al. (2007)
α/β-Hydrolase: SL signaling	*AtD14*	OsD14	*PsRMS3*	Arite et al. (2009); Chevalier et al. (2014); de Saint Germain et al. (2016)
Class I Clp ATPase: Proteolytic target(s) in SL signaling	*AtSMXL6*, *AtSMXL7*, *AtSMXL8*	OsD53, OsD53-LIKE		Zhou et al. (2013); Soundappan et al. (2015); Wang et al. (2015); Liang et al. (2016)

of the synthesis pathway, which may be more complex in other plants; rice, for example, has five genes homologous to MAX1 (Zhang et al. 2014). The CLA produced by MAX1 is converted to methyl carlactonoate (MeCLA) by an unknown enzyme (Abe et al. 2014). The MeCLA is in turn converted to a yet to be identified molecule by LATERAL BRANCHING OXIDOREDUCTASE (LBO) (Brewer et al. 2016). The remaining steps to generate the diverse A–B rings remain largely unknown.

Grafting experiments elucidated the importance of roots in SL biosynthesis. *Arabidopsis* WT rootstocks could rescue the branching excess phenotype of scions of biosynthetic mutants, *max3, max4* and *max1* (Turnbull et al. 2002; Sorefan 2003; Booker et al. 2004). However, gene expression of *MAX3, MAX4*, and *MAX1* in tissue types other than roots and the inability of the mutant rootstocks to induce a branching phenotype indicate that SL synthesis is more complicated than originally thought (Turnbull et al. 2002; Sorefan 2003). *D27* is expressed in lateral roots and crown roots of rice (Lin et al. 2009). *MAX3* expression can be observed in a number of

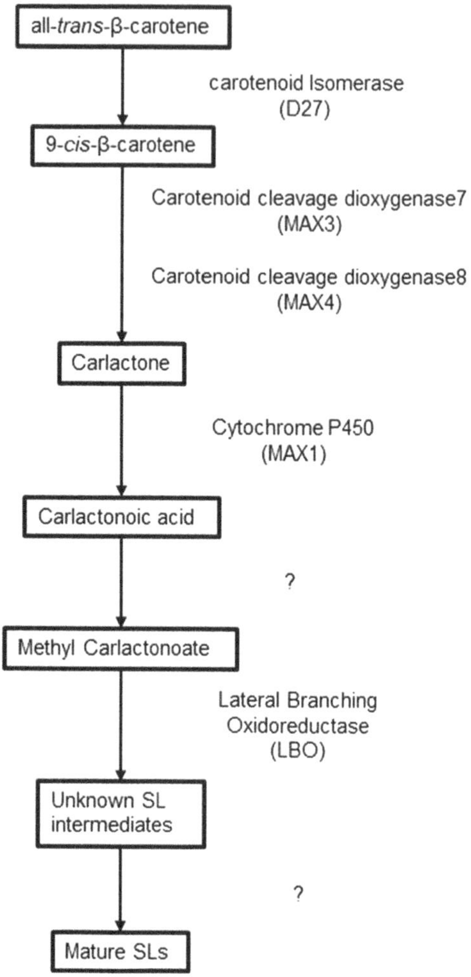

Fig. 2.2 The known SL biosynthetic pathway known so far follow a sequential action of the following enzymes: carotenoid isomerase (AtD27/OsD27), carotenoid cleavage dioxigenase7 (AtMAX3/OsD17/PsRMS5), carotenoid cleavage dioxygenase8 (AtMAX4/OsD10/PsRMS1), cytochrome P450 (AtMAX1/OsMAX1), 2-oxoglutarate-, and Fe(II)-dependent dioxygenase (AtLBO)

tissues (primary and secondary inflorescence stems, siliques). *MAX4* is mainly expressed in the primary root (PR) tip, although low levels of expression were detected in the hypocotyl, petioles, and nodal tissue (Sorefan 2003; Booker et al. 2004; Bainbridge et al. 2005). *MAX1* was demonstrated to be expressed in all tissues, with the highest concentrations in vascular tissues (Booker et al. 2005). This apparent lack of overlap in expression patterns for the various biosynthetic genes suggests that SLs and their metabolic precursors are probably mobile in the plant. However, the transport of SLs and their precursors is very poorly understood at present.

2.3 Strigolactone Transport

Data on SL movement within plants is somewhat contradictory. A mass spectrometry analysis by Kohlen et al. (2011) found SLs to be present in xylem sap of tomato and *Arabidopsis*. However, a long-distance transport study on a wide range of species contradicted these findings (Xie et al. 2015). Although it appeared that endogenous and exogenous SLs may be transported from root to shoot, SLs could not be detected in the xylem sap in this latter study (Xie et al. 2015). This might indicate that SLs are moved via active cell-to-cell transport. Evidence from petunia supports this perspective; an ATP-binding cassette transporter PLEIOTROPIC DRUG RESISTANCE1 (PDR1) was named as a key component in SL distribution and as an SL transporter (Kretzschmar et al. 2012; Sasse et al. 2015). Whether SLs are actively transported from cell to cell or are transported in the transpiration stream, once they reach their destination, they must be perceived by the target tissues. This perception and signal transduction process is still in the process of being fully elucidated.

2.4 Perception of Strigolactones

Most of the strigolactone perception mutants and biosynthetic mutants have overlapping phenotypes, particularly with regards to excessive aerial branching. However, a notable difference between the synthesis (*max3*, *max4*, and *max1*) and perception (*max2*) mutants is the inability of exogenously applied SLs to revert the branching phenotype of the perception mutants to WT levels.

Strigolactone perception begins with D14, an α-/β-fold hydrolase protein. Unlike most other hormone receptors, D14 has both enzyme and receptor functions. The protein contains a conserved Ser–His–Asp domain known as the SL catalytic triad which is crucial for SL hydrolysis and signaling (Hamiaux et al. 2012; Nakamura et al. 2013; Waters et al. 2015). Following binding to D14, SLs are hydrolyzed into two products, 5-hydroxy-3-methylbutenolide (D-OH) and a tricyclic lactone, which are not able to inhibit shoot branching (Hamiaux et al. 2012; Zhao et al. 2013; Nakamura et al. 2013). Thus, SL perception also acts as a feedback mechanism removing the hormonal signal, allowing for tighter control of the signaling process. D14 is a paralogue of KARRIKIN INSENSITIVE2 (KAI2). Interestingly, KAI2 is the ancestral gene, being found in more ancient plant lineages such as mosses which do not have D14 homologues, which suggests that strigolactone perception is a relatively recent development in higher plants (Waters et al. 2017). KAI2 appears to be involved in plant responses to karrikins from smoke. Both D14 and KAI2 interact with MAX2 (Nelson et al. 2011; Stanga et al. 2013) as the next step of their respective signaling pathways.

A presumed conformational shift, resulting either from the binding of SL by D14 or the enzymatic cleavage of D-OH from the strigolactone molecule, enables MAX2, the F-box protein in the SCF^{MAX2} E3 ubiquitin ligase complex (Stirnberg et al. 2007) to recognize and bind to D14 (Zhou et al. 2013). Hormone-activated proteolysis, resulting in a relief-of-restraint response, is a common signaling mechanism observed among plant growth regulators. Gibberellin, auxin, and jasmonate signal transduction all function on this principle. In this type of signaling, the F-box protein of a hormone-specific Skp1-Cullin-F-box (SCF) E3 ubiquitin ligase complex targets specific proteins, usually negative regulators of transcription, for poly-ubiquitination and degradation via the 26S proteasome (Devoto et al. 2002; McGinnis 2003; Tan et al. 2007). In strigolactone signaling, SCF^{MAX2} targets members of the SMXL protein families for degradation, allowing for strigolactone signaling. The SMAX1 protein was discovered in a screen for genes that suppress *max2* phenotypes in *Arabidopsis*. D53 was identified in a rice mutant that shared *d3* and *d13* phenotypes (Stanga et al. 2013; Jiang et al. 2013; Zhou et al. 2013). This further leads to the discovery of SMAX1-LIKE6 (SMXL6), SMXL7, and SMXL8. The *Arabidopsis* triple mutant *smxl6–smxl7–smxl8* could suppress all the SL-associated *max2* phenotypes, and *smax1* suppressed all the karrikin-linked phenotypes of *max2* (Soundappan et al. 2015; Wang et al. 2015). SMXL6, SMXL7, and SMXL8 primarily interact with D14 and are quickly degraded in the presence of *rac*-GR24 in a MAX2-dependent fashion (Jiang et al. 2013; Zhou et al. 2013; Umehara et al. 2015; Soundappan et al. 2015; Wang et al. 2015; Liang et al. 2016). The proteolytic degradation of D53, SMXL6, and SMXL7 can be suppressed by removing the conserved Arg–Gly–Lys–Thr motive located in the C-termini of these proteins, suggesting that this domain is critical for their recognition by MAX2 (Jiang et al. 2013; Zhou et al. 2013; Soundappan et al. 2015; Wang et al. 2015; Liang et al. 2016). The exact method of action and targets of the D53, SMAX, and SMXL protein families are still poorly understood.

MAX2 and D14 are expressed in various plant organs, with partial overlap (Stirnberg et al. 2007; Shen et al. 2007; Chevalier et al. 2014). MAX2 is localized mainly to the nucleus (Stirnberg et al. 2007; Shen et al. 2007), while D14 is found in both the nucleus and the cytoplasm (Chevalier et al. 2014). MAX2 expression is spread throughout the plant, mainly within the vascular tissue. In roots, MAX2 expression is observed in vascular, endodermal, and pericycle cells, with decreasing expression toward the base of the root (Stirnberg et al. 2007). Expression of D14 mainly coincides with MAX2 expression in the shoot; however, the root expression is mostly observed in the differentiation and elongation zones but is absent in the meristematic zone of the root tip (Chevalier et al. 2014).

2.5 Use of Synthetic Strigolactones in Research

Strigolactone research is mostly reliant on the use of synthetic SL analogues. This is mainly due to the difficulties associated with the isolation of natural SLs, which are produced in picomolar quantities *in planta* and their extremely short half-lives, due to the easy cleavage of the enol-ether bridge between the C and D rings by nucleophilic agents including water (Mangnus and Zwanenburg 1992), and the fact that natural SLs are too complex for large-scale chemical synthesis. The most common synthetic SL used to date has been GR24 (Fig. 2.2) (Johnson et al. 1981), which has a half-life of approximately 10 days in water (Akiyama et al. 2010) and is therefore considerably more stable than natural SLs. One confounding aspect of strigolactone research to date has been the use of racemic GR24 (*rac*-GR24). All natural strigolactones have an identical stereochemistry on the D ring, which is in the *2′R* configuration (Ćavar et al. 2015). Until relatively recently, most research was conducted with *rac*-GR24, which contains both *2′R* and *2′S* enantiomers in equal proportions. However, *2′S* SL compounds may be able to initiate other, non-SL signaling pathways via KAI2 (Scaffidi et al. 2014). Considering that both D14 and KAI2 operate through MAX2, some early data generated prior to the identification of these receptors and ascribed to SL activity because the use of *max2* mutants may also be due to a non-SL signaling pathway. Consequently, some care is required in interpreting early results generated using *rac* GR24, particularly in combination with *max2* but not *D14* mutants, as some of the effects observed due to the use of the racemic mix may not be true SL-mediated responses.

2.6 Roles of Strigolactones in Root Development and Architecture

Strigolactones have a wide range of outcomes in plant architecture, growth, and development both above- and belowground (Fig. 2.3). In this chapter, we focus on the effects of strigolactones on plant roots. In recent years, it became apparent that a wide range of factors play a role in SL-mediated root development, namely, hormonal cross talk between auxin, cytokinin, and ethylene, nutritional status, and symbiotic relationships with arbuscular mycorrhizal fungi (AMF). SL's effects on roots systems are well documented across different species and affect all major parts of root architecture, including primary roots, lateral roots, root hairs, and adventitious roots (Kapulnik et al. 2011a; Ruyter-Spira et al. 2011; Koltai 2011; Rasmussen et al. 2012b).

2.6.1 Primary Roots

SLs appear to be positive regulators of primary root (PR) length, although these effects may be modified by a variety of environmental factors. Under conditions with favorable phosphate and sugar levels, addition of *rac*-GR24 has been shown to increase PR length in a MAX2-dependent fashion (Jain et al. 2007). However, high concentrations (above 2.5 μM) caused a MAX2-independent reduction in PR length. It is possible that such concentrations, which are considerably in excess of

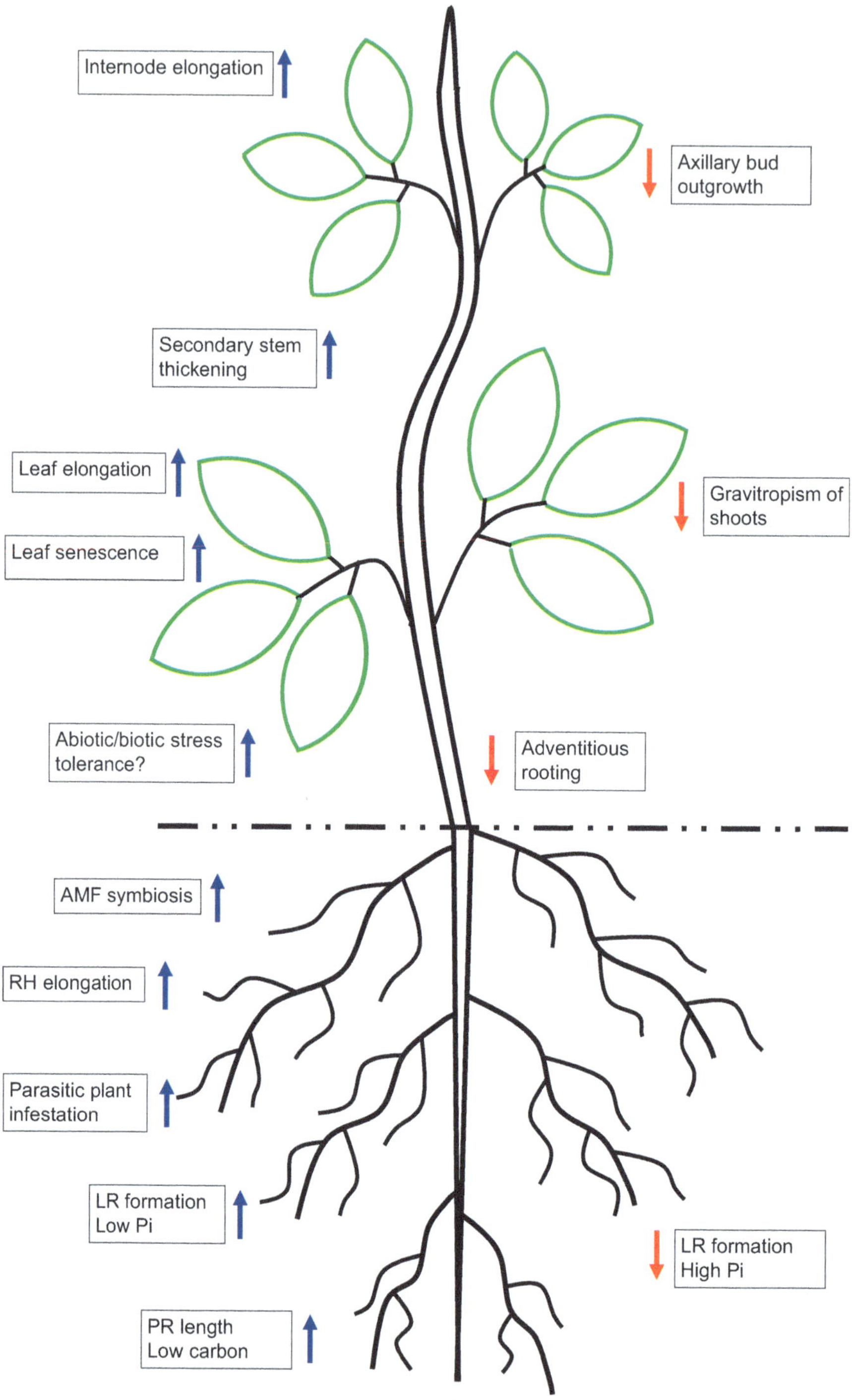
Internode elongation
Axillary bud outgrowth
Secondary stem thickening
Leaf elongation
Leaf senescence
Gravitropism of shoots
Abiotic/biotic stress tolerance?
Adventitious rooting
AMF symbiosis
RH elongation
Parasitic plant infestation
LR formation Low Pi
LR formation High Pi
PR length Low carbon

physiological levels, may be toxic to the plant (Jain et al. 2007; Ruyter-Spira et al. 2011; Shinohara et al. 2013). Carbon-limiting conditions reduce PR lengths in plants. Under these conditions, both SL biosynthetic (*max4*, *max1*) and perception (*max2*) mutants in *Arabidopsis* had shorter primary roots than WT plants. The addition of *rac*-GR24 enhanced PR length in the SL-deficient mutants, but no significant change was observed in the perception mutant (Ruyter-Spira et al. 2011). The effects on PR length in these instances appeared to be mediated through effects on cell division in the cortical cells of the PR meristem (Ruyter-Spira et al. 2011). In rice, SL biosynthesis and perception mutants had shorter seminal roots under conditions with limiting P and N; again the root lengths of SL-deficient, but not insensitive, mutants could be restored to WT phenotype upon GR24 application (Sun et al. 2014). However, GR24 application had no effect on the seminal root length of wild-type plants. In tomato and *Medicago truncatula* too, *rac*-GR24 treatment led to an increase in PR length in biosynthetic mutants but had no effect in wild-type plants (Koltai et al. 2010a; De Cuyper et al. 2015). Taken together, these results suggest that SLs are generally positive regulators of PR elongation, but this varies between species and is largely dependent on the growth conditions.

2.6.2 Lateral Roots

In contrast to the relatively subtle effects they have on PR growth, SLs have considerably more pronounced effects on lateral root (LR) development. These effects are strongly dependent on the P status of the soil. Under P deplete conditions, SLs exhibit a negative effect on LR formation in *Arabidopsis*. Accordingly, *max2*, *max3* and *max4* plants had an increase in LR number when compared to WT lines (Kapulnik et al. 2011a; Ruyter-Spira et al. 2011). Treatment with *rac*-GR24 reduced LR number to wild-type levels in *max3* and *max4* plants, while the *max2* plants showed no changes. However, when plants were grown under P-limiting conditions, *rac*-GR24 had the opposite effect, increasing LR numbers in wild-type and SL-deficient mutants but not in *max2*, suggesting that this effect is indeed mediated through MAX2 (Ruyter-Spira et al. 2011).

Fig. 2.3 Effects of strigolactones on plant growth and development. *Blue arrows* indicate an enhancement/increase, and *red arrows* indicate a reduction/inhibition in growth or development. Aboveground, SLs have a positive effect on internode elongation, secondary stem thickening, leaf elongation, and leaf senescence and appear to enhance both biotic and abiotic stress resistance. SLs have a negative effect on axillary bud outgrowth and gravitropism of shoots. Belowground, SL enhances symbiosis with arbuscular mycorrhizal fungi and the elongation of root hairs. Their effects on roots are often determined by carbon and nutrient status. Primary root length may be enhanced by SLs under carbon-limiting conditions. Depending on the phosphate status, lateral root development may either be enhanced (low P) or reduced (high P). The effects of SLs on adventitious rooting appear to be species-specific. In *Arabidopsis*, SL treatment reduces adventitious rooting, but in rice, crown root lengths are enhanced by SLs

2.6.3 Adventitious and Crown Roots

Adventitious roots (AR) develop from any non-root plant tissue. These may form during normal plant development, as is the case with crown roots in rice plants, or in response to various stress conditions such as wounding or nutrient deprivation. In both *Arabidopsis* and pea, SL-deficient and SL-insensitive mutants have higher rates of adventitious rooting than wild-type plants (Rasmussen et al. 2012a, b). Application of *rac*-GR24 to WT and biosynthetic mutants resulted in a reduction in AR, but not in the perception mutants (Rasmussen et al. 2012a, b).

Rice plants have two types of roots, the seminal roots and crown roots (Hoshikawa 1989). Seminal roots are embryonic roots which only function in the early stages of seedling development; in older plants the crown roots are the main functional roots. These are shoot-borne AR which develop from stem nodes, both above- and belowground. In contrast to *Arabidopsis* and pea, rice SL mutants have fewer and shorter crown roots than wild-type plants (Arite et al. 2012; Sun et al. 2014). This phenotype can be rescued by GR24 application. It is tempting to speculate that the reason for this difference in response to SLs may be due to the differences in function between adventitious roots in these different species. In rice, where crown roots are the main functional roots, the effects of SLs on these roots are more like the effects of SLs on PR length in plants which have a PR-based root system. It would be interesting to determine the effects of SLs on other forms of adventitious roots, such as nodal roots in strawberries and prop roots. Nonetheless, the effects of LS on root initiation from non-root tissues are strongly species-specific. SLs are negative regulators of adventitious rooting in *Arabidopsis* and pea but positive regulators of crown root growth in rice.

2.6.4 Root Hairs

In *Arabidopsis* and tomatoes, *rac*-GR24 promotes root hair (RH) elongation (Koltai et al. 2010a; Kapulnik et al. 2011a). However, neither SL-deficient nor SL-insensitive mutants have shorter RH than wild type (Kapulnik et al. 2011a; Kohlen et al. 2013; Pandya-Kumar et al. 2014), suggesting that factors other than SLs also play a role in the overall control of root hair elongation. Intriguingly, under low P conditions, both *max2* and *max4* mutants displayed a decrease in RH number and density (Mayzlish-Gati et al. 2012). This effect could be rescued by the application of high doses of rac-GR24. Consequently, SLs also appear to be positive regulators of RH elongation.

There is still considerable need and scope for further research into the effects of SLs on all four forms of root development discussed above. Equally clear is that root growth and development are end results of a complex interplay between many components, of which SLs are just one player. This is evident form the differing responses of plants to SL treatments under different nutritional states. Furthermore, other phytohormones play critical roles in the regulation of root development. It is therefore important to consider SL effects in the light of nutrient status and cross talk with other phytohormones.

2.7 Role of Strigolactones in Root Development in Association with Nutrient Status

Root development is directly linked to the plant's ability to absorb macro- and micronutrients from the soil. RHs play a pivotal role in this nutrient acquisition (Gilroy and Jones 2000). Unsurprisingly deficiency in phosphate (P), iron (Fe), and nitrogen (N) can all induce the formation and elongation of RHs (López-Bucio et al. 2003; Muller and Schmidt 2004). P levels were shown to influence the architecture of primary and lateral roots and play a part in the plant's adaptive growth processes (Hell and Hillebrand 2001; López-Bucio 2002; Osmont et al. 2007). The aluminum (Al), calcium (Ca), sulfur (S), nitrogen (N), and iron (Fe) contents of the soil all have a marked effect on PRL and LR initiation (López-Bucio et al. 2003; Osmont et al. 2007).

Phytohormones play roles in RH, PR, and LR development in association with the plant nutritional status (Mayzlish-Gati et al. 2012; Czarnecki et al. 2013; Niu et al. 2013). In *Arabidopsis* LR formation is increased when plants are grown under low P, possibly though increased sensitivity though the auxinTIR1 receptor (Perez-Torres et al. 2008; Ruyter-Spira et al. 2011). A second example is the RH elongation response to low Fe, which was shown to require two hormone pathways, ethylene and auxin. The nutrition environment can also influence hormone synthesis; for instance, plants grown under low nitrate conditions had increased auxin levels in their roots (Caba et al. 2000; Walch-Liu et al. 2006).

Since SLs are known to affect most aspects of root architecture, PR, LR, RH, and AR as discussed above, this could place SL in a key role to regulate these aspects under nutrition-restrictive conditions. SL can enhance RH elongation, and this can facilitate more efficient nutrient uptake (Kapulnik et al. 2011a). Similarly, SL increases LR formation under P limitation, and this can be a crucial way to survive under nonoptimal growth conditions (Al-Ghazi et al. 2003; Osmont et al. 2007). SL synthesis and exudation increase when P and N are below optimal concentration in the soil (Yoneyama et al. 2007; López-Ráez and Bouwmeester 2008; López-Ráez et al. 2008; Kohlen et al. 2011). The actions of phytohormones are not the only method plants use to enhance their root systems to better nutrient accumulation; symbiotic relationships with arbuscular mycorrhizal fungi (AMF) can be exploited to gain access to increased nutrient uptake.

2.7.1 Arbuscular Mycorrhizal Fungi

Most land-based plants utilize a symbiotic relationship with arbuscular mycorrhizal fungi (AMF) to increase the reach of their root systems; this is an ancient form of symbiosis and is considered to be at least partially responsible for the success of the plant kingdom (Remy et al. 1994; Smith and Read 2008). Fittingly, considering their roles in enhancing plant root systems for increased nutrient acquisition under conditions of nutrient stress, SLs also act as plant-derived signal molecules for AMF to identify their host roots (Simon et al. 1993; Bouwmeester et al. 2007; Remy et al. 1994; Smith and Read 2008).

AMF spores can germinate independently of their hosts, but the germinated spores show restricted growth and hyphal development. Extensive branching only occurs in close vicinity to the host root systems (Ganinazzi-Pearson et al. 1989; Giovannetti et al. 1996; Buee et al. 2000; Nagahashi and Douds 2000). SLs exuded from plant roots trigger a very wide range of effects, which include spore germination, hyphal architecture, mitosis, respiratory reaction, and various changes in gene expression (Akiyama et al. 2005; Besserer et al. 2006, 2008; Tisserant et al. 2012; Genre et al. 2013). SL are not absolutely critical for AMF symbiosis, but they still play a major role as SL biosynthetic mutants of pea, rice, tomato, and petunia have much lower levels of mycorrhization associated with their roots (Gomez-Roldan et al. 2008; Vogel et al. 2009; Koltai et al. 2010b; Kretzschmar et al. 2012; Gutjahr et al. 2012; Kohlen et al. 2012; Yoshida et al. 2012).

During the symbiotic phase, the AMF enter plant root epidermal cells and proliferate within the root cortical cells. The AMF remain localized in the apoplastic compartment of the root cortex (Smith and Read 2008). It is in this location where the AMF produce distinctive structures called arbuscules (Smith and Read 2008). Nutrient exchange is proposed to happen primarily in this specialized fungal structure (Bucher 2007). The AMF increase the total volume of soil plant roots can interact with; this consequently helps the plant to overcome the limitations of the nutrient depletion zone around the roots and increase mineral acquisition from the soil. Studies have shown that P uptake and availability to roots is enhanced by AMF though increasing the root absorption surface. AMF also actively import and store P in fungal vacuoles (Harrison and Buuren 1995; Maldonado-Mendoza et al. 2001; Smith et al. 2003; Benedetto et al. 2005; Bucher 2007).

In exchange for P, the fungus receives fixed carbon from its host. This can range between 4 and 20% of the total photosynthates in different plants (Bago et al. 2000). This carbon exchange can affect carbon flux and partitioning in plants (Bago et al. 2000). Despite all the signals that can be triggered in a plant by a change in carbon partitioning, the major signal molecule between plants and AMF remains SLs (Besserer et al. 2006). There is still uncertainty in the precise mechanism for SL perception in AMF, but the fungal SL receptor is highly sensitive because it is able to detect *rac*-GR24 at nanomolar concentrations (Besserer et al. 2006). Neither the gene encoding for this receptor nor its protein structure has yet been identified. For AMF to perceive SLs, these must be transported from roots to soil. In petunia this exudation is performed by an ABC transport protein named PLEIOTROPIC DRUG RESISTENCE1 (PDR1) (Kretzschmar et al. 2012). SL transporters in other species have yet to be identified. It is still uncertain whether SLs play a role in determining host specificity in AMF. Non-AMF-host plant root exudates could not trigger AMF hyphal branching (Buee et al. 2000; Nagahashi and Douds 2000), and intriguingly SLs present in AMF-host plants are also present in non-AMF-host plants *Arabidopsis, and* Lupin (Goldwasser et al. 2008; Yoneyama et al. 2008).

2.8 Hormonal Cross Talk

As noted above, SLs play a pivotal role in above- and belowground plant architecture, growth, and development. There are several other phytohormones that also have prominent roles in these processes, including auxin, cytokinin, and ethylene. It is reasonable to assume that SLs must work in concert with these phytohormones to regulate shoot and root development in plants.

2.8.1 Auxins

In terms of interaction and cross talk between SLs and other phytohormones, the interaction between SL and auxin is probably the best studied. One of the current hypotheses on cross talk between these hormones is that SL plays a more indirect role, whereby it inhibits PIN1 activity, changing the sink-source dynamic and reducing auxin export from buds (Bennett et al. 2014). Similarly, SLs may regulate RH elongation by modulation of the auxin flux in roots. Koltai et al. (2010a) proposed that SL can negatively affect the activity of auxin efflux carriers in the presence of exogenously added auxin. It was elucidated that applied *rac*-GR24 lowers the expression of *PIN1*, *PIN3*, and *PIN7* in the root tip vascular tissue (Ruyter-Spira et al. 2011). RH elongate with increased auxin concentrations in epidermal cells in a non-auxin transport-dependent fashion (Pitts et al. 1998). This auxin transport through non-hair epidermal cells is facilitated by AUX1 (Jones et al. 2009). Consequently it is feasible that SL enhancement of RH elongation functions via auxin (Kapulnik et al. 2011b).

The change in *Arabidopsis* LR density is also linked to the plant's auxin status (Ruyter-Spira et al. 2011). When concentrations of auxin were relatively low, *rac*-GR24 treatments lead to a reduction in LR density. In an environment where auxin concentrations were higher, the addition of *rac*-GR24 resulted in a higher LR density (Ruyter-Spira et al. 2011). This result could be explained by SL modulation of the auxin flux through PIN1. Under lower auxin conditions, *rac*-GR24s reduction of auxin levels allow these to drop below the optimal concentration for LR development, whereas under high auxin conditions, the fall in auxin might push the flux into the optimal range for LR development (Ruyter-Spira et al. 2011).

2.8.2 Cytokinins

Cytokinins (CKs) are phytohormones that play a prominent role in root development and its regulation. As with SLs, CKs reduce LR formation by reducing the activity of PIN auxin transporters (Vanstraelen and Benková 2012). Recently an interaction between CK signaling and SL via the AHK3/ARR1/ARR12 signaling complex that associates with GR24 to inhibit LR development was elucidated (Jiang et al. 2016). Mutants of this CK signaling complex are nonresponsive to *rac*-GR24, and LR reduction was not observed (Jiang et al. 2016).

2.8.3 Ethylene

Cross talk between SL and ethylene is interesting in the sense that SL signaling seems to be not required for an ethylene response, but the reverse seems to be true. Ethylene appears to be involved in the SL response. The *etr* and *ein* mutants involved in ethylene signaling presented a non-GR24 responsive RH phenotype. Ethylene can then be seen to be epistatic to SLs (Kapulnik et al. 2011b). Furthermore, GR24-induced expression of ACS (1-aminocyclopropane-1 carboxylate synthase, ACC synthase) genes in tomato and *Arabidopsis*, the activity of which is a critical step in ethylene biosynthesis and is strongly correlated with the amount of ethylene being produced (Yamamoto et al. 1995; Barry et al. 2000; Yamagami et al. 2003).

2.9 Concluding Perspectives

In the past decade since their identification as previously unidentified class of plant hormones, major strides have been made in unraveling the biosynthesis, signal transduction mechanism, and physiological effects of SLs. However, there are still many questions to answer. We are still unsure of all the steps in the biosynthetic pathway. We understand much of the signaling pathways, which has led to a better understanding of how this hormone functions, but we do not know how the downstream targets of MAX2, the SMXL family, control the physiological changes that we can observe in response to SL treatment. We do not yet fully understand how different species can respond in different ways to the same SL treatment, or even why this should be so. Increasing the complexity of our task in unraveling these questions is the interaction between SLs and various abiotic factors such as nutrient status. SLs appear to play an important role in the mycorrhizal symbiosis, but we do not yet understand how the fungus perceives SLs and whether these play any role in aiding the fungus to identify its host roots. Finally, as with much hormone research, the ability to understand SL functioning cannot occur in a vacuum; SL actions are modulated by their interaction with other phytohormones and their signaling pathways. Teasing these interactions apart is tedious and time-consuming but necessary. Hopefully, the answers to these questions will be quickly elucidated.

References

Abe S, Sado A, Tanaka K, Kisugi T, Asami K, Ota S, Il KH, Yoneyama K, Xie X, Ohnishi T, Seto Y, Yamaguchi S, Akiyama K, Yoneyama K, Nomura T (2014) Carlactone is converted to carlactonoic acid by MAX1 in *Arabidopsis* and its methyl ester can directly interact with AtD14 *in vitro*. Proc Natl Acad Sci USA 111:18084–18089

Akiyama K, Matsuzaki K-I, Hayashi H (2005) Plant sesquiterpenes induce hyphal branching in arbuscular mycorrhizal fungi. Nature 435:824–827

Akiyama K, Ogasawara S, Ito S, Hayashi H (2010) Structural requirements of strigolactones for hyphal branching in AM fungi. Plant Cell Physiol 51:1104–1117

Al-Ghazi Y, Muller B, Pinloche S, Tranbarger TJ, Nacry P, Rossignol M, Tardieu F, Doumas P (2003) Temporal responses of *Arabidopsis* root architecture to phosphate starvation: evidence for the involvement of auxin signalling. Plant Cell Environ 26:1053–1066

Arite T, Iwata H, Ohshima K, Maekawa M, Nakajima M, Kojima M, Sakakibara H, Kyozuka J (2007) DWARF10, an RMS1/MAX4/DAD1 ortholog, controls lateral bud outgrowth in rice. Plant J 51:1019–1029

Arite T, Umehara M, Ishikawa S, Hanada A, Maekawa M, Yamaguchi S, Kyozuka J (2009) *d14*, a strigolactone-insensitive mutant of rice, shows an accelerated outgrowth of tillers. Plant Cell Physiol 50:1416–1424

Arite T, Kameoka H, Kyozuka J (2012) Strigolactone positively controls crown root elongation in rice. J Plant Growth Regul 31:165–172

Bago B, Pfeffer PE, Shachar-Hill Y (2000) Carbon metabolism and transport in arbuscular mycorrhizas. Plant Physiol 124:949–957

Bainbridge K, Sorefan K, Ward S, Leyser O (2005) Hormonally controlled expression of the *Arabidopsis MAX4* shoot branching regulatory gene. Plant J 44:569–580

Barry CS, Llop-Tous MI, Grierson D (2000) The regulation of 1-aminocyclopropane-1-carboxylic acid synthase gene expression during the transition from system-1 to system-2 ethylene synthesis in tomato. Plant Physiol 123:979–986

Benedetto A, Magurno F, Bonfante P, Lanfranco L (2005) Expression profiles of a phosphate transporter gene (*GmosPT*) from the endomycorrhizal fungus *Glomus mosseae*. Mycorrhiza 15:620–627

Bennett T, Hines G, Leyser O (2014) Canalization: what the flux? Trends Genet 30:41–48

Besserer A, Puech-Pagès V, Kiefer P, Gomez-Roldan V, Jauneau A, Roy S, Portais J-C, Roux C, Bécard G, Séjalon-Delmas N (2006) Strigolactones stimulate arbuscular mycorrhizal fungi by activating mitochondria. PLoS Biol 4:e226

Besserer A, Becard G, Jauneau A, Roux C, Sejalon-Delmas N (2008) GR24, a synthetic analogue of strigolactones, stimulates the mitosis and growth of the arbuscular mycorrhizal fungus *Gigaspora rosea* by boosting its energy metabolism. Plant Physiol 148:108.121400

Beveridge CA, Ross JJ, Murfet IC (1996) Branching in pea - Action of genes *Rms3* and *Rms4*. Plant Physiol 110:859–865

Booker J, Auldridge M, Wills S, McCarty D, Klee H, Leyser O (2004) MAX3/CCD7 is a carotenoid cleavage dioxygenase required for the synthesis of a novel plant signaling molecule. Curr Biol 14:1232–1238

Booker J, Sieberer T, Wright W, Williamson L, Willett B, Stirnberg P, Turnbull C, Srinivasan M, Goddard P, Leyser O (2005) *MAX1* encodes a cytochrome P450 family member that acts downstream of MAX3/4 to produce a carotenoid-derived branch-inhibiting hormone. Dev Cell 8:443–449

Bouvier F, Suire C, Mutterer J, Camara B (2003) Oxidative remodeling of chromoplast carotenoids: identification of the carotenoid dioxygenase CsCCD and CsZCD genes involved in crocus secondary metabolite biogenesis. Plant Cell Online 15:47–62

Bouwmeester HJ, Roux C, Lopez-Raez JA, Bécard G (2007) Rhizosphere communication of plants, parasitic plants and AM fungi. Trends Plant Sci 12:224–230

Brewer PB, Yoneyama K, Filardo F, Meyers E, Scaffidi A, Frickey T et al (2016) Lateral branching oxidoreductase acts in the final stages of strigolactone biosynthesis in *Arabidopsis*. Proc Natl Acad Sci USA 113:6301–6306

Bucher M (2007) Functional biology of plant phosphate uptake at root and mycorrhiza interfaces. New Phytol 173:11–26

Buee M, Rossignol M, Jauneau A, Ranjeva R, Bécard G (2000) The pre-symbiotic growth of arbuscular mycorrhizal fungi is induced by a branching factor partially purified from plant root exudates. Mol Plant Microbe Interact 13:693–698

Caba JM, Centeno ML, Fernández B, Gresshoff PM, Ligero F (2000) Inoculation and nitrate alter phytohormone levels in soybean roots: differences between a supernodulating mutant and the wild type. Planta 211:98–104

Ćavar S, Zwanenburg B, Tarkowski P (2015) Strigolactones: occurrence, structure, and biological activity in the rhizosphere. Phytochem Rev 14:691–711

Chevalier F, Nieminen K, Sanchez-Ferrero JC, Rodriguez ML, Chagoyen M, Hardtke CS, Cubas P (2014) Strigolactone promotes degradation of DWARF14, an α/ß hydrolase essential for strigolactone signaling in *Arabidopsis*. Plant Cell 26:1134–1150

Cook C, Whichhard LP, Turner B, Wall ME, Egley GH (1966) Germination of witchweed (*Striga lutea* Lour.): isolation and properties of a potent stimulant. Sci New Ser 154:1189–1190

Czarnecki O, Yang J, Weston D, Tuskan G, Chen J-G (2013) A dual role of strigolactones in phosphate acquisition and utilization in plants. Int J Mol Sci 14:7681–7701

De Cuyper C, Fromentin J, Yocgo RE, De Keyser A, Guillotin B, Kunert K, Boyer F-D, Goormachtig S (2015) From lateral root density to nodule number, the strigolactone analogue GR24 shapes the root architecture of *Medicago truncatula*. J Exp Bot 66:137–146

de Saint Germain A, Clave G, Badet-Denisot M-A, Pillot J-P, Cornu D, Le Caer J-P, Burger M, Pelissier F, Retailleau P, Turnbull C, Bonhomme S, Chory J, Rameau C, Boyer F-D (2016) An histidine covalent receptor and butenolide complex mediates strigolactone perception. Nat Chem Biol 12:787–794

Devoto A, Nieto-Rostro M, Xie D, Ellis C, Harmston R, Patrick E, Davis J, Sherratt L, Coleman M, Turner JG (2002) COI1 links jasmonate signalling and fertility to the SCF ubiquitin-ligase complex in *Arabidopsis*. Plant J 32:457–466

Ganinazzi-Pearson V, Branzanti B, Gianinazzi S (1989) *In vitro* enhancement of spore germination and early hyphal growth of a vesicular-arbuscular mycorrhizal fungus by host root exudates and plant flavonoids. Symbiosis 7:243–255

Genre A, Chabaud M, Balzergue C, Puech-Pagès V, Novero M, Rey T, Fournier J, Rochange S, Bécard G, Bonfante P, Barker DG (2013) Short-chain chitin oligomers from arbuscular mycorrhizal fungi trigger nuclear Ca^{2+} spiking in *Medicago truncatula* roots and their production is enhanced by strigolactone. New Phytol 198:190–202

Gilroy S, Jones DL (2000) Through form to function: root hair development and nutrient uptake. Trends Plant Sci 5:56–60

Giovannetti M, Sbrana C, Silvia A, Avio L (1996) Analysis of factors involved in fungal recognition response to host-derived signals by arbuscular mycorrhizal fungi. New Phytol 133:65–71

Goldwasser Y, Yoneyama K, Xie X, Yoneyama K (2008) Production of Strigolactones by *Arabidopsis thaliana* responsible for *Orobanche aegyptiaca* seed germination. Plant Growth Regul 55:21–28

Gomez-Roldan V, Fermas S, Brewer PB, Puech-Pagès V, Dun EA, Pillot J-P, Letisse F, Matusova R, Danoun S, Portais J-C, Bouwmeester H, Bécard G, Beveridge CA, Rameau C, Rochange SF (2008) Strigolactone inhibition of shoot branching. Nature 455:189–194

Gutjahr C, Radovanovic D, Geoffroy J, Zhang Q, Siegler H, Chiapello M, Casieri L, An K, An G, Guiderdoni E, Kumar CS, Sundaresan V, Harrison MJ, Paszkowski U (2012) The half-size ABC transporters STR1 and STR2 are indispensable for mycorrhizal arbuscule formation in rice. Plant J 69:906–920

Hamiaux C, Drummond RSM, Janssen BJ, Ledger SE, Cooney JM, Newcomb RD, Snowden KC (2012) DAD2 is an α/β hydrolase likely to be involved in the perception of the plant branching hormone, strigolactone. Curr Biol 22:2032–2036

Harrison MJ, Buuren ML van (1995) A phosphate transporter from the mycorrhizal fungus *Glomus versiforme*. Nature 378:626–629

Hell R, Hillebrand H (2001) Plant concepts for mineral acquisition and allocation. Curr Opin Biotechnol 12:161–168

Hoshikawa K (1989) The growing rice plant: an anatomical monograph. Nobunkyo XVI, Tokyo

Ishikawa S, Maekawa M, Arite T, Onishi K, Takamure I, Kyozuka J (2005) Suppression of tiller bud activity in tillering dwarf mutants of rice. Plant Cell Physiol 46:79–86

Jain A, Poling MD, Karthikeyan AS, Blakeslee JJ, Peer WA, Titapiwatanakun B, Murphy AS, Raghothama KG (2007) Differential effects of sucrose and auxin on localized phosphate deficiency-induced modulation of different traits of root system architecture in *Arabidopsis*. Plant Physiol 144:232–247

Jiang L, Liu X, Xiong G, Liu H, Chen F, Wang L, Meng X, Liu G, Yu H, Yuan Y, Yi W, Zhao L, Ma H, He Y, Wu Z, Melcher K, Qian Q, Xu HE, Wang Y, Li J (2013) DWARF 53 acts as a repressor of strigolactone signalling in rice. Nature 504:401–405

Jiang L, Matthys C, Marquez-Garcia B, De Cuyper C, Smet L, De Keyser A, Boyer F-D, Beeckman T, Depuydt S, Goormachtig S (2016) Strigolactones spatially influence lateral root development through the cytokinin signaling network. J Exp Bot 67:379–389

Johnson AW, Gowada G, Hassanali A, Knox J, Monaco S, Razavi Z, Rosebery G (1981) The preparation of synthetic analogues of strigol. J Chem Soc Perkin Trans 1:1734

Johnson X, Brcich T, Dun EA, Goussot M, Haurogne K, Beveridge CA, Rameau C (2006) Branching genes are conserved across species. genes controlling a novel signal in pea are coregulated by other long-distance signals. Plant Physiol 142:1014–1026

Jones AR, Kramer EM, Knox K, Swarup R, Malcolm J, Lazarus CM, Leyser HMO, Grierson CS (2009) Auxin transport through non-hair cells sustain root-hair development. Plant Cell 11:78–84

Kapulnik Y, Delaux P-M, Resnick N, Mayzlish-Gati E, Wininger S, Bhattacharya C, Séjalon-Delmas N, Combier J-P, Bécard G, Belausov E, Beeckman T, Dor E, Hershenhorn J, Koltai H (2011a) Strigolactones affect lateral root formation and root-hair elongation in *Arabidopsis*. Planta 233:209–216

Kapulnik Y, Resnick N, Mayzlish-Gati E, Kaplan Y, Wininger S, Hershenhorn J, Koltai H (2011b) Strigolactones interact with ethylene and auxin in regulating root-hair elongation in *Arabidopsis*. J Exp Bot 62:2915–2924

Kohlen W, Charnikhova T, Liu Q, Bours R, Domagalska MA, Beguerie S, Verstappen F, Leyser O, Bouwmeester H, Ruyter-Spira C (2011) Strigolactones are transported through the xylem and play a key role in shoot architectural response to phosphate deficiency in nonarbuscular mycorrhizal host *Arabidopsis*. Plant Physiol 155:974–987

Kohlen W, Charnikhova T, Lammers M, Pollina T, Tóth P, Haider I, Pozo MJ, de Maagd RA, Ruyter-Spira C, Bouwmeester HJ, López-Ráez JA (2012) The tomato carotenoid cleavage dioxygenase8 (SlCCD8) regulates rhizosphere signaling, plant architecture and affects reproductive development through strigolactone biosynthesis. New Phytol 196:535–547

Kohlen W, Charnikhova T, Bours R, López-Ráez JA, Bouwmeester H (2013) Tomato strigolactones: a more detailed look. Plant Signal Behav 8:e22785

Koltai H (2011) Strigolactones are regulators of root development. New Phytol 190:545–549

Koltai H, Dor E, Hershenhorn J, Joel DM, Weininger S, Lekalla S, Shealtiel H, Bhattacharya C, Eliahu E, Resnick N, Barg R, Kapulnik Y (2010a) Strigolactones' effect on root growth and root-hair elongation may be mediated by auxin-efflux carriers. J Plant Growth Regul 29:129–136

Koltai H, Lekkala SP, Bhattacharya C, Mayzlish-Gati E, Resnick N, Wininger S, Dor E, Yoneyama K, Yoneyama K, Hershenhorn J, Joel DM, Kapulnik Y (2010b) A tomato strigolactone-impaired mutant displays aberrant shoot morphology and plant interactions. J Exp Bot 61:1739–1749

Kretzschmar T, Kohlen W, Sasse J, Borghi L, Schlegel M, Bachelier JB, Reinhardt D, Bours R, Bouwmeester HJ, Martinoia E (2012) A petunia ABC protein controls strigolactone-dependent symbiotic signalling and branching. Nature 483:341–344

Liang Y, Ward S, Li P, Bennett T, Leyser O (2016) SMAX1-LIKE7 signals from the nucleus to regulate shoot development in *Arabidopsis* via partially EAR motif-independent mechanisms. Plant Cell 28(7). https://doi.org/10.1105/tpc.16.00286

Lin H, Wang R, Qian Q, Yan M, Meng X, Fu Z, Yan C, Jiang B, Su Z, Li J, Wang Y (2009) DWARF27, an iron-containing protein required for the biosynthesis of strigolactones, regulates rice tiller bud outgrowth. Plant Cell Online 21:1512–1525

López-Bucio J (2002) Phosphate availability alters architecture and causes changes in hormone sensitivity in the *Arabidopsis* root system. Plant Physiol 129:244–256

López-Bucio J, Cruz-Ramırez A, Herrera-Estrella L (2003) The role of nutrient availability in regulating root architecture. Curr Opin Plant Biol 6:280–287

López-Ráez JA, Bouwmeester H (2008) Fine-tuning regulation of strigolactone biosynthesis under phosphate starvation. Plant Signal Behav 3:963–965

López-Ráez JA, Charnikhova T, Gómez-Roldán V, Matusova R, Kohlen W, De Vos R, Verstappen F, Puech-Pages V, Bécard G, Mulder P, Bouwmeester H (2008) Tomato strigolactones are derived from carotenoids and their biosynthesis is promoted by phosphate starvation. New Phytol 178:863–874

Maldonado-Mendoza I, Dewbre G, Harrison M (2001) A phosphate transporter gene from the extraradical mycelium of an arbuscular mycorrhizal fungus *Glomus intraradices* is regulated in response to phosphate in the environment. Mol Plant Microbe Interact 14:1140–1148

Mangnus EM, Zwanenburg B (1992) Tentative molecular mechanism for germination stimulation of *Striga* and *Orobanche* seeds by strigol and its synthetic analogs. J Agric Food Chem 40:1066–1070

Mangnus EM, Dommerholt FJ, De Jong RLP, Zwanenburg B (1992) Improved synthesis of strigol analog GR24 and evaluation of the biological activity of its diastereomers. J Agric Food Chem 40:1230–1235

Matthys C, Walton A, Struk S, Stes E, Boyer F-D, Gevaert K, Goormachtig S (2016) The whats, the wheres and the hows of strigolactone action in the roots. Planta 243:1327–1337

Matusova R (2005) The strigolactone germination stimulants of the plant-parasitic striga and orobanche spp. are derived from the carotenoid pathway. Plant Physiol 139:920–934

Mayzlish-Gati E, De-Cuyper C, Goormachtig S, Beeckman T, Vuylsteke M, Brewer PB et al. (2012) Strigolactones are involved in root response to low phosphate conditions in *Arabidopsis*. Plant Physiol 160:1329–1341

McGinnis KM (2003) The *Arabidopsis SLEEPY1* gene encodes a putative F-box subunit of an SCF E3 ubiquitin ligase. Plant Cell Online 15:1120–1130

Muller M, Schmidt W (2004) Environmentally induced plasticity of root hair development in *Arabidopsis*. Plant Physiol 134:409–419

Nagahashi G, Douds DD (2000) Partial separation of root exudate components and their effects upon the growth of germinated spores of AM fungi. Mycol Res 104:1453–1464

Nakamura H, Xue Y-L, Miyakawa T, Hou F, Qin H-M, Fukui K, Shi X, Ito E, Ito S, Park S-H, Miyauchi Y, Asano A, Totsuka N, Ueda T, Tanokura M, Asami T (2013) Molecular mechanism of strigolactone perception by DWARF14. Nat Commun 4:1–10

Nelson DC, Scaffidi A, Dun EA, Waters MT, Flematti GR, Dixon KW, Beveridge CA, Ghisalberti EL, Smith SM (2011) F-box protein MAX2 has dual roles in karrikin and strigolactone signaling in *Arabidopsis thaliana*. Proc Natl Acad Sci USA 108:8897–8902

Niu YF, Chai RS, Jin GL, Wang H, Tang CX, Zhang YS (2013) Responses of root architecture development to low phosphorus availability: a review. Ann Bot 112:391–408

Osmont KS, Sibout R, Hardtke CS (2007) Hidden branches: developments in root system architecture. Annu Rev Plant Biol 58:93–113

Pandya-Kumar N, Shema R, Kumar M, Mayzlish-Gati E, Levy D, Zemach H, Belausov E, Wininger S, Abu-Abied M, Kapulnik Y, Koltai H (2014) Strigolactone analog GR24 triggers changes in PIN2 polarity, vesicle trafficking and actin filament architecture. New Phytol 202:1184–1196

Perez-Torres C-A, Lopez-Bucio J, Cruz-Ramirez A, Ibarra-Laclette E, Dharmasiri S, Estelle M, Herrera-Estrella L (2008) Phosphate availability alters lateral root development in *Arabidopsis* by modulating auxin sensitivity via a mechanism involving the TIR1 auxin receptor. Plant Cell Online 20:3258–3272

Pitts RJ, Cernac A, Estelle M (1998) Auxin and ethylene promote root hair elongation in *Arabidopsis*. Plant J 16:553–560

Rasmussen A, Beveridge CA, Geelen D (2012a) Inhibition of strigolactones promotes adventitious root formation. Plant Signal Behav 7:694–697

Rasmussen A, Mason MG, De Cuyper C, Brewer PB, Herold S, Agusti J, Geelen D, Greb T, Goormachtig S, Beeckman T, Beveridge CA (2012b) Strigolactones suppress adventitious rooting in *Arabidopsis* and Pea. Plant Physiol 158:1976–1987

Remy W, Taylortt TN, Hass H, Kerp H (1994) Four hundred-million-year-old vesicular arbuscular mycorrhizae. Proc Natl Acad Sci USA 91:11841–11843

Ruyter-Spira C, Kohlen W, Charnikhova T, van Zeijl A, van Bezouwen L, de Ruijter N, Cardoso C et al (2011) Physiological effects of the synthetic strigolactone analog GR24 on root system architecture in *Arabidopsis*: another belowground role for strigolactones? Plant Physiol 155:721–734

Ruyter-Spira C, Al-Babili S, van der Krol S, Bouwmeester H (2013) The biology of strigolactones. Trends Plant Sci 18:72–83

Sasse J, Simon S, Gübeli C, Liu GW, Cheng X, Friml J, Bouwmeester H, Martinoia E, Borghi L (2015) Asymmetric localizations of the ABC transporter PaPDR1 trace paths of directional strigolactone transport. Curr Biol 25:647–655

Scaffidi A, Waters MT, Sun YK, Skelton BW, Dixon KW, Ghisalberti EL, Flematti GR, Smith SM (2014) Strigolactone hormones and their stereoisomers signal through two related receptor proteins to induce different physiological responses in *Arabidopsis*. Plant Physiol 165:1221–1232

Schwartz SH, Qin X, Zeevaart JAD (2001) Characterization of a novel carotenoid cleavage dioxygenase from plants. J Biol Chem 276:25208–25211

Seto Y, Sado A, Asami K, Hanada A, Umehara M, Akiyama K, Yamaguchi S (2014) Carlactone is an endogenous biosynthetic precursor for strigolactones. Proc Natl Acad Sci USA 111:1640–1645

Shen H, Luong P, Huq E (2007) The F-box protein MAX2 functions as a positive regulator of photomorphogenesis in *Arabidopsis*. Plant Physiol 145:1471–1483

Shinohara N, Taylor C, Leyser O (2013) Strigolactone can promote or inhibit shoot branching by triggering rapid depletion of the auxin efflux protein PIN1 from the plasma membrane. PLoS Biol 11:e1001474

Simon L, Bousquet J, Lévesque RC, Lalonde M (1993) Origin and diversification of endomycorrhizal fungi and coin. Nature 363:67–69

Smith SE, Read D (2008) Mycorrhizal symbiosis, 3rd edn. Elsevier, Burlington, MA, pp vii–vix

Smith SE, Smith FA, Jakobsen I, Smith SE, Smith FA, Jakobsen I (2003) Mycorrhizal fungi can dominate phosphate supply to plants irrespective of growth. Plant Physiol 133:16–20

Sorefan K (2003) *MAX4* and *RMS1* are orthologous dioxygenase-like genes that regulate shoot branching in *Arabidopsis* and pea. Genes Dev 17:1469–1474

Soundappan I, Bennett T, Morffy N, Liang Y, Stanga JP, Abbas A, Leyser O, Nelson DC (2015) SMAX1-LIKE/D53 Family members enable distinct MAX2-dependent responses to strigolactones and karrikins in *Arabidopsis*. Plant Cell 27:3143–3159

Stanga JP, Smith SM, Briggs WR, Nelson DC (2013) *SUPPRESSOR OF MORE AXILLARY GROWTH2 1* controls seed germination and seedling development in *Arabidopsis*. Plant Physiol 163:318–330

Stirnberg P, Furner IJ, Ottoline Leyser HM (2007) MAX2 participates in an SCF complex which acts locally at the node to suppress shoot branching. Plant J 50:80–94

Sun H, Tao J, Liu S, Huang S, Chen S, Xie X, Yoneyama K, Zhang Y, Xu G (2014) Strigolactones are involved in phosphate- and nitrate-deficiency-induced root development and auxin transport in rice. J Exp Bot 65:6735–6746

Tan X, Calderon-Villalobos LIA, Sharon M, Zheng C, Robinson CV, Estelle M, Zheng N (2007) Mechanism of auxin perception by the TIR1 ubiquitin ligase. Nature 446:640–645

Tisserant E, Kohler A, Dozolme-Seddas P, Balestrini R, Benabdellah K, Colard A et al (2012) The transcriptome of the arbuscular mycorrhizal fungus *Glomus intraradices* (DAOM 197198) reveals functional tradeoffs in an obligate symbiont. New Phytol 193:755–769

Turnbull CGN, Booker JP, Leyser HMO (2002) Micrografting techniques for testing long-distance signalling in *Arabidopsis*. Plant J 32:255–262

Umehara M, Cao M, Akiyama K, Akatsu T, Seto Y, Hanada A, Li W, Takeda-Kamiya N, Morimoto Y, Yamaguchi S (2015) Structural requirements of strigolactones for shoot branching inhibition in Rice and *Arabidopsis*. Plant Cell Physiol 56:1059–1072

Vanstraelen M, Benková E (2012) Hormonal interactions in the regulation of plant development. Annu Rev Cell Dev Biol 28:463–487

Vogel JT, Walter MH, Giavalisco P, Lytovchenko A, Kohlen W, Charnikhova T, Simkin AJ, Goulet C, Strack D, Bouwmeester HJ, Fernie AR, Klee HJ (2009) SlCCD7 controls strigolactone biosynthesis, shoot branching and mycorrhiza-induced apocarotenoid formation in tomato. Plant J 61:300–311

Walch-Liu P, Ivanov II, Filleur S, Gan Y, Remans T, Forde BG (2006) Nitrogen regulation of root branching. Ann Bot 97:875–881

Wang L, Wang B, Jiang L, Liu X, Li X, Lu Z, Meng X, Wang Y, Smith SM, Li J (2015) Strigolactone signaling in *Arabidopsis* regulates shoot development by targeting D53-like SMXL repressor proteins for ubiquitination and degradation. Plant Cell 27:3128–3142

Waters MT, Brewer PB, Bussell JD, Smith SM, Beveridge CA (2012) The *Arabidopsis* ortholog of rice DWARF27 acts upstream of MAX1 in the control of plant development by strigolactones. Plant Physiol 159:1073–1085

Waters MT, Scaffidi A, Moulin SLY, Sun YK, Flematti GR, Smith SM (2015) A *Selaginella moellendorffii* ortholog of KARRIKIN INSENSITIVE2 functions in *Arabidopsis* development but cannot mediate responses to karrikins or strigolactones. Plant Cell 27:1925–1944

Waters MT, Gutjahr C, Bennett T, Nelson DC (2017) Strigolactone signaling and evolution. Annu Rev Plant Biol 68:291–322

Xie X, Yoneyama K, Harada Y, Fusegi N, Yamada Y, Ito S, Yokota T, Takeuchi Y, Yoneyama K (2009) Fabacyl acetate, a germination stimulant for root parasitic plants from *Pisum sativum*. Phytochemistry 70:211–215

Xie X, Yoneyama K, Yoneyama K (2010) The strigolactone story. Annu Rev Phytopathol 48:93–117

Xie X, Yoneyama K, Kisugi T, Nomura T, Akiyama K, Asami T, Yoneyama K (2015) Strigolactones are transported from roots to shoots, although not through the xylem. J Pestic Sci 40:214–216

Yamagami T, Tsuchisaka A, Yamada K, Haddon WF, Harden LA, Theologis A (2003) Biochemical diversity among the 1-amino-cyclopropane-1-carboxylate synthase isozymes encoded by the *Arabidopsis* gene family. J Biol Chem 278:49102–49112

Yamamoto M, Miki T, Ishiki Y, Fujinami K, Yanagisawa Y, Nakagawa H, Ogura N, Hirabayashi T, Sato T (1995) The synthesis of ethylene in melon fruit during the early stage of ripening. Plant Cell Physiol 36:591–596

Yoneyama K, Xie X, Kusumoto D, Sekimoto H, Sugimoto Y, Takeuchi Y, Yoneyama K (2007) Nitrogen deficiency as well as phosphorus deficiency in sorghum promotes the production and exudation of 5-deoxystrigol, the host recognition signal for arbuscular mycorrhizal fungi and root parasites. Planta 227:125–132

Yoneyama K, Xie X, Sekimoto H, Takeuchi Y, Ogasawara S, Akiyama K, Hayashi H, Yoneyama K (2008) Strigolactones, host recognition signals for root parasitic plants and arbuscular mycorrhizal fungi, from Fabaceae plants. New Phytol 179:484–494

Yoshida S, Kameoka H, Tempo M, Akiyama K, Umehara M, Yamaguchi S, Hayashi H, Kyozuka J, Shirasu K (2012) The D3 F-box protein is a key component in host strigolactone responses essential for arbuscular mycorrhizal symbiosis. New Phytol 196:1208–1216

Zhang Y, Haider I, Ruyter-Spira C, Bouwmeester HJ (2013) Strigolactone biosynthesis and biology. In: de Bruijn FJ (ed) Molecular microbial ecology of the rhizosphere, vols 1 & 2. Wiley, Hoboken, NJ, pp 355–337

Zhang Y, van Dijk ADJ, Scaffidi A, Flematti GR, Hofmann M, al CTV (2014) Rice cytochrome P450 MAX1 homologs catalyze distinct steps in strigolactone biosynthesis. Nat Chem Biol 10:1028–1033

Zhao L-H, Zhou XE, Wu Z-S, Yi W, Xu Y, Li S et al (2013) Crystal structures of two phytohormone signal-transducing α/β hydrolases: karrikin-signaling KAI2 and strigolactone-signaling DWARF14. Cell Res 23:436–439

Zhou F, Lin Q, Zhu L, Ren Y, Zhou K, Shabek N et al. (2013) D14–SCFD3-dependent degradation of D53 regulates strigolactone signalling. Nature 504:406–410

Zou J, Zhang S, Zhang W, Li G, Chen Z, Zhai W, Zhao X, Pan X, Xie Q, Zhu L (2006) The rice HIGH-TILLERING DWARF1 encoding an ortholog of *Arabidopsis* MAX3 is required for negative regulation of the outgrowth of axillary buds. Plant J 48:687–698

Chapter 3
Root Hair Growth and Development in Response to Nutrients and Phytohormones

De-Jian Zhang, Yu-Jie Yang, Chun-Yan Liu, Fei Zhang, and Qiang-Sheng Wu

3.1 Introduction

Root hairs are tip-growing extensions from root epidermal cell (Grierson and Schiefelbein 2002). As extensions from root epidermis, root hairs can highly increase root surface area and play important roles in nutrient and water absorption, the interaction with soil microflora, such as arbuscular mycorrhizal (AM) fungi, nitrogen-fixing bacteria, etc. (Grierson and Schiefelbein 2002; Libault et al. 2010; Tanaka et al. 2014; Li et al. 2017; Vincent et al. 2017; Zou et al. 2017; Dolan 2017). Especially, it is very important for the plant to absorb scarcely mobile soil minerals, such as phosphorus (P) and potassium (K) from soil (Gahoonia et al. 1997; Gahoonia and Nielsen 1998; Zuchi et al. 2011; Brown et al. 2012; Cao et al. 2013; Jungk 2015). Root hair has been demonstrated to be important for P acquisition in a number of research papers in several crop species, particularly in low-P environment (Föhse et al. 1991; Hoffmann and Jungk 1995; Bates and Lynch 1996; Gahoonia et al. 1997; Bates and Lynch 2000; Yan et al. 2004; Brown et al. 2012). Furthermore, water channels, as well as calcium, phosphate, potassium, etc., are localized in root hair, and it has been suggested that root hair takes part in the absorption of most macro- and micronutrients in a number of crop plants (Gilroy and Jones 2000; Libault et al. 2010). Even more, Wang et al. (2016) considered that long and dense root hairs are important traits in ensuring efficient absorption of both macro- and micronutrients in the early establishment of crops in nutrient-limited soil and low nutrient input environment cropping systems.

De-Jian Zhang and Yu-Jie Yang contributed equally to this work.

D.-J. Zhang · Y.-J. Yang · C.-Y. Liu · F. Zhang · Qiang-ShengWu (✉)
College of Horticulture and Gardening, Yangtze University, Jingzhou, Hubei, China

Institute of Root Biology, Yangtze University, Jingzhou, Hubei, China

B. Giri et al. (eds.), *Root Biology*, Soil Biology 52,
https://doi.org/10.1007/978-3-319-75910-4_3

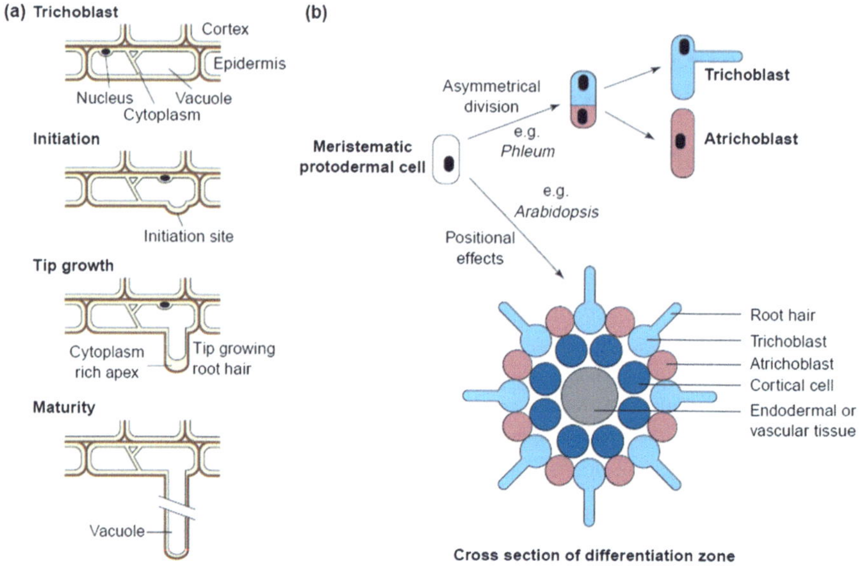

Fig. 3.1 Patterns of root hair development. (**a**) Cross section of a trichoblast (epidermal cell that will produce a root hair) during root hair development. Note the nuclear movements accompanying root hair emergence and changes in the organization of the cytoplasm at each new developmental phase. (**b**) Divergence of root epidermal cell fate to trichoblast or atrichoblast might be determined by asymmetrical division and differences in subsequent differentiation of the daughter cells or by positional effects relative to underlying cell layers (Gilroy and Jones 2000)

3.2 Root Hair Development

Root hair development consists of four stages, viz., cell fate specification, initiation, subsequent tip growth, and maturation (Gilroy and Jones 2000; Parker et al. 2000) (Fig. 3.1).

3.2.1 Cell Fate Specification

The mechanism of epidermal cell fate to root hair has lucubrated in *Arabidopsis thaliana*. Research evidence suggest genes, such as *TRANSPARENT TESTA GLABRA* (*TTG*), *WEREWOLF* (*WER*) and *GLABRA2* (*GL2*), are specific expression in the non-hair epidermis cell and are negative transcriptional regulators of root hair formation (Galway et al. 1994; Di et al. 1996; Masucci et al. 1996; Zhu et al. 2017). Conversely the MYB-like transcription factor encoded by the *CAPRICE* (*CPC*) and *TRIPTYCHON* (*TRY*) genes is thought to be specific expression in the non-hair epidermis cell and is a positive regulator of root hair fate (Wada et al. 1997; Schellmann et al. 2002; Schiefelbein 2003; Savage et al. 2008; Libault et al. 2010).

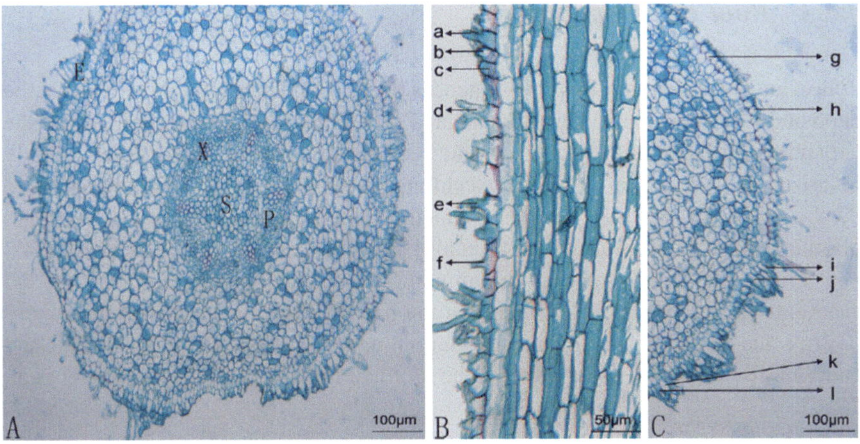

Fig. 3.2 The photos of cross and longitudinal section in the regions 2.0–3.0 cm from root apex of the taproot of citrus. (**A**) E, epidermis cell; X, xylem; P, phloem; S, pith; (**B**, **C**) a, d, k, l, root hair outgrow from epidermis cell which positioned over a single cortical cell; b, g, epidermis cell cannot initiate root hair when located over two underlying cortical cells; c, h, epidermis cell cannot form root hair when positioned over a single cortical cell; e, f, i, j, root hair form from epidermis cell which is located over two underlying cortical cells (Zhang et al. 2013)

3.2.2 *Root Hair Initiation*

Root hair initiates from root epidermal cell. The initiation patterns of root hair have been divided into three types (Clowes 2000; Kim et al. 2006a). In type 1, root hair cell can differentiate from any epidermal cell (random type). The random type pattern occurs throughout most dicots, many monocots, and most ferns' plants, such as *Soleirolia soleirolii* and *Poncirus trifoliata* (Fig. 3.2) (Clowes 2000; Zhang et al. 2013). In type 2, the root epidermis consists of two sizes of cells, long and short, but only the short cell differentiates into root hair cell (asymmetrical cell division type). The asymmetrical cell division type occurs in monocots, basal tracheophytes, and basal angiosperm families, such as *Oryza sativa* (Kim and Dolan 2011). Type 3 is position-dependent hair cell differentiation: root hair is located over epidermal cell overlying the junction of two cortical cells, whereas non-hair cell is located over a single cortical cell (positionally cued type) (Dolan et al. 1993; Galway et al. 1994). This positionally cued type has been found in Brassicaceae and other eudicot families, such as *Arabidopsis* (Cormack 1947; Clowes 2000; Dolan and Costa 2001). In *Arabidopsis*, the epidermal cell overlying the junction of two cortical cells was named as trichoblasts which could be initiated to root hair, while atrichoblast was the epidermal cell located over a single cortical cell, which cannot bulge to root hair (Duckett et al. 1994; Kwasniewski et al. 2013). Trichoblasts can be distinguished from atrichoblasts by differences in their cytoplasmic structure in the time of their formation in the meristematic zone, such as reduced vacuolation, etc. (Galway et al. 1997).

3.2.3 Root Hair Tip Growth

Following root hair initiation, it commences tip growth. During tip growth stage, the deposition of new plasma membrane and cell wall material is confined to the expanding tip, and the cytoplasm and pectic substance of the hair are highly polarized, with secretory vesicles concentrated located behind the hair tip, followed by the organelles required for the production and secretion of new cell wall and plasma membrane materials (Galway et al. 1997; Carol and Dolan 2002, 2006). Ca^{2+} and cytoskeleton, annexins, calmodulin, GTPases, and protein kinases are the candidate for the regulatory elements of the root hair tip-forced growth machinery (Gilroy and Jones 2000). In the maturation stage, ribosomes, mitochondria, and endoplasmic reticulum concentrate at the root hair tip (Cormack 1947; Nestler et al. 2014).

3.3 Root Hair Growth and Development in Response to Nutrients

Recent studies have indicated that root hair growth and development could be influenced by various environmental factors such as culture substrates (yellow soil, Hoagland solution, etc.), mineral nutrients (nitrogen, phosphorus, potassium, calcium, iron, magnesium, etc.), plant growth regulators (auxin, ethylene, jasmonic acid, ethyl jasmonate, strigolactone, etc.), soil edaphon (arbuscular mycorrhizal fungi, etc.), and so on (Zhu et al. 2006; Lee and Cho 2009; Libault et al. 2010; Kapulnik et al. 2011; Niu et al. 2011; Yang et al. 2011; Peret et al. 2011; Muday et al. 2012; Wu and He 2011; Cao et al. 2013; Vandamme et al. 2013; Zhang et al. 2013; Wu et al. 2016). Here, the effects of nutrients and phytohormones on root hair growth and development have been detailed in the Introduction.

3.3.1 Nitrogen (N)

The NH_4^+ and NO_3^- are main N-compounds in soil for plants utilization (Forde 2002). With respect to NH_4^+ and NO_3^-, there is now clear molecular and electrophysiological evidence that root hair could transport N-compounds. The expression of two putative NH_4^+ and NO_3^- transporter genes (*LeNRT1-2* and *LeAMT1*) is root hair-specific in tomato, which is regulated by an external N supply1-38. In addition, Meharg and Blatt (1995) have revealed that the high-affinity NO_3^- transporter in *Arabidopsis* root hair is greatly upregulated under NO_3^- deficiency, indicating that root hair provides adequate nutrients to the root. Direct evidence using SEM (scanning electron microscope) has suggested that split root hair were initiated in *Arabidopsis thaliana* after the addition of NH_4NO_3 to its roots (Yang et al. 2011).

Yang et al. (2011) consider that the effects of NH_4^+ and NO_3^- on split root hair may through methyl jasmonate and ethylene signaling pathway that methyl jasmonate can enhance the effect on split root hair while ethylene decrease it. Furthermore, NH_4^+ and NO_3^- may regulate calcium ion (Ca^{2+}) in root hair cell through Rho-related GTPase from plants (ROPs) and reactive oxygen species (ROS); however, Ca^{2+} gradient is indispensable for root hair tip growth (Shin et al. 2005; Bloch et al. 2011; Bai et al. 2014). Thus, N may have interactions with phytohormones, ROS, or calcium ion in regulating root hair growth.

3.3.2 *Phosphorus (P)*

Phosphorus (P) is extremely immobile in soil and is frequently growth-limiting, and it is an important macronutrient in plants, not only as a constituent of key cellular molecules such as ATP, phospholipids, and nucleic acids but also playing a pivotal role in energy conservation and metabolic regulation (Marschner 1995; Raghothama and Karthikeyan 2005; Shin et al. 2005). Research efforts are now being focused on understanding the mechanistic basis of P efficiency, to develop crops that require less input.

Gahoonia et al. (1997) confirmed that root hairs can satisfy 60% of the plant's P demand in soil (Gahoonia and Nielsen 1998). P deficiency in the soil often induces dense root hairs in plants such as *Arabidopsis* and citrus (Zhu et al. 2005; Cao et al. 2013). In P-deficient soil, the length and density of *Arabidopsis* root hairs increase significantly, expanding the root's surface area from 0.21 mm^2 mm^{-1} root under P-sufficient conditions to 1.44 mm^2 mm^{-1} roots under P-starvation conditions, with the root hairs constituting 91% of the total root's surface area (Bates and Lynch 1996). In addition, the response to P deprivation in root hairs was accompanied by an increase in ROS, which is necessary for root hair initiation and elongation through Ca^{2+} gradient (Foreman et al. 2003; Shin et al. 2005; Carol and Dolan 2006). Thus, there has interaction effect between ROS and P on root hair growth and development.

3.3.3 *Potassium (K)*

Potassium ion (K^+) is the most abundant cation in cells of higher plants, and it plays a crucial role in plant growth and development, such as leaf movements, enzyme homeostasis, photosynthesis, assimilate transport, enzyme activation, etc. (Gassmann and Schroeder 1994; Miao et al. 2010). K deficiency not only reduces crop resistance to pathogens, nutritional quality, and mechanical stability but also decreases root hair growth, such as citrus root hair (Pettigrew 2008; Cao et al. 2013). Gassmann and Schroeder (1994) considered that inward-rectifying K^+ channels in root hairs can function as both a physiologically important mechanism for low-affinity K^+ uptake and as regulators of membrane potential. This evidence is a major step toward a detailed molecular characterization of the multiple components involved in K^+ uptake, transport, and membrane potential control in root epidermal cells.

TRH1 (AtKT/AtKUP/HAK K^+ transporter family) was demonstrated in *Arabidopsis* that its mutant *trh1* has partially impaired in K^+ transport, which also blocked root hair growth and development (Rigas et al. 2001). However, the defected root hair phenotype in *trh1* could not restore when mutant seedlings were grown at high external K concentrations (Rigas et al. 2001). It may demonstrate that *TRH1* mediates K^+ transport in *Arabidopsis* roots and is responsible for specific K^+ translocation, which is essential for root hair elongation (Rigas et al. 2001).

Interestingly, the researchers found that K interacts with phytohormones on root hair growth and development, especially ethylene and auxin. When treated with K starvation, both ethylene production and the transcription of genes involved in ethylene biosynthesis were increased (Shin and Schachtman 2004; Jung et al. 2009). Ethylene signaling is a component of the plant's response to low K that stimulates the production of reactive oxygen species (ROS) and is important for changes in root hair morphology and whole plant tolerance to low-potassium conditions (Jung et al. 2009). With regard to auxin, Rigas et al. (2013) consider that auxin modulates root hair morphogenesis at the differentiation zone. *TRH1* (the potassium transporter) was cell-specific subcellular localized in stele and epidermis, which engagement in auxin transport and redistribution auxin (Rigas et al. 2013). The role of the *TRH1* in the sensing of external K^+ and the regulation of potassium-dependent root hair growth (Rigas et al. 2013). In *Arabidopsis* mutant *trh1*, auxin imbalance caused poor root hairs (Vicente-Agullo et al. 2004; Rigas et al. 2013). Hence, K may regulate root hair growth and development though the auxin signaling.

3.3.4 Calcium (Ca)

Calcium ion (Ca^{2+}) is involved in most biological processes in the plant kingdom (Weinl and Kudla 2009). It serves as a second messenger and regulates a multitude of adaptational, developmental, and physiological processes in plants, including cell division, cell expansion, cytoplasmic streaming, responses to abiotic stress, adjustment of ion homeostasis, and pathogen defense (Bush 1995; Sanders et al. 2002; White and Broadley 2003; Kim et al. 2009). Cytoplasmic Ca^{2+} is known to regulate cytoskeletal rearrangements and vesicular trafficking in tip-growing systems such as root hair development system (Hepler et al. 2001). Early research has discovered that the Ca^{2+} gradient existed in the growing root hairs, which is being most pronounced in the rapidly elongating root hairs (Fig. 3.3) (Wymer et al. 1997). With regard to the effect of Ca^{2+} on root hair growth, studies have shown that a growing root hair has a localized gradient of Ca^{2+} concentration toward the growing apex, and the intensity of this gradient correlates with the growth rate of the root hair (Schiefelbein et al.

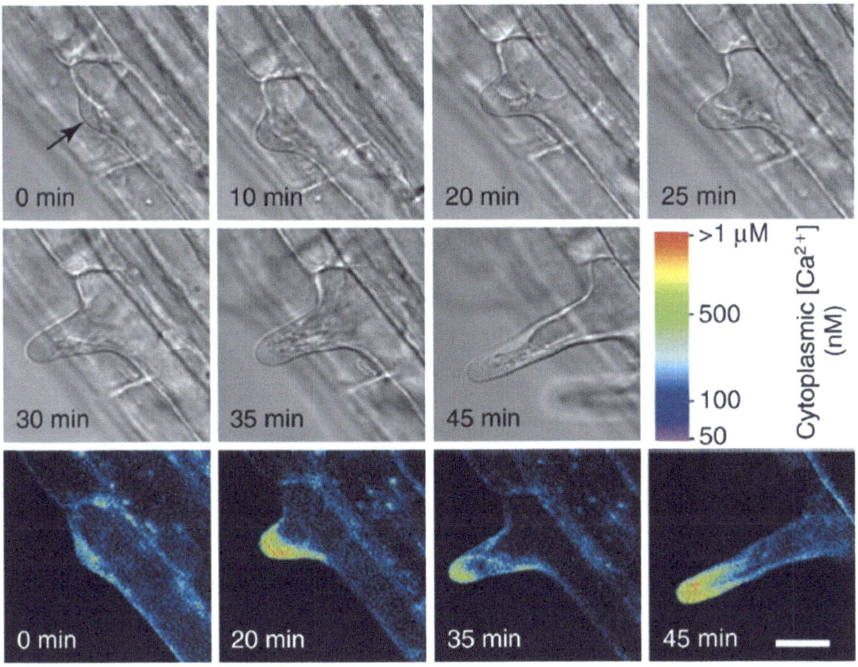

Fig. 3.3 Gradients in cytoplasmic Ca^{2+} associated with tip growth of *Arabidopsis* root hairs (Wymer et al. 1997). There are no detectable Ca^{2+} gradients in the stage of root hair initiation (0 min) but on the stage of tip growth. The photos showed the Ca^{2+} gradients focused on the elongating tip of root hair. Images are confocal ratio images of *Arabidopsis* root hairs loaded with the fluorescent Ca^{2+} sensor Indo-1 and monitored using a confocal microscope. Ca^{2+} levels in the cytoplasm have been color coded according to the *inset scale*. The *arrow* indicates the initiation point of root hair. *Scale bar* = 20 mm

1992; Pierson et al. 1996; Felle and Hepler 1997; Wymer et al. 1997; Takeda et al. 2008; Li et al. 2012). Electrophysiological studies, using the vibrating probe, showed that Ca^{2+} influx was higher at root hair tip than at the base or sides of growing root hair (Schiefelbein et al. 1992; Herrmann and Felle 1995; Jones et al. 1995). Based on confocal ratio imaging, Bibikova et al. (1997) demonstrated that a tip-focused Ca^{2+} gradient was centered at the site of active growth. If the root hair direction was changed, the Ca^{2+} gradient was also reoriented. Imposing an artificial tip-focused Ca^{2+} gradient reorients root hair growth toward the new gradient (Bibikova et al. 1997). These observations are all consistent with a model, whereby a localized increase in Ca^{2+} at the root hair apex is specifically associated with root hair growth (Gilroy and Jones 2000). Further evidence for the role of Ca^{2+}, using Ca^{2+} ionophores and channel blockers, suggested that disruption the Ca^{2+} concentration gradient result in inhibiting root hair tip growth (Miller et al. 1992; Pierson et al. 1994; Herrmann and Felle 1995; Wymer et al. 1997).

With regard to the effect of ROS and Ca^{2+} on root hair growth and development, Foreman et al. (2003) and Takeda et al. (2008) showed that ROS produced by

NADPH oxidase activated Ca^{2+} channels in the apical plasma membrane, leading to the tip-focused Ca^{2+} concentration gradient required for root hair polar growth. Additional evidence from *Arabidopsis rhd2* mutant indicated that application of exogenous ROS to in vivo root apices stimulated cell growth and Ca^{2+} influxes in root hair, which induced root hair elongation (Foreman et al. 2003). The positive effect of exogenous ROS on root hair growth and development could be blocked by pharmacological inhibitors of calcium channels (Foreman et al. 2003). Artificially increasing cytosolic Ca^{2+} by treating with calcium ionophore A23187 (mixed calcium-magnesium salt) induced production of elevated level of ROS around the root hair tip and inhibited its tip growth for that these large amounts of ROS disrupted the Ca^{2+} concentration gradient (Foreman et al. 2003).

As a consequence, Ca^{2+} concentration gradient is a requirement in sustaining root hair tip growth, and ROS can activate calcium channel-mediated influx of Ca^{2+}, which creates a Ca^{2+} concentration gradient and subsequently influences root hair elongation.

3.3.5 Iron (Fe)

Iron (Fe) is an essential nutrient for plants, which catalyzes crucial cellular functions such as photosynthesis, chlorophyll synthesis, chloroplast development, antioxidative cell protection, etc. (Siminis and Stavrakakis 2008; Curie and Mari 2017; Tsai and Schmidt 2017). Although Fe is one of the most abundant elements in soils, it mainly exists as insoluble and nonavailable to plants (Tsai and Schmidt 2017). So plants have evolved efficient strategies to increase iron solubility and absorption efficiency, such as increasing root hair (Muller and Schmidt 2004; Cao et al. 2013). The number of root hairs was increased in response to Fe deficiency in *Arabidopsis* and citrus (Muller and Schmidt 2004; Cao et al. 2013). Fe regulates root hair growth and development through plant ferredoxin-like protein (PFLP), which affects ROS content by NADPH oxidase (NOX) (Shin et al. 2011; Sundaravelpandian et al. 2013; Nestler et al. 2014; Lin et al. 2015). Further evidence suggests that Fe has interaction effect with ethylene on root hair initiation and elongation. Fe deficiency can lead to the formation of extra root hairs located in positions normally occupied by non-hair cells, which is dependent on ethylene signaling and requires functional *EIN2* and *ETR1* genes (Schmidt and Schikora 2001).

3.3.6 Magnesium (Mg)

Magnesium (Mg) is an essential mineral nutrient for plant metabolic processes and reactions, such as photophosphorylation, photooxidation in leaf tissues, chlorophyll formation, protein synthesis, photosynthetic CO_2 fixation, phloem loading, generation of ROS, and so forth (Cakmak and Yazici 2010). Mg is also involved in many critical biochemical and physiological processes in plants, thus, influencing plant growth and development (Cakmak and Yazici 2010; Gransee and Führs 2013).

Recently, the effect of interactions among Mg, ROS, and Ca^{2+} on root hair has been reported. Low Mg availability resulted in longer and denser root hair in *Arabidopsis* with greater concentrations of ROS and Ca^{2+} in the root tip and a stronger Ca^{2+} concentration gradient in the root hair tip (Niu et al. 2014). However, when treated with diphenylene iodonium (DPI, an NADPH oxidase inhibitor) or 1,2-bis(*o*-aminophenoxy) ethane-*N*,*N*,*N′*,*N′*-tetraacetic acid (BAPTA, a Ca^{2+} chelator), the Ca^{2+} concentration gradient was eliminated for the enhanced growth of root hair in low-Mg treatment (Frahry and Schopfer 1998; Yoshioka et al. 2001; Kadota et al. 2004). Instead, root hair growth was blocked in high-Mg treatment, and the inhibiting effect was restored when supplied with $CaCl_2$ or phenazine methosulfate (PMS, a ROS generator) in *Arabidopsis* (Zhang et al. 2009; Niu et al. 2014). The study outlined by Niu et al. (2014) showed that NADPH oxidase in root was positively regulated by low Mg and was inhibited by high-Mg level, which further evidenced that Mg could control ROS to regulate root hair growth and development. Therefore, the growth and development of root hair is closely related to Mg availability, which is through ROS and Ca^{2+} signaling.

3.4 Root Hair Growth and Development in Response to Phytohormones

3.4.1 Auxins (IAAs)

The phytohormone auxins are a major determinant and regulatory component important for plant growth and development (Kuhn et al. 2017). In particular, auxin is considered the main signaling molecule involved in regulating root hair growth (Pitts et al. 1998; Muday et al. 2012). Epidermal cell as cradle of root hair needed adequate auxin for its initiation (Cho et al. 2007a; Lee and Cho 2009). In *Arabidopsis*, the auxin response mutant *axr1* has fewer and shorter root hairs, but root hair growth can be dramatically promoted by exogenous auxins, such as 1-naphthylacetic acid (NAA) and indole-3-butyric acid (IBA) (Pitts et al. 1998; Rahman et al. 2002; Muday et al. 2012). Auxin improved root hair density by increased trichoblast number and the percentage of trichoblast bulged to root hair (Masucci and Schiefelbein 1996; Niu et al. 2011; Lee and Cho 2013). Bruex et al. (2012) found that 90% of genes related to root hair growth were positively regulated by auxin based on transcriptome sequencing data analysis.

Auxin transport between cells is mediated by a complex system of transporters such as *AUXIN RESISTANT 1* (*AUX1*), *LIKE AUX1* (*LAX*), Pin-formed (*PIN*), and *ATP-BINGING CASSETTE B* (*ABCB*) genes, and their localization and activity are thought to participate in root hair initiation and elongation (Rahman et al. 2002; Cho et al. 2007a; Jones et al. 2009; Strader and Bartel 2009; Ganguly et al. 2010; Kuhn et al. 2017). Auxin is primarily synthesized at the shoot apex, transferred toward the root tip by the vascular tissues of the stem, and moved basipetally to the elongation zone through peripheral root tissues (Rigas et al. 2013). Such transportation can be blocked by auxin transport inhibitors, such as 1-naphthoxyacetic acid (1-NOA), 2-naphthoxyacetic acid (2-NOA), and 3-chloro-4-hydroxyphenylacetic acid (CHPAA) which regulate overall auxin transport (LaňKová et al. 2010). In addition, auxin fluxes at the root apex are mainly controlled by various *PIN* auxin efflux carriers, in addition to *AUX1*, *LAX* auxin influx carriers, and some members of the *ABCB* transporters (auxin efflux genes) (Yang and Murphy 2009; Tromas and Perrot-Rechenmann 2010; Swarup and Péret 2012). Auxin transport and auxin synthesis are controlled by many genes, such as tryptophan aminotransferase related (*TAR*) and flavin-containing monooxygenase (*YUCs*) (Mano and Nemoto 2012). Understanding auxin biosynthesis and transport is an important factor in understanding root hair growth. For instance, overexpression of *YUCs* in *Arabidopsis* enhanced root hair growth, compared to the wild type (Zhao et al. 2001). In *Arabidopsis* mutant *ein2* (auxin efflux carrier), the auxin transportation channel from root tip to root hair zone was blocked, which inhibited root hair initiation and elongation results in short and thin root hairs (Cho et al. 2007b). In addition, *Arabidopsis* mutant *aux1* also performs similar phenotype on root hair to *ein2* (Strader et al. 2010).

3.4.2 Ethylene (ETH)

Ethylene is considered as the vital signal molecule involved in regulating development and growth of root hair (Pitts et al. 1998; Muday et al. 2012). In *Arabidopsis*, the ethylene response mutant *ctr1* (the gene encodes a Raf-like protein kinase that negatively regulates the ethylene signal transduction pathway) significantly reduced the growth of root hairs (Kieber et al. 1993). Furthermore, 1-aminocyclopropane-1-carboxylic acid (ACC, the ethylene precursor) increased root hair density and length, whereas aminoethoxyvinylglycine (AVG, an ethylene biosynthesis inhibitor) and $AgNO_3$ (Ag, an ethylene action inhibitor) reduced root hair (Masucci and Schiefelbein 1994; Tanimoto et al. 1995; Masucci and Schiefelbein 1996; Pitts et al. 1998; Dolan 2001; Moeder et al. 2002; Zhu et al. 2006; Parimalan et al. 2011; Shah et al. 2014).

In *Arabidopsis*, ethylene effect improved the percentage of trichoblast to root hair and promoted the initiation of ectopic root hair (Masucci and Schiefelbein 1996; Niu et al. 2011; Lee and Cho 2013). *EIN2* (*Arabidopsis* ethylene receptor resistant) is a positive regulator of ethylene responses, which as an ethylene receptor participated in inducing root hair formation (Lin et al. 2015; Zheng and Zhu 2016). Rahman et al. (2002) have demonstrated that *Arabidopsis* mutant *ein2* has short and thin root hairs. *ACS*, an important role in the ethylene biosynthesis pathway, could catalyze the conversion of AdoMet to ACC which is the precursor of ethylene (Moeder et al. 2002). Ribaudo et al. (2006) demonstrated that root hair grow through a mechanism of *ACS* in tomato. Thus, ethylene-regulated root hair formation may be through its biosynthesis, receptor, and signal transduction pathways.

However, the interaction between ethylene and auxin is more attractive for scientist. Bruex et al. (2012) found that 90% of genes related to root hair growth and development were upregulated by ethylene and auxin based on transcriptome sequencing data analysis. In *Arabidopsis*, ACC increased root hair growth on auxin response mutant *axr1*, while NAA could effectively relieve the inhibitory effect on ethylene signal transduction mutant *ein2* (Pitts et al. 1998; Rahman et al. 2002; Muday et al. 2012). On the one hand, Michael (2001) proposed the hypothesis of the "ethylene center in root hair development," i.e., soil environment, growth regulator, and mineral nutrients may control the effective endogenous ethylene concentration in root hair cells and further regulate root hair growth. On the other hand, endogenous auxin plays an important role in root hair growth, and its concentration determines by its synthesis and transport efficiency, which are regulated by auxin synthesis and transport genes (*TAR2* and *YUCs*; *AUX1*, *ABCBs*, and *LAXs*) (Marchant et al. 1999; Rahman et al. 2002; Cho et al. 2007a; Jones et al. 2009; Ganguly et al. 2010). Auxin could induce ethylene biosynthesis, which positively regulates root hair growth (Yoshii and Imaseki 1982; Pitts et al. 1998; Muday et al. 2012). Ethylene also could regulate auxin biosynthesis and transport-dependent auxin distribution that are important for root and root hair initiation (Rahman et al. 2002; Cho et al. 2007a; Růžička et al. 2007; Ganguly et al. 2010). Furthermore, ethylene promotes the induction by auxin of the cortical microtubule randomization required for root hair growth in *Lactuca sativa* (Takahashi et al. 2003). In citrus plant, Zhang et al. (2016) confirmed that ethylene may positively regulate root hair growth by ethylene synthesis, receptor, and signal transduction genes. Furthermore, it achieved a local activation of the auxin-signaling pathway and may regulate root hair growth by both stimulating the auxin biosynthesis and modulating the auxin transport machinery (Zhang et al. 2016). So, ethylene modulated root hair growth partially through auxin-signaling pathway.

3.4.3 Jasmonic Acid and Methyl Jasmonate (JA and MeJA)

Besides auxin and ethylene, JAs and MeJAs also participate in root hair development and growth. JA and MeJA could induce more bifurcate root hair, so they have a pronounced effect on promoting root hair formation and elongation (Zhu et al. 2006). However, the effect of MeJA and JA on root hair formation and elongation was dramatically diluted by ethylene inhibitors aminoethoxyvinylglycine (AVG) and $AgNO_3$ (Zhu et al. 2006). The enhancing effects of JA and MeJA were also diminished in ethylene-insensitive mutant *etr1*. Furthermore, the JA biosynthesis inhibitors ibuprofen and salicylhydroxamic acid (SHAM) not only decreased the enhancing effect of ethylene precursor (1-aminocyclopropane-1-carboxylic acid, ACC) on root hair initiation and elongation but also diminished the facilitating effect on root hair growth in ethylene overproducing mutant *eto1* (Zhu et al. 2006). So, there is an interaction between JA/MeJA and ethylene in the regulation of root hair growth.

Hentrich et al. (2013) demonstrated that JA-signaling pathway was linked to auxin homeostasis through the regulation of the expression of auxin signal transduction genes (*YUCCA8* and *YUCCA9*). Because auxin is a major determinant and regulatory component important for root hair growth and development (Kuhn et al. 2017). Thus, JAs and MeJAs may regulate root hair growth and development through auxin-signaling pathway.

3.4.4 Strigolactone

Strigolactones (SLs) were originally isolated from plant root exudates as germination stimulants for root parasitic plants of the family Orobanchaceae, including *Alectra* spp., Witchweeds (*Striga* spp.), and broomrapes (*Orobanche* and *Phelipanche* spp.), and so were regarded as detrimental to the producing plants (Gomezroldan et al. 2008; Xie et al. 2010; Waters et al. 2017). It has been recently shown that SLs or their metabolites are a novel class of plant hormones that regulate many aspects of shoot and root growth and development (Gomezroldan et al. 2008; Xie et al. 2010; Waters et al. 2017). GR24, as a synthetic bioactive SL, could lead to increased root hair length in *Arabidopsis* (Kapulnik et al. 2011). Additional studies have suggested cross talk between SLs and auxin in determining root hair growth. SLs might be involved in root hair tip growth via the modulation of auxin efflux in roots by regulating in *PINs* and *TIR1*. Evidence in tomato roots suggested that SLs interfered with the inhibitory effect of exogenously applied auxin on root hair elongation and auxin efflux carriers were involved in this process (Koltai et al. 2010; Kapulnik et al. 2011; Mayzlish-Gati et al. 2012). However, excess GR24 could lead to disturbances in auxin efflux and hence to an excess in cellular concentration of auxin; this, in turn, leads to reduction in the GR24 induction of root hair elongation (Kapulnik et al. 2011). Thus, SLs have an additional hormonal role in plants, acting as positive regulator of root hair growth and development.

3.4.5 Brassinosteroids

Brassinosteroids (BRs), compounds characterized as polyhydroxy steroids, are natural substances that play vital roles in the regulation of metabolic processes including respiration and in plant tolerance against abiotic stresses including water deficit and cold stress (Derevyanchuk et al. 2015; Lima and Lobato 2017). Interestingly, BRs have different effects on root hair growth and development, such as the reduction of numbers and length of root hairs in *Arabidopsis* but positively regulating in *Oryza sativa* (Kim et al. 2006b; Hardtke et al. 2007). It has been demonstrated that BR is required for cell proliferation in the root, and *AXR3/IAA17* might be involved in the BR-signaling pathway in root development (Mouchel et al. 2004, 2006; Kim et al. 2006b). In addition, the gain-of-function mutations of *AXR3/IAA17* inhibited root hair growth and development (Knox et al. 2003). In brief, it suggests that it has a cross talk of BR and auxin signaling in root hair growth and development.

3.5 Future Perspectives and Conclusions

Improving plants to make root hair high efficiency of nutrients and water uptake should reduce the environmental impact of agriculture as well as increase crop production and quality. Even abundant research papers carry on deeper studying to the effects of nutrients and phytohormones on root hair growth. But root hair morphogenesis is driven by an amalgam of interacting processes controlled by complex signaling events. Some of the major factors involved in the signaling process during root hair growth and development have been identified, such as N, Ca^{2+}, Mg^{2+}, P, and so forth. It is not clear how these signaling component interactions regulate root hair at the molecular and cellular level or how these signals interact with phytohormones to regulate root hair growth and development, such as auxin, ethylene, JAs, MeJAs, etc. Furthermore, the role of a Ca^{2+} gradient in root hair tip growth is still scarce.

More works are needed to clone the genes of additional root hair mutants and elucidate their roles, as well as add to our current knowledge of the signaling networks involving nutrient and phytohormone regulation on root hair cell fate specification, initiation, tip growth, and maturation by reverse genetics and mutant complementation studies.

Acknowledgments This study was supported by the Yangtze Youth Fund (2016cqn64) and the Yangtze Initial Fund (801190010132) of Yangtze University and the Plan in Scientific and Technological Innovation Team of Outstanding Young, Hubei Provincial Department of Education (T201604).

References

Bai L, Ma X, Zhang G, Song S, Zhou Y, Gao L, Miao Y, Song CP (2014) A receptor-like kinase mediates ammonium homeostasis and is important for the polar growth of root hairs in *Arabidopsis*. Plant Cell 26:1497–1511

Bates TR, Lynch JP (1996) Stimulation of root hair elongation in *Arabidopsis thaliana* by low phosphorus availability. Plant Cell Environ 19:529–538

Bates TR, Lynch JP (2000) The efficiency of *Arabidopsis thaliana* (Brassicaceae) root hairs in phosphorus acquisition. Am J Bot 87:964–970

Bibikova TN, Zhigilei A, Gilroy S (1997) Root hair growth in *Arabidopsis thaliana* is directed by calcium and an endogenous polarity. Planta 203:495–505

Bloch D, Monshausen G, Gilroy S, Yalovsky S (2011) Co-regulation of root hair tip growth by ROP GTPases and nitrogen source modulated pH fluctuations. Plant Signal Behav 6:426–429

Brown LK, George TS, Thompson JA, Wright G, Lyon J, Dupuy L, Hubbard SF, White PJ (2012) What are the implications of variation in root hair length on tolerance to phosphorus deficiency in combination with water stress in barley (*Hordeum vulgare*)? Ann Bot 110:319–328

Bruex A, Kainkaryam RM, Wieckowski Y, Kang YH, Bernhardt C, Xia Y, Zheng X, Wang JY, Lee MM, Benfey P et al (2012) A gene regulatory network for root epidermis cell differentiation in *Arabidopsis*. PLoS Genet 8:e1002446

Bush DS (1995) Calcium regulation in plant cells and its role in signaling. Annu Rev Plant Biol 9:95–122

Cakmak I, Yazici AM (2010) Magnesium: a forgotten element in crop production. Better Crops 11:23–25

Cao X, Chen CL, Zhang DJ, Shu B, Xiao J, Xia RX (2013) Influence of nutrient deficiency on root architecture and root hair morphology of trifoliate orange (*Poncirus trifoliata* L. Raf.) seedlings under sand culture. Scientia Horticulturae 162:100–105

Carol RJ, Dolan L (2002) Building a hair: tip growth in Arabidopsis thaliana root hairs. Philos Trans R Soc Lond B Biol Sci 357:815–821

Carol RJ, Dolan L (2006) The role of reactive oxygen species in cell growth: lessons from root hairs. J Exp Bot 57:1829–1834

Cho M, Lee SH, Cho HT (2007a) P-glycoprotein4 displays auxin efflux transporter-like action in *Arabidopsis* root hair cells and tobacco cells. Plant Cell 19:3930–3943

Cho M, Lee OR, Ganguly A, Cho HT (2007b) Auxin-signaling: short and long. J Plant Biol 50:79–89

Clowes FAL (2000) Pattern in root meristem development in angiosperms. New Phytol 146:83–94

Cormack RGH (1947) A comparative study of developing epidermal cells in white mustard and tomato roots. Am J Bot 34:310–314

Curie C, Mari S (2017) New routes for plant iron mining. New Phytol 214:521

Derevyanchuk M, Litvinovskaya R, Khripach V, Martinec J, Kravets V (2015) Effect of 24-epibrassinolide on *Arabidopsis thaliana* alternative respiratory pathway under salt stress. Acta Physiol Plant 37:1–10

Di CM et al (1996) The *Arabidopsis ATHB10 (GLABRA2)* is a HD-ZIP protein required for repression of ectopic root hair formation. Plant J 10:393–402

Dolan L (2001) The role of ethylene in root hair growth in *Arabidopsis*. J Plant Nutr Soil Sci 164:141–145

Dolan L (2017) Root hair development in grasses and cereals (*Poaceae*). Curr Opin Genet Dev 45:76–81

Dolan L, Costa S (2001) Evolution and genetics of root hair stripes in the root epidermis. J Exp Bot 52(suppl 1):413–417

Dolan L, Janmaat K, Willemsen V, Linstead P, Poethig S, Roberts K, Scheres B (1993) Cellular organisation of the *Arabidopsis thaliana* root. Development 119:71–84

Duckett CM, Grierson C, Linstead P, Schneider K, Lawson E, Dean C, Poethig S, Roberts K (1994) Clonal relationships and cell patterning in the root epidermis of *Arabidopsis*. Development 120:2465–2474

Felle HH, Hepler PK (1997) The cytosolic Ca^{2+}-concentration gradient of *Sinapis alba* root hairs as revealed by $Ca2^{+}$-selective microelectrode tests and fura-dextran ratio imaging. Plant Physiol 114:39–45

Föhse D, Claassen N, Jungk A (1991) Phosphorus efficiency of plants: II. Significance of root radius, root hairs and cation-onion balance for phosphorus influx in seven plant species. Plant Soil 132:261–272

Forde BG (2002) Local and long-range signaling pathways regulating plant responses to nitrate. Annu Rev Plant Biol 53:203–224

Foreman J, Demidchik V, Bothwell JH, Mylona P, Miedema H, Torres MA, Linstead P, Costa S, Brownlee C, Jones JD, Davies JM, Dolan L (2003) Reactive oxygen species produced by NADPH oxidase regulate plant cell growth. Nature 422:442–446

Frahry G, Schopfer P (1998) Inhibition of O_2-reducing activity of horseradish peroxidase by diphenyleneiodonium. Phytochemistry 48:223–227

Gahoonia TS, Nielsen NE (1998) Direct evidence on participation of root hair in phosphorus (^{32}P) uptake from soil. Plant Soil 198:147–152

Gahoonia TS, Care D, Nielsen NE (1997) Root hairs and phosphorus acquisition of wheat and barley cultivars. Plant Soil 191:181–188

Galway ME, Masucci JD, Lloyd AM, Walbot V, Davis RW (1994) The *TTG* gene is required to specify epidermal cell fate and cell patterning in the *Arabidopsis* root. Dev Biol 166:740–754

Galway ME, Heckman JW Jr, Schiefelbein JW (1997) Growth and ultrastructure of *Arabidopsis* root hairs: the *rhd3* mutation alters vacuole enlargement and tip growth. Planta 201:209–218

Ganguly A, Lee SH, Cho M, Lee OR, Yoo H, Cho HT (2010) Differential auxin-transporting activities of PIN-FORMED proteins in *Arabidopsis* root hair cells. Plant Physiol 153:1046–1061

Gassmann W, Schroeder JI (1994) Inward-rectifying K^+ channels in root hairs of wheat: a mechanism for aluminum-sensitive low-affinity K^+ uptake and membrane potential control. Plant Physiol 105:1399–1408

Gilroy S, Jones DL (2000) Through form to function: root hair development and nutrient uptake. Trends Plant Sci 5:56–60

Gomezroldan V, Fermas S, Brewer PB et al (2008) Strigolactone inhibition of shoot branching. Nature 455:189–194

Gransee A, Führs H (2013) Magnesium mobility in soils as a challenge for soil and plant analysis, magnesium fertilization and root uptake under adverse growth conditions. Plant Soil 368:5–21

Grierson C, Schiefelbein J (2002) Root hairs. In: Somerville C, Meyerowitz EM (eds) The *Arabidopsis* book, vol 1. American Society of Plant Biologists, Rockville

Hardtke CS, Dorcey E, Osmont KS, Sibout R (2007) Phytohormone collaboration: zooming in on auxin-brassinosteroid interactions. Trends Cell Biol. 17:485–492

Hentrich M, Böttcher C, Düchting P, Cheng Y, Zhao Y, Berkowitz O, Masle J, Medina J, Pollmann S (2013) The jasmonic acid signaling pathway is linked to auxin homeostasis through the modulation of *YUCCA8* and *YUCCA9* gene expression. Plant J 74:626–637

Hepler PK, Vidali L, Cheung AY (2001) Polarized cell growth in higher plants. Annu Rev Cell Dev Biol 17:159–187

Herrmann A, Felle HH (1995) Tip growth in root hair cells of *Sinapis alba* L.: significance of internal and external Ca21 and pH. New Phytol 129:523–533

Hoffmann C, Jungk A (1995) Growth and phosphorus supply of sugar beet as affected by soil compaction and water tension. Plant Soil 176:15–25

Jones DL, Shaff JE, Kochian LV (1995) Role of calcium and other ions in directing root hair tip growth in *Limnobium stoloniferum*. I. Inhibition of tip growth by aluminum. Planta 197:672–680

Jones AR, Kramer EM, Knox K, Swarup R, Bennett MJ, Lazarus CM, Leyser HMO, Grierson CS (2009) Auxin transport through non-hair cells sustains root-hair development. Nat Cell Biol 11:78–84

Jung JY, Shin R, Schachtman DP (2009) Ethylene mediates response and tolerance to potassium deprivation in *Arabidopsis*. Plant Cell 21:607

Jungk A (2015) Root hairs and the acquisition of plant nutrients from soil. J Plant Nutr Soil Sci 164:121–129

Kadota Y, Goh T, Tomatsu H et al (2004) Cryptogein-induced initial events in tobacco BY-2 cells: pharmacological characterization of molecular relationship among cytosolic Ca^{2+} transients, anion efflux and production of reactive oxygen species. Plant Cell Physiol 45:160–170

Kapulnik Y, Delaux PM, Resnick N et al (2011) Strigolactones affect lateral root formation and root-hair elongation in *Arabidopsis*. Planta 233:209–216

Kieber JJ, Rothenberg M, Roman G, Feldmann KA, Ecker JR (1993) *CTR1*, a negative regulator of the ethylene response pathway in *Arabidopsis*, encodes a member of the Raf family of protein kinases. Cell 72:427–441

Kim CM, Dolan L (2011) Root hair development involves asymmetric cell division in Brachypodium distachyon and symmetric division in *Oryza sativa*. New Phytol 192:601–610

Kim DW, Lee SH, Choi SB, Won SK, Heo YK, Cho M, Park YI, Cho HT (2006a) Functional conservation of a root hair cell-specific cis-element in angiosperms with different root hair distribution patterns. Plant Cell 18:2958–2970

Kim H, Park PJ, Hwang HJ, Lee SY, Oh MH, Kim SG (2006b) Brassinosteroid signals control expression of the *AXR3/IAA17* gene in the cross-talk point with auxin in root development. Biosci Biotechnol Biochem 70:768–773

Kim MC, Chung WS, Yun D, Cho MJ (2009) Calcium and calmodulin-mediated regulation of gene expression in plants. Mol Plant 2:13–21

Knox K, Grierson CS, Leyser O (2003) *AXR3* and *SHY2* interact to regulate root hair development. Development 130:5769–5777

Koltai H, Dor E, Hershenhorn J et al (2010) Strigolactones' effect on root growth and root-hair elongation may be mediated by Auxin-efflux carriers. J Plant Growth Regul 29:129–136

Kuhn BM, Nodzyński T, Errafi S et al (2017) Flavonol-induced changes in *PIN2* polarity and auxin transport in the *Arabidopsis thaliana* rol1-2 mutant require phosphatase activity. Sci Rep 7:41906

Kwasniewski M, Chwialkowska K, Kwasniewska J et al (2013) Accumulation of peroxidase-related reactive oxygen species in trichoblasts correlates with root hair initiation in barley. J Plant Physiol 170:185

LaňKová M, Smith RS, Pešek B et al (2010) Auxin influx inhibitors 1-NOA, 2-NOA, and CHPAA interfere with membrane dynamics in tobacco cells. J Exp Bot 61:3589–3598

Lee SH, Cho HT (2009) Auxin and root hair morphogenesis. Plant Cell Monogr 12:45–64

Lee RDW, Cho HT (2013) Auxin, the organizer of the hormonal/environmental signals for root hair growth. Front Plant Sci 4:448

Li S, Yu JL, Zhu M, Zhao F, Luan S (2012) Cadmium impairs ion homeostasis by altering K^+, and Ca^{2+}, channel activities in rice root hair cells. Plant Cell Environ 35:1998–2013

Li TC, Yang HY, Zhang W, Xu DF, Dong Q, Wang F, Lei YL, Liu GH, Zhou YB, Chen HJ, Li C (2017) Comparative transcriptome analysis of root hairs proliferation induced by water deficiency in maize. J Plant Biol 60:26–34

Libault M, Brechenmacher L, Cheng J, Xu D, Stacey G (2010) Root hair systems biology. Trends Plant Sci 15:641–650

Lima JV, Lobato AKS (2017) Brassinosteroids improve photosystem II efficiency, gas exchange, antioxidant enzymes and growth of cowpea plants exposed to water deficit. Physiol Mol Biol Plants 23:59–72

Lin CY, Huang LY, Chi WC, Huang TL, Kakimoto T, Tsai CR, Huang HJ (2015) Pathways involved in vanadate-induced root hair formation in *Arabidopsis*. Physiol Plant 153:137–148

Mano Y, Nemoto K (2012) The pathway of auxin biosynthesis in plants. J Exp Bot 63:2853–2872

Marchant A, Kargul J, May ST, Muller P, Delbarre A, Perrot-Rechenmann C, Bennett MJ (1999) *AUX1* regulates root gravitropism in *Arabidopsis* by facilitating auxin uptake within root apical tissues. EMBO J 18:2066–2073

Marschner A (1995) Mineral nutrition of higher plants. Academic, San Diego, CA

Masucci JD, Schiefelbein JW (1994) The *rhd6* mutation of *Arabidopsis thaliana* alters root-hair initiation through an Auxin- and Ethylene-associated process. Plant Physiol 106:1335–1346

Masucci JD, Schiefelbein JW (1996) Hormones act downstream of *TTG* and *GL2* to promote root hair outgrowth during epidermis development in the *Arabidopsis* root. Plant Cell 8:1505–1517

Masucci JD et al (1996) The homeobox gene *GLABRA2* is required for position-dependent cell differentiation in the root epidermis in *Arabidopsis thaliana*. Development 122:1253–1260

Mayzlish-Gati E, De-Cuyper C, Goormachtig S, Beeckman T, Vuylsteke M, Brewer PB, Beveridge CA, Yermiyahu U, Kaplan Y, Enzer Y, Wininger S, Resnick N, Cohen M, Kapulnik Y, Koltai H (2012) Strigolactones are involved in root response to low phosphate conditions in *Arabidopsis*. Plant Physiol 160:1329–1341

Meharg AA, Blatt MR (1995) NO_3^- transport across the plasma-membrane of *Arabidopsis thaliana* root hairs-kinetic control by pH and membrane voltage. J Membr Biol 145:49–66

Miao BH, Han XG, Zhang WH (2010) The ameliorative effect of silicon on soybean seedlings grown in potassium-deficient medium. Ann Bot 105:967

Michael G (2001) The control of root hair formation: suggested mechanisms. J Plant Nutr Soil Sci 164:111–119

Miller DD, Callaham DA, Gross DJ, Hepler PK (1992) Free Ca^{2+} gradient in growing pollen tubes of *Lilium*. J Cell Sci 101:7–12

Moeder W, Barry CS, Tauriainen AA, Betz C, Tuomainen J, Utriainen M, Grierson D, Sandermann H, Langebartels C, Kangasjärvi J (2002) Ethylene synthesis regulated by biphasic induction of 1-aminocyclopropane-1-carboxylic acid synthase and 1-aminocyclopropane-1-carboxylic acid oxidase genes is required for hydrogen peroxide accumulation and cell death in ozone-exposed tomato. Plant Physiol 130:1918–1926

Mouchel CF et al (2004) Natural genetic variation in Arabidopsis identifies BREVIS RADIX, a novel regulator of cell proliferation and elongation in the root. Genes Dev 18:700–714

Mouchel CF et al (2006) BRX mediates feedback between brassinosteroid levels and auxin signalling in root growth. Nature 443:458–461

Muday GK, Rahman A, Binder BM (2012) Auxin and ethylene: collaborators or competitors. Trends Plant Sci 17:181–195

Muller M, Schmidt W (2004) Environmentally induced plasticity of root hair development in *Arabidopsis*. Plant Physiol 134:409–419

Nestler J, Liu S, Wen TJ, Paschold A, Marcon C, Tang HM, Li D, Li L, Meeley RB, Sakai H, Bruce W, Schnable PS, Hochholdinger F (2014) Roothairless5, which functions in maize (*Zea mays* L.) root hair initiation and elongation encodes a monocot-specific NADPH oxidase. Plant J 79:729–740

Niu Y, Jin CW, Jin G, Zhou Q, Lin X, Tang C, Zhang Y (2011) Auxin modulates the enhanced development of root hairs in *Arabidopsis thaliana* (L.) Heynh. under elevated CO_2. Plant Cell Environ 34:1304–1317

Niu Y, Jin G, Yong SZ (2014) Root development under control of magnesium availability. Plant Signal Behav 9:e29720

Parimalan R, Giridhar P, Ravishankar GA (2011) Enhanced shoot organogenesis in *Bixa orellana* L: in the presence of putrescine and silver nitrate. Plant Cell Tissue Organ Cult 105:285–290

Parker JS, Cavell AC, Dolan L et al (2000) Genetic interactions during root hair morphogenesis in *Arabidopsis*. Plant Cell 12:1961–1974

Peret B, Clement M, Nussaume L, Desnos T (2011) Root developmental adaptation to phosphate starvation: better safe than sorry. Trends Plant Sci 16:442–450

Pettigrew WT (2008) Potassium influences on yield and quality production for maize, wheat, soybean and cotton. Physiologia Plantarum 133:670–681

Pierson ES, Miller DD, Callaham DA et al (1994) Pollen tube growth is coupled to the extracellular calcium ion flux and the intracellular calcium gradient: effect of BAPTA-type buffers and hypertonic media. Plant Cell 6:1815–1828

Pierson ES, Miller DD, Callaham DA, van Aken J, Hackett G, Hepler PK (1996) Tip-localized calcium entry fluctuates during pollen tube growth. Dev Biol 174:160–173

Pitts RJ, Cernac A, Estelle M (1998) Auxin and ethylene promote root hair elongation in Arabidopsis. Plant J 16:553–560

Raghothama KG, Karthikeyan AS (2005) Phosphate acquisition. Plant Soil 274:37–49

Rahman A, Hosokawa S, Oono Y, Amakawa T, Goto N, Tsurumi S (2002) Auxin and ethylene response interactions during *Arabidopsis* root hair development dissected by auxin influx modulators. Plant Physiol 130:1908–1917

Ribaudo CM, Krumpholz EM, Cassán FD, Bottini R, Cantore ML, Curá JA (2006) Azospirillum sp: promotes root hair development in tomato plants through a mechanism that involves ethylene. J Plant Growth Regul 25:175–185

Rigas S, Debrosses G, Haralampidis K, Vicente-Agullo F, Feldmann KA, Grabov A, Dolan L, Hatzopoulos P (2001) *TRH1* encodes a potassium transporter required for tip growth in *Arabidopsis* root hairs. Plant Cell 13:139–151

Rigas S, Ditengou FA, Ljung K, Daras G, Tietz O, Palme K, Hatzopoulos P (2013) Root gravitropism and root hair development constitute coupled developmental responses regulated by auxin homeostasis in the *Arabidopsis* root apex. New Phytol 197:1130–1141

Růžička R, Ljung K, Vanneste S, Podhorská R, Beeckman T, Friml J, Benková E (2007) Ethylene regulates root growth through effects on auxin biosynthesis and transport-dependent auxin distribution. Plant Cell 19:2197–2212

Sanders D, Pelloux J, Brownlee C, Harper JF (2002) Calcium at the crossroads of signaling. Plant Cell 14:S401–S417

Savage NS, Walker T, Wieckowski Y, Schiefelbein J, Dolan L, Monk NA (2008) A mutual support mechanism through intercellular movement of *CAPRICE* and *GLABRA3* can pattern the Arabidopsis root epidermis. PLoS Biol 6:e235

Schellmann S, Schnittger A, Kirik V, Wada T, Okada K, Beermann A, Thumfahrt J, Jürgens G, Hülskamp M (2002) *TRIPTYCHON* and *CAPRICE* mediate lateral inhibition during trichome and root hair patterning in *Arabidopsis*. EMBO J 21:5036–5046

Schiefelbein J (2003) Cell-fate specification in the epidermis: a common patterning mechanism in the root and shoot. Curr Opin Plant Biol 6:74–78

Schiefelbein JW, Shipley A, Rowse P (1992) Calcium influx at the tip of growing root-hair cells of *Arabidopsis thaliana*. Planta 187:455–459

Schmidt W, Schikora A (2001) Different pathways are involved in phosphate and iron stress-induced alterations of root epidermal cell development. Plant Physiol 125:2078–2084

Shah SH, Ali S, Jan SA, Din JU, Ali GM (2014) Assessment of silver nitrate on callus induction and in vitro shoot regeneration in tomato (*Solanum lycopersicum* Mill.) Pak J Bot 46:2163–2172

Shin R, Schachtman DP (2004) Hydrogen peroxide mediates plant root response to nutrient deprivation. Proc Natl Acad Sci USA 101:8827–8832

Shin R, Berg RH, Schachtman DP (2005) Reactive oxygen species and root hairs in *Arabidopsis* root response to nitrogen, phosphorus and potassium deficiency. Plant Cell Physiol 46:1350–1357

Shin LJ, Huang HE, Chang H, Lin YH, Feng TY, Ger MJ (2011) Ectopic ferredoxin I protein promotes root hair growth through induction of reactive oxygen species in *Arabidopsis thaliana*. J Plant Physiol 168:434–440

Siminis CI, Stavrakakis MN (2008) Iron induces root and leaf ferric chelate reduction activity in grapevine rootstock 140 Ruggeri. Hortscience 43:685–690

Strader LC, Bartel B (2009) The *Arabidopsis* PLEIOTROPIC DRUG RESISTANCE8/ABCG36 ATP binding cassette transporter modulates sensitivity to the auxin precursor indole-3-butyric acid. Plant Cell 21:1992–2007

Strader LC, Chen GL, Bartel B (2010) Ethylene directs auxin to control root cell expansion. Plant J 64:874–884

Sundaravelpandian K, Chandrika NNP, Schmidt W (2013) *PFT1*, a transcriptional Mediator complex subunit, controls root hair differentiation through reactive oxygen species (ROS) distribution in *Arabidopsis*. New Phytol 197:151–161

Swarup R, Péret B (2012) AUX/LAX family of auxin influx carriers-an overview. Front Plant Sci 3:225

Takahashi H, Kawahara A, Inoue Y (2003) Ethylene promotes the induction by auxin of the cortical microtubule randomization required for low-pH-induced-root hair initiation in lettuce (*Lactuca sativa* L.) seedlings. Plant Cell Physiol 44:932–940

Takeda S, Gapper C, Kaya H, Bell E, Kuchitsu K, Dolan L (2008) Local positive feedback regulation determines cell shape in root hair cells. Science 319:1241–1244

Tanaka N, Kato M, Tomioka R, Kurata R, Fukao Y, Aoyama T, Maeshima M (2014) Characteristics of a root hair-less line of *Arabidopsis thaliana* under physiological stresses. J Exp Bot 65:1497–1512

Tanimoto M, Roberts K, Dolan L (1995) Ethylene is a positive regulator of root hair development in Arabidopsis thaliana. Plant J 8:943–948

Tromas A, Perrot-Rechenmann C (2010) Recent progress in auxin biology. CR Biol 333:297–306

Tsai HH, Schmidt W (2017) Mobilization of iron by plant-borne coumarins. Trends Plant Sci 22:538–548

Vandamme E, Renkens M, Pypers P, Smolders E, Vanlauwe B, Merckx R (2013) Root hairs explain P uptake efficiency of soybean genotypes grown in a P-deficient Ferralsol. Plant Soil 369:269–282

Vicente-Agullo F, Rigas S, Desbrosses G, Dolan L, Hatzopoulos P, Grabov A (2004) Potassium carrier *TRH1* is required for auxin transport in *Arabidopsis* roots. Plant J 40:523–535

Vincent C, Rowland D, Na C, Schaffer B (2017) A high-throughput method to quantify root hair area in digital images taken in situ. Plant Soil 412:61–80

Wada T, Tachibana T, Shimura Y, Okada K (1997) Epidermal cell differentiation in *Arabidopsis* is determined by a Myb homolog, CPC. Science 227:1113–1116

Wang Y, Kristian TK, Stoumann JL, Jakob M (2016) Vigorous root growth is a better indicator of early nutrient uptake than root hair traits in spring wheat grown under low fertility. Front Plant Sci 7:865

Waters MT, Gutjahr C, Bennett T et al (2017) Strigolactone signaling and evolution. Annu Rev Plant Biol 68:291

Weinl S, Kudla J (2009) The CBL-CIPK Ca^{2+}-decoding signaling network: function and perspectives. New Phytol 184:517–528

White PJ, Broadley MR (2003) Calcium in plants. Ann Bot 92:487–511

Wu Y, He D (2011) Advances in root hairs in Gramineae and *Triticum aestivum*. Afr J Agric Res 6:1047–1050

Wu QS, Liu CY, Zhang DJ et al (2016) Mycorrhiza alters the profile of root hairs in trifoliate orange. Mycorrhiza 26:237–247

Wymer CL, Bibikova TN, Gilroy S (1997) Cytoplasmic free calcium distribution during the development of root hairs of *Arabidopsis thaliana*. Plant J 12:427–439

Xie X, Yoneyama K, Yoneyama K (2010) The strigolactone story. Annu Rev Phytopathol 48:93

Yan XL, Liao H, Beebe SE, Blair MW, Lynch JP (2004) QTL mapping of root hair and acid exudation traits and their relationship to phosphorus uptake in common bean. Plant Soil 265:17–29

Yang HB, Murphy AS (2009) Functional expression and characterization of *Arabidopsis ABCB*, *AUX1* and *PIN* auxin transporters in *Schizosaccharomyces pombe*. Plant J 59:179–191

Yang N, Zhu C, Gan L, Ng D, Xia K (2011) Ammonium-stimulated root hair branching is enhanced by methyl jasmonate and suppressed by ethylene in *Arabidopsis thaliana*. J Plant Biol 54:92–100

Yoshii H, Imaseki H (1982) Regulation of auxin-induced ethylene biosynthesis: repression of inductive formation of 1-Aminocyclopropane-1-carboxylate synthase by ethylene. Plant Cell Physiol 23:639–649

Yoshioka H, Sugie K, Park HJ et al (2001) Induction of plant gp91 *phox* homolog by fungal cell wall, arachidonic acid, and salicylic acid in potato. Mol Plant Microbe Interact 14:725–736

Zhang YY, Zhu HY, Zhang Q et al (2009) Phospholipase Dα1 and phosphatidic acid regulate NADPH oxidase activity and production of reactive oxygen species in ABA-mediated stomatal closure in *Arabidopsis*. Plant Cell 21:2357–2377

Zhang DJ, Xia RX, Cao X, Shu B, Chen CL (2013) Root hair development of *Poncirus trifoliata* grown in different growth cultures and treated with 3-indolebutyric acid and ethephon. Scientia Horticulturae 160:389–397

Zhang D, Xia R, Cao X (2016) Ethylene modulates root hair development in trifoliate orange through auxin-signaling pathway. Scientia Horticulturae 213:252–259

Zhao Y, Christensen SK, Fankhauser C, Cashman JR, Cohen JD, Weigel D et al (2001) A role for flavin monooxygenase-like enzymes in auxin biosynthesis. Science 291:306–309

Zheng Y, Zhu Z (2016) Relaying the ethylene signal: new roles for *EIN2*. Trends Plant Sci 21:2–4

Zhu J, Kaeppler SM, Lynch JP (2005) Mapping of QTL controlling root hair length in maize (*Zea mays* L.) under phosphorus deficiency. Plant Soil 270:299–310
Zhu C, Gan L, Shen Z, Xia K (2006) Interactions between jasmonates and ethylene in the regulation of root hair development in *Arabidopsis*. J Exp Bot 57:1299–1308
Zhu Y, Rong L, Luo Q et al (2017) The histone chaperone NRP1 interacts with *WEREWOLF* to activate *GLABRA2* in Arabidopsis root hair development. Plant Cell 29:260–276
Zou YN, Wang P, Liu CY, Ni QD, Zhang DJ, Wu QS (2017) Mycorrhizal trifoliate orange has greater root adaptation of morphology and phytohormones in response to drought stress. Sci Rep 7:41134
Zuchi S, Cesco S, Gottardi S, Pinton R, Römheld V, Astolfi S (2011) The root-hairless barley mutant brb used as model for assessment of role of root hairs in iron accumulation. Plant Physiol Biochem 49:506–512

Chapter 4
Morphological and Symbiotic Root Modifications for Mineral Acquisition from Nutrient-Poor Soils

A. Kleinert, V. A. Benedito, R. J. L. Morcillo, J. Dames, P. Cornejo-Rivas, A. Zuniga-Feest, Mabel Delgado, and Gastón Muñoz

4.1 Introduction: The Making of the Root System and Modulation of Its Architecture

The root system encompasses a defining organ in the plant kingdom. It has evolved to perform many diverse functions, such as support, water absorption, nutrient uptake, vegetative reproduction, hormone biosynthesis, photoassimilate storage, and the establishment of symbiotic relationships with microorganisms. This organ

A. Kleinert (✉)
Botany and Zoology Department, University of Stellenbosch, Matieland, South Africa
e-mail: kleinert@sun.ac.za

V. A. Benedito
Division of Plant & Soil Sciences, West Virginia University, Morgantown, WV, USA

R. J. L. Morcillo
Shanghai Center for Plant Stress Biology, Shanghai Institutes of Biological Sciences, Chinese Academy of Sciences, Shanghai, China

J. Dames
Department of Biochemistry and Microbiology, Rhodes University, Grahamstown, South Africa

P. Cornejo-Rivas
Departamento de Ciencias Químicas y Recursos Naturales, Universidad de La Frontera, Temuco, Chile

Centro de Investigación en Micorrizas y Sustentabilidad Agroambiental (CIMYSA), Universidad de La Frontera, Temuco, Chile

A. Zuniga-Feest · M. Delgado
Laboratorio de Biología vegetal, Instituto de Ciencias Ambientales y Evolutivas, Facultad de Ciencias, Universidad Austral de Chile, Campus Isla Teja, Valdivia, Chile

Centro de investigaciones en Suelos volcánicos (CISVo), Universidad Austral de Chile, Valdivia, Chile

B. Giri et al. (eds.), *Root Biology*, Soil Biology 52,
https://doi.org/10.1007/978-3-319-75910-4_4

has an intimate contact with its environment, the soil, and shows extreme plasticity to respond to local stimuli by dramatically shaping its form (architecture) through continually developing lateral root (LR) branching and elongation.

Root development starts very early during embryogenesis when the morphogenetic polarity of the embryo is determined during the first cell divisions of the zygote. The formation of a root–shoot axis establishes the plant body plan that defines both root and shoot apical meristems. These events are genetically orchestrated by transcription factors and other regulators of gene expression, such as miRNAs, protein kinases, signaling molecules, chromatin remodelers, and hormones, in order to ensure the acquisition of the correct identity of every cell type and tissues in each apical meristem. During germination and throughout the life of the plant, LRs develop via de novo organogenesis by successive divisions of a founder cell of the pericycle that divides successively leading to organ growth and breaking through the outer layers of the root (endodermis, cortex, and epidermis), finally culminating in the establishment of a new organ with its own meristem. Gravity, light (or lack thereof), moisture, temperature, and nutrient availability are strong stimuli that direct the direction of root growth and dynamically define the architecture of the root. The genetic network dynamics that define the root architecture have recently been assessed and reviewed by several researchers (Drapek et al. 2017; McCleery et al. 2017; Möller et al. 2017; Ötvös and Benková 2017; Walker et al. 2017). In this chapter, we reviewed some aspects of the genetics involved in root architecture remodeling (and occasionally the development of new organs) induced by low-nutrient conditions.

The ideotype of a root system that is nutrient efficient is deep reaching with lateral roots concentrating in high-nutrient patches that quickly respond genetically and epigenetically to low-nutrient status (Mi et al. 2010; Kiba and Krapp 2016). Understanding the genes involved in dynamically modulating root architecture in response to low-nutrient conditions and selecting genotypes with the best allelic combinations for higher nutrient efficiencies will potentially have a transformative impact on agriculture.

While we have not focused on the basics of root architecture in this chapter, it remains an important point of departure for studies in root responses to nutrient deficiencies. The recent reviews by Morris et al. (2017), on the shaping of the 3D root system architecture by environmental signals together with that of Shahzad and Amtmann (2017), where they review recent advances in studies of the interactive effects between two or more nutrients on root system architecture, are highly recommended. Furthermore, The Plant Cell has recently launched a new teaching tool called "Phenomics of root system architecture: measuring and analyzing root phenes" which looks at quantification of root system architecture (York and Lobet 2017). Some of the concepts covered by this teaching tool are phenes (like genes but

G. Muñoz
Laboratorio de Biología vegetal, Instituto de Ciencias Ambientales y Evolutivas, Facultad de Ciencias, Universidad Austral de Chile, Campus Isla Teja, Valdivia, Chile

Facultad de Ciencia, Universidad San Sebastián, Sede Valdivia, Valdivia, Chile

referring to the phenotype), morphology, geometry, topology, as well as the practical considerations associated with root system architecture quantification and different methods of analyzing the data obtained.

Below, we describe and review the literature about important aspects of root adaptations to low-nutrient conditions, such as development of cluster roots, mycorrhizal associations, and symbiotic nitrogen fixation. We addressed changes in organ, morphology, physiology, biochemistry, and metabolism, as well as molecular genetics and epigenetics, and pointed out our perspectives and research questions to advance our understanding in this field of study.

4.2 Cluster Roots

4.2.1 Changes in Root Morphology as a Result of Nutrient Constraints

Cluster roots (CRs) are ephemeral, adaptive root structures (described as dense clusters of fine rootlets around the main axis) that form in different parts of the root system and exude various compounds (Purnell 1960) (Fig. 4.1). CRs are present in most species belonging to the Proteaceae family, except members of Persoonieae tribe (*Persoonia*, *Garnieria*, *Toronia*, and *Acidonia*) (Lamont 1982). Additionally, outside of the Proteaceae, species that develop CR have been found in nine other families distributed in both the Northern and Southern Hemispheres in regions with significant biodiversity (Skene 2000), such as Betulaceae, Casuarinaceae, Cyperaceae, Elaeagnaceae, Fabaceae, Myricaceae, and Restionaceae (Lambers and Shane 2007).

CRs are composed by determinate rootlets that arise endogenously from opposite protoxylem poles in the pericycle of lateral roots. These rootlets grow for a limited time and remain physiologically active for some days (Skene et al. 1998). CRs increase the surface area available to absorb nutrients and can exude copious amounts of carboxylates (Lambers et al. 2002; Lamont 2003; Lambers et al. 2006; Sousa et al. 2007), acid phosphatase (Reddell et al. 1997; Gilbert et al. 1999; Neumann et al. 1999; Delgado et al. 2013), and proteases (Schmidt et al. 2003; Paungfoo-Lonhienne et al. 2008). The exuded carboxylates are important to promote P mobilization by forming stable complexes with cations bound to phosphates (e.g., Al^{3+}, Fe^{3+}, and Ca^{2+}) or displace phosphate from the soil matrix by ligand exchange (Jones and Brassington 1998; Shane and Lambers 2005). Proteaceae CRs release large amounts of carboxylates at relatively fast rates (Roelofs et al. 2001) and also exude acid phosphatase to hydrolyze organic P compounds to Pi, which is the form available for plant uptake (Bieleski 1973; Duff et al. 1994; Ryan et al. 2001). This is particularly important in P-deficient soils, as occurs in Western Australia, sub-Saharan Africa, and Brazil, in addition to young soils with high total P but low availability, such as those in southern South America (Lambers et al. 2012a).

Fig. 4.1 *Embothrium coccineum* seedling growing in hydroponics (**a**), showing abundant cluster roots (CR) of different developmental stages: *white arrow* shows mature CR; the same CR are closer in **c**; *black arrow* shows senescent CR. *Scale* indicated by a *white bar* at right side below in **a** corresponds to 1 cm. Mature CRs are shown in **b** and **c** produced in hydroponics and mature CR growing inside a volcanic rock (pumicita). *White bar* in **b** represents 1 mm (**b–d** are in the same scale)

The development of CR occurs without the intervention of microorganisms, and its formation depends mainly on low P availability at the tissue level (Shane and Lambers 2005; Delgado et al. 2014). However, CR formation has recently been linked to the presence of plant growth-promoting bacteria (PGPB) in the soil for some species of Australian Proteaceae as *Hakea laurina* and the Fabaceae as *Viminaria juncea* (Lamont et al. 2015). The indole-3-acetic acid (IAA) produced by PGPB may be involved in inducing CR formation. The Proteaceae species that form CR are non-mycorrhizal (Brundett 2009; Shane and Lambers 2005), with some exceptions, such as *Hakea verrucosa*, which is endemic from ultramafic soils in Western Australia (Boulet and Lambers 2005), and *Placospermum coriaceum* from tropical rain forest in northeastern Australia (Lambers et al. 2015).

CRs assist with the absorption of soluble forms not only of P but also other nutrients, such as Mn^{2+} (Muler et al. 2013), and in some species even the acquisition of organic N (e.g., peptides), such as *Hakea actites* (Paungfoo-Lonhienne et al. 2008). CR-exuded proteolytic enzymes aid to digest protein at the root surface (Schmidt and Steward 1997, Schmidt et al. 2003). In *H. actites* CR, the expression of small peptide transporter genes was regulated in response to N supply (Paungfoo-Lonhienne et al. 2009). Protease release has not yet been evaluated in other members of the Proteaceae family and could be relevant mainly in temperate zones of the globe, where N mineralization is limited by low temperatures.

CR formation has been extensively studied in South Western Australia, where the most severely P-impoverished soils occur, with values of P availability $\leq$1 mg kg^{-1} (Hayes et al. 2014). However, Proteaceae species that form CR do grow in soils with higher total P from southern South America (Delgado et al. 2015; Piper et al. 2013; Lambers et al. 2012a). These species also form CR in soils with low P availability (~2 mg kg^{-1}) due to Al^{3+} and Fe^{2+} complex formation (Borie and Rubio 2003). *Embothrium coccineum* can colonize ash depositions and young metamorphic rocky soils in Patagonia and are sometimes found growing in association with some cushion plants, e.g., *Acaena integerrima* (Rosaceae) (Zúñiga-Feest et al. 2010, 2014, 2015; Delgado et al. 2014). Additionally, *Euplassa cantareirae* (Proteaceae) develops CR in the Restinga biome of Brazil (de Britto et al. 2015), a habitat characterized by seasonal flooding and soils with very low pH (3.4), low-nutrient availability, and high Al^{3+} (Scarano 2002; Joly et al. 2012).

The presence of species with CR that solubilize P may also benefit neighboring plants or even soil microorganisms, since the activity of CR increases available P forms (Li et al. 2014). Therefore, combining plants with different root adaptations in agricultural systems could promote higher yields due to an increased nutrient availability in the rhizosphere of neighboring plants. In this regard, available P forms increased in the rhizosphere around *E. coccineum* CR compared with the adjacent bulk soil (Delgado et al. 2015). Similar results have been found in the CR rhizosphere of *Orites myrtoidea* (Proteaceae) growing on recent volcanic deposits (Avila 2013). Additionally, Muler et al. (2013) showed that *Banksia attenuate* (Proteaceae), a South Western Australian species, facilitate P and Mn^{2+}acquisition for neighboring species.

4.2.2 Cluster Root Morphology

CRs are classified into simple and compound (Lambers et al. 2015). Simple CR has a distinct bottlebrush-like appearance (e.g., *Leucadendron meridianum*) with diverse morphologies in the Proteaceae and other families, such as *Aspalathus linearis* (rooibos, Fabaceae) (Lambers and Shane 2007). South American Proteaceae species, such as *Gevuina avellana*, *Embothrium coccineum*, *Lomatia ferruginea*, *L. hirsuta*, and *Euplassa cantareirae*, also develop simple CR (Ramírez et al. 2005; Zúñiga-Feest et al. 2010; Lambers et al. 2012a; Delgado et al. 2013; de Britto et al. 2015). Compound CRs are essentially branched simple CR. Few genera of the Proteaceae produce compound CR (alone or in combination with simple CR). Compound CRs develop in some Australian (e.g., *Banksia* spp.) and several South African genera (e.g., *Leucadendron* and *Protea*) (Lamont 1982).

Rootlet proliferation in a cluster creates a high surface area, thereby increasing the contact of the root with the soil. For example, in *Hakea obliqua*, each cluster has a surface area (excluding root hairs) 25 times greater than that of an equivalent mass of axial root (Watt and Evans 1999). The rootlet surface of mature CR in *Hakea prostrate* represents around 2500 mm^2 cm^{-1} of root axis (Lambers et al. 2015). These CRs are temporary (ephemeral) structures of a similar formation in Proteaceae species of different geographical and phylogenetic relationships. CR develops from rootlet emergence to senescence within 25–30 days (Shane et al. 2004; Delgado et al. 2013; de Britto et al. 2015). This process occurs in different parts of the root, such as in upper horizons of the soil (Australian Proteaceae) or even 1 m deep (Chilean Proteaceae), depending on the soil profile. Senescent CRs are replaced by new CR that start to develop in other points of the root system to explore additional nutrient-rich soil patches. For example, in *Grevillea robusta*, CRs develop in predetermined areas at set distances along lateral roots (Skene et al. 1998), which has also been observed for *Lomatia ferruginea* and *L. dentata* (A. Zuñiga-Feest, personal observation).

The morphology, physiology, and functions of CR have been studied in a wide range of species, including *Lupinus albus* (Gardner et al. 1983; Johnson et al. 1994, 1996; Neumann et al. 1999; Neumann and Römheld 1999; Kihara et al. 2003), and South African and Australian Proteaceae, such as *Hakea prostrata* and *Banksia attenuata* (Lamont 2003; Shane et al. 2004; Lambers et al. 2006). Anatomical studies of the South American species, *Gevuina avellana*, revealed the presence of CR described with "claviform structures" at the end of each rootlet (Grinbergs et al. 1987; Ramírez et al. 2005). Because no other Proteaceae species exhibit this kind of root structure, their physiological function and significance remain unclear (Lambers et al. 2015).

Environmental factors such as soil texture can affect CR morphology. In fact, in *Hakea obliqua* (Dell et al. 1980), *Grevillea robusta* (Skene 1998), and *L. albus* (Watt and Evans 1999), the rootlet length is shorter when plants are grown in hydroponics, compared with seedlings grown in vermiculite or soil. In addition, when grown in the hydroponic system, root hairs are absent in *H. obliqua* or *G. Robusta*, while *E. coccineum* displays longer root hairs (A. Zuñiga-Feest, personal observation).

Other CR types have been described in monocot species, such as Cyperaceae (sedges) and Restionaceae (rushes). Cyperaceae species form a "dauciform root," a carrot-shaped cluster (Lamont 1974). These structures were first reported by Selivanov and Utemova (1969) and later found in other Cyperaceae around the world (Lamont 1982; Shane et al. 2006; Lambers et al. 2006). The most remarkable external feature is the very dense formation of long root hairs over the carrot-shaped axis. Several Restionaceae species in Australia and South Africa (the "Southern Hemisphere rushes") also develop CR with high numbers of rootlets densely covered with long root hairs called "capillaroid," due to their spongelike capability of holding water (Lamont 1982). Dauciform CR exudate carboxylate and their development are stimulated by low P supply (Lamont 1974; Shane and Lambers 2005; Playsted et al. 2006).

4.2.3 Factors Influencing Cluster Root Formation

The main factor inducing CR formation along with carboxylate and acid phosphatase exudation is low P, although its formation can occasionally be influenced by Fe deficiency (Shane and Lambers 2005), such as in the genus *Casuarina* (Zaid et al. 2003). Importantly, *Casuarina* species can also establish nitrogen-fixing actinorhizal symbiosis with *Frankia* bacteria. Recently, Delgado et al. (2013) showed that CR formation in *E. coccineum* was suppressed in hydroponic plants provided with 10 μM P. Additionally, different modulations of CR formation (without P supply) were observed when comparing seedlings of *E. coccineum* produced by seeds from different origins. In a greenhouse experiment, a southern *E. Coccineum* ecotype (Coyhaique 46°S) showed no CR suppression under full nutrient solution, unlike those observed in plants collected from temperate locations (Zúñiga-Feest et al. 2015). The genetics and evolutionary implications of the variability in this trait are intriguing.

Differences in carboxylate composition exuded by CR have been widely reported, with citrate and malate the most frequently released (Shane et al. 2004; Lambers et al. 2015). Recently, *E. coccineum* showed a higher exudation rate of citrate compared with *Hakea prostrata* under similar hydroponic conditions (Delgado et al. 2014). *E. coccineum* seedlings grown in sand exuded oxalate and succinate, depending on the origin of the population (Sepúlveda 2016). *Lomatia dentate* CR secreted mainly oxalic acid, with the highest exudation rates achieved when seedlings were exposed to P deprivation. *Gevuina avellana* exuded primarily succinic acid and, to a lesser extent, oxalic and malic acid with no differences observed under different nutritional treatments (Sepúlveda 2016). *Euplassa cantareirae* exuded mostly oxalic but also isocitric and malic acids at lower rates (de Britto et al. 2015).

CR biomass allocation in several Proteaceae species ranges between 40 and 50% of the total plant biomass. *E. coccineum* exhibits relatively low resource allocation to CR formation (8–10% plant biomass), forming small CR but with a higher carboxylate exudation rate per unit of fresh biomass (Zúñiga-Feest et al. 2010; Delgado et al. 2014). In contrast, *G. avellana*, another southern South America Proteaceae, shows a high CR biomass allocation of 25–40%, depending on nutritional conditions. In other Australian Proteaceae, *Hakea prostrata* (Shane et al. 2004) and *Grevillea crithmifolia*, CR allocation was around 25% of the total plant biomass (Shane and Lambers 2006).

Other factors influencing CR allocation were reported in *Myrica gale* and *Gymnostoma papuanum*, with soil N forms affecting size, number, and biomass distribution (Crocker and Schwintzer 1993; Racette et al. 1990).

4.2.4 Cluster Root Metabolism and Regulation

A remarkable feature of CR is their ability to strongly acidify the rhizosphere. For example, in calcareous soil (20% $CaCO_3$), CR of *Lupinus albus* acidified the rhizosphere from pH 7.5 to 4.8 (Dinkelaker et al. 1989). In some cases, rhizosphere soil pH can be decreased due to CR activity to as low as 3.6 (Li et al. 1997). This acidification is produced by exudation of organic acids, predominantly citric and malic acids (Gardner et al. 1983; Neumann et al. 1999). The C cost associated with organic acid exudation has been evaluated in *Hakea prostrata* (Shane et al. 2004) and showed threefold increases in the respiration rate at 4 days after CR initiation. These results reveal significant metabolic changes during the exudative burst in coordination with CR development. About 30% of the organic C released originates from dark CO_2 fixation by phosphoenolpyruvate carboxylase (PEPC) in P-deficient *Lupinus albus* roots (Johnson et al. 1996), while the remaining may be from photosynthesis. Much of the protons produced during the synthesis of organic acids may be released from the cell by plasma membrane H^+ ATPases. This enzyme responds to many environmental factors, such as saline stress (Niu et al. 1993), nutrient supply (Schubert and Yan 1997), Fe deficiency (Dellórto et al. 2000), and P starvation (Yan et al. 2002). Additionally, PEPC activity in *Lupinus albus* CR was threefold greater compared with P-supplemented roots. Yan et al. (2002) showed that in functioning CR cells, H^+ and organic anions are exported separately and that modification of the plasma membrane H^+ is essential to enhance CR rhizosphere acidification.

Malate and citrate exudation has been reported in CR of different species, such as *H. protrata*, *L. albus* (Kihara et al. 2003, Shane et al. 2004; Peñaloza et al. 2005), and recently in *E. coccineum* (Delgado et al. 2013). The organic acid release rate ranged between 2.4 and 7.4 mol $h^{-1}g^{-1}$ fresh weight for *Lupinus albus* CR (Keerthisinghe et al. 1998, Neumann et al. 1999). Because the rootlets are so tightly arranged, the released organic acids accumulate at high concentrations in the rhizosphere close to the CR (Yan et al. 2002). Indeed, 0.1 mmol citric acid can be produced per gram of soil by CR activity (Gerke et al. 1994; Li et al. 1997), a concentration sufficient to release P from sparingly soluble Fe and Al phosphates through ligand exchange or chelation of metal ions (Hinsinger 1998). Although several carboxylates can be exuded at the same time (Shane et al. 2004), citrate is one of the most potent extracting acids from inorganic P sources (Jones 1998, Ryan et al. 2001). The plasticity of carboxylate exudation has been observed in *Banksia grandis*, which differ its carboxylate compositions according to whether Al phosphate or Fe phosphate was supplied (Lambers et al. 2002). Factors controlling citrate exudation are important targets of research. Citrate efflux has been studied in P-starved *Lupinus albus* roots and accounted for up to 23% of the plant dry weight (Dinkelaker et al. 1989).This exudation occurred close to CR sections (Peñaloza et al. 2002). It began shortly after CR rootlets reached their final length (Watt and Evans 1999), following a transient pattern that persisted 3–9 days. The highest exudation occurred in mature CR, while young and old CR showed limited activity (Neumann et al. 1999; Watt and Evans 1999).

The metabolic source of organic acids exuded by roots is assumed to be from both the glycolysis (from sucrose translocated in the phloem) and the tricarboxylic cycle (TCA), operating in the root (Massonneau et al. 2001). An active accumulation of carboxylates before the exudative burst is a metabolic bypass occurring in the root cell TCA that leads to an accumulation of citrate and/or malate (Cramer et al. 2005). Metabolic changes in CR leading to citrate accumulation and efflux have been associated with increased PEPC activity (Neumann et al. 1999; Shane et al. 2004). Kihara et al. (2003) reported that metabolic changes in P-deprived *Lupinus albus* CR involve increased citrate biosynthesis by boosting the supply of substrates (pyruvate, oxalacetate) and blocking citrate catabolism in the cytosol. PEP phosphatase activity also increased threefold in P-deprived root apexes, which may be relevant to maintain the acetyl-CoA supply by providing pyruvate to the mitochondria. During CR development, there is also an increase in alternative oxidase (AOX) activity, while PEPC expression was relatively constant regardless of the developmental stage (Shane et al. 2004). AOX protein accumulation increased in advance to citrate and malate exudation, presumably to allow for a higher flow of electrons through the mitochondrial electron transport chain in the absence of a rapid ATP turnover. Citrate and isocitrate synthesis and accumulation contributed predominantly to the subsequent burst of citrate and malate exudation.

The enhanced exudation of organic acids from P-stressed *L. albus* roots is accompanied by increased in vitro activities of PEPC and malate dehydrogenase (MDH), which involved differential expression of the corresponding genes (Uhde-Stone et al. 2003). Peñaloza et al. (2005) isolated three PEPC cDNAs from *L. albus* CR, which expression was enhanced by Pi deficiency and showed that more than one PEPC isoform is responsible for the low Pi-induced PEPC activity in CR. However, low P-regulated gene expression is not the only mechanism to control PEPC activity in CR. Indeed, a novel control of PEPC in *Hakea prostrata* CR was described by Shane et al. (2013), involving posttranslational modifications affecting its kinetic/allosteric properties. Thus, an equivalent ratio of monoubiquitinated 110 kDa and phosphorylated 107 kDa PEPC polypeptides is present in immature proteoid roots. CR maturation correlated with PEPC activation via in vivo deubiquitination of the 110 kDa peptide followed by phosphorylation. This novel mechanism of posttranslational control results in a rapid adaptation to low P, not requiring modification of gene expression or de novo protein synthesis and contributing to the massive synthesis and excretion of organic acids that dominate the C metabolism of mature *H. prostrata* CR.

4.2.5 Cluster Root Activity Modifies the Rhizospheric Microbiota

Multiple studies showed that not only the type but also the developmental stage of a plant species significantly shape rhizospheric microbial communities (Badri et al. 2009; Chaparro et al. 2014; Jacoby et al. 2017). In the rhizosphere, most bacteria rely on root exudates as a source of C and energy. Therefore, roots are a crucial driving force for multitrophic interactions involving microorganisms and neighboring plants (Bais et al. 2006; Doornbos et al. 2012; Chaparro et al. 2014). The influence of CR and carboxylate exudation on soil microbiota has already been recorded for some *Banksia* species from Australia (Marschner et al. 2005), and *Leucospermum* spp. from South Africa (Stafford et al. 2005), showing that each species strongly influences the rhizospheric microbiota. Recently, Delgado et al. (2015) showed a higher β-glucosidase activity and fluorescein diacetate (FDA) hydrolysis in the rhizosphere of senescent *E. coccineum* CR, compared to the other developmental stages and bulk soil. These results suggest that colonizing species from recent volcanic depositions influence microbial activity mainly through CR activity.

Lupinus albus (Fabaceae) develop CR that exudate large amounts of carboxylates, especially malate and citrate but also oxalate and fumarate (Dessureault-Rompre et al. 2007; Weisskopf et al. 2008). In this species, a decrease in bacterial population was noted in mature CR and attributed to the low pH as well as carboxylate and flavonoid excretion. However, *Burkholderia* bacteria are predominant in the soil bacterial community associated around white lupin CR (Weisskopf et al. 2011). In addition to their ability to grow on citrate or malate, almost all collected *Burkholderia* isolates can use oxalate as a C source. This property, denoted as oxalotrophy, corresponds to a widespread characteristic related to plant-associated *Burkholderia*, a trait also required for root colonization in lupin and maize (Kost et al. 2014). Ubiquitous in the soil and able to colonize more than 30 plant species, some *Burkholderia* species are considered as one of the most potent plant growth-promoting rhizobacteria (Coenye and Vandamme, 2003; Compant et al. 2008; Suárez-Moreno et al. 2012).

Other carboxylates, such as malate, have been postulated to act as a signal to recruit beneficial microorganisms (Rudrappa et al. 2008). Malate is secreted from roots of *Arabidopsis thaliana* as a signal to recruit beneficial rhizobacterium isolates of *Bacillus subtilis* in a dose-dependent manner to promote binding and biofilm formation on roots (Chen et al. 2012). *Bacillus subtilis* can utilize a wide range of C sources, but glucose and malate are the two preferred substrates (Kleijn et al. 2010). Consequently, CRs from different species exude a distinct profile of C compounds and at different rates to modify rhizospheric microbial populations, by selecting or attracting specific microorganisms that may have beneficial plant growth properties. These microorganisms may also help promote soil changes at the chemical and physical level, favoring soil nutrient availability.

4.3 Root Symbiotic Modification: Ectomycorrhizal Associations

Many Basidiomycota and some Ascomycota fungi form ectomycorrhizal (ECM) relationships with roots of many perennial woody species. These include economically important forest species, such as those of the Pinaceae, Fagaceae, and Dipterocarpaceae families (Wang and Qiu 2006). An ECM association is distinguished from other mycorrhizal types by three characteristics: (1) an interwoven hyphal mantle or sheath surrounding the root tips, (2) a network of hyphae that penetrates between epidermal and cortical root cells, and (3) an extraradical mycelial network composed of emanating hyphae and rhizomorphs that extends to the soil (Smith and Read 2008). These distinctive ECM features are formed by approximately 6000 fungi species that are both structurally and functionally diverse. ECM fungi generally have a broad host range such as *Pisolithus*. In contrast, *Suillus* sp. associates only with members of the Pinaceae. Host plants can associate with many different ECM species (Brundrett 2004). Examples of ECM fungi belonging to the Basidiomycota include the genera *Amanita*, *Boletus*, *Xerocomus*, *Piloderma*, *Inocybe*, *Pisolithus*, and *Scleroderma* and the Ascomycota genera *Cenococcum*, *Choiromyces*, and *Tuber*, to name a few (Agerer 2006; Smith and Read 2008). This more generalist approach ensures colonization of seedling roots and provides access to a greater pool of nutrients (Molina et al. 1992).

4.3.1 Ectomycorrhizal Formation and Development

The presence of ECM fungi prior to colonization results in stimulation of lateral root (LR) development. Ditengou et al. (2015) reported that this stimulus was independent of the host plant because the nonhost *Arabidopsis thaliana* also developed more LRs in the presence of the ECM fungus, *Laccaria bicolor*. This indicates that the fungus releases diffusible fungal signals in their search for a host. Volatile organic compounds (VOCs) were suggested as the primary signals although other molecules may also be involved. ECM fungi, such as *Tuber melanosporum*, emit complex species-specific profiles, in this case imparting a unique aroma to these highly prized black Périgord truffles (Culleré et al. 2010). Further investigation of *L. bicolor* by Ditengou et al. (2015) identified the sesquiterpene (-)-thujopsene as responsible for LR stimulation (21–158%) of both a host plant (*Populus*) and *Arabidopsis*. The VOC also increased root hair length by 39%. The ECM ascomycete *Ceonoccum geophilum* cannot synthesize (-)-thujopsene and did not stimulate LR in *Arabidopsis*.

ECM formation of non-colonized roots or those developing from seedlings initiates with fungal propagules, either spores or ECM mycelia (Smith and Read 2008). ECM colonization of new lateral roots (LRs) may be from the same parent fungi that already colonize the Hartig net or the inner mantle of an existing colonized

LR. Recognition between ECM fungus and the plant involves root exudate signaling. Several low molecular weight compounds are thought to act as signal molecules and have been identified to include cytokinins, such as zeatin (Gogala 1991; Lagrange et al. 2001), the flavonol rutin (Lagrange et al. 2001), IAA, and hypaphorine (Gay and Debaud 1987; Ditengou et al. 2000; Ditengou and Lapeyrie 2000). The auxin-inducible protein Pp-iaa88 triggers a cascade of events resulting in ECM formation of *Pinus pinaster* (Reddy et al. 2003). The ECM symbiosis between *Eucalyptus* and *Pisolithus* is controlled by both partners through several unique symbiosis-related (SR) proteins called ectomycorrhizins (Malajczuk et al. 1990). These SR proteins encompass the fungal-encoded acid polypeptide 32 (SRAP 32) along with the hydrophobins, HydPt-2 and HydPt-3 encoded by the host (Laurent et al. 1999).

Transcriptional profiles revealed increased expression in 17% of 850 genes in 4-day-old *Eucalyptus–Pisolithus* mycorrhizae (Voiblet et al. 2001). Similar profiles have also been reported for other plant–fungus combinations, such as *Tilia–Tuber* (Polidori et al. 2002), *Pinus–Laccaria* (Podila et al. 2002), and *Betula–Paxillus* (Johansson et al. 2004). Working with different ECM plant–fungal systems, Duplessis et al. (2005) and Le Quere et al. (2005) both described similar gene expression patterns over a period of 21 days. At day 4, cell wall SR proteins and hydrophobins were expressed indicating fungal adhesion to root surfaces. Between days 4 and 7, primary metabolism genes involved in C transfer were induced, such as hexokinase, NAD-malate dehydrogenase, aspartate aminotransferase, and NADH dehydrogenase, indicating functional symbiotic interaction. Other cellular functions were induced by day 12, including protein synthesis and mitochondrial activity. Increases in stress and defense-related proteins were also noticed during this time and decreased thereafter. The role of the extraradical mycelia was studied by Morel et al. (2005) and Wright et al. (2005) using the *Betula–Paxillus* symbiosis. Sixty-five genes were differentially expressed between the external mycelium and the ECM roots. An investigation of the ECM root, rhizomorphs, and mycelia revealed differential expression of 31% of 1075 fungal genes between these three symbiotic components. Genes involved in the N cycle, amino acid and urea metabolism, NH_4^+ assimilation, and the glyoxylate cycle were strongly expressed in mycelia and rhizomorphs. A tubulin gene was expressed in rhizomorphs implicating the role of actin in long-distance motile tubular vacuole transport (Wright et al. 2005).

Root hairs proliferate behind the growing apex of non-colonized root tips, increasing the probability of contact with ECM propagules. Immediately behind the root cap, hyphae make contact within 2–4 days, and root growth continues along the root surface enveloping the root hairs and keeping pace with organ elongation. This is accompanied by an increased hyphal branching, fusion of hyphal tips, and swelling of the root tip (Massicotte et al. 1990). Adhesion to the root surface is promoted by fibrillar polymers (glycoproteins or lectins) that extend from the fungal wall. These polymers have been observed on free-living mycelium of *Laccaria bicolor* (Lei and Dexheimer 1988) and *Pisolithus tinctorius*, while lectin-binding sites have been reported on root surfaces (Lei et al. 1990). The hyphae closest to the root (the inner mantle) enter the root behind the root tip penetrating between root cells. This process continues with repeated branching in the epidermal layer to form the Hartig net. Differences are apparent in various plant–fungal interactions. In *Alnus crispa–Alpova diplophoeus* (Massicotte et al. 1986), epidermal cell wall ingrowth is reminiscent of haustoria. In *Eucalyptus–Pisolithus* mycorrhizae (Massicotte et al. 1987), a zone lacking a Hartig net is formed directly behind the root tip. This pre-Hartig net zone is accompanied by a radial swelling of epidermal cells. The Hartig net is relatively simple, but LRs are induced to develop clusters or tubercles, which results in an increased surface area (Dell et al. 1994). Primary roots of hosts such as *Eucalyptus*, *Fagus*, and *Pinus* maintain the capacity for continuous growth, and LR will abort if they do not become colonized by an ECM fungus (Smith and Read 2008).

4.3.2 Ectomycorrhizal Nutrient Acquisition Strategies

The interaction between fungus and host plant is primarily a nutrient acquisition strategy particularly for N and P. It is well known that mycorrhizal fungi enhance nutrient uptake for plant growth, and, in return, the fungi are rewarded with photosynthetically derived C compounds (Smith and Read 2008). The ECM mantle surrounds the root. Thus, all nutrients must enter the roots through the mycorrhizal pathway (Lambers et al. 2008; Smith and Read 2008). Nutrients in inorganic forms are accessible to the plant, but uptake is dependent on nutrient solubility, nutrient concentration within the root zone, and root architecture. Being fairly immobile, inorganic P can be absorbed faster by roots than it can move through the soil, resulting in a depletion zone around the roots (Lambers et al. 2008). The extraradical mycelial network is an essential component of nutrient acquisition. The mantle itself does not increase the surface area of the root and serves mainly for nutrient storage and transfer to the host (Smith and Read 2008).

Ectomycorrhizal fungi are effective scavengers of nutrients through their mycelial network that increases the volume of soil which can be explored; the absorptive hyphae can transport nutrients over long distances (up to 25 cm) back to the host plant, thus overcoming nutrient limitations (Smith and Read 2008).

Ectomycorrhizae can take up soluble inorganic nutrient sources, but they also have access to unavailable organic or other complexed forms of nutrients. For example, inorganic P is rapidly complexed with Fe^{3+} and Al^{3+} in acidic soils or with Ca^{2+} when pH is high, thus becoming a limiting nutrient. In order to access these unavailable nutrients, ECM adopted a mining strategy through the production of hydrolytic enzymes, such as phosphatases, and the synthesis of organic acids (Courty et al. 2005, 2009). Because of the symbiotic association, mobilization of these inaccessible nutrients does not depend on the C/N ratio, since unlike saprotrophs, ECM fungi obtain their C directly from the host plant (Lambers et al. 2008).

Several different ECM hydrolytic enzymes have been implicated in nutrient mobilization (Courty et al. 2005, 2009). Amylase has been reported in *Tricholoma matsutake* (Kusuda et al. 2003); protease in *Suillus variegatus* (Ramstedt and Soderhall, 1983), *Amanita muscaria*, and *Lactarius* sp. (Nygren et al. 2007); and cellulase in *Suillus bovinus* (Maijala et al. 1991). This highlights the importance of the mycelial network and its role in nutrient acquisition. This network is recognized as having, to varying degrees, hyphae that emanate from the ECM mantle and remain closely associated with the root or rhizomorphs that are more suited for long-distance transport. ECM is, therefore, classified into different exploration types that allow for contact and short-, medium-, and long-range nutrient acquisition (Agerer 2001).

Short-distance ECM roots are surrounded by extensive emanating hyphae but no rhizomorphs, e.g., *Coenococcum*. ECM fungi, such as *Tormentella*, have a smooth mantle with few emanating hyphae and ECM root tips in close contact with the substrate. Three subtypes of medium-distance exploration types are recognized: (1) the fringe subtype forms fans of emanating hyphae and rhizomorphs that ramify and frequently interconnect, and rhizomorphs are hairy and form emanating hyphae that extend further into the soil, e.g., Cortinarius; (2) the mat subtype typically covers a large area, e.g., Geastrum; and (3) the smooth subtype is recognized by a smooth mantle with no emanating hyphae but rhizomorphs that are frequently morphologically differentiated, e.g., Amanita. Long-distance exploration types have a smooth mantle with few but highly differentiate rhizomorphs, e.g., *Boletus*. These types were described in detail by Agerer (2001).

4.3.3 Ectomycorrhizal Contribution to Soil Weathering

Weathering is the physical and chemical degradation of rocks. Except for C and N, all other essential elements are made available through this process. The exploratory nature of the ECM fungal mycelial network and their ability to bridge distances and produce organic anions and proton (Gadd 1999) assist in the breakdown of weak spots in solid rock. The physical pressure exerted through hyphae allows for penetration into cracks, spaces, and mineral cleavage points. Hyphae may also expand and contract during wet/dry and freeze/thaw cycles (Hoffland et al. 2004). ECM fungi account for more than 50% of the total soil respiration with the production of CO_2 and carbonic acid. This H^+ release is responsible for the dissolution of mineral phosphates (Hoffland et al. 2004). Low molecular weight organic acids (LMWOA), such as oxalate, malate, and citrate, are produced and exuded by ECM fungi. These acids bind to metals, such as Al^{3+}, Fe^{3+}, Ca^{2+}, Mg^{2+}, Mn^{2+}, Zn^{2+}, and Cu^{2+}, and promote mineral weathering. Production of LMWOA requires a substantial cost in C that ECM fungi are provided by the host plant (Landeweert et al. 2001). Acidification due to N uptake can also contribute to weathering. ECM fungi preferentially utilize ammonium, rather than nitrate, and account for 80–90% of the cation uptake (Hoffland et al. 2004). Siderophores are Fe chelators produced by many ECM fungi (Haselwandter 1995). Fe solubilization from the mineral goethite (ferric oxyhydroxides) by the ECM fungus *Suillus granulatus* was attributed to siderophore release (Watteau and Berthelin 1994). Haselwandter and Winkelmann (2002) investigating *Cenococcumn geophilum* noted that ferricrocin was the main siderophore.

4.3.4 Succession

ECM fungal succession has been reported (Nara et al. 2003a, b; Ashkannejhad and Horton 2006), with non-colonized roots forming simple ECM by "early-stage" ECM fungal genera, e.g., *Telephora*, as a result of spore inoculum (Deacon and Fleming 1992). "Late-stage" ECM fungal genera like *Amanita* and *Russula* colonize mainly through mycelial contact. Nara et al. (2003a) observed that early-stage fungi occupied the outer root zone where root densities are low, whereas late-stage fungi colonized roots closer to the trunk where root densities are higher. In mature forests, roots may be 100% colonized with ECM fungi. Critical to maintaining the symbiosis is the colonization of new roots. ECM mycelia are more efficient at colonizing new roots as the mycelium can support the C requirements. Colonization via spores is more limited as they are reliant on conditions that favor germination (Peay et al. 2011).

The availability of non-colonized roots will depend on fine root turnover and rate of root elongation. Seedlings establishing near to the trunk may be colonized by "late-stage" fungi especially under low root densities, whereas short exploration ECM fungi types are best suited to high root density situations (Peay et al. 2011). Hobbie and Agerer (2010) further demonstrated a correlation between N^{15} values in sporocarps and ECM exploration types. These results suggested that short-distance exploration types were related to accessible labile versus non-labile (longer-distance exploration types) N sources.

4.4 Root Symbiotic Modification: Arbuscular Mycorrhizal Associations

4.4.1 Arbuscular Mycorrhizal Development and Function

Considering plant nutrition, especially unavailable P forms, root attributes are critical factors and especially its architecture. The plant P status is perhaps the most important variable influencing root architecture, and particularly a high root-to-shoot ratio, which is a characteristic of plants under P deprivation (Gruber et al. 2013). Phosphate content varies at a microspatial level in the soil leading to heterogeneous distribution. Therefore, roots need to develop continuously to reach new soil sectors that are rich in P, thus reshaping in architecture.

The topsoil layer usually has higher P availability for plants and microorganisms due to the poor P mobility in the soil. In this sense, some noticeable adaptations to access this P-rich environment are (1) the prevalence of axial roots together with a shallower angle, (2) the enhancement of adventitious rooting, and (3) a greater density and dispersion of lateral roots and root hairs (Lynch 2007). The higher density of roots in topsoil can reduce water uptake, since water is more abundant at deeper levels (Ho et al. 2005), as well as impose an increased C cost due to the development of higher density of root hairs (Zhu et al. 2010). Characteristics such as root length, root diameter, and surface-to-volume ratio are variations of root morphological traits related to P acquisition efficiency (PAE) (Lynch 2007; Miguel 2011).

The main effect attributed to AM symbiosis for enhancing PAE is the well-known increase of soil volume exploration by hyphae, beyond the space explored by lateral roots (LRs) and root hairs, both in length and in the porous space of soil. This alternative can be more efficient than the development of new roots for most AM host plants due to a lower C cost per unit of hyphal surface compared to the root surface (Gregory 2006; Schnepf et al. 2008). This explains that in general AM plants develop less root biomass and length. This effect is noticeable in plants that present a low density of root hairs and roots with large diameter (Fitter 2004; Smith and Read 2008), especially under P-limiting conditions (Schweiger et al. 1995; Jakobsen et al. 2005).

The root architecture is also influenced by some fungal exudates known as Myc factors, which stimulate colonization and can also act as plant growth regulators by modifying root development in some species (Maillet et al. 2011; Mukherjee and Ané 2011). This modification is shown in many plant species as an increase in root branching, probably to increase suitable places for AM colonization (Harrison 2005).

4.4.2 Arbuscular Mycorrhizal Development Under Limiting Nutrient Conditions

Arbuscular mycorrhizal (AM) symbiosis is an ancient microbe–plant relationship existing between higher plants (more than 80% of species) and fungi belong to the phylum Glomeromycota. This interaction provides a wide range of benefits to both partners (Smith and Read 2008; Smith and Smith 2012). The AM symbiosis emerged about 460 million years ago, in the Devonian period, being a key adaptation that allowed the early transition of plants from aquatic to terrestrial habitats (Simon et al. 1993; Schüßler 2000; Redecker et al. 2000; Heckman et al. 2001). It was the most important adaptive advantage for the nutrient uptake by first terrestrial plants. Regarding nutrition, plants provide photosynthetically derived C to the fungus, which in return supplies the plant with water and nutrients, mainly phosphate but also N (Harrison 1999; Govindarajulu et al. 2005; Javot et al. 2007). Therefore, nutrients play an important role during the symbiosis establishment. Nutrient availability is crucial for a successful association and, particularly, for root architecture modulation. In this sense, the form, availability, and concentration of P, which is delivered as phosphate by the fungus to the root through highly branched fungal structures formed within root cortical cells, named arbuscules, are critical factors for establishment and root architecture during both early and late stages of the symbiosis.

In contrast to the well-known and studied influence of phosphate in the establishment of the symbiosis and root architecture modulation, the role of other mineral nutrients, including N, the most limiting nutrient to plant growth, and other micronutrients, such as Zn^{2+} or Fe^{2+}, which are also transferred to the plant through arbuscules (Marschner and Dell 1994), remains unclear.

Moreover, the formation of the symbiosis begins with a dialogue between the plant and the fungus, in which hormones (and other associated molecules) act as messengers (Ludwig-Müller 2010; Miransari et al. 2014). Indeed, hormones play an essential role in plant development and control a large number of events, including root growth and morphogenesis under nutrient constraints (Jung and McCouch 2013). In this sense, the participation of the plant hormones in root architecture regulation during symbiotic interaction is plausible.

Phosphorus is an essential element in the establishment and development of the AM symbiosis. One of the main benefits of symbiosis is the P transfer from the fungus, which captures it from the soil, toward the plant. Therefore, it is expected that P availability and depletion to be important in regulating the process and, particularly, modulating root development. High concentrations of soluble phosphate in the soil inhibit the formation of mycorrhiza colonization, whereas phosphate starvation promotes the symbiosis establishment (McArthur and Knowles 1992; López-Ráez et al. 2008; Breuillin-Sessoms et al. 2010).

Several studies showed that P availability changes in the composition of root exudates, modifying the ability to stimulate the fungus growth. Thus, root exudates of plants cultivated in low P induce growth and branching of fungal hyphae, while exudates from plants grown at elevated P caused the opposite effect (Tawaraya et al. 1998; Nagahashi et al. 1996). Moreover, AM fungal exudates can directly induce root modifications in host plants. Exudates from germinating spores of various AM fungal species caused an increase of LR formation and total root length (Oláh et al. 2005; Mukherjee and Ané 2011), probably as part of a mechanism intended to develop a broader contact surface to ensure the establishment of the symbiosis. Although the stimulation of LR development seems to be a well-preserved mechanism during the early stages of the symbiosis, modifications on primary root and total root length diverge among the plant species in response to AM interactions (Fusconi 2014). For example, exudates from germinating spores of *Rhizophagus irregularis* (before *Glomus intraradices*) produce an increase of LR and total root length in *Medicago truncatula* without inhibiting the primary root (Oláh et al. 2005), whereas the association of *Allium porrum* with *Glomus* sp. reduced the main root development but increased LR, not affecting total root length (Berta et al. 1993).

Some of these signaling molecules released by AM fungi have a functional analogy with "Nod factors" of N_2-fixing rhizobial bacteria. For AM symbiosis, they are called "Myc factors" and are a mixture of simple sulfated and non-sulfated lipochitooligosaccharides (LCO; Maillet et al. 2011) and chitooligosaccharides (CO; Genre et al. 2013). The root response to Myc factors is partially dependent on a common symbiosis signaling pathway shared between rhizobial and AM symbioses. Similarly of Nod factors, Myc factors are recognized by plasma membrane receptors of root hairs before the physical contact between both symbionts, promoting the induction of the expression of specific genes of the host plant (Kosuta et al. 2003; Weidmann et al. 2004) and stimulating LR development (Oláh et al. 2005; Maillet et al. 2011). After the initial recognition, a signaling cascade begins with a rapid and transient increase in Ca^{2+} at cytoplasmic level (Oldroyd and Downie 2004; Chabaud et al. 2011), which is common of rhizobial and mycorrhizal responses (Kosuta et al. 2008; Sieberer et al. 2012). Ca^{2+} oscillations are transmitted to the cell nucleus by components of the nuclear pore (Kanamori et al. 2006; Saito et al. 2007) and ion channels (Ané et al. 2004), followed by an activation of transcription factors with common as well as distinct roles for each symbiosis, in order to initiate the biological processes necessary for the symbiosis establishment (Middleton et al. 2007; Gobbato et al. 2012; Sun et al. 2015). LR stimulation by Myc factors requires *DMI1* and *DMI2* in *Medicago truncatula*, but not *DMI3*, which is needed only for rhizobium–legume symbiosis (Oláh et al. 2005), while rice does not require activation of *DMI1* and *DMI3* for LR induction by germinating spore exudates (Gutjahr et al. 2009). Moreover, stimulation of LR occurs in response to Myc LCOs, but not CO4, a class of CO in *M. truncatula*, whereas both Myc LCO and CO4 are required in rice (Sun et al. 2015). This suggests that plant species respond differently to signaling molecules released by AM fungi.

One of the most evident changes in the production of exudates under different P levels is related to strigolactones. Balzergue et al. (2011) observed that pea cultivated under high P lost the ability to stimulate mycorrhiza due to a lower synthesis of strigolactones. However, supplementation with exogenous strigolactones failed to fully restore the root colonization capacity. This indicated that while strigolactones are involved in the regulation of mycorrhizal mediated by P availability, it is not the only factor involved and suggested the existence of other early signals controlling the stimulation of colonization. The metabolic profile of root exudates revealed significant changes in pea cultivated under high and low P (Laparre et al. 2011). They also reported a greater stimulation of certain ions under low P, which may be an alternative pathway of mycorrhizal regulation and LR stimulation by P availability, independent of strigolactones. In a similar trend, the *ccd8* pea mutant, which is impaired in strigolactones biosynthesis, showed comparable AM root colonization relative to that of wild-type plants under low P (Foo et al. 2013), indicating once again that strigolactones are not indispensable to AM colonization.

Typically, the inhibition of P-mediated AM colonization occurs during the presymbiotic phase, before the appressorium is formed. However, if the nutritional status of the plant is altered after the formation of the symbiosis, such as the result of an exogenous P contribution, the relationship between plant and fungus may change as part of a plant strategy to save photosynthates that would otherwise be allocated to the fungus. A study on the influence of P supplementation during the presymbiotic phase revealed that the most important effect of high P was on roots, which became unable to host mycorrhizal fungi even when appropriately stimulated (Balzergue et al. 2013). P supplementation did not disturb root Ca^{2+} spiking responses to the AM fungus, which is essential for downstream activation of transcription factors that trigger the biological processes required for the establishment of symbiosis. This suggested that high P levels did not affect the root ability to perceive the fungus (Balzergue et al. 2013). Also, the host plant can also limit the development of the fungus within the root through a mechanism mediated by P availability, regulating specific phosphate transporters responsible for transferring P from the fungus to the plant (Nagy et al. 2005). Recently, the AM-induced *Lotus japonicas* transcription factor (TF) *LjMAMI* (meristem and arbuscular mycorrhiza induced) was identified (Volpe et al. 2013b). *LjMAMI* is related to *PHR1*, a master regulator of phosphate starvation response in *Arabidopsis* (Rubio et al. 2001). LjMAMI is involved in both LR developments in response to AM symbiosis and P starvation responses in root meristems in an AM-independent manner (Volpe et al. 2013a, b). It shows a similar expression pattern than *LjPT4*, a specific phosphate transporter exclusively induced by AM symbiosis (Javot et al. 2007; Guether et al. 2009). RNAi transgenic hairy root lines of *LjMAMI* present a severe reduction in root branching, as a consequence of the gene downregulation. However, in the presence of the AM fungus, root proliferation and emission of lateral roots are enhanced (Volpe et al. 2013b), demonstrating the role of AM in lateral root development. Nevertheless, little is still known about the molecular mechanisms involved in how P starvation signaling regulated the establishment of the symbiosis, especially for root morphogenesis.

The role of N in the establishment of AM symbiosis has been largely discussed since many argue that the capacity of N uptake and transport from the AM fungus is much lower than the potential of the plant roots to directly acquire this nutrient (Hawkins et al. 2000; Hodge 2003). However, it has been demonstrated that like P, N supplementation inhibits AM root colonization (Nouri et al. 2014), and the deficiency of both P or N leads a significantly greater increase in AM colonization and arbuscule incidence in *Medicago truncatula* roots inoculated with *Rhizophagus irregularis* compared with individual deficiency of either phosphate or nitrogen (Bonneau et al. 2013). Moreover, not only P increases strigolactone amounts in root exudates in response to nutrient starvation but also N deficiency, although there are differences among plant species (Yoneyama et al. 2012). Thus, the participation of the N in both establishment of the symbiosis and regulation of root architecture during the symbiotic interaction is reasonable.

Considerable efforts have recently been made to decipher and define the role of N during the establishment of AM symbiosis. Most of the studies have been focused on N uptake, assimilation, and translocation in AM association. Several genes from both symbionts related to these biological processes have been identified and characterized (cf. Chen et al. 2018). However, little is known about the role of N in root architecture modulation during the early or late stages of the symbiotic association, probably because the importance of N in AM symbiosis seems to be linked more specifically to maintaining an adequate balance of N, P, and C photosynthates between both partners. In this sense, arbuscular degeneration produced by the absence of a functional *MtPT4*, a phosphate transporter essential for maintaining AM symbiosis in *Medicago truncatula*, is suppressed if *mtpt4* mutants are deprived of N but not of C (Javot et al. 2011; Breuillin-Sessoms et al. 2015). Nevertheless, it has been proposed that N affects strigolactone exudation through its effect on P levels (Kapulnik and Koltai 2014). Thus, it is possible that N availability influences root architecture and LR formation in early stages of AM symbiosis by modulating strigolactone responses through its impact on P, C, and N balance, even though N deficiency regulation of strigolactone biosynthesis is not required for AM colonization (Foo et al. 2013).

4.4.3 Arbuscular Mycorrhizal Physiology Under Limiting Nutrient Conditions

The role of AM fungi in plant nutrition and water uptake is well recognized but also underscores an important biotic component that increases plant tolerance to biotic and abiotic stressors through diverse mechanisms (Cornejo et al. 2017b). These include (1) increase of nutrient acquisition efficiency (Evelin et al. 2012; Krishnamoorthy et al. 2016), (2) increase of both photosynthetic activity and water use efficiency (Porcel et al. 2015; Santander et al. 2017), (3) accumulation of several compatible solutes, (4) control of reactive oxygen species (ROS) by nonenzymatic systems, (5) increase activity of antioxidant enzymes (Bárzana et al. 2015; Calvo-Polanco et al. 2016), and (6) modification of the (myco)rhizosphere (Yin et al. 2016), including accumulation of glomalin-related soil protein (GRSP) (Cornejo et al. 2008a).

The increased production of AM hyphae and spores able to sequester, adsorb, and accumulate high amounts of toxic elements (Aguilera et al. 2011; Cornejo et al. 2013; Meier et al. 2012) has been proposed as a potential mechanism to guarantee the AM symbiotic functioning under extreme environmental conditions (Fig. 4.2). In conjunction with the previously reported ability of GRSP to immobilize toxic elements (Cornejo et al. 2008a, 2017a; Seguel et al. 2015), the AM symbiosis represents a crucial component to diminish the bioavailability of these contaminants and indirectly creates better conditions for plant performance through enhanced mineral nutrition. Based on the above, it is necessary to broaden the biotechnological knowledge and the possibilities to manage AM symbiosis, with the aim of developing strategies and bioproducts for sustainable plant production under nutrient-limiting conditions.

Fertilizer applications are expensive, and most soils are deficient in some nutrients, mainly P and N. Nitrogen supplementation in sustainable agriculture is best managed by using legumes in crop rotation or the use of free-living microorganisms (*Azotobacter*, *Azospirillum*, among others) as bioinoculants (Cornejo et al. 2017b). This is not the case for P because its availability is usually scarce due to its low solubility nature and high anion fixation capacity of acidic soils, in special (Seguel et al. 2013, 2017). This low P availability leads to the continuous application of high

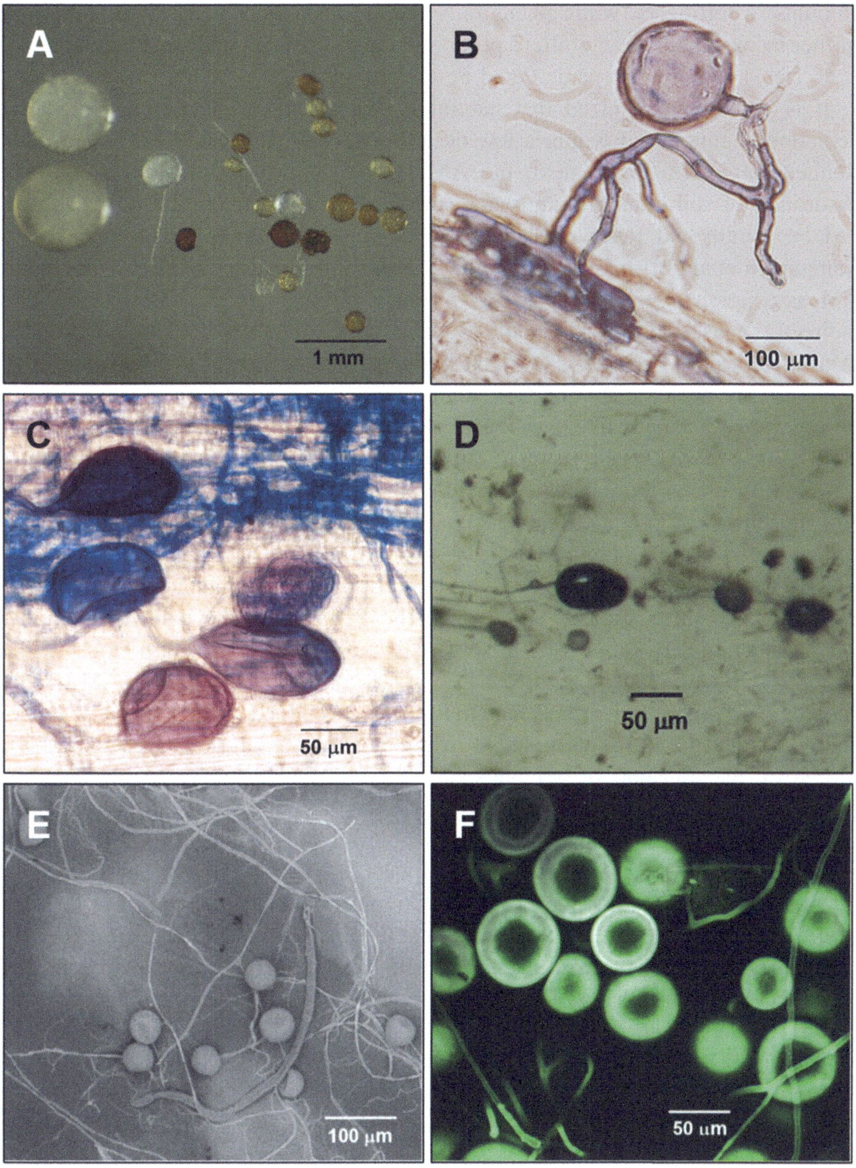

Fig. 4.2 Different structures of arbuscular mycorrhizal (AM) fungi and the generated symbiosis with plant roots. (**a**) Different AM resistance spores extracted from a low-nutrient multi-contaminated soil in Central Chile. Note the high size of the resistance spores, in some cases higher than 1 mm. (**b**) The first stage of AM formation represented for the appressorium formation in the root cells of *Rosmarinus officinalis* by the AM fungus *Funneliformis constrictum*. (**c**) Intraradical mycelium and spores of the AM fungus *Rhizophagus intraradices* in roots of *Lavandula officinalis*. (**d**) Vesicles and intraradical mycelium of unidentified AM fungi in roots of *Triticum aestivum* obtained from soil with high levels of aluminum (Al) and stained with hematoxylin. This stain reveals the presence of phytotoxic Al related to P deficiency in acid allophanic soils and represents the "barrier effect" of AM fungal structures to the entry of Al to the root cell. (**e**) Scanning electron microscopy of spores and extraradical hyphae of *R. intraradices* associated with transformed roots

amounts of fertilizers, while plants only require a fraction of the total applied P, further increasing the immobilized portion in the soil (Borie and Rubio 2003; Seguel et al. 2017), mainly as organic P and Fe–Al phosphates.

It is globally recognized that natural P sources are decreasing and the total depletion is anticipated within a few decades (Cordell and White 2011; Elser and Bennett 2011). In this context, the AM symbiosis is especially critical for the absorption of soil-immobile nutrients, such as P as well as Zn, Cu, Mn, Ca, Mg, and Fe (Smith and Read 2008; Cornejo et al. 2008b, 2009; Kaya et al. 2009; Campanelli et al. 2012; Evelin et al. 2012; Garg and Bhandari 2016). Additionally, AM association also can improve N nutrition (Cornejo et al. 2008b, 2009) and indirectly cause changes in root exudates that affect microbial communities (Marschner and Timonen 2005), resulting, for example, in the increase of ammonifying bacteria, nitrobacters, denitrifying bacteria, and phosphobacteria to improve urease and alkaline phosphatase activity in the (myco)rhizosphere (Ye et al. 2015). The latter is especially important in P solubilization and increased absorption by AM host plants (Fernández-Bidondo et al. 2012). Additionally, the extraradical AM mycelium can produce acid phosphatases that contribute to P solubilization, as demonstrated by Sato et al. (2015) in monoxenic cultures of *Rhizophagus clarus*.

The role of AM fungi improving P nutrition is so fundamental that some AM-colonized hosts dramatically reduce the P direct uptake pathway of roots, which is completely suppressed in some cases, as reported by Smith et al. (2004) in flax and tomato plants, in which 100% of P is taken up via AM. The AM fungus absorbs P via high-affinity phosphate transporters (Maldonado-Mendoza et al. 2001) and accumulates it in the vacuoles of extraradical mycelium (Viereck et al. 2004). Moreover, the transport of polyphosphate through fungal hyphae is mediated by aquaporins (Kikuchi et al. 2016).

AM symbiosis also plays an influential role favoring N absorption in deficient environments. Indeed, AM increases the uptake of NO_3^-, NH_4^+, and even some N organic sources (Hodge et al. 2001, 2010; Govindarajulu et al. 2005; López-Pedrosa et al. 2006). Bago et al. (2001) proposed that the uptake, transfer, and subsequent N absorption by the plant are related processes associated with the urea cycle and transport of polyphosphates. This hypothesis was confirmed by Govindarajulu et al. (2005). In this AM-mediated pathway, NO_3^- is absorbed by the extraradical mycelium coupled to H^+ symport and subsequently converted to NH_4^+ by nitrate reductase or directly taken up as NH_4^+. Next, via the GS/GOGAT cycle, NH_4^+ is assimilated into arginine and transported to intraradical mycelium associated with

---

Fig. 4.2 (continued) of *Daucus carota* revealing the comparatively small diameter of AM hypha, which allows the profuse colonization of small pores in the soil by fungal structures connected to the root. (**f**) Confocal laser scanning microscopy of hyphae and spores of AM fungi extracted from agricultural acidic soils with high Al saturation. The adsorption and accumulation of Al in extraradical fungal structures are shown by the fluorescence generated and suggest the presence of a physical barrier to the entry of Al beyond the roots (photographs **d**, **e**, and **f** courtesy of Dr. Paula Aguilera, Universidad de La Frontera)

the transport of polyphosphates (Subramanian and Charest 1998; Bago et al. 2001; Govindarajulu et al. 2005). Several NH_4^+ and NO_3^- transporters have been identified in extra- and intraradical mycelium of AM fungi (López-Pedrosa et al. 2006; Tisserant et al. 2012), which explains that in some AM plants 21–75% of the total N is absorbed through the AM pathway (Govindarajulu et al. 2005; Tanaka and Yano 2005; Smith and Read 2008; Kobae et al. 2010).

Regarding other nutrients, it is noticeable the role AM fungi play as a primary barrier and in selecting ions (Daei et al. 2009; Santander et al. 2017). For example, in saline soils, AM reduce the translocation of Na^+ ions to plant tissues avoiding accumulation to toxic levels, because of the ability to retain these ions in vacuoles of intra- and extraradical mycelium and vesicles (Al-Karaki 2006; Mardukhi et al. 2011). Furthermore, AM fungi selectively absorb minerals such as K^+, Mg^{2+}, and Ca^{2+}, thereby maintaining high ratios of K^+/Na^+, Ca^{+2}/Na^+, and Mg^{+2}/Na^+ (Hammer et al. 2011). This mechanism is regulated at the molecular level through membrane transporters, such as for *R. irregularis*, which genes encoding for K^+ and small conductance Ca-activated K^+ ion channels (Casieri et al. 2013) were selectively induced to balance the K^+/Na^+ ratio. Taking into account the obligatory nature of AM fungi biotrophy, study of the molecular mechanisms by which this particular type of fungi contributes to the plant nutrition under adverse conditions is particularly challenging. However, there is a consensus about the crucial role of this association in modifying the plant physiology and contributing to stress coping mechanisms. Therefore, more research is needed to advance our understanding of the underlying mechanisms involved in this association and whether the presumably different degree of functional fungus/host compatibilities could be used as a tool to design "à la carte" bioproducts in a sustainable plant production systems.

4.5 Root Symbiotic Modification: Nodule Development and Function Under Limiting Nutrient Conditions

Nitrogen is one of the most abundant elements on earth (78% of the atmosphere) while simultaneously being a critical limiting element for most plants due to its unavailability as available forms (Valentine et al. 2011). Plants can assimilate N from both soils and the atmosphere (Vance 2001). Soil nitrogen can be originated from commercial fertilizer, organic manure, or the mineralization of organic matter, while atmospheric N is assimilated through symbiotic N_2 fixation (Vance 2001). Up to 80% of this fixed N_2 is the product of symbioses between species of *Rhizobium*, *Bradyrhizobium*, *Sinorhizobium*, *Azorhizobium*, *Mesorhizobium*, and *Allorhizobium* and their leguminous plant hosts (Graham and Vance 2000). Herein, they are collectively referred to as rhizobia.

4.5.1 Nodule Development and Nutrient Transfer Between Symbionts

Nodule development is herein described in broad terms; for an in-depth review on the developmental biology of legume nodulation, readers are referred to Gage (2004). Rhizobia sense secreted compounds, such as flavonoids and betaines originating from host roots, and in turn induce the expression of *nod* genes. These bacterial *nod* genes encode approximately 25 proteins essential for synthesis and export of Nod factors, which are lipooligosaccharides responsible for the initiation of many developmental changes observed early on in the nodulation process of host plants. These developmental changes include root hair deformation, membrane depolarization, intracellular Ca^{2+} oscillations, and initiation of cell division in the root cortex, which establishes a meristem and nodule primordium. During the initial stages of symbiosis, rhizobia migrate from the root surface to the inner root tissue to populate cells in the developing nodule. A tubule known as the infection thread grows down the inside of a root hair. Rhizobia proliferate inside the infection thread, thereby keeping the tubule filled with bacteria. The infection thread fuses with the distal cell wall of the epidermal cell, which enables the bacteria to reach the intercellular space between the epidermal cell and the underlying cell layer. Furthermore, the infection thread branches as it grows through the root and enters the nodule primordium. This branching increases the number of sites where bacteria can enter nodule cells to ensure sufficient colonization. Once inside the nodule cells, bacteria continue to differentiate and synthesize proteins required for N_2 fixation and symbiosis maintenance (Gage 2004).

After successful initiation and development of nodules, the specialized N_2-fixing bacteria (now referred to as bacteroids) are enclosed by plant-derived membranes inside the plant cytosol (White et al. 2007). These organelle-like structures are known as symbiosomes. The plant-derived membrane is called the symbiosome or peribacteroid membrane with a symbiosome/peribacteroid space occurring between the symbiosome and symbiosome membrane. The peribacteroid solution is acidified and contains high levels of sugars, such as inositols (Tejima et al. 2003), which together with other molecules initiate the differentiation of rhizobia into bacteroids (Ohkama-Ohtsu et al. 2015). It is possible for one infected nodule cell to contain thousands of symbiosomes, with one or a few bacteroids in each, depending on the species. N_2 fixation requires coordinated nutrient exchanges whereby the plant provides reduced C and all nutrients to bacteroids, while in turn the plant is supplied with reduced N from the bacteroids (White et al. 2007). In order for nutrient exchange to take place, molecules must cross both the symbiosome and bacteroid membranes, and these pathways of exchange can follow more than one pattern and often are mediated by membrane transporters.

There are fundamental differences between the two types of nodules, namely, determinate and indeterminate in both their development and carbon and nitrogen metabolism (White et al. 2007) (Fig. 4.3). Indeterminate nodules are found in all three legume subfamilies of, whereas determinate nodules develop in legumes

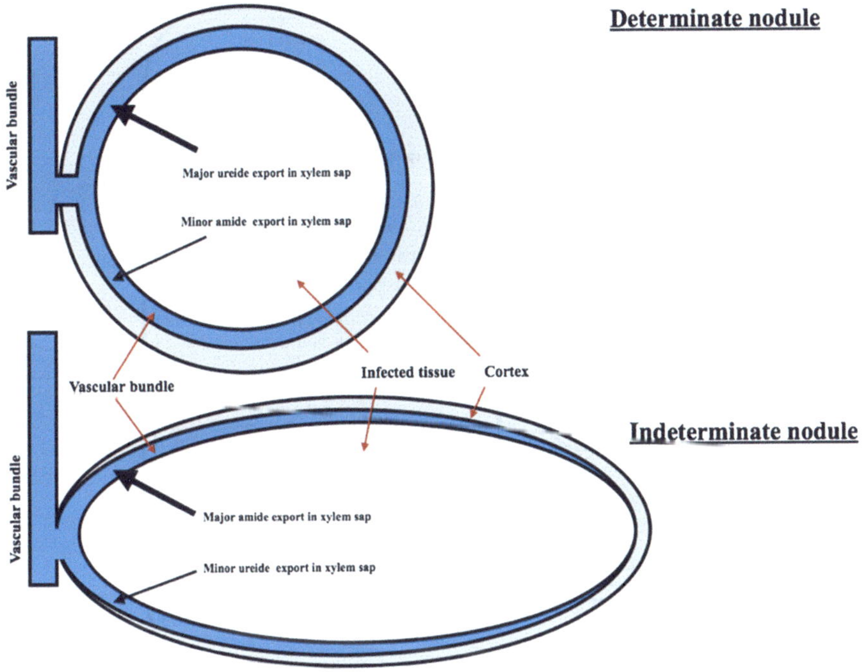

Fig. 4.3 The anatomical and functional differences between legumes with either determinate or indeterminate Nodules

belonging to Phaseoleae and Loteae tribes (Sprent and James 2008). The most remarkable metabolic difference between the two types of nodule is that indeterminate nodules mainly export assimilated N as amides, such as glutamine and asparagine, whereas determinate nodules primarily export ureides, such as allantoin and allantoic acid (Streeter 1992; Tajima and Kouchi 1997).

The biochemical conditions inside the symbiosome of mature nodules provide the optimal environment for maintaining rhizobial respiration and permit the oxygen-sensitive nitrogenase enzyme complex to convert unreactive atmospheric N_2 to ammonia (Oldroyd et al. 2001). The symbiosome and bacteroid membranes are sites where all C and N molecules, in addition to ions and O_2, must move across, which underlines their importance in the establishment and maintenance of symbiosis. Symbiotic nitrogen fixation (SNF) is an energy-intensive process that consumes 16–18 ATP per N_2 fixed (Dixon and Kahn 2004). The plant provides a C source to the bacteroids to fuel N_2 fixation in exchange for secretion of reduced N. The derived ammonia is converted to asparagine, glutamine, or similar N products and exported to the legume host (Prell and Poole 2006). The C provided to the bacteroid is derived from photosynthesis, which is translocated as sucrose via the phloem (Fig. 4.4) (Gordon et al. 1999; White et al. 2007). It is accepted that C_4-dicarboxylates such as malate and succinate are the main products of sucrose degradation supplied to

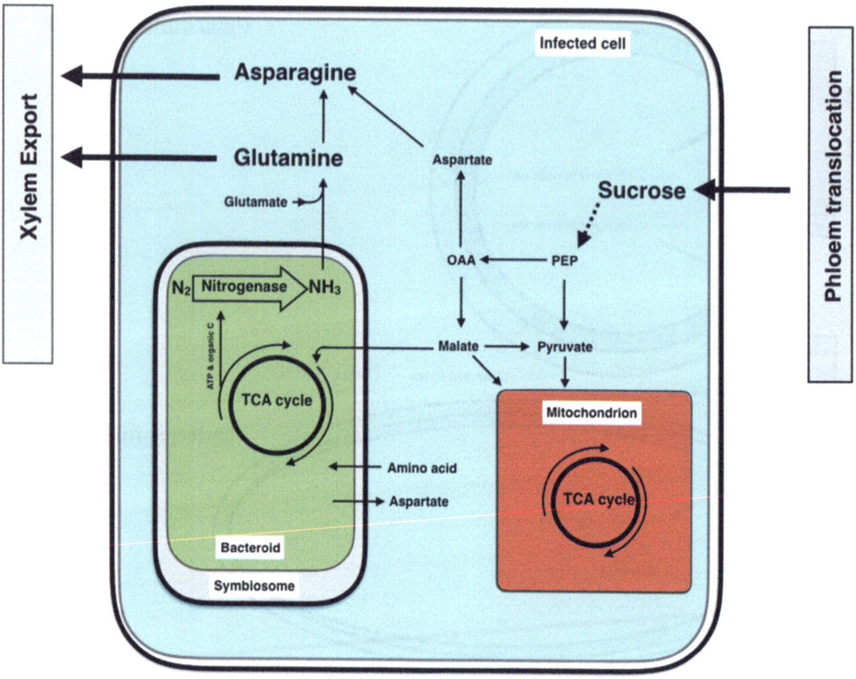

Fig. 4.4 Organic supply to nodules and N metabolism and export from nodules

bacteroids in most legumes (Udvardi and Day 1997). This energetically expensive process is inhibited by the presence of N compounds such as NO_3^- at various stages of the nodulation process, such as rhizobial infection, and also inhibits nitrogenase activity in established nodules, triggering early nodule senescence (Carroll and Matthews 1990).

Bacteroid respiration requires a high flux of O_2 that must be achieved in an environment of very low free oxygen. The answer to this paradox is reached with the action of leghemoglobin (White et al. 2007). The O_2 concentration in the infection zone of nodules is maintained at approximately 18 nmol (Layzell and Hunt 1990). Although the mechanism of O_2 control in legumes is not entirely understood yet, the regulation appears to comprise of three levels (Udvardi and Poole 2013): (1) bacteroids and plant mitochondria have high respiration rates that increase O_2 consumption; (2) leghemoglobin has a high binding affinity for O_2 in the cytoplasm and delivers it to the infected cells for consumption by bacteroids and mitochondria; and (3) a barrier exists in the outer cell layers of nodules that limits O_2 diffusion into the infected zone (Fig. 4.5).

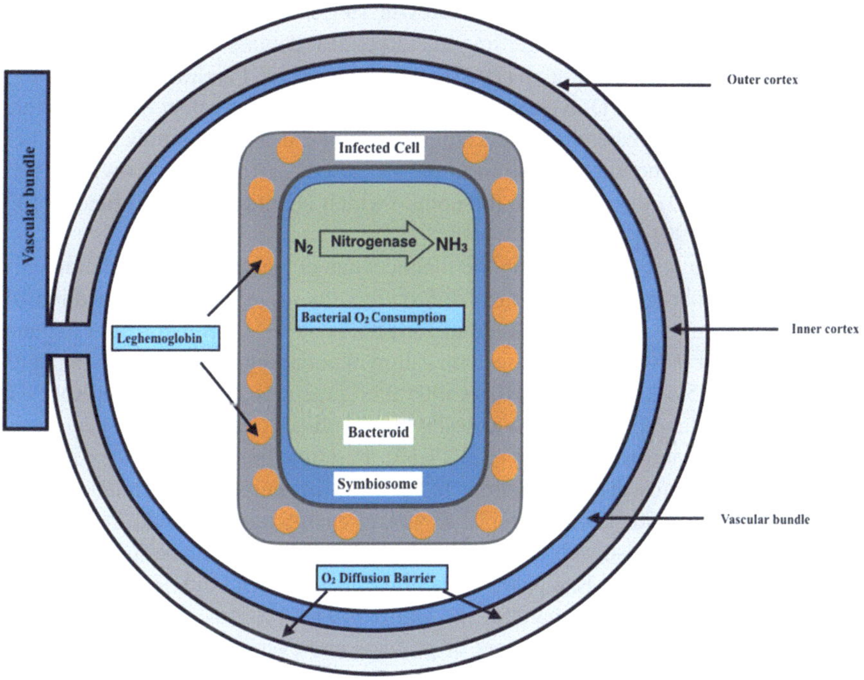

Fig. 4.5 Proposed mechanisms of O_2 regulation in determinate nodules

4.5.2 Nutrient Status Affects Nodule Development

The number of nodule that develops on legume roots is highly regulated by a mechanism called autoregulation, by which previously formed or forming nodules will suppress formation of subsequent nodules (Schulze and Kondorosi 1998). Legumes preferentially take up mineral N instead of forming symbiotic nodules, and N application suppresses nodule formation in legumes (Carroll and Matthews 1990). Biomass crop production is negatively influenced by environmental stresses, such as drought, salinity, low pH, extreme temperatures, heavy metals, as well as the low-nutrient availability. Among the nutritional stresses, P deficiency has received the most attention (Sulieman and Tran 2015). The reliance on SNF to fulfil N needs requires higher levels of P, K, and S than legumes relying solely on mineral N (Israel 1987; Sulieman et al. 2013). The levels of these nutrients can influence nodule growth, formation, and functioning (Duke et al. 1980; Almeida et al. 2000; Varin et al. 2010).

Phosphorus plays a role in many metabolic processes such as energy generation, nucleic acid synthesis, photosynthesis, respiration, glycolysis, membrane synthesis and integrity, activation/inactivation of enzymes, redox reactions, signaling, carbohydrate metabolism, as well as nitrogen fixation (Vance et al. 2003). P availability is critical for nodule development and also nodule activity due to the high ATP requirement of nitrogenase (Ribet and Drevon 1995; Al-Niemi et al. 1997). There have been conflicting reports regarding the effect of P deficiency on nodule number per unit shoot mass. Schulze et al. (2006) found an increase, Pereira and Bliss (1989) reported a decrease, and Drevon and Hartwig (1997) found no effect of P on nodule number. The strategy whereby P-deficient plants invest in a larger number of smaller nodules increases the surface/volume ratio of nodules (Ribet and Drevon 1995) which facilitates O_2 diffusion into the nodule which is critical for SNF functioning (Layzell et al. 1990).

Potassium plays a crucial role in the maintenance of electrical potential gradients across cell membranes, generation of turgor in cells, as well as the activation of numerous enzymes. Also, K is essential for photosynthesis, protein synthesis, and stomatal control as well as being the main cation maintaining ionic balances in plants (Marschner 1995, Zhang et al. 2010). Cultivation practices, cultivated crop species, and environmental influences such as soil type and climatic conditions all play a role in influencing K nutrition (Zhang et al. 2010). K status directly influences SNF by affecting nodule growth and function, metabolism of ammonia assimilation, conversions of amino acids, carbon supply, as well as energy transfer (Duke et al. 1980). Nodule number and fresh weight of nodules per plant together with nodule weight all increased with increasing K supply, while nitrogenase activity did not increase with an increase in K supply (Premaratne and Oertli 1994).

Sulfur is essential as a component of proteins and cysteine-containing peptides, such as glutathione and secondary metabolites (Varin et al. 2010). S deficiency impacted negatively on nodule development, nitrogenase, and leghemoglobin, leading to a drastic decline in N_2 fixation (Varin et al. 2010). S deficiency also led to decreased ferredoxin levels as well as limitations on energy supply (Scherer et al. 2008). Apart from the direct impact of P, K, and S on C and N metabolism, constrains on SNF appear as a result of host plant growth (Almeida et al. 2000; Hogh-Jensen 2003; Varin et al. 2010). The deleterious effect of nutrient deficiency on plant growth triggers an N feedback mechanism which downregulates nodule development and activity. This mechanism also impedes SNF during other stresses such as drought, salinity, toxic metals, and pathogen attack (Lea et al. 2007).

4.5.3 Nodule Functioning Under Nutrient Deficiency

The P requirements of nodulated plants are higher than those of non-nodulated plants supplied with a mineral N form (Leidi and Rodríguez-Navarro 2000) due to the role of P in the nodule's energy requirements (Sulieman and Tran 2015). This is supported by the significant correlation between N_2 fixation and nodule P content. The growth of N_2-fixing legumes is negatively affected by P deficiency due to the constraints on N_2 fixation in bacteroids as well as on ammonium assimilation into amino acids and ureides, which are not sufficient to support plant growth. P requirements for the synthesis of mitochondrial and symbiosome membranes add a further burden to P demands of N_2-fixing legumes (Sulieman and Tran 2015). In this token, legumes developed a diverse suite of responsive and adaptive strategies to conserve the P supply and maintain high SNF rates during periods of P stress. These strategies are (1) maintenance of high nodular P concentrations compared to other organs, (2) adjustments in root morphology thereby enhancing root exudation and P uptake mechanisms, (3) improved N_2 fixation per unit of nodule mass to compensate for reduction in nodule number, and (4) higher nodule permeability, which is linked to increased O_2 consumption per unit of reduced N_2 (Sulieman and Tran 2015).

4.5.3.1 P Is Preferentially Partitioned to Nodules

P is preferentially partitioned to nodules for maintenance of SNF during P deficiency, often at the expense of plant growth (Hogh-Jensen et al. 2002). Nodules are strong sinks for P even under adequate P supply (Drevon and Hartwig 1997). This is compounded during P deficiency when often nodules exhibit even higher P content compared to roots and shoots (Drevon and Hartwig 1997). Nodulated *L. albus* plants supplied with sufficient P showed increased nodule biomass coinciding with decreased cluster root production (Thuynsma et al. 2014), but without a concomitant increase in SNF. The unchanged SNF efficiencies during sufficient P supply indicate optimal functioning of nodules under low P supply, as found in *L. angustifolius* where no changes in SNF efficiencies or %NDFA were observed during short-term P deprivation (Le Roux et al. 2006, 2009). Nodules are strong P sinks able to obtain P from the host, but not readily releasing it to the host root during P-deficient conditions (Al-Niemi et al. 1998). The suggestion that nodules act as a strong P sinks is supported by Vadez et al. (1996) who found threefold higher P levels in nodules than other tissues (Valentine et al. 2011). Furthermore, several legume species studied to date, such as *M. truncatula* (Cabeza et al. 2014), *L. luteus* (Kleinert et al. 2014), and *Virgilia divaricata* (Vardien et al. 2014), are able to regulate P influx from host cells and do not release these P reserves to other organs during periods of P stress (Sulieman and Tran 2015). The nodule response to P stress is dependent on the period of P starvation (Valentine et al. 2011) with the nodule response to P stress slower compared to that of other organs. Nodules of *Lupinus angustifolius* experienced no P stress after 14 days of P starvation (Le Roux et al. 2006) and showed a decline in P*i* levels only after 25 days of the onset of P starvation (Le Roux et al. 2008). Once P stress was exhibited, a concomitant decrease in nodule weight and nodular P concentrations was observed (Valentine et al. 2011).

Roots experience P deficiency more severely than nodules during short-term P stress (Le Roux et al. 2006) with fast decreases in cellular P*i* and ATP/ADP ratio (Valentine et al. 2011), whereas nodules maintain P*i* and energy levels. Nodules appear to function optimally at low P*i* levels (Al-Niemi et al. 1997, 1998, Colebatch et al. 2004), and bacteroids, which consistently operate at low P concentrations, can scavenge P from the host cells to fulfil their metabolic requirements (Al-Niemi et al. 1997, 1998; Valentine et al. 2011). Physiological studies have shown that P was unevenly distributed in the nodular tissue under P deficiency, with the bacteroid fraction preferentially accumulating higher levels of P required for nitrogenase activity (Sulieman and Tran 2015). P resupply to P-deficient *L. albus* plants led to metabolic changes that resulted in preferential P allocation to the symbiotic fraction (Thuynsma et al. 2014).

The critical role of nodule P homeostasis during P stress led to research on how P enters into the nodules. Two main pathways exist to facilitate P flow into symbiotic tissues: (1) a direct route whereby P is absorbed from the surrounding soil or solution by the nodule surface and (2) an indirect pathway whereby P is translocated from the roots of the plant to the nodule via the vascular tissue (Sulieman and Tran 2015). The high-affinity plasma membrane-localized phosphate transporter of soybean, *GmPT5*, is associated with phosphate entry into soybean nodules (Qin et al. 2012) and appears to transport P into nodules from host-plant roots primarily during periods of P limitation. The upregulation of high-affinity phosphate transporter genes may be one of the strategies used for maintaining P homeostasis, thereby conferring adaptation to P stress (Sulieman and Tran 2015).

4.5.3.2 Effect of P Deficiency on Nodule Carbon Metabolism

Prolonged P deficiency leads to a drastic decline in intracellular levels of cytoplasmic P*i*, which in turn reduces intracellular levels of ATP, ADP, and associated nucleoside – Ps. Despite P*i* starvation, plants continue producing energy and C skeletons for core metabolic pathways (Plaxton and Podestá 2006). Severe stress hampers the metabolic functioning of P*i* or adenylate-dependent enzymes in the glycolytic pathway (Vance et al. 2003). P*i* starvation induces a cluster of adenylate-independent enzymes that serve as alternatives to the P*i*-dependent glycolytic and mitochondrial routes. The alternative routes bypass the requirements for adenylates and P*i* that is usual during P-sufficient conditions (Le Roux et al. 2008). These alternative enzymes promote P*i* recycling and synthesis of organic acids with P*i* as a by-product of their reactions (Vardien et al. 2014). Phosphoenolpyruvate carboxylase (PEPC, EC 4.1.1.31) together with malate dehydrogenase (MDH, EC 1.1.1.37) and the mitochondrial malic enzyme (ME, EC 1.1.1.38) function as an alternative route to generate pyruvate, which is formed during normal conditions via the ADP-dependent cytosolic pyruvate kinase (PK, EC 2.7.1.40) (Plaxton 2004). The functioning of this metabolic bypass during P*i* stress possibly ensures a continued pyruvate supply to the TCA cycle, and at the same time, it releases P*i* back into the metabolic pool (Plaxton 2004). Le Roux et al. (2006, 2014) reported increased activities of PEPC and MDH in P*i*-deficient *L. angustifolius* nodules, which support the hypothesis that the alternative pathway exists to increase pyruvate synthesis and P*i* recycling during P-limited conditions.

4.5.3.3 Recycling of Phosphorus by Acid Phosphatases

The remobilization of P inside the plant is an adaptive strategy by which P is translocated from the aboveground organs to nodules to ensure SNF (Sulieman and Tran 2015). Studies with *Medicago truncatula* found that nodules can sustain constant P levels until the P concentrations in leaves reached critical points that any further reductions would render the leaves nonviable (Cabeza et al. 2014). At this critical level, the nodules are capable of switching to alternative internal P resources to support the symbiotic activity.

Organic phosphate presents a large internal P pool for P-stressed nodules. Several genes involved in P recycling from organic substances are upregulated during P deficiency (Hernandez et al. 2009), of which the most important genes are those encoding acid phosphatases (Sulieman and Tran 2015). Induction of both intracellular and secreted acid phosphatases during P stress seems to be a widespread response (Duff et al. 1994). Increased nodular phytase and phosphoenolpyruvate phosphatase during P deficiency provide further evidence in support of the assumption that acid-phosphatases are another component in the vast array of legume adaptations to low P conditions. Furthermore, acid phosphatases also play a role in nodule carbon metabolism, in regulation of O_2 permeability, as well as in oxidative stress (Sulieman and Tran 2015). The acid phosphatase group (EC 3.1.3.2, orthophosphoric monoester phosphohydrolases) are enzymes that hydrolyze and mobilize inorganic orthophosphate from organic P substances (Bargaz et al. 2012, Sulieman and Tran 2015). Intracellular acid phosphatases break down P nucleotides, sugar phosphates, as well as phosphate monoesters and to recycle the released P for utilization in amino acid biosynthesis and nodule metabolism during P stress (Pennheiter et al. 1997, Vardien et al. 2014).

Acid phosphatases are found in nuclei, cell walls and intracellular spaces, amyloplasts, mitochondria, and Golgi bodies. They are also present in the endoplasmic reticulum to a lesser degree (Vardien et al. 2014; Lazali et al. 2013). These enzymes facilitate P recycling at an intracellular level and thereby improve P utilization (Vance et al. 2003). The physiological role of these P*i* mobilizing enzymes in nodules of N_2-fixing legumes is not yet fully understood. Fructose-1,6-bisphosphatase (FBPase, EC 3.1.3.11) is an essential regulatory enzyme in gluconeogenesis and converts fructose 1,6-bisphosphate to fructose 6-phosphate, which is an important precursor in various biosynthetic pathways. Increased FBPase transcription in the nodule cortex correlated with increased enzyme activity under P deficiency and improved P use efficiency in nodules (Lazali et al. 2016). Fructose 6-phosphate serves as a regulating molecule in plant sugar metabolism and development (Gupta et al. 1998). The in situ localization of FBPase transcripts in the cortex and infected zone under P deficiency is indicative of its role in nodule permeability to O_2 diffusion (Lazali et al. 2016). As mentioned previously, P deficiency increases nodule O_2 permeability (Drevon et al. 2015), and this notion is further supported by the positive correlation between phytase activity and nodule permeability to O_2 diffusion. Although this is an interesting correlation, how the precise mechanism operates is not currently known.

Nodular acid phosphatase activity increases during low P supply but is relatively stable at increased P levels (Araujo et al. 2008). Expression levels of nodule purple acid phosphatase (PAP) genes were induced during P stress in soybean (Li et al. 2010). Acid phosphatases occur at the symbiosome membrane, which is indicative of the importance of these enzymes in P acquisition and recycling and maintaining high levels of nodular P during P stress (Pennheiter et al. 1997; Bargaz et al. 2013). Another source of internal P exists as phospholipids of cell membranes. Intracellular phosphohydrolases mediate the replacement of phospholipids with sulfolipids to release P*i* to the cell (Lambers et al. 2012b). Vardien et al. (2014) found a decline in membrane phospholipids of P-stressed nodules, but did not investigate changes in sulfolipid composition.

4.5.3.4 Leghemoglobin Functioning Under P Deficiency

Leghemoglobin is found within the cytosol of infected nodule cells and plays the crucial role of regulating O_2 supply to the plant mitochondria and bacteroids (Scherer et al. 2008). Leghemoglobin has a significant buffering capacity against free oxygen in the nodule interior and maintains free O_2 levels at low nM concentrations. This ensures a constant flow of oxygen for the high-affinity bacterial cytochrome oxidases for sufficient respiration (Liese et al. 2017b). Several studies reported increased O_2 permeability correlating with reduced SNF during P deficiency (Ribet and Drevon 1995; Drevon and Hartwig 1997; Schulze and Drevon 2005; Le Roux et al. 2009), making the role of leghemoglobin even more significant when the O_2 diffusion barrier is reduced during P stress.

P deficiency had no effect on leghemoglobin concentration in *G. max* nodules even though reduced SNF was observed (Miao et al. 2007). Leghemoglobin may affect the levels of P-containing compounds such as ATP. Scherer et al. (2008) found that S deficiency reduced leghemoglobin levels, which was associated with declined ATP levels. The ATP decrease was ascribed to the reduced leghemoglobin supply of O_2 to mitochondria, where ATP production is directly coupled to O_2. Concerted downregulation of leghemoglobin transcripts under P deficiency suggests that O_2 buffering is not the main function of leghemoglobin. Interestingly, lower expression of leghemoglobin might be a mechanism whereby nodule and bacteroid respiration are decreased with the result that less ATP is produced, with potentially a negative impact on nitrogenase activity. Loss of function of the hemoglobin gene was accompanied by a near total loss of nitrogenase protein in the nodule (Liese et al. 2017b).

4.6 Genetic Regulation of Root Form and Function During Nutrient Constraints

Ultimately, the anatomy and metabolism of an organism are the expression of a sophisticated network of gene activities. The availability of genome sequences and gene annotations for many species, as well as large and detailed sets of gene expression data, along with user-friendly analytical tools allows for understanding the mechanisms by which genes are turned on and off and how the intricate genetic grids are interconnected to advance developmental programs and control metabolism in the plant. Nevertheless, much still needs to be learned about the spatial–temporal orchestration of molecular functions that essential genes play in these contexts and especially in developmental programs of root structures in response to symbioses and nutritional homeostasis in adverse environments.

Under low-nutrient availability, roots initially undergo a limited spurt in root growth to increase their rhizosphere space and better explore mineral resources in the soil. An important question that has not been fully answered yet is about the very basal mechanisms by which the initial cues of low-nutrient conditions ultimately trigger the transcriptional regulation of the main genes involved in nutrient homeostasis. A possible scenario is that the cellular environment with low-nutrient concentrations could induce allosteric modifications of transcription factors (TFs) that alter their DNA-binding affinity to *cis* elements in target gene promoters, thus inducing or inhibiting gene expression. Although these studies have never been carried out in eukaryotes, there is evidence that such mechanism exists in prokaryotes (Zhao et al. 1993; Weaver et al. 2001; Beckett 2009; Popovych et al. 2009; Reichheld et al. 2009). Despite being a long-reaching goal, a more comprehensive understanding of how low-nutrient conditions initially trigger the signaling cascade leading to root modifications, both developmentally (architecture) and biochemically (metabolism and gene expression), to enhance nutrient uptake will be instrumental in creating highly efficient genotypes of crops that grow better in poor soils. For example, by identifying allosteric pockets or which amino acid residues undergo posttranslational modifications in transcription factors linked with nutritional homeostasis, one could identify or produce better alleles that recognize low-nutrient cues earlier to enhance nutrient uptake efficiency.

Much of the nutrient use efficiency mechanisms in plant physiology and metabolism are ultimately dependent upon the primary sequence of gene products (e.g., optimal amino acid sequences of enzymes and membrane transporters for ideal rates of catalysis or transport under adverse conditions). Additionally, this trait is also dependent on the best spatial–temporal gene expression activity program (i.e., whether the genes are expressed in the best place and time, possibly via sensitive recognition of cues for low-nutrient levels) that evolved under natural systems. Therefore, an early sensing or signaling mechanism of low-nutrient levels triggering gene expression along with an optimized activity of transport (uptake, translocation, intracellular storage), as well as enzymes for assimilation and metabolism, are essential components of high-nutrient use efficiency.

Most of the knowledge we have amassed on the molecular genetics controlling root architecture has been gained mostly from plant model species, such as *Arabidopsis thaliana*. Despite the usefulness of these models to unveil the intricacies of molecular genetics in plants, *Arabidopsis* roots, for example, do not establish common symbiotic associations, such as mycorrhization (which occur in ~90% of all land plants), nor nodulation for nitrogen fixation (a trait that is mostly limited to legumes and a few other rosid species). Therefore, additional models (such as tomato, *Medicago truncatula*, rice, sorghum, and maize) are essential to reveal gene functions and the molecular networks involved in root development and uptake systems (e.g., membrane transporters, assimilation enzymes, cell signaling components, transcription factors, miRNAs, etc.) that are necessary to recognize and react to low-nutrient conditions in order to guarantee proper plant growth and development.

4.6.1 Nitrogen

In plants, N is the most critical and energetically expensive element to acquire. While most plants go to great lengths to obtain N from soil (Han et al. 2015) or via mycorrhizal associations (Chen et al. 2018), some species are able to develop a symbiotic relationship with prokaryotes to reduce atmospheric N_2. In most crops, N fertilizer is the most expensive and energy-demanding input in the cropping enterprise, the most limiting nutrient for high yields, and often a significant contributor to water and air pollution. Therefore, a better understanding of the genetics involved in defining root architecture to improve N uptake can help create breeding tools to create more efficient genotypes.

Roots are stimulated to branch in patches of soil with a higher concentration of N upon a local stimulus of nitrate, ammonium (Lima et al. 2010), or glutamate (Singh et al. 2016). The perception of a local patch with high N in the soil will trigger a signaling cascade to stimulate the pericycle to start dividing to form a new LR as well as stimulate root elongation.

4.6.1.1 Regulation of Root Architecture Under Low N Conditions: Low Nitrate Conditions

In addition to its value as a nutrient, nitrate is also a potent signaling molecule that can dramatically alter root architecture (Sun et al. 2017). Localized nitrate applications stimulate local root growth and branching. Likewise, high nitrate levels reduce lateral root formation and growth (Joshi et al. 2016). In general, low N availability will ultimately induce and activate high-affinity nitrogen (NRT and AMT) transporters, whereas a mild N limitation will promote the growth of the primary and lateral root system via master regulators of root architecture, such as MDR4, TAR2, CBL7, and WAK4. On the other hand, a severe N limitation will repress root developmental overall, by inducing receptor kinases that act as repressors of root growth, such as CLAVATA 1 (CLV1) and the ARABIDOPSIS CRINKLY 4 (ACR4) and AXR5 (Kiba and Krapp 2016).

In *Arabidopsis*, key elements orchestrating the genetic network under low nitrate conditions are the MADS-box transcription factor ARABIDOPSIS NITRATE REGULATED 1 (ANR1/AGL44) and its antagonist miR444a (Yan et al. 2014), as well as ARF8/miR167, AFB3/miR393, CLE/CLV1, AHA2, TAR2, NRT1.1, and NRT2.1. For example, ANR1 expression in founder pericycle cells initiates LR development that culminates in the establishment of a meristem of the lateral root. Low nitrate induces ANR1 expression in lateral roots via the dual-affinity nitrate transporter and sensor NITRATE TRANSPORTER 1.1 (NRT1.1/CHL1). Arabidopsis *nrt1.1* mutants are more drought tolerant but show a severely decreased lateral root system, which makes them unable to modify the architecture of their root systems to better explore nitrate-rich patches in the soil.

4.6.1.2 Regulation of Root Architecture Under Low N Conditions: Low Ammonium Conditions

Ammonium seems to complement nitrate in shaping the root architecture. In *Arabidopsis* growing under low ammonium conditions as the sole N source, LR branching is primarily induced by the ammonium transporter AMT1;3 (Lima et al. 2010). The regulatory gene networks involved in this mechanism have yet to be discovered.

4.6.1.3 Nodulation and Symbiotic Nitrogen Fixation in Legumes

Nitrogen fixation (i.e., the reduction of N_2 to NH_3) is necessarily a reaction carried out by diazotrophic prokaryotes often in symbiotic association with eukaryotic cells. In legume roots, free-living rhizobia associate with roots to induce the development of nodules and establish an endocellular association that allows the microorganisms to switch their genetic program to fix atmospheric nitrogen. In exchange for reduced nitrogen (NH_3), plants provide rhizobia with all nutrients needed for the bacteria to thrive and a mutualistic entity, including a source of energy (TCA dicarboxylates), branched-chain amino acids (Leu, Ile, and Val), and minerals (Benedito et al. 2006, 2010).

Legumes have co-opted the first steps of an ancient symbiotic pathway that evolved for the establishment of mycorrhizal symbiosis. Therefore, these initial steps are known as the common symbiotic pathway. Legume roots do not form nodules under N-sufficient conditions. N-sufficient shoots send unidentified inhibitory signals to the roots through a mechanism called autoregulation of nodulation (AON), which inhibits nodule formation. However, under N deficiency, roots exude flavonoids that are perceived by rhizobia, triggering the synthesis and release of lipochitooligosaccharides (LCOs) called nodulation (Nod) factors (NF), which are recognized by NF receptors located in the root hairs. This recognition induces root hair curling, forming a structure called shepherd's hook that traps the rhizobia and causes the formation of an infection thread (IT) toward the inner cortex of the root. Meanwhile, the cortex recognizes the invasion and starts dividing, giving rise to the nodule meristem. Depending on the legume species, this meristem can be persistent in the indeterminate nodules of phaseoloid legumes (alfalfa, pea, clover, vetch) as well as *Lotus japonicus* with tissues in sequentially distinctive differentiation stages in the same mature nodule. Alternatively, in galegoid legumes (e.g., beans, soybean), meristem formation, cell infection, bacteroid differentiation, N_2 fixation, and, at last, senescence occur sequentially in time.

Not surprisingly, a significant number of genes are involved in this process. For example, 473 genes in the *Medicago truncatula* genome were identified as exclusively expressed during nodule development (Benedito et al. 2008). There is a negative correlation between nodulation and LR development. The LRR-RLK receptor kinase COMPACT ROOT ARCHITECTURE 2 (CRA2, Medtr3g110840) represses LR formation from local signals in *Medicago truncatula* while it induces nodulation systemically. *cra2* knockout mutants develop a short but dense root system with many lateral roots that develop fewer nodules upon rhizobial inoculation (Huault et al. 2014). More recently, CRA2 was proposed as the likely receptor of C-TERMINALLY ENCODED PEPTIDE 1 (MtCEP1), which mediates root and nodule organogenesis (Imin et al. 2013; Mohd-Radzman et al. 2016). Intriguingly, while the CRA2/CEP1 complex promotes nodulation via ethylene signaling (EIN2/SKL), inhibition of LR development occurs via an alternate pathway. Comparative RNA-Seq analyses of nodulating hairy roots overexpressing CEP1 revealed 89 and 116 genes induced and repressed, respectively, making them potential targets of the complex.

On the other hand, the miR390/TAS3 module represses nodulation and promotes LR development in *Medicago truncatula* (Hobecker et al. 2017). AGO7 activity mediates this process via trans-acting small interference RNAs created by TAS3 cleavage that targets auxin response factors. In soybean, the cell wall β-expansin GmEXPB2 is also involved in both processes, nodulation and root architecture (Li et al. 2015). GmEXPB2 overexpression increases root hair density and induces longer roots, along with a higher number of nodules.

These examples underscore the trade-off between nodulation and LR development in defining the architecture of legume roots under rhizobial symbiosis. Most genetic mechanisms seem to be locked in a way that a gene that promotes one pathway will inhibit the other. Given that N is the most limiting element for plant growth, it seems logical that legumes evolved a mechanism to restrict growth of their root system to allocate energy to nodule development and sustain symbiotic nitrogen fixation.

4.6.2 Phosphate

Plants acquire P from the soil as phosphate anion (PO_4^{3-} or P*i*) through secondary membrane transporters. Given the imminent P crisis due to the global exhaustion of phosphate rock mines for fertilizer production, it is important that we create crop varieties with improved P use efficiency capable of growing with reduced P inputs.

In *Arabidopsis*, genes involved with remodeling the root system under low P include PIP5K, PRD, SIZ1, and WRKY75. In soybean, besides its role in nodulation, as mentioned earlier, GmEXPB2 has also been implicated with modification of root architecture under low P (Guo et al. 2011). As a major factor in defining root architecture, some plants evolved effective strategies to improve P uptake by either associating with mycorrhizal fungi or dramatically changing their root architecture (e.g., brushlike proteoid roots) and exudating phosphatases from roots.

4.6.2.1 Development of Cluster Roots

The dramatic development of cluster roots (a.k.a. hairbrush or proteoid roots) evolved particularly in the Proteaceae family (27 of its 83 genera, including macadamia) but also independently in other 28 species from 6 dicot families [Betulaceae, Casuarinaceae, Elaeagnaceae, Fabaceae (e.g., lupins), Moraceae, and Myricaceae]. Correspondingly, less studied examples of monocot Poales develop dauciform roots in the Cyperaceae (sedges) and Restionaceae (restiads), which are structurally different but functionally analogous to the dicot cluster roots (Lambers et al. 2006; Shane et al. 2006). Very interestingly, all dicot species capable of developing cluster roots can also establish symbiosis for nitrogen fixation (either rhizobial or actinorhizal), and most do not form mycorrhizal associations (Watt and Evans 1999; Dinkelaker et al. 1995).

The cluster root developmental program and phosphatase exudation are typically activated by internal or external cues of low P and high N (Güsewell 2017). Given the convenient availability in the *Lupinus* genus of species with contrasting phenotypes for the development of cluster roots, lupins are ideal comparative models for understanding the genetics of this trait. Initial efforts in this regard were the publication of a comparative transcriptome analysis in white lupin under sufficient vs. low P conditions (O'Rourke et al. 2013) followed by the identification of regulatory gene networks associated with the cluster root developmental program (Secco et al. 2014; Wang et al. 2014). Currently, the key questions are as follows: What are the genetic triggers that initiate this developmental program? How could this trait be potentially introduced to crops? Albeit a largely overlooked topic in plant biology research, given the impending P supply crisis that may affect food security in the midterm (Mew 2016), funding for genetics research on cluster roots should be highly prioritized.

4.6.2.2 Mycorrhization

Mycorrhizal fungi associate with roots of ~80% of land plants. This symbiosis expands broadly the reach of exploration of the rhizosphere through the fungi hyphae. Some species with limited root systems are largely dependent on this association to thrive in P-poor soils, including crops like garlic, flax, and maize (Bona et al. 2016). The association is based on the exchange of sugars provided by the plant to the fungus for minerals (particularly P but also N and micronutrients) to the host. This mutualistic association can be ectomycorrhizal (such as in orchids and many temperate woody species) or endomycorrhizal (such as arbuscular mycorrhiza fungi). However, most members of the Brassicaceae family cannot establish mycorrhizal associations, including the ubiquitous plant model *Arabidopsis thaliana*, making them vulnerable to cultivation in poor soils.

The genetics of mycorrhization involves initially a common symbiotic pathway (co-opted by legumes for rhizobial perception and triggering nodule development), which evolved just after plants conquered land (circa 400 million years ago) and has been retained ever since. Mycorrhization starts with root efflux of strigolactones to the rhizosphere to induce spore germination of mycorrhizal fungi. Germinating spores synthesize and release Myc factors (LCO) that are perceived by (still elusive) Myc receptors in root cells to allow hyphal invasion, hyphae penetration between the cell wall and symplast, and the formation of arbuscules, which are extra-symplastic structures that immensely increase the contact between the plasma membranes of both symbionts. This intimate contact allows for efficient exchanges of molecules and ions through specific membrane transporters and signals.

Mycorrhizal associations lead to an increase of root branching and crown root length independently of the plant species colonized (Liese et al. 2017a; Yu et al. 2016; Maherali 2014). In the model legume *Lotus japonicus*, the MYB transcription factor LjMAMI is independently involved in root growth and AM symbiosis, although via a yet unknown mechanism (Volpe et al. 2013a, b). It is also becoming clear that mycorrhization has profound implications in the aboveground tissues and dramatically affect the plant physiology, including the proteome (Bona et al. 2016). Beyond this point, the complex molecular networks involving AM associations and root architecture remodeling under low P remain to be unveiled.

4.6.3 Other Nutrients

The depletion of macroelements (K, Ca, Mg, and S) as well as micronutrients (especially Fe, B, Mn, Zn, Cu, and Mo) in heavily cultivated soils often requires fertilization. Ideally, plant nutrition would originate from organic matter that releases nutrients slowly during mineralization while also avoiding the eventual soil salinization to toxic levels, especially of adjuvant anions such as Cl^-. Not much is known about the molecular genetics of architecture remodeling of symbiotic roots in low levels of nutrients other than N and P. Only a few genes controlling root architecture under low-nutrient, non-symbiotic conditions have been identified. As one of the few examples, the nitrate transporter NRT1.5/NPF7.3 controls LR development under low K in *Arabidopsis* (Zheng et al. 2016). Although some physiological mechanisms and downstream genes have been identified to alter root architecture under low Fe in *Arabidopsis*, the triggering master regulatory genes remain to be discovered (Jeong et al. 2017).

4.6.4 Epigenetics of Root Architecture Remodeling

Unsurprisingly, one of the molecular changes in a plant cell under nutrient stress is at the epigenetic level, including histone modifications, partial replacement of chromatin constitution by histone variants, and DNA methylation (Secco et al. 2017). A recent review has been published on the epigenetic regulation of root architecture under mineral deficiency (Sirohi et al. 2016). Identification of epigenetic mutants related to root architecture remodeling under low-nutrient availability in crop species will be particularly interesting to unveil the role of epigenetics in this process.

4.6.5 Perspectives on Understanding the (Epi)genetics of Architecture Remodeling of Symbiotic Roots Under Low-Nutrient Conditions

The genetics of root development under low-nutrient conditions, especially for N and P in model species, have been relatively well studied. Less is known, however, about remodeling of root architecture under symbiotic conditions (e.g., mycorrhization, nodulation, other soil microorganisms) and in particular under low levels of nutrients other than N and P. In the same token, we need more detailed information on root biology of crop species. Currently, with the technological advances of genome and transcriptome sequencing and analyses that allow non-model species to be studied, relevant questions regarding molecular genetics of non-model crops can be addressed. Elaborate experiments considering nutrient levels and symbioses in specific root zones and tissues will lead to a better understanding of the genetic networks involved in low-nutrient adaptations and root architecture remodeling leading to improved nutrient uptake and use efficiency.

Identification of homologue genes in crop species to critical root architecture genes identified in model species and identification of generation of mutants (e.g., via CRISPR genome editing) will be helpful to understand the molecular mechanisms by which nutrients ultimately affect root architecture under symbiotic conditions. Mathematical modeling of root architecture development under different nutrient levels and types coupled with gene transcription and construction of regulatory gene networks considering not whole root systems but rather specific tissues and cell types (e.g., using laser-capture microdissection) will lead to advances in the field.

4.7 Conclusions

The diversity of the alterations in root anatomy and physiology for the purpose of nutrient acquisition is staggering. As a consequence of their intimate contact with the immediate environment in the soil rhizosphere, roots have evolved exquisite adaptations to efficiently extract minerals from this environment. These modifications range from the fascinating alterations in root anatomy which can form cluster roots, to the truly sublime adaptations, which can accommodate the symbioses with soilborne fungi and bacteria in form and function, to produce nitrogen-fixing nodules and mycorrhizas. In this regard, roots display perhaps the most innovative and exotic adaptations of all the plant organs, owing to their capacity to diversify their functional architectures for nutrient acquisition. This is clear in the plasticity of cluster roots to acidify the rhizosphere with different carboxylate compositions, in the abilities of certain mycorrhizas to not only benefit host nutrient acquisition but also contribute to soil formation by the weathering of rock, and in the different types of nodules to interact with their host legumes.

As more information becomes available regarding the genes and molecular regulation that underpin these adaptations in roots, it should remain a goal for future research to increase the elucidations of these root mechanisms for nutrient acquisition. Since most of the information on genetic regulation of these roots come from the model plant, *Arabidopsis thaliana*, there is a limitation in that this species does not form mycorrhizal associations or symbiotic nodules with soilborne microbes. Therefore, more resources should be invested into the establishment of appropriate model plants. Although this endeavor is certainly of great fundamental interest, the survival of the human race may also depend on it. Agricultural crop production in a world of declining natural resources will ultimately have to consider crop modifications that enable roots to acquire essential minerals more efficiently, from soil environments which are already damaged or threatened by global climate change. The grand challenge would be to impart to crop plants a combination of these diverse root strategies for nutrient acquisition.

Acknowledgements The authors would like to acknowledge the various research agencies, whose funding supported much of the work, upon which this chapter is based. P Cornejo, A Zuniga-Feest and M Delgado were funded by the following grants—Fondecyt grant postdctoral N° 3150187 (MD), Fondecyt iniciation grant N° 11170368 (MD), Fondecyt N° 1130440 (AZF) and FONDECYT grant N° 1170264 (PC) (CONICYT-Chile). Research at V Benedito's laboratory is funded by the USDA National Institute of Food and Agriculture (Hatch Project WVA00686). A Kleinert is funded by the National Research Foundation of South Africa. The authors would like to thank Alexander Valentine for Figs. 4.3–4.5.

References

Agerer R (2001) Exploration types of ectomycorrhizae—a proposal to classify ectomycorrhizal mycelial systems according to their patterns of differentiation and putative ecological importance. Mycorrhiza 11:107–114

Agerer R (2006) Fungal relationships and structural identity of their ectomycorrhizae. Mycol Prog 5:67–10

Aguilera P, Borie F, Seguel A, Cornejo P (2011) Fluorescence detection of aluminum in arbuscular mycorrhizal fungal structures and glomalin by using confocal laser scanning microscopy. Soil Biol Biochem 43:2417–2431

Al-Karaki GN (2006) Nursery inoculation of tomato with arbuscular mycorrhizal fungi and subsequent performance under irrigation with saline water. Sci Horticult 109:1–7

Almeida JPF, Hartwig UA, Frehner M, Nosberger J, Luscher A (2000) Evidence that P deficiency induces N feedback regulation of symbiotic N2 fixation in white clover (*Trifolium repens* L.) J Exp Bot 51:1289–1297

Al-Niemi TS, Kahn ML, McDermott TR (1997) P metabolism in the bean-*Rhizobium tropici* symbiosis. Plant Physiol 113:1233–1242

Al-Niemi TS, Kahn ML, McDermott TR (1998) Phosphorus uptake by bean nodules. Plant Soil 198:71–78

Ané JM, Kiss GB, Riely BK, Penmetsa RV, Oldroyd GE, Ayax C, Levy J, Debelle F, Baek JM, Kalo P, Rosenberg C, Roe BA, Long SR, Denarie J, Cook DR (2004) *Medicago truncatula DMI1* required for bacterial and fungal symbioses in legumes. Science 303:1364–1367

Araujo AP, Plassard C, Drevon JJ (2008) Phosphatase and phytase activities in nodules of common bean genotypes at different levels of phosphorus supply. Plant Soil 312:129–138

Ashkannejhad S, Horton TR (2006) Ectomycorrhizal ecology under primary succession on coastal sand dunes: interactions involving *Pinus contorta*, suilloid fungi and deer. New Phytol 169:345–354

Avila A (2013) Evaluación de la capacidad de extracción de fósforo en Proteaceas chilenas. Tesis para optar al grado de Magister en Ciencias vegetales. Escuela de Postgrado en Ciencias Agrarias. Universidad Austral de Chile. Valdivia, Chile

Badri DV, Weir TL, Van der Lelie D, Vivanco JM (2009) Rhizosphere chemical dialogues: plant–microbe interactions. Curr Opin Biotechnol 20(6):642–650

Bago B, Pfeffer P, Shachar-Hill Y (2001) Could the urea cycle be translocating nitrogen in the arbuscular mycorrhizal symbiosis? New Phytol 149:4–8

Bais HP, Weir TL, Perry LG, Gilroy S, Vivanco JM (2006) The role of root exudates in rhizosphere interactions with plants and other organisms. Annu Rev Plant Biol 57:233–266

Balzergue C, Puech-Pags V, Bécard G, Rochange SF (2011) The regulation of arbuscular mycorrhizal symbiosis by phosphate in pea involves early and systemic signalling events. J Exp Bot 62:1049–1060

Balzergue C, Chabaud M, Barker DG, Bécard G, Rochange SF (2013) High phosphate reduces host ability to develop arbuscular mycorrhizal symbiosis without affecting root calcium spiking responses to the fungus. Front Plant Sci 4:426

Bargaz A, Ghoulam C, Amenc L, Lazali M, Faghire M, Drevon JJ (2012) Phosphoenol pyruvate phosphatase is induced in the root nodule cortex of *Phaseolus vulgaris* under phosphorus deficiency. J Exp Bot 63:4723–4730

Bargaz A, Faghire M, Farissi M, Drevon J-J, Ghoulam C (2013) Oxidative stress in the root nodules of *Phaseolus vulgaris* is induced under conditions of phosphorus deficiency. Acta Physiol Plant 35:1633–1644

Bárzana G, Aroca R, Ruiz-Lozano JM (2015) Localized and non-localized effects of arbuscular mycorrhizal symbiosis on accumulation of osmolytes and aquaporins and on antioxidant systems in maize plants subjected to total or partial root drying. Plant Cell Environ 38:1613–1627

Beckett D (2009) Regulating transcription regulators via allostery and flexibility. Proc Natl Acad Sci USA 106(52):22035–22036

Benedito VA, Dai X, He J, Zhao PX, Udvardi MK (2006) Functional genomics of plant transporters in legume nodules. Funct Plant Biol 33(8):731–736

Benedito VA, Torres-Jerez I, Murray JD, Andriankaja A, Allen S, Kakar K, Wandrey M, Verdier J, Zuber H, Ott T, Moreau S, Niebel A, Frickey T, Weiller G, He J, Dai X, Zhao PX, Tang Y, Udvardi MK (2008) A gene expression atlas of the model legume *Medicago truncatula*. Plant J 55(3):504–513

Benedito VA, Li H, Dai X, Wandrey M, He J, Kaundal R, Torres-Jerez I, Gomez SK, Harrison MJ, Tang Y, Zhao PX, Udvardi MK (2010) Genomic inventory and transcriptional analysis of *Medicago truncatula* transporters. Plant Physiol 152(3):1716–1730

Berta G, Fusconi A, Trotta A (1993) VA mycorrhizal infection and the morphology and function of root systems. Environ Exp Bot 33:159–173

Bieleski RL (1973) Phosphate pools, phosphate transport, and phosphate availability. Annu Rev Plant Physiol 24:225–252

Bona E, Scarafoni A, Marsano F, Boatti L, Copetta A, Massa N, Gamalero E, D'Agostino G, Cesaro P, Cavaletto M, Berta G (2016) Arbuscular mycorrhizal symbiosis affects the grain proteome of *Zea mays*: a field study. Sci Rep 6:26439

Bonneau L, Huguet S, Wipf D, Pauly N, Truong HN (2013) Combined phosphate and nitrogen limitation generates a nutrient stress transcriptome favorable for arbuscular mycorrhizal symbiosis in *Medicago truncatula*. New Phytol 199(1):188–202

Borie F, Rubio R (2003) Total and organic phosphorus in Chilean volcanic soils. Gayana Bot 60:69–78

Boulet F, Lambers H (2005) Characterisation of arbuscular mycorrhizal fungi colonization in cluster roots of shape *Hakea verrucosa* F. Muell (Proteaceae), and its effects on grown and nutrient acquisition in ultramafic soils. Plant Soil 269:357–367

Breuillin-Sessoms F, Schramm J, Hajirezaei M, Ahkami A, Favre P, Druege U, Hause B, Bucher M, Kretzschmar T, Bossolini E, Kuhlemeier C, Martinoia E, Franken P, Scholz U, Reinhardt D (2010) Phosphate systemically inhibits development of arbuscular mycorrhiza in Petunia hybrida and represses genes involved in mycorrhizal functioning. Plant J 64:1002–1017

Breuillin-Sessoms F, Floss DS, Gomez SK, Nathan Pumplin N, Ding Y, Levesque-Tremblay V, Noar RD, Daniels DA, Bravo A, Eaglesham JB, Benedito VA, Udvardi MK, Harrisona MJ (2015) Suppression of arbuscule degeneration in *Medicago truncatula* phosphate transporter4 Mutants is dependent on the ammonium transporter 2 family protein AMT2;3. Plant Cell 27 (4):1352–1366

Brundett MC (2009) Mycorrhizal associations and other means of nutrition of vascular plants: understanding the global diversity host plants by resolving conflicting information and developing reliable means of diagnosis. Plant Soil 320:37–77

Brundrett M (2004) Diversity and classification of mycorrhizal associations. Biol Rev Camb Philos Soc 79:473–495

Cabeza RA, Liese R, Lingner A, von Stieglitz I, Neumann J, Salinas-Riester G, Pommerenke C, Dittert K, Schulze J (2014) RNA-seq transcriptome profiling reveals that *Medicago truncatula* nodules acclimate N_2 fixation before emerging P deficiency reaches the nodules. J Exp Bot 65:6035–6048

Calvo-Polanco M, Sánchez-Castro I, Cantos M, García JL, Azcón R, Ruiz-Lozano JM, Beuzón CR, Aroca R (2016) Effects of different arbuscular mycorrhizal fungal backgrounds and soils on olive plants growth and water relation properties under well-watered and drought conditions. Plant Cell Environ 39:2498–2514

Campanelli A, Ruta C, De Mastro G, Morone-Fortunato I (2012) The role of arbuscular mycorrhizal fungi in alleviating salt stress in *Medicago sativa* L. var. icon. Symbiosis 59(2):65–76

Carroll BJ, Matthews A (1990) Nitrate inhibition of nodulation in legumes. In: Gresshoff PM (ed) Molecular biology of symbiotic nitrogen fixation. CRC, Boca Raton, FL, pp 159–180

Casieri L, Lahmidi NA, Doidy J, Veneault-Fourrey C, Migeon A, Bonneau L, Courty PE, Garcia K, Charbonnier M, Delteil A, Brun A, Zimmermann S, Plassard C, Wipf D (2013) Biotrophic transportome in mutualistic plant fungal interactions. Mycorrhiza 23:597–625

Chabaud M, Genre A, Sieberer BJ, Faccio A, Fournier J, Novero M, Barker DG, Bonfante P (2011) Arbuscular mycorrhizal hyphopodia and germinated spore exudates trigger Ca2+ spiking in the legume and nonlegume root epidermis. New Phytol 189:347–355

Chaparro JM, Badri DV, Vivanco JM (2014) Rhizosphere microbiome assemblage is affected by plant development. ISME J 8:790–803

Chen Y, Cao S, Chai Y, Clardy J, Kolter R, Guo JH, Losick R (2012) A *Bacillus subtilis* sensor kinase involved in triggering biofilm formation on the roots of tomato plants. Mol Microbiol 85:418–430

Chen A, Gu M, Wang S, Chen J, Xu G (2018) Transport properties and regulatory roles of nitrogen in arbuscular mycorrhizal symbiosis. Semin Cell Dev Biol 74:80–88. https://doi.org/10.1016/j.semcdb.2017.06.015

Coenye T, Vandamme P (2003) Diversity and significance of *Burkholderia* species occupying diverse ecological niches. Environ Microbiol 5:719–729

Colebatch G, Desbrosses G, Ott T, Krussell L, Montanari O, Kloska S, Kopka J, Udvardi M (2004) Global changes in transcription orchestrate metabolic differentiation during symbiotic nitrogen fixation in *Lotus japonicus*. Plant J 39:487–512

Compant S, Nowak J, Coenye T, Clément C, Ait Barka E (2008) Diversity and occurrence of *Burkholderia* spp. in the natural environment. FEMS Microbiol Rev 32:607–626

Cordell D, White S (2011) Peak phosphorus: clarifying the key issues of a vigorous debate about long-term phosphorus security. Sustainability 3:2027–2049

Cornejo P, Meier S, Borie G, Rillig M, Borie F (2008a) Glomalin-related soil protein in a Mediterranean ecosystem affected by a copper smelter and its contribution to Cu and Zn sequestration. Sci Total Environ 406:154–160

Cornejo P, Rubio R, Castillo C, Azcón R, Borie F (2008b) Mycorrhizal effectiveness on wheat nutrient acquisition in an acidic soil from Southern Chile as affected by nitrogen sources. J Plant Nutr 31(9):1555–1569

Cornejo P, Rubio R, Borie F (2009) Mycorrhizal propagule persistence in a succession of cereals in a disturbed and undisturbed andisol fertilized with two nitrogen sources. Chilean J Agric Res 69 (3):426–434

Cornejo P, Pérez-Tienda J, Meier S, Valderas A, Borie F, Azcón-Aguilar C, Ferrol N (2013) Copper compartmentalization in spores as a survival strategy of arbuscular mycorrhizal fungi in Cu-polluted environments. Soil Biol Biochem 57:925–928

Cornejo P, Meier S, García S, Ferrol N, Durán P, Borie F, Seguel A (2017a) Contribution of inoculation with arbuscular mycorrhizal fungi to the bioremediation of a copper polluted soil using *Oenothera picensis*. J Soil Sci Plant Nutr 17:14–21

Cornejo P, Seguel A, Aguilera P, Meier S, Larsen J, Borie F (2017b) Arbuscular mycorrhizal fungi improves tolerance of agricultural plants to cope abiotic stress conditions. In: Singh DP (ed) - Plant-microbe interactions in agro-ecological perspectives. Vol. 2. Microbial interactions and agro-ecological impacts. Springer, Singapore. ISBN: 978-981-10-6592-7

Courty PE, Pritsch K, Schloter M, Hartmann A, Garbaye J (2005) Activity profiling of ectomycorrhiza communities in two forest soils using multiple enzymatic tests. New Phytol 167:309–319

Courty PE, Hoegger PJ, Kilaru S, Kohler A, Buee M, Garbaye J, Martin F, Kues U (2009) Phylogenetic analysis, genomic organisation, and expression analysis of multi-copper oxidases in the ectomycorrhizal basidiomycete *Laccaria bicolor*. New Phytol 182:736–750

Cramer MD, Shane MW, Lambers H (2005) Physiological changes in white lupin associated with variation in root-zone CO_2 concentration and cluster-root P mobilization. Plant Cell Environ 28:1203–1217

Crocker LJ, Schwintzer CR (1993) Factors affecting formation of cluster roots in Myrica gale seedlings in water culture. Plant Soil 152:287–298

Culleré L, Ferreira V, Chevret B, Venturini ME, Sánchez-Gimeno AC, Blanco D (2010) Characterisation of aroma active compounds in black truffles (*Tuber melanosporum*) and summer truffles (*Tuber aestivum*) by gas chromatography–olfactometry. Food Chem 122:300–306

Daei G, Ardakani M, Rejali F, Teimuri S, Miransari M (2009) Alleviation of salinity stress on wheat yield, yield components, and nutrient uptake using arbuscular mycorrhizal fungi under field conditions. J Plant Physiol 166:617–625

Deacon JW, Fleming LV (1992) Interactions of ectomycorrhizal fungi. In: Allen MF (ed) Mycorrhizal functioning: an integrative plant-fungal process. Chapman and Hall, New York, pp 249–300

de Britto P, Abrahão A, Gorne Viani RA, Santin Brancalion PH, Lambers H, Frankand Sawaya CH, Oliveira R (2015) Cluster-root formation and carboxylate release in *Euplassa cantareirae* (Proteaceae) from a neotropical biodiversity hotspot. Plant Soil. https://doi.org/10.1007/s11104-015-2630-2

Delgado M, Zúñiga-Feest A, Alvear M, Borie F (2013) The effect of phosphorus on cluster root formation and functioning of Embothrium coccineum (R. et J. Forst.) Plant Soil 373:765–773. https://doi.org/10.1007/s11104-013-1829-3

Delgado M, Suriyagoda L, Zúñiga-Feest A, Borie F, Lambers H, Field K (2014) Divergent functioning of Proteaceae species: the South American *Embothrium coccineum* displays a combination of adaptive traits to survive in high-phosphorus soils. Funct Ecol 28:1356–1366

Delgado M, Zúñiga-Feest A, Almonacid L, Lambers H, Borie F (2015) Cluster roots of *Embothrium coccineum* (Proteaceae) affect enzyme activities and phosphorus lability in rhizosphere soil. Plant Soil 395:189–200

Dell B, Kuo J, Thompson GJ (1980) Development of proteoid roots in *Hakea oblique* R. BR. (Proteaceae) grown in water culture. Aust J Bot 28:27–37

Dell J, Malajczuk N, Bougher NL, Thomson G (1994) Development and function of *Pisolithus* and *Scleroderma* ectomycorrhizas formed *in vivo* with *Allocasuarina*, *Casuarina* and *Eucalyptus*. Mycorrhiza 5:129–138

Dellórto M, Santi S, de Nisi P, Cesco S, Varanini Z, Zocchi G, Pinton R (2000) Development of Fe-deficiency responses in cucumber (*Cucumis sativus* L.) roots: involvement of plasma membrane H+-ATPase activity. J Exp Bot 51:695–701

Dessureault-Rompre J, Nowack B, Schulin R, Luster J (2007) Spatial and temporal variation in organic acid anion exudation and nutrient anion uptake in the rhizosphere of *Lupinus albus* L. Plant Soil 301:123–134

Dinkelaker B, Romheld V, Marschner H (1989) Citric acid excretion and precipitation of calcium citrate in the rhizosphere of white lupin (*Lupinus albus* L.) Plant Cell Environ 12:285–292

Dinkelaker B, Hengeler C, Marschner H (1995) Distribution and function of proteoid roots and other root clusters. Bot Acta 108:183–200

Ditengou FA, Lapeyrie F (2000) Hypaphorine from the ectomycorrhizal fungus *Pisolithus tinctorius* counteracts activities of indole-3-acetic acid and ethylene but not synthetic auxins in eucalypt seedlings. Mol Plant Microbe Interact 13:151–158

Ditengou FA, Beguiristain T, Lapeyrie F (2000) Root hair elongation is inhibited by hypaphorine, the indole alkaloid from the ectomycorrhizal fungus *Pisolithus tinctorius*, and restored by indole-3-acetic acid. Planta 211:722–728

Ditengou FA, Müller A, Rosenkranz M, Felten J, Lasok H, van Doorn MM, Legué V, Palme K, Schnitzler J-P, Polle A (2015) Volatile signalling by sesquiterpenes from ectomycorrhizal fungi reprogammes root architecture. Nat Commun 6:6279. https://doi.org/10.1038/ncomms7279

Dixon R, Kahn D (2004) Genetic regulation of biological nitrogen fixation. Nat Rev Microbiol 2:621–631

Doornbos RF, Van Loon LC, Bakker P (2012) Impact of root exudates and plant defense signaling on bacterial communities in the rhizosphere. A review. Agron Sustain Dev 32:227–243

Drapek C, Sparks EE, Benfey PN (2017) Uncovering gene regulatory networks controlling plant cell differentiation. Trends Genet 33(8):529–539

Drevon JJ, Hartwig UA (1997) Phosphorus deficiency increases the argon-induced decline of nodule nitrogenase activity in soybean and alfalfa. Planta 201:463–469

Drevon JJ, Abadie J, Alkama N, Andriamananjara A, Amenc L, Bargaz A, Carlssonn G, Jaillard B, Lazali M, Ghoulam C, Ounane SM (2015) Phosphorus use efficiency for N_2 fixation in the rhizobial symbiosis with legumes. In: de Bruijn FJ (ed) Biological Nitrogen Fixation. Wiley, Hoboken, NJ, pp 455–464

Duff SMG, Sarath G, Plaxton WC (1994) The role of acid phosphatases in plant phosphorus metabolism. Physiol Plant 90:791–800

Duke SH, Collins M, Soberalske RM (1980) Effects of potassium fertilization on nitrogen fixation and nodules enzymes of nitrogen metabolism in alfalfa. Crop Sci 20:213–219

Duplessis S, Courty PE, Tagu D, Martin F (2005) Transcript patterns associated with ectomycorrhizal development in *Eucalyptus globulus* and *Pisolithus microcarpus*. New Phytol 165:599–611

Elser J, Bennett E (2011) Phosphorus cycle: a broken biogeochemical cycle. Nature 478:29–31

Evelin H, Giri B, Kapoor R (2012) Contribution of *Glomus intraradices* inoculation to nutrient acquisition and mitigation of ionic imbalance in NaCl-stressed *Trigonella foenum-graecum*. Mycorrhiza 22:203–217

Fernández-Bidondo L, Bompadre J, Pergola M, Silvani V, Colombo R, Bracamonte F, Godeas A (2012) Differential interaction between two *Glomus intraradices* strains with different extraradical mycelium architecture and a phosphate solubilizing bacterium in maize rhizosphere. Pedobiologia 55(12):227–232

Fitter AH (2004) Magnolioid roots—hairs, architecture and mycorrhizal dependency. New Phytol 164:15–16

Foo E, Yoneyama K, Hugill CJ, Quittenden LJ, Reid JB (2013) Strigolactones and the regulation of pea symbioses in response to nitrate and phosphate deficiency. Mol Plant 6(1):76–87

Fusconi A (2014) Regulation of root morphogenesis in arbuscular mycorrhizae: what role do fungal exudates, phosphate, sugars and hormones play in lateral root formation? Ann Bot 113:19–33

Gadd GM (1999) Fungal production of citric and oxalic acid: importance of metal speciation, physiology and biogeochemical processes. Adv Microb Physiol 41:48–92

Gage DJ (2004) Infection and invasion of roots by symbiotic, nitrogen-fixing rhizobia during nodulation of temperate legumes. Microbiol Mol Biol Rev 68:280–300

Gardner WK, Barber DA, Parbery DG (1983) The acquisition of phosphorus by *Lupinus albus* L. III. The probable mechanism by which phosphorus, movement in the soil/root interface is enhanced. Plant Soil 70:107–124

Garg N, Bhandari P (2016) Silicon nutrition and mycorrhizal inoculations improve growth, nutrient status, K^+/Na^+ ratio and yield of *Cicer arietinum* L .genotypes under salinity stress. Plant Growth Regul 78(3):371–387

Gay G, Debaud JC (1987) Genetic study on indole-3-acetic acid production by ectomycorrhizal *Hebeloma* species: inter- and intraspecific variability in homo- and dikaryotic mycelia. Appl Microbiol Biotechnol 26:141–146

Genre A, Chabaud M, Balzergue C, Puech-Pagès V, Novero M, Rey T, Fournier J, Rochange S, Bécard G, Bonfante P, Barker DG (2013) Short-chain chitin oligomers from arbuscular mycorrhizal fungi trigger nuclear Ca2+ spiking in Medicago truncatula roots and their production is enhanced by strigolactone. New Phytol 198:190–202

Gerke J, Römer W, Jungk A (1994) The excretion of citric and malic acid by proteoid roots of Lupinus albus L.: effects on soil solution concentration of phosphate, iron, and aluminum in the proteoid rhizosphere in samples of an oxisol and a luvisol. Z Pflanzen Bodenk 157:289–294

Gilbert GA, Knight JD, Vance CP, Allan DL (1999) Acid phosphatase activity in phosphorus-deficient white lupin roots. Plant Cell Environ 22:801–810

Gobbato E, Marsh JF, Vernié T, Wang E, Maillet F, Kim J, Miller JB, Sun J, Bano SA, Ratet P, Mysore KS, Dénarié J, Schultze M, Oldroyd GE (2012) A GRAS-type transcription factor with a specific function in mycorrhizal signaling. Curr Biol 22:2236–2241

Gogala N (1991) Regulation of mycorrhizal infection by hormonal factors produced by hosts and fungi. Experientia 47:331–340

Gordon AJ, Minchin FR, James CL, Komina O (1999) Sucrose synthase in legume nodules is essential for nitrogen fixation. Plant Physiol 120:867–878

Govindarajulu M, Pfeffer PE, Jin H, Abubaker J, Douds DD, Allen JW, Bücking H, Lammers PJ, Shachar-Hill Y (2005) Nitrogen transfer in the arbuscular mycorrhizal Simbiosis. Nature 435:819–823

Graham PH, Vance CP (2000) Nitrogen fixation in perspective: an overview of research and extension needs. Field Crop Res 65:93–106

Gregory PJ (2006) Plant roots: growth, activity and interaction with soils. Blackwell, Oxford

Grinbergs J, Valenzuela E, Ramírez C (1987) Formación y desarrollo de raíces proteiformes en plántulas de Gevuina avellana Mol. Agro Sur 15(1):1–9

Gruber BD, Giehl RFH, Friedel S, von Wirén N (2013) Plasticity of the Arabidopsis root system under nutrient deficiencies. Plant Physiol 163:161–179

Guether M, Balestrini R, Hannah M, He J, Udvardi MK, Bonfante P (2009) Genome-wide reprogramming of regulatory networks, transport, cell wall and membrane biogenesis during arbuscular mycorrhizal symbiosis in Lotus japonicus. New Phytol 182:200–212

Guo WB, Zhao J, Li XX, Qin L, Yan XL, Liao H (2011) A soybean β-expansin gene *GmEXPB2* intrinsically involved in root system architecture responses to abiotic stresses. Plant J 66:541–552

Gupta AK, Kaur V, Kaur N (1998) Appearance of different phosphatase forms and phosphorus partitioning in nodules of chickpea (Cicer arietinum L.) during development. Acta Physiol Plant 20:369–374

Güsewell S (2017) Regulation of dauciform root formation and root phosphatase activities of sedges (*Carex*) by nitrogen and phosphorus. Plant Soil 415(1-2):57–72

Gutjahr C, Casieri L, Paszkowski U (2009) *Glomus intraradices* induces changes in root system architecture of rice independently of common symbiosis signaling. New Phytol 182:829–837

Hammer EC, Nasr H, Pallon J, Olsson PA, Wallander H (2011) Elemental composition of arbuscular mycorrhizal fungi at high salinity. Mycorrhiza 21:117–129

Han M, Okamoto M, Beatty PH, Rothstein SJ, Good AG (2015) The genetics of nitrogen use efficiency in crop plants. Annu Rev Genet 49:269–289

Harrison MJ (1999) Molecular and cellular aspects of the arbuscular mycorrhizal symbiosis. Annu Rev Plant Physiol Plant Mol Biol 50:361–389

Harrison MJ (2005) Signaling in the arbuscular mycorrhizal symbiosis. Annu Rev Microbiol 59:19–42

Haselwandter K (1995) Mycorrhizal fungi: siderophore production. Crit Rev Biotechnol 15:287–291

Haselwandter K, Winkelmann G (2002) Ferricrocin—an ectomycorrhizal siderophore of *Cenoccoccum geophilum*. Biometals 15:73–77

Hawkins H-J, Johansen A, George E (2000) Uptake and transport of organic and inorganic nitrogen by arbuscular mycorrhizal fungi. Plant Soil 226:275–285

Hayes P, Turner BL, Lambers H (2014) Foliar nutrient concentrations and resorption efficiency in plants of contrasting nutrient-acquisition strategies along a 2-million-year dune chronosequence. J Ecol 102:396–410

Heckman DS, Geiser DM, Eidell BR, Stauffer RL, Kardos NL, Hedges SB (2001) Molecular evidence for the early colonization of land by fungi and plants. Science 293:1129–1133

Hernandez G, Valdes-Lopez O, Ramirez M, Goffard N, Weiller G, Aparicio-Fabre R, Fuentes SI, Erban A, Kopka J, Udvardi MK, Vance CP (2009) Global changes in the transcript and metabolic profiles during symbiotic nitrogen fixation in phosphorus-stressed common bean plants. Plant Physiol 151:1221–1238

Hinsinger P (1998) How do plant roots acquire mineral nutrients? Chemical properties involved in the rhizosphere. Adv Agron 64:225–265

Ho MD, Rosas JC, Brown KM, Lynch JP (2005) Root architectural tradeoffs for water and phosphorus acquisition. Funct Plant Biol 32:737–748

Hobbie EA, Agerer R (2010) Nitrogen isotopes in ectomycorrhizal sporocarps correspond to belowground exploration types. Plant Soil 327:71–83

Hobecker KV, Reynoso MA, Bustos-Sanmamed P, Wen J, Mysore KS, Crespi M, Blanco FA, Zanetti ME (2017) The microRNA390/TAS3 pathway mediates symbiotic nodulation and lateral root growth. Plant Physiol 174(4):2469–2486

Hodge A (2003) Plant nitrogen capture from organic matter as affected by spatial dispersion, interspecific competition and mycorrhizal colonization. New Phytol 157:303–314

Hodge A, Campbell CD, Fitter AH (2001) An arbuscular mycorrhizal fungus accelerates decomposition and acquires nitrogen directly from organic material. Nature 413:297–299

Hodge A, Helgason T, Fitter AH (2010) Nutritional ecology of arbuscular mycorrhizal fungi. Fungal Ecol 3(4):267–273

Hoffland E, Kuyper TW, Wallander H, Plassard C, Gorbushina AA, Haselwandter K, Holström S, Landeweert R, Lundström US, Rosling A, Sen R, Smits MM, van Hess PAW, van Breemen N (2004) The role of fungi in weathering. Front Ecol Environ 2:258–264

Hogh-Jensen H (2003) The effect of potassium deficiency on growth and N2-fixation in *Trifolium repens*. Physiol Plant 119:440–449

Hogh-Jensen H, Schjoerring JK, Soussana JF (2002) The influence of phosphorus deficiency on growth and nitrogen fixation of white clover plants. Ann Bot 90:745–753

Huault E, Laffont C, Wen J, Mysore KS, Ratet P, Duc G, Frugier F (2014) Local and systemic regulation of plant root system architecture and symbiotic nodulation by a receptor-like kinase. PLoS Genet 10(12):e1004891

Imin N, Mohd-Radzman NA, Ogilvie HA, Djordjevic MA (2013) The peptide-encoding CEP1 gene modulates lateral root and nodule numbers in *Medicago truncatula*. J Exp Bot 64(17):5395–5409

Israel DW (1987) Investigation of the role of phosphorus in symbiotic dinitrogen fixation. Plant Physiol 84:835–840

Jacoby R, Peukert M, Succurro A, Koprivova A, Kopriva S (2017) The role of soil microorganisms in plant mineral nutrition—current knowledge and future directions. Front Plant Sci 8:1617

Jakobsen I, Leggett ME, Richardson AE (2005) Rhizosphere microorganisms and plant phosphorus uptake. In: Sims JT, Sharpley AN (eds) Phosphorus, agriculture and the environment. American Society for Agronomy, Madison, pp 437–494

Javot H, Pumplin N, Harrison M (2007) Phosphate in arbuscular mycorrhizal symbiosis: transporter properties and regulatory role. Plant Cell Environ 30:310–322

Javot H, Penmetsa RV, Breuillin F, Bhattarai KK, Noar RD, Gomez SK, Cook DR (2011) *Medicago truncatula mtpt4* mutants reveal a role for nitrogen in the regulation of arbuscule degeneration in arbuscular mycorrhizal symbiosis. Plant J 68:954–965

Jeong J, Merkovich A, Clyne M, Connolly EL (2017) Directing iron transport in dicots: regulation of iron acquisition and translocation. Curr Opin Plant Biol 39:106–113

Johansson T, Le Quéré A, Ahren D, Söderström B, Erlandsson R, Lundeberg J, Uhlén M, Tunlid A (2004) Transcriptional responses of *Paxillus involutus* and *Betula pendula* during formation of ectomycorrhizal root tissue. Mol Plant Microbe Interact 17:202–215

Johnson JF, Allan D, Vance CP (1994) Phosphorus stress-induced proteoid roots show altered metabolism in *Lupinus albus*. Plant Physiol 104:657–665

Johnson JF, Allan D, Vance CP, Weiblen G (1996) Root carbon dioxide fixation by phosphorus-deficient *Lupinus albus*. Contribution to organic acid exudation by proteoid roots. Plant Physiol 112:19–30

Joly C, Assis M, Bernacci L (2012) Florística e fitossociologia em parcelas permanentes da Mata Atlântica do sudeste do Brasil ao longo de um gradiente altitudinal. Biota Neotrop 12:125–145

Jones D (1998) Organic acids in the rhizosphere—a critical review. Plant Soil 205:25–44

Jones DL, Brassington DS (1998) Sorption of organic acids in acid soils and its implications in the rhizosphere. Eur Soil Sci 49:447–455

Joshi M, Fogelman E, Belausov E, Ginzberg I (2016) Potato root system development and factors that determine its architecture. J Plant Physiol 205:113–123

Jung JKH, McCouch S (2013) Getting to the roots of it: genetic and hormonal control of root architecture. Front Plant Sci 4:186

Kanamori N, Madsen LH, Radutoiu S, Frantescu M, Quitsgaard EMH, Miwa H, Downie JA (2006) A nucleoporin is required for induction of Ca2+ spiking in legume nodule development and essential for rhizobial and fungus symbiosis. Proc Natl Acad Sci USA 103:359–364

Kapulnik Y, Koltai H (2014) Strigolactone involvement in root development, response to abiotic stress, and interactions with the biotic soil environment. Plant Physiol 166(2):560–569

Kaya C, Ashraf M, Sonmez O (2009) The influence of arbuscular mycorrhizal colonisation on key growth parameters and fruit yield of pepper plants grown at high salinity. Sci Hortic 121:1–6

Keerthisinghe G, Hocking PJ, Ryan PR, Delhaize E (1998) Effect of phosphorus supply on the formation and function of proteoid roots of lupin white (*Lupinus albus* L.) Plant Cell Environ 21:467–478

Kiba T, Krapp A (2016) Plant nitrogen acquisition under low availability: regulation of uptake and root architecture. Plant Cell Physiol 57(4):707–714

Kihara T, Wada T, Suzuki Y, Hara T, Koyama H (2003) Alteration of citrate metabolism in cluster roots of white lupin. Plant Cell Physiol 44:901–908

Kikuchi Y, Hijikata N, Ohtomo R, Handa Y, Kawaguchi M, Saito K, Masuta C, Ezawa T (2016) Aquaporin-mediated long-distance polyphosphate translocation directed towards the host in arbuscular mycorrhizal symbiosis: application of virus-induced gene silencing. New Phytol 211:1202–1208

Kleijn RJ, Buescher JM, Le Chat L, Jules M, Aymerich S, Sauer U (2010) Metabolic fluxes during strong carbon catabolite repression by malate in *Bacillus subtilis*. J Biol Chem 285:1587–1596

Kleinert A, Venter M, Kossmann J, Valentine A (2014) The reallocation of carbon in P deficient lupins affects biological nitrogen fixation. J Plant Physiol 171:1619–1624

Kobae Y, Tamura Y, Takai S, Banba M, Hata S (2010) Localized expression of arbuscular mycorrhiza-inducible ammonium transporters in soybean. Plant Cell Physiol 51:1411–1415

Kost T, Stopnisek N, Agnoli K, Eberl L, Weisskopf L (2014) Oxalotrophy, a widespread trait of plant-associated *Burkholderia* species, is involved in successful root colonization of lupin and maize by *Burkholderia phytofirmans*. Front Microbiol 4:421

Kosuta S, Chabaud M, Lougnon G, Gough C, Denaire J, Baker DG, Becard G (2003) Diffusible factor from arbuscular mycorrhizal fungi induces symbiosis-specific *MtENOD11* expression in roots of *Medicago truncat*ula. Plant Physiol 131:952–962

Kosuta S, Hazledine S, Sun J, Miwa H, Morris RJ, Downie JA, Oldroyd GED (2008) Differential and chaotic calcium signatures in the symbiosis signaling pathway of legume. Proc Natl Acad Sci USA 105:9823–9828

Krishnamoorthy R, Kim K, Subramanian P, Senthilkumar M, Anandham R, Sa T (2016) Arbuscular mycorrhizal fungi and associated bacteria isolated from salt-affected soil enhances the tolerance of maize to salinity in coastal reclamation soil. Agric Ecosyst Environ 231:233–239

Kusuda M, Nagai M, Ueda THM, Terashita T (2003) Purification and some properties of α-amylase from an ectomycorrhizal fungus, *Tricholoma matsutake*. Mycoscience 44:311–317

Kusuda M, Ueda M, Konishi Y, Yamanaka K, Miyatake K (2007) Effects of carbohydrate substrate on the vegetative growth of an ectomycorrhizal mushroom, *Tricholoma matsutake*, isolated from *Quercus*. Mycoscience 48:358–364

Lagrange H, Jay-Allgmand C, Lapeyrie F (2001) Rutin, the phenolglycoside from eucalyptus root exudates, stimulates *Pisolithus* hyphal growth at picomolar concentration. New Phytol 149:349–355

Lambers H, Shane MW (2007) Role of root clusters in phosphorus acquisition and increasing biological diversity in agriculture. In: JHJ S, Struik PC, van Laar HH (eds) Scale and complexity in plant systems research: gene-plant-crop relations. Springer, Dordrecht, pp 237–250

Lambers H, Juniper D, Cawthray GR, Veneklaas EJ, Martínez-Ferri E (2002) The pattern of carboxylate exudation in *Banksia grandis* (Proteaceae) is affected by the form of phosphate added to the soil. Plant Soil 238:111–122

Lambers H, Shane WM, Cramer MD, Pearse SJ, Veneklaas EJ (2006) Root structure and functioning for efficient acquisition of phosphorus: matching morphological and physiological traits. Ann Bot 98:693–713

Lambers H, Raven JA, Shaver GR, Smith SE (2008) Plant nutrient-acquisition strategies change with soil age. Trends Ecol Evol 23:95–103

Lambers H, Bishop J, Hopper S, Laliberté E, Zuñiga-Feest A (2012a) Phosphorus mobilization ecosystem engineering: the roles of cluster roots and carboxylate exudation in young P-limited ecosystems. Ann Bot 110:329–348

Lambers H, Cawthray G, Giavalisco P, Kuo J, Laliberte E, Pearse S, Scheibke W, Stitt M, Teste F, Turner BL (2012b) Proteaceae from severely phosphorus-impoverished soils extensively replace phospholipids with galactolipids and sulfolipids during leaf development to achieve a high photosynthetic phosphorus-use-efficiency. New Phytol 196:1098–1108

Lambers H, Clode P, Hawkins HJ, Oliveira RS, Reddell P, Shane SW, Weston P (2015) Chapter 11. Metabolic adaptations of the non-mycotrophic Proteaceae to soil with a low phosphorus availability. In: Plaxton WC, Lambers H (eds) Phosphorus metabolism in plants in the post-genomic era: from gene to ecosystem. Wiley-Blackwell, Chicester

Lamont B (1974) The biology of dauciform roots in the sedge *Cyathochaeta avenacea*. New Phytol 73:985–996

Lamont BB (1982) Mechanisms for enhancing nutrient uptake in plants, with particular reference to Mediterranean South Africa and Western Australia. Bot Rev 48:597–689

Lamont BB (2003) Structure, ecology and physiology of root clusters—a review. Plant Soil 248:1–19

Lamont BB, Pérez-Fernández M, Rodríguez-Sánchez J (2015) Soil bacteria hold the key to root cluster formation. New Phytol 206(3):1156–1162

Landeweert R, Hoffland E, Finlay RD, Kuyper TW, Van Breemen N (2001) Linking plants to rocks: ectomycorrhizal fungi mobilize nutrients from minerals. Trends Ecol Evol 16 (5):248–254

Laparre J, Balzergue C, Rochange S, Ludwiczak P, Letisse F, Portais JC, Bécard G, Puech-Pages V (2011) Metabolite profiling of pea roots in response to phosphate availability. Plant Signal Behav 6:837–839

Laurent P, Voiblet C, Tagu D, De Carvalho D, Nehls U, De Bellis R, Balestrini R, Bauw G, Bonfante P, Martin F (1999) A novel class of ectomycorrhiza-regulated cell wall polypeptides in Pisolithus tinctorius. Mol Plant Microbe Interact 12:862–871

Layzell DB, Hunt S (1990) Oxygen and the regulation of nitrogen fixation in legume nodules. Physiol Plant 80:322–327

Layzell DB, Hunt S, Palmer GR (1990) Mechanisms of nitrogenase inhibition in soybean nodules. Pulse-modulated spectroscopy indicates that nitrogenase activity is limited by O_2. Plant Physiol 92:1101–1107

Lazali M, Zaman-Allah M, Amenc L, Ounane G, Abadie J, Drevon J-J (2013) A phytase gene is overexpressed in root nodules cortex of *Phaseolus vulgaris*–rhizobia symbiosis under phosphorus deficiency. Planta 238:317–324

Lazali M, Bargaz A, Brahimi S, Amenc L, Abadie J, Drevon JJ (2016) Expression of a phosphate-starvation inducible fructose-1,6-bisphosphatase gene in common bean nodules correlates with phosphorus use efficiency. J Plant Physiol 205:48–56

Le Quere A, Wright DP, Söderström B, Tunlid A, Johansson T (2005) Global patterns of gene regulation associated with the development of ectomycorrhiza between birch (*Betula pendula* Roth.) and *Paxillus involutus* (Batsch) Fr. Mol Plant Microbe Interact 18(7):659–673

Le Roux MR, Ward CL, Botha FC, Valentine AJ (2006) Routes of pyruvate synthesis in phosphorous-deficient lupin roots and nodules. New Phytol 169:399–408

Le Roux MR, Kahn S, Valentine AJ (2008) Organic acid accumulation inhibits N2–fixation in P-stressed lupin nodules. New Phytol 177(4):956–964

Le Roux MR, Khan S, Valentine AJ (2009) Nitrogen and carbon costs of soybean and lupin root systems during phosphate starvation. Symbiosis 48:102–109

Le Roux M, Phiri E, Khan W, Şakiroglu M, Valentine A, Khan S (2014) Expression of novel cytosolic malate dehydrogenases (cMDH) in lupinus angustifolius nodules during phosphorus starvation. J Plant Physiol 171(17):1609–1618

Lea PJ, Sodek L, Parry MAJ, Shewry PR, Halford NG (2007) Asparagine in plants. Ann Appl Biol 150:1–26

Lei J, Dexheimer J (1988) Ultrastructural localisation of ATPase activity in *Pinus sylvestris/ Laccaria laccata* mycorrhizal association. New Phytol 108:329–334

Lei J, Lapeyrie F, Malajczuk N, Dexheimer J (1990) Infectivity of pine and eucalypt isolates of *Pisolithus tinctorius* Pers. Coker and Couch on roots of *Eucalyptus urophylla* S.T. Blake *in vitro*. New Phytol 116:115–122

Leidi EO, Rodríguez-Navarro DN (2000) Nitrogen and phosphorus availability limit N_2 fixation in bean. New Phytol 147:337–346

Li M, Shinano T, Tadona T (1997) Distribution of exudates of lupin roots in the rhizosphere under phosphorus deficient conditions. Soil Sci Plant Nutr 43:237–245

Li C, Giu S, Yang T, Walk T, Wang X, Liao H (2010) Identification of soybean purple acid phosphatase genes and their expression response to phosphorus availability and symbiosis. Ann Bot 109:275–285

Li L, Tilman D, Lambers H, Zhang FS (2014) Plant diversity and overyielding: insights from belowground facilitation of intercropping in agriculture. New Phytol 203(1):63–69

Li X, Zhao J, Tan Z, Zeng R, Liao H (2015) GmEXPB2, a cell wall β-expansin, affects soybean nodulation through modifying root architecture and promoting nodule formation and development. Plant Physiol 169(4):2640–2653

Liese R, Alings K, Meier IC (2017a) Root branching is a leading root trait of the plant economics spectrum in temperate trees. Front Plant Sci 8:315

Liese R, Schulze J, Cabeza RA (2017b) Nitrate application or P deficiency induce a decline in Medicago truncatula N2-fixation by similar changes in the nodule transcriptome. Sci Rep 7:46264. https://doi.org/10.1038/srep46264

Lima JE, Kojima S, Takahashi H, von Wirén N (2010) Ammonium triggers lateral root branching in Arabidopsis in an AMMONIUM TRANSPORTER1;3-dependent manner. Plant Cell 22 (11):3621–3633

López-Pedrosa A, González-Guerrero M, Valderas A, Azcón-Aguilar C, Ferrol N (2006) GintAMT1 encodes a functional high-affinity ammonium transporter that is expressed in the extraradical mycelium of *Glomus*. Fungal Genet Biol 43:102–110

López-Ráez JA, Charnikhova T, Gómez-Roldán V, Matusova R, Kohlen W, De Vos R, Verstappen F, Puech-Pages V, Bécard G, Mulder P, Bouwmeester H (2008) Tomato strigolactones are derived from carotenoids and their biosynthesis is promoted by phosphate starvation. New Phytol 178:863–879

Ludwig-Müller J (2010) Hormonal responses in host plants triggered by arbuscular mycorrhizal fungi. In: Koltai HY, Kapulnik Y (eds) Arbuscular mycorrhizas: physiology and function. Springer, Amsterdam, pp 169–190

Lynch JP (2007) Roots of the second green revolution. Aust J Bot 55:1–20

Maherali H (2014) Is there an association between root architecture and mycorrhizal growth response? New Phytol 204(1):192–200

Maijala P, Fagerstedt KV, Raudaskoski M (1991) Detection of extracellular cellulolytic and proteolytic activity in ectomycorrhizal fungi and *Heterobasidion annosum* (Fr.) Bref. New Phytol 117:643–648

Maillet F, Poinsot V, André O (2011) Fungal lipochitooligosaccharide symbiotic signals in arbuscular mycorrhiza. Nature 469:58–63

Malajczuk N, Lapeyrie F, Garbaye J (1990) Infectivity of pine and eucalypt isolates of *Pisolithus tinctorius* on the roots of *Eucalyptus urophylla in vitro*. New Phytol 114:627–631

Maldonado-Mendoza IE, Dewbre GR, Harrison MJ (2001) A phosphate transporter gene from the extra-radical mycelium of an arbuscular mycorrhizal fungus *Glomus intraradices* is regulated in response to phosphate in the environment. Mol Plant Microbe Interact 14(10):1140–1148

Mardukhi B, Rejali F, Daei G, Ardakani MR, Malakouti M, Miransari M (2011) Arbuscular mycorrhizas enhance nutrient uptake in different wheat genotypes at high salinity levels under field and greenhouse conditions. Com Rend Biol 334:564–571

Marschner H (1995) Mineral nutrition of higher plants, 2nd edn. Academic, London

Marschner H, Dell B (1994) Nutrient uptake in mycorrhizal symbiosis. Plant Soil 159:89–102

Marschner P, Timonen S (2005) Interactions between plant species and mycorrhizal colonization on the bacterial community composition in the rhizosphere. Appl Soil Ecol 28:23–36

Marschner P, Grierson PF, Rengel Z (2005) Microbial community composition and functioning in the rhizosphere of three *Banksia* species in native woodland in Western Australia. Appl Soil Ecol 28:191–201

Massicotte HB, Peterson RL, Ackerley CA, Piché Y (1986) Structure and ontogeny of *Alnus crispa-Alpova diplophloeus* ectomycorrhizae. Can J Bot 64:177–192

Massicotte HB, Peterson RL, Ackerley CA, Ashford AE (1987) Ontogeny of *Eucalyptus pilularis–Pisolithus tinctorius* ectomycorrhizae. II. Transmission electron microscopy. Can J Bot 65:1940–1947

Massicotte HB, Peterson RL, Ackerley CA, Melville LH (1990) Structure and ontogeny of *Betula alleghaniensis-Pisolithus tinctorius* ectomycorrhizae. Can J Bot 68:579–593

Massonneau A, Langlade N, Leon S, Smutny J, Vogt E, Neumann G, Martinoia E (2001) Metabolic changes associated with cluster root development in white lupin (*Lupinus albus* L.): relationship between organic acid excretion, sucrose metabolism and energy status. Planta 213:534–542

McArthur DAJ, Knowles NR (1992) Resistance responses of potato to vesicular arbuscular Mycorrhizal fungi under varying abiotic phosphorus levels. Plant Physiol 100:341–351

McCleery WT, Mohd-Radzman NA, Grieneisen VA (2017) Root branching plasticity: collective decision-making results from local and global signalling. Curr Opin Cell Biol 44:51–58

Meier S, Borie F, Bolan N, Cornejo P (2012) Phytoremediation of metal-polluted soils by arbuscular mycorrhizal fungi. Crit Rev Environ Sci Technol 42:741–775

Mew MC (2016) Phosphate rock costs, prices and resources interaction. Sci Total Environ 542:1008–1012

Mi G, Chen F, Wu Q, Lai N, Yuan L, Zhang F (2010) Ideotype root architecture for efficient nitrogen acquisition by maize in intensive cropping systems. Sci China Life Sci 53 (12):1369–1373

Miao SJ, Han XZ, Liu XB, Qiao YF (2007) Seedling treatments and phosphorus solution concentrations affect nodulation and nodule functions in soybean (*Glycine max* L.) Plant Soil Environ 53:65–71

Middleton PH, Jakab J, Penmetsa RV, Starker CG, Doll J, Kaló P, Prabhu R, Marsh JF, Mitra RM, Kereszt A, Dudas B, VandenBosch K, Long SR, Cook DR, Kiss GB, Oldroyd GE (2007) An ERF transcription factor in Medicago truncatula that is essential for Nod factor signal transduction. Plant Cell 19:1221–1234

Miguel MA (2011) Functional role and synergistic effect of root traits for phosphorus acquisition efficiency and their genetic basis in common bean (*Phaseolus vulgaris* L.). PhD thesis. Pennsylvania State University, University Park

Miransari M, Abrishamchi A, Khoshbakht K, Niknam V (2014) Plant hormones as signals in arbuscular mycorrhizal symbiosis. Crit Rev Biotechnol 34(2):123–133

Mohd-Radzman NA, Laffont C, Ivanovici A, Patel N, Reid D, Stougaard J, Frugier F, Imin N, Djordjevic MA (2016) different pathways act downstream of the CEP peptide receptor CRA2 to regulate lateral root and nodule development. Plant Physiol 171(4):2536–2548

Molina R, Massicotte H, Trappe JM (1992) Specificity phenomena in mycorrhizal symbiosis: community ecological consequences and practical application. In: Allen MF (ed) Mycorrhizal functioning. Chapman and Hall, New York, pp 357–423
Möller BK, Xuan W, Beeckman T (2017) Dynamic control of lateral root positioning. Curr Opin Plant Biol 35:1–7
Morel M, Jacob C, Kohler A, Johansson T, Martin F, Chalot M, Brun A (2005) Identification of genes differentially expressed in extraradical mycelium and ectomycorrhizal roots during *Paxillus involutus-Betula pendula* ectomycorrhizal symbiosis. Appl Environ Microbiol 71:382–391
Morris EC, Griffiths M, Golebiowska A, Mairhofer S, Burr-Hersey J, Goh T, von Wangenheim D, Atkinson B, Sturrock CJ, Lynch JP, Vissenberg K, Ritz K, Wells DM, Mooney SJ, Bennett MJ (2017) Shaping 3D root system architecture. Curr Biol 27(17):R919–R930
Mukherjee A, Ané JM (2011) Germinating spore exudates from arbuscular mycorrhizal fungi: molecular and developmental responses in plants and their regulation by ethylene. Mol Plant Microbe Interact 24:260–270
Muler AL, Oliveira RS, Lambers H, Veneklaas EJ (2013) Does cluster-root activity benefit nutrient uptake and growth of co-existing species? Oecologia 174:23–31
Nagahashi G, Douds DD, Abney GD (1996) Phosphorus amendment inhibits hyphal branching of the VAM fungus *Gigaspora margarita* directly and indirectly through its effect on root exudation. Mycorrhiza 6:403–408
Nagy R, Karandashov V, Chague V, Kalinkevich K, Tamasloukht B, Xu G, Jakobsen I, Levy AA, Amrhein N, Bucher M (2005) The characterization of novel mycorrhiza-specific phosphate transporters from *Lycopersicon esculentum* and *Solanum tuberosum* uncovers functional redundancy in symbiotic phosphate transport in solanaceous species. Plant J 42:236–250
Nara K, Nakaya H, Hogetsu T (2003a) Ectomycorrhizal sporocarp succession and production during early primary succession on Mount Fuji. New Phytol 158:193–206
Nara K, Nakaya H, Wu BY, Zhou ZH, Hogetsu T (2003b) Underground primary succession of ectomycorrhizal fungi in a volcanic desert on Mount Fuji. New Phytol 159:743–756
Neumann G, Römheld V (1999) Root excretion of carboxylic acids and protons in phosphorus-deficient plants. Plant Soil 211:121–130
Neumann G, Massonneau A, Martinoia E, Römheld V (1999) Physiological adaptations to phosphorus deficiency during proteoid root development in White lupin. Planta 208:373–382
Niu X, Narasimhan ML, Salzman RA, Bressan RA, Hasegawa PM (1993) NaCl regulation of plasma membrane H+-ATPase gene expression in a glycophyte and a halophyte. Plant Physiol 103:713–718
Nouri E, Breuillin-Sessoms F, Feller U, Reinhardt D (2014) Phosphorus and nitrogen regulate arbuscular mycorrhizal symbiosis in *Petunia hybrid*. PLoS One 9(6):e90841
Nygren CMR, Edqvist J, Elfstrand M, Heller G, Taylor AFS (2007) Detection of extracellular protease activity in different species and genera of ectomycorrhizal fungi. Mycorrhiza 17:241–248
Ohkama-Ohtsu N, Ichida S, Yamaya H, Ohwada T, Itakura M, Hara Y, Mitsui H, Kaneko T, Tabata S, Tejima K, Saeki K, Omori H, Hayashi M, Maekawa T, Murooka Y, Tajima S, Simomura K, Nomura M, Uchiumi T, Suzuki A, Shimoda Y, Abe M, Minamisawa K, Arima Y, Yokoyama T (2015) Peribacteroid solution of soybean root nodules partly induces genomic loci for differentiation into bacteroids of free-living *Bradyrhizobium diazoefficiens* cells. Soil Sci Plant Nutr 61:461–470
Oláh B, Briere C, Becard G, Denarie J, Gough C (2005) Nod factors and a diffusible factor from arbuscular mycorrhizal fungi stimulate lateral root formation in *Medicago truncatula* via the *DMI1/DMI2* signalling pathway. Plant J 44:195–207
Oldroyd GE, Engstrom EM, Long SR (2001) Ethylene inhibits the Nod factor signal transduction pathway of Medicago truncatula. Plant Cell 13:1835–1849
Oldroyd GED, Downie JA (2004) Calcium, kinases and nodulation signalling in legumes. Nat Rev Mol Cell Biol 5:566–576

O'Rourke JA, Yang SS, Miller SS, Bucciarelli B, Liu J, Rydeen A, Bozsoki Z, Uhde-Stone C, Tu ZJ, Allan D, Gronwald JW, Vance CP (2013) An RNA-Seq transcriptome analysis of orthophosphate-deficient white lupin reveals novel insights into phosphorus acclimation in plants. Plant Physiol 161(2):705–724

Ötvös K, Benková E (2017) Spatiotemporal mechanisms of root branching. Curr Opin Genet Dev 45:82–89

Paungfoo-Lonhienne C, Lonhienne TGA, Rentsch D, Robinson M, Christie M, Webb R, Gamage H, Carroll B, Schenk P, Schmidt S (2008) Plants can use protein as a nitrogen source without assistance from other organisms. Proc Natl Acad Sci USA 105:4524–4529

Paungfoo-Lonhienne C, Schenk P, Lonhienne TGA, Brackin R, Meier S, Rentsch D, Schmidt S (2009) Nitrogen affects cluster root formation and expression of putative peptide transporters. J Exp Bot 60:2665–2676

Peay KG, Kennedy PG, Bruns TD (2011) Rethinking ectomycorrhizal succession: are root density and hyphal exploration types drivers of spatial and temporal zonation? Fungal Ecol 4:233–240

Peñaloza E, Corcuera LJ, Martinez J (2002) Spatial and temporal variation in citrate and malate exudation and tissue concentration as affected by P stress in roots of white lupin. Plant Soil 241:209–221

Peñaloza E, Muñoz G, Salvo-Garrido H, Silva H, Corcuera L (2005) Phosphate deficiency regulates phosphoenolpyruvate carboxylase expression in proteoid root clusters of white lupin. J Exp Bot 56:145–153

Pennheiter AR, Duff SMG, Sarath G (1997) Soybean root nodule acid phosphatase. Plant Physiol 114:597–604

Pereira PAA, Bliss FA (1989) Selection of common bean (*Phaseolus vulgaris* L.) for N_2 fixation at different levels of available phosphorus under field and environmentally-controlled conditions. Plant Soil 115:75–82

Piper F, Baeza G, Zúñiga-Feest A, Fajardo A (2013) Soil nitrogen, and not phosphorus promotes cluster root formation in a South American Proteaceae. Am J Bot 100(12):2328–2338

Plaxton WC (2004) Biochemical adaptations of phosphate starved plants. In: Goodman R (ed) Encyclopedia of plant and crop sciences. Marcel Dekker, New York, pp 976–980

Plaxton WC, Podestá FE (2006) The functional organization and control of plant respiration. CRC Crit Rev Plant Sci 25:159–198

Playsted CWS, Johnston ME, Ramage CM (2006) Functional significance of dauciform roots: exudation of carboxylates and acid phosphatase under phosphorus deficiency in Caustis blakei (Cyperaceae). New Phytol 170(3):491–500

Podila GK, Zheng J, Balasubramanian S (2002) Fungal gene expression in early symbiotic interactions between *Laccaria bicolor* and red pine. Plant Soil 244:117–128

Polidori E, Agostini D, Zeppa S, Potenza L, Palma F, Sisti D, Stocchi G (2002) Identification of differentially expressed cDNA clones in *Tilia platyphyllos–Tuber borchii* ectomycorrhizae using a differential screening approach. Mol Genet Genomics 266:858–864

Popovych N, Tzeng SR, Tonelli M, Ebright RH, Kalodimos CG (2009) Structural basis for cAMP-mediated allosteric control of the catabolite activator protein. Proc Natl Acad Sci USA 106 (17):6927–69232

Porcel R, Redondo-Gómez S, Mateos-Naranjo E, Aroca R, García R, Ruiz-Lozano JM (2015) Arbuscular mycorrhizal symbiosis ameliorates the optimum quantum yield of photosystem II and reduces non-photochemical quenching in rice plants subjected to salt stress. J Plant Physiol 185:75–83

Prell J, Poole P (2006) Metabolic changes of rhizobia in legume nodules. Trends Microbiol 14:161–168

Premaratne KP, Oertli JJ (1994) The influence of potassium on nodulation, nitrogenase activity and nitrogen accumulation of soybean (Glycine max L. Merrill) grown in nutrient solution. Fertil Res 38:95–99

Purnell H (1960) Studies of the family Proteaceae. I. Anatomy and morphology of the roots of some Victoria species. Aust J Bot 8:38–50

Qin L, Zhao J, Tian J, Chen L, Sun Z, Guo Y, Lu X, Gu M, Xu G, Liao H (2012) The high-affinity phosphate transporter GmPT5 regulates phosphate transport to nodules and nodulation in soybean. Plant Physiol 159(4):1634–1643

Racette S, Louis I, Torrey JG (1990) Cluster root formation by *Gymnostoma papuanum* (Casuarinaceae) in relation to aeration and mineral nutrient availability in water culture. Can J Bot 68(2):2564–2570

Ramírez C, Valenzuela E, San Martín C (2005) Nuevos antecedentes sobre desarrollo temprano, morfología y anatomía de las raíces proteiformes de Gevuina avellana. Agro Sur 32:33–44

Ramstedt M, Soderhall K (1983) Protease, phenoloxidase and pectinase activities in mycorrhizal fungi. Trans Br Mycol Soc 81:157–161

Reddell P, Yun Y, Warren A (1997) Cluster roots and mycorrhizae in Casuarina cunninghamiana: their occurrence and formation in relation to phosphorus supply. Aust J Bot 45:41–51

Reddy SM, Pandey AK, Melayah D, Marmeisse R, Gay G (2003) The auxin responsive gene-*C61* is up-regulated in *Pinus pinaster* roots following inoculation with ectomycorrhizal fungi. Plant Cell Environ 26:681–691

Redecker D, Morton JB, Bruns TD (2000) Ancestral lineages of arbuscular mycorrhizal fungi (Glomales). Mol Phylogenet Evol 14:276–284

Reichheld SE, Yu Z, Davidson AR (2009) The induction of folding cooperativity by ligand binding drives the allosteric response of tetracycline repressor. Proc Natl Acad Sci USA 106 (52):22263–22268

Ribet J, Drevon JJ (1995) Increase in permeability to oxygen and in oxygen uptake of soybean nodules under limiting phosphorus nutrition. Physiol Plant 94:298–304

Roelofs RFR, Rengel Z, Cawthra YGR, Dixon KW, Lambers H (2001) Exudation of carboxylates in Australian Proteaceae: chemical composition. Plant Cell Environ 24:891–903

Rubio V, Linhares F, Solano R, Martin AC, Iglesias J, Leyva A, Paz-Ares J (2001) A conserved MYB transcription factor involved in phosphate starvation signaling both in vascular plants and in unicellular algae. Genes Dev 15:2122–2133

Rudrappa T, Czymmek KJ, Pare PW, Bais HP (2008) Root-secreted malic acid recruits beneficial soil bacteria. Plant Physiol 148:1547–1556

Ryan PR, Delhaize E, Jones DL (2001) Function and mechanism of organic anion exudation from plant roots. Annu Rev Plant Physiol Plant Mol Biol 52:527–560

Saito K, Yoshikawa M, Yamo K, Miwa H, Uchide H, Asamizu E, Sato S et al (2007) NUCLEOPORIN85 is required for calcium spiking, fungal and bacterial symbiosis, and seed production in *Lotus japonicus*. Plant Cell 19:610624

Santander C, Aroca R, Ruiz-Lozano JM, Olave J, Borie F, Cornejo P (2017) Arbuscular mycorrhiza effects on plant performance under osmotic stress. Mycorrhiza 27:639–657

Sato T, Ezawa T, Cheng W, Tawaraya K (2015) Release of acid phosphatase from extraradical hyphae of arbuscular mycorrhizal fungus *Rhizophagus clarus*. Soil Sci Plant Nutr 61 (2):269–274

Scarano FR (2002) Structure, function and floristic relationships of plant communities in stressful habitats marginal to the Brazilian Atlantic Rainforest. Ann Bot 90:517–524

Scherer HW, Pacyna S, Spoth KR, Schulz M (2008) Low levels of ferredoxin, ATP, and leghemoglobin contribute to limited N_2 fixation of peas (*Pisum sativum* L.) and alfalfa (*Medicago sativa* L.) under S deficiency conditions. Biol Fertil Soils 44:909–916

Schmidt S, Steward GR (1997) Waterlogging and fire impacts on nitrogen availability and utilization in subtropical wet heathland (Wallum). Plant Cell Environ 20:1231–1241

Schmidt S, Masonm M, Sangtiean T, Stewart GR (2003) Do cluster roots of *Hakea actities* (Proteaceae) acquire complex organic nitrogen? Plant Soil 248:157–165

Schnepf A, Roose T, Schweiger P (2008) Impact of growth and uptake patterns of arbuscular mycorrhizal fungi on plant phosphorus uptake—a modelling study. Plant Soil 312:85–99

Schubert S, Yan F (1997) Nitrate and ammonium nutrition of plants: effects on acid/base balance and adaptation of root cell plasmalemma H+ ATPase. Z Planzen Bodenk 160:275–281

Schulze J, Drevon J-J (2005) P-deficiency increases the O_2 uptake per N_2 reduced in alfalfa. J Exp Bot 56:1779–1784

Schulze M, Kondorosi A (1998) Regulation of symbiotic root nodule development. Ann Rev Genet 32:33–57

Schulze J, Temple G, Temple SJ, Beschow H, Vance CP (2006) Nitrogen fixation by white lupin under phosphorus deficiency. Ann Bot 98:731–740

Schüβler A (2000) *Glomus claroideum* forms an arbuscular mycorrhiza-like symbiosis with the hornwort Anthoceros punctatus. Mycorrhiza 10:15–21

Schweiger PF, Robson AD, Barrow NJ (1995) Root hair length determines beneficial effect of a Glomus species on shoot growth of some pasture species. New Phytol 131:247–254

Secco D, Shou H, Whelan J, Berkowitz O (2014) RNA-seq analysis identifies an intricate regulatory network controlling cluster root development in white lupin. BMC Genomics 15:230

Secco D, Whelan J, Rouached H, Lister R (2017) Nutrient stress-induced chromatin changes in plants. Curr Opin Plant Biol 39:1–7

Seguel A, Cumming J, Klugh-Stewart K, Cornejo P, Borie F (2013) The role of arbuscular mycorrhizas in decreasing Al phytotoxicity in acidic soils: a review. Mycorrhiza 23:167–183

Seguel A, Barea JM, Cornejo P, Borie F (2015) Role of arbuscular mycorrhizal symbiosis in phosphorus-uptake efficiency and aluminium tolerance in barley growing in acid soils. Crop Pasture Sci 66:696–705

Seguel A, Cornejo P, Ramos A, Von Baer E, Cumming J, Borie F (2017) Phosphorus acquisition by three wheat cultivars contrasting in aluminum tolerance growing in an aluminum-rich Andisol. Crop Pasture Sci 68:305–316

Selivanov IA, Utemova LD (1969) Root anatomy of sedges in relation to their mycotrophy. Trans Perm State Pedagogical Inst 68:45–55

Sepúlveda F (2016) Estudio de la colimitación por déficit de Nitrógeno y Fósforo en el funcionamiento de las Proteaceas nativas chilenas: *Embothrium coccineum*, *Gevuina avellana* y *Lomatia dentata*. In: Seminario de Titulación Licenciatura en Ciencias. Facultad de Ciencias, Universidad Austral de Chile, Valdivia, Chile

Shahzad Z, Amtmann A (2017) Food for thought: how nutrients regulate root system architecture. Curr Opin Plant Biol 39:80–87

Shane MW, Lambers H (2005) Cluster roots: a curiosity in context. Plant Soil 274:101–125

Shane MW, Lambers H (2006) Systemic suppression of cluster-root formation and net P-uptake rates in Grevillea crithmifolia at elevated P supply: a proteacean with resistance for developing symptoms of 'P toxicity'. J Exp Bot 57:413–423

Shane MW, Cramer MD, Funayama-Noguchi S, Cawthray GR, Millar AH, Day DA, Lambers H (2004) Developmental physiology of cluster-root carboxylate synthesis and exudation in Harsh Hakea. Expression of phosphoenolpyruvate carboxylase and the alternative oxidase. Plant Physiol 135:549–560

Shane MW, Cawthray GR, Cramer MD, Kuo J, Lambers H (2006) Specialized 'dauciform' roots of Cyperaceae are structurally distinct, but functionally analogous with 'cluster' roots. Plant Cell Environ 29:1989–1999

Shane MW, Fedosejevs ET, Plaxton WC (2013) Reciprocal control of anaplerotic phospho*enol*-pyruvate carboxylase by *in vivo* monoubiquitination and phosphorylation in developing proteoid roots of phosphate deficient *Hakea prostrata* RBr. Plant Physiol 161:1634–1644

Sieberer BJ, Chabaud M, Fournier J, Timmers AC, Barker DG (2012) A switch in Ca2+ spiking signature is concomitant with endosymbiotic microbe entry into cortical root cells of *Medicago truncatula*. Plant J 69:822–830

Simon L, Bousquet J, Levesque RC, Lalonde M (1993) Origin and diversification of endomycorrhizal fungi and coincidence with vascular plants. Nature 363:67–69

Singh SK, Chien CT, Chang IF (2016) The Arabidopsis glutamate receptor-like gene GLR3.6 controls root development by repressing the Kip-related protein gene KRP4. J Exp Bot 67 (6):1853–1869

Sirohi G, Pandey BK, Deveshwar P, Giri J (2016) Emerging trends in epigenetic regulation of nutrient deficiency response in plants. Mol Biotechnol 58(3):159–171
Skene KR (1998) Cluster root: some ecological considerations. J Ecol 86:1060–1064
Skene KR (2000) Pattern formation in cluster roots: some developmental and evolutionary considerations. Ann Bot 85:901–908
Skene KR, Raven JA, Sprent JI (1998) Cluster roots development in Grevillea robusta (Proteaceae). I. Xylem, pericycle, cortex and epidermis development in a determinate root. New Phytol 138:725–732
Smith SE, Read DJ (2008) Mycorrhizal symbiosis, 3rd edn. Academic, London
Smith SE, Smith FA (2012) Fresh perspectives on the roles of arbuscular mycorrhizal fungi in plant nutrition and growth. Mycologia 104:1–13
Smith SE, Smith FA, Jakobsen I (2004) Functional diversity in arbuscular mycorrhizal (AM) symbioses: the contribution of the mycorrhizal P uptake pathway is not correlated with mycorrhizal responses in growth or total P uptake. New Phytol 162:511–524
Sousa MF, Façanha AR, Tavares RM, Lino-Neto T, Gerós H (2007) Phosphate transport by proteoid roots of *Hakea sericea*. Plant Sci 173:550–558
Sprent JI, James EK (2008) Legume-rhizobia symbiosis: an anorexic model. New Phytol 179:3–5
Stafford WH, Baker GC, Brown SA, Burton SG, Cowan DA (2005) Bacterial diversity in the rhizosphere of Proteaceae species. Environ Microbiol 7(11):1755–1768
Streeter JG (1992) Transport and Metabolism of carbon and nitrogen in legume nodules. Adv Bot Res 18:130–187
Suárez-Moreno ZR, Caballero-Mellado J, Coutinho BG, Mendonça-Previato L, James EK, Venturi V (2012) Common features of environmental and potentially beneficial plant-associated *Burkholderia*. Microb Ecol 63(2):249–266
Subramanian KS, Charest C (1998) Arbuscular mycorrhizae and nitrogen assimilation in maize after drought and recovery. Physiol Plantarum 102:285–296
Sulieman S, Tran L-SP (2015) Phosphorus homeostasis in legume nodules as an adaptive strategy to phosphorus deficiency. Plant Sci 239:36–43
Sulieman S, Ha CV, Schulze J, Tran LSP (2013) Growth and nodulation of symbiotic *Medicago truncatula* at different levels of phosphorus availability. J Exp Bot 64:2701–2712
Sun J, Miller JB, Granqvist E et al (2015) Activation of symbiosis signaling by arbuscular mycorrhizal fungi in legumes and rice. Plant Cell 27(3):823–838
Sun CH, Yu JQ, Hu DG (2017) Nitrate: a crucial signal during lateral roots development. Front Plant Sci 8:485
Tajima S, Kouchi H (1997) Metabolism and compartmentation of carbon and nitrogen in legume nodules. In: Stacey G, Keen NT (eds) Plant-microbe interactions, vol 2. Chapman & Hall, New York, pp 27–60
Tanaka Y, Yano K (2005) Nitrogen delivery to maize via mycorrhizal hyphae depends on the form of N supplied. Plant Cell Environ 28:1247–1254
Tawaraya K, Hashimoto K, Wagatsuma T (1998) Effect of root exudate fractions from P-deficient and P-sufficient onion plants on root colonization by the arbuscular mycorrhizal fungus *Gigaspora margarita*. Mycorrhiza 8:67–70
Tejima K, Arima Y, Yokoyama T, Sekimoto H (2003) Composition of amino acids, organic acids, and sugars in the peribacteroid space of soybean root nodules. Soil Sci Plant Nutr 49 (2):239–247. https://doi.org/10.1080/00380768.2003.10410003
Thuynsma R, Valentine AJ, Kleinert A (2014) Phosphorus deficiency affects the allocation of below-ground resources to combined cluster roots and nodules in *Lupinus albus*. J Plant Physiol 17:285–291
Tisserant E, Kohler A, Dozolme-Seddas P et al (2012) The transcriptome of the arbuscular mycorrhizal fungus *Glomus intraradices* (DAOM 197198) reveals functional tradeoffs in an obligate symbiont. New Phytol 193:755–769
Udvardi MK, Day DA (1997) Metabolite transport across symbiotic membranes of legume nodules. Annu Rev Plant Physiol Plant Mol Biol 48:493–523

Udvardi M, Poole PS (2013) Transport and metabolism in legume-rhizobia symbioses. Annu Rev Plant Biol 64:781–805

Uhde-Stone C, Gilbert G, Johnson JM, Litjens R, Zinn KE, Temple SJ, Allan DL (2003) Acclimation of white lupin to phosphorus deficiency involves enhanced expression of genes related to organic acid metabolism. Plant Soil 248(1–2):99–116

Vadez V, Beck DP, Lasso JH, Drevon J-J (1996) Utilization of the acetylene reduction assay to screen for tolerance of symbiotic N_2 fixation to limiting P nutrition in common bean. Physiol Plant 99:227–232

Valentine AJ, Benedito VA, Kang Y (2011) Legume nitrogen fixation and soil abiotic stress: from physiology to genomics and beyond. Annu Plant Rev 42:207–248

Vance CP (2001) Symbiotic nitrogen fixation and phosphorus acquisition. Plant nutrition in a world of declining renewable resources. Plant Physiol 127:390–397

Vance CP, Uhde-Stone C, Allan DL (2003) Phosphorus acquisition and use: critical adaptations by plants for securing a non-renewable resource. New Phytol 157:423–447

Vardien W, Mesjasz-Pryzbylowicz J, Pryzbylowicz WJ, Wang Y, Steenkamp ET, Valentine AJ (2014) Nodules from Fynbos legume *Virgilia divaricata* have high functional plasticity under variable P supply levels. J Plant Physiol 171:1732–1739

Varin S, Cliquet JB, Personeni E, Avice JC, Lemauviel-Lavenant S (2010) How does sulphur availability modify N acquisition of white clover (*Trifolium repens* L). J. Exp Bot 61:225–234

Viereck N, Hansen PE, Jakobsen I (2004) Phosphate pool dynamics in the arbuscular mycorrhizal fungus *Glomus intraradices* studied by in vivo 31P NMR spectroscopy. New Phytol 162:783–794

Voiblet C, Duplessis S, Encelot N, Martin F (2001) Identification of symbiosis-regulated genesin *Eucalyptus globulus-Pisolithus tinctorius* ectomycorrhiza by differential hybridization of arrayed cDNAs. Plant J 25:181–191

Volpe V, Dell'Aglio E, Bonfante P (2013a) The *Lotus japonicus MAMI* gene links root development, arbuscular mycorrhizal symbiosis and phosphate availability. Plant Signal Behav 8(3): e23414

Volpe V, Dell'Aglio E, Giovannetti M, Ruberti C, Costa A, Genre A, Guether M, Bonfante P (2013b) An AM-induced, MYB-family gene of *Lotus japonicus* (*LjMAMI*) affects root growth in an AM-independent manner. Plant J 73(3):442–455

Walker L, Boddington C, Jenkins D, Wang Y, Grønlund JT, Hulsmans J, Kumar S, Patel D, Moore JD, Carter A, Samavedam S, Bomono G, Hersh DS, Coruzzi GM, Burroughs NJ, Gifford ML (2017) Root architecture shaping by the environment is orchestrated by dynamic gene expression in space and time. Plant Cell 29:2393–2412

Wang B, Qiu YL (2006) Phylogenetic distribution and evolution of mycorrhizas in land plants. Mycorrhiza 16:299–363

Wang Z, Straub D, Yang H, Kania A, Shen J, Ludewig U, Neumann G (2014) The regulatory network of cluster-root function and development in phosphate-deficient white lupin (*Lupinus albus*) identified by transcriptome sequencing. Physiol Plant 151(3):323–338

Watt M, Evans JR (1999) Proteoid roots. Physiology and development. Plant Physiol 121 (2):317–324

Watteau F, Berthelin J (1994) Microbial dissolution of iron and aluminium from soil minerals: efficiency and specificity of hydroxamate siderophores compared to aliphatic acids. Eur J Soil Biol 30:1–9

Weaver LH, Kwon K, Beckett D, Matthews BW (2001) Corepressor-induced organization and assembly of the biotin repressor: a model for allosteric activation of a transcriptional regulator. Proc Natl Acad Sci U S A 98(11):6045–6050

Weidmann S, Sanchez L, Descombin J, Chatagnier O, Gianinazzi F, Gianinazzi-Pearson V (2004) Fungal elicitation of signal transduction-related plant genes precedes mycorrhiza establishment and requires the *dmi3* gene in *Medicago truncatula*. Mol Plant Microbe Interact 17:1385–1393

Weisskopf L, Le Bayon RC, Kohler F, Page V, Jossi M, Gobat JM, Aragno M (2008) Spatio-temporal dynamics of bacterial communities associated with two plant species differing in organic acid secretion: a one-year microcosm study on lupin and wheat. Soil Biol Biochem 40 (7):1772–1780

Weisskopf L, Heller S, Eberl L (2011) *Burkholderia* species are major inhabitants of white lupin cluster roots. Appl Environ Microbiol 77(21):7715–7720

White J, Prell J, James EK, Poole P (2007) Nutrient sharing between symbionts. Plant Physiol 144:604–614

Wright DP, Johansson T, Le Quere A, Söderström B, Tunlid A (2005) Spatial patterns of gene expression in the extramatrical mycelium and mycorrhizal root tips formed by the ectomycorrhizal fungus *Paxillus involutus* in association with birch (*Betula pendula*) seedlings in soil microcosms. New Phytol 167:579–596

Yan F, Zhu Y, Möller C, Zörb C, Schubert S (2002) Adaptation of H+-pumping and plasma membrane H+ ATPase activity in proteoid roots of white lupin under phosphate deficiency. Plant Physiol 129:50–63

Yan Y, Wang H, Hamera S, Chen X, Fang R (2014) miR444a has multiple functions in the rice nitrate-signaling pathway. Plant J 78(1):44–55

Ye S, Yanga Y, Xin G, Wang Y, Ruan L, Ye G (2015) Studies of the Italian ryegrass–rice rotation system in southern China: arbuscular mycorrhizal symbiosis affects soil microorganisms and enzyme activities in the *Lolium multiflorum* L. rhizosphere. Appl Soil Ecol 90:26–34

Yin N, Zhang Z, Wang L, Qian K (2016) Variations in organic carbon, aggregation, and enzyme activities of gangue-fly ash-reconstructed soils with sludge and arbuscular mycorrhizal fungi during 6-year reclamation. Envi Sci Pollut Res 23(17):17840–17849

Yoneyama K, Xie X, Kim HI, Kisugi T, Nomura T, Sekimoto H, Yokota T, Yoneyama K (2012) How do nitrogen and phosphorus deficiencies affect strigolactone production and exudation? Planta 235(6):1197–1207

York L, Lobet G (2017) Phenomics of root system architecture: measuring and analyzing root phenes. Teaching tools in plant biology: lecture notes. Plant Cell (Online). https://doi.org/10.1105/tpc.117.tt0917

Yu P, Gutjahr C, Li C, Hochholdinger F (2016) Genetic control of lateral root formation in cereals. Trends Plant Sci 21(11):951–961

Zaid H, Arahou M, Diem H, El Morabet R (2003) Is Fe deficiency rather than P deficiency the cause of cluster root formation in *Casuarina* species? Plant Soil 248:229–235

Zhang F, Niu J, Zhang W, Chen X, Li C, Yuan L, Xie J (2010) Potassium nutrition of crops under varied regimes of nitrogen supply. Plant Soil 335:21–34

Zhao D, Arrowsmith CH, Jia X, Jardetzky O (1993) Refined solution structures of the *Escherichia coli trp* holo- and aporepressor. J Mol Biol 229(3):735–746

Zheng Y, Drechsler N, Rausch C, Kunze R (2016) The Arabidopsis nitrate transporter NPF7.3/NRT1.5 is involved in lateral root development under potassium deprivation. Plant Signal Behav 11(5):e1176819

Zhu JM, Brown KM, Lynch JP (2010) Root cortical aerenchyma improves the drought tolerance of maize (*Zea mays* L.) Plant Cell Environ 33:740–749

Zúñiga-Feest A, Delgado M, Alberdi M (2010) The effect of phosphorus on growth and cluster-root formation in the Chilean Proteaceae: *Embothrium coccineum* (R. et J. Forst.) Plant Soil 334:113–121

Zúñiga-Feest A, Delgado M, Bustos A (2014) Cluster roots. In: Varma A, Morte A (eds) Root engineering, Soil biology series. Springer, Heidelberg, pp 353–367. http://www.springer.com/series/5138

Zúñiga-Feest A, Delgado M, Bustos-Salazar A, Ochoa V (2015) The southern South American Proteaceae, Embothrium coccineum exhibits intraspecific variation in growth and cluster-root formation depending on climatic and edaphic origins. Plant Soil 396:201–213. https://doi.org/10.1007/s11104-015-2574-6

Chapter 5
Root Exudates and Microbial Communities Drive Mineral Dissolution and the Formation of Nano-size Minerals in Soils: Implications for Soil Carbon Storage

Guanghui Yu

5.1 Introduction

Soils are the largest stable terrestrial carbon (C) pool, and they are often assumed to be a major sink for future C storage (Schmidt et al. 2011; Stockmann et al. 2013). Recent studies indicate that microbial accessibility to substrates rather than chemical complexity of organic C dominantly controls long-term C stability in soils (Schmidt et al. 2011; Sulman et al. 2014; Kleber et al. 2015; Lehmann and Kleber 2015) and that a significant proportion of stable soil organic matter (SOM) is derived from simple C rather than resistant compounds (Kelleher and Simpson 2006; Lehmann et al. 2008; Schmidt et al. 2011). Such stable SOM mainly results from physical disconnection in microaggregates and chemical sorption in organo-mineral complexes (Baldock and Skjemstad 2000; Koegel-Knabner et al. 2008; Schmidt et al. 2011; Jones and Singh 2014). However, the majority of studies take organo-mineral complexes as "biogeochemical black boxes," where inputs and outputs of organics and minerals are estimated but the underlying mechanisms controlling C stabilization and storage are rarely explored (Johnson et al. 2015).

This situation is partly due to the inherent physical and biogeochemical complexity of soil systems, fluctuation of environmental conditions (Schmidt et al. 2011), and the existence of nano-size (~1–100 nm) minerals that may dominate C binding (Hochella et al. 2008). These nano-size minerals are composed of nanominerals and mineral

G. Yu (✉)
Institute of Surface-Earth System Science, Tianjin University, Tianjin, China

College of Resources & Environmental Sciences, Nanjing Agricultural University, Nanjing, China
e-mail: yuguanghui@njau.edu.cn

B. Giri et al. (eds.), *Root Biology*, Soil Biology 52,
https://doi.org/10.1007/978-3-319-75910-4_5

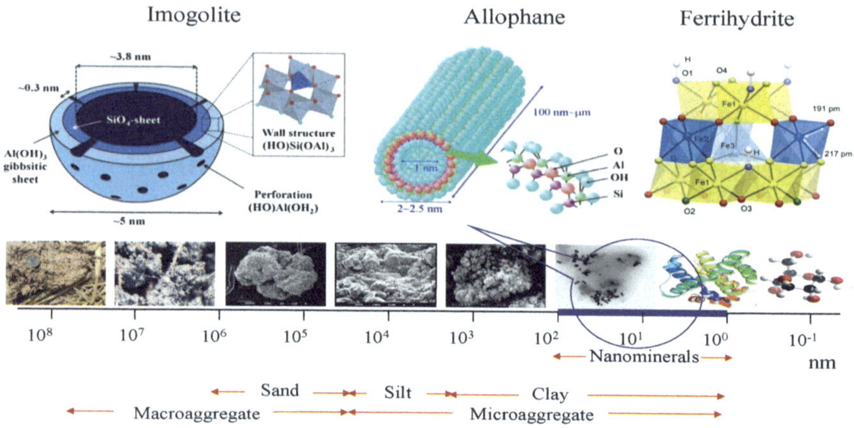

Fig. 5.1 The focused scales in soils

nanoparticles (Hochella et al. 2008). Nanominerals are defined here as minerals that only exist in this size range; that is, one will not find their equivalent at sizes larger than this, e.g., ferrihydrite, allophane, and imogolite. These nanominerals are also well known as short-range ordered (SRO) minerals. Mineral nanoparticles are minerals that can also exist in larger sizes, and these probably include most of all minerals (Hochella et al. 2008). What has been generally recognized more recently is that nanominerals and mineral nanoparticles commonly behave differently as a function of their size within the nano-size range. Recently, some investigations showed that management practices, e.g., organic amendments, can transfer bulk minerals to nano-size minerals for further C binding (Wen et al. 2014a; Xiao et al. 2016; Yu et al. 2017). These nano-size minerals behave different with other soil components, e.g., sand, silt, clay, macroaggregate, and microaggregate, which are shown in Fig. 5.1.

Two major mechanisms critically control the formation of nano-size minerals for C binding. First, soil physiochemical conditions, such as pH (Dakora and Phillips 2002; Scheel et al. 2007; Mimmo et al. 2014) and redox potential (Schwertmann and Cornell 2007; Colombo et al. 2014; Mimmo et al. 2014), and dissolution–precipitation processes regulate the release of mineral elements from primary minerals (Colombo et al. 2014). Second, both plants and microbes also affect the formation of biogenic minerals through their exudates and metabolic compounds (Colombo et al. 2014; Mimmo et al. 2014; Li et al. 2016). By delivering a continuous supply of individual exudate solutions through an artificial root into unperturbed soil, low molecular weight (LMW) acids have been shown to have strong metal-complexing abilities, increasing mineral dissolution by promoting the formation of nano-size minerals (Keiluweit et al. 2015).

In soil environments, the transformation of minerals (especially for iron minerals) may be dominated by the aging–rejuvenation cycle (Raiswell 2011). During mineral weathering, they first formed the young minerals (e.g., ferrihydrite), and then aggregated and/or attached young minerals will age and become increasingly crystalline such as hematitic and/or goethitic and non-bioavailable (Fig. 5.2). By

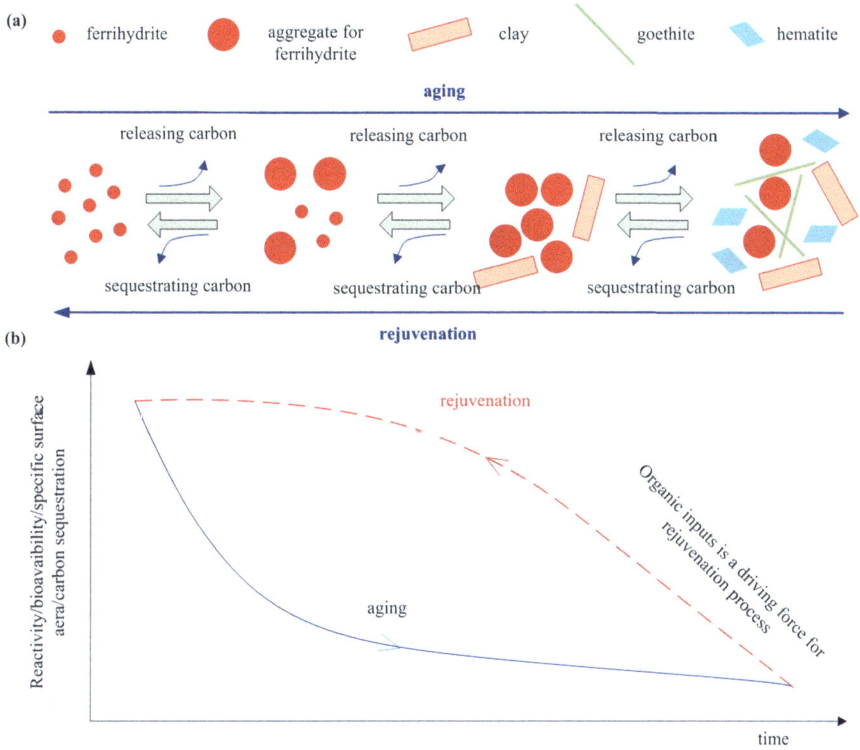

Fig. 5.2 The aging–rejuvenation transformations of minerals in the soils

contrast, the aged minerals can also be rejuvenated by physical and biological reworking. This recycling produces nano-size mineral again, which is initially labile and bioavailable, from more crystalline iron oxides in a process that can be termed rejuvenation. In this aging–rejuvenation cycle, root exudates and microbial communities in soils may play a critical role.

In this chapter, organic acids resulting from long-term organic amendments were shown to increase soil mineral dissolution/availability and the formation of SRO minerals and that SRO minerals facilitate C retention. Meanwhile, long-term chemical and organic fertilization regimes are shown to have a distinct effect on the Fe redox microbial community shifts and the Fe mineralogy in soils.

5.2 Root Exudates Promoted the Formation of Nano-size Minerals

The long-term field studies demonstrated that organic amendments significantly increased the availability of Al and Fe minerals, particularly their SRO phases (Wen et al. 2014a; Xiao et al. 2016; Yu et al. 2017). To explore the factors that

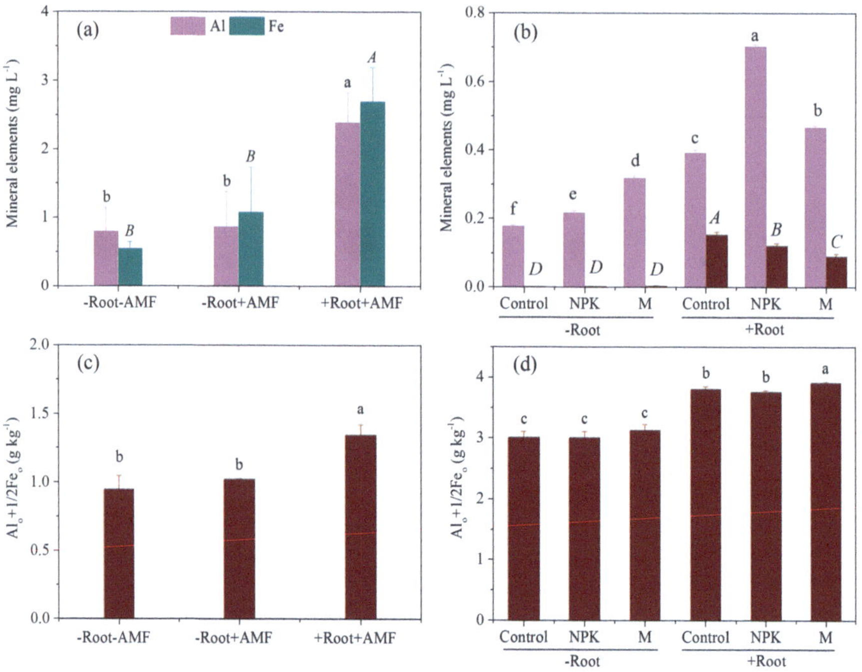

Fig. 5.3 Influence of roots and organic amendments on mineral dissolution and the formation of SRO minerals. (**a**) Effect of roots on the mobilized minerals. (**b**) Effect of roots and organic amendments on the mobilized minerals. (**c**) Effect of roots on the formation of SRO minerals. (**d**) Effect of roots and organic amendments on the formation of SRO minerals. –Root-AMF, not allowing either AMF hyphae or roots growing into the TEST compartments; –Root+AMF, allowing AM hyphae but not roots growing into the TEST compartments; +Root+AMF, allowing both AMF hyphae and roots growing into the TEST compartments; Control, no fertilizer inputs; NPK, chemical fertilizer inputs; M, manure inputs ($n = 3$) (Modified from Yu et al. 2017)

increase mineral availability, two microcosm studies were conducted, which allowed us to investigate the contribution of root and arbuscular mycorrhiza fungi (AMF) exudates as well as fertilizers on mineral availability and SRO mineral formation (Fig. 5.3). The presence of roots increased the release of Al and Fe from soils ($P < 0.05$) over two times for mean values with or without the application of fertilizers (Fig. 5.3a, b), but AMF had no significant impact on Al and Fe release ($P > 0.05$) (Fig. 5.3a). Interestingly, both microcosm studies demonstrated that the presence of roots also markedly increased the concentration of SRO minerals (Fig. 5.3c, d).

Compared to chemical fertilizers, organic amendments significantly decreased mineral mobilization (Fig. 5.3b) but increased (>20%, $P < 0.05$) the concentration of SRO minerals from 3.7 to 3.9 g kg^{-1} in the presence of roots (Fig. 5.3d). These results indicate that roots, in concert with organic amendments, may be responsible for mineral availability and the formation of SRO minerals.

Mineral availability driven by microbes or plants is believed as an important step (Vorhies and Gaines 2009). The microcosm experiments showed that plant roots and their exudates may play a bigger role than microbes (e.g., AMF) in the development of mineral availability and subsequent formation of SRO minerals (Fig. 5.3). This result also challenges the long-standing conceptual view that the weathering of minerals and the formation of SRO minerals are very slow processes and cannot be detected in a short-term system (Whalen et al. 2000; Colombo et al. 2014). For this point, it is suggested that the formation of SRO minerals can be accelerated or regulated by plant roots and some agricultural practices (e.g., organic amendments). In addition, decreased mineral mobility under organic amendments was probably attributable to the combination of mobilized minerals with organic materials as soil structures.

Addition of LMW organic acids (e.g., citric acid—one of the most abundant exudate classes) to soils benefits the formation of SRO minerals (Keiluweit et al. 2015). C 1s near edge X-ray absorption fine structure (NEXAFS) spectroscopy was used to identify the composition of organic C in soil colloids from all three field experiments (Figs. 5.4a and 5.5). Carboxyl C (1–π* transition of COOH) was dominant in soluble organic C, accounting for approximately 61% of the organic C at the Qiyang Experiment, while aromatic C (1s–π* transition of conjugated C=C) only constituted less than 5% of the organic C in the three field experiments (Fig. 5.4a, Table 5.1). As for the composition of organic C at the Park Grass and Broadbalk experiments, carboxyl C and aromatic C accounted for 35–44% and approximately 3% of the organic C, respectively (Fig. 5.5, Table 5.1). The other C forms were present as phenolic C, alkyl C, O-alkyl C, and carbonyl C (Table 5.1). In addition, long-term organic amendments also markedly increased the concentration of dissolved organic C when compared to chemical fertilization in all of the three experiments (Yu et al. 2017). LMW organic acids have in other experiments shown to consist approximately 0.5–5% of C in soil solution (Fox and Comerford 1990; van Hees et al. 2000). Therefore, long-term organic amendments may increase production of organic acids, especially with LMW components, in soils.

Furthermore, the mineral transformation experiments provided direct evidence that the formation of SRO minerals is promoted by LMW organic acids (Fig. 5.4), which may be produced by roots or the degradation of organic amendments. To test the critical role of root exudates in the formation of SRO minerals, we designed a simulated study by adding citric acid—one of the most abundant exudate classes—to soil colloids. The colloids were derived from soils with long-term organic amendments at the Qiyang Experiment. Iron k^3-weighted Extended X-Ray Absorption Fine Structure (EXAFS) spectra (Fig. 5.4b) showed that two peaks at $k = 5.7-6.0$ Å and $8.0-8.8$ Å, respectively, were observed in the raw soil colloids and those with 10 mg L^{-1} citrate addition but disappeared in the soil colloids with 100 mg L^{-1} citrate addition. These two peaks could be observed in goethite mineral standards but were not present in ferrihydrite (Fig. 5.4b). Linear combination fitting (LCF) results (Table 5.2) of the Fe k^3-weighted EXAFS spectra further demonstrated that incubation of soil colloids with citric acid at a concentration of 10 and 100 mg L^{-1} for 1 day could decrease goethite from 27.6 to 13% and 5.1% of the total Fe mineralogy but increase ferrihydrite from 39.4 to 49.9% and 74.6% of the iron mineralogy, respectively. Because ferrihydrite is more mobile and has a higher specific surface area than goethite (Raiswell 2011), these results provide spectroscopic evidence that

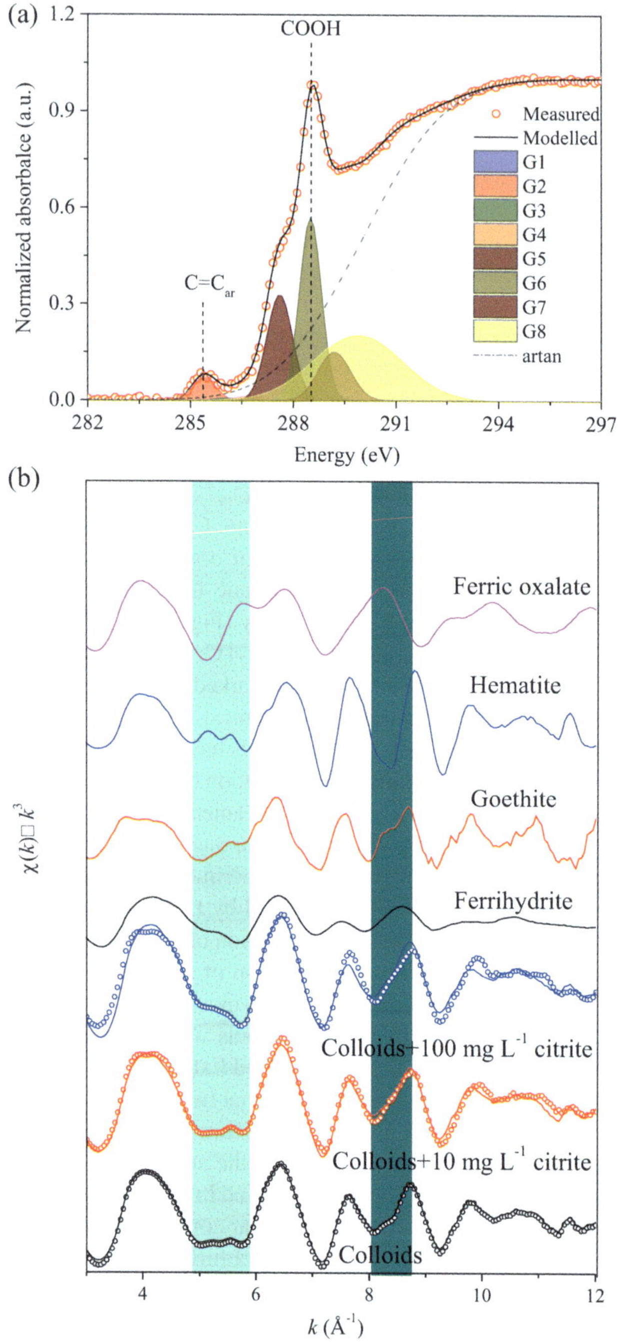

Fig. 5.4 Typical C 1s NEXAFS spectra for soil colloids (**a**) and the formation of SRO minerals following addition of citric acid (**b**) at the Qiyang Experiments. Note that G1–G8 represent eight Gaussian curves. Artan represents an arctangent step function. The specific C forms of G1–G8 are given in Table 5.1. *Open circles* indicate experimental data, and *solid lines* show the model fits (Modified from Yu et al. 2017)

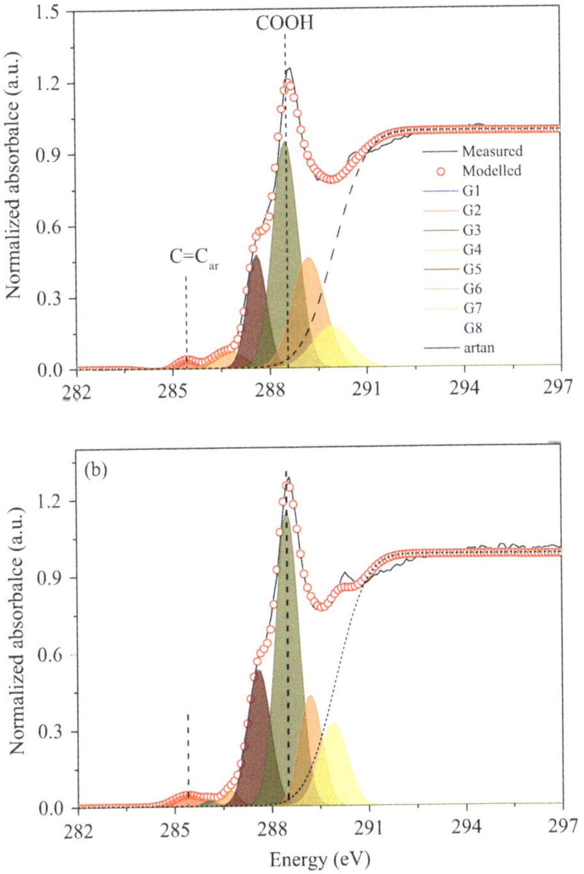

Fig. 5.5 Carbon 1s NEXAFS spectra for soil colloids from the Qiyang (**a**) and Broadbalk (**b**) experiments, respectively (Modified from Yu et al. 2017). Note that G1–G8 represent eight Gaussian curves. Artan represents an arctangent step function. The specific C forms of G1–G8 are given in Table 5.1

Table 5.1 Deconvolution results for C 1s NEXAFS on soil colloids from the long-term organic amendments at the Qiyang, Broadbalk, and Park Grass Experiments

	Proportion of absorption regions (%)					
	Aromatic C	Phenolic C	Alkyl C	Carboxylic C	O-alkyl C	Carbonyl C
	(263–286.1 eV)	(286.2–287.5 eV)	(287.6–288.3 eV)	(288.4–289.1 eV)	(289.2–289.8 eV)	(289.9–290.2 eV)
Experiment sites	G1+G2 +G3	G4	G5	G6	G7	G8
Qiyang	4.7	0.7	11.4	61.1	7.0	15.0
Broadbalk	2.8	1.7	20.2	44.2	16.0	15.0
Park Grass	3.3	1.5	13.7	34.9	11.0	35.6

Modified from Yu et al. (2017)

Table 5.2 Linear combination fit (LCF) results of Fe k^3-weighted EXAFS spectra of soil colloids in the stimulated studies (data are given as % values)

Samples	Ferrihydrite	Goethite	Hematite	Ferric oxalate
Soil colloids	39.4	27.6	24.2	8.8
Soil colloids + 10 mg L^{-1} citric acid	49.9	13.0	23.3	13.8
Soil colloids + 100 mg L^{-1} citric acid	74.6	5.1	15.9	4.5

Modified from Yu et al. (2017)

citric acid can increase mineral mobilization and promote transformation of goethite to ferrihydrite, the most reactive SRO iron (oxyhydr)oxide.

Similarly, oxalate, another common root exudates or intermediate of soil microbes, was also found to have the same effect on the dissolution of goethite (Cheah et al. 2003; Mimmo et al. 2014) and Al minerals (Li et al. 2006). Although these organic acids only account for a small percentage of soil soluble C (van Hees et al. 2000), they represent the most reactive forms of organic matter and exist widely in soils, especially in the rhizosphere (Li et al. 2006). Since ferrihydrite is more bioavailable than goethite (Raiswell 2011), this transformation is particularly important because the absence of iron in an available form limits C storage in many soils. Also, further investigations should be done on the effect of organic amendments on Fe mineral species in less Fe-rich soils. The newly formed SRO minerals may adsorb or precipitate on soil aggregates and promote soil aggregation (Zhou et al. 2013). The aggregation role of newly formed SRO minerals is also supported by the results from high-resolution transmission electron microscopy (HRTEM) combined with energy-dispersive X-ray spectroscopy (EDS) analysis that Al and Fe are enriched on the surface of soil particles with long-term organic amendments (Yu et al. 2012; Wen et al. 2014b; Xiao et al. 2015). The driving force for this aggregation may be the decrease in surface energy that appears to be low enough for SRO minerals (Waychunas et al. 2005). This increased soil aggregation lowers rates of respiration per unit of soil C, one of the main mechanisms of soil C storage and preservation (Scheel et al. 2007).

Except for microbes or plants, other parameters, i.e., pH, complexation, or, most importantly, redox variation, also affected mineral availability (Scheel et al. 2007; Schwertmann and Cornell 2007) and the formation of SRO minerals (Wen et al. 2014b). Several previous results from the Qiyang Experiment indicated that compared with chemical fertilization, organic fertilization significantly ($P < 0.05$) increased soil pH and the concentration of Al and Fe, and amorphous Al but decreased exchangeable Al (Yu et al. 2012; Wen et al. 2014a).

These available Al and Fe decrease the C mineralization and benefit for the SOC sequestration (Scheel et al. 2007; Schwertmann and Cornell 2007). The percentage of SRO minerals in organic-amended soil was higher than that of no fertilizer and chemical fertilizer inputs based on the previous results achieved from selective extraction methods (Wen et al. 2014a) and TEM analysis (Yu et al. 2012; Wen et al. 2014b) at the Qiyang Experiment. However, selective extraction method gives only an operational defined pool of SROs and suffers from intrinsic limitations owning to artifacts associated with reagent selectivity and inability to differentiate specific SROs (Kaiser and Zech 1996). X-ray absorption fine structure (XAFS) spectroscopy is a good addition to sequential extraction techniques because it provides direct identification of important SROs (Li et al. 2014; Yu et al. 2017). These SRO minerals possess structural defects, high specific surface area and charge density, and variably charged surfaces, enabling them to bind and thereby potentially chemically stabilize organic matter (Torn et al. 1997; Rasmussen et al. 2006). Although the importance of SRO minerals in protecting soil C has increasingly been recognized (Masiello et al. 2004; Rasmussen et al. 2006; Keiluweit et al. 2015), the information about their regulation is still very limited. Having a well-controlled long-term field system and advanced technologies allows investigators to identify weathering of soil primary minerals and the formation of SRO minerals.

5.3 Microbial Communities Determined the Consumption of Ferrihydrite

Soils under four long-term (23 years) different fertilizations served as the inoculum (1% vol: vol) to artificial groundwater (AGW) and incubated for 6 weeks. Similar patterns of Fe(II) accumulation and consumption were observed in all soils during the 6-week incubation (Fig. 5.6). During the Fe(III) reduction stage, Fe(II) was formed and increased in all treatments, but the extent of this increase in the NPK treatment incubation (by 1.1 g L^{-1}) was much higher than that in the organic fertilization (i.e., M and NPKM) (by ca. 0.5 g L^{-1}). Here, NPK represents inorganic nitrogen, phosphorus, and potassium fertilization, whereas NPKM denotes NPK plus swine manure fertilization, and M means swine manure fertilization. For the NO_3^--dependent Fe(II) oxidation stage, addition of NO_3^- at day 21 resulted in the rapid oxidation of Fe(II); Fe(II) increased quickly with the fast rate of oxidation rate observed in M and NPKM treatments and relatively slow oxidation rate in NPK treatment (Fig. 5.6). Total extractable Fe in all soils showed little difference within different treatments.

Nitrate was consumed during the initial 4 days of incubation, resulting in transient accumulation of NO_2^- and production of approximately NH_4^+ (Fig. 5.6). Addition of NO_3^- at day 21 resulted in the rapid decrease of NO_3^- and the production of NH_4^+ and NO_2^-, with lager rate observed in M and NPKM treatments, which is consistent with the changes in Fe(II) (Fig. 5.6).

To explore the bacterial community responsible for Fe(III) reduction and Fe (II) oxidation during the incubation experiment, the soil bacterial communities were examined at the initial stage of soil slurry, the end of Fe(III) reduction stage, and the end of Fe(II) oxidation stage using Illumina MiSeq sequencing. The dominant phyla across all samples were *Proteobacteria*, *Acidobacteria*, *Chloroflexi*, *Firmicutes*, *Bacteroidetes*, and *Actinobacteria*, accounting for more than 85% of the bacterial sequences from each soil (Fig. 5.7).

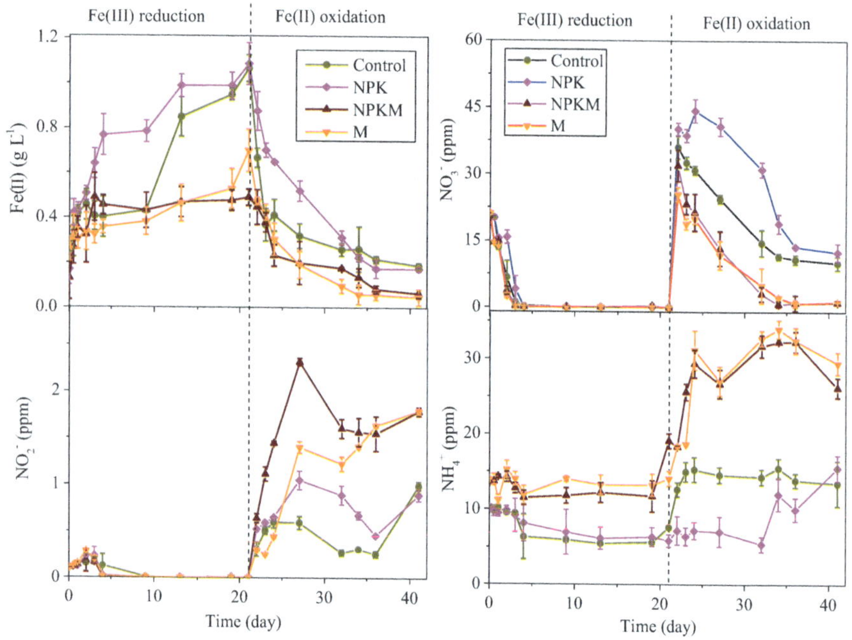

Fig. 5.6 Changes in 0.5 M HCl-extractable Fe(II) (A), NO_3^- (B), NO_2^- (C), and NH_4^+ (D) concentrations over time in the soil enrichment culture (Modified from Wen et al. 2018). *Error bars* indicate standard error of triplicate cultures

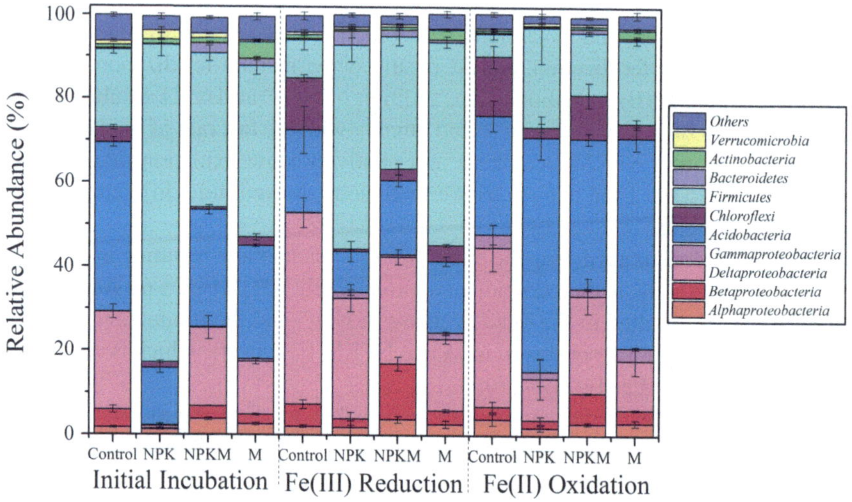

Fig. 5.7 Relative abundance of the dominant bacteria phyla (*proteobacterial* classes) at the initial incubation stage, the end of the Fe reduction stage, and the end of the Fe(II) oxidation stage in each sample. Relative abundances are based on the proportional frequencies of those DNA sequences that could be classified at the phylum (class) level

Furthermore, bacterial richness and diversity variation among different samples were examined using common biotic indices. Statistically significant differences in richness and diversity ($P < 0.05$) were observed for OTUs, ACE, Chao1, and Shannon (Table 5.3). Both the bacterial diversity indexes (i.e., the Shannon index) and richness indexes (i.e., ACE and Chao1) were significantly larger in organic fertilized soils (i.e., M and NPKM) than those of chemical fertilized soils (NPK) during the different incubation stages.

In addition, the variations of microbial composition were also determined among the different samples using unweighted UniFrac distance matrixes. Nonmetric multidimensional scaling (NMDS) plot (Fig. 5.8) revealed that soil samples from the same fertilization regime were clustered closely together but those from the different fertilization regime (i.e., chemical fertilization and organic fertilization) was clearly divided in bacterial community composition. Furthermore, no significant difference was found along the incubation time for the control, NPKM, and M fertilization. However, the bacterial communities from NPK-treated soils were well differentiated across the incubation time along NMDS axi1.

Five genera related to Fe(III) reduction bacteria were found during the microcosm experiment and stimulated in Fe(III) reduction stage, including *Geobacter*, *Clostridium*, *Desulfovibrio*, *Desulfosporosinus*, *Desulfitobacterium*, and *Bacillus* (Fig. 5.9a). The relative abundances of *Geobacter* increased significantly during Fe(III) reduction stage, indicating that *Geobacter* plays an important role in Fe(III) reduction. Compared with control, NPKM, and M, this extent increased much more in NPK. During Fe(II) oxidation stage, three genera related to Fe(II) oxidation bacteria were enriched (Fig. 5.9b) and were shown to have significantly higher relative abundance in organic fertilization treatments than chemical fertilization at the end of the Fe (II) oxidation stage.

Furthermore, the oxidation state and structure of Fe during the microcosm experiment were analyzed using Fe K-edge EXAFS. Qualitatively, the first shell of Fe in all examined samples was similar to that of standard ferrihydrite, but the second shell showed some differences (Fig. 5.10a). Specifically, the peak at 2.5 $Å^{-1}$ was higher in NPK at both the reduction and oxidation stages (i.e., NPK-R and NPK-O) than that of other samples and the standard ferrihydrite. This peak was more similar to that from lepidocrocite and goethite, indicating a lager extent transformation of ferrihydrite in NPK fertilization regime during the incubation experiment. To confirm this result, we next do the linear combination fitting (LCF) of Fe k^3-weighted spectra to determine the amount of ferrihydrite transformation (Table 5.4). The LCF results showed that at the end of Fe (III) reduction stage, approximately 41% of ferrihydrite was transformed for chemical fertilization regime and only less than 29% of ferrihydrite was transformed for organic fertilization regimes. Meanwhile, more unstable Fe species (i.e., lepidocrocite) were formed under chemical fertilization regime than organic fertilization regimes (i.e., goethite). During the Fe oxidation stage, the ferrihydrite was produced in different rate, with only 65% under chemical fertilization regime but over 81% under organic fertilization regimes.

Table 5.3 Shifts in OTUs, richness, and diversity over the incubation time in different fertilization regimes

			Richness[b]		Diversity[c]
Incubation stage	Fertilization regime	OTUs[a]	ACE	Chao1	Shannon
Initial incubation	Control	1265 ± 51 e	1590 ± 65 cd	1595 ± 78 c	6.61 ± 0.11 bc
	NPK	568 ± 36 g	736 ± 39 e	764 ± 40 d	4.93 ± 0.20 f
	NPKM	1785 ± 117 b	2253 ± 92 b	2262 ± 94 b	7.34 ± 0.16 a
	M	2096 ± 310 a	2517 ± 188 a	2515 ± 218 a	7.32 ± 0.15 a
Fe(III) reduction	Control	1392 ± 91 cd	1782 ± 51 c	1788 ± 91 c	6.11 ± 0.18 cd
	NPK	1244 ± 172 e	1599 ± 172 cd	1626 ± 184 c	6.19 ± 0.59 cd
	NPKM	1531 ± 140 c	2185 ± 121 b	2111 ± 124 b	6.46 ± 0.26 bc
	M	2176 ± 74 a	2598 ± 73 a	2609 ± 108 a	6.69 ± 0.21 b
Fe(II) oxidation	Control	1258 ± 108 e	1720 ± 134 cd	1758 ± 163 c	5.95 ± 0.35 de
	NPK	1006 ± 119 f	1549 ± 163 cd	1532 ± 149 c	5.07 ± 0.53 f
	NPKM	1455 ± 137 cd	2134 ± 151 b	2105 ± 198 b	6.16 ± 0.28 cd
	M	1552 ± 232 c	2166 ± 201 b	2149 ± 270 b	5.56 ± 0.19 e

Note: Significant differences between the fertilization treatments were determined using one-way ANOVA followed by Duncan's multiple range test at $P < 0.05$, in which the conditions of normality and homogeneity of variance were met. The data are shown as the means ± SD ($n = 4$). Abbreviations: Control, no fertilization; NPK, inorganic nitrogen, phosphorus, and potassium fertilization; NPKM, NPK plus swine manure; M, swine manure

[a]OTUs: operational taxonomic units (97% similarity)

[b]Based on the Chao1 and abundance-based coverage estimator (ACE) richness indices

[c]Based on Shannon diversity indices

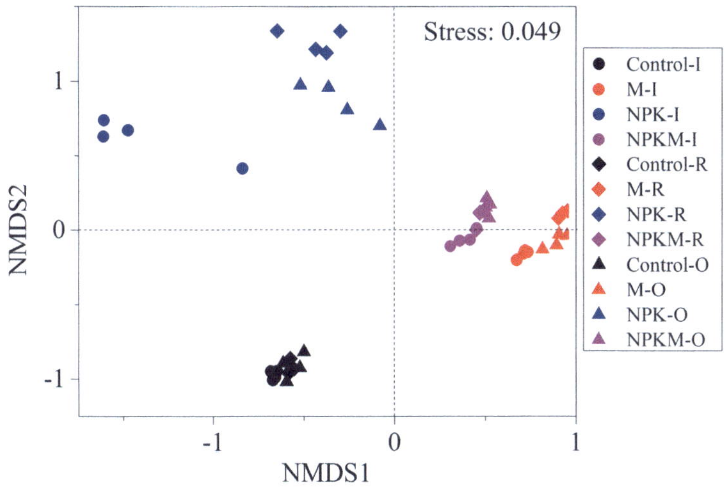

Fig. 5.8 Changes in bacterial community composition, as assessed with OTUs using high-throughput sequencing and analyzed by nonmetric multidimensional scaling (NMDS) of phylogenetic similarity (UniFrac)

It is initially hypothesized that long-term contrasting fertilization regimes have a distinct effect on the Fe redox microbial community shifts and the Fe redox cycles. The results from the incubation experiment demonstrated that during the anaerobic incubation stage, significant Fe(III) consumption and formation were found in the presence of ferrihydrite in all treatments but apparently affected in opposite way between the NPK fertilization and organic fertilizations (i.e., M and NPKM) when compared with control (Figs. 5.6, 5.7; Table 5.3). Chemical fertilization enhanced Fe (III) reduction, which is similar to a previous research in paddy soils (Ding et al. 2015). However, for the first time, it is demonstrated that organic fertilization inhibited the Fe(III) reduction significantly (Figs. 5.6, 5.7; Table 5.3) and played a relatively positive role in Fe(II) oxidation process. This result may be attributable to the shift in microbial community as suggested by 16S rRNA-based pyrosequencing (Table 5.3). Specifically, microbes had a higher abundance and diversity in organic fertilization treatments than those in chemical fertilization treatment (Table 5.3). At genus level, we found that *Geobacter*, a well-known branch of dissimilatory Fe reducers (Lovley and Phillips 1988; Childers et al. 2002), was apparently enriched and showed significantly higher abundance in chemical fertilization treatment during the Fe(III) reduction stage (Fig. 5.8). In addition, organic fertilization treatments have significantly higher concentration of *Pseudomonas* (Fig. 5.9), which has been reported as a kind of Fe oxide (Straub et al. 1996). These results provided direct evidence to support our hypothesis. Shift of Fe redox cycling microbe communities responding to long-term fertilization regime may be related to changes of soil characteristics. Numerous studies have reported that proton concentrations, soil dissolved organic carbon (DOC), could alter soil microbe communities (Högberg et al. 2006; Reeve et al. 2010; Hartmann et al. 2015). However, it needs more future

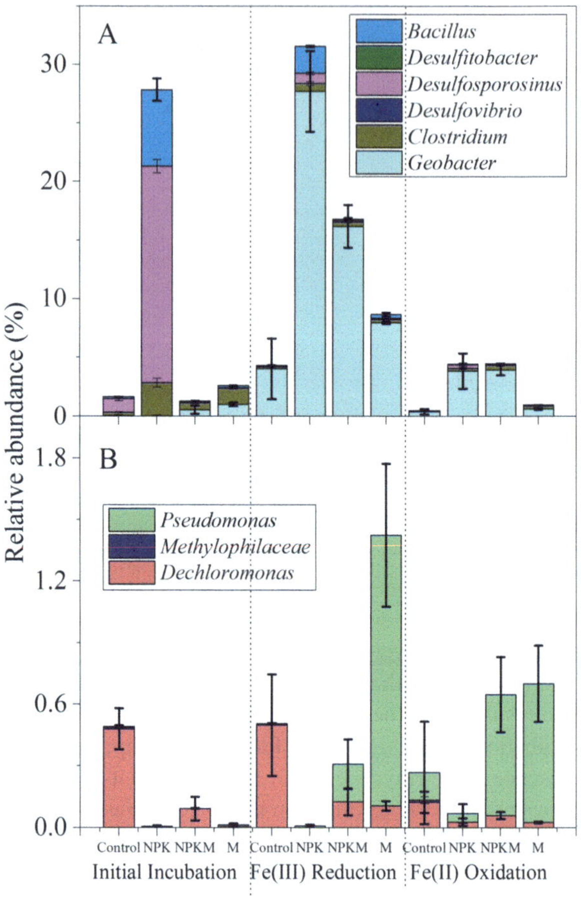

Fig. 5.9 Relative abundance of putative dissimilatory Fe(III)-reducing bacterial (**a**) and Fe (II)-oxidizing bacterial (**b**) community structures at the genus level at the initial incubation stage, the end of the Fe(III) reduction stage, and the end of the Fe (II) oxidation stage in each sample. Abundance is expressed as the average percentage of targeted sequences out of the total high-quality bacterial sequences of samples from triplicate plots from each fertilization treatment. *Error bars* indicate standard error of triplicates

work to identify the relationship between shift of Fe redox bacteria community and the soil characteristics driven by long-term fertilization.

The EXAFS data from the incubation experiment suggested that during the anaerobic incubation stage, compared to control, chemical (NPK), and organic (i.e., M and NPKM) fertilizations, regimes had an opposite influence on Fe mineralogy, that is, after a 6-week incubation, more poorly ordered ferrihydrite were formed in organic fertilization treatments (i.e., M and NPKM), whereas more crystalline Fe minerals were found in NPK treatment. It is well established that transient redox fluctuations have a significant impact on Fe(III) oxide mineralogy (Ferris 2005; Thompson et al. 2006; Coby et al. 2011; Posth et al. 2014). The biogenic Fe (hydr) oxides, produced as a result of Fe oxidation, are always poorly crystalline forms of ferrihydrite (Ferris 2005; Emerson et al. 2010). However, some studies showed that some environmental factors, for example, geochemical conditions of the medium and the rate of Fe(II) oxidation, can subtly alter the mineral structure, mineral surface area, and overall reactivity (Chan et al. 2004; Mikutta et al. 2008; Emerson et al. 2010; Larese-Casanova et al. 2010; Baumgartner and Faivre 2015). Chemical fertilization

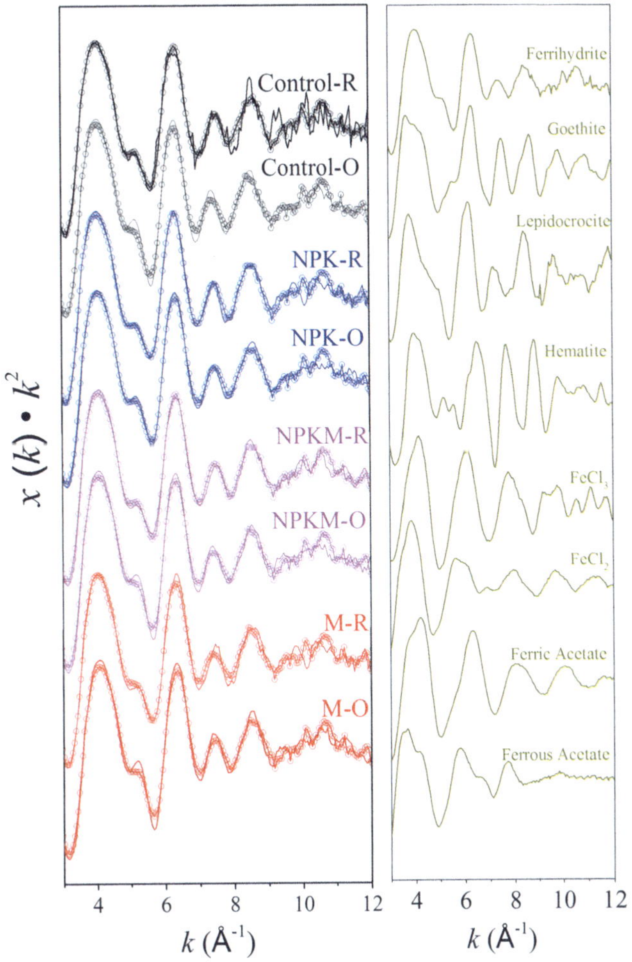

Fig. 5.10 Fits of the Fe k^2-weighted spectra of samples at the end of the Fe(III) reduction stage (-R) and the end of the Fe(II) oxidation stage (-O) and standards used in the fitting process

consumed more ferrihydrite at the end of Fe(III) reduction stage (Table 5.4), which might be due to the influence of *Geobacter*, as significantly more abundance of *Geobacter* was detected in NPK treatment after the Fe(III) was reduced (Fig. 5.10). In addition, more poorly ordered ferrihydrite yielded in organic fertilization treatments (i.e., M and NPKM) at the end of Fe(II) oxidation stage, probably owing to higher abundance of *Pseudomonas* activated by the addition of NO_3^-. During the incubation experiment, the main factor that influenced the Fe(III) (hydr)oxide mineralogy is the rate of Fe(II) oxidation, which is dominated by the *Pseudomonas* in the tested soils. Similar to our results, Senko et al. also found that in cultures of nitrate-reducing and Fe(II)-oxidizing bacteria, a stronger goethite signal was identified, and a larger proportion of Fe(III) was in the crystalline form as the Fe (II) oxidation rate decreased (Senko et al. 2005). Therefore, the different abundance

Table 5.4 Linear combination fitting (LCF) of the EXAFS models of the adsorbed Fe spectra with the contributions (in %) of the various components required to achieve the best fit[a]

Incubation stage	Fertilization regime	Ferrihydrite	Goethite	Lepidocrocite	Hematite	$Fe_2(SO_4)_3$	Ferrous oxalate	*R* factor ($\times 10^{-3}$)
Fe reduction	Control	69.0	7.3	23.7				0.235
	NPK	58.6		29.4	6.6		5.4	0.117
	NPKM	70.9	18.8			10.3		0.233
	M	74.2	16.1			9.7		0.100
Fe oxidation	Control	80.4		17.0		1.5	1.1	0.085
	NPK	65.0		26.2		5.9	2.9	0.379
	NPKM	81.2	8.4			7.6	2.8	0.403
	M	82.6	9.0			7.8	0.6	0.063

[a]The fits were conducted over K-range 3–12 $Å^{-1}$, based on the spectra of ferrihydrite, goethite, lepidocrocite, hematite, $Fe_2(SO_4)_3$, and ferrous oxalate. Individual fractions normalized to a sum of 100% are reported, together with the effective sum of the fitted fractions. Reference and sample spectra and the LCF fits are shown in Fig. 5.5. This table provides the proportion of the newly formed nano-precipitates for each sample. Abbreviations: Control, no fertilization; NPK, inorganic nitrogen, phosphorus, and potassium fertilization; NPKM, NPK plus swine manure; M, swine manure

of *Geobacter* and *Pseudomonas* during our incubation experiment are responsible for the different Fe(III) (hydr)oxide mineralogies in chemical and organic fertilization treatments. Besides, it is well established that over time, amorphous Fe phase could gradually transform to more stable and more crystalline phases, which could be affected by soil characteristics, for example, pH, SOC, and the presence of solutes (Schwertmann 1966; Das et al. 2010; Lalonde et al. 2012; Shimizu et al. 2013). High concentrations of SOC might stabilize amorphous Fe minerals by incorporating into the network structure of SRO minerals and thus inhibiting further growth of reactive Fe minerals to their crystalline counterpart (Cai et al. 2011; Yu et al. 2012; Huang et al. 2016). This might be another reason that soils under organic fertilizations have more amorphous Fe as the application of manure to soils results in increased SOC concentrations significantly. Also, decreased soil pH under NPK fertilization could reduce the nutrient availability, especially the P availability in red soil, which might stimulate the transformation of ferrihydrite (Weber et al. 2006).

Higher level of amorphous Fe(III) oxide in organic fertilization treatment might favor Fe-reducing bacterial growth because of higher substrate and nutrient availability (Ding et al. 2015), whereas we tested a low abundance of *Geobacter* in the organic fertilization during the Fe(III) reduction stage. That might be owing to the fact that some other abiotic factors under field conditions control the Fe(III) reducer's growth more than the substrate concentration, but it need more evidence in the future work.

To date, most studies focus on the Fe redox cycling process in hydromorphic environment, i.e., freshwater and marine sediments (Weber et al. 2006; Duckworth et al. 2009; Luef et al. 2013) and paddy soils (Li et al. 2011; Yi et al. 2013; Ding et al. 2015), in which periodic redox changes a lot. However, our controlled laboratory incubation experiments provided a valuable insight into the microbial reduction of Fe(III) and oxidation of Fe(II) in the red soil with a typical cropping rotation system of wheat with corn. The microbial-dominant Fe redox cycling occurs probably due to the considerable amount of Fe in the examined soils and the anoxic microclimate gradient created by the interaction among plants roots, soils, and the aggregation of soil particles. In summary, our findings demonstrated that long-term fertilization has the potential to regulate the Fe redox bacteria community and results in different Fe (III) mineralogies in these soils. These findings may be meaningful for other coupling elements cycling of Fe in soils, e.g., carbon and nitrogen cycles.

5.4 Retention of Carbon by Nano- or Submicron-size Minerals

To verify the strong retention capability for C by the mobilized Al and Fe minerals, an isotopic labeling experiment (using ^{13}C-labeled amino acid) combined with nanoscale secondary ion mass spectrometry (nanoSIMS) observation was designed (Fig. 5.11). After 24 h of incubation with a ^{13}C amino acid mixture, the composite nanoSIMS image

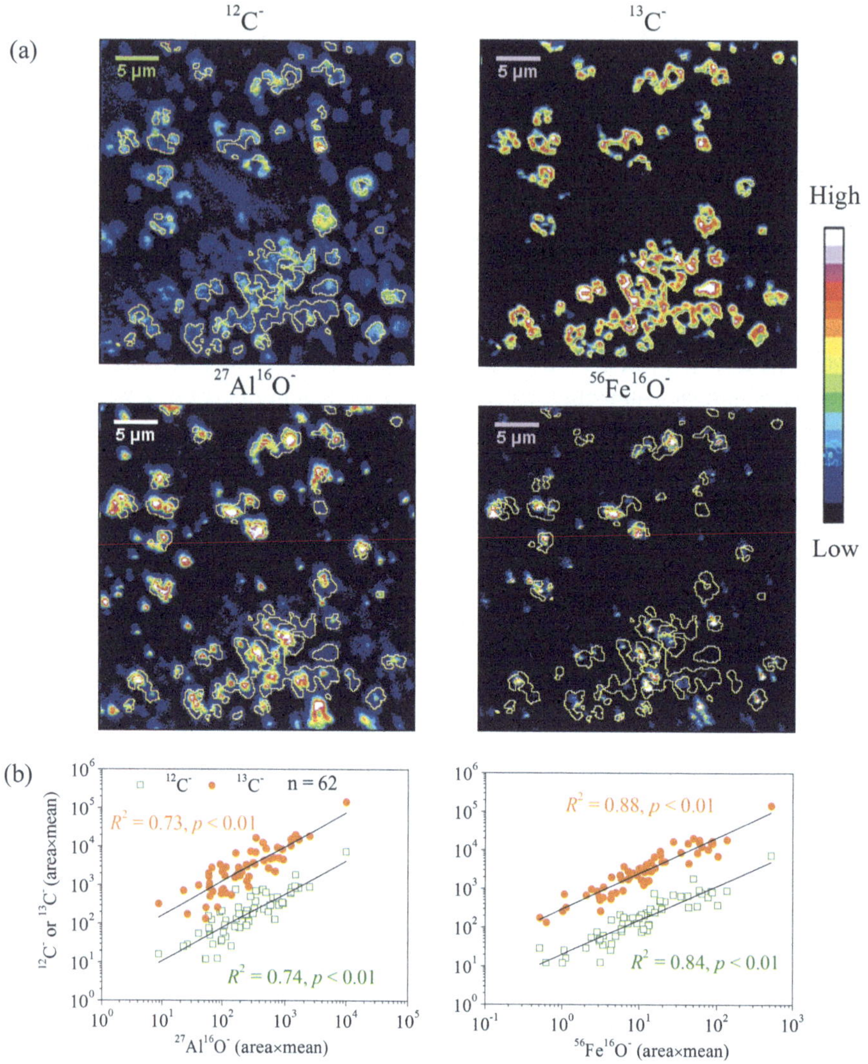

Fig. 5.11 Isotopic labeling experiment illustrated retention of C by Al and Fe minerals. (**a**) Element distribution map of $^{12}C^-$, $^{13}C^-$, $^{27}Al^{16}O^-$, and $^{56}Fe^{16}O^-$ following 24 h incubation of ^{13}C-labeled amino acid with soil colloids from long-term organic amendment samples of the Qiyang Experiment. (**b**) Linear relationship between $^{12}C^-$, $^{13}C^-$, and $^{27}Al^{16}O^-$, $^{56}Fe^{16}O^-$ ion counts from nanoSIMS secondary ion analyses

showed a profound enrichment of newly added $^{13}C^-$ on $^{27}Al^{16}O^-$ and $^{56}Fe^{16}O^-$, which served as "nuclei" for the retention of $^{13}C^-$. Line profiles further indicated that the distribution patterns of $^{13}C^-$, $^{12}C^-$, $^{27}Al^{16}O^-$, and $^{56}Fe^{16}O^-$ were similar.

It is hypothesized that the mobilized mineral particles in soils with organic amendments have a strong capability to retain C in soils. To test this hypothesis,

Yu et al. (2017) examined the distribution patterns of native C, newly added C, and minerals in soil colloids to support the C storage potential of the mobilized mineral particles in soil colloids (Yu et al. 2017). The nanoSIMS results indicate that native C and newly added C are co-localized best with minerals (Fig. 5.11). By contrast, it has recently been shown that only a limited proportion (<19%) of the clay-sized surfaces contributes to organic C stabilization (Vogel et al. 2014), indicating that using the amount of clay as a proxy to predict the storage potential of soil C is not sufficient. Meanwhile, some authors demonstrated that the particle surface area covered by SOM decreased with increasing fraction density, as the proportion of aggregated particles decreased (Hatton et al. 2015). Together, our nanoSIMS results are supporting our hypothesis that the mobilized mineral particles in soil colloids have a strong capability to retain labile C in soils. By combining long-term data from a grassland biodiversity experiment and radiocarbon (^{14}C) modeling, some investigators demonstrated that the increase in soil C storage is mainly limited by the integration of new C into soil and less by the decomposition of pre-existing soil C (Lange et al. 2015), suggesting that the protection of new C plays a major role in soil C storage. For example, this mineral binding labile C has recently been shown to markedly contribute to the formation of SOM (Cotrufo et al. 2015). This binding of labile C on mineral surfaces has also been described in terms of a layer-by-layer "onion" model (Chasse et al. 2015). Due to the surface reactivity of mineral particles varying as a function of particle size (Hochella et al. 2008), soil colloids, composed of the mobilized submicron- and nanoscale mineral particles with high reactivity, deserve more attention.

The micro-scale binding sites and distribution of mineral elements and different organic functional groups within soil colloid particles were characterized jointly using micro-X-ray fluorescence (μ-XRF) spectromicroscopies, electron probe microanalysis (EPMA), and synchrotron radiation (SR)-based Fourier transform infrared (SR-FTIR) (Fig. 5.12). In general, mineral elements of interest (i.e., Fe, Al, and Si) and functional groups were heterogeneously distributed at the micro-scale. Specifically, μ-XRF spectromicroscopy demonstrated that a high concentration of Fe was distributed on the edge of soil particles. This result is further supported by EPMA mapping, which showed an Fe spatial distribution pattern similar to that obtained from μ-XRF spectromicroscopy. In addition, EPMA mapping also showed that Fe had a similar distribution pattern with Al but not Si at the micro-scale.

SR-FTIR spectromicroscopy provided information about the distribution pattern of functional groups in the soil particles. For the small particle (Fig. 5.12, Region a) and the edge of the large soil particle (Fig. 5.12, Region b), all the functional groups had the same distribution pattern with Fe, revealing that all of these functional groups can contribute to the binding of Fe. However, certain C forms (but not all) showed clearly discernible spatial patterns with Fe in the edge of the large soil particle (Fig. 5.12, Region b). For instance, aromatic C (1613 cm^{-1}) and aliphatic C (1465 cm^{-1}) had a discernible spatial relationship with Fe. And only clay O-H, as well as mineral elements (Silicates Si-O, 1100 cm^{-1}; clay Si-O, 965 cm^{-1}; Al-O, 900 cm^{-1}), had a similar distribution pattern with Fe.

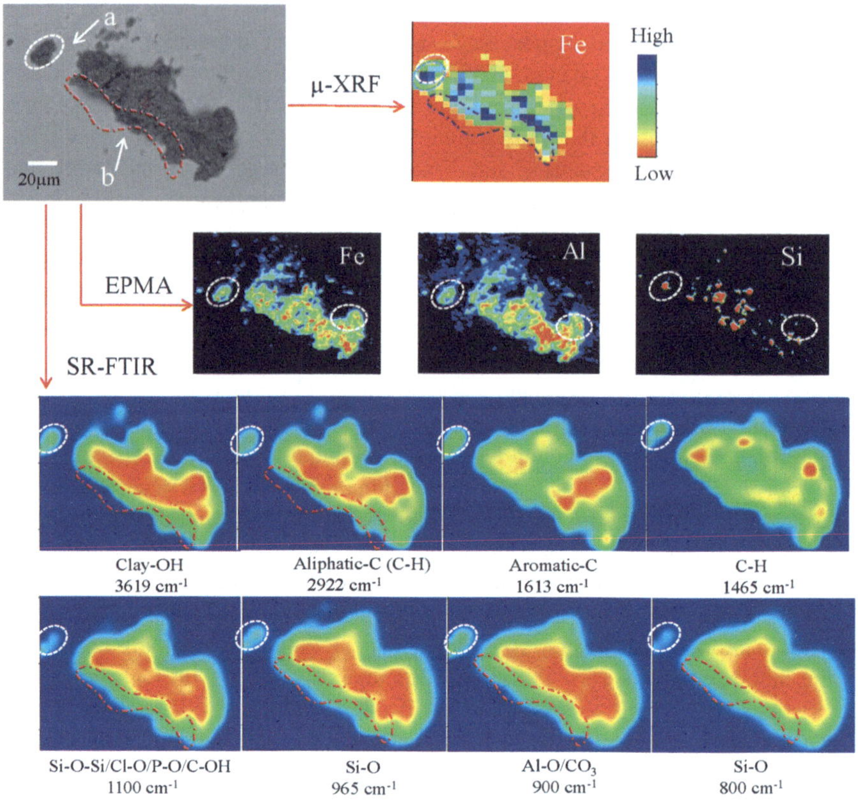

Fig. 5.12 Correlative SR-FTIR, μ-XRF, and EPMA analysis of the thin section from the manure-treated soil with the addition of Cd. (a, b) represent the selected regions of interest (ROI) (Modified from Sun et al. 2017). The *color scale* is a relative scale for each peak height and does not allow quantitative comparisons between peaks

5.5 Conclusions and Future Perspectives

Taken together, this chapter illustrates linkages among organic acids from root exudates and organic inputs, SRO minerals, and soil C stability within field and incubation experiments. Continuous organic amendments initialize a positive feedback loop, in which high organic inputs liberate minerals that can promote C sequestration in soils. The liberated minerals in the soil colloids, and hence the high content of SRO minerals formed by organically growth-limited precipitation, are therefore expected to be key factors that control the storage of soil C. More importantly, this chapter also provides a pathway for regulating mineral availability and the formation of SRO minerals in the field, which will be beneficial for managing the global C cycle. Therefore, organic amendments may represent practical tools for managing and increasing global terrestrial C stocks. Meanwhile, fertilization regimes were linked with the microbially mediated Fe redox reactions in soils. This chapter demonstrates that long-term chemical and organic fertilization regimes have a distinct influence on the shifts of Fe redox microbial community and the corresponding Fe mineralogy in soils. More amorphous Fe minerals are presented in soils under organic fertilizations, whereas more crystalline Fe minerals are presented in soil under NPK fertilization, which to a large extent is a result of shifts in Fe redox bacteria community. Long-term NPK fertilization enhanced the acetate-assimilating Fe(III)-reducing bacteria and promotes Fe(III) reduction in the incubation experiment, resulting in the more consumption of ferrihydrite. Meanwhile, soils under long-term organic fertilization have more abundance of Fe (II) oxides, producing relatively more amorphous Fe oxides, which is closely interacted with soil C cycling and contributes to SOC storage. In summary, these findings may prove vital in our understanding of C cycling in a changing climate.

Acknowledgments We thank Xiangzhi Zhang and Lijuan Zhang for the help and support at the BL08U1 beamline, Jingyuan Ma at the BL14W1 beamline, Jichao Zhang at the BL15U1 beamline, and Jiajia Zhong, Xiaojie Zhou, and Yuzhao Tang at the BL01B1 of the Shanghai Synchrotron Radiation Facility (SSRF). This work was funded by the National Key Research and Development Program of China (2016YFD0200301) and the National Natural Science Foundation of China (41371248).

References

Baldock JA, Skjemstad JO (2000) Role of the soil matrix and minerals in protecting natural organic materials against biological attack. Org Geochem 31:697–710

Baumgartner J, Faivre D (2015) Iron solubility, colloids and their impact on iron (oxyhydroxide) formation from solution. Earth Sci Rev 150:520–530

Cai P, He X, Xue A, Chen H, Huang Q, Yu J et al (2011) Bioavailability of methyl parathion adsorbed on clay minerals and iron oxide. J Hazard Mater 185:1032–1036

Chan CS, De Stasio G, Welch SA, Girasole M, Frazer BH, Nesterova MV et al (2004) Microbial polysaccharides template assembly of nanocrystal fibers. Science 303:1656–1658

Chasse AW, Ohno T, Higgins SR, Amirbahman A, Yildirim N, Parr TB (2015) Chemical force spectroscopy evidence supporting the layer-by-layer model of organic matter binding to Iron (oxy)hydroxide mineral surfaces. Environ Sci Technol 49:9733–9741

Cheah S-F, Kraemer SM, Cervini-Silva J, Sposito G (2003) Steady-state dissolution kinetics of goethite in the presence of desferrioxamine B and oxalate ligands: implications for the microbial acquisition of iron. Chem Geol 198:63–75

Childers SE, Ciufo S, Lovley DR (2002) Geobacter metallireducens accesses insoluble Fe(III) oxide by chemotaxis. Nature 416:767–769

Coby AJ, Picardal F, Shelobolina E, Xu H, Roden EE (2011) Repeated anaerobic microbial redox cycling of iron. Appl Environ Microbiol 77:6036–6042

Colombo C, Palumbo G, He J-Z, Pinton R, Cesco S (2014) Review on iron availability in soil: interaction of Fe minerals, plants, and microbes. J Soils Sediments 14:538–548

Cotrufo MF, Soong JL, Horton AJ, Campbell EE, Haddix ML, Wall DH, Parton WJ (2015) Formation of soil organic matter via biochemical and physical pathways of litter mass loss. Nat Geosci 8:776–779

Dakora F, Phillips D (2002) Root exudates as mediators of mineral acquisition in low-nutrient environments. Plant Soil 245:35–47

Das S, Hendry MJ, Essilfie-Dughan J (2010) Transformation of two-line ferrihydrite to goethite and hematite as a function of pH and temperature. Environ Sci Technol 45:268–275

Ding LJ, Su JQ, Xu HJ, Jia ZJ, Zhu YG (2015) Long-term nitrogen fertilization of paddy soil shifts iron-reducing microbial community revealed by RNA-(13)C-acetate probing coupled with pyrosequencing. ISME J 9:721–734

Duckworth OW, Holmström SJM, Peña J, Sposito G (2009) Biogeochemistry of iron oxidation in a circumneutral freshwater habitat. Chem Geol 260:149–158

Emerson D, Fleming EJ, McBeth JM (2010) Iron-oxidizing bacteria: an environmental and genomic perspective. Annu Rev Microbiol 64:561–583

Ferris F (2005) Biogeochemical properties of bacteriogenic iron oxides. Geomicrobiol J 22:79–85

Fox TR, Comerford NB (1990) Low-Molecular-Weight organic acids in selected forest soils of the Southeastern USA. Soil Sci Soc Am J 54:1139–1144

Hartmann M, Frey B, Mayer J, Mader P, Widmer F (2015) Distinct soil microbial diversity under long-term organic and conventional farming. ISME J 9:1177–1194

Hatton PJ, Remusat L, Zeller B, Brewer EA, Derrien D (2015) NanoSIMS investigation of glycine-derived C and N retention with soil organomineral associations. Biogeochemistry 125:303–313

Hochella MF Jr, Lower SK, Maurice PA, Penn RL, Sahai N, Sparks DL, Twining BS (2008) Nanominerals, mineral nanoparticles, and Earth systems. Science 319:1631–1635

Högberg MN, Högberg P, Myrold DD (2006) Is microbial community composition in boreal forest soils determined by pH, C-to-N ratio, the trees, or all three? Oecologia 150:590–601

Huang C, Liu S, Li R, Sun F, Zhou Y, Yu G (2016) Spectroscopic evidence of the improvement of reactive iron mineral content in red soil by long-term application of Swine manure. PLoS One 11:e0146364

Johnson K, Purvis G, Lopez-Capel E, Peacock C, Gray N, Wagner T et al (2015) Towards a mechanistic understanding of carbon stabilization in manganese oxides. Nat Commun 6:7628

Jones E, Singh B (2014) Organo-mineral interactions in contrasting soils under natural vegetation. Front Environ Sci 2:2

Kaiser K, Zech W (1996) Defects in estimation of aluminum in humus complexes of podzolic soils by pyrophosphate extraction. Soil Sci 16:452–458

Keiluweit M, Bougoure JJ, Nico PS, Pett-Ridge J, Weber PK, Kleber M (2015) Mineral protection of soil carbon counteracted by root exudates. Nat Climate Change 5:588–595

Kelleher BP, Simpson AJ (2006) Humic substances in soils: are they really chemically distinct? Environ Sci Technol 40:4605–4611

Kleber M, Eusterhues K, Keiluweit M, Mikutta C, Mikutta R, Nico PS (2015) Mineral-organic associations: formation, properties, and relevance in soil environments. In: Donald LS (ed) Advances in agronomy. Academic, Waltham, pp 1–140

Koegel-Knabner I, Guggenberger G, Kleber M, Kandeler E, Kalbitz K, Scheu S et al (2008) Organo-mineral associations in temperate soils: integrating biology, mineralogy, and organic matter chemistry. J Plant Nutr Soil Sci 171:61–82

Lalonde K, Mucci A, Ouellet A, Gélinas Y (2012) Preservation of organic matter in sediments promoted by iron. Nature 483:198–200

Lange M, Eisenhauer N, Sierra CA, Bessler H, Engels C, Griffiths RI et al (2015) Plant diversity increases soil microbial activity and soil carbon storage. Nat Commun 6:6707

Larese-Casanova P, Haderlein SB, Kappler A (2010) Biomineralization of lepidocrocite and goethite by nitrate-reducing Fe(II)-oxidizing bacteria: effect of pH, bicarbonate, phosphate, and humic acids. Geochim Cosmochim Acta 74:3721–3734

Lehmann J, Kleber M (2015) The contentious nature of soil organic matter. Nature 528:60–68

Lehmann J, Solomon D, Kinyangi J, Dathe L, Wirick S, Jacobsen C (2008) Spatial complexity of soil organic matter forms at nanometre scales. Nat Geosci 1:238–242

Li J, Xu R, Tiwari D, Ji G (2006) Mechanism of aluminum release from variable charge soils induced by low-molecular-weight organic acids: kinetic study. Geochim Cosmochim Acta 70:2755–2764

Li H, Peng J, Weber KA, Zhu Y (2011) Phylogenetic diversity of Fe(III)-reducing microorganisms in rice paddy soil: enrichment cultures with different short-chain fatty acids as electron donors. J Soils Sediments 11:1234–1242

Li W, Joshi SR, Hou G, Burdige DJ, Sparks DL, Jaisi DP (2014) Characterizing phosphorus speciation of Chesapeake bay sediments using chemical extraction, P NMR, and X-ray absorption fine structure spectroscopy. Environ Sci Technol 49:203–211

Li H, Hu SJ, Polizzotto ML, Chang XL, Shen QR, Ran W, Yu GH (2016) Fungal biomineralization of montmorillonite and goethite to short-rangeordered minerals. Geochim Cosmochim Acta 191:17–31

Lovley DR, Phillips EJ (1988) Novel mode of microbial energy metabolism: organic carbon oxidation coupled to dissimilatory reduction of iron or manganese. Appl Environ Microbiol 54:1472–1480

Luef B, Fakra SC, Csencsits R, Wrighton KC, Williams KH, Wilkins MJ et al (2013) Iron-reducing bacteria accumulate ferric oxyhydroxide nanoparticle aggregates that may support planktonic growth. ISME J 7:338–350

Masiello CA, Chadwick OA, Southon J, Torn MS, Harden JW (2004) Weathering controls on mechanisms of carbon storage in grassland soils. Global Biogeochem Cycles 18:GB4023

Mikutta C, Mikutta R, Bonneville S, Wagner F, Voegelin A, Christl I, Kretzschmar R (2008) Synthetic coprecipitates of exopolysaccharides and ferrihydrite. Part I: Characterization. Geochim Cosmochim Acta 72:1111–1127

Mimmo T, Del Buono D, Terzano R, Tomasi N, Vigani G, Crecchio C et al (2014) Rhizospheric organic compounds in the soil-microorganism-plant system: their role in iron availability. Eur J Soil Sci 65:629–642

Posth NR, Canfield DE, Kappler A (2014) Biogenic Fe(III) minerals: from formation to diagenesis and preservation in the rock record. Earth Sci Rev 135:103–121

Raiswell R (2011) Iron transport from the continents to the open ocean: the aging-rejuvenation cycle. Elements 7:101–106

Rasmussen C, Southard RJ, Horwath WR (2006) Mineral control of organic carbon mineralization in a range of temperate conifer forest soils. Global Change Biol 12:834–847

Reeve JR, Schadt CW, Carpenter-Boggs L, Kang S, Zhou J, Reganold JP (2010) Effects of soil type and farm management on soil ecological functional genes and microbial activities. ISME J 4:1099–1107

Scheel T, Dörfler C, Kalbitz K (2007) Precipitation of dissolved organic matter by aluminum stabilizes carbon in acidic forest soils. Soil Sci Soc Am J 71:64–74

Schmidt MW, Torn MS, Abiven S, Dittmar T, Guggenberger G, Janssens IA et al (2011) Persistence of soil organic matter as an ecosystem property. Nature 478:49–56

Schwertmann U (1966) Inhibitory effect of soil organic matter on the crystallization of amorphous ferric hydroxide. Nature 212:645–646

Schwertmann U, Cornell RM (2007) Iron oxides in the laboratory: preparation and characterization. Wiley-VCH, New York, pp 121–134

Senko JM, Dewers TA, Krumholz LR (2005) Effect of oxidation rate and Fe(II) state on microbial nitrate-dependent Fe(III) mineral formation. Appl Environ Microbiol 71:7172–7177

Shimizu M, Zhou J, Schröder C, Obst M, Kappler A, Borch T (2013) Dissimilatory reduction and transformation of ferrihydrite-humic acid coprecipitates. Environ Sci Technol 47:13375–13384

Stockmann U, Adams MA, Crawford JW, Field DJ, Henakaarchchi N, Jenkins M et al (2013) The knowns, known unknowns and unknowns of sequestration of soil organic carbon. Agric Ecosyst Environ 164:80–99

Straub KL, Benz M, Schink B, Widdel F (1996) Anaerobic, nitrate-dependent microbial oxidation of ferrous iron. Appl Environ Microbiol 62:1458–1460

Sulman BN, Phillips RP, Oishi AC, Shevliakova E, Pacala SW (2014) Microbe-driven turnover offsets mineral-mediated storage of soil carbon under elevated CO_2. Nat Climate Change 4:1099–1102

Sun FS, Polizzotto ML, Guan DX, Wu J, Shen QR, Ran W et al (2017) Exploring the interactions and binding sites between Cd and functional groups in soil DOM using two-dimensional correlation spectroscopy and synchrotron radiation based spectromicroscopies. J Hazard Mater 326:18–25

Thompson A, Chadwick OA, Rancourt DG, Chorover J (2006) Iron-oxide crystallinity increases during soil redox oscillations. Geochim Cosmochim Acta 70:1710–1727

Torn MS, Trumbore SE, Chadwick OA, Vitousek PM, Hendricks DM (1997) Mineral control of soil organic carbon storage and turnover. Nature 389:170–173

van Hees PAW, Lundström US, Giesler R (2000) Low molecular weight organic acids and their Al-complexes in soil solution-composition, distribution and seasonal variation in three podzolized soils. Geoderma 94:173–200

Vogel C, Mueller CW, Hoschen C, Buegger F, Heister K, Schulz S et al (2014) Submicron structures provide preferential spots for carbon and nitrogen sequestration in soils. Nat Commun 5:2947

Vorhies JS, Gaines RR (2009) Microbial dissolution of clay minerals as a source of iron and silica in marine sediments. Nat Geosci 2:221–225

Waychunas GA, Kim CS, Banfield JF (2005) Nanoparticulate iron oxide minerals in soils and sediments: unique properties and contaminant scavenging mechanisms. J Nanopart Res 7:409–433

Weber KA, Urrutia MM, Churchill PF, Kukkadapu RK, Roden EE (2006) Anaerobic redox cycling of iron by freshwater sediment microorganisms. Environ Microbiol 8:100–113

Wen YL, Xiao J, Li H, Shen QR, Ran W, Zhou QS et al (2014a) Long-term fertilization practices alter aluminum fractions and coordinate state in soil colloids. Soil Sci Soc Am J 78:2083–2089

Wen YL, Li H, Xiao J, Wang C, Shen QR, Ran W et al (2014b) Insights into complexation of dissolved organic matter and Al(III) and nanominerals formation in soils under contrasting fertilizations using two-dimensional correlation spectroscopy and high resolution-transmission electron microscopy techniques. Chemosphere 111:441–449

Wen YL, Xiao J, Liu FF, Goodman BA, Li W, Jia ZJ et al (2018) Contrasting effects of chemical and organic fertilization regimes on shifts in the Fe redox bacterial community. Soil Biol Biochem 117:56–67

Whalen JK, Chang C, Clayton GW, Carefoot JP (2000) Cattle manure amendments can increase the pH of acid soils. Soil Sci Soc Am J 64:962–966

Xiao J, Wen YL, Li H, Hao JL, Shen QR, Ran W et al (2015) In situ visualisation and characterisation of the capacity of highly reactive minerals to preserve soil organic matter (SOM) in colloids at submicron scale. Chemosphere 138:225–232

Xiao J, He XH, Hao JL, Zhou Y, Zheng LY, Ran W et al (2016) New strategies for submicron characterization the carbon binding of reactive minerals in long-term contrasting fertilized soils: implications for soil carbon storage. Biogeosciences 13:3607–3618

Yi W, You J, Zhu C, Wang B, Qu D (2013) Diversity, dynamic and abundance of Geobacteraceae species in paddy soil following slurry incubation. Eur J Soil Biol 56:11–18

Yu GH, Wu MJ, Wei GR, Luo YH, Ran W, Wang BR et al (2012) Binding of organic ligands with Al(III) in dissolved organic matter from soil: implications for soil organic carbon storage. Environ Sci Technol 46:6102–6109

Yu GH, Xiao J, Hu SJ, Polizzotto ML, Zhao FJ, McGrath SP et al (2017) Mineral availability as a key regulator of soil carbon storage. Environ Sci Technol 51:4960–4969

Zhou H, Peng X, Perfect E, Xiao T, Peng G (2013) Effects of organic and inorganic fertilization on soil aggregation in an Ultisol as characterized by synchrotron based X-ray micro-computed tomography. Geoderma 195–196:23–30

Chapter 6
Root Exudates Dominate the Colonization of Pathogen and Plant Growth-Promoting Rhizobacteria

Jun Yuan, Waseem Raza, and Qirong Shen

6.1 Introduction

The plant roots perform a multitude of functions, including anchorage, absorption of water and nutrients, storage of food reserves, and secretion of photosynthates as root exudates. The root–soil interface is the site of maximum biological and chemical activities in soil and is called rhizosphere. The term rhizosphere was first described by Lorenz Hiltner (1904) over a century ago and later was redefined by Pinton et al. (2001) as the soil zone that is influenced by plant roots along with the root tissues colonized by microorganisms. The rhizosphere is divided into three distinct zones: (1) rhizoplane, (2) endorhizosphere, and (3) ectorhizosphere (Lynch 1987). The plant roots excrete 5–21% of their photosynthates as root exudates in the form of soluble sugars, amino acids, or secondary metabolites (Fig. 6.1) (Badri et al. 2013b; Chaparro et al. 2013). Microorganisms use root exudates as substrates resulting in higher microbial biomass and activity around roots also called as rhizosphere effect. Generally, the root exudates are grouped into low molecular weight compounds such as amino acids, vitamins, organic acids, sugars, phenolic compounds, and other secondary metabolites and high molecular weight compounds, such as polysaccharides and proteins (Bais et al. 2006; Badri et al. 2009a).

The actual composition of root exudates varies with the type of plant cultivar, species, and developmental stages and environmental factors including soil type, pH, temperature, and microbial communities (Badri et al. 2009a; Uren 2000). In a study,

J. Yuan (✉) · W. Raza · Q. Shen
Jiangsu Provincial Key Lab for Organic Solid Waste Utilization, National Engineering Research Center for Organic-based Fertilizers, Jiangsu Collaborative Innovation Center for Solid Organic Waste Resource Utilization, Nanjing Agricultural University, Nanjing, China
e-mail: junyuan@njau.edu.cn

B. Giri et al. (eds.), *Root Biology*, Soil Biology 52,
https://doi.org/10.1007/978-3-319-75910-4_6

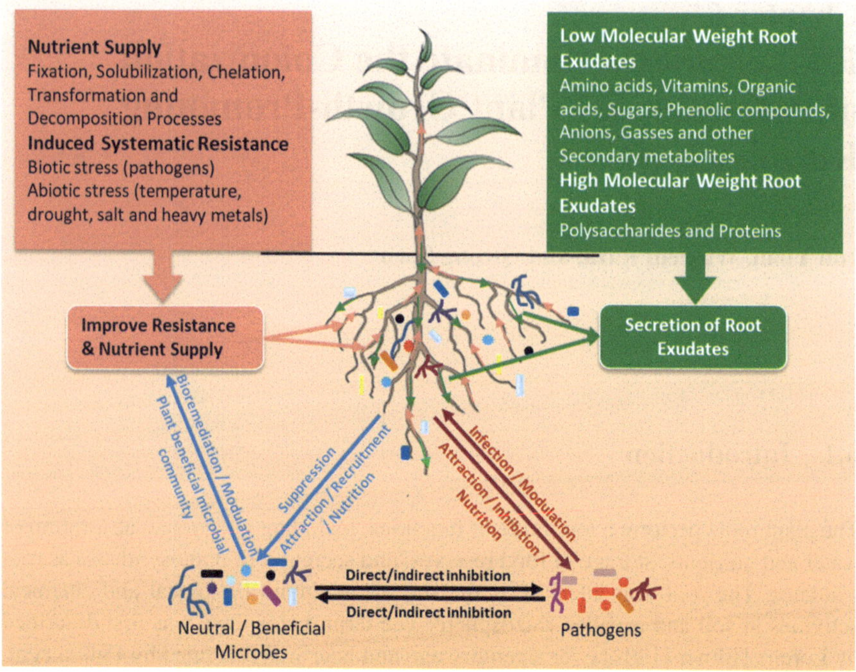

Fig. 6.1 Plant roots secrete low molecular weight exudates containing amino acids, vitamins, organic acids, sugars, phenolic compounds, anions, gasses, and other secondary metabolites, and high molecular weight exudates contain polysaccharides and proteins (Badri et al. 2009a; Kamilova et al. 2006b). The root exudates, as a food source, attract neutral and beneficial microbes and pathogens toward the rhizosphere (Badri et al. 2013a) and, in response, are modulated in composition, which causes the recruitment of beneficial microbes and suppression/inhibition of not greeted microbes and pathogens to avoid infection and non-beneficial microbial community (Zhang et al. 2009). The recruitment of microbes develops a beneficial microbial community in the rhizosphere, which affects plants in different ways. The beneficial microbes improve nutrient supply to plants by fixing nitrogen, solubilizing and chelating unavailable nutrients, decomposing organic matter, and transforming nutrients (Kiers et al. 2011; Zahran 1999). Further, the beneficial microbes directly inhibit pathogen growth and virulence traits by producing antibiotics, hydrolytic enzymes, and volatile organic compounds and indirectly inhibit pathogens by competition for nutrients and space to colonize plant roots and by inducing systematic resistance in plants (Benhamou et al., 2000; Sturz and Nowak 2000). In addition, beneficial microbes which induced systematic resistance also help plants to tolerate abiotic stresses like temperature, drought, salt, and heavy metals (Raaijmakers et al. 2009). The pathogens also try to compete with beneficial microbes for nutrients, root colonization, and survival by producing toxins, hydrolytic enzymes, etc. (Mercado-Blanco and Bakker 2007)

Kamilova et al. (2006a, b) found that tomato, cucumber, and sweet pepper root exudates contained higher amounts of organic acids mainly citric acid, succinic acid, and malic acid compared to sugars. The major sugars were found to be fructose and glucose. The tomato roots secreted higher levels of total organic acids when inoculated with a bacterial biocontrol agent *Pseudomonas fluorescens* WCS365, while the secretion of succinic acid was decreased (Kamilova et al. 2006b). On the other hand, when tomato roots were inoculated with a foot and root rot-causing pathogenic fungus *Fusarium oxysporum* f. sp. *radicis-lycopersici*, roots secreted less amount of citric acid, while the secretion of succinic acid was increased compared to uninoculated control (Kamilova et al. 2006b). When both biocontrol agent WCS365 and pathogen *F. oxysporum* were inoculated, then the disease severity was reduced along with the secretion of succinic acid. There are many studies that reported similar results indicating that the composition and availability of nutrients for microorganisms as root exudates are highly dynamic and not only microbial community affect the composition of root exudates but also the composition of root exudates affects the microbial community around roots. The differences in root exudates define and generate microbial communities having a certain degree of specificity for each plant species in the rhizosphere.

The plant roots secrete exudates by using both passive and active transport systems. The transport of small polar and uncharged molecules is conducted via direct passive diffusion and is dependent on membrane permeability, cytosolic pH, and polarity of the exuded compounds (Badri et al. 2009a). On the other hand, secondary metabolites, polysaccharides, and proteins are released by plant root cells using different membrane-bound transporter proteins, including the ATP-binding cassette (ABC) transporters (Loyola-Vargas et al. 2007; Sugiyama et al. 2008), the multidrug and toxic compound extrusion (MATE) family (Yazaki 2005), the major facilitator superfamily (Reddy et al. 2012), and the aluminum-activated malate transporter family (Weston et al. 2012). There is not much information available about the functions of these membrane-bound transport proteins; however, those have been reported to be associated with a wide range of compounds in the rhizosphere. For example, 25 ABC transporter genes were reported to be significantly overexpressed in the *Arabidopsis thaliana* (L.) Heynh. roots and played important roles in these secretion processes (Badri et al. 2008, 2009b). Other than ABC transporters, many substrates, like plant-derived alkaloids, toxic compounds, antibiotics, citrate anions, and phenolic compounds, are released by root cells using the electrochemical gradient of ions by MATEs active transporters (Weston et al. 2012). This significant role of MATEs active transporters has been reported for *Arabidopsis* (Li et al. 2002; Liu et al. 2009), sorghum (Magalhaes et al. 2007), barley (Furukawa et al. 2007), and rice (Ishimaru et al. 2011). In addition, a wide variety of plants possess specialized root cells containing many mitochondria, Golgi stacks, and Golgi-derived vesicles that are indicative of active secretion of metabolites. Other than that, root boarder cells are detached from the roots and become the part of mucilage surrounding the root surface (Hawes et al. 1998). The root border cells have been reported to be involved in many functions, including attraction and recruitment of beneficial microorganisms, reduction of sensitivity to heavy metals, and entrapment of pathogenic bacteria and nematodes in the mucilage surrounding the roots (Hawes et al. 1998; Miyasaka and Hawes 2001).

The root exudate-mediated interactions between plant roots and microbial communities in the rhizosphere have been extensively studied in the last decade (Badri et al. 2013a; Broeckling et al. 2008; Chaparro et al. 2013) (Fig. 6.1). The interactions of plant root-secreted phytochemicals vary from neutral to beneficial or deleterious and can be plant–plant, plant–microbe, and plant–fauna (Mercado-Blanco and Bakker 2007; Raaijmakers et al. 2009). All the components of root exudates play different roles but serve the same purpose to better plant growth and health. Phenolic and aldonic acids secreted by N_2-fixing legumes serve as major signals to *Rhizobiaceae* bacteria to initiate a symbiotic association, especially under nitrogen-limiting conditions (Coronado et al. 1995; Zhang et al. 2009). Some of the same compounds affect the development of mycorrhizal fungi that are crucial for phosphate uptake. These interactions are based on an equal supply of nutrients for both plant and mycorrhiza. In a study with *Medicago truncatula*, mycorrhiza provided the plant with more phosphorous, when it was provided with more carbon (Kiers et al. 2011). This fair trade was also observed with respect to N, as the mycorrhiza only provides the plant with N when it receives plant carbon (Fellbaum et al. 2012).

Under nutrient-limited environments, the plant roots excrete exudates in a way other than as symbiotic signals to attract soil microbes involved in nutrient procurement. The organic acids and extracellular enzymes present in root exudates release soil Ca, Fe, P, and Al phosphates from organic compounds, increase their solubility, and improve their availability through chelation (Dinkelaker and Marschner 1992; Goldstein et al. 1988; Lee 1988). Some plant nutrients also maintain electronic neutrality by releasing an excess of anions, including hydroxyl ions, especially nitrate. Legumes benefit from the reduced nitrogen in the root nodules and secrete a net excess of protons, which can markedly lower rhizosphere pH. This decrease in pH not only affects the growth and activity of several soil microbes but also decreases the availability of several nutrients. A few plants such as *Aspalathus linearis* L. actively modify their rhizosphere pH by extruding OH^- and HCO_3^- to facilitate growth in low-pH soils (pH 3–5). The microbes can transition from pathogenic to symbiotic depending upon the environmental conditions (Newton et al. 2010). Symbiotic nitrogen-fixing rhizobacteria can have a symbiotic or neutral interaction with plant roots depending upon the N levels of soil (Davidson and Robson 1986; Zahran 1999). Similarly, under N-limiting conditions, roots of legume plants secrete more flavones and flavonols to attract N-fixing rhizobacteria to initiate symbiosis (Coronado et al. 1995; Zhang et al. 2009). Strigolactones have been reported to be present in the root exudates of a wide range of different plants. Some studies indicated that strigolactones are specific signals for AMF but not general plant signals for fungi.

The microbial population can undergo temporary variations in their structure due to the selective pressure exerted by the plant root exudates and environmental conditions via exchange within local population and migration between distinct populations. Previous studies demonstrated that plants select their rhizosphere community by secreting root exudates to establish a habitat favorable for the plant (Broeckling et al. 2008; Houlden et al. 2008; Rudrappa et al. 2008). Smalla et al. (2001) observed plant-specific abundance and composition of rhizosphere bacterial community of strawberry, oilseed rape, and potato for 2 consecutive years, and differences became more pronounced in the second year. Similarly, Lemanceau et al. (1995) reported that flax and tomato roots affected *Pseudomonas* populations differentially, and rhizosphere populations differed from those in bulk soil. In addition, the plant development stages affect the production and then secretion of root exudates (Hamlen et al., 1972), which employ selection and repulsion of microbes in the rhizosphere (Burr et al. 1984; Miller et al. 1989) that varies in function of time due to plant age. Consequently, the development of more adapted microorganisms at different plant growth stages may be favored. Di Cello et al. (1997) found that *Burkholderia cepacia* populations associated with maize roots decreased significantly during plant development. Moreover, several studies on the rhizosphere microbial community structure of wheat and maize revealed that fast-growing bacteria (r-strategist) were predominant on young immature roots, whereas slow-growing bacteria (k-strategist) became predominant on mature roots (De Leij et al. 1995; Nacamulli et al. 1997). Young and Kucharek (1977) reported the increase of *Trichoderma viride* and *Fusarium moniliforme* counts in roots from the silking stage through subsequent growth stages. On the other hand, Windham and King (1983) found that fungi generally were isolated more frequently from roots of seedling stages than from roots of plants at silking stage. The capacity of bacteria and fungi to colonize internal tissues of plants could confer an ecological advantage over others that can only colonize plants epiphytically (Hallmann et al. 1997). Rhizosphere is widely accepted to be an important source of root endophytes, which play an important role in promoting plant growth and protecting plant against pathogens (Benhamou et al. 2000; Sturz and Nowak 2000). Frommel et al. (1993) reported that the high density of *Pseudomonas* spp. on and in potato roots was associated with the growth and yield stimulation of potato. Hinton and Bacon (1995) found that a corn endophyte *Enterobacter cloacae* was associated with the control of corn pathogen systematically.

6.2 Root Exudates Mediated Pathogen–Plant Interaction

There is an immense diversity of microbial life in the soil and rhizosphere, so the assessment of the abundance and diversity of microbial populations and their relationship with the ecological functions is very important (Kent and Triplett 2002; Torsvik and Øvreås 2002). The root colonization with soil microbes leads to multiple types of physical and chemical interactions between plants and microbes, which can be beneficial or neutral on one side and deleterious on other side if pathogen is involved (Mercado-Blanco and Bakker 2007; Raaijmakers et al. 2009). In addition, microorganisms can transit between pathogenic and symbiotic states depending on environmental conditions (Newton et al. 2010). The interactions between plant roots, soil, and microbes significantly alter soil physical and chemical properties, which in turn alter the microbial population in the rhizosphere (Nihorimbere et al. 2011).

The microbial population in and around roots includes bacteria, fungi, yeasts, and protozoa. The colonization of roots by microbes is an important step in the interaction between beneficial microbes and the host plant, as the PGPRs unable to colonize plant roots are often not useful under field conditions (Benizri et al. 2007; Bloemberg and Lugtenberg 2001). Similarly, rhizosphere colonization is also important as the first step in pathogenesis of soilborne pathogens. The root exudates not only attract beneficial microbes but also pathogens as a general signal. Ruan et al. (1995) reported that flavonoids present in the root exudates of legumes stimulate the germination of macroconidia of a pea and bean pathogen, *Fusarium solani.* According to Steinkellner et al. (2007), flavonoids seem not to be specific compounds in stimulating tomato wilt pathogen *F. oxysporum*. On the other hand, Straney et al. (2002) found the highest germination of *F. solani* f. sp. *pisi* by the pea root exudates and a partial stimulation with bean root exudates, while root exudates of soybean and non-legume crops did not show stimulatory activity. These results showed the host-specific effects linked with the presence of flavonoids in root exudates as the macroconidia germination of pathogen was not correlated with the nutrient levels. However, this specificity is not the general characteristic, as Steinkellner et al. (2005) found that not only tomato root exudates stimulated the microconidia germination of tomato wilt pathogen *F. oxysporum* but also the nonhost plants such as sweet pepper, bean, barley, tobacco, and cucumber stimulated microconidia germination, suggesting general signals for *F. oxysporum* stimulation in root exudates. Another function of flavonoids on fungal root pathogens has been documented for *F. oxysporum* f. sp. *dianthi*. Curir et al. (2001) isolated a kaempferide triglycoside in carnation stems and roots, involved in the resistance of carnation *to F. oxysporum* f. sp. *dianthi*, suggesting this flavonol as a typical phytoalexin, responsive to *F. oxysporum* pathogenic to carnation (Curir et al. 2010).

6.3 Root Exudates Mediated PGPR–Plant Interaction

The plant roots have been reported to recruit beneficial microbes. These beneficial microbes not only promote plant growth and protect plants from pathogens (Bent 2006; Loon et al. 2006) but also increase tolerance to abiotic stresses such as drought, nutrient deficiency (Yang et al. 2009), and heavy metal toxicity (Zhuang et al. 2007). Rudrappa et al.'s (2008) results showed that the infection of a bacterial leaf pathogen *Pseudomonas syringae* pv. *Tomato* (DC3000) in *Arabidopsis* results in the recruitment of the biocontrol strain *Bacillus subtilis* FB17 to the roots. The authors further showed that infected plant roots secrete higher amounts of malic acid which is a chemoattractant for biocontrol strain. These results showed that diseased plants signal for help in the rhizosphere. In another study, the causal agent of take-all in wheat *Gaeumannomyces graminis* var. *tritici* was suppressed after monoculturing of wheat for several years. The further experiments showed that this suppression was associated with a specific strain of fluorescent *Pseudomonas* spp. in the wheat rhizosphere that produced the antibiotic 2,4-diacetylphloroglucinol (DAPG) and successfully suppressed *Ggt* (Weller 2007; Kwak et al. 2012).

Further, the root exudates can also secrete antimicrobial defensive compounds. For example, root cultures of *Ocimum basilicum* infected with *Pythium ultimum* produce rosmarinic acid, an antimicrobial agent against a number of soilborne pathogens (Bais et al. 2002). In another study, it was shown that root-derived antimicrobial metabolites from *Arabidopsis* confer resistance to a variety of *P. syringae* pathovars (Bais et al. 2005). These secreted antimicrobial compounds might also influence rhizosphere microbial communities (Glandorf et al. 1997). Liu et al. (2017) reported that raffinose and tryptophan positively affect the root colonization of *F. oxysporum* f. sp. *cucumerinum* and *Bacillus amyloliquefaciens* SQR9, respectively. These results indicate that cucumber roots colonized by *F. oxysporum* f. sp. *cucumerinum* or SQR9 increase root secretion of tryptophan to strengthen further colonization of SQR9. In contrast, these colonized cucumber roots reduce raffinose secretion to inhibit root colonization of *F. oxysporum* f. sp. *cucumerinum*. The above-described results showed that root exudates can attract both beneficial microbes and pathogens but plant roots also systematically respond to the presence of beneficial microbes and attract those for betterment of plant health while suppressing pathogens by secreting antimicrobial compounds and recruiting beneficial microbes.

6.4 Root Exudates Mediated Pathogen–PGPR–Host Interaction

Root exudates as the inducer excite to irritate the bioactivities in the rhizospheric area, which affect the plant–microbe interaction. Root exudates were the medium taking part in the dynamic and co-evolutionary process between the plant root and soil microbe (Morgan et al. 2005). In our former experiment, *B. amyloliquefaciens* NJN-6 strain can better colonize on the banana root surface when the plant is grown in vermiculite under sterilized condition (Jun Yuan et al. 2013). Further investigation found that low molecular weight OAs in banana root exudates especially oxalic acid, malic acid, and fumaric acid take an important role in attracting PGPRs and inducing their colonization on their host root (Yuan et al. 2015a).

Root exudates can also be used as a tool to attract PGPRs when plant host is attacked by pathogens. Tritrophic interactions between beneficial microorganisms, plants, and pathogens in the rhizosphere have been more and more focused on, such as Rudrappa et al. (2008) reporting that *Arabidopsis* selectively secrete L-malic acid to enhance biofilm formation of the beneficial rhizobacterium *Bacillus subtilis* FB17 on the root when infected by a foliar pathogen. A similar result was also found in cucumber roots selectively secreting citric and fumaric acids to attract *B. amyloliquefaciens* SQR-9 colonization on the root when faced with the cucumber pathogen *Fusarium oxysporum* f. sp. *cucumerinum* (Liu et al. 2014). Recent studies found that when banana infected by *Fusarium oxysporum*, the abundance of phenolic acid including phthalic acid, salicylic acid, and cinnamic acid in root exudate was highly increased. Further evaluation indicated these phenolic acids could upregulate genes involved in biofilm formation and antifungal compound production of PGPR strain *B. amyloliquefaciens* NJN-6 (Yuan et al. unpublished data). Root exudates as

Fig. 6.2 Plant can release distinct compounds to recruit PGPRs when faced with pathogens' attack

the useful strategy take the important roles in host plant defense process by altering their component to recruit various functional microbes (Fig. 6.2).

6.5 Root Exudates Mediated Rhizosphere Community Assembling

Rhizosphere soil is a hotspot area in pedosphere. Root exudates cultured millions of soil microbes by supporting substrates and energy. The rhizosphere soil microbial community development is highly correlated with host plant root exudates (Fig. 6.3). It is wildly reported that rhizosphere soil microbial communities were significantly distinguished among plant species (Edwards et al. 2015). An important reason is that the root exudates are diverse with different genotypes of host plant. Researchers found that composition of root exudates differed among development points in even certain genotype, which definitely led to significantly different rhizosphere soil microbial community patterns (Chaparro et al. 2014; Yuan et al. 2015b). As known, rhizosphere is a thinner layer between roots and bulk soil (Andrade et al. 1997). The spatial position determined that the rhizosphere was affected by both roots and bulk soil. The microbial communities living in the rhizosphere would be also determined by host plant roots and bulk soil microbial community: bulk soil microbial communities as the microbial reservoir support the "seeds," and roots select and culture what they want to form in the rhizosphere soil microbial

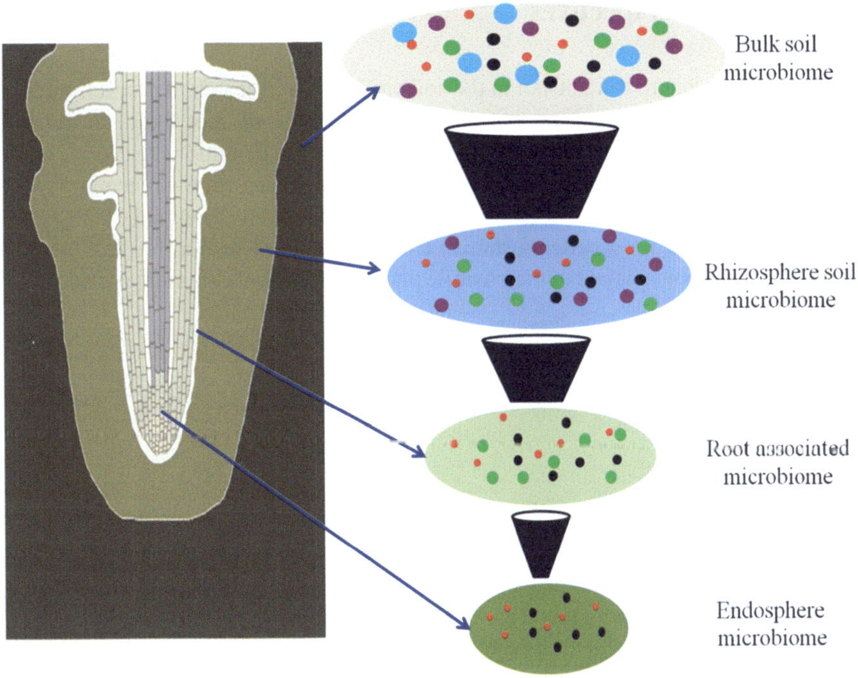

Fig. 6.3 Plant roots and its exudates showed the filtration effect (decrease the diversity) when assembling their microbial community

communities. In the selected and cultured process, as mentioned above, the root exudates take vital roles in the rhizosphere soil microbial community assembling.

Based on the knowledge, rhizosphere soil microbial community, closely linked to plant growth, can be modified by alteration of root exudate secretion. Scientists have found that changing the root exudate patterns by modifying ABC transporter could affect the rhizosphere microbiome (Badri et al. 2009a, b; Chaparro et al. 2013). Based on these experimental results, breeding scientists tended to culture disease-resistant cultivar not by altering enhance-resistant gene expression but by influencing the root exudate pattern in the recruitment of PGPRs (Wei and Jousset 2017). In fact, there are other strategies to alter root exudate pattern, such as inducement of PGPRs. Application of PGPRs and/or agents containing PGPRs, such as bioorganic fertilizers, can change the component of root exudate after colonization by PGPRs on the host roots and then modify the assembling of the rhizosphere soil microbiome (Fu et al. 2016; Wang et al. 2017; Xiong et al. 2017). Even colonization of pathogens could also affect rhizosphere soil microbiome dynamic; an outstanding scientific phenomenon is the development of natural disease-suppressive soil microbial community (Weller et al. 2002). Although lots of studies are working on the mechanism of the development of natural disease-suppressive soil, there are still many concerns unclear, though root exudates might take a vital role in the process.

6.6 Future Perspectives

Although rhizosphere is a hotspot in the soil and/or plant research, it is still limitedly known in many aspects. First of all, the components of root exudates remain indefinitely clear, especially when plant is under stress conditions. Second, compared with low molecular weight substrates of root exudates, high molecular weight compounds need to be paid more attention on their function among the rhizosphere interactions. Third, for a better sustainable development of modern agriculture, the role of root exudates played in the positive plant–soil feedback should be made clear in the near future.

References

Andrade G, Mihara KL, Linderman RG, Bethlenfalvay GJ (1997) Bacteria from rhizosphere and hyphosphere soils of different arbuscular-mycorrhizal fungi. Plant Soil 192(1):71–79

Badri DV, Loyola-Vargas VM, Broeckling CD, De-La-Peña C, Jasinaski M, Santelia D et al (2008) Altered profile of secondary metabolites in the root exudates of Arabidopsis ATP-binding cassette transporter mutants. Plant Physiol 146(2):762–771

Badri DV, Quintana N, El Kassis EG, Kim HK, Choi YH, Sugiyama A et al (2009a) An ABC transporter mutation alters root exudation of phytochemicals that provoke an overhaul of natural soil microbiota. Plant Physiol 151(4):2006–2017

Badri DV, Weir L, Lelie DVD, Vivanco JM (2009b) Rhizosphere chemical dialogues: plant–microbe interactions. Curr Opin Biotechnol 20(6):642–650

Badri DV, Chaparro JM, Zhang R, Vivanco JM (2013a) Application of natural blends of phytochemicals derived from the root exudates of Arabidopsis to the soil reveal that phenolic-related compounds predominantly modulate the soil microbiome. J Biol Chem 288(42):4502–4512

Badri DV, Zolla G, Bakker MG, Manter DK, Vivanco JM (2013b) Potential impact of soil microbiomes on the leaf metabolome and on herbivore feeding behavior. New Phytol 198(1):264–273

Bais HP, Walker TS, Schweizer HP, Vivanco JM (2002) Root specific elicitation and antimicrobial activity of rosmarinic acid in hairy root cultures of *Ocimum basilicum*. Plant Physiol Biochem 40(11):983–995

Bais HP, Prithiviraj B, Jha AK, Ausubel FM, Vivanco JM (2005) Mediation of pathogen resistance by exudation of antimicrobials from roots. Nature 434(7030):217–221

Bais HP, Weir TL, Perry LG, Gilroy S, Vivanco JM (2006) The role of root exudates in rhizosphere interactions with plants and other organisms. Annu Rev Plant Biol 57:233–266

Benhamou N, Gagné S, Le QD, Dehbi L (2000) Bacterial-mediated induced resistance in cucumber: beneficial effect of the endophytic bacterium serratia plymuthica on the protection against infection by Pythium ultimum. Phytopathology 90(1):45

Benizri E, Nguyen C, Piutti S, Slezack-Deschaumes S, Philippot L (2007) Additions of maize root mucilage to soil changed the structure of the bacterial community. Soil Biol Biochem 39(5): 1230–1233

Bent E (2006) Induced systemic resistance mediated by plant growth-promoting rhizobacteria (PGPR) and fungi (PGPF). In: Tuzun A, Bent E (eds) Multigenic and induced systemic resistance in plants. Springer, New York, pp 225–258

Bloemberg GV, Lugtenberg BJ (2001) Molecular basis of plant growth promotion and biocontrol by rhizobacteria. Curr Opin Plant Biol 4(4):343–350

Broeckling CD, Broz AK, Bergelson J, Manter DK, Vivanco JM (2008) Root exudates regulate soil fungal community composition and diversity. Appl Environ Microbiol 74(3):738

Burr TJ, Caesar A, Schrolh MN (1984) Beneficial plant bacteria. Crit Rev Plant Sci 2(1):1–20
Cello FD, Bevivino A, Chiarini L, Fani R, Paffetti D, Tabacchioni S et al (1997) Biodiversity of a Burkholderia cepacia population isolated from the maize rhizosphere at different plant growth stages. Appl Environ Microbiol 63(11):4485
Chaparro JM, Badri DV, Bakker MG, Sugiyama A, Manter DK, Vivanco JM (2013) Root exudation of phytochemicals in Arabidopsis follows specific patterns that are developmentally programmed and correlate with soil microbial functions. PLoS One 8(2):525–534
Chaparro JM, Badri DV, Vivanco JM (2014) Rhizosphere microbiome assemblage is affected by plant development. ISME J 8(4):790
Coronado C, Zuanazzi J, Sallaud C, Quirion JC, Esnault R, Husson HP et al (1995) Alfalfa root flavonoid production is nitrogen regulated. Plant Physiol 108(2):533–542
Curir P, Dolci M, Lanzotti V, Taglialatela-Scafati O (2001) Kaempferide triglycoside: a possible factor of resistance of carnation (Dianthus caryophyllus) to Fusarium oxysporum, f. sp. dianthi. Phytochemistry 56(7):717–721
Curir P, Dolci M, Galeotti F (2010) A phytoalexin-like flavonol involved in the carnation (Dianthus caryophyllus)-Fusarium oxysporum f.sp. dianthi pathosystem. J Phytopathol 153(2):65–67
Davidson IA, Robson MJ (1986) Effect of contrasting patterns of nitrate application on the nitrate uptake, N2-fixation, nodulation and growth of white clover. Ann Bot 57(3):331–338
De Leij LF, Sutton EJ, Whipps JM, Fenlon JS, Lynch JM (1995) Impact of field release of genetically modified pseudomonas fluorescens on indigenous microbial populations of wheat. Appl Environ Microbiol 61(9):3443
Dinkelaker B, Marschner H (1992) In vivo demonstration of acid phosphatase activity in the rhizosphere of soil-grown plants. Plant Soil 144(2):199–205
Edwards J, Johnson C, Santosmedellín C, Lurie E, Podishetty NK, Bhatnagar S et al (2015) Structure, variation, and assembly of the root-associated microbiomes of rice. Proc Natl Acad Sci U S A 112(8):E911
Fellbaum CR, Gachomo EW, Beesetty Y, Choudhari S, Strahan GD, Pfeffer PE et al (2012) Carbon availability triggers fungal nitrogen uptake and transport in arbuscular mycorrhizal symbiosis. Proc Natl Acad Sci U S A 109(7):2666–2671
Frommel MI, Nowak J, Lazarovits G (1993) Treatment of potato tubers with a growth promoting Pseudomonas sp.: Plant growth responses and bacterium distribution in the rhizosphere. Plant Soil 150(1):51–60
Fu L, Ruan Y, Tao C, Shen Q (2016) Continous application of bioorganic fertilizer induced resilient culturable bacteria community associated with banana Fusarium wilt suppression. Sci Rep 6: 27731
Furukawa J, Yamaji N, Wang H, Mitani N, Murata Y, Sato K et al (2007) An aluminum-activated citrate transporter in barley. Plant Cell Physiol 48(8):1081–1091
Glandorf DCM, Bakker PAHM, Loon LCV (1997) Influence of the production of antibacterial and antifungal proteins by transgenic plants on the saprophytic soil microflora. Plant Biol 46(1): 85–104
Goldstein AH, Baertlein DA, McDaniel RG (1988) Phosphate starvation inducible metabolism in Lycopersicon esculentum. I. Excretion of acid phosphatase by tomato plants and suspension-cultured cells. Plant Physiol 87:711–715
Hallmann J, Mahaffee WF, Kloepper JW, Quadthallmann A (1997) Bacterial endophytes in agricultural crops. Can J Microbiol 43(10):895–914
Hamlen RA, Lukezic FL, Bloom JR (1972) Influence of age and stage of development on the neutral carbohydrate. Can J Plant Sci 52(4):633–642
Hawes MC, Brigham LA, Wen F, Woo HH, Zhu Y (1998) Function of root border cells in plant health: pioneers in the rhizosphere. Annu Rev Phytopathol 36(1):311–327
Hiltner L (1904) Uber neuer erfahrungen und probleme auf dem gebiet der bodenbakteriologie unter besonderer berücksichtigung der gründüngung und brache. Arbeiten der Deutschen Landwirtschaftlichen Gesellschaf 32:1405–1417

Hinton DM, Bacon CW (1995) Enterobacter cloacae is an endophytic symbiont of corn. Mycopathologia 129(2):117

Houlden A, Timmswilson TM, Day MJ, Bailey MJ (2008) Influence of plant developmental stage on microbial community structure and activity in the rhizosphere of three field crops. FEMS Microbiol Ecol 65(2):193

Ishimaru Y, Kakei Y, Shimo H, Bashir K, Sato Y, Sato Y et al (2011) A rice phenolic efflux transporter is essential for solubilizing precipitated apoplasmic iron in the plant stele. J Biol Chem 286(28):24649

Kamilova F, Kravchenko LV, Shaposhnikov AI, Makarova N, Lugtenberg B (2006a) Effects of the tomato pathogen Fusarium oxysporum f. sp. radicis-lycopersici and of the biocontrol bacterium Pseudomonas fluorescens WCS365 on the composition of organic acids and sugars in tomato root exudate. Mol Plant Microbe Interact 19(10):1121–1126

Kamilova F, Kravchenko LV, Shaposhnikov AI, Azarova T, Makarova N, Lugtenberg B (2006b) Organic acids, sugars, and L-tryptophane in exudates of vegetables growing on stonewool and their effects on activities of rhizosphere bacteria. Mol Plant Microbe Interact 19(3):250

Kent AD, Triplett EW (2002) Microbial communities and their interactions in soil and rhizosphere ecosystems. Annu Rev Microbiol 56(1):211–236

Kiers ET, Duhamel M, Beesetty Y, Mensah JA, Franken O, Verbruggen E et al (2011) Reciprocal rewards stabilize cooperation in the mycorrhizal symbiosis. Science 333(6044):880–882

Kwak YS, Bonsall RF, Okubara PA, Paulitz TC, Thomashow LS, Weller DM (2012) Factors impacting the activity of 2,4-diacetylphloroglucinol-producing Pseudomonas fluorescens against take-all of wheat. Soil Biol Biochem 54(6):48–56

Lee RB (1988) Phosphate influx and extracellular phosphatase activity in barley roots and rose cells. New Phytol 109(2):141–148

Lemanceau P, Corberand T, Gardan L, Latour X, Laguerre G, Boeufgras J et al (1995) Effect of two plant species, flax (Linum usitatissinum L.) and tomato (Lycopersicon esculentum Mill.), on the diversity of soilborne populations of Fluorescent Pseudomonads. Appl Environ Microbiol 61(3):1004–1012

Li L, He Z, Pandey GK, Tsuchiya T, Luan S (2002) Functional cloning and characterization of a plant efflux carrier for multidrug and heavy metal detoxification. J Biol Chem 277(7):5360

Liu J, Magalhaes JV, Shaff J, Kochian LV (2009) Aluminum-activated citrate and malate transporters from the MATE and ALMT families function independently to confer Arabidopsis aluminum tolerance. Plant J Cell Mol Biol 57(3):389

Liu Y, Zhang N, Qiu M, Feng H, Vivanco JM, Shen Q et al (2014) Enhanced rhizosphere colonization of beneficial Bacillus amyloliquefaciens SQR9 by pathogen infection. FEMS Microbiol Lett 353(1):49

Liu Y, Chen L, Wu G, Feng H, Zhang G, Shen Q et al (2017) Identification of root-secreted compounds involved in the communication between cucumber, the beneficial bacillus amyloliquefaciens, and the soil-borne pathogen Fusarium oxysporum. Mol Plant Microbe Interact 30(1):53

Loon LCV, Geraats BPJ, Linthorst HJM (2006) Ethylene as a modulator of disease resistance in plants. Trends Plant Sci 11(4):184–191

Loyola-Vargas VM, Broeckling CD, Badri D, Vivanco JM (2007) Effect of transporters on the secretion of phytochemicals by the roots of Arabidopsis thaliana. Planta 225(2):301–310

Lynch JM (1987) The rhizosphere. Wiley Interscience, Chichester

Magalhaes JV, Liu J, Guimarães CT, Lana UG, Alves VM, Wang YH et al (2007) A gene in the multidrug and toxic compound extrusion (MATE) family confers aluminum tolerance in sorghum. Nat Genet 39(9):1156–1161

Mercado-Blanco J, Bakker PAHM (2007) Interactions between plants and beneficial Pseudomonas spp.: exploiting bacterial traits for crop protection. Antonie Van Leeuwenhoek 92(4):367–389

Miller HJ, Henken G, Veen JAV (1989) Variation and composition of bacterial populations in the rhizospheres of maize, wheat, and grass cultivars. Can J Microbiol 35(6):656–660

Miyasaka SC, Hawes MC (2001) Possible role of root border cells in detection and avoidance of aluminum toxicity. Plant Physiol 125(4):1978

Morgan JA, Bending GD, White PJ (2005) Biological costs and benefits to plant-microbe interactions in the rhizosphere. J Exp Bot 56(417):1729–1739

Nacamulli C, Bevivino A, Dalmastri C, Tabacchioni S, Chiarini L (1997) Perturbation of maize rhizosphere microflora following seed bacterization with Burkholderia cepacia MCI7. FEMS Microbiol Ecol 23(3):183–193

Newton AC, Fitt BDL, Atkins SD, Walters DR, Daniell TJ (2010) Pathogenesis, parasitism and mutualism in the trophic space of microbe-plant interactions. Trends Microbiol 18(8):365–373

Nihorimbere V, Ongena M, Smargiassi M, Thonart P (2011) Beneficial effect of the rhizosphere microbial community for plant growth and health. Biotechnologie Agronomie Société Et Environnement 15(2):327–337

Pinton R, Varanini Z, Nannipieri P, Pinton R, Varanini Z, Nannipieri P (2001) The rhizosphere. Biochemistry and organic substances at the soil-plant interface. Soil Sci Soc Am J 72(6): 339–353(15)

Raaijmakers JM, Paulitz TC, Steinberg C, Alabouvette C, Moënne-Loccoz Y (2009) The rhizosphere: a playground and battlefield for soilborne pathogens and beneficial microorganisms. Plant Soil 321(1–2):341–361

Reddy VS, Shlykov MA, Castillo R, Sun EI, Saier MH Jr (2012) The major facilitator superfamily (MFS) revisited. FEBS J 279(11):2022–2035

Ruan Y, Kotraiah V, Straney DC (1995) Flavonoids stimulate spore germination in Fusarium solani pathogenic in legumes in a manner sensitive to inhibitors of cAMP-dependent protein kinase. Mol Plant Microbe Interact 8(6):929–938

Rudrappa T, Czymmek KJ, Paré PW, Bais HP (2008) Root-secreted malic acid recruits beneficial soil bacteria. Plant Physiol 148(3):1547–1556

Smalla K, Wieland G, Buchner A, Zock A, Parzy J, Kaiser S et al (2001) Bulk and rhizosphere soil bacterial communities studied by denaturing gradient gel electrophoresis: plant-dependent enrichment and seasonal shifts revealed. Appl Environ Microbiol 67(10):4742–4751

Steinkellner S, Mammerler R, Vierheilig H (2005) Microconidia germination of the tomato pathogen Fusarium oxysporum in the presence of root exudates. J Plant Interact 1(1):23–30

Steinkellner S, Lendzemo V, Langer I, Schweiger P, Khaosaad T, Toussaint JP et al (2007) Flavonoids and strigolactones in root exudates as signals in symbiotic and pathogenic plant-fungus interactions. Molecules 12(7):1290–1306

Straney D, Khan R, Tan R, Bagga S (2002) Host recognition by pathogenic fungi through plant flavonoids. Adv Exp Med Biol 505:9

Sturz AV, Nowak J (2000) Endophytic communities of rhizobacteria and the strategies required to create yield enhancing associations with crops. Appl Soil Ecol 15(2):183–190

Sugiyama A, Shitan N, Yazaki K (2008) Signaling from soybean roots to rhizobium: an ATP-binding cassette-type transporter mediates genistein secretion. Plant Signal Behav 3(1): 38–40

Torsvik V, Øvreås L (2002) Microbial diversity and function in soil: from genes to ecosystems. Curr Opin Microbiol 5(3):240–245

Uren N (2000) Types, amounts, and possible functions of compounds released into the rhizosphere by soil-grown plants. In: Pinton R, Varanini Z, Nannipiero P (eds) The rhizosphere: biochemistry and organic substances at the soil-plant interface. Marcel Dekker, New York. English 1–21

Wang L, Li J, Yang F, E Y, Raza W, Huang Q et al (2017) Application of bioorganic fertilizer significantly increased apple yields and shaped bacterial community structure in orchard soil. Microb Ecol 73(2):1–13

Wei Z, Jousset A (2017) Plant breeding goes microbial. Trends Plant Sci 22(7):555

Weller DM (2007) Pseudomonas biocontrol agents of soilborne pathogens: looking back over 30 years. Phytopathology 97(2):250

Weller DM, Raaijmakers JM, Gardener MS, Thomashow LS (2002) Microbial populations responsible for specific soil suppressiveness to plant pathogens1. Annu Rev Phytopathol 40(1):309

Weston LA, Ryan PR, Watt M (2012) Mechanisms for cellular transport and release of allelochemicals from plant roots into the rhizosphere. J Exp Bot 63(9):3445

Windham MT, King SB (1983) Mycoflora of roots of maize plants at seedling and silking stages in Mississippi. Plant Dis 67(12):1366–1368

Xiong W, Guo S, Jousset A, Zhao Q, Wu H, Li R et al (2017) Bio-fertilizer application induces soil suppressiveness against Fusarium wilt disease by reshaping the soil microbiome. Soil Biol Biochem 114:238–247

Yang J, Kloepper JW, Ryu CM (2009) Rhizosphere bacteria help plants tolerate abiotic stress. Trends Plant Sci 14(1):1–4

Yazaki K (2005) Transporters of secondary metabolites. Curr Opin Plant Biol 8(3):301–307

Young TR, Kucharek TA (1977) Succession of fungal communities in roots and stalks of hybrid field corn grown in Florida. J Phys Soc Jpn 21(20):1866–1872

Yuan J, Ruan Y, Wang B, Zhang J, Waseem R, Huang Q et al (2013) Plant growth-promoting rhizobacteria strain Bacillus amyloliquefaciens NJN-6-enriched bio-organic fertilizer suppressed Fusarium wilt and promoted the growth of banana plants. J Agric Food Chem 61(16):3774–3780

Yuan J, Chaparro JM, Manter DK, hang R, Vivanco JM, Shen Q (2015a) Roots from distinct plant developmental stages are capable of rapidly selecting their own microbiome without the influence of environmental and soil edaphic factors. Soil Biol Biochem 89:206–209

Yuan J, Zhang N, Huang Q, Raza W, Li R, Vivanco JM et al (2015b) Organic acids from root exudates of banana help root colonization of PGPR strain *Bacillus amyloliquefaciens* NJN-6. Sci Rep 5(13438):121–127

Zahran HH (1999) Rhizobium-legume symbiosis and nitrogen fixation under severe conditions and in an arid climate. Microbiol Mol Biol Rev 63(4):968–989

Zhang J, Subramanian SG, Yu O (2009) Flavones and flavonols play distinct critical roles during nodulation of Medicago truncatula by Sinorhizobium meliloti. Plant J Cell Mol Biol 57(1):171

Zhuang X, Chen J, Shim H, Bai Z (2007) New advances in plant growth-promoting rhizobacteria for bioremediation. Environ Int 33(3):406–413

Chapter 7
Biocontrol of Soilborne Root Pathogens: An Overview

Pratibha Thakur and Ishwar Singh

7.1 Introduction

Conventional agricultural practices are now being replaced world over with modern practices for higher crop yields. For the same reason, the use of chemical fertilizers and pesticides has taken a big boost in agricultural sector. However, the excessive use of these synthetic chemicals is causing serious environmental and health threats. It is expected that in the long run, the negative impacts of applying these chemicals especially pesticides in agriculture would be far greater than the short-term benefits and much more severe hazardous consequences would be apparent in the near future. It has been reported that about three billion kilograms of pesticides are being used worldwide annually and are suspected to be responsible for over 220,000 deaths per year (Pimentel 2009). Therefore, some of the major challenges in twenty-first century are to reduce the chemical pollution in agriculture ecosystem, to produce healthy and sustainable crops and to enhance agricultural production for feeding growing human population.

One of the promising solutions which scientists are looking nowadays is the biological control that not only safeguards crops from the hazardous effects of various plant pathogens but also safeguards the environment and human health (Handelsman and Stabb 1996). This chapter is intended to provide an overview of different aspects of biocontrol methods emphasizing especially on the control of soilborne pathogens which cause root diseases in various plants of economic importance.

P. Thakur · I. Singh (✉)
Department of Botany, Hansraj College, University of Delhi, Delhi, India

B. Giri et al. (eds.), *Root Biology*, Soil Biology 52,
https://doi.org/10.1007/978-3-319-75910-4_7

7.2 Biological Control or Biocontrol

Biological control or Biocontrol term implies to the use of microbial antagonists to suppress various diseases. Biocontrol can be defined as the practice of introducing or managing populations of natural or genetically modified organisms against undesirable or harmful pests and pathogens of a plant in order to reduce their populations or at least slow down their reproductive rate (Stirling 2014). It is now being used as a promising strategy for pathogen management in order to maintain sustainable agricultural production while simultaneously reducing the harmful impacts of chemical pesticides in agricultural environment.

7.3 Why Go for Biocontrol Methods?

Reducing the dependence of growers on chemical pesticides by promoting biocontrol methods is one of the positive approaches toward integrated pest management (IPM). Biocontrol effects are not only characterized by suppression of pathogens but also by an increased tolerance and/or resistance against disease, higher growth, and higher yield of inoculated plants. Biological control using antagonistic microbes has many advantages in disease management such as little or no side effects, rare chances of resistance, long-term control on pests, best alternative to completely or substantially eliminate the use of synthetic pesticides, favorable cost/benefit ratio, nontoxic or harmless to plants and beneficial organisms, prevention of secondary diseases, and a vital part of integrated disease management (Guédez et al. 2008). It is an environment-friendly approach for the management of various pathogens, soil-borne pathogens in particular, and reducing plant diseases, which is already showing broad perspectives in agriculture.

The first and foremost requirement for successful pest management is the fertile soil-field, managed and maintained likely through good agricultural practices such as soil quality maintenance, crop rotations, and improved irrigation methods. The next line of defense is the use of healthy and disease-resistant varieties which could be developed through conventional breeding methods or advanced molecular techniques. Furthermore, some additional measures are still needed to grow productive and economically sustainable crops which are resistant to various diseases (Ponmurugan and Baby 2007). For this, biocontrol methods can be very promising as they do not have negative effects on environment and nontarget organisms (Guetsky et al. 2001, 2002; Jacobsen et al. 2004; Kergunteuil et al. 2016).

7.4 Biocontrol Agents

The significant role that biocontrol methods can play in reducing the activity and spread of phytopathogens is possible because of tremendous progress made in the field of plant–microbe interaction in the past few decades (Hussain et al. 2016). In any biocontrol method, selection of effective antagonistic organisms (BA) is the fundamental step. While selecting a biocontrol agent (BA), various parameters, namely, anti-pathogenic efficacy, adverse effect (if there is any) on the host or other beneficial organisms, mechanisms of inhibition involved, large-scale production, formulation, preservation conditions, shelf life, and application methods, are taken into consideration (Deketelaere et al. 2017).

The BAs include microorganisms such as viruses, bacteria, fungi, and nematodes which help in crop protection (Mehta et al. 2014). As compared to chemical insecticides/pesticides, BAs are cost-effective and most importantly environment-friendly. Additionally, pest resistance to BAs has been barely reported. However, the insecticidal markets are still dominated by chemical pesticides and biopesticides represent only 2% of that insecticidal market (Bravo et al. 2011). The importance of BAs against plant pathogens is hardly known as relatively very few products containing BAs are available in the market (Table 7.1), and therefore, the benefits of biocontrol methods are beyond the reach of farmers.

7.5 Formulation and Application of Biocontrol Agents

In order to have effective suppression of diseases, BAs can be applied majorly at three levels: first, direct application of biocontrol microorganisms to the specific (infection) site, at high population to obliterate the plant pathogen(s) (inundate application); second, application at one place but at lower proportions which on multiplication will then increase in population and gradually get spread to other plant parts and provide protection against pathogens (augmentative application); and third, one time or occasional application of biocontrol agent that maintains population of pathogens below threshold levels (Heydari and Pessarakli 2010; Nega 2014). Biocontrol microorganisms have been mostly used at soil or seed level. Some microorganisms have also been used in the form of composts in some plants.

7.5.1 Formulation

Formulation is one of the important techniques in commercial success of a biocontrol agent against plant pathogen. A formulated microbial product constitutes a product of one or more biological control agents (consortium) mixed with ingredients to improve its survival and effectiveness (Schisler et al. 2004). It requires the

Table 7.1 Some commercially available biopesticides registered in the USA

Trade name	Active biocontrol agent	Target disease/ organism	Crop	Manufacturer
Actino-Iron	*Streptomyces lydicus* WYEC108	*Pythium*, *Fusarium*, *Rhizoctonia solani*	Vegetables and ornamentals in green houses	Novozymes/ Monsanto BioAg
AtEze	*Pseudomonas chlororaphis* strain 63-28	*Pythium* spp., *Rhizoctonia solani*, and *Fusarium oxysporum*	Ornamental and vegetable crops	Agrium U.S. Inc.
BIO-TAM	*Trichoderma asperellum* strain ICC 012 and *Trichoderma gamsii* strain ICC 080	Various soilborne diseases by *Armillaria* spp., *Fusarium* spp., *Phytophthora* spp., *Pythium* spp., *Rhizoctonia* spp., *Rosellinia* spp., *Sclerotinia* spp., *Sclerotium rolfsii*, *Thielaviopsis basicola*, and *Verticillium* spp.	Alfalfa, barley, oats, rye, triticale, wheat, clover, cotton, ginseng grass, peanut, sunflower, tobacco	Bayer
DiTera WDG	*Myrothecium verrucaria* strain AARC-0255	Plant parasitic nematodes	Food, fiber and ornamental crops	Valent USA
Double Nickel LC	*Bacillus amyloliquefaciens* strain D747	Damping off, seedling blights, and root or crown diseases caused by *Pythium*, *Rhizoctonia*, *Fusarium*, *Phytophthora*, or *Verticillium*	Vegetable, fruits	Certis USA
MeloCon WG	*Paecilomyces lilacinus* strain 251	Plant parasitic nematodes	Fruits, ornamentals, vegetables	Certis USA
Mycostop Biofungicide	*Streptomyces griseoviridis* strain K61	Seed rot, root and stem rot and wilt caused by *Fusarium*, *Alternaria*, and *Phomopsis*	Vegetables, herbs, and ornamentals	Verdera
Prima stop soil guard	*Gliocladium catenulatum* strain JI446	Soilborne pathogens	Vegetables, herbs, spices	Kemira Agro Oy, Finland
Pro-Mix Biofungicide	*Bacillus subtilis* strain MBI 600	*Pythium*, *Fusarium*, *Rhizoctonia*, *Alternaria* and *Aspergillus*	Vegetable, fruits	Premier Tech Horticulture

(continued)

Table 7.1 (continued)

Trade name	Active biocontrol agent	Target disease/ organism	Crop	Manufacturer
Rot Stop C	*Phlebiopsis gigantea* strain VRA 1992	Root and butt rot (species of the *Heterobasidion annosum* complex including *H. irregulare*)	Conifers	BioForest Technologies Inc.
Serenade Garden	*Bacillus subtilis* strain QST 713	Black Root Rot/Black Crown Rot (*Alternaria* spp.), Early Blight (*Alternaria solani*)	Broccoli, onion, melon, pepper, tomato, lettuce, apple, pear, carrot, walnut	Bayer Crop Science
SoilGard	*Gliocladium virens* strain GL-21	Damping off and Root rot by *Pythium*, *Rhizoctonia*	Fruits, ornamentals	Certis USA

Environmental Protection Agency (EPA) for soilborne pathogens

blending of active ingredients (e.g., fungal spores) with an inert material (diluents, surfactants, etc.) in order to alter the physical properties into a more desirable form while simultaneously preserving the viability and virulence of the strain used (Junaid et al. 2013). Carriers are defined as the inert ingredients used during formulation to hold or dilute the microorganisms to the required concentration that improves coverage and distribution (Burges 1998). To maximize the potential for successfully developing and deploying a biocontrol product formulation, it is required to begin with a thorough microbial screening procedure, followed by development of mass production protocols that optimize quantity and quality and finally developing a formulation that aids in product delivery, enhances bioactivity, and also preserves shelf life (Schisler et al. 2004). While developing formulation, emphasis should be given on target coverage, target adhesion, biomass survival, and enhancing biocontrol efficacy after delivery to the target (Rahman 2016).

An effective formulation is based on thorough and detailed study of the biocontrol organism, plant pathogen, environment, and their interactions (Leggett et al. 2011). The development of a commercially viable formulation must include selection of a potent stain, mass multiplication, compatibility, shelf life, storage, quality control, biosafety, application technology, and registration (Junaid et al. 2013).

According to Jones and Burges (1998), the four basic functions of formulation of biocontrol agents include:

1. To stabilize the biocontrol organism throughout production, distribution and storage
2. To help in proper handling and application of the biocontrol product such that it is delivered to the target in the most appropriate form
3. To protect the biocontrol agent from harmful environmental factors, for that reason increasing its persistence at the target site

4. To enhance the activity of biocontrol organism at the target site by increasing its activity, reproduction, and interaction with the target organism

7.5.2 Application

The formulated microbial products can be applied to the soil in the form of fluid suspensions, powder formulations, and granules and through spray application (Singh et al. 2016). Depending upon the compatibility of the biocontrol strain with different mineral and organic carriers, various types of formulations include dry products, such as wettable powders, dusts, granules, and dry flakes, and liquid products (aqueous suspensions and flowable), such as cell biomass suspensions in water, oils, and emulsions (Schisler et al. 2004).

Wettable powder is a powdered formulation consisting of the active ingredient in a finely ground state combined with wetting agents. Wettable powders are formulated to be mixed with water (diluted prior to application) to be applied as a dilute suspension through liquid spraying equipment (Keswani et al. 2016). Wettable powders contain 5–95% (usually 50% or more) active ingredients by weight.

Dusts are formulated by adding an active ingredient on fine dry mineral powder such as clay, chalk, and talc, with particle size ranging from 50 to 100 μm. Dusts are applied dry, directly to the target either manually or mechanically, generally to the seeds or foliage. Dusts usually contain a low percentage of active ingredients, <10% of microorganisms by weight (Keswani et al. 2016).

Granular formulations are similar to dust formulations except that granular particles are heavier and larger than dust. Concentration of active ingredient in granules usually ranges from 5 to 20%. Granular formulations are available in three types in general: (1) the microorganism is sprayed onto a rotating granular carrier without a sticker, (2) the microorganism is adhered to the outer surface of a granular carrier by a sticker in a rotating drum, and (3) the microorganism is incorporated into a paste or powdered carrier that sets as a matrix (Keswani et al. 2016).

Encapsulation of potential biocontrol agent within crosslinked polymers such as alginate and carrageen is also one of the formulation techniques known as microencapsulation (Cho and Lee 1999).

Botanigard, Mycotrol, Biostat, Mycotal, Trianum, and Met52 are the major commercially available fungal preparations used against root-knot nematode—*Meloidogyne* spp. (Tranier et al. 2014). Some of *Bacillus*-based plant disease biocontrol products with their formulation types are Serenade (WP, aqueous suspension), EcoGuard (flowable), Kodiak (wettable powder concentrate, i.e., WP (conc.), flowable), Yield Shield (WP (conc.)), Bio-Yield (dry flake), Subtilex (WP (conc.)), and HiStick L + Subtilex (flowable) (Schisler et al. 2004); most of these are active against fungal pathogens.

It is imperative to develop formulations of biocontrol microorganisms with high degree of stability and survival before commercializing, as their performances may vary under different environmental conditions. Moreover, the commercially available formulations should be cost-effective in order to effectively integrate their use in crop protection programs.

7.6 Root Environment

Root is an important part of a plant body that not only provides anchorage and support to the plant system but also helps in absorption (mainly occurs through root hairs) and conduction of water, oxygen, and nutrients. The other plant-specific important functions, like storage of food, incorporation of absorbed nitrogen into amino acids, and production of growth hormones, also have vital roles in plant growth and development. The extent of underground root expansion determines the limitation in plant growth, as the greater root surface area enhances ability of a plant to access more nutrients from soil and indirectly promotes better plant growth (Vessey 2003). A healthy root system can build a plant to tolerate and cope up the adverse environmental conditions. The factors that affect root system as well as soilborne pathogens in soil include soil type, temperature, moisture, aeration, pore size, level of carbon dioxide, pH, mineral elements, and salt concentration (Bot and Benites 2005; Pegg and Manners 2014a).

Anaerobic conditions (hypoxia) suffocate root system resulting into restricted root growth, reduced root hairs, and browning or blackening of root tips. This lack of aeration would ultimately affect the nutrient and water uptake, particularly in aerial parts of plants. Reduced aeration and inadequate pore space cause poor drainage and make soil highly saturated. Poor drainage in turn restricts leaching process, leading to accumulation of salt. Soil salinity causes osmotic stress and nutrient deficiency, imposes ion toxicity and high osmotic stress on plants, and limits water uptake from soil. All these factors interfere with plant growth and ultimately result into reduced crop production, low economic returns, and soil erosion (Shrivastava and Kumar 2015).

Water-saturated, oxygen-deficient, and highly saline conditions in soil stress plants and make them prone to attack by pathogens, particularly by water molds such as *Phytophthora* and *Pythium*. Moreover, oxygen-deprived roots leak more of soluble metabolites that attract zoospores of such pathogens (Pegg and Manners 2014a). Symptoms caused by various soilborne root pathogens vary depending on the pathogen, host plant, and the extent of root rot. Usually, the infected root tissue becomes soft as well as dark brown or black in color. There may be decay of root tips and absence of secondary and tertiary roots at later stages. Plants may as a result appear stunted and slow growing followed by yellowing of leaves, wilting, and necrosis.

7.7 Root Pathogens in Rhizosphere

The rhizosphere is an interactive zone where the plant root releases metabolites such as sugars, amino acids, and many other organic compounds that may be utilized by diverse range of soil microorganisms (Dobbelaere et al. 2003; Singh et al. 2004; Lambers et al. 2009). The interactions in rhizosphere include root–root, root–insect, and root–microbe interactions. Because of the occurrence of plant-secreted chemicals, the rhizosphere remains ecologically dynamic as more number of microorganisms is attracted toward this region in comparison to surroundings. As a result, the rhizosphere harbors a unique microbial population called rhizomicroflora which comprises of neutral, beneficial, and pathogenic microorganisms.

Weller and Thomashow (1994) described root colonization as the multistage process of introducing rhizobacteria on seeds, vegetatively propagated plant parts, or into the soil which then become distributed along roots in soil, multiply, and continue to exist for longer time in the presence of indigenous soil microflora. Colonization of the rhizosphere, rhizoplane, and/or inside the root describes the root colonization and the relative root-colonizing ability as rhizosphere competence. The nutritious environment of the rhizosphere harbors a large population, but less diversity, of bacteria than the bulk soil (Lugtenberg and Kamilova 2009). The bacteria occupying the rhizosphere are together termed rhizobacteria, show immense effects on plant growth and health. Rhizobacteria have been classified as neutral, deleterious, and beneficial (Antoun and Prevost 2006). The colonization of roots is under the influence of various factors, viz., bacterial strains, root exudates, environmental factors, etc., in the rhizosphere (Benizri et al. 2001).

7.8 Classes of Root Pathogens

Plant root diseases are generally caused by a wide variety of fungi, and by some bacteria and nematodes such as *Aphanomyces euteiches*, *Criconemella xenoplax*, *Fusarium oxysporum*, *Thielaviopsis basicola*, *Phytophthora cinnamomi*, *Gaeumannomyces graminis* var. *tritici*, *Phytophthora infestans*, *Pythium splendens*, *Pythium ultimum*, *Plasmodiophora brassicae*, *Rhizoctonia solani*, *Streptomyces scabies*, *Ralstonia solanacearum*, *Heterodera avenae*, *Heterodera schachtii*, and *Meloidogyne* spp. (Weller et al. 2002). Root pathogens in absence of suitable host remain dormant in soil due to limited resources and get active once conditions become favorable. Generally, root infection impairs water and nutrient uptake in plants, resulting into symptoms like wilting, leaf fall, and death of branches and, in severe cases, also the death of plant. In contrast to foliar diseases, root diseases caused by soilborne pathogens are often more destructive as they are difficult to detect before substantial damage had already occurred (Koike et al. 2003). Therefore, in general, root diseases are recognized late in comparison to easily visualized foliar diseases like rusts, smuts, and mildews on plants. Further, it is difficult to distinguish the causal agent(s) of a disease as a primary pathogen when the root is also infected by some other associated pathogens. Hence, a keen assessment of the surrounding microorganisms of root is essential to investigate and identify the existing primary root pathogen(s) in soil to safeguard plants from disease.

7.8.1 Fungal Pathogens

Fungal pathogens affecting underground parts of different plants mainly belong to groups *Plasmodiophoromycetes*, *Chytridiomycetes*, *Oomycetes*, *Ascomycetes*, and *Basidiomycetes*. The major soilborne fungal pathogens causing root disease in plants pertain to various genera such as *Aphanomyces*, *Chalara*, *Cylindrocladium*, *Fusarium*, *Phytophthora*, *Pythium*, *Rhizoctonia*, *Sclerotinia*, *Sclerotium*, and *Verticillium*. Few of these have been described below.

Phytophthora Various species of genus *Phytophthora* are hemibiotrophs as these derive energy initially from living cells and later from dead cells. They are regarded as "plant destroyer" (Greek phyton, plant; phthora, destruction), known to be one of the most widespread and destructive plant pathogens (Webster and Weber 2007). *Phytophthora* infection usually starts at the root tip that rapidly spread to other underground parts of the plant. It destroys the entire root system of mature plants and causes trunk cankers near the soil surface. *Phytophthora* spp. over-seasons through oospores and/or chlamydospores in soil which serve as weapons of the plant destroyer and contribute to their pathogenic success (Judelson and Blanco 2005). Among the most common species, *P. cinnamomi*, *P. colocasiae*, and *P. cactorum* show a very wide host range in contrast to *P. infestans*, which exhibits a narrow host range. *Phytophthora cinnamomi* is known to have the widest host range of all species and is being capable of infecting over 1000 plants (Zentmyer 1980).

Pythium *Pythium* is worldwide in distribution, occurring as saprophyte in water, soil, or decaying organic matter or as parasite on plants and rarely on animals. Many species of *Pythium* are facultative parasites causing root rot, fruit rot, seed rot, preemergence killing, and damping-off diseases. In older plants, generally, they attack undifferentiated feeder roots and disrupt water and nutrients uptake. The most common species are *P. debaryanum*, *P. aphanidermatum*, *P. ultimum*, *P. butleri*, and *P. graminicolum*. Most *Pythium* species have a wider host range than *Phytophthora* and infect a wide variety of plants such as papaya, mustard, wheat, ginger, tobacco, and tomato. *Pythium* survives in the soil as oospores produced through sexual reproduction (Webster and Weber 2007).

Fusarium *Fusarium* species are common soilborne organisms, capable of surviving in the form of chlamydospores and as mycelium in plant and soil for long periods. It is a root-infecting fungal pathogen which targets a number of plants such as tomato, cotton, banana, and *Arabidopsis*. *F. oxysporum* is one of the most commonly isolated species from soil, which is often found to grow in association with plant roots (Webster and Weber 2007). *Fusarium* root rot is one such important disease of leguminous soybean that is caused by a species complex of *Fusarium*, of which *F. solani* and *F. oxysporum* are the most frequent (Diaz Arias et al. 2013). In severe root rot infection, the infected plants may also develop foliar symptoms such as stunting, marginal or whole leaf chlorosis, wilting, and defoliation.

The genus also causes wilt disease on large number of plant species. The wilt and root rot-inducing strains of *F. oxysporum* have been reported to cause serious losses of many ecologically and economically important agricultural crops worldwide (Fravel et al. 2003).

Rhizoctonia *Rhizoctonia* is another root-infecting fungal pathogen having an enormous host range and distribution. The fungus was named so because it rapidly attacks and kills the roots of plants (root killer). The genus comprises of a highly divergent group of fungi that are effective saprophytes (aggressively colonizes organic debris) and can survive in soil for longer periods with the help of sclerotia and hyphae that it produces in soil and plant residues (Webster and Weber 2007). Several pathogenic species of *Rhizoctonia* are known to infect and cause serious diseases in plants. This versatile fungus can infect and cause a myriad of diseases including root rot disease, seed decay, post- and preemergence damping-off disease, stem cankers, fruit decay, and foliar blights. The most important disease-causing species of this genus is *R. solani*. The species infects root crops like sugar beets and carrots. The symptoms of *Rhizoctonia* infection include rusty-brown, dry sunken lesions on stems and roots (Pegg and Manners 2014b). It also causes decay of lateral roots. In soybean, it causes root and stem rot that typically causes most damage to seedlings. *Rhizoctonia* isolates have been recovered from soybean seedlings with damping-off, root rot, and hypocotyl rot symptoms, of which maximum isolates (~80) were confirmed to be of *R. solani* (Ajayi-Oyetunde and Bradley 2017).

Verticillium The soilborne fungus *Verticillium* causes serious vascular disease in a wide variety of annual crops and woody perennials in different parts of the world. The genus *Verticillium* contains a relatively small number of soilborne ascomycete fungi. Several of them cause wilt disease on a variety of plant hosts which is most prevalent in temperate and subtropical regions and rare in tropical regions (Deketelaere et al. 2017). Currently, there are ten species, viz., *Verticillium albo-atrum*, *V. alfalfa*, *V. dahliae*, *V. isaacii*, *V. klebahnii*, *V. longisporum*, *V. nonalfalfae*, *V. nubilum*, *V. tricorpus*, and *V. zaregamsianum*, recognized within the genus, of those *V. dahliae* has the broadest host range and infects over 200 plant species (Inderbitzin and Subbarao 2014). *Verticillium* spp. produce long-lasting resting structures such as microsclerotia, chlamydospores, and mycelium in dead or dying plant tissues that serve as the primary inoculum. Hyphae developed from primary inoculum penetrate the roots and fungus colonizes the xylem vessels of the vascular system of the host (Deketelaere et al. 2017). Symptoms associated with *Verticillium* wilt are stunting, chlorosis, wilting, vascular discoloration, and early senescence. However, symptoms can differ considerably among hosts (Fradin and Thomma 2006).

7.8.2 Bacterial Pathogens

In comparison to fungal pathogens, relatively few species of bacteria such as species belonging to genera *Erwinia*, *Rhizomonas*, and *Streptomyces* are known to cause root infections (Koike et al. 2003). Majority of the bacterial pathogens in soil enter along with infected plant debris and persist as long as debris resists decomposition due to their limited competitive saprophytic ability; only few bacterial pathogens are actual soil-dwelling species. Bacterial pathogens, *Agrobacterium tumefaciens*, *Ralstonia solanacearum*, and *Streptomyces scabies*, described below are some of the common examples of root disease-causing agents.

Ralstonia solanacearum An extremely damaging soilborne bacterial pathogen has a wide geographical distribution and a very broad host range, infecting over 200 plant species in 50 families (Mansfield et al. 2012). It has been reported to affect different crops such as potato, tomato, eggplant, tobacco, peanut, pepper, and banana and be responsible for potato brown rot; bacterial wilt of tomato, tobacco, eggplant, and some ornamentals; and Moko disease of banana (Chandrasekaran et al. 2016). World over, on potato alone, an estimated US$1 billion loss has been reported to occur each year by this pathogen (Elphinstone 2005). The entry of pathogen in hosts takes place through wounds and root tips/cracks at the sites of lateral root emergence. The bacterium consequently colonizes the root cortex, invades xylem vessels, and reaches the stem and aerial plant parts through vascular system; hence, it is also regarded as vascular pathogen. It is known to multiply rapidly in the xylem up to very high cell densities, causing wilting symptoms and plant death.

Streptomyces scabies Members of genus *Streptomyces* are gram-positive, soil-inhibiting, saprophytic, mycelial bacteria with high GC content (Loria et al. 2006). Majority of these are harmless and known to produce a large number of antibiotics which are commonly used in the treatment of various infectious diseases of plants and animals including human beings. However, some of these like *Streptomyces scabies*, *S. turgidiscabies*, *S. acidiscabies*, *S. stelliscabiei*, *S. luridiscabiei*, *S. puniciscabiei*, *S. niveiscabiei*, and *S. ipomoeae* are pathogenic and cause scab disease in potato, carrot, radish, beet, peanut, and other tap root crops (Loria et al. 2006; Bignell et al. 2010; Zhang and Loria 2017). These pathogenic species secrete a phytotoxic class of secondary metabolites, known as the thaxtomins that induce a variety of phenotypic changes in the plant host including cell hypertrophy, root and shoot stunting, tissue necrosis, inhibition of cellulose synthesis, alterations in plant Ca^{2+} and H^{+} ion influx, programmed cell death, and production of the antimicrobial plant phytoalexin scopoletin (Bignell et al. 2010). Among various pathogenic species, *S. scabies* is the most notorious as it is widely distributed and causes common scab disease in potato crop besides other root crops. Common scab disease is an economically important disease of potato and is characterized by the presence of necrotic lesions with a corky texture on the potato tuber surface. The lesions can be superficial (russet), erumpent (raised), or deep pitted that can remain small and round or coalesce to cover significant areas of the tuber surface (Zhang and Loria

2017). Pathogen survives in the soil as spores in infected tissue and spreads through water, infected plant material, and wind-blown soil. Entry of *S. scabies* may be direct in young tissues (like developing tubers) or through wounds and natural opening in older tissues.

Agrobacterium tumefaciens *Agrobacterium* spp. are ubiquitous components of the soil microflora, the vast majority of them are saprophytes which survive primarily on decaying organic matter. However, several agrobacteria cause neoplastic diseases in plants, including *Agrobacterium rhizogenes* (hairy root disease), *A. rubi* (cane gall disease), *A. tumefaciens* (crown gall disease), and *A. vitis* (crown gall of grape) (Escobar and Dandekar 2003). Agrobacteria are aerobic, bacillus, gram-negative microorganisms which do not form endospores and exhibit motility due to the presence of one to six peritrichous flagella. *A. tumefaciens* is a cosmopolitan soilborne pathogen that affects various dicotyledonous plants in more than 60 different plant families and causes crown gall disease in host plants. Although crown gall disease is of wide occurrence, it is economically important only on a relatively small number of young, rapidly growing plants that include almond, apple, apricot, blackberry, cherry, cottonwood, crabapple, euonymus, fig, grape, honeysuckle, nectarine, peach, pecan, pear, plum, prune, poplar, pyracantha, raspberry, rose, sugar beet, turnip, walnut, and willow (DeCleene and DeLey 1976; Kennedy 1980; Otten et al. 2008). Crown gall manifests itself initially as small swellings on the root or stem near the soil line and occasionally on aerial portions of the plant. Young tumors, which often resemble the callus tissue that results from wounding, are soft, somewhat spherical, and white to cream colored. As tumors become older, their shape becomes quite irregular, and they turn brown or black in color. Additional symptoms include stunting, chlorotic leaves, and plants may also become more susceptible to adverse environmental conditions and secondary infection.

Pathogenic strains of *A. tumefaciens* may live saprophytically in soil for up to 2 years. When a nearby host plant is wounded near the soil line by insect feeding, transplant injury or any other means; the bacterium chemotactically moves into the wound site and causes infection.

7.8.3 Nematodes

Plant-parasitic nematodes (PPN) are root pests of a wide range of important agricultural crops and, on the basis of feeding strategy, have been classified into different groups, viz., ectoparasites, endoparasites, migratory endoparasites, and sedentary parasites (Perry and Moens 2011). The direct PPN-mediated damages can be further aggravated by secondary infections by other pathogens; moreover, some PPN such as *Xiphinema* spp. can transmit plant viruses also (Hao et al. 2012). Sedentary endoparasitic nematodes such as cyst and root-knot nematodes (RKNs) are considered to be the most damaging pests of agricultural crops worldwide (Bartlem et al. 2014) which become sedentary with the onset of feeding in the vascular cylinder and continuously derive nourishment from adjacent cells and subsequently produce galls or knots (Gheysen and Mitchum 2011). For instance, root knots caused by nematodes, *M. graminicola* and *M. javanica* in rice (Anita and Samiyappan 2012) and mung bean (Ahmed et al. 2009), respectively, damage the vascular system of roots and affect the water and nutrient uptake in plants, making the root system susceptible to other pathogens like fungi and bacteria (Rahman 2003; Ralmi et al. 2016).

RKNs are the major pests of several crops in agriculture that attack nearly all crops resulting into significant losses in crop yield (Kiewnick and Sikora 2006). This group includes various species of genus *Meloidogyne* like *M. hapla*, *M. incognita*, *M. arenaria*, *M. javanica*, *M. graminicola*, and *M. naasi* which are highly adaptable, obligate, and polyphagous plant parasites (Bird et al. 2009; García and Sánchez-Puerta 2012). *Meloidogyne* nematodes are distributed worldwide and parasitize most flowering plants including economically relevant crops (Moens et al. 2009). Among various species of *Meloidogyne*, *M. incognita* and *M. javanica* are notorious for causing severe damages in tobacco, tomato, sunflower, and pepper (Wesemael et al. 2011). The two important species of root-knot nematodes are being discussed below:

Meloidogyne hapla forms smaller galls on host plants—alfalfa, asters, barberry, beans, cabbage, carrot, cauliflower, *Delphinium*, and eggplant—and, to some extent, is less destructive than other root-knot nematode species such as *M. incognita*. It usually causes excessive root branching just behind the growing point of the root, inhibiting root growth.

M. incognita produces large sized root galls, and much more severe symptoms of stunting, yellowing, and wilting are observed; death of young plants is more common in contrast to *M. hapla*. It causes damage to a number of crops including alfalfa, *Asparagus*, beans, cabbage, carrot, clovers, corn, cotton, cucumber, eggplant, grape, lettuce, etc. (Moens et al. 2009).

RKNs often interact with other soil-inhabiting plant pathogens to form disease complexes. In such cases, the resulting disease is much more severe than the individuals of the complex would cause alone. For example, *Meloidogyne* species are known to interact with both *Verticillium* and *Fusarium* fungi, which cause wilt diseases of tomatoes, pepper, potatoes, and some other plants. In certain conditions, the nematode has been responsible for breaking disease resistance to *Fusarium* wilt. Disease complexes often result into death of young plants, while the nematodes alone rarely cause such a severe effect.

7.9 Action—Mechanism of Biocontrol

The foremost facet of implicating any biocontrol method is the selection of BAs and study of various mechanisms employed by them in disease control. Different strategies adapted by antagonistic microorganisms including BAs can be clubbed in two broad categories: first, where a shielding effect is extended to the plants by BAs (consequently pathogens do not get sufficient opportunities to have successful

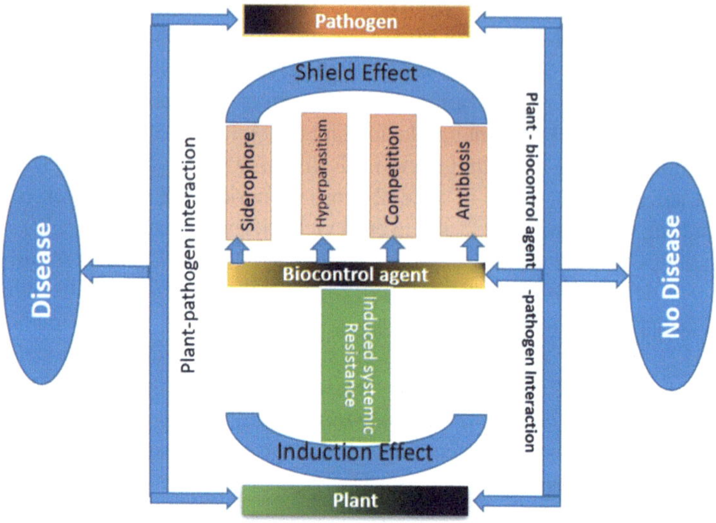

Fig. 7.1 Mechanisms of action exhibited by different biocontrol agents against root pathogens

plant–pathogen interaction that leads to appearance of the disease), and second, where an inducing effect is generated in the plants by BAs (during that, the defense system of the plants is activated, which makes plants resistant to diseases) (Fig. 7.1). The most important mechanisms belonging to mentioned strategies and adopted by prominent groups of the BAs against root pathogens in controlling plant diseases are antibiosis, competition for space and nutrients, iron chelation (siderophores), hyperparasitism, and induced systemic resistance (ISR).

7.9.1 Antibiosis

Antibiosis is a condition in which single to many metabolites are produced by one organism (biocontrol agent) to harm or kill others (pathogens) (Haas and Défago 2005). The production of antimicrobials is one of the major mechanisms for biological control of plant root diseases (Handelsman and Stabb 1996). Chemicals synthesized as a part of antibiosis may be specific or nonspecific metabolites, lytic agents, enzymes, volatile compounds, or other toxic substances (Fravel 1988; Mehta et al. 2014) and are released by various BAs including bacteria and fungi. Antibiotics are an important group of antimicrobials, which are secreted by a variety of antagonistic microbes in rhizosphere and help in controlling plant diseases by suppressing pathogens (Glick 1995).

Several species of *Pseudomonas* and *Bacillus* have been reported to synthesize a broad spectrum of antimicrobial metabolites that suppress the growth of pathogens of several important crops and improve plant growth (Weller and Cook 1983; Haas

Table 7.2 Biocontrolling activities of antibiotics produced by different species of genera *Bacillus*, *Gliocladium*, and *Pseudomonas*

Antibiotics	Producer	Biocontrol activity	References
2,4-Diacetylphloro-glucinol	*Pseudomonas protegens*, *P. fluorescens*	Antifungal	Hassan et al. (2011), Zhang et al. (2016)
Bacitracin	*Bacillus amyloliquefaciens*	Antibacterial	Sabaté et al. (2017)
Cerexin	*Bacillus mycoides*, *B. cereus*	Antibacterial	Shoji et al. (1975), Cochrane et al. (2015)
Circulin	*Bacillus circulans*	Antibacterial and Antifungal	Katz and Demain (1977)
Colistin	*Bacillus polymyxa*	Antibacterial	Katz and Demain (1977)
Difficidin	*Bacillus subtilis*, *B. amyloliquefaciens*	Antibacterial	Wu et al. (2015)
Fengycin	*B. amyloliquefaciens*	Antifungal	Sabaté et al. (2017)
Gliotoxin	*Gliocladium virens*	Antifungal and antibacterial	Lumsden et al. (1992)
Gliovirin	*G. virens*	Antifungal	Lumsden et al. (1992)
Gramicidin	*Bacillus brevis*	Broad spectrum	Berditsch et al. (2007)
Iturins	*B. amyloliquefaciens*	Antifungal	Sabaté et al. (2017)
Kurstakin	*B. amyloliquefaciens*	Antibacterial and antifungal	Sabaté et al. (2017)
Laterosporin	*Bacillus laterosporus*	Antifungal and antibacterial	Barnes (1949), Castillo et al. (2013)
Mycobacillin	*B. subtilis*	Antifungal	Castillo et al. (2013)
Mycosubtilin	*B. subtilis*	Antifungal and anti-*Micrococcus* spp.	Castillo et al. (2013)
Phenazine-1-carboxylic acid	*Pseudomonas fluorescens* and *P. aureofaciens*, *P. chlororaphis*	Antifungal	Labuschagne et al. (2010)
Polymyxin	*B. polymyxa*, *B. subtilis*	Antibacterial	Katz and Demain (1977)
Prodigiosin	*Serratia plymuthica*	Broad spectrum	Kalbe et al. (1996)
Pyocyanin	*Pseudomonas aeruginosa*	Antibacterial and antifungal	Jayaseelan et al. (2014)
Pyoluteorin	*Pseudomonas protegens*, *P. fluorescens*, *P. putida*	Antifungal	Hassan et al. (2011), Zhang et al. (2016)
Pyrrolnitrin	*P. fluorescens*, *B. cepacia*, *Enterobacter agglomerans*, *Serratia* spp.	Antifungal	Burkhead et al. (1994), Chernin et al. (1996), Ligon et al. (2000)
Surfactins	*B. amyloliquefaciens*	Antifungal and Antibacterial	Sabaté et al. (2017)
Viscosinamide	*P. fluorescens*	Antifungal	Thrane et al. (2000)
Zwittermicin	*B. cereus*, *B. thuringiensis*	Broad spectrum	Silo-Suh et al. (1998)

and Keel 2003; Kumar et al. 2005) (Table 7.2). An antimicrobial compound 2,4 diacetylphloroglucinol produced by fluorescent pseudomonads has been used for protecting plant roots against pathogenic fungi (Shirifi et al. 1998). Hanlon et al. (1994) have shown that *B. subtilis* inhibited phytopathogenic fungi by producing antibiotic lipopeptide substance. These antibiotics work as antagonists of pathogens and suppress their growth, longevity, spore germination, etc. and thus reduce their populations. Bacteriocins are another class of antibiotics produced by different bacteria. These are very specific in action and effective against other related strains of same producer species (Sindhu et al. 2009). Several hydrolytic enzymes such as β-1,3-glucanase, chitinase, cellulase, and protease are secreted to exert a direct inhibitory effect on the hyphal growth of fungal pathogens (Labuschagne et al. 2010). Chitinase and β-1,3-glucanase degrade chitin, a major component of the fungal cell wall. The β-1,3-glucanase synthesized by strains of *Paenibacillus* and *Streptomyces* spp., and *Bacillus cepacia* lyses fungal cell walls of pathogenic *F. oxysporum*, *R. solani*, *P. ultimum*, and *S. rolfsii* (Compant et al. 2005). Similarly, potential BAs with chitinolytic activities include *B. licheniformis*, *B. cereus*, *B. circulans*, *B. thuringiensis*, *Serratia marcescens*, *Enterobacter agglomerans*, *Pseudomonas aeruginosa*, and *P. fluorescens* (Sadfi et al. 2001; Nielsen and Sorensen 1999).

Volatile organic compounds (VOCs) are low molecular weight organic compounds such as terpenoids, phenylpropanoids, and fatty acid derivatives that possess antimicrobial properties in addition to other activities and often synthesized by different BAs. VOCs, namely, 2-nonanone, 2-undecanone, decanal, dimethyl disulfide, and benzeneacetaldehyde, produced by *Bacillus megaterium* have been demonstrated to exhibit strong nematicidal activities against *M. incognita* (Huang et al. 2010). Kai et al. (2007) reported various degrees of inhibition of *Rhizoctonia solani* by VOCs released by bacteria, *Stenotrophomonas maltophilia*, *Serratia plymuthica*, *Stenotrophomonas rhizophila*, *Serratia odorifera*, *Pseudomonas trivialis*, *S. plymuthica*, and *Bacillus subtilis*. Similarly, inhibitory activities of VOCs of PGPR, *Paenibacillus polymyxa* (*B. polymyxa*) and *Bacillus amyloliquefaciens* against *F. oxysporum* and *Pseudomonas fluorescens* against *Ralstonia solanacearum*, have also been reported by Yuan et al. (2012) and Raza et al. (2016a, b).

7.9.2 *Competition for Space and Nutrients*

The rhizosphere is a region of high microbial activity as it provides richer nutritional environment to both beneficial and pathogenic microorganisms (Bot and Benites 2005). Therefore, to prevent roots from infections and diseases, successful biocontrol agents should be able to maintain a higher population in order to provide protection to plant roots from rhizosphere-inhabiting pathogens. It is believed that soilborne pathogens (e.g., *Fusarium*, *Pythium*) that infect through mycelial contact are more subjected to competition by root- as well as plant-associated microbes than foliar pathogens that germinate directly on plant surfaces and infect through appressoria and infection pegs (Nega 2014).

Competition for space and nutrients is one of the mechanisms to restrict/control the disease incidence and severity caused by pathogenic species. Biological control agents are largely antagonists that occupy similar niches of pathogenic organisms, and in due course either naturally or through manipulations of root environment, they gradually outcompete the pathogens in their niches (Handelsman and Stabb 1996). Initial competition for occupancy is followed by competition for limiting nutrients or minerals resulting into depletion of resources and nutritional requirements of root pathogens through their biocontrol activity. Poor biocontrol performance of a biocontrol agent has been correlated with ineffective root colonization by using mutants of *Pseudomonas* strains, which had lost their biocontrol potential. The production of plant glycoprotein agglutinin was associated with effective root colonization by *Pseudomonas putida*, lacking which it showed reduced capacity to colonize the rhizosphere and subsequently limited potentiality of suppressing *Fusarium* wilt disease in cucumber (Tari and Anderson 1988).

Van Dijk and Nelson (2000) provided direct evidence where fatty acid competition aids in suppression of *Pythium ultimum* by *Enterobacter cloacae* and thus prevents *P. ultimum*-induced seed infections in cotton. Though there were lack of direct evidences supporting competition as a mechanism of suppressing plant diseases microbiologically, unequivocal evidences now suggest that competitive interaction between plant-associated microbes and plant pathogens plays an important role in plant growth and development and suppression of plant diseases.

7.9.3 Siderophores

Siderophores play a significant role in plant iron nutrition and biocontrol of soilborne plant diseases (Loper and Buyer 1991). These are low molecular weight, high affinity compounds that works as iron (III) chelators and aid in iron transport into the bacterial cells (Leong 1986). Two major types of siderophoric compounds produced by microorganisms were reported by Carson et al. (2000), namely, hydroxamate and catechol. Hydroxamate siderophores generally contain 6-*N*-hydroxyornithine as the ligand which is involved in the chelation of iron. Siderophore systems of bacteria are composed of ferric-specific ligands and their membrane receptors as chelating agents (Neilands 1989). Since the limited supply of iron (III) is being sequestered by the siderophores in the rhizosphere, the limited iron availability to pathogens ultimately results into their suppressed growth (Schroth et al. 1984; Handelsman and Stabb 1996). Siderophore-mediated suppression of rice fungal pathogens *Rhizoctonia solani* and *Pyricularia oryzae* has been reported by Battu and Reddy (2009).

7.9.4 Hyperparasitism

In hyperparasitism, different BAs suppress plant diseases by parasitizing the root pathogen as a part of direct antagonism. This mode of action mechanism has been reported in viruses, bacteria, and fungi. In general, there are four major classes of hyperparasites: obligate bacterial pathogens, hypoviruses, facultative parasites, and predators. *Pasteuria penetrans* is an obligate bacterial pathogen of root-knot nematodes that affects a wide range of nematodes (Chen and Dickson 1998). Mycoviruses are fungi-infecting viruses which cause hypovirulence, a reduction in disease-producing capacity of the pathogen (Nuss 2005). Fungi exhibiting hyperparasitism may possess special structures like clamps and appresoria to capture the pathogen. Fungi, e.g., mycoparasites, may also secrete certain enzymes such as chitinase and β-1,3-glucanase that decrease the biomass of soilborne pathogens, e.g., *Fusarium oxysporum* (Chen et al. 2015). *Trichoderma harzianum* exhibits excellent mycoparasitic activity against *Rhizoctonia solani* hyphae (Altomare et al. 1999). *Acremonium alternatum*, *Acrodontium crateriforme*, *Ampelomyces quisqualis*, and *Gliocladium virens* are few of the fungi that have the capacity to parasitize powdery mildew pathogens (Kiss 2003). Fungi such as *Purpureocillium lilacinum* (syn. *Paecilomyces lilacinus*) and *Pochonia chlamydosporia* (syn. *Verticillium chlamydosporium*) are common soil inhabitants that parasitize nematodes (Timper 2014).

7.9.5 Induced Systemic Resistance

Induced systemic resistance (ISR) is a plant immune response mediated by the root microbiome. It is an important defensive mechanism in plants in which particular plant growth-promoting bacteria and fungi or the beneficial microbes in the rhizosphere directs the whole plant body for an enhanced level of protection against a broad spectrum of pathogens (Handelsman and Stabb 1996). Beneficial microbial population can trigger ISR by producing different microbe-associated molecular patterns (MAMPs) and elicitors (Pieterse et al. 2014).

Induced resistance is expressed not only locally at the site of induction but also systemically in other plant parts which are spatially separated from the inducer, hence the term ISR. Induced resistance is regulated by a network of interconnected signaling pathways in which plant hormones such as jasmonic acid and ethylene play a chief regulatory role (Pineda et al. 2010). A large number and variety of root-associated microbes such as *Pseudomonas*, *Bacillus*, *Trichoderma*, and mycorrhizal species sensitize the immune system in plants for synthesis of defense chemicals such as chitinase, peroxidase, and pathogenesis-related proteins (Pozo and Azcon-Aguilar 2007; Contreras-Cornejo et al. 2009; Jung et al. 2012; Cameron et al. 2013; Zamioudis et al. 2013). Priming (preparation/sensitization of the whole plant to better combat pathogen attack) for enhanced defensive capacity, rather than direct activation of resistance, is a general attribute of systemic immunity elicited by beneficial microbes (Pieterse et al. 2003, 2014).

7.10 Groups of Biocontrol Agents

The most common species employed as BAs for suppression of root pathogens and inducing systemic resistance in plants for their healthy growth and development, particularly in agricultural sector, are comprised of bacteria, fungi, arbuscular mycorrhizal fungi (AMF), and nematodes (Table 7.3). Some groups of microorganisms having importance in biocontrol methods are mentioned below:

7.10.1 Bacteria

Plant growth-promoting rhizobacteria (PGPRs) are the beneficial bacteria that colonize roots and promote plant growth as well as help in disease control by a variety of mechanisms (Yan et al. 2003; Ashrafuzzaman et al. 2009). These include a diverse array of microorganisms belonging to genera such as *Acetobacter*, *Actinoplanes*, *Agrobacterium*, *Alcaligenes*, *Arthrobacter*, *Azospirillum*, *Azotobacter*, *Bacillus*, *Bradyrhizobium*, *Cellulomonas*, *Clostridium*, *Enterobacter*, *Erwinia*, *Flavobacterium*, *Pasteuria*, *Pseudomonas*, *Rhizobium*, *Serratia*, and *Xanthomonas* (Sindhu et al. 2009). Of these, biological control is usually achieved the most by various species of *Pseudomonas* and *Bacillus*. However, in comparison to *Bacillus* spp., *Pseudomonas* spp. are extensively used for biocontrol of soilborne diseases (Silo-Suh et al. 1994; Tu et al. 2016). Both *Bacillus* and especially *Pseudomonas* bacteria possess different anti-pathogenic properties such as antibiotic production and competition for iron and other nutrients that are helpful in suppression of root-infecting pathogens in soil. Further, these bacterial organisms also exhibit high efficiency for colonization of host root as well as production of growth metabolites leading to improved growth and subsequently higher crop yield (Fiddman and Rossall 1993; Haas and Défago 2005).

Pseudomonas The genus *Pseudomonas* contains a group of ubiquitous microorganisms that can be found in diverse ecological habitats such as soils, foliage, freshwater, sediments, and seawater (Von Graevenitz 1977). *Pseudomonas* bacteria are gram-negative, rod-shaped bacteria, characterized by their motility due to presence of one or more polar flagella, aerobic respiration, metabolic versatility, and a high (59–68%) genomic GC content (Haas and Défago 2005; Ramadan et al. 2016). *Pseudomonas* species are known to be effective root colonizers and BAs. Members of the genus *Pseudomonas* include both fluorescent and nonfluorescent species. Fluorescent pseudomonads can be visually distinguished from other *Pseudomonas* spp. by their ability of producing a water-soluble yellow-green fluorescent pigment,

Table 7.3 Biocontrol agents effective against root pathogens of various plants

Biocontrol agent	Plant root pathogen	Host plant	References
Pseudomonas fluorescens	*Rhizoctonia solani*	Rapeseed	Dahiya et al. (1988)
P. fluorescens CHA0	*Meloidogyne javanica*	Tomato	Siddiqui et al. (2006)
P. fluorescens strains UP61, UP143, and UP148	*Pythium ultimum*, *Rhizoctonia solani*	*Lotus corniculatus*	Bagnasco et al. (1998)
P. fluorescens 2-79RN(10)	*Aphanomyces euteiches* f. sp. *pisi*	Pea	Dandurand and Knudsen (1993)
P. fluorescens WCS417r	*Fusarium oxysporum*	Tomato	Duijff et al. (1998)
P. fluorescens MKB 100 and MKB 249	*Fusarium culmorum*	Wheat, barley	Khan et al. (2006)
Pseudomonas strains BTP1 and M3	*Pythium ultimum*	Cucumber	Ongena et al. (1999)
Bacillus subtilis and *B. cereus*	*Gaeumannomyces graminis* var. *tritici*	Wheat	Ryder et al. (1999)
Bacillus amyloliquefaciens	*Fusarium verticillioides*	Maize	Pereira et al. (2009)
Bacillus amyloliquefaciens AK0	*Cylindrocarpon destructans*	Ginseng	Kim et al. (2017)
B. amyloliquefaciens B14	*Macrophomina phaseolina*	Common bean	Sabaté et al. (2017)
Pasteuria penetrans	*Meloidogyne incognita, M. javanica*	Tobacco	Weibelzahl-Fulton et al. (1996)
Paecilomyces lilacinus strain 251 (PL251)	*Meloidogyne incognita*	Tomato	Kiewnick and Sikora (2006)
Entrophospora infrequens	*Pseudomonas syringae* pv. *glycinea*	Soybean	Malik et al. (2016)
Trichoderma harzianum	*Macrophomina phaseolina*	Soybean	Barari and Foroutan (2016), Khaledi and Taheri (2016)
Trichoderma spp.	*Fusarium oxysporum* f. sp. *phaseoli*, *Fusarium solani* f. sp. *phaseoli*, and *Macrophomina phaseolina*	Common bean	Pierre et al. (2016)
Non-pathogenic *Fusarium oxysporum*	Pathogenic *Fusarium oxysporum* f. sp. *radicis-cucumerinum*	Cucumber	Abeysinghe (2009)
Glomus etunicatum	*Rhizoctonia solani*, *Pythium ultimum*	Potato, Cucumber	Rosendahl and Rosendahl (1990), Yao et al. (2002)
Glomus intraradices	*Xiphinema index*, *Radopholus similis*, *Pratylenchus coffeae*, *Nacobbus aberrans*, *Xanthomonas campestris* pv. *alfalfae*, *Phytophthora nicotianae*	Grape, banana, tomato, *Medicago*	Liu et al. (2007), Lioussanne et al. (2008), Elsen et al. (2008), Hao et al. (2012), Marro et al. (2014)
Glomus mosseae	*Tylenchulus semipenetrans*, *Meloidogyne incognita*, *M. javanica*, *Ralstonia solanacearum*, *Phytophthora capsici*	Citrus, tomato, banana, pepper	O'Bannon et al. (1979), Elsen et al. (2002), Tahat et al. (2012), Vos et al. (2013), Pereira et al. (2016)

pyoverdine or pseudobactin. The heterogeneous group of fluorescent *Pseudomonas* comprises, most notably, *P. aeruginosa*, *P. putida*, *P. fluorescens*, and pathogenic *P. syringae* (Scarpellini et al. 2004).

Mode of action Rhizosphere-inhabiting fluorescent pseudomonads are rapid root colonizers which possess biocontrolling (Keel and Défago 1997) and plant growth and yield-promoting properties (Schippers et al. 1987; Couillerot et al. 2009). Growth promotion and disease suppression mechanism of fluorescent pseudomonads in rhizosphere involves direct and indirect mechanisms that include production of diverse array of metabolites (auxin, cytokinin, and gibberellins), antibiotics, and volatile compounds, e.g., hydrogen cyanide (HCN) (Thomashow and Weller 1996), competition for space, competition for iron and other nutrients via siderophore production (Ahemad and Kibret 2014), and ISR (Pieterse et al. 2003, 2014), or they may interfere specifically with fungal pathogenicity factors (Lucy et al. 2004; Adesemoye et al. 2008).

Antibiotic compounds such as phenazine-1-carboxylic acid, pyocyanin, pyrrolnitrin, pyoluteorin, and 2,4-diacetylphloroglucinol (Phl) have been isolated from fluorescent pseudomonads (Howell and Stipanovic 1979, 1980; Dahiya et al. 1988; Thomashow and Weller 1996; Thomashow et al. 1997). Phenazine-producing *Pseudomonas* have been shown to help in natural suppression of *Fusarium* wilt in soils (Mazurier et al. 2009) and possess strong inhibitory activity against *Rhizoctonia solani* in wheat roots, both in vitro and in vivo (Mavrodi et al. 2012).

Active motility as well as chemotaxis provides these bacteria a competitive edge over pathogens in grabbing space and nutrients as these can reach root surfaces faster than non-motile pathogens (de Weert et al. 2002). Siderophore-assisted iron acquisition ability of fluorescent *Pseudomonas* further assures a competitive advantage over other microorganisms. Siderophores bind to ferric iron in soil or the root zone, thus making iron unavailable to phytopathogens. These siderophores have been reported to be involved in the suppression of pathogenic *F. oxysporum* (Baker et al. 1986). The pathogens are known either to lack the ability of producing siderophores or if produce, their affinity for iron is lower as compared to beneficial microorganisms (Kloepper et al. 1988), which lead to their suppression or eradication.

Suppression of root rot of soybean caused by *Phytophthora* (Lifshitz et al. 1987), control of tobacco black root rot (Keel et al. 1989), fungal diseases of citrus roots (Gardner et al. 1984), and some ornamental plants (Yuen and Schorth 1986) are few examples exhibiting biocontrol activity of fluorescent *Pseudomonas* strains. Chandrasekaran et al. (2016) used meta-analytical approach to quantitatively review the results of 650 studies on biological control efficacy for controlling *Ralstonia* wilt disease caused by *Ralstonia solanacearum*. This analysis showed that the extent of disease suppression by biological control agents varied widely among studies. The disease incidence and severity were significantly decreased on average by 53.7% and 49.3%, respectively. Mean effect sizes for genus *Pseudomonas* spp. as biological control agents were higher than for genus

Bacillus spp. Among antagonists tested, *P. fluorescens* was found to be most effective in disease reduction.

Bacillus The genus *Bacillus* is characterized by gram-positive, rod-shaped bacteria that may be obligate or facultative aerobes. *Bacillus* species are abundant in various ecological niches, viz., soil, water, and air (Zhang et al. 2009; Logan and Halket 2011) and recognized as one of the most valuable biological control agents due to the growth-inhibiting properties of pathogens (Sid et al. 2003; Schisler et al. 2004; Shafi et al. 2017). Bacilli form endospores in response to adverse growth conditions which can resist environmental stress, for instance, high temperatures and current chemical disinfectants (Collins and Jacobsen 2003; Kumar et al. 2011). This endospore-forming characteristic of *Bacillus* may function as potential biocontrol inoculant as the endospore can resist heat and desiccation ensuring the stability of formulation for a longer time (Emmert and Handelsman 1999). The species most commonly used in biocontrol process are motile, having peritrichous flagella (Driks 2004).

Mode of action The chief action mechanism of *Bacillus* is antibiosis. It produces extracellular hydrolytic enzymes that decompose polysaccharides, nucleic acids, and antibiotics such as bacitracin, polymyxin, and gramicidin (Guillén-Cruz et al. 2006; Hussein and AL-Janabi 2006; Li et al. 2009). *B. subtilis* is regarded to be a safe biocontrol agent that may be due to production of peptide antibiotics, viz., iturin A and surfactin (Chang and Kommedahl 1968; Hiraoka et al. 1992). Important antibiotics produced by various bacilli have been mention in Table 7.2. Another species, *B. thuringiensis* represents about 95% of microorganisms used in biological control of agricultural pests (Schünemann et al. 2014).

The other mechanisms include competition with pathogens to occupy an ecological niche. Unlike *Pseudomonas*, the genus *Bacillus* is considered to be a non-rhizosphere competent, but, some strains may also be rhizosphere competent, given that rhizospheric competency is strain-dependent (Kumar et al. 2011). *Bacillus* also acts by metabolizing root exudates on pathogens affecting their growth and activity (Doornbos et al. 2012). It also induces plants to produce phytoalexins, when installed in roots and leaves, and elicits resistance against pathogenic attack by fungi, bacteria, and nematodes in plants (Kloepper et al. 2004). *Bacillus* species also produce different kinds of metabolites, cell wall-degrading enzymes, and volatile compounds (Priest 1977; Pelletier and Sygusch 1990; Fiddman and Rossall 1993; Pleban et al. 1997) which elicit disease resistance mechanisms in plants (Kloepper et al. 2004; Kumar et al. 2011). Bacterial inoculation of seeds with cells of *Pseudomonas* or *Bacillus* was found antagonistic to *Rhizoctonia solani* (Georgakopoulos et al. 2002). In an another study, *B. subtilis* RB14 coated seeds were used as biological control agent against *Rhizoctonia solani* to suppress damping off disease in tomato plants (Nawar 2016).

Many species of *Bacillus*, such as *B. subtilis*, *B. licheniformis*, *B. pumilus*, *B. amyloliquefaciens*, *B. cereus*, *B. mycoides*, and *B. thuringiensis*, have been reported to be effective in disease control by suppressing growth of several pathogenic fungi including *Macrophomina*, *Rhizoctonia*, *Fusarium*, *Pythium*, *Phytophthora*, *Sclerotinia*, *Gaeumannomyces*, *Sclerotium*, *Nectria*, and *Verticillium*

(Carrillo et al. 2003; Nalisha et al. 2006; Zhang et al. 2009; Khan et al. 2011; Basurto-Cadena et al. 2012; Sabaté et al. 2017). The beneficial role of *B. cereus* in crop health improvement in soybean (increase in yield and nodulation), suppression of damping off of tomato and alfalfa (Benizri et al. 2001), plant development, and protection of plant roots from devastating effects of phytopathogens has been well stated by Kazmar et al. (2000).

7.10.2 Actinomycetes

Actinomycetes are important sources of various antimicrobial metabolites (Terkina et al. 2006). Actinomycetes strains, such as species of *Micromonospora*, *Streptomyces*, *Streptosporangium*, and *Thermobifida*, have been evaluated as very effective root colonizers, showing an immense potential to function as biocontrol agents against many root pathogenic fungi (Franco-Correa et al. 2010). The use of endophytic actinobacteria in control of *Rhizoctonia solani* and *Pseudomonas solanacearum* in tomato and *Colletotrichum musae* in banana has been reported (Taechowisan et al. 2003).

Streptomyces species have been known as rhizosphere-colonizing bacteria and important antifungal BAs that have been helpful in controlling fungal root diseases. Their properties of producing antibiotics, siderophores, and plant growth-promoting hormones make them beneficial microbes for the plants (Rothrock and Gottlieb 1984; Miller et al. 1990). *Streptomycetes* has been reported as potential biocontrol agent of *Fusarium* and *Armillaria* pine rot and as plant growth promoter of *Pinus taeda* (de Vasconcellos and Cardoso 2009). Antagonistic activity of *Streptomyces griseorubiginosus* against *F. oxysporum* f. sp. cubense has been described by Cao et al. (2004). Naturally occurring populations of *Bacillus subtilis* and saprophytic *Streptomyces* species suppress *Streptomyces scabies*, the causal organism of potato scab.

Pasteuria species are gram-positive, mycelial, endospore-forming bacteria that have shown great potential as a biological control agent of root-knot nematodes. *Pasteuria* spp. are distributed worldwide and have been reported from 323 nematode species belonging to 116 genera of free-living, predatory, plant-parasitic, and entomopathogenic nematodes (Chen and Dickson 1998). The biological control potential of *Pasteuria* spp. have been demonstrated on 20 crops against host nematodes, *Belonolaimus longicaudatus, Heterodera* spp., *Meloidogyne* spp., and *Xiphinema diversicaudatum.* Among various species, *P. penetrans* has been studied most extensively, and its efficacy as biocontrol agent has been examined against different PPNs including *Meloidogyne* spp. such as *M. arenaria* and *M. incognita* (Jonathan et al. 2000; Kariuki and Dickson 2007).

7.10.3 Fungi

The fungi that have shown biocontrolling potentialities against various plant pathogens include both free-living and symbiotic heterotrophic mycelial organisms. Among various fungal biocontrol agents, *Trichoderma* spp. and mycorrhizae, which are also the most widely studied agents, have been described below:

Trichoderma Species belonging to the genus *Trichoderma* are free soil- and rhizosphere-dwelling, saprophytic, and imperfect fungi of worldwide occurrence (Motlagh and Samimi 2013). Being highly interactive fungi of root, soil, and foliar environments, *Trichoderma* species are considered effective biological means of reducing the incidence of diseases caused by plant pathogenic fungi, particularly management of soilborne diseases (Howell 2003). It readily colonizes plant roots and has evolved various mechanisms to attack other fungi and enhance plant growth (Datnoff et al. 1995; Cotxarrera et al. 2002; Yigit and Dikilitas 2007).

Mode of action Different *Trichoderma* strains act against most of the pathogenic fungi by suppressing the growth of root pathogen population through competition and production of antibiotics and toxins such as trichothecin and sesquiterpene as a part of their interactions with pathogens and plants (Vinale et al. 2014). Consequently, different root pathogens cause milder disease in plants in which the roots are colonized by *Trichoderma* (Harman et al. 2004). The other mechanisms used against pathogens include mycoparasitism, hyphal interactions, and enzyme secretion. *Trichoderma* produces enzymes, cellulase, and chitinase which break down cellulose in plant cell walls and chitin in the fungal cell wall (Harman et al. 2004). The syntheses of antagonistic compounds such as proteins, enzymes, and antibiotics and micronutrients in, e.g., vitamins, hormones, and minerals enhance their biocontrol efficiency (Vinale et al. 2014).

Trichoderma enhances yield, quality, germination rate, and root and shoot length, promotes healthy growth in early stages of crop, and increases dry matter production through various means. The species has been used widely as plant growth-promoting fungi (PGPF), owing to its ability to produce siderophores, phosphate-solubilizing enzymes, and phytohormones (Ellis and Roberts 1981).

Trichoderma has been successfully applied against various pathogenic fungi belonging to different genera such as *Fusarium*, *Phytophthora*, and *Sclerotium*. *Trichoderma* isolates have shown significant reduction in tomato wilt disease caused by pathogenic *Fusarium oxysporum* (Ghazalibiglar et al. 2016). It has shown clear antagonistic effects against *Fusarium solani*, *F. oxysporum*, and *Macrophomina phaseolina*, the causative agents of root rot/wilt disease in *Phaseolus vulgaris* (Pierre et al. 2016). Further, some species, like *T. longibrachiatum* T6, exhibit nematicidal properties against nematodes, *Heterodera avenae* and *M. incognita*, by parasitizing the egg and second-stage juvenile of the nematode (Zhang et al. 2017).

The two most common species of *Trichoderma* used in biocontrol are *T. harzianum* and *T. viride*; *T. harzianum* is the most common species, which is used as a biofungicide for foliar application, seed treatment, and soil treatment to inhibit activity of disease-causing fungal pathogens (Yigit and Dikilitas 2007).

7.10.4 Mycorrhiza

Mycorrhizae are the symbiotic associations between non-pathogenic fungi and roots which promote growth of plants. The fungal partner may occur inside or on the surface of host roots, extending hyphae in soil to enhance availability of phosphate and other nutrients in addition to increasing water uptake to the host plant.

Arbuscular mycorrhizal fungi (AMF) especially in association with plant growth-promoting rhizobacteria (PGPR) have shown immense potential to control soilborne diseases including PPN (Hussain et al. 2016). Dual or individual inoculation of AMF and PGPR has shown to enhance plant growth and reduce root-knot nematodes infection in tomato, indicating their biocontrol potential for the management of nematodes (*Meloidogyne incognita*) (Sharma and Sharma 2017). Mycorrhizal roots also elicit PGPR for increased production of anti-pathogenic compounds; Siasou et al. (2009) reported that rhizospheric fluorescent *Pseudomonas* strains produce increasing amounts of antibiotic 2,4-diacetylphloroglucinol in the presence of AMF, *Glomus intraradices*, which confers plant protection against *Gaeumannomyces graminis* var. *tritici*. Similarly, AMF-colonized roots synthesize root volatiles, for instance, cyanides and isothiocyanates (the volatile products resulting from glucosinolate or cyanogenic glycoside conversion), that have been found to be harmful or toxic to a wide range of belowground herbivores and pathogens (Potter et al. 1998; Hopkins et al. 2009; Kissen et al. 2009).

Three species of AMF (*Glomus mosseae*, *Scutellospora* sp., and *Gigaspora margarita*) showed significant potential in controlling tomato bacterial wilt caused by *Ralstonia solanacearum* under glasshouse conditions (Tahat et al. 2012). There are reports suggesting the role of AMF in disease control in plants, e.g., role of mycorrhizal fungus *Pisolithus tinctorius* where the thick symbiont sheath forms a barrier to infection by pathogen *Phytophthora cinnamomi* attacking eucalyptus trees, but their action mechanism against root pathogens needs to be investigated. Though, there are several studies suggesting the important role of mycorrhizal fungi in promoting plant growth, studies on their biocontrol activity are limited (Azcon-Aguilar and Barea 1996) and still need further research.

Mode of action Mycorrhizal fungi suppress plant diseases especially those caused by soil-dwelling pathogens by adapting a number of strategies, namely, improved plant nutrition, damage compensation, direct competition for colonization sites or root exudates, alteration in the root morphology, changes in rhizosphere microbial populations, biochemical changes associated with plant defense mechanisms, and the activation of plant defense mechanisms (Jung et al. 2012). Fungal colonization in mycorrhiza promotes branching of roots which subsequently causes higher absorption of water and nutrients (Harrison 2005); additionally, these fungi make insoluble minerals available to plant and simultaneously deprive nutrition to pathogens by competing with them for available resources in the rhizosphere (Whipps 2004). This enhances the plant growth as well as makes mycorrhizal plant robust to resist disease-causing microbes.

Mycorrhizal roots in comparison to non-mycorrhizal roots have fewer infection sites to soil pathogens as shown in studies related to fungal pathogens and PPNs (Cordier et al. 1996; Schouteden et al. 2015). Furthermore mycorrhizae modify the chemical composition of root exudates that attract PGPR such as *Azotobacter chroococcum* and *Pseudomonas fluorescens* (Sood 2003) and repel plant pathogens like *Phytophthora nicotianae* (Lioussanne et al. 2008). It has been observed that, in response to mycorrhizal colonization, though weak or very local and transient, activation of specific plant defense compounds such as phytoalexins, chitinases, β-1,3-glucanases, pathogenesis-related proteins, callose, hydroxyproline-rich glycoproteins, phenolic compounds, and enzymes of the phenylpropanoid pathway takes place which elicits specific defense reactions and makes the plants proactive against attackers/pathogens (Gianinazzi-Pearson et al. 1994; Azcon-Aguilar and Barea 1996). The chemicals produced under plant defense mechanism are diversified groups of biochemicals which are synthesized mostly as secondary metabolites (Singh 2017). Mycorrhizal fungi also exhibit the ability to reduce the impacts of plant pathogens through induced resistance which may be localized as well as systemic (Pozo and Azcon-Aguilar 2007).

7.10.5 Nematodes

Entomopathogenic nematodes (EPNs) of families Steinernematidae and Heterorhabditidae are the most promising biocontrol agents of root pests reported so far (Kaya and Gaugler 1993). EPNs are being currently used as classical, conservational, and augmentative biological control agents. Several factors made them most suitable candidates for commercialization such as broad host range, rapid killing of insect host, active searching behavior using olfactory cues, easy mass production both in vivo and in vitro, and application potential in integrated pest management (Lacey and Georgis 2012; Lacey et al. 2015). However, their formulation, shelf life, and application optimization exceed the production cost of chemical pesticides. Further limitation to their use as effective biocontrol agent is their sensitivity to abiotic factors such as low humidity, high UV radiation, high soil salinity, and high or low pH; additionally, EPNs are also known to be quite sensitive to several pesticides, nematicides and fumigants (Lacey and Georgis 2012). To conquer these constraints related to sensitivity, several possible ways out have been proposed, for instance, encapsulating them in biocompatible and biodegradable natural polymers in order to provide them physical protection against abiotic (low humidity, high UV radiation, high soil salinity, and high or low pH) and biotic (fungi and bacteria) factors (John et al. 2011; Vemmer and Patel 2013; Kim et al. 2015). The EPN-based capsules made by co-encapsulation of EPNs with other nutritional components can enhance the efficacy of biocontrol of root pests which may divert the insects feeding from roots toward eating EPN-based capsules (Hiltpold et al. 2012).

The life cycle of an entomopathogenic nematode comprises of an egg stage, four juvenile stages, and an adult stage. Among these, only the third juvenile stage has potentiality as biocontrol agent as it is the infective, free-living stage that can survive in soil for long periods, before infecting a new host (Poinar 1990).

7.11 Conservation and Management

Conservation of beneficial antagonists can be achieved either by preserving the already existing microbe population which attack the pathogens or outcompete them or by enhancing the prevalent conditions for their survival and reproduction at the expense of pathogenic organisms. Further, soil environment may be enhanced for desirable organisms by soil amendments, adding organic matter (Van Driesche and Bellows 1996) and by avoiding practices which may negatively affect them, for instance, soil treatments with fungicides.

Biocontrol through augmentation (inoculation of soils or plants with specific desirable microorganisms) of natural enemy populations based on mass-culturing antagonistic species is applicable where they are low in numbers or present in locations different than the desired. The purpose of augmentation is to increase their population or modify their distribution. In the context of plant pathology, it is sometimes termed as "introduction," i.e., adding them to the system (Andrews 1992; Cook 1993), where the antagonist microbes are usually found in local ecosystem and are not introduced from another region.

7.12 Conclusion

Biocontrol is a sustainable alternative to suppress pathogenic activity and controlling plant diseases over chemical control. Chemical control of pathogens using synthetic pesticides pollutes soil as well as shows harmful effects on living organisms including humans and, thus, threatens global diversity (Evangelista-Martinez 2014). Despite the documented applications and environment-friendly approach of biocontrol methods for longer than a century, it still remained an underexplored area of insect/pest management. In the present scenario, the environmental contamination caused by excessive use of chemical pesticides has increased the scope of biocontrol applications in integrated pest management, where chemical pesticides are substituted by biopesticides to control plant pests and diseases. Based on the overwhelmingly positive features of biological control and increasing interest and public concerns, it is certainly one of the various safe and useful strategies of utilizing natural antagonists for controlling noxious organisms.

7.13 Suggestions

Application of microbial agents for biocontrol has been widely known especially in agricultural field. The focus of their effective utilization should now be based on several undermentioned important aspects which need more information.

- Effective application strategies should be developed to enhance the efficacy of biocontrol mechanisms.
- More research is needed for selection of highly effective strains or strain variants.
- Genetically modified microbes can be used to improve biocontrol application methods.
- Development of proper formulations of microbial agents is required to enhance their biocontrol activities.
- Identification of genes and gene products in the genotypes of microorganisms and plants that govern efficient rhizosphere colonization and improve biocontrol potential.
- Root colonization by AM fungi and root infection by parasitic fungus can be studied using molecular mechanisms.

The ultimate goal of biological control of plant diseases is to assist the farmers in combating and controlling plant pathogens for healthy crop production. Thus, it is pertinent to practically integrate these biocontrol methods in agriculture without imposing further harmful effects on crops which are generally caused by chemical pesticides.

References

Abeysinghe S (2009) Use of nonpathogenic *Fusarium oxysporum* and rhizobacteria for suppression of *Fusarium* root and stem rot of *Cucumis sativus* caused by *Fusarium oxysporum* f. sp. *radicis – cucumerinum*. Arch Phytopathol Plant Protect 42(1):73–82

Adesemoye AO, Obini M, Ugoji EO (2008) Comparison of plant growth-promotion with *Pseudomonas aeruginosa* and *Bacillus subtilis* in three vegetables. Braz J Microbiol 39:423–426

Ahemad M, Kibret M (2014) Mechanisms and applications of plant growth promoting rhizobacteria: current perspective. J King Saud Univ Sci 26(1):1–20

Ahmed N, Abbasi MW, Shaukat SS, Zaki MJ (2009) Physiological changes in leaves of mungbean plants infected with *Meloidogyne javanica*. Phytopathol Mediterr 48:262–268

Ajayi-Oyetunde OO, Bradley CA (2017) Identification and characterization of *Rhizoctonia* species associated with soybean seedling disease. Plant Dis 101(4):520–533

Altomare C, Norvell WA, Bjorkman T, Harman GE (1999) Solubilization of phosphate and micro nutrients by the plant growth promoting fungus *Trichoderma harzianum* Rifai 1295-22. Appl Environ Microbiol 65(7):2926–2933

Andrews JH (1992) Biological control in the phyllosphere. Annu Rev Phytopathol 30:603–635

Anita B, Samiyappan R (2012) Induction of systemic resistance in rice by *Pseudomonas fluorescens* against rice root knot nematode *Meloidogyne graminicola*. J Biopest 5:53–59

Antoun H, Prevost D (2006) Ecology of plant growth promoting rhizobacteria. In: Siddiqui ZA (ed) PGPR: biocontrol and biofertilization. Springer, Dordrecht, pp 1–38

Ashrafuzzaman M, Hossen FA, Razi Ismail M, Anamul Hoque MD, Zahurul Islam M, Shahidullah SM, Meon S (2009) Efficiency of plant growth-promoting rhizobacteria (PGPR) for the enhancement of rice growth. Afr J Biotechnol 8(7):1247–1252

Azcon-Aguilar C, Barea JM (1996) Arbuscular mycorrhizas and biological control of soil borne plant pathogens: an overview of the mechanisms involved. Mycorrhiza 6:457–464

Bagnasco P, De La Fuente L, Gualtieri G, Noya F, Arias A (1998) Fluorescent *Pseudomonas* spp. as biocontrol agents against forage legume root pathogenic fungi. Soil Biol Biochem 30 (10-11):1317–1322

Baker PA, Weisheek PJ, Schippers B (1986) The role of siderophores in plant growth stimulation by fluorescent *Pseudomonas* sp. Med Fac Landboucow Rijksumiv Gent 51(31):1357–1362

Barari H, Foroutan A (2016) Biocontrol of soybean charcoal root rot disease by using *Trichoderma* spp. Cercetări Agronomice în Moldova 49(2):41–51

Barnes EM (1949) Laterosporin A and Laterosporin B antibiotics produced by *B. laterosporus*. Br J Exp Pathol 30(2):100–104

Bartlem DG, Jones MGK, Hammes UZ (2014) Vascularization and nutrient delivery at root-knot nematode feeding sites in host roots. J Exp Bot 65:1789–1798

Basurto-Cadena MGL, Vázquez-Arista M, García-Jiménez J, Salcedo-Hernández R, Bideshi DK, Barboza-Corona JE (2012) Isolation of a New Mexican strain of *Bacillus subtilis* with antifungal and antibacterial activities. Sci World J 2012:384978. https://doi.org/10.1100/2012/384978

Battu PR, Reddy MS (2009) Siderophore mediated antibiosis of rhizobacterial fluorescent pseudomonads against rice fungal pathogens. Int J Pharm Tech Res 1:227–229

Benizri E, Baudoin E, Guckert A (2001) Root colonization by inoculated plant growth-promoting rhizobacteria. Biocontrol Sci Technol 11:557–574

Berditsch M, Afonin S, Ulrich AS (2007) The ability of *Aneurinibacillus migulanus* (*Bacillus brevis*) to produce the antibiotic gramicidin S is correlated with phenotype variation. Appl Environ Microbiol 73(20):6620–6628

Bignell DR, Huguet-Tapia JC, Joshi MV, Pettis GS, Loria R (2010) What does it take to be a plant pathogen: genomic insights from Streptomyces species. Antonie Van Leeuwenhoek 98 (2):179–194

Bird DMK, Opperman CH, Williamson VM (2009) Plant infection by root-knot nematode. In: Berg RH, Taylor CG (eds) Cell biology of plant nematode parasitism, Plant Cell Monograph, vol 15. Springer, Berlin

Bot A, Benites J (2005) The importance of soil organic matter- Key to drought-resistant soil and sustained food production, FAO Soils Bulletin 80. Food and Agriculture Organization of the United Nations, Rome, pp 1–80

Bravo A, Cristina del Rincon-Castro M, Ibarra JE, Soberon M (2011) Towards a healthy control of insect pests: potential use of microbial insecticides. In: Lopez O, Fernandez Bolanos JG (eds) Green trends in insect control. Royal Society of Chemistry, London, pp 266–299

Burges HD (1998) Formulation of microbial biopesticides: beneficial microorganisms, nematodes, and seed treatments. Kluwer Academic, Dordrecht, p 412. https://doi.org/10.1007/978-94-011-4926-6

Burkhead KD, Schisler DA, Slininger PJ (1994) Pyrrolnitrin production by biological control agent *Pseudomonas cepacia* b37w in culture and in colonized wounds of potatoes. Appl Environ Microbiol 60:2031–2039

Cameron DD, Neal AL, Van Wees SCM, Ton J (2013) Mycorrhiza-induced resistance: more than the sum of its parts? Trends Plant Sci 18:539–545

Cao L, Qiu Z, Dai X, Tan H, Lin Y, Zhou S (2004) Isolation of endophytic actinomycetes from roots and leaves of banana (*Musa acuminata*) plants and their activities against *Fusarium oxysporum* f. sp. *cubense*. World J Microbiol Biotechnol 20:501–504

Carrillo C, Teruel JA, Aranda FJ, Ortiz A (2003) Molecular mechanism of membrane permeabilization by the peptide antibiotic surfactin. Biochim Biophys Acta 16(11):91–97

Carson KC, Meyer JM, Dilworth MJ (2000) Hydroxamate siderophores of root nodule bacteria. Soil Biol Biochem 32:11–21

Castillo HFD, Reyes CF, Morales GG, Herrera RR, Aguilar C (2013) Biological control of root pathogens by plant-growth promoting *Bacillus* spp. https://doi.org/10.5772/54229

Chandrasekaran M, Subramanian D, Yoon E, Taehoon Kwon CS (2016) Meta-analysis reveals that the genus *Pseudomonas* can be a better choice of biological control agent against bacterial wilt disease caused by *Ralstonia solanacearum*. Plant Pathol J 32:216–227

Chang I, Kommedahl T (1968) Biological control of seedling blight of corn by coating kernels with antagonistic microorganisms. Phytopathology 58:1395–1401

Chen ZX, Dickson DW (1998) Review of *Pasteuria penetrans*: biology, ecology, and biological control potential. J Nematol 30(3):313–340

Chen Y, Xu H, Zhou M, Wang Y, Wang S, Zhang J (2015) Salecan enhances the activities of β-1,3-glucanase and decreases the biomass of soilborne fungi. PLoS One 10(8):e0134799. https://doi.org/10.1371/journal.pone.0134799

Chernin L, Brandis A, Ismailov Z, Chet I (1996) Pyrrolnitrin production by an *Enterobacter agglomerans* strain with a broad spectrum of antagonistic activity towards fungal and bacterial phytopathogens. Curr Microbiol 32:208–212

Cho C-F, Lee W-C (1999) Formulation of a biocontrol agent by entrapping biomass of *Trichoderma viride* in gluten matrix. J Biosci Bioeng 87(6):822–824

Cochrane SA, Surgenor RR, Khey KMV, Vederas JC (2015) Total synthesis and stereochemical assignment of the antimicrobial lipopeptide cerexin A. Org Lett 17(21):5428–5431

Collins DP, Jacobsen BJ (2003) Optimizing a *Bacillus subtilis* isolate for biocontrol of sugar beet *Cercospora* leaf spot. Biol Control 26(2):153–161

Compant S, Duffy B, Nowak J, Clèment C, Barka EA (2005) Use of plant growth-promoting bacteria for biocontrol of plant diseases: principles, mechanisms of action, and future prospects. Appl Environ Microbiol 71:4951–4959

Contreras-Cornejo HA, Macias-Rodriguez L, Cortes-Penagos C, Lopez-Bucio J (2009) *Trichoderma virens*, a plant beneficial fungus, enhances biomass production and promotes lateral root growth through an auxin-dependent mechanism in *Arabidopsis*. Plant Physiol 149:1579–1592

Cook RJ (1993) Making greater use of introduced microorganisms for biological control of plant pathogens. Annu Rev Phytopathol 31:53–80

Cordier C, Gianinazzi S, Gianinazzi-Pearson V (1996) Colonisation patterns of root tissues by fungus *Glomus mosseae* in tomato plants. Plant Soil 185:199–209

Cotxarrera L, Trillas-Gay M, Steinberg C, Alabouvette C (2002) Use of sewage sludge compost and *Trichoderma asperellum* isolates to suppress *Fusarium* wilt of tomato. Soil Biol Biochem 34 (4):467–476

Couillerot O, Prigent-Combaret C, Caballero-Mellado J, Moënne-Loccoz Y (2009) *Pseudomonas fluorescens* and closely related fluorescent pseudomonads as biocontrol agents of soilborne phytopathogens. Lett Appl Microbiol 48(5):505–512

Dahiya JS, Woods DL, Tewari JP (1988) Control of *Rhizoctonia solani*, causal agent of brown girdling root rot of rapeseed, by *Pseudomonas fluorescens*. Bot Bull Acad Sinica 29:135–142

Dandurand LM, Knudsen GR (1993) Influence of *Pseudomonas fluorescens* on hyphal growth and biocontrol activity of *Trichoderma harzianum* in the spermosphere and rhizosphere of pea. Phytopathology 83:265–270

Datnoff L, Nemec S, Pernezny K (1995) Biological control of *Fusarium* crown and root rot of tomato in Florida using *Trichoderma harzianum* and *Glomus intraradices*. Biol Control 5 (3):427–431

de Vasconcellos RLF, Cardoso EJBN (2009) Rhizospheric streptomycetes as potential biocontrol agents of *Fusarium* and *Armillaria* pine rot and as PGPR for *Pinus taeda*. BioControl 54:807–816

de Weert S, Vermeiren H, Mulders IH, Mulders IH, Kuiper I, Hendrickx N, Bloemberg GV, Vanderleyden J, De Mot R, Lugtenberg BJ (2002) Flagella driven chemotaxis towards exudate components is an important trait for tomato root colonization by *Pseudomonas fluorescens*. Mol Plant Microbe Interact 15(11):1173–1180

DeCleene M, DeLey J (1976) The host range of crown gall. Bot Rev 42:389–466

Deketelaere S, Tyvaert L, França SC, Höfte M (2017) Desirable traits of a good biocontrol agent against *Verticillium* Wilt. Front Microbiol 8:1186. https://doi.org/10.3389/fmicb.2017.01186

Diaz Arias MM, Leandro LF, Munkvold GP (2013) Aggressiveness of *Fusarium* species and impact of root infection on growth and yield of soybean. Phytopathology 103(8):822–832

Dobbelaere S, Vanderleyden J, Okon Y (2003) Plant growth-promoting effects of diazotrophs in the rhizosphere. Crit Rev Plant Sci 22:107–149

Doornbos RF, Loon LC, Bakker PAHM (2012) Impact of root exudates and plant defense signaling on bacterial communities in the rhizosphere. Agron Sustain Dev 32:227–243

Driks A (2004) The *Bacillus* spore coat. Phytopathology 94(11):1249–1251

Duijff BJ, Pouhair D, Olivain C, Alabouvette C, Lemanceau P (1998) Implication of systemic induced resistance in the suppression of *Fusarium* wilt of tomato by *Pseudomonas fluorescens* WCS417r and by nonpathogenic *Fusarium oxysporum* Fo47. Eur J Plant Pathol 104:903–910

Ellis RA, Roberts EH (1981) The quantification of ageing and survival in orthodox seeds. Seed Sci Technol 9:373–409

Elphinstone JG (2005) The current bacterial wilt situation: a global overview. In: Allen C, Prior P, Hayward AC (eds) Bacterial wilt disease and the *Ralstonia solanacearum* species complex. APS Press, St Paul, pp 9–28

Elsen A, Declerck S, De Wasele D (2002) Effects of three arbuscular mycorrhizal fungi on root knot nematode (*Meloidogyne* spp.) infection of *Musa*. Infomusa 11:21–23

Elsen A, Gervacio D, Swennen R, De Waele D (2008) AMF-induced biocontrol against plant-parasitic nematodes in *Musa* sp.: a systemic effect. Mycorrhiza 18:251–256

Emmert EA, Handelsman J (1999) Biocontrol of plant diseases: a (Gram-) positive perspective. FEMS Microbiol Lett 171:1–9

Escobar MA, Dandekar AM (2003) *Agrobacterium tumefaciens* as an agent of diseases. Trends Plant Sci 8:380–385

Evangelista-Martinez Z (2014) Isolation and characterization of soil *Streptomyces* species as potential biological control agents against fungal plant pathogens. World J Microbiol Biotechnol 30:1639–1647

Fiddman PJ, Rossall S (1993) The production of antifungal volatiles by *Bacillus subtilis*. J Appl Bacteriol 74:119–126

Fradin EF, Thomma BP (2006) Physiology and molecular aspects of *Verticillium* wilt diseases caused by *V. dahliae* and *V. albo-atrum*. Mol Plant Pathol 7:71–86

Franco-Correa M, Quintana A, Duque C, Suarez C, Rodriguez MX, Barea JM (2010) Evaluation of actinomycete strains for key traits related with plant growth promotion and mycorrhiza helping activities. Appl Soil Ecol 45:209–217

Fravel D (1988) Role of antibiosis in the biocontrol of plant diseases. Annu Rev Phytopathol 26:75–91

Fravel D, Olivain C, Alabouvette C (2003) *Fusarium oxysporum* and its biocontrol. New Phytol 157(3):493–502

García LE, Sánchez-Puerta MV (2012) Characterization of a root-knot nematode population of *Meloidogyne arenaria* from Tupungato (Mendoza, Argentina). J Nematol 44(3):291–301

Gardner JM, Chandler L, Feldman AW (1984) Growth promotion and inhibition by antibiotics producing fluorescent pseudomonads on citrus root. Plant Soil 77:103–113

Georgakopoulos DG, Fiddaman P, Leifert C, Malathrakis NE (2002) Biological Control of cucumber and sugar beet damping-off caused by *Pythium ultimum* with bacterial and fungal antagonists. J Appl Microbiol 92:1078–1086

Ghazalibiglar H, Kandula DRW, Hampton JG (2016) Biological control of *Fusarium* wilt of tomato by *Trichoderma* isolates. NZ Plant Prot 69:57–63

Gheysen G, Mitchum MG (2011) How nematodes manipulate plant development pathways for infection. Curr Opin Plant Biol 14:415–421

Gianinazzi-Pearson V, Gollotte A, Dumas-Gaudot E, Franken P, Gianinazzi S (1994) Gene expression and molecular modifications associated with plant responses to infection by arbuscular mycorrhizal fungi. In: Daniels M, Downic JA, Osbourn AE (eds) Advances in molecular genetics of plant-microbe interactions. Kluwer, Dordrecht, pp 179–186

Glick BR (1995) The enhancement of plant growth promotion by free living bacteria. Can J Microbiol 41(2):109–117

Guédez C, Castillo C, Cañizales L, Olivar R (2008) Biological control a tool for sustaining and sustainable development. Control Biol 7(13):50–74

Guetsky R, Shtienberg D, Elad Y, Dinoor A (2001) Combining biocontrol agents to reduce the variability of biological control. Phytopathology 91:621–627

Guetsky R, Shtienberg D, Elad Y, Fischer E, Dinoor A (2002) Improving biological control by combining biocontrol agents each with several mechanisms of disease suppression. Phytopathology 92:976–985

Guillén-Cruz R, Hernández-Castillo FD, Gallegos-Morales G, Rodríguez-Herrera R, Aguilar-González CN, PadrónCorral E, Reyes-Valdés MH (2006) *Bacillus* spp. como biocontrol en un suelo infestado con *Fusarium* spp., *Rhizoctonia solani* Kühn y *Phytophthora capsici* Leonina y su efecto en el desarrollo y rendimiento del cultivo de chile (*Capsicum annuum* L.) Rev Mex Fitopatol 24:105–114

Haas D, Défago G (2005) Biological control of soilborne pathogens by fluorescent pseudomonads. Nat Rev Microbiol 3:307–319

Haas D, Keel C (2003) Regulation of antibiotic production in root-colonizing *Pseudomonas* spp. and relevance for biological control of plant disease. Annu Rev Phytopathol 41:117–153

Handelsman J, Stabb EV (1996) Biocontrol of soilborne plant pathogens. Plant Cell 8 (10):1855–1869

Hanlon DW, Rosario MML, Ordal GW, Venema G, Van Sinderen D (1994) Identification of TipC, a novel 62 kDa MCP-like protein from *Bacillus subtilis*. Microbiologica 140:1847–1856

Hao Z, Fayolle L, van Tuinen D, Chatagnier O, Xiaolin L, Gianinazzi S, Gianinazzi-Pearson V (2012) Local and systemic mycorrhiza-induced protection against the ectoparasitic nematode *Xiphinema index* involves priming of defence gene responses in grapevine. J Exp Bot 63:3657–3672

Harman GE, Howell CR, Viterbo A, Chet I, Lorito M (2004) *Trichoderma* species – opportunistic, avirulent plant symbionts. Nat Rev Microbiol 2:43–56

Harrison MJ (2005) Signaling in the arbuscular mycorrhizal symbiosis. Annu Rev Microbiol 59:19–42

Hassan MN, Afghan S, Hafeez FY (2011) Biological control of red rot in sugarcane by native pyoluteorin-producing *Pseudomonas putida* strain NH-50 under field conditions and its potential modes of action. Pest Manag Sci 67(9):1147–1154

Heydari A, Pessarakli M (2010) A review on biological control of fungal plant pathogens using microbial antagonists. J Biol Sci 10(4):273–290

Hiltpold I, Hibbard B, French BW, Turlings TJ (2012) Capsules containing entomopathogenic nematodes as a Trojan horse approach to control the western corn rootworm. Plant Soil 358:11–25

Hiraoka H, Asaka O, Ano T, Shoda M (1992) Characterization of *Bacillus subtilis* RB14, coproducer of peptide antibiotics Iturin A and Surfactin. J Gen Appl Microbiol 38:635–640

Hopkins RJ, van Dam NM, van Loon JJA (2009) Role of glucosinolates in insect-plant relationships and multitrophic interactions. Annu Rev Entomol 54:57–83

Howell C (2003) Mechanisms employed by *Trichoderma* species in the biological control of plant diseases: the history and evolution of current concepts. Plant Dis 87(1):4–10

Howell CR, Stipanovic RD (1979) Control of *Rhizoctonia solani* on cotton seedlings with *Pseudomonas fluorescens* and with an antibiotic produced by the bacterium. Phytopathology 69:480–482

Howell CR, Stipanovic RD (1980) Suppression of *Pythium ultimum*-induced damping off of cotton seedlings by *Pseudomonas fluorescens* and its antibiotic pyoluteorin. Phytopathology 70 (8):712–715

Huang Y, Xu C, Ma L, Zhang K, Duan C, Mo M (2010) Characterization of volatiles produced from *Bacillus megaterium* YFM 3.25 and their nematicidal activity against *Meloidogyne incognita*. Eur J Plant Pathol 126:417–422

Hussain I, Alam SS, Khan I, Shah B, Naeem A, Khan N, Ullah W, Adnan M, Shah SRA, Junaid K, Ahmed N, Iqbal M (2016) Medicinal plants rhizosphere exploration for the presence of potential biocontrol fungi. J Entomol Zool Stud 4(3):108–113

Hussein AA, AL-Janabi S (2006) Identification of bacitracin produced by local isolate of *Bacillus licheniformis*. Afr J Biotechnol 18:1600–1601

Inderbitzin P, Subbarao KV (2014) *Verticillium* systematics and evolution: how confusion impedes *Verticillium* wilt management and how to resolve it. Phytopathology 104:564–574

Jacobsen BJ, Zidack NK, Larson BJ (2004) The role of bacillus-based biological control agents in integrated pest management systems: plant diseases. Phytopathology 94:1272–1275

Jayaseelan S, Ramaswamy D, Dharmaraj S (2014) Pyocyanin: production, applications, challenges and new insights. World J Microbiol Biotechnol 30(4):1159–1168

John RP, Tyagi RD, Brar SK, Surampalli RY, Prevost D (2011) Bio-encapsulation of microbial cells for targeted agricultural delivery. Crit Rev Biotechnol 31:211–226

Jonathan EI, Barker KR, Abdel-Alim FF, Vrain TC, Dickson DW (2000) Biological control of *Meloidogyne incognita* on tomato and banana with rhizobacteria, actinomycetes, and *Pasteuria penetrans*. Nematropica 30:231–240

Jones KA, Burges HD (1998) Technology of formulation and application. In: Burges HD (ed) Formulation of microbial pesticides – beneficial microorganisms, nematodes and seed treatments. Springer, Dordrecht, pp 7–30

Judelson HS, Blanco FA (2005) The spores of *Phytophthora*: weapons of the plant destroyer. Nat Rev Microbiol 3:47–58

Junaid JM, Dar NA, Bhat TA, Bhat AH, Bhat MA (2013) Commercial biocontrol agents and their mechanism of action in the management of plant pathogens. Int J Mod Plant Anim Sci 1 (2):39–57

Jung SC, Martinez-Medina A, Lopez-Raez JA, Pozo MJ (2012) Mycorrhiza-induced resistance and priming of plant defenses. J Chem Ecol 38:651–664

Kai M, Effmert U, Berg G, Piechulla B (2007) Volatiles of bacterial antagonists inhibit mycelial growth of the plant pathogen *Rhizoctonia solani*. Arch Microbiol 187:351–360

Kalbe C, Marten P, Berg G (1996) Members of the genus *Serratia* as beneficial rhizobacteria of oilseed rape. Microbiol Res 151:433–439

Kariuki GM, Dickson DW (2007) Transfer and development of *Pasteuria penetrans*. J Nematol 39 (1):55–61

Katz E, Demain AL (1977) The peptide antibiotics of *Bacillus*: chemistry, biogenesis, and possible functions. Bacteriol Rev 41(2):449–474

Kaya K, Gaugler R (1993) Entomopathogenic nematodes. Annu Rev Entomol 38:181–206

Kazmar ER, Goodman RM, Grau CR, Johnson DW, Norrdheim EV, Undersander KJ, Handelsman J (2000) Regression analyses for evaluating the influence of *Bacillus cereus* on alfalfa yield under variable disease intensity. Phytopathology 90:657–665

Keel C, Défago G (1997) Interactions between beneficial soil bacteria and root pathogens: mechanisms and ecological impact. In: Gange AC, Brown VK (eds) Multitrophic interactions in terrestrial system. Blackwell Science, Oxford, pp 27–47

Keel C, Voisard C, Berling CH, Kadr G, Defago G (1989) Iron sufficiency, a prerequisite for the suppression of tobacco root rot by *Pseudomonas fluorescens* strain CHAO under gnotobiotic conditions. Phytopathology 79:584–589

Kennedy BW (1980) Estimates of U.S. crop losses to prokaryote plant pathogens. Plant Dis 647:674–676

Kergunteuil A, Bakhtiari M, Formenti L, Xiao Z, Defossez E, Rasmann S (2016) Biological control beneath the feet: a review of crop protection against insect root herbivores. Insects 7(70):1–22. https://doi.org/10.3390/insects7040070

Keswani C, Bisen K, Singh V, Sarma BK, Singh HB (2016) Formulation technology of biocontrol agents: present status and future prospects. In: Arora NK, Mehnaz S, Balestrini R (eds) Bioformulations: for sustainable agriculture. Springer, New Delhi, p 299. https://doi.org/10.1007/978-81-322-2779-3

Khaledi N, Taheri P (2016) Biocontrol mechanisms of *Trichoderma harzianum* against soybean charcoal rot caused by *Macrophomina phaseolina*. J Plant Prot Res 56(1):21–31

Khan MR, Fischer S, Egan D, Doohan FM (2006) Biological control of *Fusarium* seedling blight disease of wheat and barley. Phytopathology 96:386–394

Khan AAH, Naseem RL, Prathibha B (2011) Screening and potency evaluation of antifungal from soil isolates of *Bacillus subtilis* on selected fungi. Adv Biotechnol 10(7):35–37

Kiewnick S, Sikora RA (2006) Biological control of the root-knot nematode *Meloidogyne incognita* by *Paecilomyces lilacinus* strain 251. Biol Control 38:179–187

Kim J, Jaffuel G, Turlings TJ (2015) Enhanced alginate capsule properties as a formulation of entomopathogenic nematodes. Biol Control 60:527–535

Kim YS, Balaraju K, Jeon YH (2017) Biological characteristics of *Bacillus amyloliquefaciens* AK-0 and suppression of ginseng root rot caused by *Cylindrocarpon destructans*. J Appl Microbiol 122:166–179

Kiss L (2003) A review of fungal antagonists of powdery mildews and their potential as bioagents. Pest Manag Sci 59:475–483

Kissen R, Rossiter JT, Bones AM (2009) The "mustard oil bomb": not so easy to assemble?! Localization, expression and distribution of the components of the myrosinase enzyme system. Phytochem Rev 8:69–86

Kloepper JW, Lifshitz R, Novacky A (1988) *Pseudomonas* inoculation to benefit plant production. Anim Plant Sci 8:60–64

Kloepper JW, Ryu C-M, Zhang S (2004) Induced systemic resistance and promotion of plant growth by *Bacillus* spp. Phytopathology 94(11):1259–1266

Koike ST, Subbarao KV, Davis RM, Turini TA (2003) Vegetable diseases caused by soilborne pathogens. University of California, Publication 8099. http://anrcatalog.ucdavis.edu

Kumar RS, Ayyadurai N, Pandiaraja P, Reddy AV, Venkateswarlu Y, Prakash O, Sakthivel N (2005) Characterization of antifungal metabolite produced by a new strain *Pseudomonas aeruginosa* PUPa3 that exhibits broad spectrum antifungal activity and biofertilizing traits. J Appl Microbiol 98(1):145–154

Kumar A, Prakash A, Johri BN (2011) *Bacillus* as PGPR in crop ecosystem. In: Maheshwari DK (ed) Bacteria in agrobiology: Crop ecosystem. Springer, Berlin, pp 37–59

Labuschagne N, Pretorius T, Idris AH (2010) Plant growth promoting rhizobacteria as biocontrol agents against soilborne plant diseases. In: Maheshwari DK (ed) Plant growth and health promoting bacteria Microbiology Monographs 18. https://doi.org/10.1007/978-3-642-13612-2_9

Lacey LA, Georgis R (2012) Entomopathogenic nematodes for control of insect pests above and below ground with comments on commercial production. J Nematol 44:218–225

Lacey LA, Grzywacz D, Shapiro-Ilan DI, Frutos R, Brownbridge M, Goettel MS (2015) Insect pathogens as biological control agents: back to the future. J Invertebr Pathol 132:1–41

Lambers H, Mougel C, Jaillard B, Hinsinger P (2009) Plant-microbe-soil interactions in the rhizosphere: an evolutionary perspective. Plant Soil 321:83–115

Leggett M, Leland J, Kellar K, Epp B (2011) Formulation of microbial biocontrol agents – an industrial perspective. Can J Plant Pathol 33(2):101–107

Leong J (1986) Siderophores: their biochemistry, and possible role in the biocontrol of plant pathogens. Annu Rev Phytopathol 24:187–209

Li J, Yang Q, Zhao L, Zhang S, Wang Y, Zhao X (2009) Purification and characterization of a novel antifungal protein from *Bacillus subtilis* strain B29. J Zhejiang Univ Sci B 10(4):264–272

Lifshitz R, Kloepper JW, Mozlowski M, Simonson C, Carlson J, Tipping EM, Zaleska I (1987) Growth promotion of canola (rapeseed) seedlings by a strain of *Pseudomonas putida* under gnotobiotic conditions. Can J Microbiol 33:390–395

Ligon JM, Hill DS, Hammer PE, Torkewitz NR, Hofmann D, Kempf H-J, van Pée K-H (2000) Natural products with antifungal activity from *Pseudomonas* biocontrol bacteria. Pest Manag Sci 56:688–695. https://doi.org/10.1002/1526-4998(200008)56:8<688::AID-PS186>3.0.CO;2-V

Lioussanne L, Jolicoeur M, St-Arnaud M (2008) Mycorrhizal colonization with *Glomus intraradices* and development stage of transformed tomato roots significantly modify the chemotactic response of zoospores of the pathogen *Phytophthora nicotianae*. Soil Biol Biochem 40(9):2217–2224

Liu J, Maldonado-Mendoza I, Lopez-Meyer M, Cheung F, Town CD, Harrison MJ (2007) Arbuscular mycorrhizal symbiosis is accompanied by local and systemic alterations in gene expression and an increase in disease resistance in the shoots. Plant J 50:529–544

Logan NA, Halket G (2011) Developments in the taxonomy of aerobic, endospores-forming bacteria. In: Longan N, Vos P (eds) Endospore-forming soil bacteria. Soil biology, vol 27 Springer, Berlin, pp 1–29

Loper JE, Buyer JW (1991) Siderophores in microbial interactions on plant surfaces. Mol Plant-Microbe Int 4:5–13

Loria R, Kers J, Joshi M (2006) Evolution of plant pathogenicity in *Streptomyces*. Annu Rev Phytopathol 44:469–487

Lucy M, Reed E, Glick BR (2004) Applications of free living plant growth-promoting rhizobacteria. Antonie van Leeuwenhoek Int J Gen Mol Microbiol 86:1–25

Lugtenberg B, Kamilova F (2009) Plant-growth-promoting rhizobacteria. Annu Rev Microbiol 63:541–556

Lumsden RD, Rldout CJ, Vendemia ME, Harrlson DJ, Waters RM, Walter JF (1992) Characterization of major secondary metabolites produced in soilless mix by a formulated strain of the biocontrol fungus *Gliocladium virens*. Can J Microbiol 38:1274–1280

Malik RJ, Dixon MH, Bever JD (2016) Mycorrhizal composition can predict foliar pathogen colonization in soybean. Biol Control 103:46–53

Mansfield J, Genin S, Magori S, Citovsky V, Sriariyanum M, Ronald P, Dow M, Verdier V, Beer SV, Machado MA, Toth I, Salmond G, Foster GD (2012) Top 10 plant pathogenic bacteria in molecular plant pathology. Mol Plant Pathol 13:614–629

Marro N, Lax P, Cabello M, Doucet ME, Becerra AG (2014) Use of the arbuscular mycorrhizal fungus *Glomus intraradices* as biological control agent of the nematode *Nacobbus aberrans* parasitizing tomato. Braz Arch Biol Technol 57(5):668–675

Mavrodi DV, Mavrodi OV, Parejko JA, Bonsall RF, Kwak Y-S, Paulitz TC, Thomashow LS, Weller DM (2012) Accumulation of the antibiotic phenazine-1-carboxylic acid in the rhizosphere of dryland cereals. Appl Environ Microbiol 78(3):804–812

Mazurier S, Corberand T, Lemanceau P, Raaijmakers JM (2009) Phenazine antibiotics produced by fluorescent pseudomonads contribute to natural soil suppressiveness to *Fusarium* wilt. ISME J 3 (8):977–991

Mehta CM, Palni U, Franke-Whittle IH, Sharma AK (2014) Compost: its role, mechanism and impact on reducing soilborne plant diseases. Waste Manag 34:607–622

Miller JR, Hare EW, Wu J (1990) Quantitative characterization of the vegetation red edge reflectance. I. An inverted-Gaussian reflectance model. Int J Rem Sens 11:1755–1773

Moens M, Perry RN, Starr JL (2009) *Meloidogyne* species – a diverse group of novel and important plant parasites. In: Perry RN, Moens M, Starr JL (eds) Root knot nematodes. CAB International, Oxfordshire

Motlagh MRS, Samimi Z (2013) Evaluation of *Trichoderma* spp., as biological agents in some of plant pathogens. Ann Biol Res 4(3):173–179

Nalisha I, Muskhazli M, Nor Farizan T (2006) Production of bioactive compounds by *Bacillus subtilis* against *Sclerotium rolfsii*. Mal J Microbiol 2(2):19–23

Nawar LS (2016) Interactions between vascular arbuscular mycorrhizal fungi and *Streptomyces* as biocontrol agent for tomato damping-off disease caused by *Rhizoctonia Solani* Kuhn. IOSR-JPBS 11(6):89–96

Nega A (2014) Review on concepts in biological control of plant pathogens. J Biol Agri Healthcare 4(27):33–54
Neilands JB (1989) Siderophore systems of bacteria and fungi. In: Beveridge TJ, Doyle RJ (eds) Metal ions and bacteria. Wiley, Somerset, pp 141–164
Nielsen MN, Sorensen J (1999) Chitinolytic activity of *Pseudomonas fluorescens* isolates from barley and sugar beet rhizosphere. FEMS Microbiol Ecol 30(3):217–227
Nuss DL (2005) Hypovirulence: mycoviruses at the fungal plant interface. Nat Rev Microbiol 3:632–642
O'Bannon J, Inserra R, Nemec S, Vovlas N (1979) The influence of *Glomus mosseae* on *Tylenchulus semipenetrans*-infected and uninfected *Citrus limon* seedlings. J Nematol 11:247–250
Ongena M, Daayf F, Jacques P, Thonart P, Benhamou N, Paulitz TC, Cornelis P, Koedam N, Belanger RR (1999) Protection of cucumber against *Pythium* root rot by fluorescent pseudomonads: predominant role of induced resistance over siderophores antibiosis. Plant Pathol 48:66–76
Otten L, Burr T, Szegedi E (2008) *Agrobacterium*: a disease-causing bacterium. In: Tzfira T, Citovsky V (eds) *Agrobacterium*: From biology to biotechnology. Springer, New York, pp 1–46. https://doi.org/10.1007/978-0-387-72290-0_1
Pegg K, Manners A (2014a) Soil borne root pathogens in production nurseries. Nursery production plant health & biosecurity project, Plant health biosecurity, risk management and capacity building for the nursery industry. Agri-science Queensland, Department of Agriculture, Fisheries and Forestry (DAFF)
Pegg K, Manners A (2014b) *Rhizoctonia* – A variable and versatile nursery pathogen. Nursery production plant health & biosecurity project, Plant health biosecurity, risk management and capacity building for the nursery industry. Agri-science Queensland, Department of Agriculture, Fisheries and Forestry (DAFF)
Pelletier A, Sygusch J (1990) Purification and characterization of three chitosanase activities from *Bacillus megaterium* P1. Appl Environ Microbiol 56(4):844–848
Pereira P, Nesci A, Etcheverrg MG (2009) Efficacy of bacterial seed treatments for the control of *Fusarium verticillioides* in maize. Biol Control 54:103–111
Pereira JAP, Vieira IJC, Freitas MSM, Prins CL, Martins MA, Rodrigues R (2016) Effects of arbuscular mycorrhizal fungi on *Capsicum* spp. J Agric Sci 154(5):828–849
Perry RN, Moens M (2011) Introduction to plant-parasitic nematodes; modes of parasitism. In: Jones JT, Gheysen L, Fenoll C (eds) Genomics and molecular genetics of plant–nematode interactions. Springer, Heidelberg, pp 3–20
Pierre E, Louise NW, Marie TKR, Valère TFP, Arc-en-ce JM, Fekam BF (2016) Integrated assessment of phytostimulation and biocontrol potential of endophytic *Trichoderma* spp against common bean (*Phaseolus vulgaris* L.) root rot fungi complex in centre region, Cameroon. Int J Pure App Biosci 4(4):50–68
Pieterse CMJ, van Pelt JA, Verhagen BWM, Ton J, van Wees ACM, Léon-Kloosterziel KM, van Loon LC (2003) Induced systemic resistance by plant growth-promoting rhizobacteria. Symbiosis 35(1–3):39–54
Pieterse CMJ, Zamioudis C, Berendsen RL, Weller DM, Van Wees SCM, Bakker PAHM (2014) Induced systemic resistance by beneficial microbes. Annu Rev Phytopathol 52:347–375
Pimentel D (2009) Environmental and economic costs of the application of pesticides primarily in the United States. In: Peshin R, Dhawan AK (eds) Integrated pest management: innovation-development process, vol 1. Springer, Dordrecht, pp 89–111
Pineda A, Zheng SJ, van Loon JJA, Pieterse CMJ, Dicke M (2010) Helping plants to deal with insects: the role of beneficial soilborne microbes. Trends Plant Sci 15:507–514
Pleban S, Chernin L, Shet I (1997) Chitinolytic activity of an endophytic strain of *Bacillus cereus*. Lett Appl Microbiol 25(1):284–288

Poinar GO (1990) Taxonomy and biology of Steinernematidae and Heterorhabditidae. In: Gaugler R, Kaya HK (eds) Entomopathogenic nematodes in biological control. CRC Press, Boca Raton

Ponmurugan P, Baby UI (2007) Evaluation of fungicides and biocontrol agents against Phomopsis canker of tea under field conditions. Aust Plant Pathol 36:68–72

Potter MJ, Davies K, Rathjen AJ (1998) Suppressive impact of glucosinolates in *Brassica* vegetative tissues on root lesion nematode *Pratylenchus neglectus*. J Chem Ecol 24:67–80

Pozo MJ, Azcon-Aguilar C (2007) Unraveling mycorrhiza-induced resistance. Curr Opin Plant Biol 10(4):393–398

Priest FG (1977) Extracellular enzyme synthesis in the genus *Bacillus*. Bacteriol Rev 41 (3):711–753

Rahman L (2003) Root knot diseases and its control. Agfact AB1, 3rd edn, pp 1–10

Rahman M (2016) *Bacillus* spp.: a promising biocontrol agent of root, foliar and postharvest diseases of plants. In: Islam MT, Rahman MM, Pandey P, Jha CK, Aeron A (eds) Bacilli and agrobiotechnology. Springer, New York, p 416. https://doi.org/10.1007/978-3-319-44409-3_6

Ralmi NHAA, Khandaker MM, Mat N (2016) Occurrence and control of root knot nematode in crops: a review. Aust J Crop Sci (AJCS) 10(12):1649–1654

Ramadan EM, AbdelHafez AA, Hassan EA, Saber FM (2016) Plant growth promoting rhizobacteria and their potential for biocontrol of phytopathogens. Afr J Microbiol Res 10 (15):486–504

Raza W, Ling N, Yang L, Huang Q, Shen Q (2016a) Response of tomato wilt pathogen *Ralstonia solanacearum* to the volatile organic compounds produced by a biocontrol strain *Bacillus amyloliquefaciens* SQR-9. Sci Rep 6:24856. https://doi.org/10.1038/srep24856

Raza W, Ling N, Liu D, Wei Z, Huang Q, Shen Q (2016b) Volatile organic compounds produced by *Pseudomonas fluorescens* WR-1 restrict the growth and virulence traits of *Ralstonia solanacearum*. Microbiol Res 192:103–113

Rosendahl CN, Rosendahl S (1990) The role of vesicular arbuscular mycorrhizal fungi in controlling damping-off and growth reduction in cucumber caused by *Pythium ultimum*. Symbiosis 9:363–366

Rothrock CS, Gottlieb D (1984) Role of antibiosis in antagonism of *Streptomyces hygroscopicus* var *geldanus* to *Rhizoctonia solani* in soil. Can J Microbiol 30:1440–1447

Ryder MH, Yan Z, Terrace TE, Rovira AD, Tang W, Correll RL (1999) Use of *Bacillus* isolated in China to suppress take-all and rhizoctonia root rot, and promote seedling growth of glass house grown wheat in Australian soils. Soil Biol Biochem 31:19–29

Sabaté DC, Brandan CP, Petroselli G, Erra-Balsells R, Audisio MC (2017) Decrease in the incidence of charcoal root rot in common bean (*Phaseolus vulgaris* L.) by *Bacillus amyloliquefaciens* B14, a strain with PGPR properties. Biol Control 113:1–8

Sadfi N, Chérif M, Fliss I, Boudabbous A, Antoun H (2001) Evaluation of *Bacillus* isolates from salty soils and *Bacillus thuringiensis* strains for the biocontrol of *Fusarium* dry rot of potato tubers. J Plant Pathol 83:101–118

Scarpellini M, Franzetti L, Galli A (2004) Development of PCR assay to identify *Pseudomonas fluorescens* and its biotype. FEMS Microbiol Lett 236:257–260

Schippers B, Bakker AW, Bakker PAHM (1987) Interactions of deleterious and beneficial rhizosphere microorganisms and the effect of cropping practices. Annu Rev Phytopathol 25 (1):339–358

Schisler DA, Slininger PJ, Behle RW, Jackson MA (2004) Formulation of *Bacillus* spp. for biological control of plant diseases. Phytopathology 94(11):1267–1271

Schouteden N, De Waele D, Panis B, Vos CM (2015) Arbuscular mycorrhizal fungi for the biocontrol of plant-parasitic nematodes: a review of the mechanisms involved. Front Microbiol 6:1280. https://doi.org/10.3389/fmicb.2015.01280

Schroth MN, Loper JE, Hildebrand DC (1984) Bacteria as biocontrol agents of plant disease. In: Klug MJ, Reddy CA (eds) Current perspectives in microbial ecology. American Society for Microbiology, Washington, DC, pp 362–369

Schünemann R, Knaak N, Fiuza LM (2014) Mode of action and specificity of *Bacillus thuringiensis* toxins in the control of caterpillars and stink bugs in soybean culture. ISRN Microbiol pages12 ID135675. https://doi.org/10.1155/2014/135675

Shafi J, Tian H, Ji M (2017) *Bacillus* species as versatile weapons for plant pathogens: a review. Biotechnol Biotechnol Equip 31(3):446–459

Sharma IP, Sharma AK (2017) Physiological and biochemical changes in tomato cultivar PT-3 with dual inoculation of mycorrhiza and PGPR against root-knot nematode. Symbiosis 71:175–183

Shirifi A, Zala M, Natsch A, Dejaga G (1998) Biocontrol of soil borne fungi plant diseases by 2,4-diacetlyphloroglucinol producing *Pseudomonas fluorescens*. Eur J Plant Pathol 104:631–643

Shoji J, Hinoo H, Wakisaka Y, Koizumi K, Mayama M, Matsuura S, Matsumoto K (1975) Isolation of two new related peptide antibiotics, cerexin A and B. J Antibiot 28:56–59

Shrivastava P, Kumar R (2015) Soil salinity: a serious environmental issue and plant growth promoting bacteria as one of the tools for its alleviation. Saudi J Biol Sci 22:123–131

Siasou E, Standing D, Killham K, Johnson D (2009) Mycorrhizal fungi increase biocontrol potential of *Pseudomonas fluorescens*. Soil Biol Biochem 41:1341–1343

Sid AA, Ezziyyani M, Pérez-Sanchez C, Candela ME (2003) Effect of chitin on biological control activity of *Bacillus* spp. and *Trichoderma harzianum* against root rot disease in pepper (*Capsicum annuum*) plants. Eur J Plant Pathol 109(6):633–637

Siddiqui IA, Shaukat SS, Sheikh IH Khan A (2006) Role of cyanide production by *Pseudomonas fluorescens* CHA0 in the suppression of root-knot nematode, *Meloidogyne javanica* in tomato. World J Microbiol Biotechnol 22(6):641–650

Silo-Suh LA, Lethbridge BJ, Raffel SJ, He H, Clardy J, Handelsman J (1994) Biological activities of two fungistatic antibiotics produced by *Bacillus cereus* UW85. Appl Environ Microbiol 60:2023–2030

Silo-Suh LA, Stabb EV, Raffel SJ, Handelsman J (1998) Target range of zwittermicin A, an aminopolyol antibiotic from *Bacillus cereus*. Curr Microbiol 37(1):6–11

Sindhu SS, Rakshia YS, Sahu G (2009) Biological control of soil borne plant pathogens with rhizosphere bacteria. Pest Control 3:10–21

Singh I (2017) Antimicrobials in higher plants: classification, mode of action and bioactivities. Chem Bio Lett 4(1):48–62

Singh BK, Millard P, Whiteley AS, Murrell JC (2004) Unravelling rhizosphere microbial interactions: opportunities and limitations. Trends Microbiol 12:386–393

Singh DP, Singh HB, Prabha R (2016) Microbial inoculants in sustainable agricultural productivity: functional applications, vol 2. Springer, New York, p 308

Sood GS (2003) Chemotactic response of plant-growth-promoting bacteria towards roots of vesicular-arbuscular mycorrhizal tomato plants. FEMS Microbiol Ecol 45:219–227

Stirling GR (2014) Biological control of plant-parasitic nematodes: soil ecosystem management in sustainable agriculture. Biological Crop Protection/CAB International, Brisbane/Oxfordshire

Taechowisan T, Peberdy JF, Lumyong S (2003) Isolation of endophytic actinomycetes from selected plants and their antifungal activity. World J Microbiol Biotechnol 19:381–385

Tahat MM, Sijam K, Othman R (2012) The potential of endomycorrhizal fungi in controlling tomato bacterial wilt *Ralstonia solanacearum* under glasshouse conditions. Afr J Biotechnol 11 (67):13085–13094

Tari PH, Anderson AJ (1988) *Fusarium* wilt suppression and agglutinability of *Pseudomonas putida*. Appl Environ Microbiol 54(8):2037–2041

Terkina IA, Parfenova VV, Ahn TS (2006) Antagonistic activity of actinomycetes of Lake Baikal. Appl Biochem Microbiol 42:173–176

Thomashow LS, Weller DM (1996) Current concepts in the use of introduced bacteria for biological control: mechanisms and antifungal metabolites. In: Stacey G, Keen NT (eds) Plant-microbe interactions, vol 1. Chapman & Hall, New York, pp 187–235

Thomashow LS, Bonsall RF, Weller DM (1997) Antibiotic production by soil and rhizosphere microbes in situ. In: Hurst CJ, Knudsen GR, McInervey MJ, Stetzenbach LD, Walter MV (eds) Manual of environmental microbiology. ASM Press, Washington, DC, pp 493–499

Thrane C, Harder Nielsen T, Neiendam Nielsen M, Sørensen J, Olsson S (2000) Viscosinamide-producing *Pseudomonas fluorescens* DR54 exerts a biocontrol effect on *Pythium ultimum* in sugar beet rhizosphere. FEMS Microbiol Ecol 33(2):139–146

Timper P (2014) Conserving and enhancing biological control of nematodes. J Nematol 46 (2):75–89

Tranier M-S, Pognant-Gros J, Quiroz RDC, González CNA, Mateille T, Roussos S (2014) Commercial biological control agents targeted against plant-parasitic root-knot nematodes. Braz Arch Biol Technol 57(6):831–841

Tu L, He Y, Shan C, Wu Z (2016) Preparation of microencapsulated *Bacillus subtilis* SL-13 seed coating agents and their effects on the growth of cotton seedlings. Biomed Res Int 1–7. https://doi.org/10.1155/2016/3251357

Van Dijk K, Nelson EB (2000) Fatty acid competition as a mechanism by which *Enterobacter cloacae* suppresses *Pythium ultimum* sporangium germination and damping-off. Appl Environ Microbiol 66:5340–5347

Van Driesche RG, Bellows TS Jr (1996) Methods for biological control of plant pathogens. Biological control. Springer, Boston, pp 235–256

Vemmer M, Patel AV (2013) Review of encapsulation methods suitable for microbial biological control agents. Biol Control 67:380–389

Vessey JK (2003) Plant growth promoting rhizobacteria as biofertilizers. Plant Soil 255:571–586

Vinale F, Sivasithamparam K, Ghisalberti EL, Woo SL, Nigro M, Marra R, Lombardi N, Pascale A, Ruocco M, Lanzuise S, Manganiello G, Lorito M (2014) *Trichoderma* secondary metabolites active on plants and fungal pathogens. Open Mycol J 8:127–139

Von Graevenitz A (1977) The role of opportunistic bacteria in human disease. Annu Rev Microbiol 31:447–471

Vos C, Schouteden N, Tuinen D, Chatagnier O, Elsen A, De Waele D, Panis B, Gianinazzi-Pearson V (2013) Mycorrhiza-induced resistance against the root-knot nematode *Meloidogyne incognita* involves priming of defense gene responses in tomato. Soil Biol Biochem 60:45–54

Webster J, Weber R (2007) Introduction to fungi, 3rd edn. Cambridge University Press, Cambridge

Weibelzahl-Fulton E, Dickson DW, Whitty EB (1996) Suppression of *Meloidogyne incognita* and *M. javanica* by *Pasteuria penetrans* in field soil. J Nematol 28:43–49

Weller DM, Cook RJ (1983) Suppression of take-all of wheat by seed treatments with fluorescent pseudomonads. Phytopathology 73:463–469

Weller DM, Thomashow LS (1994) Current challenges in introducing beneficial microorganisms into the rhizosphere. In: O'Gara F, Dowling DN, Boesten B (eds) Molecular ecology of rhizosphere microorganisms: biotechnology and the release of GMOs. Wiley, Weinheim. https://doi.org/10.1002/9783527615810.ch1

Weller DM, Raajimakers JM, Gardener BBM, Thomashow LS (2002) Microbial populations responsible for specific soil suppressiveness to plant pathogens. Annu Rev Phytopathol 40:309–348

Wesemael WML, Viaene N, Moens M (2011) Roor-knot nematodes (*Meloidogyne* spp.) in Europe. Nematology 13:3–16

Whipps JM (2004) Prospects and limitations for mycorrhizas in biocontrol of root pathogens. Can J Bot 82:1198–1227

Wu L, Wu H, Chen L, Yu X, Borriss R, Gao X (2015) Difficidin and bacilysin from *Bacillus amyloliquefaciens* FZB42 have antibacterial activity against *Xanthomonas oryzae* rice pathogens. Sci Rep 5:12975. https://doi.org/10.1038/srep12975

Yan Z, Reddy MS, Kloepper JW (2003) Survival and colonization of rhizobacteria in tomato transplant system. Can J Microbiol 49:383–389

Yao M, Tweddell R, Désilets H (2002) Effect of two vesicular-arbuscular mycorrhizal fungi on the growth of micropropagated potato plantlets and on the extent of disease caused by *Rhizoctonia solani*. Mycorrhiza 12:235–242

Yigit F, Dikilitas M (2007) Control of *Fusarium* wilt of tomato by combination of fluorescent *Pseudomonas*, non-pathogen *Fusarium* and *Trichoderma harzianum* T-22 in greenhouse conditions. Plant Pathol J 6(2):159–163

Yuan J, Raza W, Shen Q, Huang Q (2012) Antifungal activity of *Bacillus amyloliquefaciens* NJN-6 volatile compounds against *Fusarium oxysporum* f. sp. *Cubense*. Appl Environ Microbiol 78 (16):5942–5944

Yuen GY, Schroth MN (1986) Interaction of *Pseudomonas fluorescens* strains E6 with ornamental plants and its effect on the composition of root colonization microflora. Phytopathology 76:176–179

Zamioudis C, Mastranesti P, Dhonukshe P, Blilou I, Pieterse CMJ (2013) Unraveling root developmental programs initiated by beneficial *Pseudomonas* spp. bacteria. Plant Physiol 162:304–318

Zentmyer GA (1980) *Phytophthora cinnamomi* and the diseases it causes. Monograph No. 10. APS Press, St. Paul

Zhang Y, Loria R (2017) Emergence of novel pathogenic *Streptomyces* species by site-specific accretion and cis-mobilization of pathogenicity islands. Mol Plant Microbe Interact (MPMI) 30 (1):72–82

Zhang JX, Xue AG, Tambong JT (2009) Evaluation of seed and soil treatments with novel *Bacillus subtilis* strains for control of soybean root rot caused by *Fusarium oxysporum* and *F. graminearum*. Plant Dis 93(12):1317–1323

Zhang Q, Ji Y, Xiao Q, Chng S, Tong Y, Chen X, Liu F (2016) Role of Vfr in the regulation of antifungal compound production by *Pseudomonas fluorescens* FD6. Microbiol Res 188–189:106–112. https://doi.org/10.1016/j.micres.2016.04.013

Zhang S, Gan Y, Ji W, Xu B, Hou B, Liu J (2017) Mechanisms and characterization of *Trichoderma longibrachiatum* T6 in suppressing nematodes (*Heterodera avenae*) in wheat. Front Plant Sci 8. https://doi.org/10.3389/fpls.2017.01491

Chapter 8
Biological Control of Root-Knot and Cyst Nematodes Using Nematophagous Fungi

Geeta Saxena

8.1 Introduction

Plant-parasitic nematodes are important pests of crops which have a direct economic impact in reducing crop yield. They are non-segmented tiny eelworms usually measuring about 100–1000 μm in length. The plant-feeding nematodes are sustained by photosynthetic activity of plants and food supply from roots in the form of exudates and exfoliates. Along with other microorganisms, plant-parasitic nematodes are abundant in rhizosphere soil, as they are source of high-quality nutrients. Estimated annual crop loss caused by nematodes worldwide each year is over $100 billion (Chitwood 2003). On a worldwide basis, the ten most damaging genera of plant-parasitic nematodes are *Heterodera*, *Globodera*, *Meloidogyne*, *Tylenchulus*, *Pratylenchus*, *Ditylenchus*, *Rotylenchus*, *Helicotylenchus*, *Xiphinema* and *Radopholus* (Sasser and Freckman 1987). In general root symptoms vary widely but can include cysts, galls, lesions, stunting and decay. Aboveground symptoms include wilting, yellowing and loss of foliage. Majority of plant parasitic forms enter the root completely, feed, mature, and lay eggs within the root or attached to it.

Biological check mechanisms operate in the soil to control high population densities of plant-parasitic nematodes. Nematophagous fungi also known as nematode-destroying fungi or predaceous fungi are most fascinating soil organisms because of their spectacular predaceous ability to capture nematodes thus suppress the population of plant-parasitic nematodes. They are present throughout the world in all types of climate. Their habitats include soil, cultivated lands, decaying plant materials, decaying woods, dung, garden compost, leaf litter, moss cushions (Mittal et al. 1988, 1989; Saxena and Mukerji 1991; Saxena and Lysek 1993; Saxena 2008), and permanent pasture (Bailey and Gray 1989). They have also been reported to come from freshwater and marine habitats (Hao et al. 2005), brackish

G. Saxena (✉)
Department of Botany, Swami Shraddhanand College, University of Delhi, Delhi, India

B. Giri et al. (eds.), *Root Biology*, Soil Biology 52,
https://doi.org/10.1007/978-3-319-75910-4_8

water (Johnson and Autsery 1961), mangroves (Swe et al. 2008, 2009), and polluted soil (Mo et al. 2006, 2008). In rhizosphere soil, a high number of nematophagous fungi have been recorded (Persmark and Jansson 1997; McSorley et al. 2006). Nematophagous fungi have been divided into four main groups based on their mode of capture and infection mechanism: (1) nematode-trapping fungi, (2) endoparasitic fungi, (3) egg- and cyst-parasitic fungi, and (4) toxin-producing fungi.

8.2 Nematode-Trapping Fungi

Nematode-trapping fungi also known as predaceous or predatory fungi have an extensive hyphal development in the substratum. They possess a comparatively good saprophytic ability, while endoparasitic fungi are often obligate parasites. They capture nematodes either by adhesive or nonadhesive trapping devices (Barron 1977). The presence of nematodes or nematode secretions triggers the formation of these traps. Nematode-trapping fungi were formerly classified into a number of genera based on the shape, size and septa of conidia and branching and modifications of the conidiophores. They were *Arthrobotrys*, *Dactylaria*, *Dactylella*, *Drechslerella*, and *Monacrosporium* (Drechsler 1933, 1937; Duddington 1951; Subramanian 1963; Cooke and Godfrey 1964). Based on the phylogenetic analysis of 18S and ITS rDNA sequences (Pfister 1997), nematode-trapping *Hyphomycetes* fungi are now placed within Orbiliaceae of Ascomycota. Yu et al. (2014) assigned orbilaceous members in three asexual genera, namely, *Arthrobotrys* (54 species), *Dactylellina* (28 species) and *Drechslerella* (14 species). Apart from this, *Stylopage* and *Cystopage* of Zygomycota (Zoopagales) and *Nematoctonus* of *Basidiomycota* (*Pleurotaceae*) also form traps.

8.2.1 Adhesive Trapping Devices

8.2.1.1 Adhesive Hyphae

This trapping device is most primitive type, characteristics of *Stylopage* and *Cystopage*. Nematodes are captured by means of adhesive coated along the entire coenocytic hyphae or by adhesive produced at any point of hyphae. Following capture, nematode struggles to set itself free, becomes exhausted, and is finally penetrated by the fungus through appressorium-like structure. The nematode body gets filled with absorptive hyphae. After consuming nematode contents, fungal protoplasts retrieve back after forming cross wall. Outside the nematode body, conidiophores bearing single conidium are formed.

8.2.1.2 Adhesive Branches

Adhesive branches also called adhesive columns are morphologically simplest capture organs. They are mostly a few cells in height as they grow erect from prostrate hyphae. They usually anastomose forming scalariform two-dimensional network. They are characteristic trapping devices of *Monacrosporium cionopagum* and *M. gephyrophagum.*

8.2.1.3 Adhesive Knobs

These are globose to subglobose distinct cells coated with a film of adhesive. The knobs are either sessile or produced on a short stalk. After a nematode adheres to a knob, it gets caught by several knobs during struggle, leading to penetration by the fungus. After that a globose infection bulb is produced from which assimilative hyphae arise and digest body contents of the nematode (Barron 1977).

8.2.1.4 Adhesive Nets

Net-forming fungi are ubiquitous and very aggressive predators. Nets vary from 2-D loops as in *Arthrobotrys musiformis* to complex 3-D multibranched network as in *Arthrobotrys oligospora*. The adhesive is highly effective, and the prey is held fast. Violent movement of the prey leads to its further entanglement in the network. All the protoplast from the corpse is translocated for growth and reproduction of the fungus. The cluster of two-celled conidia is produced at the apex of each conidiophore. In *A. oligospora*, adhesion leads to piercing of nematode cuticle which is composed mainly of proteins including collagen. Proteases isolated from nematophagous fungi can hydrolyze protein of the cuticle. After obtaining data from sequencing, it has been found that they have a high homology to subtilisin type of serene proteases. Large amount of lectins are accumulated in the trophic hyphae inside the nematode body, which are transported to other parts of mycelium for further growth (Ahman et al. 1996, 2002; Nordbring-Hertz et al. 2006). Other hydrolytic enzymes include chitinases, glucanases, or collagenases (Yang et al. 2007).

8.2.2 Non-adhesive Trapping Devices

8.2.2.1 Non-constricting Rings

Non-constricting rings are three-celled rings arising from prostrate septate hyphae on a slender support stalk. They are passive in action. Ring breaks off often during struggle of the prey. Detached rings are viable; the nematode is penetrated by the

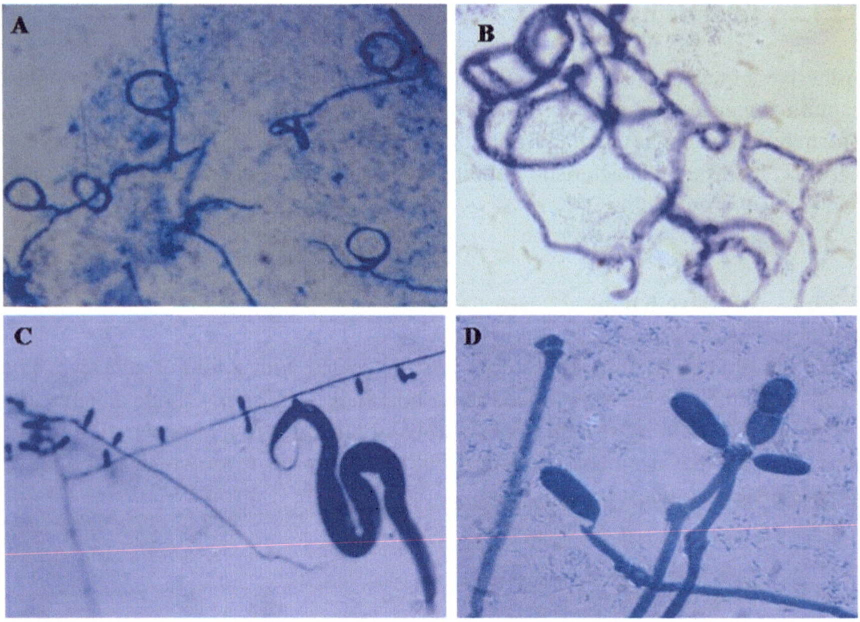

Fig. 8.1 (**a**) Several constricting rings produced on the hyphae of *Arthrobotrys dactyloides*. (**b**) Adhesive network produced by *Arthrobotrys oligospora*. (**c**) Nematode trapped by adhesive branches of *Monacrosporium cionopagum* produced on hyphae. Many one-celled adhesive branches are visible. (**d**) Tip of the conidiophores and conidia of *Arthrobotrys* sp. are visible

fungus. Fungi producing non-constricting rings often form adhesive knobs as in *Dactylaria candida* and *D. lysipaga* (Drechsler 1937).

8.2.2.2 Constricting Rings

The constricting rings are also three-celled rings produced on short and stout support stalk. When a nematode enters the ring, the friction of the body stimulates ring cells to rapidly swell inward to grasp the nematode allowing no chance for escape. The reaction is very rapid (about 1/10 of a second), is irreversible, and leads to almost complete closure of the aperture of the ring. The mechanism of action of constricting ring has been studied by Muller (1958) and Rudek (1975). Chen et al. (2001) found that in *Arthrobotrys dactyloides* signal transduction pathway is involved in the inflation of the ring cells. Nordbring-Hertz et al. (2006) suggest that pressure exerted by a nematode on the ring activates G-proteins in the ring cells leading to an increase in cytoplasmic Ca^{2+} activation of calmodulin and finally the opening of water channels (Fig. 8.1).

8.3 Endoparasitic Fungi

Endoparasites are obligate parasites with a broad host range. In the soil they do not produce extensive mycelia but exist as viable conidia which either adhere to the surface of the host or are ingested by the nematode. Second category is of fungi-producing zoospores which attach to the host cuticle.

8.3.1 Conidia-Producing Endoparasites

Some endoparasitic species exclusively belonging to *Harposporium* have palatable conidia. These conidia are morphologically adapted to special shapes which aid in getting them lodged in the digestive tract of the prey. Stylet-bearing nematodes cannot ingest spores of *Harposporium*. Adhesive conidia are found in genera like *Drechmeria*, *Hirsutella*, *Haptocillium*, *Pochonia*, *Purpureocillium* in Hypocreales of Ascomycota (Quandt et al. 2014), and *Meristacrum* of Zygomycota. *Nematoctonus* of Basidiomycota is an exceptional genus as it produces adhesive knobs to behave as predator and adhesive spores to behave as endoparasite. The conidia, after attached by adhesive, germinate inside the nematode body. Then conidiophores break out through the cuticle, which produce phialides and conidia.

8.3.2 Zoospores Producing Endoparasites

The zoopores are produced by *Catenaria* (*Chytridiomycota*) and members of *Oomycota* (Straminipila) including *Myzocytium*, *Haptoglossa* and *Nematophthora*. *Catenaria anguillulae* produces zoospore, which has single posterior whiplash flagellum. The zoospores are attracted and then get attached to a specific site on the nematode. They encyst and germinate to form a narrow outgrowth tube which swell to form vesicle and narrow rhizoids. The vesicles develop into sporangia; their contents cleave to form zoospores (Deacon and Saxena 1997). In *Myzocytium* biflagellate zoospores come out through evacuation tubes produced by the sporangia present inside the host (Fig. 8.2).

8.4 Parasites of Cyst and Root-Knot Nematodes

The third group is of parasites of cyst and root-knot nematodes, which attack eggs and females of these nematodes by in-growth of vegetative hyphae. These fungi do not have any morphological adaptations but have some measures of specifications enabling them to exploit nematodes (Morgan-Jones and Rodriguez-Kabana 1987).

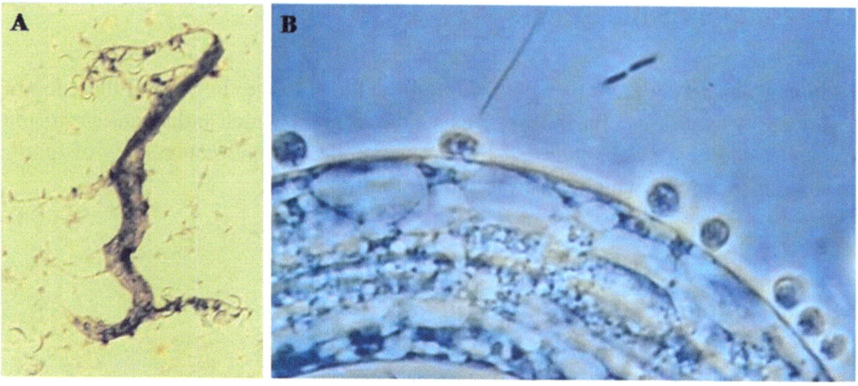

Fig. 8.2 (**a**) Nematode infected with *Harposporium anguillulae*, several conidiophores breaking out through the host cuticle, and crescent-shaped conidia are seen. (**b**) Seven zoospores of *Catenaria anguillulae* are seen attached to the nematode's middle part of the body. One of the zoospores showcases its flagellum

They are of two types, obligate and opportunistic parasitic fungi. These fungi infect eggs by forming appressoria where certain proteases bring about enzymatic disruption of egg shell which is three layered (Lopez-Llorca et al. 2008; Gortari and Hours 2008). Penetration is facilitated by extracellular enzymes such as chitinases and proteases (Yang et al. 2007). The major proteases identified from nematophagous fungi belong to the proteinase K family of subtilases (from peptidase S8 subtilase family) (Morton et al. 2004). Lipases have been implicated in the infection of *Heterodera schachtii* eggs; fungi appear to degrade inner lipid layers (Perry and Trett 1986). The fungus proliferates endogenously, ultimately destroying the larval contents. Common examples are *Lecanicillium lecanii*, *Pochonia chlamydosporia*, *P. rubescens* etc. infecting eggs and females by appressoria and zoospores (*Catenaria auxiliaris*) (Fig. 8.3).

8.5 Toxin-Producing Fungi

These fungi attack their prey by secreting diffusible toxins, even before any physical contact between host and the fungus (Lopez-Llorca et al. 2008). Giuma and Cooke (1971) demonstrated that nematotoxins are produced by *Nematoctonus haptocladus* and *N.concurrens*, *Pleurotus*, and *Coprinus*. Thorn and Barron (1984) reported that the *Pleurotus* species secrete toxins and *P. ostreatus* produced tiny droplets of toxins. Kwok et al. (1992) isolated the toxin trans-2-decenedioic acid from *P. ostreatus*. Luo et al. (2007) isolated nematicidal compounds from *Coprinus comatus*, which were active against *Meloidogyne arenaria* and *Panagrellus redivivus*. Some predatory fungi, besides producing trapping organs, produce nematicidal compounds at the same time trapping the nematodes (Li and Zhang 2014). An aliphatic compound

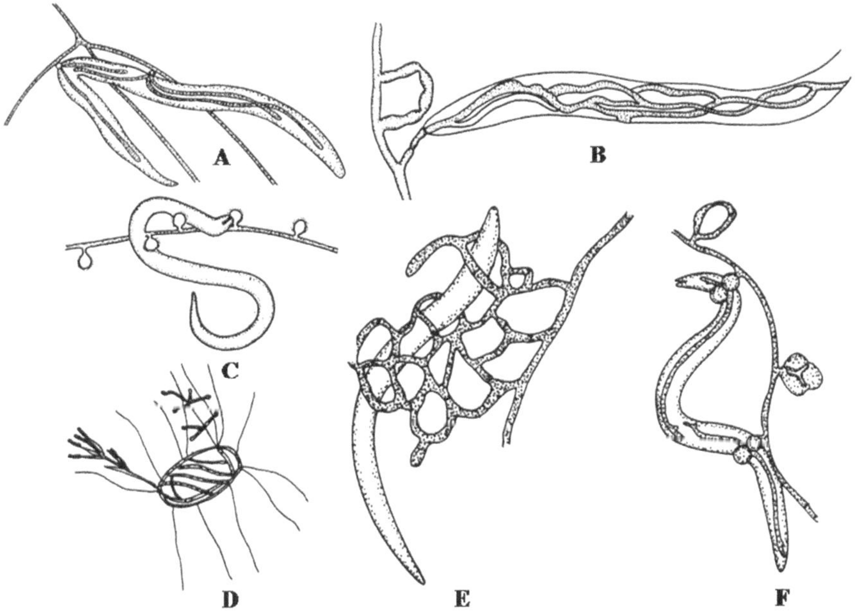

Fig. 8.3 (**a**) Nematode attached to the adhesive hypha of *Stylopage hadra* by means of globular protuberances and the body filled with trophic hyphae. (**b**) Adhesive branches fused to form scalariform network in *Monacrosporium gephyrophagum*. Nematode filled with assimilative hyphae attached to the adhesive branch is seen. (**c**) Nematode captured by adhesive knobs of *Monacrosporium parvicollis*. (**d**) Root-knot nematode egg parasitized by *Purpureocillium lilacinum*. (**e**) Freshly captured nematode in 3-D adhesive network of *Arthrobotrys conoides*. (**f**) Nematode captured in constricting rings of *Arthrobotrys brochopaga* at its head and tail region. An inflated ring is also seen (All figures are drawn with the aid of camera lucida at ×400 magnification except part **d**)

linoleic acid was detected in *Arthrobotrys brochopaga*, *A. conoides*, *A. dactyloides*, *A. oligospora*, *Dactylella candida*, and *Monacrosporium doedycoides* (Stadler et al. 1993; Anke et al. 1995).

8.6 Commonly Used Nematophagous Fungi in Biological Control of Nematodes

Biological control involves the action of nematophagous fungi to reduce nematode populations. Root-knot nematodes such as *Meloidogyne* spp. enter the root completely, feed, mature, and lay eggs within the root or on the surface. Root galling induced by *Meloidogyne* is a well-known host response. Abnormally large, multinucleate cells in the vascular tissues of susceptible plants are formed (Sasser 1989). *Meloidogyne incognita*, *M. javanica* and *M. arenaria* are mostly found in tropical and subtropical climates. Some of the most economically important crops such as potato, tomato, soybean, sugar beet, carrot and tobacco are common hosts of various *Meloidogyne* spp. They trigger susceptible response in the host by altering signaling pathways and genetic expression (Moens et al. 2009).

8.6.1 Purpureocillium lilacinum *syn.* Paecilomyces lilacinus

Purpureocillium lilacinum is ubiquitously present in soil and different types of organic materials in mostly tropical and subtropical soils. It penetrates the egg shell of root-knot nematodes, occasionally through the formation of appressoria. Ultrastructural studies have shown that it is capable of destroying both eggs and larvae of *Meloidogyne* (Morgan-Jones et al. 1984). Once this fungus is applied to the soil, nematode population begins to decline. It has great potential among all the parasites of cyst- and egg-parasitizing fungi, hence the most widely researched biocontrol agent of nematodes all over the world (Jatala et al. 1981; Jatala 1985; Kerry 1987; Saxena et al. 1991; Mittal et al. 1999; Sun et al. 2006; Saxena 2004, 2007; Kiewnick 2010; Stirling 2011; Abd-Elgawad 2016). Jatala et al. (1979) first reported the use of *P. lilacinum* for the control of *Meloidogyne incognita acrita* and *Globodera pallida* in Peru, and interest developed subsequently in the use of this fungus. The potential of *P. lilacinum* as biocontrol agent against *M. incognita* on different crops has been well worked out by several workers (Mittal et al. 1995; Noe and Sasser 1995). It has also proved to be an efficient biocontrol agent in controlling *M. arenaria and M. javanica* (Culbreath et al. 1986; Khan and Saxena 1997; Stirling 2013). Rawat et al. (1999) during a microflora study in the rhizosphere of Okra showed the presence of *P. lilacinus*, *P. variotii* and *P. fusisporum* during all stages of plant growth.

Effectiveness of *P. lilacinum* in biocontrol of other phytonematodes, viz., *Heterodera*, *Globodera*, *Radopholus*, *Tylenchus* and *Pratylenchus*, has also been demonstrated successfully in various greenhouse and field trials by several workers (Gapasin 1995; Davide and Zorilla 1995; Wang et al. 1997). Major progress was made when water-dispersible granules and wettable powders having high concentrations of *P. lilacinus* strain 251 (PL 251) spores were manufactured. They are available in many countries under trade names such as BioAct ™ and MeloCon ™ (Kiewnick 2010). Kiewnick et al. (2011) found that application of PL 251 (2×10^5 cfu g^{-1} soil) reduced galling and the number of egg masses by 45% and 69%, respectively. Hussain et al. (2017) determined the efficacy of six fungi on *Meloidogyne incognita* eggs and second-stage juveniles. *Lecanicillium muscarium* (79%), *L. psalliotae* (65%), and *P. lilacinum* (64%) parasitized eggs.

8.6.2 Pochonia chlamydosporia

Pochonia chlamydosporia (syn. *Verticillium chlamydosporium*) is a member of order Hypocreales in Ascomycota (Zare et al. 2001). It parasitizes egg masses attached on the root surface by hyphae and forms appressoria that penetrate the eggs. Segers et al. (1996) found that several enzymes are involved in dissolving the egg shell. The enzyme VCP1, an alkaline serine protease, breaks down protein layer in the egg shell of *Meloidogyne*. Larriba et al. (2014) sequenced and assembled 41Mb genomic DNA of *P. chlamydosporia* and found that it contains genes encoding hydrolytic enzymes.

It is a parasite of both root-knot nematodes (*Meloidogyne* spp.) and cyst nematodes *Globodera* and *Heterodera* like *H. schachtii* and *H. avenae* (Kerry 1981; Crump et al. 1983, Kerry and Crump 1998). It maintains the nematode population below the economic threshold (Kerry and Jaffee 1997). Its growth and survival in the rhizosphere soil is dependent on soil environment, nematode density and the host status of the plant. Luambano et al. (2015) found that pre-decomposed organic materials added to soil increased saprophytic growth of the fungus, but high C:N ratio did not affect parasitism of *M. incognita* eggs. It grows well in organic soil; optimum temperature for growth is 25–30°C and pH range 4.0–7.0 (Karakas et al. 2012). It produces chlamydospores to survive in the soil and does not germinate unless they are triggered by nutrients exuding from roots or by the presence of organic matter (de Leij et al. 1993; Hallman et al. 2009).

Populations of *P. chlamydosporia* increase markedly when egg masses are extruded on the galled root surface (Bourne et al. 1996). Isolates of *P. chlamydosporia* vary significantly in their pathogenicity (Yang et al. 2012; Dallemole-Giaretta et al. 2012). Isolates from cyst nematodes often are not pathogenic to root-knot nematodes and vice versa. Mauchline et al. (2002) observed increases in fungal growth rate of *P. chlamydosporia* isolate Jersey in the rhizosphere of potato cyst nematode infested plants but not in the rhizosphere of root-knot infested plants, after 14 weeks using a competitive PCR (cPCR) assay. There is considerable genetic diversity among isolates from different locations and hosts (Morton et al. 2004). Wang et al. (2015) found that a nematicidal strain of *P. chlamydosporia* isolated from root knots of tobacco yielded a total yellow pigment aurovertin. Aurovertin D showed strong toxicity toward *Meloidogyne incognita*. *P. chlamydosporia* is a potential biocontrol agent, generally as a component of an integrated nematode management. Its commercial formulation is available in Cuba and Brazil (Dallemole-Giaretta et al. 2011).

8.6.3 Arthrobotrys oligospora

Arthrobotrys oligospora is another potential predatory biocontrol agent which captures nematodes by adhesive network. It has been successful in controlling nematode infections in all forms of formulations. Strains of *A. oligospora* (ORS 18692 57 and ORS 18692 95) were found effective against *Meloidogyne* in both in vitro and in vivo experiments. About 98% juveniles were trapped by the fungus within 48 h in vitro experiments. It reduced nematode population and stimulated seedling growth in pot and field trials during summer and winter season tomato crops (Duponnois et al. 1995). Singh (2012) tested the potential of *A. oligospora* against *Meloidogyne graminicola.* They tested five isolates and found traps to be induced within 24 h after inoculation in in vitro studies. In pot culture too, reduction in number of galls was observed in rice. Degenkolb and Vilcinskas (2016) reviewed nematicidal activities of metabolites from *A. oligospora* CBS 115.81. Three colorless oils oligosporon, oligosporol A, and oligosporol B had nematicidal effect. Persmark and Jansson (1997), in a field experiment on barley, pea and white mustard, found greater densities of nematode-trapping fungi in the rhizosphere soil. *A. oligospora* was the most common species in both rhizosphere and soil. In pea rhizosphere, up to 780 propagules of nematode-trapping fungi g^{-1} were found which were 19 times higher than in the root-free soil.

The genomics and proteomics of *A. oligospora* are being researched (Yang et al. 2011).

Castaneda-Ramirez et al. (2016) sequenced five isolates of *A. oligospora* and *Monacrosporium eudermatum* using ITS4 and ITS5 primers. They observed polymorphisms between *Arthrobotrys* and *Monacrosporium* indicating differences in predatory activity.

8.6.4 Hirsutella rhossiliensis

Hirsutella rhossiliensis is an endoparasitic fungus, a member of Ascomycota, which is extensively used as biocontrol agent of cyst nematode *Heterodera.* Conidia surrounded by adhesive attach to passing nematode. The germ tube penetrates the cuticle; infection bulb is formed inside the host, from which assimilative hyphae grow and proliferate within about 3 days. The fungus produces new conidia on phialides. The phialides of *H. rhossiliensis* are essential for transmission because detached conidia do not adhere to passing nematode (McInnis and Jaffee 1989). *Heterodera schachtii* is an important pest of sugar beet and other crops. Vermiform juveniles hatch from eggs within cysts (the dead bodies of the adult female), move through soil pores, and penetrate into plant roots. Within roots, juveniles establish a feeding site.

Jaffee and Muldoon (1989) showed that natural populations had the potential to suppress *Heterodera schachtii.* In soil infested with this fungus, penetration of cabbage roots by the nematode was reduced by 50–77%. *H. rhossiliensis* parasitized second-stage juveniles in soil from all fields, but high level of parasitism was found in a field that had been in soybean monoculture for 35 years. Levels of suppression for *Heterodera schachtii* were not very high (Jaffee et al. 1992). Density-dependent parasitism was observed, as there was an increase in percentage parasitism with greater density of second-stage juveniles of *H. schachtii* (Jaffee et al. 1993). Alginate pellets (both dried and non-dried) of vegetative colonies of *H. rhossiliensis* were prepared. Addition of dried pellets to soil suppressed invasion of roots by *H. schachtii* (Lackey et al. 1993). Velvis and Kamp (1995) collected soil from five experimental plots which had annual crops of potatoes for 9 years. The infection of second-stage juveniles of potato cyst nematode *Globodera pallida* by fungal parasites was studied. *H. rhossiliensis* was the only fungus found, which infected 10–60% juveniles. It is also known to parasitize second-stage juveniles of soybean cyst nematode (*Heterodera glycines*). In two experiments, the level of parasitism was 11–53% in one field and 43% in another field (Chen 1997; Liu and Chen 2000). Formulations of the fungus such as corn grits, liquid suspensions containing mycelial fragments, and spores were evaluated. This reduced nematode population, but the application rates required to obtain control were not cost-effective in soybean production (Liu and Chen 2005).

8.6.5 Nematophthora gynophila

Nematophthora gynophila, an Oomycete, is an obligate parasite common in cereal-grown soils of England and infects *Heterodera* females by means of biflagellate zoospores. Hyphae inside the nematode form sporangia, in which zoospores are formed. They are rapidly released through discharge tubes outside the body and initiate new infections. Thick-walled spores are produced on hyphal segments. The fungus completes its life cycle in the cereal cyst nematode *Heterodera avenae* within 5 days at 13°C (Kerry and Crump 1980). *N. gynophila* invaded females as they emerged from roots of cereals and destroyed them completely within 1 week (Kerry 1975). *N. gynophila* reduces the number of females forming cysts and thus the number of viable eggs that these cysts contained.

The activity of *N. gynophila* is influenced by soil moisture, density of females and spores in soil. The thick-walled oospores have longevity of at least 5 years in the absence of nematodes (Kerry 1984). The fungus initiates infection by overwintering resting spores and is then spread by zoospores that have a limited capacity to move in the soil. Distribution of females on roots is more important than their average density in regulating nematode populations (Kerry et al. 1982). Despite the fact that free-draining soils and dry seasons limit the activity of *N. gynophila*, it provides significant control of *H. avenae* population in Europe. This indicates that once suppressiveness of cereal cyst nematode is established, its population can be permanently maintained at levels below the economic threshold.

8.7 Conclusions

Over the last two decades, considerable progress has been made in biological control of plant-parasitic nematodes. Some predatory and parasitic fungi like *Pochonia chlamydosporia* and *Purpureocillium lilacinum* have been mass produced on a commercial scale and being marketed. For these fungi, achieving high levels of control of nematodes is challenging as they have to compete with other microorganisms already present in soil. The solution is integrated nematode management which utilizes multiple management tactics such as host plant resistance, better farming system, soil amendments, and combination with other antagonists which has generated mixed results. The consistency of biological control may be improved by combining different antagonists of nematodes. *Pasteuria penetrans* and *Bacillus firmus* are effective biocontrol agents. The combination of nematophagous fungi with these bacteria will be beneficial.

Organic amendments are useful; their successive inputs at low application rates are economically beneficial. Organic amendments including chitin, organic manure, neem cake, etc. have been used to suppress plant-parasitic nematodes. It can involve release of toxic compounds, stimulate antagonistic organisms, maintain nutrient and labile carbon levels, and enhance the survival and proliferation of biological control organisms. Since arbuscular mycorrhizal fungi promote better plant health, combining them with nematophagous fungi gives a better nematode control. With the advancement of molecular technologies, our understanding of interactions between nematodes and their natural antagonists enhances day by day. When used collectively, these practices will increase soil organic matter, thus enhancing the suppressiveness of field soils to nematodes and other soil-borne diseases. What still remains a challenge are active implementations of pest management practices at farmers' level. The adoption of these practices instead of exploitive use of chemicals is where the future of farming lies.

Acknowledgments The author is thankful to her family members for constant encouragement and support throughout the preparation of this chapter.

References

Abd-Elgawad MMM (2016) Biological control agents of plant-parasitic nematodes. Egypt J Biol Pest Control 26(2):423–429

Ahman J, Ek B, Rask L, Tunlid A (1996) Sequence analysis and regulation of a cuticle degrading serine protease from the nematophagous fungus *Arthrobotrys oligospora*. Microbiology 142:1605–1616

Ahman J, Johansson T, Olsson M, Peter JP, Cees AMJJ, Hondel v, Tunlid A (2002) Improving the pathogenicity of a nematode-trapping fungus by genetic engineering of a subtilisin with nematotoxic activity. Appl Environ Microbiol 68:3408–3415

Anke H, Stadler M, Mayer A, Sterner O (1995) Secondary metabolites with nematicidal and antimicrobial activity from nematophagous fungi and Ascomycetes. Can J Bot 73(1):932–939

Bailey F, Gray NF (1989) The comparison of isolation techniques for nematophagous fungi from soil. Ann Appl Biol 114:125–132

Barron GL (1977) The nematode destroying fungi. Topics in mycobiology I. Canadian Biological Publications, Guelph, Canada, pp 140

Bourne JM, Kerry BR, De Leij FAAM (1996) The importance of the host plant on the interaction between root-knot nematodes (*Meloidogyne* spp.) and the nematophagous fungus *Verticillium chlamydosporium* Goddard. Bicontrol Sci Technol 6:539–548

Castaneda-Ramirez GS, Mendoza-de-Gives P, Aguilar-Marcelino L, Lopez-Arellano ME, Hernandez-Romano J (2016) Phylogenetic analysis of nucleotide sequence from the ITS region and biological characterization of nematophagous fungi from Morelos, Mexico. J Mycol. Article ID 8502629, pp 13

Chen S (1997) Infection of *Heterodera glycines* by *Hirsutella rhossiliensis* in a Minnesota soybean field. J Nematol 29:573

Chen TH, Hsu CS, Tsai PJ, Ho YF, Lin NS (2001) Heterotrimeric G-protein and signal transduction in the nematode-trapping fungus *Arthrobotrys dactyloides*. Planta 212:858–863

Chitwood DJ (2003) Research on plant parasitic nematode biology conducted by the United States Department of Agriculture-Agricultural Research Service. Pest Manag Sci 59:748–753

Cooke RC, Godfrey BES (1964) A key to the nematode-destroying fungi. Trans Br Mycol Soc 47:61–74

Crump DH, Sayre RM, Young LD (1983) Occurrence of nematophagous fungi in cyst nematode populations. Plant Dis 67:63–64

Culbreath AK, Rodriguez-Kabana R, Morgan-Jones G (1986) Chitin and *Paecilomyces lilacinus* for control of *Meloidogyne arenaria*. Nematropica 16:153–166

Dallemole-Giaretta R, de Freitas LG, Lopes EA, Ferraz S, de Poderta GS, Agnes EL (2011) Cover crops and *Pochonia chlamydosporia* for the control of *Meloidogyne javanica*. Nematology (13):919–926

Dallemole-Giaretta R, de Freitas LG, Lopes EA, Pereira OL, Zooca RJF, Ferraz S (2012) Screening of *Pochonia chlamydosporia* Brazilian isolates as biocontrol agents of *Meloidogyne javanica*. Crop Prot 42:102–107

Davide RG, Zorilla RA (1995) Farm trials by potato growers on the use of *Paecilomyces lilacinus* fungus for the biological control of potato cyst nematode and other species. Biocontrol 1:63–75

De Leij FAAM, Kerry BR, Dennehy JA (1993) *Verticillium chlamydosporium* as a biological control agent for *Meloidogyne incognita* and *M. hapla* in pot and microplot test. Nematologica 39:115–126

Deacon JW, Saxena G (1997) Oriented zoospore attachment and cyst germination in *Catenaria anguillulae*, facultative endoparasite of nematodes. Mycol Res 101(5):513–522

Degenkolb T, Vilcinskas A (2016) Metabolites from nematophagous fungi and nematicidal natural products from fungi as an alternative for biological control. Part I: metabolites from nematophagous ascomycetes. Appl Microbiol Biotechnol 100:3799–3812

Drechsler C (1933) Morphological diversity among fungi capturing and destroying nematodes. J Wash Acad Sci 23:138–141

Drechsler C (1937) Some hyphomycetes that prey on free-living terricolous nematodes. Mycologia 29:447–552

Duddington CL (1951) The ecology of predaceous fungi I. Trans Br Mycol Soc 34:322–338

Duponnois R, Mateille T, Gueye M (1995) Biological characteristics and effect of two strains of *Arthrobotrys oligospora* from Senegal on *Meloidogyne* species parasitizing tomato plants. Biocontrol Sci Technol 5:517–525

Gapasin RM (1995) Evaluation of *Paecilomyces lilacinus* (Thom) Samson for the control of *Pratylenchus* sp. in corn. Biocontrol 1:35–39

Giuma AY, Cooke RC (1971) Nematoxin production by *Nematoctonus haptocladus* and *N. concurrens*. Trans Br Mycol Soc 56:89–94

Gortari MC, Hours RA (2008) Fungal chitinases and their biological role in antagonism onto nematode eggs. Annu Rev Mycol Prog 7:221–238

Hallman J, Davies KG, Sikora R (2009) Biological control using microbial pathogens, endophytes and antagonists. In: Perry RN, Moens M, Starr JL (eds) Root-knot nematodes. CAB, Wallingford

Hao Y, Mo M, Su H, Zhang KQ (2005) Ecology of aquatic nematode-trapping hyphomycetes in southwestern China. Aquat Microb Ecol 40:175–181

Hussain M, Zouhar M, Rysamek P (2017) Effects of nematophagous fungi on viability of eggs and juveniles of *Meloidogyne incognita*. J Anim Plant Sci 27(1):252–258

Jaffee BA, Muldoon AE (1989) Suppression of cyst nematode by a natural infestation of a nematophagous fungus. J Nematol 21:505–510

Jaffee BA, Phillips R, Muldoon AE, Mangel M (1992) Density-dependent host-pathogens dynamics in soil microcosms. Ecology 73:495–506

Jaffee BA, Tedford EC, Muldoon AE (1993) Tests for density-dependent parasitism of nematodes by nematode-trapping and endoparasitic fungi. Biol Control 3:329–326

Jatala P (1985) Biological control of nematodes. In: Sasser JN, Carter C (eds) An advanced treatise on *Meloidogyne*, biology and control, vol I. North Carolina State Graphics, Raleigh, NC, pp 303–308

Jatala P, Kaltenbach R, Bocangel M (1979) Biological control of *Meloidogyne incognita acrita* and *Globodera pallida* on potatoes. J Nematol 11:303

Jatala P, Salas R, Kaltenbach R, Bocangel M (1981) Multiple application and long term effect of *Paecilomyces lilacinus* in controlling *Meloidogyne incognita* under field conditions. J Nematol 13:445

Johnson TW, Autsery CL (1961) An *Arthrobotrys* from brackish water. Mycologia 53:432–433

Karakas M, Benli M, Cebesoy S (2012) Effects of some cultural conditions on the growth of nematophagous fungus *Pochonia chlamydosporia* (Fungi: Clavicipitaceae) isolated from *Meloidogyne incognita* eggs. J Anim Vet Adv 11(24):4644–4647

Kerry BR (1975) Fungi and the decrease of cereal cyst-nematode population in cereal monoculture. EPPO Bull 5:353–361

Kerry BR (1981) Progress in the use of biological agents for control of nematodes. In: Papavizas GC (ed) Biological control in crop production. BARC symposium no. 5. Allenheld and Osmum, Totowa, NJ, pp 79–90

Kerry BR (1984) Nematophagous fungi and the regulation of nematode population in soil. Helminthol Abstr Ser B 53:1–14

Kerry BR (1987) Biological control. In: Brown RH, Kerry BR (eds) Principles and practice of nematode control in crops. Academic, New York, pp 233–263

Kerry BR, Crump DH (1980) Two fungi parasitic on females of cyst nematodes (*Heterodera* spp.) Trans Br Mycol Soc 74:119–125

Kerry BR, Crump DH (1998) The dynamics of the decline of the cereal cyst nematode *Heterodera avenae* in four soils under intensive cereal production. Fundam Appl Nematol 21:617–625

Kerry BR, Jaffee BA (1997) Fungi as biological control agents for plant parasitic nematodes. In: Wicklow DT, Soderstrom B (eds) The mycota IV: Environmental and microbial relationships. Springer, Berlin, pp 203–218

Kerry BR, Crump DH, Mullen LA (1982) Studies of the cereal cyst-nematode *Heterodera avenae* under continuous cereals 1974-1978 I, Plant growth and nematode multiplication. Ann Appl Biol 100:477–487

Khan TA, Saxena SK (1997) Integrated management of root-knot nematode *Meloidogyne javanica* infecting tomato using organic materials and *Paecilomyces lilacinus*. Bioresource Technol 61:247–250

Kiewnick S (2010) Importance of multitrophic interactions for successful biocontrol of plant parasitic nematodes with *Paecilomyces lilacinus* strain 251. In: Gisi U, Chet I, Lodvica Gullino M (eds) Recent developments in management of plant diseases, plant pathology in the 21st century. Springer, Dordrecht, pp 81–92

Kiewnick S, Neumann S, Sikora RA, Frey JE (2011) Effect of *Meloidogyne incognita* inoculum density and application rate of *Paecilomyces lilacinus* strain 251 on biocontrol efficacy and colonization of egg masses analyzed by real-time quantitative PCR. Phytopathology 101:105–112

Kwok OCH, Plattner R, Weisleder D, Wicklow DT (1992) A nematicidal toxin from *Pleurotus ostreatus* NRRL 3526. J Chem Ecol 18:127–135

Lackey BA, Muldoon AE, Jaffee BA (1993) Alginate pellet formulation of *Hirsutella rhossiliensis* for biological control of plant-parasitic nematodes. Biol Control 3:155–160

Larriba E, Jaime M, Carbonell-Caballero J, Conesa A, Dopazo J, Nislow C, Martni-Nieto J, Lopez-Llorca LV (2014) Sequencing and functional analysis of the genome of a nematode egg-parasitic fungus *Pochonia chlamydosporia*. Fungal Genet Biol 65:69–80

Li GH, Zhang KQ (2014) Nematode toxic fungi and their nematicidal metabolites. In: Zhang KQ, Hyde KD (eds) Nematode-trapping fungi, Fungal diversity research series. Mushroom Research Foundation 2014. Springer, Dordrecht, pp 313–375

Liu XZ, Chen SY (2000) Parasitism of *Heterodera glycines* by *Hirsutella* spp. in Minnesota soybean fields. Biol Control 19:161–166

Liu S, Chen SY (2005) Efficacy of the fungi *Hirsutella minnesotensis* and *H. rhossiliensis* from liquid culture for control of the soybean cyst nematodes *Heterodera glycines*. Nematology 7:149–157

Lopez-Llorca LV, Macia-Vicente JG, Jansson HB (2008) Mode of action and interactions of nematophagous fungi. In: Ciancio A, Mukerji KG (eds) Integrated management and biocontrol of vegetable and grain crops nematodes. Springer, Dordrecht, pp 51–76

Luambano ND, Manzanilla-Lopez RH, Kimenju JW, Powers SJ, Narla RD, Wanjohi WJ, Kerry BR (2015) Effect of temperature, pH, carbon and nitrogen ratios on the parasitic activity of *Pochonia chlamydosporia* on *Meloidogyne incognita*. Biol Control 80:23–29

Luo H, Liu YJ, Fang L, Li X, Tang NH, Zhang KQ (2007) *Coprinus comatus* damages nematode cuticles mechanically with spiny balls and produces potent toxins. Appl Environ Microbiol 73:3916–3923

Mauchline TH, Kerry BR, Hirsch PR (2002) Quantification in soil and the rhizosphere of the nematophagous fungus *Verticillium chlamydosporium* by competitive PCR and comparison with selective plating. Appl Environ Microbiol 68(4):1846–1853

McInnis TM, Jaffee BA (1989) An assay for *Hirsutella rhossiliensis* spores and the importance of phialides for nematode inoculation. J Nematol 21:229–234

McSorley R, Wang KH, Kokalis-Burelle N, Church G (2006) Effects of soil type and steam on nematode biological control potential of the rhizosphere community. Nematropica 36:197–214

Mittal N, Saxena G, Mukerji KG (1988) Few predators from Delhi soils. J Rec Adv Appl Sci 3:383–388

Mittal N, Saxena G, Mukerji KG (1989) Ecology of nematophagous fungi: distribution in Delhi. J Phytol Res 2(1):31–37

Mittal N, Saxena G, Mukerji KG (1995) Integrated control of root-knot disease in three crop plants using chitin and *Paecilomyces lilacinus*. Crop Prot 14(8):647–651

Mittal N, Saxena G, Mukerji KG (1999) Biological control of root-knot nematode by nematode-destroying fungi. In: Singh J, Aneja KR (eds) From ethnomycology to fungal biotechnology: exploiting fungi from natural resources for novel product. Kluwer, Plenum, New York, pp 163–172

Mo MH, Chen WM, Su HY, Zhang KQ, Duan CQ, He DM (2006) Heavy metal tolerance of nematode-trapping fungi in lead-polluted soils. Appl Soil Ecol 31:11–19

Mo MH, Chen WM, Yang HR, Zhang KQ (2008) Diversity and metal tolerance of nematode-trapping fungi in Pb-polluted soils. J Microbiol 46:16–22

Moens M, Perry RN, Starr JL (2009) *Meloidogyne* species—a diverse group of novel and important plant parasites. In: Perry RN, Moens M, Starr JL (eds) Root-knot nematodes. CAB, Wallingford, pp 1–17

Morgan-Jones G, Rodriguez-Kabana R (1987) Fungal biocontrol for the management of nematodes. In: Veech JA, Dickson DW (eds) Vistas on nematology. Society Nematologists, Hyattsville, MD, pp 94–99

Morgan-Jones G, White JF, Rodriguez-Kabana R (1984) Phytonematode pathology: ultrastructural studies 2. Parasitism of *Meloidogyne arenaria* eggs and larvae by *Paecilomyces lilacinus*. Nematropica 16:21–31

Morton CO, Hirsch PR, Kerry BR (2004) Infection of plant-parasitic nematodes by nematophagous fungi—a review of the application of molecular biology to understand infection processes and to improve biological control. Nematology 6:161–170

Muller HG (1958) The constricting ring mechanism of two predacious Hyphomycetes. Trans Br Mycol Soc 41:341–364

Noe JP, Sasser JN (1995) Evaluation of *Paecilomyces lilacinus* as an agent for reducing yield losses due to *Meloidogyne incognita*. Biocontrol 1:57–67

Nordbring-Hertz B, Jansson HB, Tunlid A (2006) Nematophagous fungi. In: Encyclopedia of life sciences. Wiley, Chichester, pp 1–11. https://doi.org/10.1038/npg.els.0004293

Perry RN, Trett MW (1986) Ultrastructure of the eggshell of *Heterodera schachtii* and *Heterodera glycines* (Nematode, Tylenchida). Rev Nematol 9:399–403

Persmark L, Jansson HB (1997) Nematophagous fungi in the rhizosphere of agricultural crops. FEMS Microbiol Ecol 22:303–312

Pfister DH (1997) Castor, Pollux and life histories of fungi. Mycologia 89:1–23

Quandt CA, Kepler RM, Gams W, Evans HC, Hudghes D, Humber R, Hymel-Jones N, Li Luangsa-Ard JJ, Rehner SA, Sato M, Shrestha B, Sung GH, Yao YJ, Zara R, Spatafora JW (2014) Phylogenetic-based nomenclatural proposals for *Ophiocordycipitaceae (Hypocreales)* with new combination in *Tolypocladium*. IMA Fungus 5:121–134

Rawat R, Pandey A, Saxena G, Mukerji KG (1999) Rhizosphere biology of root-knot diseased *Abelmoschus esculentus* in relation to its biocontrol. In: Singh J, Aneja KR (eds) From ethnomycology to fungal biotechnology: exploiting fungi from natural resources for novel products. Plenum, New York, pp 173–184

Rudek WT (1975) The construction of the trapping rings in *Dactylaria brochopaga*. Mycopathologia 53:193–197

Sasser JN (1989) Plant-parasitic nematodes: the farmer's hidden enemy. Department of Plant Pathology, North Carolina State University and the Consortium for International Crop Protection, pp 115

Sasser JN, Freckman DW (1987) World perspective on Nematology: the role of the society. In: Veech JA, Dickson DW (eds) A commemoration of the twenty-fifth anniversary of the society of nematologists. Society of Nematologists, Hyattsville, MD, pp 7–14

Saxena G (2004) Biocontrol of nematode-borne diseases in vegetable crops. In: Mukerji KG (ed) Fruit and vegetable diseases. Disease management of fruits and vegetables, vol I. Kluwer, Dordrecht, pp 387–450

Saxena G (2007) Nematophagous fungi as biocontrol agents of plant-parasitic nematodes. In: Saxena G, Mukerji KG (eds) Management of nematode and insect-borne plant diseases. Haworth, Taylor and Francis, New York, pp 165–194

Saxena G (2008) Observations on the occurrence of nematophagous fungi in Scotland. Appl Soil Ecol 39:352–357

Saxena G, Lysek G (1993) Observation of nematophagous fungi in natural soils by fluorescence microscopy and correlation with isolation. Mycol Res 97(8):1005–1011

Saxena G, Mukerji KG (1991) Distribution of nematophagous fungi in Varanasi, India. Nova Hedwigia 52(3-4):487–497

Saxena G, Mittal N, Mukerji KG, Arora DK (1991) Nematophagous fungi in biological control of nematodes. In: Arora DK, Ajello L, Mukerji KG (eds) Handbook of applied mycology: humans, animals and insects, vol II. Marcel Dekker, New York, pp 707–733

Segers R, Butt TM, Kerry BR, Bechett A, Peberdy JF (1996) The role of the proteinase VCP1 produced by the nematophagous *Verticillium chlamydosporium* in the infection process of nematode eggs. Mycol Res 100:421–428

Singh UB, Sahu A, Singh RK, Singh DP, Meena KK, Srivastava JS, Renu, Manna MC (2012) Evaluation of biocontrol potential of *Arthrobotrys oligospora* against *Meloidogyne graminicola* and *Rhizoctonia solani* in Rice (*Oryza sativa* L). Biol Control 60:262–270

Stadler M, Anke H, Sterner O (1993) New biologically active compounds from the nematode-trapping fungus *Arthrobotrys oligospora*. Fres Z Naturforsch 48:843–850

Stirling GR (2011) Biological control of plant-parasitic nematodes: an ecological perspective, a review of progress and opportunities for further research. In: Davies K, Spiegel Y (eds) Biological control of plant-parasitic nematodes: building coherence between microbial ecology and molecular mechanisms. Springer, Dordrecht, pp 1–38

Stirling GR (2013) Integration of organic amendments, crop rotation, residue retention and minimum tillage into a subtropical vegetable farming system enhances suppressiveness to root-knot nematode (*Meloidogyne incognita*). Australas Plant Pathol 42:625–637

Subramanian CV (1963) *Dactylella*, *Monacrosporium* and *Dactylina*. J Ind Bot Soc 42:291–300

Sun MH, Gao L, Shi YX, Li BJ, Liu XZ (2006) Fungi and actinomycetes associated with *Meloidogyne* spp. eggs and females in China and their biocontrol potential. J Invertebr Pathol 93:22–28

Swe A, Jeewon R, Hyde KD (2008) Nematode-trapping fungi from mangrove habitats. Cryptogamie Mycologie 29(4):333–354

Swe A, Jeewon R, Pointing SB, Hyde KD (2009) Diversity and abundance of nematode-trapping fungi from decaying litter in terrestrial and mangrove habitats. Biodivers Conservat 18 (6):1695–1714

Thorn RG, Barron GL (1984) Carnivorous mushrooms. Science 224:76–78

Velvis H, Kamp P (1995) Infection of second stage juveniles of potato cyst nematodes by the nematophagous fungus *Hirsutella rhossiliensis* in Dutch potato fields. Nematologica 41:617–627

Wang CJ, Song CY, Zhang XD, Xie YQ, Xing ZL, Wang MZ (1997) Sustainable control efficacy of *Paecilomyces lilacinus* against *Heterodera glycines*. J Nematol 13:26–28

Wang YL, Li LF, Li DX, Wang B, Zhang K, Niu X (2015) Yellow pigment aurovertins mediate interactions between the pathogenic fungus *Pochonia chlamydosporia* and its nematode host. J Agric Food Chem 63:6577–6587

Yang Y, Yang E, An Z, Liu X (2007) Evolution of nematode-trapping cells of predatory fungi of the Orbiliaceae based on evidence from rRNA-encoding DNA and multiprotein sequences. Proc Natl Acad Sci USA 104:8379–8384

Yang J, Wang L, Ji XL, Feng Y, Li XM, Zou CG, Xu JP, Ren Y, Mi QL, Wu JL, Liu SQ, Liu Y, Huang XW, Wang HY, Niu XM, Li J, Liang LM, Luo YL, Ji KF, Zhou W, Yu ZF, Li GH, Liu YJ, Li L, Qiao M, Feng L, Zhang KQ (2011) Genomic and proteomic analyses of the fungus *Arthrobotrys oligospora* provide insights into nematode trap formation. PLoS Pathog 7: e1002179

Yang JI, Loffredo A, Borneman J, Jo B (2012) Biocontrol efficacy among strains of *Pochonia chlamydosporia* obtained from the root-knot suppressive soil. J Nematol 44:67–71

Yu ZF, Mo MH, Zhang Y, Zhang KQ (2014) Taxonomy of nematode-trapping fungi from Orbiliaceae, Ascomycota. In: Zhang KQ, Hyde KD (eds) Nematode-trapping fungi, Fungal diversity research series, vol 23. Springer, Dordrecht, pp 41–210

Zare R, Gams W, Evans HC (2001) A revision of *Verticillium* section Prostrata V. The genus *Pochonia* with notes on *Rotiferophthora*. Nova Hedwigia 73:51–86

Chapter 9
Optimizing Growth and Tolerance of Date Palm (*Phoenix dactylifera* L.) to Drought, Salinity, and Vascular *Fusarium*-Induced Wilt (*Fusarium oxysporum*) by Application of Arbuscular Mycorrhizal Fungi (AMF)

Abdelilah Meddich, Mohamed Ait El Mokhtar, Widad Bourzik, Toshiaki Mitsui, Marouane Baslam, and Mohamed Hafidi

9.1 Introduction

The date palm, *Phoenix dactylifera* L., is one of the most economically important perennial plants in arid and semiarid areas of the Near East and the North Africa, where it is widely cultivated for food and many other commercial purposes. The palm is considered the symbolism of life in the arid zone, being one of the oldest domesticated trees with multifold socioeconomic roles (Zohary and Hopf 2000; Chao and Krueger 2007).The fruits of date palm are a good source of essential nutrients, such as sugars, proteins, fibers, minerals (selenium, potassium, calcium, magnesium, and iron), etc., and form an important part of the human daily diet (Benmeddour et al. 2013). Over the last century, however, several constraints face

A. Meddich (✉) · M. Ait El Mokhtar
Department of Biology Biotechnology and Plant Physiology Unit, Cadi Ayyad University, Faculty of Science Semlalia, Marrakesh, Morocco
e-mail: a.meddich@uca.ma

W. Bourzik
Environment Service, Wilaya of Marrakesh-Safi Region, Marrakesh Prefecture, Marrakesh, Morocco

T. Mitsui
Faculty of Agriculture, University of Niigata, Niigata, Japan

Graduate School of Science and Technology, University of Niigata, Niigata, Japan

M. Baslam
Department of Applied Biological Chemistry, Faculty of Agriculture, University of Niigata, Niigata, Japan

M. Hafidi (✉)
Department of Biology Ecology and Environment Unit, Cadi Ayyad University, Faculty of Science Semlalia, Marrakesh, Morocco

B. Giri et al. (eds.), *Root Biology*, Soil Biology 52,
https://doi.org/10.1007/978-3-319-75910-4_9

date palm expansion cultivation. In North Africa, particularly in Morocco, the culture and production of *P. dactylifera* are adversely affected by biotic (i.e., bayoud disease) and abiotic stresses (i.e., drought and salinity) (Oihabi 2001; Meddich et al. 2015).

Bayoud, caused by the soilborne fungus *Fusarium oxysporum* f. sp. *albedinis* (Foa), is the most serious disease of date palm (Saaidi 1990). The impact of this pathogen is very serious in North Africa, where losses are increasing and its rapid expansion may be a threat for date-growing areas around the world (Daayf et al. 2003). It has destroyed over 12 million trees in Morocco and more than 3 million in Algeria, having accelerated desertification and advancing steadily (Djerbi 1998). A number of factors influence the spread of the pathogen, being the extent of human intervention the most important, by transporting infested young or adult date palm trees from contaminated to unaffected areas (Meddich and Boumezzough 2017).

Abiotic stress conditions such as drought, high and low temperature, and salinity are known to influence the occurrence and spread of pathogens, insects, and weeds (Peters et al. 2014). Drought is one of the main factors negatively affecting the productivity of agricultural and natural ecosystems (Passioura 2007; Ciais et al. 2005) and the diversity of plant species (Engelbrecht et al. 2007). Desertification processes in arid, semiarid, and dry subhumid areas are the result of various factors including water scarcity. Changes in plant physiology, nutrient acquisition, and metabolism induced by drought are highly limiting factors for plant growth and yield (Evelin et al. 2009). Additionally, soil salinity is another problem of grave concern because it adversely affects growth and development of plants especially in arid and semiarid regions (Pitman and Läuchli 2004). Actually, salinization is one of the most important agricultural and eco-environmental problems nowadays, which is increasing steadily in many parts of the world (Evelin et al. 2009; Porcel et al. 2012). It has been estimated that more than 7% of the arable land is salinized, and it is expected to increase up to 50% during the twenty-first century (Ruiz-Lozano et al. 2012). Soil salinity is a major constraint to food production because it restricts the use of previously uncultivated lands. Moreover, it dramatically limits agricultural yield over than 20% as it negatively affects plant growth and development (Porcel et al. 2012). Plants growing in saline soils are subjected to different physiological stresses that induce nutrient imbalance, damage cell organelles, and disrupt photosynthesis and respiration (Juniper and Abbott 1993; Evelin et al. 2013). Nevertheless, the majority of studies on plant responses to abiotic stresses focus on short and extreme single stresses, with plant survival or recovery rates as a measurement of plant stress tolerance.

Arbuscular mycorrhizal fungi (AMF) are considered an integral component of plant communities in both natural and agricultural ecosystems (Smith and Read 1997; Redecker et al. 2000). AMF confers numerous benefits to host plants in improving plant growth and mineral nutrition and tolerance to diverse biotic and abiotic stresses (Oihabi and Meddich 1996; Harrison 1997; Al-Karaki 2000; Vigo et al. 2000; Borowicz 2001; Cantrell and Linderman 2001; Dell'Amico et al. 2002; Al-Karaki et al. 2004; Asghari et al. 2005; Sannazzaro et al. 2006; Giri et al. 2007; Fan et al. 2011; Meddich et al. 2015; Giri 2017). Many crops could potentially benefit from mycorrhizal symbiosis, although the magnitude of symbiotic benefits conferred to crops might depend on the mycorrhizal dependency of a plant species and composition of AMF communities (Janos 2007; Oruru and Njeru 2016). In the case of date palm, the limited development of the root system (root hairs at low densities), along with the field observations of higher levels of mycorrhizal colonization, suggests its potential benefit from this mutualistic symbiosis under harsh environmental conditions.

The aims of the current study were to characterize the morphophysiological acclimations of date palm under combinations of biotic (drought and salinity) and abiotic (*Fusarium oxysporum* f. sp. *albedinis* (Foa)) stresses by using indigenous AMF to improve its tolerance. Integrating AMF for the development of date palm might be considered as an appropriate strategy to reverse the land degradation trend and encourage sustainable patterns for the development of arid and semiarid zones or date-growing zones.

9.2 Materials and Methods

9.2.1 Biological Material and Experimental Design

Arbuscular mycorrhizal fungi were obtained from different soils after trapping and multiplication on the host plant and are as follows: *Glomus monosporus* (reference strain from INRA Dijon, France), *Glomus clarum* and *Glomus deserticola* (both were selected from the Laboratory of Biotechnology at the University of Yaoundé in Cameroon), and the mycorrhizal Aoufous consortium (MAC) (obtained from the palm grove of Tafilalet). The MAC contains a mixture of indigenous species *Glomus* sp. (15 spores g^{-1} soil), *Sclerocystis* sp. (9 spores g^{-1} soil), and *Acaulospora* sp. (1 spore g^{-1} of soil) (Meddich 2001).The inoculum was therefore used in the form of plants of barley (*Hordeum vulgare* L.) mycorrhized by the mentioned AM fungi. These fungi came from a pot culture with barley (*Hordeum vulgare* L.) as a host plant, and the inoculum consisted of soil with alfalfa root fragments, spores, and hyphae. Briefly, barley seeds were disinfected and placed in germinating condition within the vermiculite (previously sterilizedat 200°C for 3 h) and watered with sterilized distilled water. After a week of germination, the barley seedlings were transferred into plastic pots (13 × 09 cm) containing soils mixed with the fungi to be tested. These plants were watered regularly with distilled water with a 30 ml weekly intake of a modified nutrient solution of Long Ashton (Plenchette et al. 1982). After 3 months of culture, the mycorrhizal roots of barley were disinfected for 10 min (Strullu 1986), rinsed three times with distilled water, and cut into fragments of 1–2 mm long. In all cases, the frequency of infection (F) of barley root was determined by the technique described by Trouvelot et al. (1986). An average of over 82% frequency of infection was maintained for the prepared inocula (Aoufous consortium, 100%; *G. monosporus*, 95.55%; *G. clarum*, 86.67%; and *G. deserticola*, 82.22%).

The cultures of date palm were performed in 5 L black plastic buckets (16 cm in diameter and 20 cm height), equipped with a drainage device for removing excess water and allow to determine the soil field capacity. The methodology for the application of water stress is described by Tobar et al. (1994) and Meddich et al. (2000). The seeds of an elite variety of date palm cv. Bouffgouss were disinfected, germinated in plastic bowls containing a sterile sandy substrate, and incubated for 3 weeks at 38°C. Two-month palm seedlings (leaf stage) are then transplanted into the plastic bucket containing 4 kg of sand–peat mixture (2:1 v/v) previously sterilized for 3 h at 180°C. The inoculation by symbiotic fungi was carried out supplying with 2.8 g (fresh weight) of mycorrhizal barley roots (Strullu 1986) near to palm date root system. The physicochemical parameters of the sand–peat mixture used are water pH 7.31, total phosphorus 0.041%, organic carbon 0.82%, total nitrogen 0.13%, and conductivity 148 μs cm^{-1}. The water treatments were applied at transplanting inoculated date palm seedlings (8 weeks after germination). The plants were weekly (30 ml) irrigated with Long Ashton amended nutrient solution (Plenchette et al. 1982). The experimental buckets were placed in a greenhouse under natural light (average temperature 24.5°C, relative humidity average of 69.12%, and light 330 μm^{-2} s^{-1}). Ten treatments (non-inoculated control and uninfected by Foa, non-mycorrhized and infected by Foa, Aoufous consortium, Aoufous consortium with Foa, *G. monosporus*, *G. monosporus* with Foa, *G. deserticola*, *G. deserticola* with Foa, *G. clarum*, and *G. clarum* with Foa) were performed. For each treatment, two water treatments (75% and 25% of field capacity) were applied. The combination of each water and fungal treatment consisted of ten repetitions of six plants, being a total of 1200 used plants.

The used fungal pathogen Foa (Fig. 9.1) was isolated from Bayoud-infected palms in the growing-palm of Drâa. Its preservation was performed on the sand, and its pathogenicity was regularly confirmed by the infection of young plants grown from the seed of the susceptible cultivar JHL Bayoud. The inoculum consists of a 10-day Foa spore suspension obtained by successive washings with distilled water and cultured in solid nutrient agar medium. Four months after date palm inoculation, 5 ml per plant of Foa isolate was provided in the form of a spore suspension at the concentration of 2.10^6 spores ml^{-1} to the roots of young palms.

9.2.2 Salt Stress Application to Date Palm Cultivation

The inoculum was multiplicated using the same method as described previously, and the mycorrhizal fungi were obtained from the Aoufous complex cited above. The seeds of the same variety of date palm (cv. Bouffgouss) were used for this assay. Two months later, date palm seedlings were transplanted to a 5 L black plastic bucket containing 4 kg of sterilized sand (Fig. 9.2).

Inoculation by symbiotic fungi was performed by supplying 3 g of mycorrhizal barley root fragments, spores, and hyphae near to date palm root system. The buckets

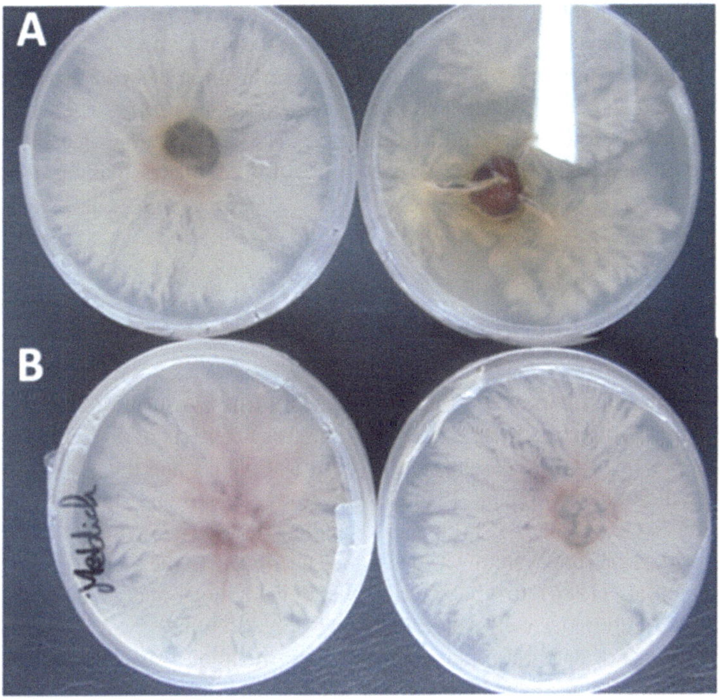

Fig. 9.1 Photographs of *Fusarium oxysporum* f. sp. *albedinis* (Foa) isolated from rachis tissues of diseased adult date palms (**a**) and from roots of diseased young palms trees (**b**) (not associated with AMF)

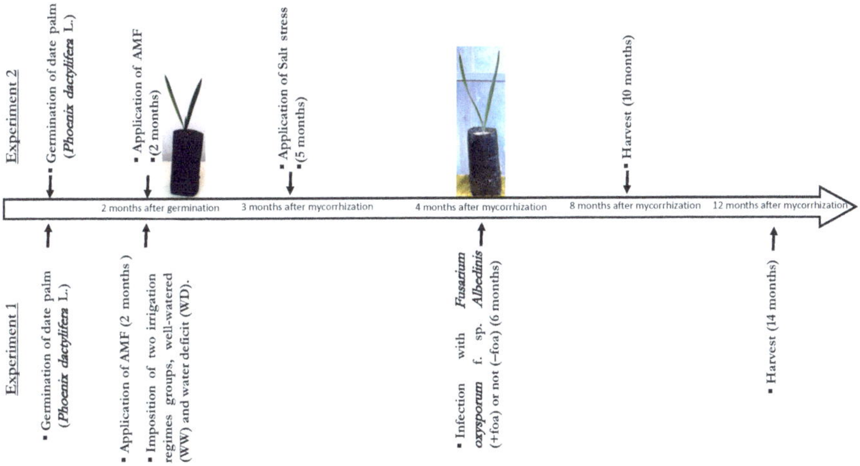

Fig. 9.2 Experimental design

were then placed in the greenhouse under natural light (average temperature 24.5°C, relative humidity average of 69.12%, and light 330 $\mu m^{-2}\ s^{-1}$).

The salt stress was applied 3 months after mycorrhization by irrigation with saline solution 0 and 240 mM NaCl solutions. Two fungal treatments (non-inoculated (control) and Aoufous consortium) were used. The combination of each salt and fungal treatment consisted of 20 repetitions of tree plants making a total of 240 plants.

9.2.3 Measured Parameters

Concerning the first experiment, after 48 weeks of mycorrhization, samples from 30 date palms per treatment were analyzed to assess the effect of drought on growth, water features of mycorrhizal and non-mycorrhizal plants, and the quantification of fungi growth and development. Plant roots were processed and stained with trypan blue 0.01% in lactoglycerol (Phillips and Hayman 1970). The review of the status of the mycorrhization root system was performed according to the method described by Trouvelot et al. (1986) to characterize the development and aggressiveness of mycorrhizal fungi under well-water and water-deficit regimes.

The response of date palm plants to the mycorrhization was estimated by determining the number of formed leaves, leaf area, and biomass production. Dry mass (DM) was determined after drying the plant material in the oven at 80°C until the weight was constant. Leaf area was measured on the leaf of the same rank for all treatments. The relative water content (RWC) was also measured on the same level of leaf in plants and for all treatments using the following equation:

$$\text{RWC\%} = \frac{\text{FM-DM}}{\text{FMsat-DM}} \times 100$$

where FMsat corresponds to the saturated fresh materials

In the second experiment, after 32 weeks of mycorrhization, aerial and shoot dry mass, water content (WC), stomatal conductance (SC), and leaf water potential (ψ_h) were carried out. Leaf water potential was measured by the method of the pressure chamber developed by Scholander et al. (1965). The stomatal conductance was determined in well-developed leaf samples of the same rank using LI-1600 porometer (LI-COR Inc, Lincoln, Nebraska).

9.2.4 Statistical Analysis

All results were analyzed statistically with the CO-STAT software (Statistical Software, New Style Anova). The study includes an analysis of variance followed by Newman–Keuls test at the 5%. All values shown in the figures are mean ± SE.

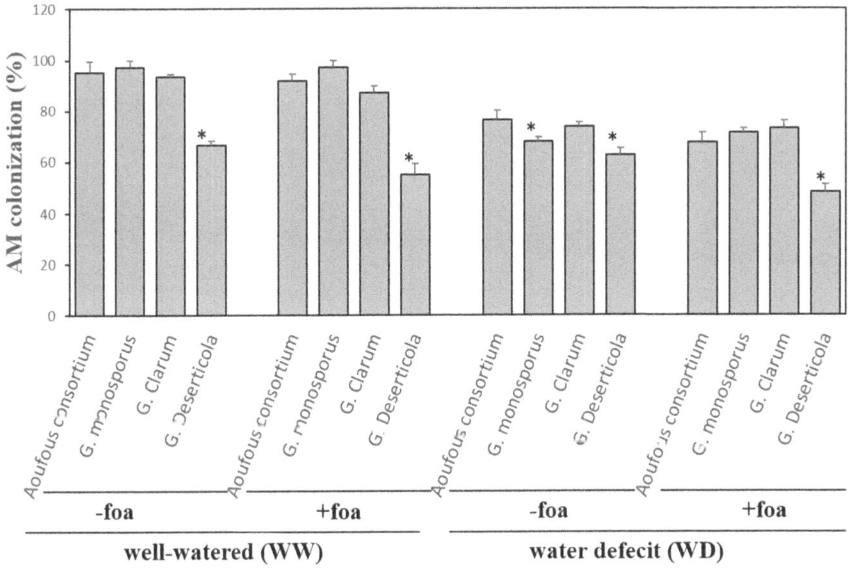

Fig. 9.3 Mycorrhizal colonization of date palm associated with Aoufous consortium, *Glomus monosporus*, *Glomus clarum*, *Glomus deserticola*, inoculated (+Foa) or not (–Foa) with *Fusarium oxysporum* f. sp. *albedinis* and subjected to well-watered (WW) or water-deficit (WD) conditions. Values represent the mean ± SD ($n = 10$ plants). *Asterisks* indicate significant differences between treatments according to Student's *t*-test ($p \leq 0.05$)

9.3 Results

9.3.1 Effect of Water Availability and Fungal Foa Disease on the Colonization of Date Palm Roots by AMF

The percentage of colonization of the date palm root system by AMF is slightly affected by soil water deficiency during 48 weeks of mycorrhization (Fig. 9.3). This frequency of mycorrhization is not affected by Foa with the exception of *G. deserticola*. For the two studied water regimes (75 and 25% FC), mycorrhizal frequency remained high (>48%) for all mycorrhizal fungi tested. Mycorrhizal isolates from Aoufous consortium, *G. monosporus*, and *G. clarum* were the most infectious ($F > 67\%$) fungi compared with *G. deserticola*, even in the presence of Foa and irrespective of soil moisture regime.

The intensity of colonization of the date palm roots by the different AMF decreases under the severe water regime (25% FC) (Fig. 9.4). In contrast, in plant subjected to well-watered conditions 75% FC, the mycorrhization intensity was significantly reduced in the presence of Foa, independently of the type of AMF used. Generally, Aoufous complex, *G. monosporus*, and *G. clarum* were the most effective to colonize date palm root regardless of water supply conditions, the positive effect being especially evident when the plants were inoculated with the

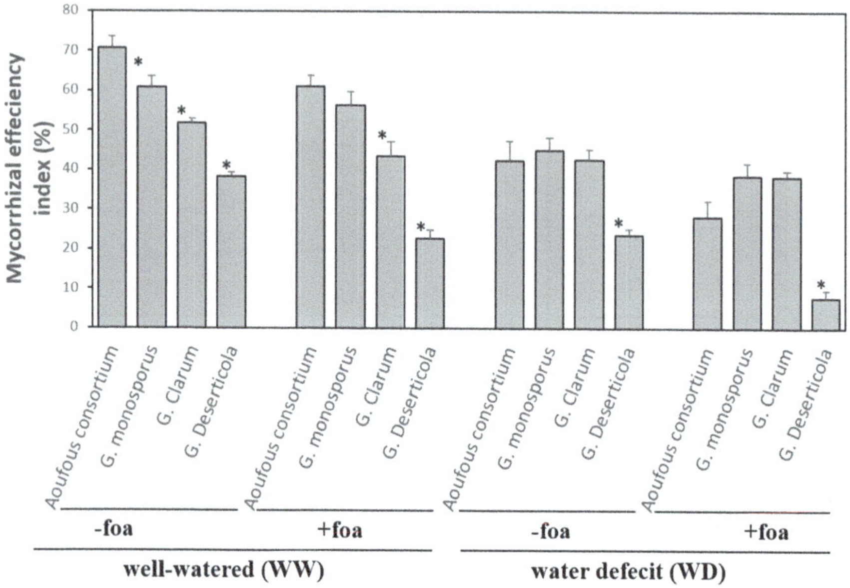

Fig. 9.4 Mycorrhizal efficiency index (MEI) of date palm plants associated with Aoufous consortium, *Glomus monosporus*, *Glomus clarum*, *Glomus deserticola*, inoculated (+Foa) or not (–Foa) with *Fusarium oxysporum* f. sp. *albedinis* and subjected to well-watered (WW) or water-deficit (WD) conditions. Values represent the mean ± SD ($n = 10$ plants). *Asterisks* indicate significant differences between treatments according to Student's *t*-test ($p \leq 0.05$)

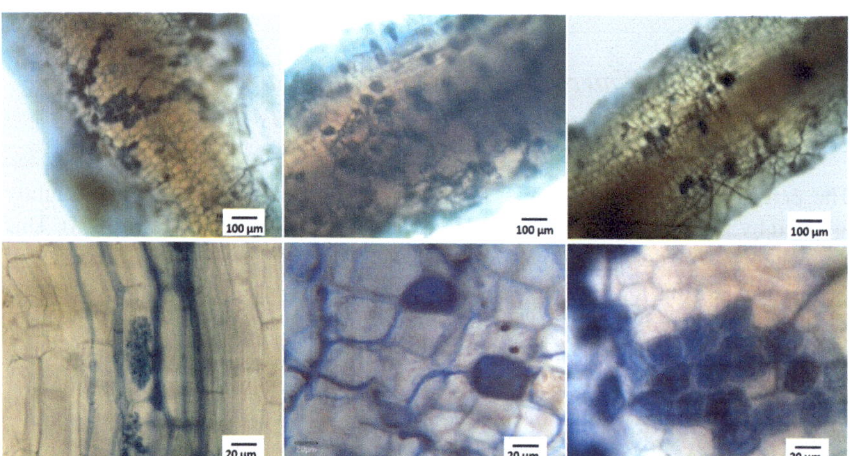

Fig. 9.5 AMF structures developed by the Aoufous consortium in date palm (*Phoenix dactylifera*)

indigenous community inoculant (Aoufous complex). Figure 9.5 shows AMF structures developed by the symbiosis Aoufous consortium and date palm roots. Besides, *Glomus deserticola* showed the low rate of mycorrhizal intensity (8%) under water stress and Foa infection conditions,

9.3.2 *Effect of AMF on the Growth of Date Palm Subjected to Water Deficiency and Attacks by Foa*

Mycorrhization of the date palm trees allowed the formation of a number of leaves significantly higher than their respective non-mycorrhizal controls under both well-watered and drought conditions (Table 9.1). The reduction in the number of leaves

Table 9.1 Number of leaves (NL), leaf area (LA), and relative water content (RWC) of date palm associated with Aoufous consortium, *Glomus monosporus*, *Glomus clarum*, *Glomus deserticola*, inoculated or not (control) under well-watered or drought-stressed conditions and inoculated (+Foa) or not (–Foa) with *Fusarium oxysporum* f. sp. *albedinis*

	Treatment	75%	25%
NL	*Aoufous consortium*	6.00 ± 0.000a	5.00 + 0.000bc
	Aoufous consortium + *Foa*	4.00 ± 0.000cd	4.00 ± 0.000cd
	Glomus monosporus	5.00 ± 0.000b	4.00 ± 0.000bc
	Glomus monosporus + *Foa*	4.33 ± 0.577bc	3.67 ± 0.577cd
	Glomus clarum	5.00 ± 0.000bc	4.33 ± 0.577bc
	Glomus clarum + *Foa*	4.67 ± 0.577bc	4.00 ± 0.000d
	Glomus deserticola	4.33 ± 0.577cd	2.67 ± 0.577e
	Glomus deserticola + *Foa*	3.33 ± 0.577cd	2.33 ± 0.577f
	Control	4.00 ± 0.000cd	3.00 ± 1.000f
	Control + *Foa*	3.67 ± 0.577d	2.33 ± 0.577f
LA (cm^2)	*Aoufous consortium*	50.25 ± 1.391a	35.50 ± 1.299de
	Aoufous consortium + *Foa*	38.25 ± 1.044b	28.00 ± 1.500e
	Glomus monosporus	48.00 ± 1.464a	33.33 ± 1.474bc
	Glomus monosporus + *Foa*	31.75 ± 1.147cd	27.00 ± 1.000de
	Glomus clarum	49.00 ± 1.00bc	31.00 ± 0.790de
	Glomus clarum + *Foa*	35.25 ± 1.070de	28.75 ± 1.035f
	Glomus deserticola	29.00 ± 1.732de	23.45 ± 1.149f
	Glomus deserticola + *Foa*	26.54 ± 1.164de	19.75 ± 1.205g
	Control	27.00 ± 1.447de	22.00 ± 0.460fg
	Control + *Foa*	20.05 ± 0.926fg	19.25 ± 1.089g
RWC (%)	*Aoufous consortium*	83.43 ± 1.974b	71.79 ± 1.043def
	Aoufous consortium + *Foa*	75.44 ± 2.158c	70.88 ± 1,141f
	Glomus monosporus	82.42 ± 1.035a	68.92 ± 1.545d
	Glomus monosporus + *Foa*	70.26 ± 2.576de	68.74 ± 1.573f
	Glomus clarum	76.88 ± 1.764c	69.23 ± 1.985g
	Glomus clarum + *Foa*	70.15 ± 1.592ef	66.01 ± 1.779h
	Glomus deserticola	71.29 ± 0.551d	60.47 ± 1.410g
	Glomus deserticola + *Foa*	50.10 ± 1.831i	38.26 ± 1.210j
	Control	75.20 ± 2.585bc	61.15 ± 1.001g
	Control + *Foa*	53.45 ± 1.552h	47.77 ± 1.948i

The values followed by the same letter are not significantly different $P < 0.05$ (Newman and Keuls test)

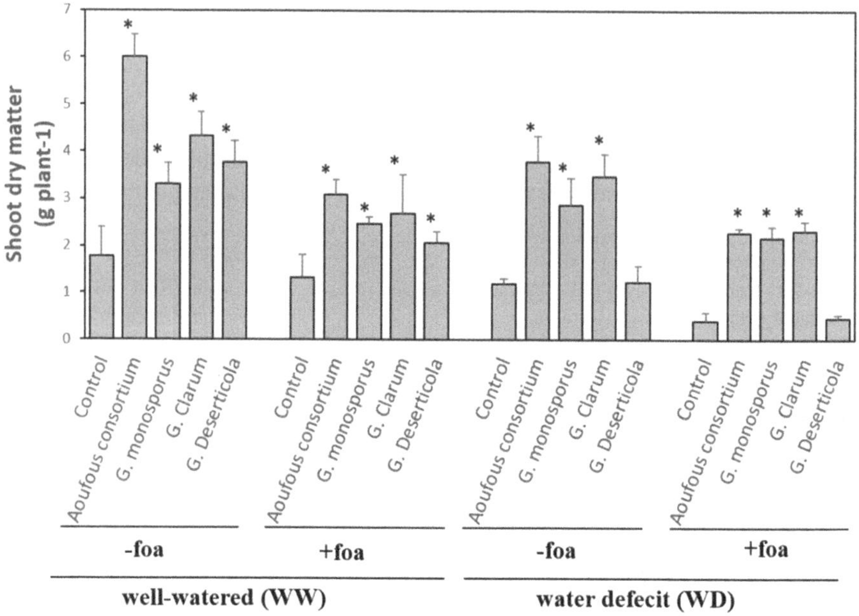

Fig. 9.6 Shoot dry matter of date in mycorrhizal (with Aoufous consortium, *Glomus monosporus*, *Glomus clarum*, *Glomus deserticola*) plants under well-watered or drought-stressed conditions and inoculated (+Foa) or not (–Foa) with *Fusarium oxysporum* f. sp. *albedinis*. Values represent the mean ± SD ($n = 10$ plants). *Asterisks* indicate significant differences between treatments according to Student's *t*-test ($p \leq 0.05$)

formed is clearly significant in the case of non-mycorrhized control plants infected with Foa and subjected to water deficiency.

Plants inoculated with Aoufous consortium and *G. monosporus* showed significant improvement in their leaf area compared to control plants or those mycorrhized with *G. deserticola*. Infection with Foa induces a significant decrease in the leaf area of mycorrhizal and non-mycorrhizal plants. The number of leaves formed and leaf surface of date palm decreased significantly when the soil water availability decreases.

In general, the non-mycorrhizal date palm plants have been more sensitive with respect to changes in soil water than mycorrhizal plants. The inoculation of date palm plants with Aoufous consortium and *G. monosporus* maintains a higher relative water content (RWC) for both water regimes (Table 9.1). The application of Foa significantly reduced RWC in non-mycorrhizal and mycorrhizal plants by *G. deserticola* reaching values of 48 and 38%, respectively, under water-deficit condition (25% FC).

Moreover, production of aerial dry matter increased significantly by the colonization of the palm plants with different mycorrhizal fungi regardless of the water regime (Fig. 9.6). The Aoufous complex, *G. monosporus*, and *G. clarum* allowed the production of a larger aerial dry mass to face the severe water regime.

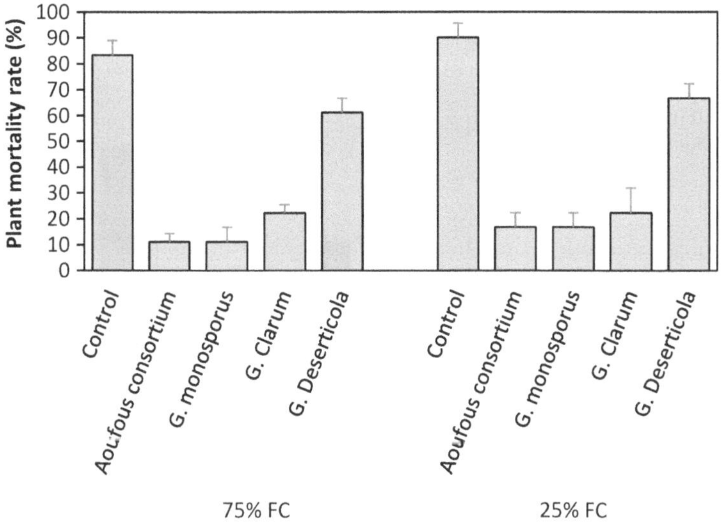

Fig. 9.7 Plant mortality rate of date palm in mycorrhizal (with Aoufous consortium, *G. monosporus*, *G. clarum*, *G. deserticola*) plants under well-watered or drought-stressed conditions after 8 months of inoculation with Foa. Values represent the mean ± SD ($n = 10$ plants). *Asterisks* indicate significant differences between treatments according to Student's *t*-test ($p \leq 0.05$)

Non-mycorrhizal plants or those inoculated with *G. deserticola* showed low weight dry matter either in the presence or absence of the pathogen Foa at the severe water regime (25% FC).

The mortality of date palm infected with Foa in terms of tested AMF and soil water regimes was evaluated after a period of 32 weeks (Fig. 9.7). Regardless of the water conditions, plant mortalities stabilized between 11 and 22% in plants inoculated with Aoufous consortium, *G. monosporus*, and *G. clarum*. After 8 months of Foa infection, the mortality rates were 61 and 83% in *G. deserticola* mycorrhized and non-mycorrhized plants, respectively, under the well-watered regime (75% FC) (Fig. 9.7). Furthermore, the mortality rates under water deficiency conditions increased exponentially in non-mycorrhizal and plants inoculated with *G. deserticola*, reaching values of 90% and 67%, respectively.

9.3.3 Potential of Arbuscular Mycorrhizal Fungi on the Mitigation of Salt Stress Negative Effects on Date Palm

In the absence of salt stress, the mycorrhizal date palm plants showed an aerial and root dry matters three times higher than their control plants (Figs. 9.8 and 9.9). In

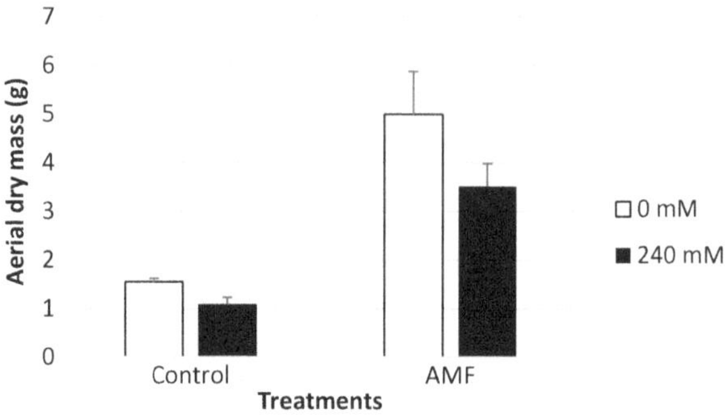

Fig. 9.8 Effect of salt stress (during 5 months) on the aerial dry matter production in control and mycorrhizal date palm (8 months of mycorrhization)

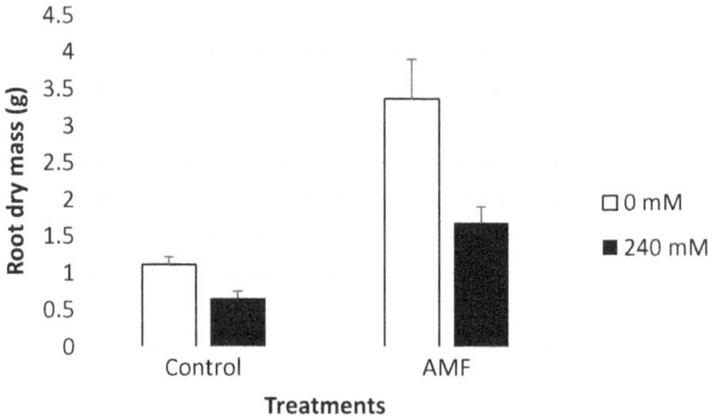

Fig. 9.9 Effect of salt stress (during 5 months) on the root dry matter production in both mycorrhized and control date palm after 8 months of mycorrhization

Table 9.2 Effect of salt stress on water parameters of date palm after 8 months of mycorrhization (5 months under salt stress)

	Treatments	0 mM	240 mM
Water content (WC) ($gH_2O\ gDM^{-1}$)	AMF	11.15 ± 1.40a	7.44 ± 0.31b
	Control	3.31 ± 0.15c	1.83 ± 0.39d
Stomatal conductance (SC) ($mmol\ m^{-2}\ s^{-1}$)	AMF	45.77 ± 4.31a	23.00 ± 1.20b
	Control	17.63 ± 1.66c	10.87 ± 1.72d
Water potential (ψ_h) (bar)	AMF	–15.96 ± 1.44a	-21.97 ± 1.06b
	Control	–23.70 ± 0.98c	-31.47 ± 2.81d

Mean followed by the same letters are not significantly different $P < 0.05$ (Neumans and Keuls test)

contrast, the application of salt stress decreased the aerial and root dry matters of date palm plants by 30% and 40%, respectively, whereas in the presence of AMF, the dry matter of seedlings was enhanced three times for the aerial dry matter and twice for the root dry matter compared to control plants under salt stress.

The water parameters were improved by the application of AMF in the absence or presence of salt stress (Table 9.2). Under the non-salt stress condition, the mycorrhizal date palm plants showed three times higher water content and stomatal conductance compared with the control plants. Also, the inoculation with AMF allowed a higher water potential (–15.96 bars). The application of salt stress decreases water content, stomatal conductance, and water potential by 45%, 38%, and 33%, respectively, in control plants. In the same conditions, the application of AMF improved the tolerance of date palm plants, since it enhances the water content to 400 %, the stomatal conductance to 230 %, and the water potential to 70% compared to stressed control plants.

9.4 Discussion

9.4.1 Enhanced Drought Stress Tolerance by the Arbuscular Mycorrhizal Symbiosis in Date Palm Plants

The fungal isolates of the Aoufous consortium and the selected *Glomus* showed a greater colonization of date palm roots. The severe water regime (25% FC) has slightly affected the frequency and the colonization of these AMF. In our previous works (Meddich et al. 2000; Meddich 2001), we observed a significant reduction in the infectivity and colonization parameters of clover and barley roots subjected to water deficit and inoculation with Aoufous consortium and *G. monosporus*. This suggests the presence of variability in the AMF infectivity parameters depending on the host plant involved and the environmental conditions. Thus, the existence of signals exchange and cross talk between both the host plant and the AMF partners exist prior to the establishment of a functional symbiosis. On the plant side, the organic compounds contained in the former root-sucking buds influence the development of arbuscular mycelia (Koske 1982; Gianinazzi-Pearson et al. 1996). Low molecular weight compounds and proteins necessary for the establishment of mycorrhizal symbiosis have been also identified in angiosperms (Delaux et al. 2013). On the fungi side, Maillet et al. (2011) have shown that *G. intraradices* secrete symbiotic signals in the form of a mixture of lipochitooligosaccharides, which stimulate the formation of mycorrhizal symbiosis in certain plant families, including the Fabaceae, Asteraceae, and Umbelliferae. Subsequently, the mycelia of the AMF isolates colonize the cortical cells and give rise to the fungal arbuscules representing the preferred site of the metabolic exchanges between the fungus and the host plant (Gianinazzi-Pearson and Gianinazzi 1986, 1988; Gianinazzi-Pearson et al. 1996).

9.4.2 Effect of Mycorrhiza and Water and Salinity Stresses on Date Palm Growth and Physiological Traits

Mycorrhization by the Aoufous consortium and *Glomus* has a positive effect on date palm plant growth. Thus, the inoculation with these AMF increased the number and area of date palm leaves under the well-watered and drought-stressed regimes. Foa infection affects slightly these parameters in mycorrhizal plants. The reduction of plant growth was remarkable in both non-mycorrhizal and plants inoculated with *G. deserticola*, subjected to the severe water regime. Similarly, AMF improved significantly the dry matter production compared to non-mycorrhizal plants. Similar responses were reported in other plants inoculated with *G. mosseae* and *G. intraradices* (Kothari et al. 1990; Tobar et al. 1994; Sheng et al. 2008; Baslam and Goicoechea 2012). The positive effects of mycorrhizal symbiosis on the growth and health of date palms have been reported (Al-Karaki 2013; Baslam et al. 2014; Meddich et al. 2015). Studies have revealed that (1) the AMF has promoted the growth of date palm seedlings in nursery conditions (Shabbir et al. 2011) compared to controls treated with chemical fertilizers (Symanczik et al. 2014), (2) increased the availability of nutrients in soil cultures (Al-Karaki et al. 2007), and (3) improved the absorption of water and nutrients in saline conditions (Bearden and Petersen 2000).

Also, mycorrhizal date palm showed a high stomatal conductance compared to control plants in absence or presence of salinity stress. The high stomatal conductance in mycorrhizal plants could improve the CO_2 fixation in mesophyll (Brown and Bethelenfalvay 1987), which contributes to an increase in plant photosynthesis rate (Lawlor 1987; Zuccarini and Okurowska 2008). Plants inoculated with the Aoufous consortium showed a higher RWC under water stress and maintained a high level of WC and ψh under salt stress compared to control plants. This reflects the ability of AMF to maintain well-hydrated host plant tissues. AM symbiosis may enhance osmotic adjustment in plants which could contribute in maintaining higher leaf water status in AM plants during drought or salinity and keeps the plants protected against oxidative stress.These cumulative effects increase the tolerance of plants to biotic and abiotic stresses. The plant genetic analyses carried out by our team (Zézé et al. 2007, 2008) revealed the expression of three types of aquaporin genes, intrinsic proteins, in clover roots mycorrhized by the Aoufous consortium and *G. monosporus* and subjected to water stress (30% FC), which would contribute to a better distribution of water movement along plant tissues. Aquaporins provide a low resistance pathway for the movement of water across a membrane. Furthermore, because aquaporins can be gated, this provides greater control for wáter loss from the root. The better distribution of water circulation in the plant may explain, in part, the tolerance of plants in the presence of AMF.

The use of mycorrhizal fungi to cope with water and salt stresses (Sharifi et al. 2007; Jahromi et al. 2008; Fini et al. 2011; Navarro et al. 2011; Sheng et al. 2011; Baslam and Goicoechea 2012; Augé et al. 2015; Taffouo et al. 2014; Zhang et al. 2014; Meddich et al. 2015) could be a key component and a promising biological strategy in increasing drought and osmotic resistance. It is worth noting that the indigenous strains (Aoufous complex) have been effective in improving the tolerance of the host plant under water and salt stresses. The indigenous and adapted fungal isolates to specific environmental conditions could be an important tool in protecting plants against harmful effects of biotic and abiotic stresses.

9.4.3 *Interactions Between AMF and Biotic Stressors (Foa) of Date Palm*

The mortality rates in palm trees infected with Foa remained lower in mycorrhizal plants than in non-mycorrhizal plants after 32 weeks of infection. The effect of prior mycorrhization by AMF is remarkable on the expression of bayoud disease. Indeed, Oihabi (1991) found that the simultaneous inoculation of *G. mosseae* and the pathogen does not allow the expression of a protective effect in the young date plants cultivated under the calcined clay. On the other hand, Caron et al. (1986) observed a reduction in root necrosis of the tomato due to *Fusarium oxysporum* f. sp. *radicis-lycopersici* when *G.* intraradices was applied 5 weeks before the pathogen. In addition, Bartschi et al. (1981) reported that the inoculation of *Chamaecyparis lawsonia* by a natural AMF mixture 6 months prior to the infection with *Phytophthora cinnamomi*, a root rot agent, greatly reduced the mortality rate of the host, whereas the simultaneous inoculation showed no effect. Jalali and Tharija (1981) showed that the mycorrhization allowed a 53% reduction in the incidence of chickpea *fusarium* wilt. In this study, the application of Foa 4 months after the establishment of mycorrhization by the Aoufous consortium and *G. clarum* affected slightly the growth parameters of the young plants, in both water regimes applied. In contrast, non-mycorrhizal plants subjected to water stress and Foa attack showed higher mortality rates. Thus, the effectiveness of prior mycorrhization may be related to the colonization and protection of the root system by AMF prior to the movement of the pathogen into the plant tissues, showing the preventive measures of AMF to curb the spread of the pathogen. In this line, Ismail and Hijri (2012) have shown that potato inoculation by *G.irregulare* significantly reduces the negative effects of *Fusarium sambucinum* on biomass and tuber production. These favorable answers were attributed to the role of AMF in the positive regulation of the expression of the majority of the defense genes (i.e., ChtA3, gluB, CEVI16, and PR-1) at host plant root level. Similarly, Ismail et al. (2011, 2013) showed that *G. irregulare* significantly inhibited the growth of *F. sambucinum* while modulating the expression of genes producing toxins (trichothecenes) by the biotic stressor. Other studies have shown the beneficial role of AMF in the protection of other host plants toward several diseases (Linderman 1994; Azcon-Aguilar and Barea 1996; Thygesen et al. 2004; Jung et al. 2012; Xiao et al. 2014). Mycorrhizal date palm could be better prepared to overcome attacks of pathogens than non-mycorrhizal ones. AMF can induce structural, physiological, and/or biochemical changes in plants in response to *F. oxysporum*. The overall data here show that AMF can be used as biocontrol agents in triggering date palm defense against the pathogenic *F. oxysporum* and a preventive and eradication measure to curb the spread of the biotic stressors under abiotic conditions.

9.5 Conclusions

Our study reveals that the association of date palm with AMF benefited growth under well-watered conditions, being the native Aoufous consortium the most effective in increasing the morphophysiological traits. Under stressful conditions, both the indigenous community and exotic AMF (*G. monosporus* and *G. clarum*) appeared as the most beneficial fungus for improving plant growth. After 48 weeks of mycorrhization, Aoufous consortium and selected *Glomus* are more aggressive colonization, even under conditions of water stress. Interestingly, plant colonized with these AMF showed restriction of Foa fungal growth in the roots allowing a marked improvement in growth parameters under water scarcity. Only plants colonized by *G. deserticola* were more susceptible to *F. oxysporum* attack under water stress. Further, the indigenous Aoufous strains improve significantly the tolerance of date palm to the combination of water and *F. oxysporum* stresses. Overall, the AMF indigenous community provided better protection of date palm against salt stress by increasing the biomass and other water and physiological parameters. The use of indigenous fungi adapted to unfavorable conditions could constitute an integrated solution to improve the plant defense and alleviate the constraints of combined stresses such as salinity, drought, and pathogenic agents, and this could provide useful tools to reduce losses in crop species too.

Acknowledgments This research was partially supported by Grant for Promotion of KAAB Projects (Niigata University) from the Ministry of Education, Culture, Sports, Science, and Technology, Japan.

References

Al-Karaki GN (2000) Growth of mycorrhizal tomato and mineral acquisition under salt stress. Mycorrhiza 10:51–54

Al-Karaki GN (2013) Application of mycorrhizae in sustainable date palm cultivation. Emir J Food Agric 25:854–862

Al-Karaki GN, McMichael B, Zak J (2004) Field response of wheat to arbuscular mycorrhizal fungi and drought stress. Mycorrhiza 14:263–269

Al-Karaki GN, Othman Y, Al-Ajmi A (2007) Effects of mycorrhizal fungi inoculation on landscape turf establishment under Arabian Gulf region conditions. Arab Gulf J Sci Res 25:147–152

Asghari HR, Marschner P, Smith SE, Smith FA (2005) Growth response of *Atriplex nummularia* to inoculation with arbuscular mycorrhizal fungi at different salinity levels. Plant Soil 273:245–256

Augé RM, Toler HD, Saxton AM (2015) Arbuscular mycorrhizal symbiosis altersstomatal conductance of host plants more under drought than under amply watered conditions: a meta-analysis. Mycorrhiza 25:13–24

Azcon-Aguilar C, Barea JM (1996) Arbuscular mycorrhizas and biological control of soil-borne plant pathogens—an overview of the mechanisms involved. Mycorrhiza 6:457–464

Bartschi H, Gianinazzi-Pearson V, Vegh I (1981) Vesicular arbuscular mycorrhiza formation and root disease (*Phytophthora cinnamomi*) development in *Chamaecyparis lawsoniana*. J Phytopathol 102:213–218

Baslam M, Goicoechea N (2012) Water deficit improved the capacity of arbuscular mycorrhizal fungi (AMF) for inducing the accumulation of antioxidant compounds in lettuce leaves. Mycorrhiza 22:347–359

Baslam M, Qaddoury A, Goicoechea N (2014) Role of native and exotic mycorrhizal symbiosis to develop morphological, physiological and biochemical responses coping with water drought of date palm, *Phoenix dactylifera*. Trees 28:161–172

Bearden B, Petersen L (2000) Influence of arbuscular mycorrhizal fungi on soil structure and aggregate stability of a vertisol. Plant Soil 218:173–183

Benmeddour Z, Mehinagic E, Le Meurlay D, Louaileche H (2013) Phenolic composition and antioxidant capacities of ten Algerian date (*Phoenix dactylifera* L.) cultivars: a comparative study. J Funct Foods 5:346–354

Borowicz VA (2001) Do arbuscular mycorrhiza fungi alter plant–pathogen relations? Ecology 82:3057–3068

Brown MS, Bethelenfalvay GJ (1987) Glycine Glomus-Rhizobium symbiosis. Plant Physiol 85:120–123

Cantrell IC, Linderman RG (2001) Preinoculation of lettuce and onion with VA fungi reduces deleterious effects of soil salinity. Plant Soil 233:269–281

Caron M, Fortin A, Richard C (1986) Effect of inoculation sequence on the interaction between *Glomus intraradices* and *Fusarium oxysporum* f. sp. *radicis-lycopersici* in tomatoes. Can J Plant Pathol 8(1):12–16. https://doi.org/10.1080/07060668609501835

Chao CCT, Krueger RR (2007) The date palm (*Phoenix dactylifera* L.), overview of biology, uses, and cultivation. Hortscience 42(5):1077–1082

Ciais P, Reichstein M, Viovy N, Granier A, Ogee J, Allard V (2005) Europe-wide reduction in primary productivity caused by the heat and drought in 2003. Nature 437:529–533

Daayf F, El Bellaj M, El Hassni M, Jaiti F, El Hadrami I (2003) Elicitation of soluble phenolics in date palm (*Phoenix dactylifera*) callus by *Fusarium oxysporum* f. sp. *albedinis* culture medium. Environ Exp Bot 49:41–47

Delaux PM, Séjalon-Delmas N, Bécard G, Ané JM (2013) Evolution of the plant-microbe symbiotic 'toolkit'. Trends Plant Sci 18(6):298–304

Dell'Amico J, Torrecillas A, Rodriguez P, Morte A, Sanchez-Blanco M (2002) Responses of tomato plants associated with the arbuscular mycorrhizal fungus *Glomus clarum* during drought and recovery. J Agric Sci 138:387–393

Djerbi M (1998) Diseases of the date palm: present status and future prospects. In: Proceedings of the international conference on integrated pest management, Muscat, Sultanate of Oman, 23–25 Feb 1998, Sultan-Qaboos-University. J Sci Res Agric Sci 3:103–114

Engelbrecht BMJ, Comita LS, Condit R, Kursar TA, Tyree MT, Turner BL (2007) Drought sensitivity shapes species distribution patterns in tropical forests. Nature 447:80–82

Evelin H, Kapoor R, Giri B (2009) Arbuscular mycorrhizal fungi in alleviation of salt stress: a review. Ann Bot 104:1263–1280

Evelin H, Giri B, Kapoor R (2013) Ultrastructural evidence for AMF mediated salt stress mitigation in *Trigonellafoenum-graecum*. Mycorrhiza 23:71–86

Fan L, Dalpé Y, Fang C, Dubé C, Khanizadeh S (2011) Influence of arbuscular mycorrhizae on biomass and root morphology of selected strawberry cultivars under salt stress. Botany 89:397–403

Fini A, Frangi P, Amoroso G (2011) Effect of controlled inoculation with specific mycorrhizal fungi from the urban environment on growth and physiology of containerized shade tree species growing under different water regimes. Mycorrhiza 21:703–719

Gianinazzi-Pearson V, Gianinazzi S (1986) The physiology of improved phosphate nutrition in mycorrhizal plants. In: Les Mycorhizes, Physiologie et Génétique. INRA, Editions, Paris, pp 101–109

Gianinazzi-Pearson V, Gianinazzi S (1988) Morphological integration and functional compatibility between symbionts in vesicular arbuscular mycorrhizal associations. In: Scannerini S, Smith D,

Bonfante-Fasolo P, Gianinazzi-Pearson V (eds) Cell to cell signals in Plant, animal and microbial symbiosis. Springer, Berlin, pp 73–84
Gianinazzi-Pearson V, Dumas-Gaudot E, Gollotte A, Tahiri-Alaoui A, Gianinazzi S (1996) Cellular and molecular defense-related root response to invasion by arbuscular mycorrhizal fungi. New Phytol 133:45–57
Giri B (2017) Mycorrhizal dependency and growth response of *Gliricidia sepium* (Jacq.) Kunth ex Walp. under saline condition. Plant Sci Today 4(4):154–160
Giri B, Kapoor R, Mukerji KG (2007) Improved tolerance of *Acacia nilotica* to salt stress by arbuscular mycorrhiza, *Glomus fasciculatum* may be partly related to elevated K/Na ratios in root and shoot tissues. Microb Ecol 54:753–760
Harrison MJ (1997) The arbuscular mycorrhizal symbiosis. In: Stancey G, Keen NT (eds) Plant-microbe interactions, vol 3. Chapman and Hall, New York, pp 1–34
Ismail Y, Hijri M (2012) Arbuscular mycorrhization with Glomus irregulare induces expression of potato PR homologues genes in response to infection by *Fusarium sambucinum*. Funct Plant Biol 39(3):236–245
Ismail Y, McCormick S, Hijri M (2011) A fungal symbiont of plant-roots modulates mycotoxin gene expression in the pathogen *Fusarium sambucinum*. PLoS One 6(3):e17990
Ismail Y, McCormick S, Hijri M (2013) The arbuscular mycorrhizal fungus, *Glomus irregulare*, controls the mycotoxin production of *Fusarium sambucinum* in the pathogenesis of potato. FEMS Microbiol Lett 348(1):46–51. https://doi.org/10.1111/1574-6968.12236. PMID: 23964970
Jahromi F, Aroca R, Porcel R, Ruiz-Lozano JM (2008) Influence of salinity on the in vitro development of *Glomus intraradices* and on the in vivo physiological and molecular responses of mycorrhizal lettuce plants. Microb Ecol 55:45–53
Jalali BL, Tharija ML (1981) Suppression of Fusarium wilt of chickpea in V.A.M. inoculated soils. International Chickpea Newsletter. Haryana. Agric Univ Hissar India 4:21–24
Janos DP (2007) Plant responsiveness to mycorrhizas differs from dependence upon mycorrhizas. Mycorrhiza 17:75–91
Jung SC, Ainhoa MM, Lopez-Raez JA, Pozo MJ (2012) Mycorrhiza-induced resistance and priming of plant defenses. J Chem Ecol 38:651–664
Juniper S, Abbott L (1993) Vesicular-arbuscular mycorrhizas and soil salinity. Mycorrhiza 4:45–58
Koske RE (1982) Evidence for a volatile attractant from plant roots affecting germ tubes of a VA mycorrhizal fungus. Trans Br Mycol Soc 79:305–310
Kothari SK, Marschner H, George E (1990) Effect of VA mycorrhizal fungi and rhizosphere microorganisms on root and shoot morphology growth and water relations in maize. New Phytol 116:303–311
Lawlor DW (1987) Stress metabolism: its implication in breeding programmes. In: Srivastava JP, Porceddu E, Acevedo E, Varma S (eds) Drought tolerance in winter cereals. ICARDA, Wiley, Chichester, pp 227–240
Linderman RG (1994) Role of VAM fungi in biocontrol. In: Pfleger FL, Linderman RG (eds) Mycorrhizae and plant health. American Phytopathological Society, St. Paul, MN, pp 1–25
Maillet F, Poinsot V, André O, Puech-Pagès V, Haouy A, Gueunier M et al (2011) Fungal lipochitooligosaccharide symbiotic signals in arbuscular mycorrhiza. Nature 469:58–63. https://doi.org/10.1038/nature09622. PMID: 21209659
Meddich A (2001) Rôle des endomycorhizes à vésicules et à arbuscules des Palmeraies Marocaines dans la tolérance des plantes de zones arides au stress hydrique. Thèse de Doctorat National. Département de Biologie, Université Cadi Ayyad de Marrakech, Faculté des Sciences Semlalia, Marrakech, Maroc, 234 pages
Meddich A, Boumezzough A (2017) First detection of *Potosia opaca* larva attacks on *Phoenix dactylifera* and *P. canariensis* in Morocco: focus on pests control strategies and soil quality of prospected palm groves. J Entomol Zool Stud 5(4):984–991
Meddich A, Oihabi A, Abbass Y, Bizid E (2000) Rôle des champignons mycorhiziens à arbuscules de zones arides dans la résistance du trèfle (*Trifolium alexandrinum L.*) au déficit hydrique. Agronomie 20:283–295

Meddich A, Jaiti F, Bourzik W, El Asli A, Hafidi M (2015) Use of mycorrhizal fungi as a strategy for improving the drought tolerance in date palm (*Phoenix dactylifera L.*) Scientia Horticulturae 192:468–474
Navarro AG, Del Pilar Bañón Árias S, Morte A, Sánchez-Blanco MJ (2011) Effects of nursery preconditioning through mycorrhizal inoculation and drought in Arbutus unedo L. plants. Mycorrhiza 21:53–64
Oihabi A (1991) Effet des endomycorhizes V.A sur la croissance et la nutrition minérale du palmier dattier. Thèse de Doctorat d'Etat, Univ. Cadi Ayyad Marrakech, Maroc/Univ. Bourgogne Dijon France, p 117
Oihabi A (2001) Date palm in North Africa: trumps and problems. In: The second international conference on Date Palms, Department of Arid Land Agriculture and Department of Agriculture and Livestock College of Food Systems, UAE University, Ai-Ain, UAE, March, 25–27
Oihabi A, Meddich A (1996) Effet des mycorhizes à arbuscules sur la croissance et la composition minérale du trèfle (*Trifolium alexandrinum*). Cahiers Agric 5:382–386
Oruru MB, Njeru EM (2016) Upscaling arbuscular mycorrhizal symbiosis and related agroecosystems services in smallholder farming systems. Biomed Res Int 2016
Passioura J (2007) The drought environment: physical, biological and agricultural perspectives. J Exp Bot 58:113–117
Peters RS, Meusemann K, Petersen M, Mayer C, Wilbrandt J, Ziesmann T, Donath A et al (2014) The evolutionary history of holometabolous insects inferred from transcriptome-based phylogeny and comprehensive morphological data. BMC Evol Biol 14(1):52
Phillips JM, Hayman DS (1970) Improved procedures for clearing roots and staining parasitic and vesicular-arbuscular mycorrhizal fungi for rapid assessment of infection. Trans Br Mycol Soc 55:158–161
Pitman M, Läuchli A (2004) Global impact of salinity and agricultural ecosystems. In: Läuchli A, Lüttge U (eds) Salinity: environment-plants-molecules. Springer, Dordrecht, pp 3–20
Plenchette C, Furlan V, Fortin JA (1982) Effects of different endomycorrhizal fungi on five host plants grown on calcined montmorillonite clay. J Am Soc Hortic Sci 107:535–538
Porcel A, Aroca R, Ruiz-Lozano JM (2012) Salinity stress alleviation using arbuscular mycorrhizal fungi. Agron Sustain Dev 32:181–200
Redecker D, Morton JB, Bruns TD (2000) Ancestral lineage of arbuscular mycorrhizal fungi (Glomales). Mol Phylogenet Evol 14:276–284
Ruiz-Lozano JM, Porcel R, Azcon R, Aroca R (2012) Regulation by arbuscular mycorrhizae of the integrated physiological response to salinity in plants. New challenges in physiological and molecular studies. J Exp Bot 63:4033–4044. https://doi.org/10.1093/jxb/ers126
Saaidi M (1990) Amélioration génétique du palmier dattier critères de sélection, techniques et résultats. Options Méditerranéennes 11:133–154
Sannazzaro AI, Ruíz OA, Alberto EO, Menendez AB (2006) Alleviation of salt stress in *Lotus glaber* by *Glomus intraradices*. Plant Soil 285:279–287
Scholander PF, Hammel HT, Bradstreet ED, Hemmingzen EA (1965) Sap pressure in vascular plants. Science 148:339–346
Shabbir G, Dakheel AJ, Brown GM, Rillig MC (2011) Potential of arbuscular mycorrhizal technology in date palm production. In: Jain SM, Al-Khayri JM, Johnson DV (eds) Date palm biotechnology. Springer, Dordrecht, pp 449–476
Sharifi M, Ghorbanli M, Ebrahimzadeh H (2007) Improved growth of salinity-stressed soybean after inoculation with pre-treated mycorrhizal fungi. J Plant Physiol 164:1144–1151
Sheng M, Tang M, Chen H, Yang B, Zhang F, Huang Y (2008) Influence of arbuscular mycorrhizae on photosynthesis and water status of maize plants under salt stress. Mycorrhiza 18:287–296
Sheng M, Tang M, Zhang F, Huang Y (2011) Influence of arbuscular mycorrhiza on organic solutes in maize leaves under salt stress. Mycorrhiza 21:423–430
Smith SE, Read DJ (1997) Mycorrhizal symbiosis. Academic, London

Strullu DG (1986) Micropropagation of chesnut and conditions of mycorrhizal synthesis in vitro. New Phytol 102:95–101

Symanczik S, Błaszkowski J, Chwat G, Boller T, Wiemken A, Al-Yahya'ei MN (2014) Three new species of arbuscular mycorrhizal fungi discovered at one location in a desert of Oman: Diversispora omaniana, *Septoglomus nakheelum* and *Rhizophagus arabicus*. Mycologia 106:243–259

Taffouo VD, Benard Ngwene B, Akoa A, Franken P (2014) Influence of phosphorus application and arbuscular mycorrhizal inoculation on growth, foliar nitrogen mobilization, and phosphorus partitioning in cowpea plants. Mycorrhiza 24:361–368

Thygesen K, Larsen J, Bodker L (2004) Arbuscular mycorrhizal fungi reduce development of pea root-rot caused by Aphanomyces euteiches using oospores as pathogen inoculum. Eur J Plant Pathol 110:411–419

Tobar R, Azcon R, Barea JM (1994) Improved nitrogen uptake and transport from15N-labelled nitrate by external hyphae of arbuscular mycorrhiza underwater-stressed conditions. New Phytol 126:119–122

Trouvelot A, Kouch J, Gianinazzi-Pearson V (1986) Mesure du taux demycorhization VA d'un système radiculaire: Recherche de méthodesd'estimation ayant une signification fonctionnelle. In: Gianinazzi S (ed) Les Mycorhizes: Physiologie et Génétique, 1er Séminaire Europeen sur les Mycorhizes, Dijon. INRA, Paris, pp 217–221

Vigo C, Norman JR, Hooker JE (2000) Biocontrol of the pathogen *Phytophthoraparasitica* by arbuscular mycorrhizal fungi is a consequence of effects on infection loci. Plant Pathol 49:509–514

Xiao LS, Ai-Rong L, Yan C, Kai-Yun G, Lu Z, Yan YL (2014) Arbuscular mycorrhizal fungi: potential biocontrol agents against the damaging root hemiparasite *Pedicularis kansuensis*. Mycorrhiza 24:187–195

Zézé A, Brou Y, Meddich A, Marty F (2007) Molecular characterisation of a mycorrhizal inoculant that enhances *Trifolium alexandrium* resistance underwater stress conditions. Afr J Biotechnol 6 (13):1524–1528

Zézé A, Brou Y, Meddich A, Marty F (2008) Molecular identification of MIP genes expressed in the roots of an arbuscular mycorrhizal *Trifolium alexandrium* L. under water stress. Afr J Agric Res 3(1):078–083

Zhang HS, Fei FQ, Qin P, Pan SM (2014) Evidence that arbuscular mycorrhizal and phosphate-solubilizing fungi alleviate NaCl stress in the halophyte Kosteletzkya virginica: nutrient uptake and ion distribution within root tissues. Mycorrhiza 24:383–395

Zohary D, Hopf M (2000) The domestication of the plants in the old world: the origin and spread of cultivated plants in West Asia, Europe and Nile Valley, 3rd edn. Oxford University Press, Oxford

Zuccarini P, Okurowska P (2008) Effects of mycorrhizal colonization and fertilization on growth and photosynthesis of sweet basil under salt stress. J Plant Nutr 31:497–513

Chapter 10
Improvement of Salt Tolerance in Rice Plants by Arbuscular Mycorrhizal Symbiosis

Juan Manuel Ruiz-Lozano, Rosa Porcel, Mónica Calvo-Polanco, and Ricardo Aroca

10.1 Introduction

Soil salinization is a serious ecological and agronomical problem. It can be due to natural causes such as contamination from the parental rocks, salty raining, waters around the coasts, and oceanic salts, but inadequate cultivation practices worldwide have exacerbated the concentration of salts in the rhizosphere (Mahajan and Tuteja 2005). Thus, salinization of arable land is getting widespread throughout the world, and estimations indicate that it will result in 30% land loss within the next 25 years and up to 50% within the next 40 years (Wang et al. 2003; Porcel et al. 2012).

Soil salinity causes important yield losses since it affects the establishment, growth, and development of plants (Evelin et al. 2009). In fact, when plants grow under excessive salinity, several aspects of plant physiology are negatively affected. Salinity decreases nutrient uptake and/or transport to the shoot, thus inducing nutrient imbalance in the plant (Marschner 1995; Evelin et al. 2009). In addition, the toxic effects of specific ions such as sodium and chloride inhibit protein synthesis, damage cell organelles, disrupt the structure of enzymes, and uncouple

J. M. Ruiz-Lozano (✉) · R. Aroca
Departamento de Microbiología del Suelo y Sistemas Simbióticos, Estación Experimental del Zaidín (CSIC), Granada, Spain
e-mail: juanmanuel.ruiz@eez.csic.es

R. Porcel
Departamento de Microbiología del Suelo y Sistemas Simbióticos, Estación Experimental del Zaidín (CSIC), Granada, Spain

Instituto Universitario de Conservación y Mejora de la Agrodiversidad Valenciana (COMAV), Universidad Politécnica de Valencia, Valencia, Spain

M. Calvo-Polanco
Biochimie et Physiologie Moléculaire des Plantes, SupAgro/INRA UMR 5004, Montpellier Cedex 2, France

B. Giri et al. (eds.), *Root Biology*, Soil Biology 52,
https://doi.org/10.1007/978-3-319-75910-4_10

photosynthesis and respiration. Moreover, the accumulation of salts in the soil lowers the soil osmotic potentials and hinders the uptake of water by roots, producing a physiological drought in the plant (Ruiz-Lozano et al. 2012).

Rice (*Oryza sativa* L.) is a species native to tropical regions which has been consumed by humans for nearly 9000 years (Callaway 2014). Indeed, rice is the most important source of food for more than half of the world population, with salinity having a remarkable negative impact on its productivity worldwide (Kumar et al. 2013). World population continues to grow, and FAO (2005) has estimated that by the middle of twenty-first century, world agriculture should produce 70% more food in order to nourish the growing population. However, rice productivity is strongly reduced by salinity (Kumar et al. 2013). Therefore, an important challenge for researchers is to achieve different strategies to make rice plants more tolerant and to improve its productivity under salinity, in order to cope with reduced food production due to soil salinization (Ruiz-Lozano et al. 2012, Augé et al. 2014).

As a consequence of the toxic, nutritional, and osmotic effects of salinity, salt stress affects the major plant metabolic processes, such as photosynthesis, growth, energy and lipid metabolisms, and protein synthesis (Ramoliya et al. 2004). However, plants have evolved physiological, biochemical, and molecular mechanisms to cope with the negative effects of salinity. These include the regulation of genes with a role in the transport or compartmentation of Na^+ and/or K^+ ions, known by their important role during ionic homeostasis (Munns 2005). Mechanisms of water/osmotic homeostasis are intended to restore the cellular ion or water content to levels similar to those present under unstressed conditions. This depends of the action of genes involved in solute biosynthesis and water channels (aquaporins). In addition, protection and damage repair mechanisms attempt to prevent or repair cellular damage caused by altered ion, water content, or reactive oxygen species (ROS) under stress (Ruiz-Lozano et al. 2012).

Several studies have shown that the arbuscular mycorrhizal (AM) symbiosis can alleviate salt stress in different host plant species (for reviews, see Evelin et al. 2009; Ruiz-Lozano et al. 2012; Augé et al. 2014). In this chapter we will summarize results obtained in relation to mechanisms involved in amelioration of salt stress tolerance by the arbuscular mycorrhizal symbiosis in a crop of such importance for human nourishment as rice.

10.2 Amelioration of Photosynthetic Performance by AM

Soil salinity leads to a decrease in crop production due to inhibition of photosynthetic processes (Pitman and Läuchli 2002). Indeed, salinity inhibits specific enzymes involved in the synthesis of photosynthetic pigments, causing a reduction in plant chlorophyll content (Giri and Mukerji 2004; Sheng et al. 2008). In addition, salinity affects photosynthetic CO_2 assimilation because the osmotic component of salt stress reduces stomatal conductance. This, in turn, results in low CO_2 supply to RuBisCo. In a second phase, salinity might cause biochemical and photochemical effects on photosynthesis (Duarte et al. 2013). Thus, the impairment in CO_2 assimilation induces accumulation of excess energy, which if not quenched may lead to excess electron accumulation from the photochemical phase in thylakoid membranes, particularly in the presence of high light intensity. This effect may lead to over-reduction of the reaction centers of PSII, causing damage to the photosynthetic apparatus (Redondo-Gómez et al. 2010). Indeed, the light energy absorbed by chlorophyll molecules can be used either to drive photosynthesis, it can be re-emitted as light-chlorophyll fluorescence or the excess energy can be dissipated as heat (Lima-Neto et al. 2014). These three processes occur in a competitive way, so that any increase in the efficiency of one will decrease the yield of the other two (Harbinson 2013). Thus, the ability of the plant to dissipate or not the excess energy can be quantified by measuring the chlorophyll *a* fluorescence. Another key determinant of plant photosynthetic efficiency is the activity of enzymes involved in carbon assimilation such as the ribulose-1,5-bisphosphate carboxylase/oxygenase (RuBisCo) (Masumoto et al. 2005; Goicoechea et al. 2014).

Improvements in photosynthetic activity or water use efficiency have been reported in AM plants growing under salt stress (Sheng et al. 2008; Zuccarini and Okurowska 2008; Hajiboland et al. 2010) or under drought stress (Birhane et al. 2012; Liu et al. 2015). Nevertheless, so far, it is not clear how the AM symbiosis affects the plant photosynthetic capacity, particularly the efficiency of photosystem II in plants subjected to salinity.

Recently, a study addressed the influence of AM fungi on rice PSII performance and on RuBisCo activity and gene expression under salt stress (Porcel et al. 2015). Thus, AM and non-AM rice plants were cultivated under non-saline conditions or subjected to 75 mM NaCl and 150 mM NaCl during 4 weeks. The inoculation of rice plants with the AM fungus *Clareidoglomus etunicatum* produced higher shoot fresh and dry biomass than non-AM plants, specially under 150 mM NaCl. Moreover, salinity decreased the shoot biomass production in non-AM plants, while in AM plants shoot biomass only decreased transiently at 75 mM NaCl, but it was not significantly reduced at 150 mM NaCl (Porcel et al. 2015). These results agree with several reports on salt stress alleviation by AM symbiosis (For reviews see Evelin et al. 2009; Porcel et al. 2012; Ruiz-Lozano et al. 2012).

AM rice plants maintained higher stomatal conductance, transpiration rate, and net photosynthetic rate than non-AM plants both under non-saline conditions and under 150 mM NaCl (Fig. 10.1A–D). The increase of plant gas exchange by the AM symbiosis has been related to alterations of host plant hormonal levels and with the enhanced uptake and translocation of water (Goicoechea et al. 1997; Sheng et al. 2008; Ruiz-Lozano and Aroca 2010) and would translate into increased photosynthesis (Birhane et al. 2012).

Chlorophyll content is a key factor for plant photosynthesis and closely reflects the photosynthetic ability of plants such as rice (Takai et al. 2010). In the study with rice, a higher chlorophyll *a* concentration was found in AM plants subjected to 150 mM NaCl (Porcel et al. 2015). Enhanced chlorophyll content in AM plants has been related to increased P and Mg uptake (Zhu et al. 2014). AM rice plants also displayed higher RuBisCo activity under all salt levels (Porcel et al. 2015),

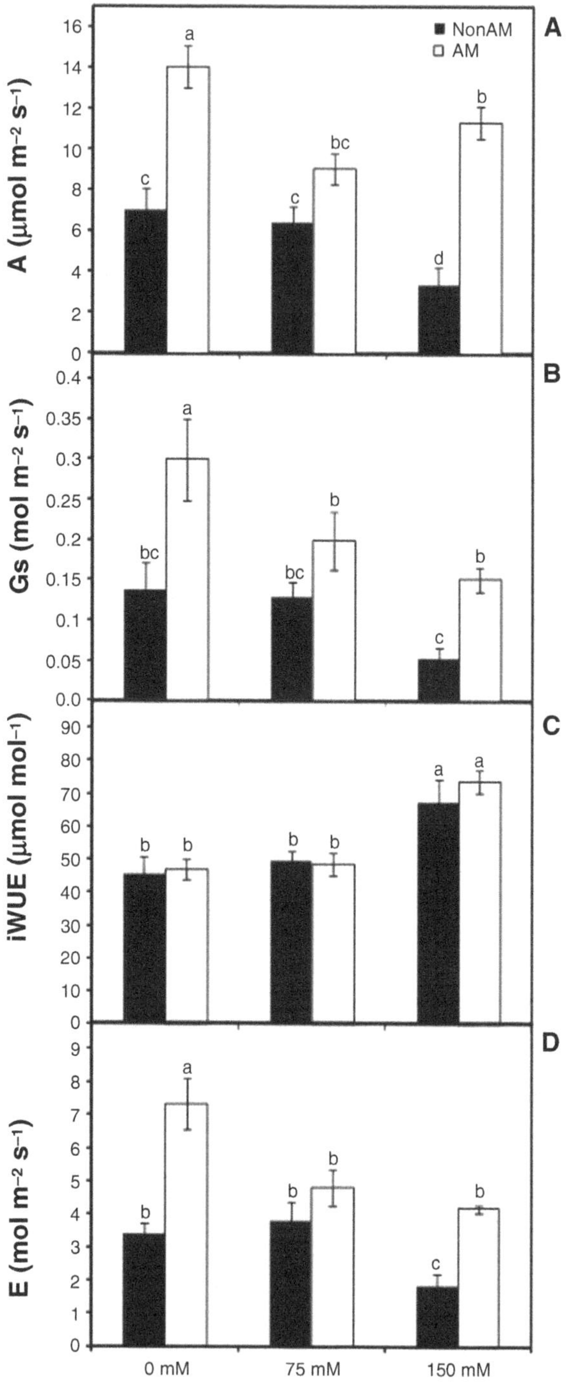

Fig. 10.1 (**A**) Net photosynthetic rate, (**B**) stomatal conductance (Gs), (**C**) intrinsic water use efficiency (iWUE) and (**D**) transpiration rate (E) in rice plants cultivated under non-saline conditions (0 mM) or subjected to 75 mM or 150 mM NaCl. Plants remained as uninoculated controls (*black columns*) or were inoculated with the arbuscular mycorrhizal fungus *Claroideoglomus etunicatum* (*white columns*). *Bars* represent mean ± standard error. *Different letters* indicate significant differences ($p < 0.05$) ($n = 10$). Reproduced from Porcel et al. (2015) with kind permission from Elsevier

demonstrating a lower metabolic limitation of photosynthesis than non-AM plants (Lima-Neto et al. 2014). RuBisCo activity is a parameter that is well correlated with CO_2 assimilation (Sanz-Sáez et al. 2013). Thus, the enhanced net photosynthetic activity of AM plants could be also due to non-stomatal factors such as higher chlorophyll *a* content and RuBisCo activity (Chen et al. 2014).

PSII activity has been widely used to study response and adaptation to stress by plants (Strasser et al. 2000). When the metabolism of a plant is disturbed by biotic or abiotic stresses, redundant energy hast to be dissipated via non-photochemical processes like heat or chlorophyll fluorescence in order to avoid damage of plant tissues (Pinior et al. 2005). The maximum quantum yield of primary photochemistry (Fv/Fm) reflects the potential quantum efficiency of PSII and is used as an index of plant photosynthetic performance, with optimal values for most plant species of around 0.83 (Björkman and Demmig 1987). In the study with rice, values of Fv/Fm ranged from 0.68 to 0.78 (Fig. 10.2A) in different treatments (Porcel et al. 2015), which suggests that the performance of the photosynthetic apparatus was not at the optimum level (Bagheri et al. 2011). However, data also showed that under salinity, the actual quantum yield of PSII photochemistry (Φ_{PSII}) was enhanced by AM symbiosis (Fig. 10.2B). The increase in Φ_{PSII} due to mycorrhization was by 25% and 34% at 75 and 150 mM NaCl, respectively. At the same time, Φ_{NPQ} was significantly lower in AM plants under both saline levels (Fig. 10.2C). The reduction ranged from 30% under 75 mM NaCl to 40% under 150 mM NaCl. Moreover, Φ_{NPQ} increased in non-AM plants as consequence of salinity but resulted unaffected by salinity in AM plants. Normally, Φ_{NPQ} increases as a mechanism to protect the leaf from light-induced damage, but this means that photochemical processes are reduced proportionally (Baker 2008; Lazár 2015). Data by Porcel et al. (2015) indicate that AM rice plants had a higher photochemical efficiency for CO_2 fixation and solar energy utilization and that AM inoculation would reduce the light-induced damage due to salinity (Zhu et al. 2014). As a result, the AM symbiosis increases salt tolerance in rice plants by preventing the injury to the photosystems reaction centers and by allowing a better utilization of light energy in photochemical processes (Pinior et al. 2005; Bagheri et al. 2011), reducing at the same time light energy dissipation as heat. These effects of the AM symbiosis enhancing Φ_{PSII} and reducing Φ_{NPQ} may be related to the sink stimulation of AM symbiosis. Indeed, Kaschuk et al. (2009) showed that the carbon sink strength due to the fungal presence in the plant root stimulates the host plant, increasing the photosynthetic rate.

10.3 Improved Ion Homeostasis by AM

Soil salinity produces nutrient imbalance due to decreased nutrient uptake and/or transport to the shoot (Munns and Tester 2008; Ruiz-Lozano et al. 2012). Thus, in plants subjected to salt stress, accumulation of Na^+ and impairment of K^+ nutrition are typical characteristics (Chen et al. 2007). Na^+ is a major cation present in saline

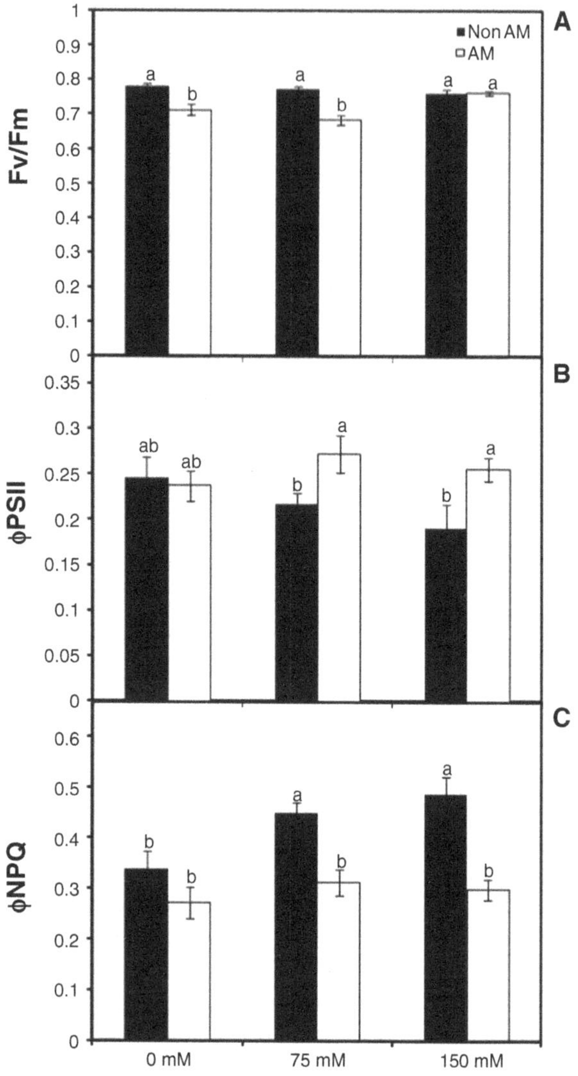

Fig. 10.2 (**A**) Maximum efficiency of photosystem II (Fv/Fm), (**B**) actual quantum yield of PSII photochemistry (Φ_{PSII}) and (**C**) quantum yield of non-photochemical quenching (Φ_{NPQ}) in rice plants cultivated under non-saline conditions (0 mM) or subjected to 75 mM or 150 mM NaCl. Plants remained as uninoculated controls (*black columns*) or were inoculated with the arbuscular mycorrhizal fungus *Claroideoglomus etunicatum* (*white columns*). *Bars* represent mean ± standard error. *Different letters* indicate significant differences ($p < 0.05$) ($n = 10$). Reproduced from Porcel et al. (2015) with kind permission from Elsevier

soils, but it is not an essential mineral nutrient for most plants, while many cytosolic enzymes are activated by K^+ and inhibited by Na^+ (Shi et al. 2002).

Plants have, thus, evolved biochemical and molecular mechanisms for the regulation of the uptake, transport, or compartmentation of Na^+ and/or K^+, which are responsible for the adequate ionic homeostasis in the plant. The first one is restricting Na^+ entry to plant cells by selective ion uptake. The second one is maximizing the efflux of Na^+ back to the growth medium or to apoplastic spaces. Finally, plants can also restrict the transfer of Na^+ to the shoot by sequestering the internalized Na^+ into vacuoles (Cuin et al. 2011). In any case, the two last mechanisms seem to be more important to control Na^+

accumulation in plants (Cuin et al. 2011; Cabot et al. 2014). Na^+ efflux from cytosol to the growth medium or to apoplastic spaces is catalyzed by a plasma membrane Na^+/H^+ antiporter SOS1 in several plants, including rice (Kumar et al. 2013). The sequestration of Na^+ into vacuoles is catalyzed by vacuolar (Na^+, K^+)/H^+ antiporters NHXs (Cuin et al. 2011), of which four genes (*OsNHX1-4*) have been reported in rice (Fukuda et al. 2011, Kumar et al. 2013). Moreover, recirculation of Na^+ from photosynthetic organs to roots and unloading of Na^+ from the xylem have also been described as strategies for salt tolerance (Davenport et al. 2007), with high-affinity HKT transporters being involved in these processes in several plants, including rice (Garciadeblás et al. 2003; Ren et al. 2005).

Mycorrhizal colonization has been shown to enhance K^+ absorption under saline conditions while preventing Na^+ translocation to shoot tissues (Giri et al. 2007; Talaat and Shawky 2011). Thus, mycorrhizal plants often have a higher K^+/Na^+ ratio under salinity and a lower shoot Na^+ concentration than non-mycorrhizal plants (Sannazzaro et al. 2006; Estrada et al. 2013), preventing the disruption of various enzymatic processes and inhibition of protein synthesis.

Na^+ uptake and distribution within the plant are major determinants for the sensitivity of a plant to salinity. Prevention of Na^+ entry into the root, transport to and allocation within the leaf, and sequestration into the vacuole are strategies by which plants cope with excessive soil salinity (Ruiz-Lozano et al. 2012). A study was conducted in rice to elucidate the effects of the AM symbiosis on the expression of several rice transporters involved in Na^+/K^+ homeostasis and to measure the Na^+ and K^+ contents and their ratios in the different plant tissues (Porcel et al. 2016). The results obtained showed a decrease in shoot K^+ concentration, while Na^+ accumulation was increased in roots of AMF-inoculated plants. This study also showed that the shoot Na^+ to root Na^+ ratio was consistently lower in AM plants than in non-AM plants (Fig. 10.3), suggesting that the translocation of Na^+ ions from roots to shoots was restricted in AM plants as a strategy to limit the accumulation of this toxic ion in

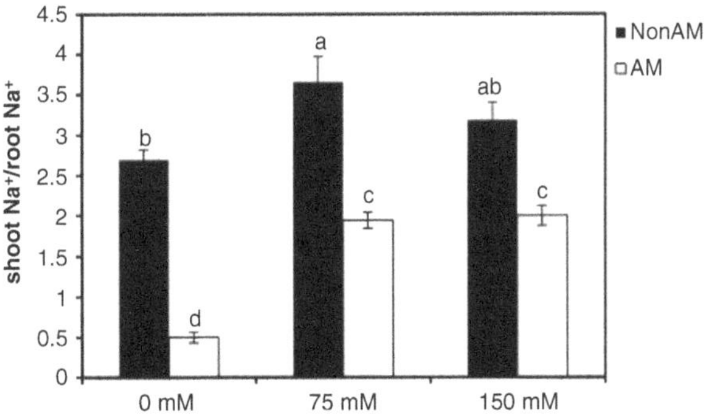

Fig. 10.3 Shoot Na^+/root Na^+ ratio in rice plants cultivated under non-saline conditions (0 mM) or subjected to 75 or 150 mM NaCl. Plants remained as uninoculated controls (*black columns*) or were inoculated with the arbuscular mycorrhizal fungus *Claroideoglomus etunicatum* (*white columns*). *Bars* represent mean ± standard error. *Different letters* indicate significant differences ($p < 0.05$) ($n = 5$). Reproduced from Porcel et al. (2016) with kind permission from Springer

photosynthetic tissues (Zhu et al. 2016). It has been proposed that in AM-inoculated plants, Na^+ might be kept inside root cell vacuoles, vesicles, or in intraradical fungal hyphae to prevent the allocation of Na^+ to the shoots (Hammer et al. 2011; Evelin et al. 2013). This would indicate that AM fungi induce a regulatory effect on the translocation of Na^+ to the aerial parts (Evelin et al. 2012). However, the molecular bases of these protective mechanisms are unknown so far.

In the study with rice, AM plants accumulated more Na^+ in root tissues than non-AM plants and had a lower shoot Na^+ to root Na^+ ratio (Porcel et al. 2016). Thus, it was studied if the AM symbiosis regulated the gene expression of well-known transporters involved in ion homeostasis. These include the following:

1. Transporter type NHX, a Na^+/H^+ antiporter system localized in the vacuole. It is expressed in roots and leaves and sequesters Na^+ into the vacuole (Munns 2005).
2. Transporter type SOS1, a plasma membrane Na^+/H^+ antiporter involved in Na^+ efflux from cytosol to the growth medium or to apoplastic spaces. It may also participate in the redistribution of Na^+ between roots and shoots, being related with the ability of plants to prevent Na^+ from reaching the photosynthetic tissues (Olías et al. 2009) and in Na^+ extrusion from the root to the external medium (Munns 2005).
3. Transporter type HKT, which allows the transport of Na^+ and K^+ and is involved in recirculation of Na^+ from photosynthetic organs to roots and unloading of Na^+ from the xylem (Davenport et al. 2007).

The expression of *OsNHX3* gene remained constant in shoots of non-AM plants, regardless of the salt level applied (Fig. 10.4A). In contrast, in AM plants, *OsNHX3* was considerably upregulated under both saline levels as compared to non-AM plants. The shoot Na^+ concentration was similar in AM and non-AM plants at both saline levels. However, at the molecular level, the upregulation of this gene suggests that in AM plants, part of the Na^+ translocated to shoots may be excluded from cytosol by sequestering it into the vacuole. In this way, AM rice plants would maintain a high K^+/Na^+ ratio within the metabolically active cytosol, which is crucial for plant tolerance to salinity (Cuin et al. 2011). Contrarily, in the non-AM plants, the low expression of this gene would mean that the Na^+ remains in the cytosol where it may damage cell metabolism (Porcel et al. 2016).

The shoot expression of *OsSOS1* and *OsHKT2;1* genes was consistently upregulated by AM fungal presence at both saline levels, especially at 150 mM NaCl. In contrast, the expression was low and remained constant in non-AM plants (Fig. 10.4B, C). This suggests that in AM plants, the enhanced expression of these two genes may contribute to a Na^+ efflux from cytosol to apoplastic spaces (via SOS1), where Na^+ is less toxic, as well as to an increased Na^+ unloading from the xylem and recirculation from photosynthetic organs to roots (via HKT). Indeed, it has been suggested that the transport functions of SOS1 and HKT systems may be coordinated to achieve Na^+ homeostasis and partitioning between plant organs (Pardo et al. 2006; Olías et al. 2009). Besides, it has been recently shown in wheat that *HKT1;4* and *HKT1;5* genes affect the activity and expression levels of the SOS1 Na^+/H^+ exchanger in both root cortical and stelar tissues (Zhu et al. 2016). In the

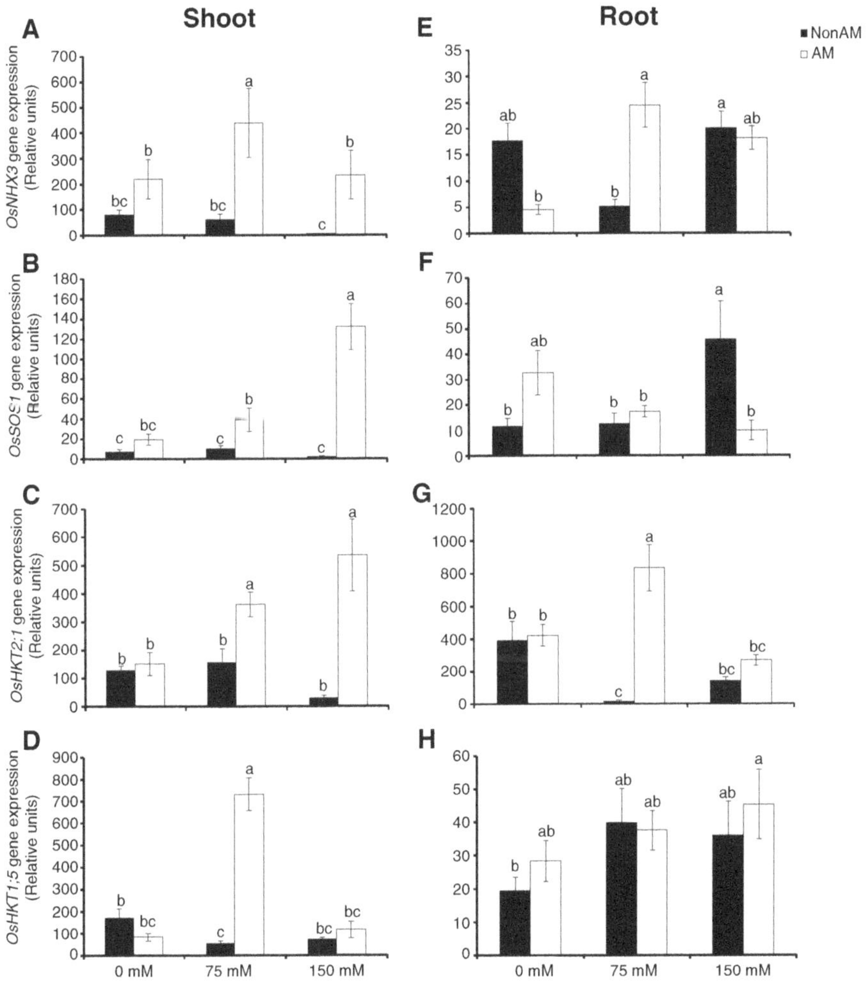

Fig. 10.4 Expression of genes *OsNHX3*, *OsSOS1*, *OsHKT2;1*, and *OsHKT1;5* in shoots (**A–D**) and roots (**E–H**) of rice plants cultivated under non-saline conditions (0 mM) or subjected to 75 or 150 mM NaCl. Plants remained as uninoculated controls (*black columns*) or were inoculated with the arbuscular mycorrhizal fungus *Claroideoglomus etunicatum* (*white columns*). *Bars* represent mean ± standard error. *Different letters* indicate significant differences ($p < 0.05$) ($n = 5$). Reproduced from Porcel et al. (2016) with kind permission from Springer

study with rice, both genes seem to be coordinated since they showed a similar expression pattern in shoots (Porcel et al. 2016). Indeed, at the highest salt level (150 mM NaCl), the expression of *OsSOS1* gene in roots resulted considerably lower in AM plants than in non-AM plants. This may be important in order to restrict the transport of Na^+ to shoots at this high salt level since SOS1 also has a role controlling the long distance Na^+ transport from root to the shoot via xylem (Olías et al. 2009). Moreover, in wheat it has been proposed that HKT1;4 and HKT1;5 confer two complementary mechanisms for reducing the xylem Na^+ content. One enhances the Na^+ retrieval from the xylem via its direct action, while the other reduces the rate of Na^+ loading into the xylem via SOS1 (Zhu et al. 2016). All these effects would also explain the higher Na^+ concentration in roots of AM plants.

In shoots, the expression of another HKT gene, *OsHKT1;5*, was upregulated by sevenfold in AM plants at 75 mM NaCl (Fig. 10.4D). However, at increasing salinity (150 mM NaCl) the expression of the gene was similar for AM and non-AM plants and comparable to non-saline conditions. Different HKT members may have different transport properties. The rice OsHKT1;5 has been shown to transport Na^+, and it was hypothesized to control shoot Na^+ recirculation by withdrawing Na^+ from the xylem stream into the xylem parenchyma cells (Ren et al. 2005). This mechanism may be operated in AM plants at low salt levels (75 mM NaCl) but avoided at higher salt levels due to the prevalence of the other homeostasis mechanisms provided by OsNXH3, OsSOS1, and OsHKT2;1, which remained upregulated in AM plants at the highest salt level (Porcel et al. 2016).

The effects of the AM symbiosis on the expression of these transporter genes were more evident in shoot tissues than in root tissues. Indeed, in root tissues only a transient upregulation of *OsNHX3* and *OsHKT2;1* genes by AM fungal presence at 75 mM NaCl was observed (Fig. 10.4E–H). At the highest salt level (150 mM NaCl), the expression of these genes was similar to that in non-AM plants. These data suggest that the effect of the AM symbiosis seems to be directed preferentially to protect the photosynthetic tissues from the detrimental effects of Na^+ rather than the root tissues. The ability of a plant to exclude Na^+ from photosynthetic tissues is considered to be a crucial feature of salinity tolerance in glycophytes (Munns and Tester 2008; Cuin et al. 2011). Moreover, it has been proposed that the regulation of the rate of Na^+ transport to the shoot over time is critical for plant salinity tolerance (Maathuis 2014). Moreover, Wu et al. (2013) showed that the ability of a plant to maintain a high K^+/Na^+ ratio resides in photosynthetically active tissues, and this determines its photosynthetic capacity (and hence growth and yield) under saline conditions. The improved plant growth in AM treatments in the study with rice seems to be related to the preferential protection of photosynthetic tissues by the AM symbiosis.

10.4 Root Water Transport Capacity and Regulation of Aquaporins by AM

One of the primary responses of plants to soil salinity is inhibition of their root water uptake capacity (i.e., root hydraulic conductivity) due to reduced soil water potential. The exact mechanism by which salinity reduces the hydraulic conductance in cells and roots is still unknown. However, it has been suggested that it could be due to changes either in the aquaporin function or in the amount of this protein present in the membrane (Carvajal et al. 2000; Martínez-Ballesta et al. 2000; Sade et al. 2010).

Aquaporins are a group of membrane intrinsic proteins that facilitate and regulate the passive movement of water molecules down a water potential gradient (Maurel et al. 2008). In plants, aquaporins comprise a large and diverse protein family composed by 31–71 different genes, depending on the plant species and are subdivided in five subgroups based on their amino acid sequence similarity. These are plasma membrane intrinsic proteins (PIPs), tonoplast intrinsic proteins (TIPs), nodulin-like intrinsic proteins (NIPs), small and basic intrinsic proteins (SIPs), and X intrinsic proteins (XIPs), a recently described group (Gupta and Sankararamakrishnan 2009), which has been shown to transport a variety of uncharged substrates (Bienert et al. 2011), including water (López et al. 2013).

In plants, aquaporin discovery has caused a significant change in the understanding of plant–water relations. High levels of aquaporin expression have been shown in tissues with high water fluxes across membranes but also in roots where water uptake occurs (Otto and Kaldenhoff 2000). Thus aquaporins seem to play an important role in controlling transcellular water transport in plant tissues (Javot and Maurel 2002; Zhao et al. 2008), including rice plants (Grondin et al. 2016). However, the relationship that exists between aquaporins and plant responses to water deficit still remains elusive and with contradictory results (Aharon et al. 2003; Lian et al. 2004).

During the AM symbiosis, plant root cells must undergo extensive morphological alterations in order to accommodate the presence of an endophytic symbiont, and most of these changes concern vacuolar or cytoplasmic membrane systems. Thus, Krajinski et al. (2000) hypothesized a variation of expression affecting genes that encode membrane-associated proteins such as the aquaporins. In addition, the AM symbiosis results in altered rates of water movement into, through, and out of the host plants (Augé 2001) and also modifies the root hydraulic conductivity (L) (Aroca et al. 2007; Ruiz-Lozano et al. 2009; Bárzana et al. 2012, 2014). Aquaporins provide a low-resistance pathway for the movement of water across a membrane, and PIPs and TIPs isoforms have been recognized as central pathways for intracellular and transcellular water transport (Maurel et al. 2008). Furthermore, since aquaporins can be gated, this provides a mechanism to control the movement of water along plant tissues. Thus, it seems likely that mycorrhizal symbiosis causes significant changes in aquaporin activity of host plants (Uehlein et al. 2007; Bárzana et al. 2014), and some of the plant aquaporins might be important for the mycorrhizal responses.

When the expression of aquaporin genes has been analyzed in salt-stressed AM and non-AM plants, the results obtained were not conclusive. Hence, results by Ouziad et al. (2006), Aroca et al. (2007), and Jahromi et al. (2008) suggest that each aquaporin gene responds differently to AM colonization depending on the nature of the stress imposed.

In the studies with rice plants subjected to two salt levels, our research group analyzed for the first time the possible regulation of rice aquaporins by the AM symbiosis in this plant species, as well as the effects on L under salt stress (Calvo-Polanco et al., unpublished). Results showed that in the absence of salinity in the growing medium, L was similar in AM and non-AM plants (Fig. 10.5). However, when plants were subjected either to 75 mM NaCl or to 150 mM NaCl, L was

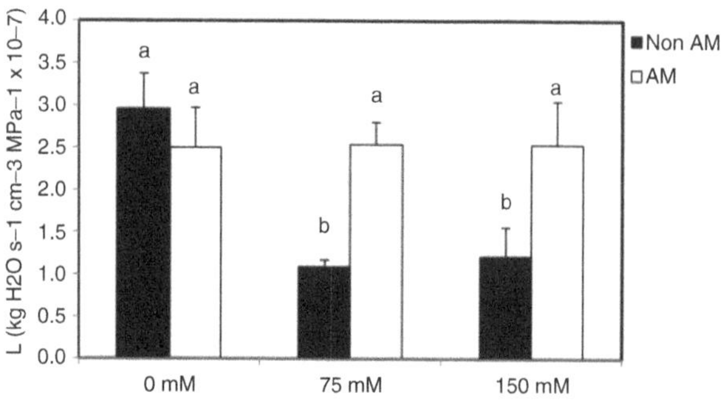

Fig. 10.5 Root hydraulic conductivity (L) in rice plants cultivated under non-saline conditions (0 mM) or subjected to 75 or 150 mM NaCl. Plants remained as uninoculated controls (*black columns*) or were inoculated with the arbuscular mycorrhizal fungus *Claroideoglomus etunicatum* (*white columns*). *Bars* represent mean ± standard error. *Different letters* indicate significant differences ($p < 0.05$) ($n = 5$)

considerably reduced in non-AM plants, while in AM plants, L was kept unaltered due to salinity. Thus, under saline conditions, AM plants had values of 109% of L higher than non-AM plants.

PIPs and TIPs rice aquaporins were analyzed in roots and shoots of these plants (Calvo-Polanco et al. unpublished). Results showed that among the 11 PIP genes analyzed in rice roots, *OsPIP1;3*, *OsPIP2;1*, *OsPIP2;2*, *OsPIP2;3*, and *OsPIP2;4* were inhibited by mycorrhization both in the absence and in the presence of salinity in the growing medium. Only *OsPIP2;8* was induced by mycorrhization, regardless of the salt level applied. The rest of genes did not show significant differences. In shoots, genes *OsPIP1;1*, *OsPIP1;3*, *OsPIP2;1*, *OsPIP2;4*, and *OsPIP2;7* were also inhibited by mycorrhization, and only *OsPIP2;6* was consistently induced, mainly under saline conditions. The rest of genes did not show significant differences.

In the case of ten TIP genes analyzed in rice roots, results showed inhibition of *OsTIP1;1* and *OsTIP4;1* in AM plants, while *OsTIP2;2* and *OsTIP5;1* were induced by the AM fungal presence. The rest of the genes did not show significant differences. In rice shoots the AM symbiosis inhibited the gene *OsTIP1;2* and induced the expression of genes *OsTIP2;1*, *OsTIP2;2*, *OsTIP3;1*, *OsTIP4;2*, and *OsTIP4;3*. The rest of the genes did not show significant differences.

Thus, it is clear that the enhanced L values in AM rice plants is correlated with a low number of aquaporin genes induced in roots, while most of the aquaporin genes were negatively correlated with L, since they were inhibited by mycorrhization. This lack of correlation is not surprising and has been observed in other studies (Boursiac et al. 2005; Aroca et al. 2007; Ruiz-Lozano et al. 2009) since symplastic movement of water via plasmodesmata may also contribute significantly to hydraulic conductivity (Galmés et al. 2007), and aquaporin regulation occurs at both transcriptional and posttranscriptional levels (Zelazny et al. 2007). In addition, it must be also taken into account that the L values measured in this study include both apoplastic and symplastic water flow and that in rice the apoplastic flow seems to be relatively larger than in other plant species (Ranathunge et al. 2004; Grondin et al. 2016). In this regard, AM fungus within the roots may have greatly contributed to the increase of the apoplastic water flow as previously reported (Lehto and Zwiazek 2011; Bárzana et al. 2012). Increased water uptake by mycorrhizal plants under drought has been related to the increased absorbing surface of growing hyphae and mycorrhizal ability to take up water from soil pores inaccessible to roots, as AM hyphae represent a low-resistance way for water movement until root cells (Ruiz-Lozano 2003; Allen 2009; Lehto and Zwiazek 2011). Hence, under such conditions water movement through AM fungal hyphae may be critical to improve the water supply to the plant, increasing, thus, the cell-to-cell and apoplastic pathways (Bárzana et al. 2012). On the other hand, AM fungal aquaporins have been related to water transport in the extraradical mycelium and in the periarbuscular membrane (Li et al. 2013). Thus, in AM plants, the enhanced root hydraulic conductivity could be also due to the activity of the own fungal aquaporins (Bárzana et al. 2014, 2015). This aspect should be the object of future studies.

10.5 Amelioration of Antioxidant Capacity by AM

During osmotic stresses such as drought or salinity, several metabolic pathways are uncoupled, and electrons are transferred to molecular oxygen to form reactive oxygen species (ROS). Indeed, under salinity the availability of atmospheric CO_2 is restricted because of increased stomatal closure, and consumption of NADPH by the Calvin cycle is reduced. When ferredoxin is over-reduced during photosynthetic electron transfer, electrons may be transferred from PSI to oxygen to form superoxide radicals ($O_2^{\bullet-}$) by the process called Mehler reaction, which initiates chain reactions that produce more harmful oxygen radicals. These include singlet oxygen ($^1O_2$), hydrogen peroxide (H_2O_2), or hydroxyl radicals ($OH^\bullet$) (Miller et al. 2010).

ROS can first act as signaling molecules for stress responses, and generation of ROS is an early event in plant response to stress. However, ROS, such as $^1O_2$, H_2O_2, $O_2^{\bullet-}$, and $HO^\bullet$, are toxic molecules capable of causing oxidative damage to proteins, DNA, and lipids (Miller et al. 2010). Thus, plants have evolved several antioxidants mechanisms to avoid oxidative damage linked to stressful conditions. These mechanisms can be divided as enzymatic or nonenzymatic. Enzymatic antioxidants include superoxide dismutase (SOD), glutathione reductase (GR), catalase (CAT), ascorbate- or thiol-dependent peroxidases, and the enzymes of the ascorbate-glutathione pathway. Nonenzymatic mechanisms include compounds able to scavenge directly several ROS as are ascorbic acid (AsA), glutathione (GSH), or α-tocopherol (Miller et al. 2010; Scheibe and Beck 2011).

In the study with rice plants subjected to two salt levels, the oxidative damage to lipids was measured in roots and shoot tissues (Porcel et al. unpublished). Results showed that in roots, salinity increased the oxidative damage to lipids in non-AM plants, especially at 150 mM NaCl, with an increase of 63% as compared to non-AM plants in the absence of salinity (Fig. 10.6). On the contrary, AM plants maintained

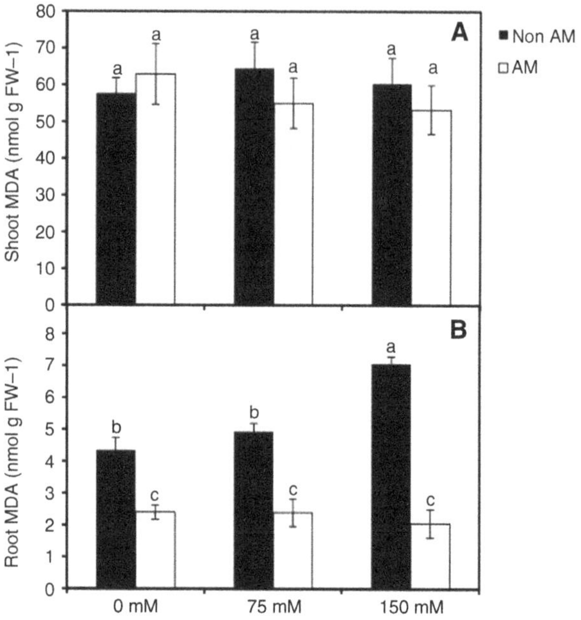

Fig. 10.6 Oxidative damage to lipids (expressed as equivalents of malondialdehyde, MDA) in shoots (**A**) and roots (**B**) of rice plants cultivated under non-saline conditions (0 mM) or subjected to 75 or 150 mM NaCl. Plants remained as uninoculated controls (*black columns*) or were inoculated with the arbuscular mycorrhizal fungus *Claroideoglomus etunicatum* (*white columns*). *Bars* represent mean ± standard error. *Different letters* indicate significant differences ($p < 0.05$) ($n = 5$)

steady-state levels of oxidative damage regardless of the saline medium, which were similar to levels found in non-AM plants in the absence of salinity. In shoots no significant differences were found.

The SOD, APX, CAT, and GR activities were evaluated, but results did not show a clear trend for each antioxidant enzymatic activity in response to the salt applied and the presence of the AM fungus (Porcel et al. unpublished). Thus, CAT and GR activities increased considerably in roots and shoots of AM plants subjected to salinity, while non-AM plants did not show significant changes in these activities. In contrast, the APX activity was lower in AM plants than in non-AM plants both in the presence and in the absence of salinity in the growing medium. For SOD activity no significant differences were found between AM and non-AM plants. Several studies have shown that AM plants possess enhanced activity of various antioxidant enzymes, but the response of the individual enzymes has been shown to vary with respect to the fungal species, the host plant, and the type of environmental stress imposed (Porcel et al. 2012). This variation may also depend on the micronutrients available to some of the enzymes. For instance, CAT, APX, and SOD are metalloenzymes whose activities can be determined by the availability of the metals they utilize. This suggests that the effect of the AM symbiosis inducing activities of several antioxidant enzymes may be the indirect result of the mycorrhizal effects on host plant growth and acquisition of P or N (Alguacil et al. 2003; Evelin et al. 2009).

The influence of the AM symbiosis on the accumulation of nonenzymatic antioxidants in the host plant has also been studied. In a study with rice subjected to drought stress, Ruíz-Sánchez et al. (2011) found an increase of AsA content in AM plants. In the study with rice plants subjected to two salt levels, the amount of GSH

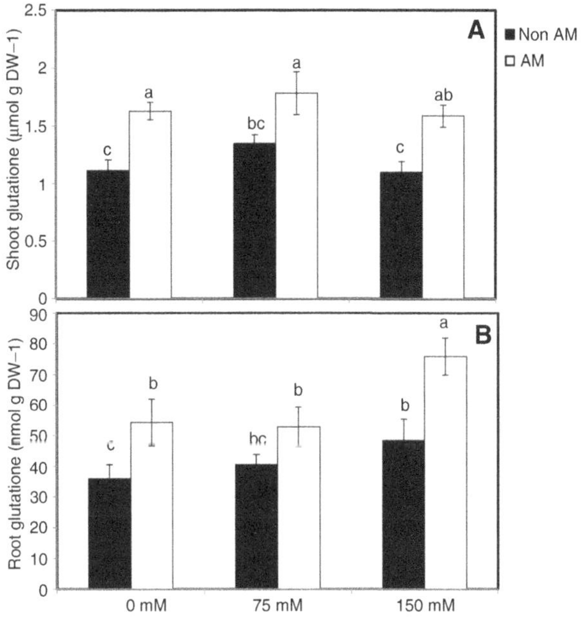

Fig. 10.7 Accumulation of reduced glutathione in shoots (**A**) and roots (**B**) of rice plants cultivated under non-saline conditions (0 mM) or subjected to 75 or 150 mM NaCl. Plants remained as uninoculated controls (*black columns*) or were inoculated with the arbuscular mycorrhizal fungus *Claroideoglomus etunicatum* (*white columns*). *Bars* represent mean ± standard error. *Different letters* indicate significant differences ($p < 0.05$) ($n = 5$)

accumulated in shoot and root tissues was always higher in AM plants than in non-AM plants in both tissues (Fig. 10.7) (Porcel et al. unpublished). In the case of AsA, it was also higher in AM plants than in non-AM plants only at the highest salt level applied (150 mM NaCl). For the rest of salt levels, the differences were not significant. Ascorbic acid is an important nonenzymatic antioxidant compound since it is involved in the removal of H_2O_2 by ascorbate peroxidases, which use AsA as electron donor and is closely related to GSH in the ascorbate-glutathione cycle (Noctor et al. 2016). An increase in GSH content in AM plants has also been found in rice plants (Ruíz-Sánchez et al. 2010), concomitantly with a reduced oxidative damage to lipids. GSH not only has the function of scavenging peroxides or regenerating AsA pool, but it also regulates the expression of photosynthetic genes and may keep the cell pools of reducing power (NADPH) under necessary conditions for plant living cells (Noctor et al. 2016). Moreover, the ratio of GSH to its oxidized form, GSSG, plays an important role in maintaining redox equilibrium in the cell during H_2O_2 degradation and other processes (Shao et al. 2008).

10.6 Concluding Remarks

Results in rice plants subjected to salinity suggest that AM plants had a higher photochemical efficiency for CO_2 fixation and solar energy utilization, and this increases plant salt tolerance by preventing the injury to the photosystems reaction centers and by allowing a better utilization of light energy in photochemical processes, reducing at the same time light energy dissipation as heat. All these processes translated into a higher photosynthetic and RuBisCo activities in AM rice plants and improved plant biomass production under salinity.

On the other hand, in aerial plant tissues, the AM symbiosis may favor Na^+ extrusion from cytoplasm, its sequestration into the vacuole, the unloading of Na^+ from the xylem, and its recirculation from photosynthetic organs to roots through regulation of *OsSOS1*, *OsNHX3*, *OsHKT2;1*, and *OsHKT1;5*. As a result there is a decrease of Na^+ root-to-shoot distribution and an increase of Na^+ accumulation in rice roots which seems to enhance the plant tolerance to salinity and allows AM rice plants to maintain their growing processes under salt stress conditions.

The AM symbiosis has been shown to improve root hydraulic conductivity under saline conditions, and this may be due to improved symplastic and apoplastic radial water flow in rice roots, mediated by the induction of specific aquaporin isoforms and also by the additional water absorbing-surface of fungal hyphae and their ability to take up water from soil pores inaccessible to roots.

In addition to the above effects, the antioxidant capacity of AM rice plants seems to be also enhanced by the symbiosis, as evidenced by the reduced oxidative damage to lipids in AM plants subjected to salinity and by the enhanced CAT or GR activities or accumulation of antioxidant compounds, mainly glutathione.

Acknowledgments This work was financed by Project P11-CVI-7107 supported by Junta de Andalucía (Spain) and Project AGL2014-53126-R financed by Ministerio de Economía y Competitividad.

References

Aharon R, Shahak Y, Wininger S, Bendov R, Kapulnik Y, Galili G (2003) Overexpression of a plasma membrane aquaporins in transgenic tobacco improves plant vigour under favourable growth conditions but not under drought or salt stress. Plant Cell 15:439–447

Alguacil MM, Hernández JA, Caravaca F, Portillo B, Roldán A (2003) Antioxidant enzyme activities in shoots from three mycorrhizal shrub species afforested in a degraded semi-arid soil. Physiol Plant 118:562–570

Allen MF (2009) Bidirectional water flows through the soil-fungal-plant mycorrhizal continuum. New Phytol 182:290–293

Aroca R, Porcel R, Ruiz-Lozano JM (2007) How does arbuscular mycorrhizal symbiosis regulate root hydraulic properties and plasma membrane aquaporins in Phaseolus vulgaris under drought, cold or salinity stresses? New Phytol 173:808–816

Augé RM (2001) Water relations, drought and vesicular-arbuscular mycorrhizal symbiosis. Mycorrhiza 11:3–42

Augé RM, Toler HD, Saxton AM (2014) Arbuscular mycorrhizal symbiosis and osmotic adjustment in response to NaCl stress: a meta-analysis. Front Plant Sci 5:562

Bagheri V, Shamshiri MH, Shirani H, Roosta HR (2011) Effect of mycorrhizal inoculation on ecophysiological responses of pistachio plants grown under different water regimes. Photosynthetica 49:531–538

Baker NR (2008) Chlorophyll fluorescence: a probe of photosynthesis in vivo. Annu Rev Plant Biol 59:89–113

Bárzana G, Aroca R, Paz JA, Chaumont F, Martinez-Ballesta MC, Carvajal M, Ruiz-Lozano JM (2012) Arbuscular mycorrhizal symbiosis increases relative apoplastic water flow in roots of the host plant under both well-watered and drought stress conditions. Ann Bot 109:1009–1017

Bárzana G, Aroca R, Bienert P, Chaumont F, Ruiz-Lozano JM (2014) New insights into the regulation of aquaporins by the arbuscular mycorrhizal symbiosis in maize plants under drought stress and possible implications for plant performance. Mol Plant Microbe Interact 27:349–363

Bárzana G, Aroca R, Ruiz-Lozano JM (2015) Localized and non-localized effects of arbuscular mycorrhizal symbiosis on accumulation of osmolytes and aquaporins and on antioxidant systems in maize plants subjected to total or partial root drying. Plant Cell Environ 38:1613–1627

Bienert GP, Désirée-Bienert M, Jahn TP, Boutry M, Chaumont F (2011) Solanaceae XIPs are plasma membrane aquaporins that facilitate the transport of many uncharged substrates. Plant J 66:306–317

Birhane E, Sterck FJ, Fetene M, Bongers F, Kuyper TW (2012) Arbuscular mycorrhizal fungi enhance photosynthesis, water use efficiency, and growth of frankincense seedlings under pulsed water availability conditions. Oecologia 169:895–904

Björkman O, Demmig B (1987) Photon yield of O_2 evolution and chlorophyll fluorescence characteristics at 77 K among vascular plants of diverse origins. Planta 170:489–504

Boursiac Y, Chen S, Luu D-T, Sorieul M, van den Dries N, Maurel C (2005) Early effects of salinity on water transport in Arabidopsis roots. Molecular and cellular features of aquaporin expression. Plant Physiol 139:790–805

Cabot C, Sibole JV, Barceló J, Poschenrieder C (2014) Lessons from crop plants struggling with salinity. Plant Sci 226:2–13

Callaway E (2014) Domestication: the birth of rice. Nature 514:S58

Carvajal M, Cerdá A, Martínez V (2000) Does calcium ameliorate the negative effect of NaCl on melon root water transport by regulating water channel activity? New Phytol 145:439–447

Chen Z, Zhou MX, Newman IA, Mendham NJ, Zhang GP, Shabala S (2007) Potassium and sodium relations in salinized barley as a basis of differential salt tolerance. Funct Plant Biol 34:150–162

Chen YY, Hu CY, Xiao JX (2014) Effects of arbuscular mycorrhizal inoculation on the growth, zinc distribution and photosynthesis of two citrus cultivars grown in low-zinc soil. Trees 28:1427–1436

Cuin TA, Bose J, Stefano G, Jha D, Tester M, Mancuso S, Shabala S (2011) Assessing the role of root plasma membrane and tonoplast Na^+/H^+ exchangers in salinity tolerance in wheat: in planta quantification methods. Plant Cell Environ 34:947–961

Davenport RJ, Muñoz-Mayor A, Jha D, Essah PA, Rus A, Tester M (2007) The Na^+ transporter AtHKT1;1 controls retrieval of Na^+ from the xylem in Arabidopsis. Plant Cell Environ 30:497–507

Duarte B, Santos D, Marques JC, Cacador I (2013) Ecophysiological adaptations of two halophytes to salt stress: photosynthesis, PS II photochemistry and anti-oxidant feedback—implications for resilience in climate change. Plant Physiol Biochem 67:178–188

Estrada B, Aroca R, Maathuis FJM, Barea JM, Ruiz-Lozano JM (2013) Arbuscular mycorrhizal fungi native from a Mediterranean saline area enhance maize tolerance to salinity through improved ion homeostasis. Plant Cell Environ 36:1771–1182

Evelin H, Kapoor R, Giri B (2009) Arbuscular mycorrhizal fungi in alleviation of salt stress: a review. Ann Bot 104:1263–1280

Evelin H, Giri B, Kapoor R (2012) Contribution of Glomus intraradices inoculation to nutrient acquisition and mitigation of ionic imbalance in NaCl-stressed *Trigonella foenum-graecum*. Mycorrhiza 22:203–217

Evelin H, Giri B, Kapoor R (2013) Ultrastructural evidence for AMF mediated salt stress mitigation in *Trigonella foenum-graecum*. Mycorrhiza 23:71–86

FAO (2005) Global network on integrated soil management for sustainable use of salt-affected soils. FAO Land and Plant Nutrition Management Service, Rome, Italy. http://www.fao.org/ag/agl/agll/spush

Fukuda A, Nakamura A, Hara N, Toki S, Tanaka Y (2011) Molecular and functional analyses of rice NHX-type Na+/H+ antiporter genes. Planta 233:175–188
Galmés J, Medrano H, Flexas J (2007) Photosynthetic limitations in response to water stress and recovery in Mediterranean plants with different growth forms. New Phytol 175:81–93
Garciadeblás B, Senn ME, Banuelos MA, Rodriguez-Navarro A (2003) Sodium transport and HKT transporters: the rice model. Plant J 34:788–801
Giri B, Mukerji KG (2004) Mycorrhizal inoculant alleviates salt stress in *Sesbania aegyptiaca* and *Sesbania grandiflora* under field conditions: evidence for reduced sodium and improved magnesium uptake. Mycorrhiza 14:307–312
Giri B, Kapoor R, Mukerji KG (2007) Improved tolerance of Acacia nilotica to salt stress by arbuscular mycorrhiza, Glomus fasciculatum may be partly related to elevated K/Na ratios in root and shoot tissues. Microb Ecol 54:753–760
Goicoechea N, Antolin MC, Sánchez-Díaz M (1997) Gas exchange is related to the hormone balance in mycorrhizal or nitrogen-fixing alfalfa subjected to drought. Physiol Plant 100:989–997
Goicoechea N, Baslam M, Erice G, Irigoyen JJ (2014) Increased photosynthetic acclimation in alfalfa associated with arbuscular mycorrhizal fungi (AMF) and cultivated in greenhouse under elevated CO_2. J Plant Physiol 171:1774–1781
Grondin A, Mauleon R, Vadez V, Henry A (2016) Root aquaporins contribute to whole plant water fluxes under drought stress in rice (*Oryza sativa* L.) Plant Cell Environ 39:349–365
Gupta AB, Sankararamakrishnan R (2009) Genome-wide analysis of major intrinsic proteins in the tree plant *Populus trichocarpa*: characterization of XIP subfamily of aquaporins from evolutionary perspective. BMC Plant Biol 9:134
Hajiboland R, Aliasgharzadeh N, Laiegh SF, Poschenrieder C (2010) Colonization with arbuscular mycorrhizal fungi improves salinity tolerance of tomato (*Solanum lycopersicum* L.) plants. Plant Soil 331:313–327
Hammer E, Nasr H, Pallon J, Olsson P, Wallander H (2011) Elemental composition of arbuscular mycorrhizal fungi at high salinity. Mycorrhiza 21:117–129
Harbinson J (2013) Improving the accuracy of chlorophyll fluorescence measurements. Plant Cell Environ 36:1751–1754
Jahromi F, Aroca R, Porcel R, Ruiz-Lozano JM (2008) Influence of salinity on the in vitro development of Glomus intraradices and on the in vivo physiological and molecular responses of mycorrhizal lettuce plants. Microbial Ecol 55:45–53
Javot H, Maurel C (2002) The role of aquaporins in root water uptake. Ann Bot 90:301–313
Kaschuk G, Kuyper TW, Leffelaar PA, Hungria M, Giller KE (2009) Are the rate of photosynthesis stimulated by the carbon sink strength of rhizobial and arbuscular mycorrhizal symbioses? Soil Biol Biochem 41:1233–1244
Krajinski F, Biela A, Schubert D, Gianinazzi-Pearson V, Kaldenhoff R, Franken P (2000) Arbuscular mycorrhiza development regulates the mRNA abundance of Mtaqp1 encoding a mercury-insensitive aquaporin of *Medicago truncatula*. Planta 211:85–90
Kumar K, Kumar M, Kim S-R, Ryu H, Cho Y-G (2013) Insights into genomics of salt stress responses in rice. Rice 6:27
Lazár D (2015) Parameters of photosynthetic energy partitioning. J Plant Physiol 175:131–147
Lehto T, Zwiazek JJ (2011) Ectomycorrhizas and water relations of trees: a review. Mycorrhiza 21:71–90
Li T, Hu Y-J, Hao Z-P, Li H, Wang Y-S, Chen B-D (2013) First cloning and characterization of two functional aquaporin genes from an arbuscular mycorrhizal fungus *Glomus intraradices*. New Phytol 197:617–630
Lian H-L, Yu X, Ye Q, Ding X-S, Kitagawa Y, Kwak S-S, Su W-S, Tang Z-C (2004) The role of aquaporin RWC3 in drought avoidance in rice. Plant Cell Physiol 45:481–489
Lima-Neto MC, Lobo AKM, Martins MO, Fontenele AV, Albenisio J, Silveira G (2014) Dissipation of excess photosynthetic energy contributes to salinity tolerance: a comparative study of salt-tolerant *Ricinus communis* and salt-sensitive *Jatropha curcas*. J Plant Physiol 171:23–30

Liu T, Sheng M, Wang CY, Chen H, Li Z, Tang M (2015) Impact of arbuscular mycorrhizal fungi on the growth, water status, and photosynthesis of hybrid poplar under drought stress and recovery. Photosynthetica 53:250–258

López D, Venisse J-S, Fumanal B, Chaumont F, Guillot E, Daniels MJ, Gousset-Dupont A (2013) Aquaporins and leaf hydraulics: poplar sheds new light. Plant Cell Physiol 54:1963–1975

Maathuis FJM (2014) Sodium in plants: perception, signalling, and regulation of sodium fluxes. J Exp Bot 65:849–858

Mahajan S, Tuteja N (2005) Cold, salinity and drought stresses: an overview. Arch Biochem Biophys 444:139–158

Marschner H (1995) Mineral nutrition of higher plants, 2nd edn. Academic, London

Martínez-Ballesta MC, Martínez V, Carvajal M (2000) Regulation of water channel activity in whole roots and in protoplasts from roots of melon plants grown under saline conditions. Aust J Plant Physiol 27:685–691

Masumoto C, Ishii T, Hatanaka T, Uchida N (2005) Mechanisms of high photosynthetic capacity in BC2F4 lines derived from a cross between *Oryza sativa* and wild relatives O-rufipogon. Plant Prod Sci 8:539–545

Maurel C, Verdoucq L, Luu DT, Santoni V (2008) Plant aquaporins: membrane channels with multiple integrated functions. Annu Rev Plant Biol 59:595–624

Miller G, Suzuki N, Ciftci-Yilmaz S, Mittler R (2010) Reactive oxygen species homeostasis and signalling during drought and salinity stresses. Plant Cell Environ 33:453–467

Munns R (2005) Genes and salt tolerance: bringing them together. New Phytol 167:645–663

Munns R, Tester M (2008) Mechanisms of salinity tolerance. Annu Rev Plant Biol 59:651–681

Noctor G, Mhamdi A, Foyer CH (2016) Oxidative stress and antioxidative systems: recipes for successful data collection and interpretation. Plant Cell Environ 39:1140–1160

Olías R, Eljakaoui Z, Li J, Alvarez De Morales P, Marín-Manzano MC, Pardo JM, Belver A (2009) The plasma membrane Na^+/H^+ antiporter SOS1 is essential for salt tolerance in tomato and affects the partitioning of Na^+ between plant organs. Plant Cell Environ 32:904–916

Otto B, Kaldenhoff R (2000) Cell-specific expression of the mercury-insensitive plasma-membrane aquaporin NtAQP1 from *Nicotiana tabacum*. Planta 211:167–172

Ouziad F, Wilde P, Schmelzer E, Hildebrandt U, Bothe H (2006) Analysis of expression of aquaporins and Na+/H+ transporters in tomato colonized by arbuscular mycorrhizal fungi and affected by salt stress. Environ Exp Bot 57:177–186

Pardo JM, Cubero B, Leidi EO, Quintero FJ (2006) Alkali cation exchangers: roles in cellular homeostasis and stress tolerance. J Exp Bot 57:1181–1199

Pinior A, Grunewaldt-Stöcker G, von Alten H, Strasser RT (2005) Mycorrhizal impact on drought stress tolerance of rose plants probed by chlorophyll *a* fluorescence, proline content and visual scoring. Mycorrhiza 15:596–605

Pitman M, Läuchli A (2002) Global impact of salinity and agricultural ecosystems. In: Läuchli A, Lüttge U (eds) Salinity: environment-plants-molecules. Kluwer, Dordrecht, pp 3–20

Porcel R, Aroca R, Ruiz-Lozano JM (2012) Salinity stress alleviation using arbuscular mycorrhizal fungi. A review. Agron Sustain Dev 32:181–200

Porcel R, Redondo-Gómez S, Mateos-Naranjo E, Aroca R, Garcia R, Ruiz-Lozano JM (2015) Arbuscular mycorrhizal symbiosis ameliorates the optimum quantum yield of photosystem II and reduces non-photochemical quenching in rice plants subjected to salt stress. J Plant Physiol 185:75–83

Porcel R, Aroca R, Azcón R, Ruiz-Lozano JM (2016) Regulation of cation transporter genes by the arbuscular mycorrhizal symbiosis in rice plants subjected to salinity suggests improved salt tolerance due to reduced Na^+ root-to-shoot distribution. Mycorrhiza 26:673–684

Ramoliya PJ, Patel HM, Pandey AN (2004) Effect of salinization of soil on growth and macro- and micro-nutrient accumulation in seedlings of *Salvadora persica* (Salvadoraceae). Forest Ecol Manag 202:181–193

Ranathunge K, Kotula L, Steudle E, Lafitte R (2004) Water permeability and reflection coefficient of the outer part of young rice roots are differently affected by closure of water channels (aquaporins) or blockage of apoplastic pores. J Exp Bot 55:433–447

Redondo-Gómez S, Mateos-Naranjo E, Figueroa ME, Davy AJ (2010) Salt stimulation of growth and photosynthesis in an extreme halophyte, *Arthrocnemum macrostachyum*. Plant Biol 2010 (12):79–87

Ren Z-H, Gao J-P, Li L-G et al (2005) A rice quantitative trait locus for salt tolerance encodes a sodium transporter. Nat Genet 37:1141–1146

Ruiz-Lozano JM (2003) Arbuscular mycorrhizal symbiosis and alleviation of osmotic stress. New perspectives for molecular studies. Mycorrhiza 13:309–317

Ruiz-Lozano JM, Aroca R (2010) Host response to osmotic stresses: stomatal behaviour and water use efficiency of arbuscular mycorrhizal plants. In: Koltai H, Kapulnik Y (eds) Arbuscular mycorrhizas: physiology and function, 2nd edn. Springer, Dordrecht, pp 239–256

Ruiz-Lozano JM, Alguacil MM, Bárzana G, Vernieri P, Aroca R (2009) Exogenous ABA accentuates the differences in root hydraulic properties between mycorrhizal and non mycorrhizal maize plants through regulation of PIP aquaporins. Plant Mol Biol 70:565–579

Ruiz-Lozano JM, Porcel R, Azcón R, Aroca R (2012) Regulation by arbuscular mycorrhizae of the integrated physiological response to salinity in plants: new challenges in physiological and molecular studies. J Exp Bot 63:4033–4044

Ruíz-Sánchez M, Aroca R, Munoz Y, Polon R, Ruíz-Lozano JM (2010) The arbuscular mycorrhizal symbiosis enhances the photosynthetic efficiency and the antioxidative response of rice plants subjected to drought stress. J Plant Physiol 167:862–869

Ruíz-Sánchez M, Armada E, Munoz Y, García de Salamone I, Aroca R, Ruíz-Lozano JM, Azcón R (2011) *Azospirillum* and arbuscular mycorrhizal colonization enhanced rice growth and physiological traits under well-watered and drought conditions. J Plant Physiol 168:1031–1037

Sade N, Gebretsadik M, Seligmann R, Schwartz A, Wallach R, Moshelion M (2010) The role of tobacco aquaporin1 in improving water use efficiency, hydraulic conductivity, and yield production under salt stress. Plant Physiol 152:245–254

Sannazzaro AI, Ruiz OA, Alberto EO, Menendez AB (2006) Alleviation of salt stress in *Lotus glaber* by *Glomus intraradices*. Plant Soil 285:279–287

Sanz-Sáez A, Erice G, Aranjuelo I, Aroca R, Ruiz-Lozano JM, Aguirreolea J et al (2013) Photosynthetic and molecular markers of CO_2-mediate photosynthetic downregulation in nodulated alfalfa. J Integr Plant Biol 55:721–734

Scheibe R, Beck E (2011) Drought, desiccation, and oxidative stress. In: Lüttge U et al (eds) Plant desiccation tolerance, Ecological studies, vol 215. Springer, Berlin, pp 209–231

Shao HB, Chu LY, Shao MA, Jaleel CA, Mi HM (2008) Higher plant antioxidants and redox signaling under environmental stresses. Comptes Redue Biolol 331:433–441

Sheng M, Tang M, Chen H, Yang B, Zhang F, Huang Y (2008) Influence of arbuscular mycorrhizae on photosynthesis and water status of maize plants under salt stress. Mycorrhiza 18:287–296

Shi HZ, Quintero FJ, Pardo JM, Zhu JK (2002) The putative plasma membrane Na^+/H^+ antiporter SOS1 controls long-distance Na^+ transport in plants. Plant Cell 14:465–477

Strasser RJ, Srivastava A, Tsimilli-Michael M (2000) The fluorescence transient as a tool to characterize and screen photosynthetic samples. In: Yunus M (ed) Probing photosynthesis: mechanisms, regulation and adaptation. Taylor & Francis, London, pp 445–483

Takai T, Kondo M, Yano M, Yamamoto T (2010) A quantitative trait locus for chlorophyll content and its association with leaf photosynthesis in rice. Rice 3:172–180

Talaat NB, Shawky BT (2011) Influence of arbuscular mycorrhizae on yield, nutrients, organic solutes, and antioxidant enzymes of two wheat cultivars under salt stress. J Plant Nutr Soil Sci 174:283–291

Uehlein N, Fileschi K, Eckert M, Bienert GP, Bertl A, Kaldenhoff R (2007) Arbuscular mycorrhizal symbiosis and plant aquaporin expression. Phytochemistry 68:122–129

Wang WX, Vinocur B, Altman A (2003) Plant responses to drought, salinity and extreme temperatures: towards genetic engineering for stress tolerance. Planta 218:1–14

Wu H, Shabala L, Barry K, Zhou M, Shabala S (2013) Ability of leaf mesophyll to retain potassium correlates with salinity tolerance in wheat and barley. Physiol Plant 149:515–527

Zelazny E, Borst JW, Muylaert M, Batoko H, Hemminga MA, Chaumont F (2007) FRET imaging in living maize cells reveals that plasma membrane aquaporins interact to regulate their subcellular localisation. Proc Natl Acad Sci USA 104:12359–12364

Zhao C-X, Shao H-B, Chu L-Y (2008) Aquaporin structure-function relationships: water flow through plant living cells. Col Surf B 62:163–172

Zhu XQ, Wang CY, Chen H, Tang M (2014) Effects of arbuscular mycorrhizal fungi on photosynthesis, carbon content, and calorific value of black locust seedlings. Photosynthetica 52:247–252

Zhu M, Shabala L, Cuin TA, Huang X, Zhou M, Munns R, Shabala S (2016) Nax loci affect SOS1-like Na+/H+ exchanger expression and activity in wheat. J Exp Bot 67:835–844

Zuccarini P, Okurowska P (2008) Effects of mycorrhizal colonization and fertilization on growth and photosynthesis of sweet basil under salt stress. J Plant Nutr 31:497–513

Chapter 11
Bioprotection of Soybean Plants from Drought Stress by Application of Bacterial and Fungal Endophytes

Dipanti Chourasiya, Richa Agnihotri, Anil Prakash, Kamal K. Pal, and Mahaveer P. Sharma

11.1 Introduction

Soybean [*Glycine max* (L.) Merrill] is one of the major oilseed crops in the world which contains 18–20% oil and 35–40% of good-quality protein (Fatima et al. 2006). Soybean is being used in human and animal nutrition and in the production of biodiesel, disinfectants, lubricants, soap, and cosmetics, among other uses (Sediyama et al. 2009). In the recent agriculture scenario, climate change and food security are the two prominent challenges faced by scientists to cater the needs of burgeoning Indian population. The top five soybean-producing countries are the USA, Brazil, Argentina, China, and India; in Southeast Asia, soybean productivity highly relies on rainfall. In India, the erratic monsoon pattern, incidence of disease and pest, and long dry spell (drought) because of climate change have given rise to uncertainty in the production of soybean in the past few years. Moisture stress decreases the significant yield of soybean every year (Joshi and Bhatia 2003). According to Grover et al. (2011), nearly two-thirds area from parts of arid and semiarid ecosystem in India are affected by drought or soil moisture stresses. Moisture stress is among the most destructive abiotic stresses that increased in intensity over the past decades affecting the world's food security. It affects different growth stages of soybean, for example, the reproductive stages were affected more

D. Chourasiya · R. Agnihotri · M. P. Sharma (✉)
ICAR-Indian Institute of Soybean Research, Indore, India

A. Prakash
Faculty of Life Sciences, Department of Microbiology, Barkatullah University, Bhopal, India

K. K. Pal
Directorate of Groundnut Research (DGR-ICAR), Junagadh, India

B. Giri et al. (eds.), *Root Biology*, Soil Biology 52,
https://doi.org/10.1007/978-3-319-75910-4_11

severely as compared to vegetative growth (Sionet and Kramer 1977). Under drought conditions, planting soybeans hinders germination, leading to low plant population and reduction in yield significantly.

To cope with the moisture stress, there is an emerging need to develop strategies. The strategy of breeding varieties tolerant to drought is time-consuming and is a long-term approach. Therefore, the use of biological resources is advocated which would have minimal damage to the environment and sustain the soybean productivity. The inoculation of plants with beneficial microorganisms such as soybean rhizobia and AM fungi has been used as an alternative strategy to ameliorate plants from abiotic stresses and help in sustaining the productivity of soybean with reduced uses of chemical fertilizers (Sharma et al. 2016). Soybean belongs to the family Leguminosae which forms root nodules harbouring from symbiotic bacteria and many endophytes and therfore relys less on N fertilizer (Zhao and Lai 2017). Under drought conditions, accumulation of free radicals occurred due to protein changes, restricted enzyme efficiency, and changes in electron transport (Berard et al. 2015), resulted into protein denaturation and lipid peroxidation leading to cell lysis (Potts 1999). During oxidative stress, antioxidant defenses and reactive oxygen species (ROS) such as superoxide radicals, hydrogen peroxide, and hydroxyl radicals are affected by free radicals which induce lipid peroxidation and membrane deterioration and degrade proteins, lipids, and nucleic acids in plants (Hendry 2005; Nair et al. 2008). The reduced precipitation and changed rainfall patterns are causing the frequent onset of droughts around the world (Lobell et al. 2011) and is associated with reduction in photosynthesis, results due to decrease in leaf expansion, impaired photosynthetic machinery, premature leaf senescence (Wahid and Rasul 2005). Drought stress in soybean decreases total seed yield and the branch seed yield (Frederick et al. 2001). Hence, it becomes imperative to modulate plant's physiology through inoculation of bacterial and fungal endophytes with a capability to maintain growth and survive during drought stress (Chaves et al. 2003). Endophytic microbes are known to enhance growth and yield of plants by fixing atmospheric nitrogen; solubilization of phosphorus, potassium, zinc, etc.; production of phytohormones (cytokinin, auxin, gibberellins, etc.), ammonia, hydrogen cyanide, and siderophore; and production of ACC deaminase while mitigating the level of ethylene in plants during drought (Suman et al. 2016). In this chapter, we provided a holistic view on types of bacterial and fungal endophytes inhabiting soybean and other crops and how these endophytes are significantly important in conferring the tolerance in soybean to moisture deficit stress conditions. The key mechanisms involved and the factors influencing the efficacy of these endophytes for mitigating the stress have been dealt with in the subsequent subsections.

11.2 Distribution of Bacterial and Fungal Endophytes

Hallmann et al. (1997) defined endophytes as "microbes that are residing in plant tissues without causing harmful effects to the host plant." Endophytic microorganisms found in plants include bacteria, actinomycetes, and fungi (Zinniel et al. 2002; Ji et al. 2014; Tenguria and Firodiya 2013; Dalal and Kulkarni 2012). Endophytes occurring in roots, stems, and leaves was first studied in soybean and maize cultivars in Argentina (Russo et al. 2016), and these can be isolated by the surface sterilization technique from the parts of the plant that has no harm to the plant (Gaiero et al. 2013). Endophytes inhabit in the apoplast, as well as intercellular spaces of the cell walls and xylem vessels of the roots, stems, leaves, fruits, tubers, ovules, internodal regions and also in root nodules which all increases the growth of soybean plant (Hallmann et al. 1997; Kuklinsky Sobral et al. 2004; Pimentel et al. 2006). Dalal and Kulkarni (2012) isolated a comparatively higher number of endophytic actinomycetes during vegetative stages than during reproductive stages in soybean cultivated under vertisol.

In general, all the bacterial endophytes are described as common habitants of the rhizosphere; therefore, endophyte microbiome is suggested as the subpopulation of the rhizosphere-inhabiting bacteria (Marquez-Santacruz et al. 2010; Germida et al. 1998). There are two types of endophytic bacteria mainly categorized based on types of activity, viz., growth promotion and disease control (Bacon and White 2000), which Hardoim et al. (2008) further subdivided into four types:

1. Facultative endophytes which live inside the plants and in other habitats
2. Obligate endophytes which are strictly bound inside a plant during their entire lifespan and that do not possess life stages outside the plant, except for plant-to-plant and plant-to-insect-to-plant transmission
3. Opportunistic endophytes which occasionally enter into the plants and benefit from the plant's internal environment (nutrient availability, protection, and lack of competition) and show particular root colonization characteristics (e.g., a chemotactic response, which enables them to colonize the rhizoplane and then invade the internal plant tissues through cracks formed at the sites of lateral root emergence and root tips)
4. Passenger endophytes that enter the plant by accident in the absence of selective forces maintaining it in the internal tissue of the plant, i.e., such endophytes might become endophytic by chance (e.g., via colonization of natural wounds or following root invasion by nematodes)

The agricultural practices maintain natural endophytic bacterial diversity in plants (Brandl 2006). Interestingly, gram-negative bacteria isolates were found low in *G. soja*, and gram-positive isolates were more in *G. max* (Stoltzfus et al. 1997; Elbeltagy et al. 2000). On the other hand, Zinniel et al. (2002) reported an equal occurrence of gram-negative and gram-positive bacteria. Dalal and Kulkarni (2014) suggested that the occurrence of different endophyte species depends mostly on growth stages of plant, bacteria genotype, and biotic and abiotic environmental conditions. They showed that vegetative and reproductive stages of soybean facilitate diversity and distribution of endophytic fungi (Dalal and Kulkarni 2014). Some factors like age of the plant and type of cultivation environment such as open fields and greenhouses also contribute to the diversity of endophytic fungi in soybean (Pimentel et al. 2006).

Very recently, about 187 species of fungal endophytes from different parts of soybean was reported, which belong to different taxonomic groups including

Ascomycota and *Basidiomycota* (De Souza Leite et al. 2013). Different types of endophytic bacteria, e.g., *Bacillus*, *Enterobacter*, and *Pseudomonas*, inhabit the soybean (Egamberdieva et al. 2016). Two types of fungal isolates were reported in soybean based on culture-dependent (CD) and culture-independent (CI) methods (Miller and Roy 1982; Pimentel et al. 2006). CD method detected greater endophyte diversity ($H' = 2.12$) than the CI method ($H' = 0.66$) (Impullitti and Malvick 2013). In general, diversity analysis of endophytes in soybean revealed that the phylum *Proteobacteria*, including the classes *Alphaproteobacteria*, *Betaproteobacteria*, and *Gammaproteobacteria*, were found to be dominant besides the members of the *Firmicutes* and *Actinobacteria*, which were consistently found as endophytes, whereas other classes such as *Bacteroidetes*, *Planctomycetes*, *Verrucomicrobia*, and acid bacteria are less commonly found as endophytes. The most commonly found genera of bacterial endophytes are *Pseudomonas*, *Bacillus*, *Burkholderia*, *Stenotrophomonas*, *Micrococcus*, *Pantoea*, and *Microbacterium* (Sun et al. 2009; Romero et al. 2014; Hallmann et al. 1997; Marquez-Santacruz et al. 2010; Shi et al. 2014). Fungal endophytes such as *Fusarium oxysporum* and *Phoma* sp. are predominately found in roots, whereas *Colletotrichum gloeosporioides* (12.29%) and *Ampelomyces* sp. (9.09%) are found in leaves (Fernandes et al. 2015). A regularly isolated species in all soybean cultivars was *Fusarium graminearum*, and the least probable isolated one was *Scopulariopsis brevicaulis*. It has been revealed that soybean leaves were somewhat richer in fungal endophytes than in roots (Fernandes et al. 2015). A list of bacterial and fungal endophytes recovered from soybean or used in soybean is provided in Table 11.1.

11.3 Interaction of Fungal and Bacterial Endophytes

Bacterial and fungal endophytes have distinctive roles in plants. They facilitate the distribution or production of biologically active metabolites, such as enzymes, biofunctional chemicals, phytohormones, nutrients, and minerals (Schulz et al. 2002). Co-inoculation of AMF and PGPR shows positive effect in drought tolerance (Kohler et al. 2009). It has been shown that *Bradyrhizobium* when co-inoculated with AM species, e.g., *G. mosseae* and *G. deserticola*, had a positive and synergistic influence on soybean. However, comparatively *G. mosseae* was found to be more compatible and effective than *G. deserticola*. Overall, AM symbiosis has typically increased water use efficiency (Simpson and Daft 1990) by modulating hormonal regulation (Levy and Krikun 1980), capturing more soil water (Sieverding 1981) through improved soil/root contacts (Reid 1979). AM also helps in stimulating gas exchange by increased sink strength (e.g., Allen et al. 1981; Snellgrove et al. 1982) and osmotic adjustments (Allen and Boosalis 1983) contributed more water absorption through extended soil hyphae (Allen 1982).

Table 11.1 List of potential bacterial and fungal endophytes isolated from soybean plants

Soybean	Endophytes strain	References
Seed	*Alternaria*, *Aspergillus*, *Cladosporium*, *Colletotrichum*, *Fusarium*, *Paecilomyces*, and *Penicillium*	Miller and Roy (1982)
	Ampelomyces, *Annulohypoxylon*, *Guignardia*, *Leptospora*, *Magnaporthe*, *Ophiognomonia*, *Paraconiothyrium*, *Phaeosphaeriopsis*, *Rhodotorula*, *Sporobolomyces*, and *Xylaria*	De Souza Leite et al. (2013)
Root	*Bacillus thuringiensis*, *Bacillus subtilis*	Pimentel et al. (2006)
	Paenibacillus sp.	Annapurna et al. (2013)
	Fusarium oxysporum, *Fusarium solani* and *Fusarium* sp., *F. graminearum*	Fernandes et al. (2015)
	Bacillus endoradicis sp.	Zhang et al. (2012)
	Pseudomonaceae, *Burkholderiacea*, and *Enterobacteriaceae* groups	Kuklinsky Sobral et al. (2004)
Leaves	*Ampelomyces* sp., *Cladosporium cladosporioides*, *Colletotrichum gloeosporioides*, *Diaporthe helianthi*, *Guignardia mangiferae*, and *Phoma* sp.	Fernandes et al. (2015)
	Alternaria, *Aspergillus*, *Cladosporium*, *Fusarium*, *Paecilomyces*, and *Penicillium from soybean leaves* and *seeds*	Miller and Roy (1982)
	Dematiaceous fungi, *Fusarium*, *Aspergillus*, *Penicillium* and *Mycelia sterile*, *Colletotrichum*, *Acremonium*, *Paecilomyces*	Pimentel et al. (2006)
	Ampelomyces, *Annulohypoxylon*, *Guignardia*, *Leptospora*, *Magnaporthe*, *Ophiognomonia*, *Paraconiothyrium*, *Phaeosphaeriopsis*, *Rhodotorula*, *Sporobolomyces*, and *Xylaria*	De Souza Leite et al. (2013)
Stem	*Colletotrichum*, *Acremonium*	Pimentel et al. (2006)
	Moraxellaceae	Kuklinsky Sobral et al. (2004)
Nodule	*Bacillus subtilis*, *Bacillus thuringiensis*, and *Bacillus pumilus*	Bai et al. (2002), Li et al. (2008)

11.4 Role of Bacterial and Fungal Endophytes in Alleviation of Stress

Plant growth promotion (PGP) activity has been observed in many endophytic bacteria (Zachow et al. 2010; Gasser et al. 2011; Malfanova et al. 2011). Endophytic symbiosis with host plants especially in roots can regulate and change the uptake of mineral nutrients, secretion of plant hormones, and exudation of defensive metabolites from roots (Khan et al. 2013; Bashan et al. 2014). Direct PGP mediation by endophytes provides essential nutrients to the plants and regulates phytohormones. Several mechanisms conferring drought tolerance to plants by plant growth-promoting bacteria

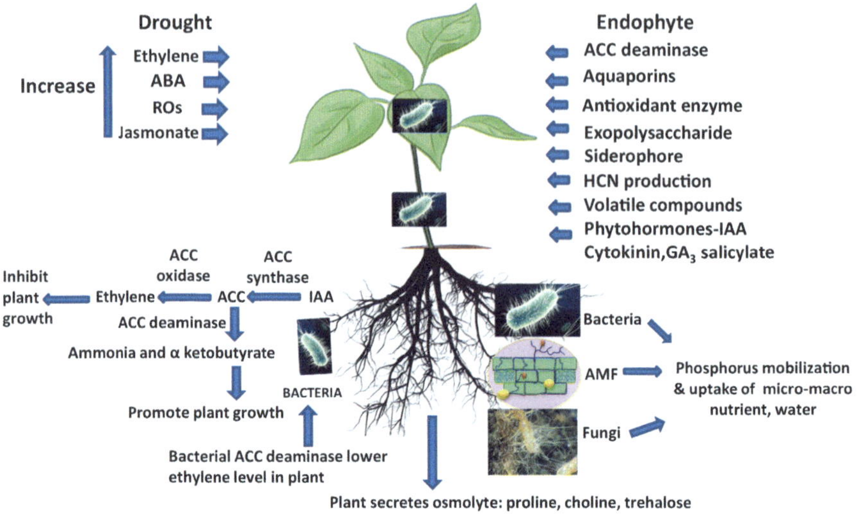

Fig. 11.1 Prospects of endophytes in the alleviation of drought stress in plants and underlying mechanisms

(PGPB) have been proposed by several works (Fig. 11.1). PGPB may influence plant growth either directly or indirectly by:

1. Indirect mechanism through the production of antibiotics, cell wall-degrading enzymes, induced systemic resistance (ISR) through decreasing iron abundance available to pathogens, and synthesis of pathogen-inhibiting volatile compounds (Yang et al. 2009). The PGPB facilitates the gaining of nutrient resources from environment including nitrogen (N), phosphorous (P), and iron (Fe), e.g., via nitrogen fixation, phosphate solubilization (Wakelin et al. 2004), iron chelation, and siderophore production (Loon et al. 1998);
2. Direct promotion by regulating various plant hormones including auxin and cytokinin (Madhaiyan et al. 2006) or by producing the enzyme 1-aminocyclopropane-1-carboxylate (ACC) deaminase, which lowers plant ethylene levels and modulates plant growth.

While analyzing the phenotypic and genotypic diversity in the bacterial endophytes of two species of soybean, viz., *Glycine max* and *G. soja*, Hung and Annapurna (2004) reported that a total of 65 bacterial endophytes were isolated from three tissues, stem, root, and nodules, and screened for hydrolytic enzymes where more than 70% isolates were found to secrete pectinase and cellulase. Sandhya et al. (2009) found the role of volatile organic compounds (VOCs) released by PGPB that played a role in IST response. An endophytic fungus *Piriformospora indica* has been widely evaluated for its role in abiotic stress tolerance and plant growth promotion (Varma et al. 2012). This fungus belongs to *Sebacinaceae* family, which colonizes the root of plants and is reported to alleviate plants from drought stress conditions (Sahay and Varma 1999; Shahollari et al. 2005). The fungal endophytes may help in decrease plant disease infestation and enhance plant growth and have widely been worked out by many workers (Hardoim et al. 2008; Marquez et al. 2007). *Cladosporium*, the widespread endophytic fungal genus detected in the vascular and pith tissues of soybean (Impullitti and Malvick 2013), has been known for plant growth promotion. *Cladosporium sphaerospermum* was found to produce gibberellin-like compounds which increased soybean biomass and height (Hamayun et al. 2009). The AM inoculation has been shown to enhance tolerance to plant by adjustments in osmoregulation of solutes and improve upon water uptake through their extended hyphal network (Mathimaran et al. 2017) that help indirectly plants in maintaining osmotic adjustments, facilitates membrane stability, protects from electrolytes leakage and maintain water potential gradient in the roots (Evelin et al. 2012).

Arbuscular mycorrhiza forms symbiotic association with soybean plants for improved P nutrition and gas exchange and also affects plant–water relation under drought condition (Safir et al. 1971). Further, induced production of growth-promoting hormones and antioxidative enzymes (Zhu et al. 2011) regulates the plant aquaporin gene (GintAQPF1and GintAQPF2) expression (Li et al. 2013). Besides improving plant growth, AMF also influence indirectly for improved RWC, LWP, K, and N in shoots that play a role in osmotic adjustment and stomata behavior and relatively increased shoot and seed dry weights. Soybean plant inoculated with either single or mixed AMF under drought conditions had significantly higher shoot biomass. However, the magnitude of response varies where mixtures of AMF isolates did not perform than single strain inoculum, excluding complementarily effects and suggesting selection of effective AMF for alleviating drought stress in soybean (Grumberg et al. 2015). AMF inoculation changes in soluble sugar, proline levels, and hydrogen peroxide in drought-stressed plants promote exchange of carbon and nitrogen required for drought adaptation of the host plants (Rapparini and Penuelas 2014). The best single strain inoculum, PGPB scavenge toxic compounds ROS by enzymatic and nonenzymatic antioxidant responses (Wang et al. 2012). Bilal et al. (2017) reported that inoculation with *Paecilomyces formosus* LHL10 significantly increased plant biomass and growth attributes as compared to non-inoculated control plants with or without Ni contamination. LHL10 enhanced the translocation of Ni from the root to the shoot as compared to the control. In addition, *P. formosus* LHL10 modulated the physiochemical apparatus of soybean plants during Ni contamination by reducing lipid peroxidation and the accumulation of reactive oxygen species.

11.4.1 Bacterial and Fungal Endophytes (Key Examples)

Many bacterial endophytes (more than 65 isolates), viz., *Alphaproteobacteria* (*Agrobacterium* sp.), *Gammaproteobacteria* (*Erwinia* sp., *Pseudomonas citronellolis*, *P. oryzihabitans*, *P. straminea*, *K. pneumoniae*, *K. oxytoca*, *Enterobacter* sp., *Pantoea* sp., and *P. agglomerans*), and *Firmicutes* (*Bacillus fastidiosus*, etc.), have been reported from stems, leaves, roots, and root nodules of soybean (Hung and Annapurna 2004). Bai et al. (2002) reported the role of nonsymbiotic endophytic bacterial strains (NEB, mainly bacilli), viz., NEB4, NEB5, and NEB17, in the growth promotion of soybean when only co-inoculated (either of one isolate) with symbiotic *Bradyrhizobium japonicum* under nitrogen-free conditions compared with plants inoculated with *B. japonicum* alone. In the absence of *B. japonicum*, these isolates neither nodulated soybean nor affected soybean growth. In case of fungal endophytes, *Fusarium graminearum* and *Colletotrichum* (from leaves) species were mainly isolated from soybean (Pimentel et al. 2006; Russo et al. 2016).

Long back Miller and Roy (1982) have also isolated fungal endophytes, viz., *Alternaria*, *Aspergillus*, *Cladosporium*, *Colletotrichum*, *Fusarium*, *Paecilomyces*, and *Penicillium*, from soybean leaves and seeds where *Cladosporium sphaerospermum* were found to increase the plant height and biomass. Recently besides AM fungi, the endophytic fungi *Piriformospora indica* which belong to the *Sebacinaceae* family plays a greater role in the alleviation of abiotic stress (Gill et al. 2016) and is gaining importance. A detailed list and examples of potential bacterial endophytes in the growth and bioprotection of soybean plants are provided in Table 11.2.

11.5 Mechanisms Involved for Conferring Drought Tolerance

11.5.1 Phosphate Solubilization and Siderophore Production

Phosphorus element is supplied through animal manure, plant residues, green manure, and chemical fertilizers (Rao 1982). Phosphate (P) solubilization is an essential microbial process required for enhancing the availability of P nutrition in crops (Vassileva et al. 1998). Different soil bacteria and fungi belong to *Bacillus*, *Pseudomonas*, and *Penicillium citrinum* are involved in conversion of insoluble phosphate to soluble phosphate in soil through secretion of organic acid such as, acetic, formic, fumaric, glycolic, and lactic. In general endophytic bacteria frequently occurs during vegetative (V6) stage of soybean (Kuklinsky Sobral et al. 2004) where Enterobacteriaceae and species of Pseudomonas and Ralstonia involves in phosphate solubilization (Kuklinsky Sobral et al. 2004) and improve P nutrition and thus can help in the reduction of phosphate fertilizers.

Plant growth-promoting endophytes promote plant growth by production of siderophores (Costa and Loper 1994). Siderophores are high-affinity iron-chelating compounds, soluble Fe^{3+}-binding agents, help in scavenging iron during iron deficiency (Neilands 1995), and play a role in disease control (Duffy and Defago 1999). Bacteria and fungi secrete siderophores. Senthilkumar et al. (2009) showed that out of 137 bacterial isolates recovered from various parts of soybean, only nine bacterial endophytes (mainly *Bacillus* and only one *Paenibacillus*) were found to be potential for controlling soybean phytopathogen, e.g., *Rhizoctonia bataticola*, *Macrophomina phaseolina*, *Fusarium udam*, and *Sclerotium rolfsii*. Based on

Table 11.2 Role of potential bacterial endophytes recovered from soybean and other plants on soybean

Bacterial endophyte	Host and niche from which isolated	Inference	References
Azoarcus sp. BH72	Rice	Nitrogen fixation	Krause et al. (2006)
Azospirillum lipoferum 4B	Rice, maize, wheat	Nitrogen fixation, phytohormone secretion	Wisniewski-Dye et al. (2011)
Azospirillum sp. B510	Rice	Nitrogen fixation, phytohormone secretion	Kaneko et al. (2010)
Burkholderia phytofirmans PsJN	Potato, tomato, maize, barley, onion, canola, grapevine	IAA synthesis, ACC deaminase	Weilharter et al. (2011)
Burkholderia spp. KJ006	Rice	ACC deaminase, *nif* gene cluster, antifungal action (indirect PGP)	Kwak et al. (2012)
Enterobacter sp. 638	Poplar	Siderophore, IAA, acetoin and 2,3-butanediol synthesis, antifungal action (indirect PGP)	Taghavi et al. (2009)
Gluconacetobacter diazotrophicus PaI5	Sugarcane, rice, coffee, tea	Nitrogen fixation, auxin synthesis	Bertalan et al. (2009)
Klebsiella pneumoniae 342	Maize, wheat	Nitrogen fixation	Fouts et al. (2008)
Pseudomonas putida W619	Poplar	IAA synthesis, ACC deaminase	Taghavi et al. (2009)
Pseudomonas stutzeri A1501	Rice	Nitrogen fixation	Yan et al. (2008)
Serratia proteamaculans 568	Soybean	IAA synthesis, ACC deaminase, acetoin and 2,3-butanediol synthesis	Taghavi et al. (2009)
Paenibacillus sp. and *Bacillus* sp.	Soybean	Biocontrol potential	Senthilkumar et al. (2009)
Sphingomonas sp.	Soybean (leaf and stem)	Plant growth promotion	Asaf et al. (2017)
Stenotrophomonas maltophilia R551-3	Poplar	IAA synthesis, ACC deaminase	Taghavi et al. (2009)
Cladosporium sphaerospermum	Soybean (root)	Increased plant height and biomass	Hamayun et al. (2009), Arnold et al. (2003)
Piriformospora indica	Wheat	Increase biomass	
Ampelomyces, *Chaetomium*, and *Phoma glomerata*	Soybean	Biocontrol of pathogen	De Souza Leite et al. (2013)
Paecilomyces formosus	Soybean	Increase biomass	Bilal et al. (2017)
Sphingomonas sp.	Soybean	Decrease JA and ABA	Asaf et al. (2017)
Bacillus amyloliquefaciens	Soybean	Reduced charcoal rot infestation	Torres et al. (2016)

assessment, they found only two most efficient strains as biocontrol isolates, i.e., *Paenibacillus* sp. HKA-15 (HKA-15) and *Bacillus* sp. HKA-121 (HKA-121), and two strains (HKA-72 and HKA-113) were found to be strong siderophore producers but ineffective in charcoal rot diseases suppression. Very recently Nhu and Diep (2017) recovered five bacterial endophytes from soybean plants identified as *Enterobacter cloacae* TSR1A, *Enterobacter cloacae* CPR1A, *Bacillus* sp. OSR12, *Bacillus subtilis* TST10c, and *Acinetobacter* sp. TGN1, and all these strains were found to be strong siderophore-producing candidates with multiple beneficial characteristics.

11.5.2 Production of Phytohormones

Plant growth and development are under the control of plant growth regulators and several phytohormones and organic compound, viz., auxins, gibberellins (GAs), cytokinins (CKs), ethylene (ET), and abscisic acid (ABA) (Farooq et al. 2009). GAs and CKs help in plant growth, whereas ethylene and abscisic acid inhibit growth (Taiz and Zeiger 2010). Drought stress leads to an increase of the inhibitory concentration of plant hormones, which inhibit plant growth, thus allowing the plants to regulate their water resources (Farooq et al. 2009). Endophytes are able to promote the growth of plants under drought by modifying the phytohormone content (Dodd et al. 2010) such as alteration of ET production (Glick et al. 1998; Belimov et al. 2009) and changing the balance of CKs and ABA (Figueiredo et al. 2008; Cohen et al. 2009) or IAA signaling etc. (Contesto et al. 2010). A brief account on the role of hormones in drought stress tolerance and regulation by endophytic bacteria has been given below in the following sub-heads.

11.5.3 Auxins

Auxins are an important regulator of plant growth and development, such as IAA, which controls different cellular functions as well as differentiation of vascular tissues, initiation of lateral and adventitious roots, stimulation of cell division, elongation of stems and roots, and orientation of root and shoot growth in response to light and gravity (Glick 1995). Auxins plays in plant development by regulating cell division and elongation, for example, *Sphingomonas* sp. (LK11), *Serratia marcescens* (TP1) bacterial endophytes produces IAA, which improve physiological characteristics such as shoot/root length, fresh/dry weight, and chlorophyll contents. Similarly, Enterobacter hormaechei isolated from S. selanica plants produces higher IAA and found compatible to support the soybean growth evaluated in a glasshouse experiment (Asaf et al. 2017).

11.5.4 Gibberellic Acid and Abscisic Acid

Gibberellic acid (GA) is known to regulate plant adaptation to drought stress. A rapid decline in levels of endogenous GA was observed in plants subjected to drought stress, resulting in growth inhibition. For example, secretion of gibberellins from *P. putida* improved soybean plant growth under drought (Kang et al. 2014). Inoculation of soybean with *Sphingomonas* sp. (LK11) and *Serratia marcescens* (TP1) significantly increased the level of GA (155.43–146.94 ng/g) and abscisic acid as compared to control (113.76 ng/g). However, inoculation of these strains decreased jasmonic acid content. Anjum et al. (2011) reported that application of methyl jasmonate (MeJA) did not directly alleviate the drought stress in soybean but rather helped in improving the drought tolerance of soybean by modulating the membrane lipid peroxidation and antioxidant activities.

Abscisic acid (ABA) performs important roles in various physiological processes in plants and is vital for the response to environmental stresses such as drought (Porcel et al. 2014; Cohen et al. 2015) and is involved in processes, e.g., inhibition of germination, restriction of shoot and root growth, and stomata closure (Daszkowska-Golec 2016). Under the drought conditions, plants secrete ABA for survival under unfavorable conditions that is synthesized in the roots and translocated to leaves, which induces stomata closure and conserves water loss (Zhu 2002).

11.5.5 Ethylene and ACC Deaminase

Biotic and abiotic stresses synchronized by ethylene levels and plant activities are regulated by biosynthesis of ethylene (Hardoim et al. 2008). In the biosynthetic pathway of ethylene, *S*-adenosylmethionine (SAM) is converted by 1-aminocyclopropane-1-carboxylate synthase (ACS) to 1-aminocyclopropane-1-carboxylate (ACC), the immediate precursor of ethylene. Under stress conditions, the plant hormone ethylene endogenously regulates plant homoeostasis resulting in reduced root and shoot growth. The ACC deaminase-producing plant growth-promoting endophytic bacteria sequester plant ACC and degrade or break it down into alpha ketobutyrate, thereby reducing the deleterious effects of ethylene, ameliorating plant stress, and promoting plant growth (Glick 2005). It has been well established that endophytic bacteria such as *Alcaligenes* sp., *Bacillus* sp., *Ochrobactrum* sp., *Burkholderia*, etc. have ACC deaminase enzyme which break down 1-aminocyclopropane-1-carboxylate (ACC) and reduce the ethylene levels in plants to survive under stress (Arshad et al. 2008; Onofre-Lemus et al. 2009). Hence, by application of ACC deaminase-producing bacteria, we can stop ethylene signaling and alleviate higher levels of ethylene during stress. Bacterial ACC deaminase can be divided into two groups, based on high or low enzymatic activity (Glick 2005). In soybean, the low ACC deaminase-producing bacteria found in tissue part have less ability to reduce ethylene level as compared to high ACC deaminase-producing bacteria under stress conditions, for example, *Serratia proteamaculans* is reported to be a high ACC deaminase-producing bacteria (Glick 2005; Taghavi et al. 2009).

11.5.6 Antioxidant Defenses

Exposure of plants to drought stress leads to the generation of reactive oxygen species (ROS), including superoxide anion radicals (O_2^-), hydrogen peroxide (H_2O_2), hydroxyl radicals (OH), singlet oxygen (O_{12}), and alkoxy radicals (RO). ROS react with proteins, lipids, and deoxyribonucleic acid causing oxidative damage and impairing the normal functions of plant cell. In order to overcome these effects, plants develop antioxidant defense systems comprising both enzymatic and nonenzymatic components that serve to prevent ROS accumulation and alleviate the oxidative damage occurring during drought stress (Miller et al. 2010). Antioxidant enzymes include superoxide dismutase (SOD), catalase (CAT), ascorbate peroxidase (APX), and glutathione reductase (GR) and nonenzymatic component constituents hold cysteine, glutathione, and ascorbic acid maintains antioxidant defense system in the plant cell (Gong et al. 2005; Kaushal and Wani 2016). However, plants inoculated with *P. formosus* had significantly ($p < 0.05$) increased levels of reduced glutathione (GSH) contents.

11.5.7 Osmotic Adjustments

Osmotic stress induced by drought can hinder the growth and yield of plants. Osmotic adjustment is the drought-tolerant mechanism found in the cell, which helps plants to cope with drought stress. Such adaptation strategies in plants to drought stress are coupled with metabolic adjustments which lead to the accumulation of several compatible solute/osmolytes like proline, sugars, polyamines, betaines, quaternary ammonium compounds, polyhydric alcohols, and other amino acids and water stress proteins like dehydrins (Yancey et al. 1982; Close 1996).

Chen et al. (2007) correlated proline accumulation with drought and salt tolerance in plants. Introduction of proBA genes derived from *Bacillus subtilis* into *A. thaliana* resulted in the production of higher levels of free proline resulting in increased tolerance to osmotic stress in the transgenic plants. Increased levels of proline along with decreased electrolyte leakage, maintenance of relative water content of leaves and selective uptake of Kions resulted in salt tolerance in *Zea mays* co-inoculated with *Rhizobium* and *Pseudomonas* (Bano and Fatima 2009). Proline protects membranes and proteins against the adverse effects of high concentration of inorganic ions and temperature extremes. It also functions as a protein-compatible hydrotope and as a hydroxyl radical scavenger (Smirnoff and Cumbs 1989). Accumulation of proline buffers cellular redox potential under environmental stresses (Jain et al. 2001; Wahid and Close 2007).

11.5.8 Exopolysaccharides Secretion

Endophytes are known to produce exopolysaccharide (EPS) which protect plants from stress and enhance their survival. Meneses et al. (2011) reported that EPS-producing strain *Gluconacetobacter diazotrophicus* (PAL5) consist a gum D gene that involved in EPS biosynthesis, which required for biofilm formation and plant colonization. Grover et al. (2011) showed that EPS can help in binding ions to cations including Na, thus making it unavailable to plants under stress. It is noteworth to mention that EPS also helps in binding soil particles to form microaggregate and macroaggregates which facilitate plant roots and fungal hyphae fit in the micro- and macropores and thus stabilize macroaggregates and also facilitate in making connection and colonizing the plant roots colonize the roots (Bashan et al. 2004). For example, endophytic fungi *Fusarium solani* and *Aspergillus* sp. are known as a backbone of the EPS produced by these species (Mahapatra and Banerjee 2013; Chen et al. 2011). Therefore, inoculating plants with EPS-producing bacteria enhanced resistance to water stress by improving the soil structure (Sandhya et al. 2009).

11.5.9 Aquaporins

Aquaporins are also known as water channels and are integral membrane proteins from a larger family of major intrinsic proteins present in all living organisms mainly in cell membranes of microbes which mainly help in facilitating transport of water between cells (Agre 2006). Cooper (2009) reported that through aquaporins the water could flow more rapidly into and out of the cell than by diffusing through the phospholipid bilayer. Ruth et al. (2011) showed that AM fungi facilitated transfer of water from soil to root and regulate root hydraulic nature via an aquaporin, and during drought AM also helped the regulation of plant aquaporin gene which regulates plant water status (Aroca et al. 2007, 2013).

11.6 Conclusion and Future Perspective

Globally, drought stress has become a major challenge for sustaining the productivity of crops including soybean. The conventional approaches involving breeding for developing drought resistance/tolerant varieties though is a viable solution but is time taking hence a long-term approach. On the other hand, modern agronomic practices, e.g., anti-transparent, growth hormones, soil conditioners, organic compounds, etc., are beyond affordable to farmers and also pose adverse effects to soil environments and hence are discouraged. Therefore, the adoption of biological interventions involving plant growth-promoting microbes in the alleviation of stress and sustaining the growth of plants is a short-term and eco-friendly solution. For the past 20–30 years, a lot of emphasis was given on PGPR research where plant growth-promoting microbes (PGPM) (mainly inhabiting in the soil rhizosphere) were used in soybean. Although nitrogen-fixing rhizobia were also used, those were mainly targeted for nitrogen management.

To harness and further improve the potential of microbes, recent impetus is to be given on exploitation of microbial endophytes in alleviation of stress and confer tolerance to plants, which opens a new dimension of PGPM research. The composition of the endosphere microbial populations depends mostly on plant and bacteria genotype and biotic and abiotic environmental factors. Endophytic species have been mostly reported throughout *Alphaproteobacteria*, *Betaproteobacteria*, and *Gammaproteobacteria* subgroups, and the latter is the most diverse and dominant group. The genera of *Bacillus* and *Pseudomonas* are identified as frequently occurring in agricultural crops including soybean. However, according to Glick group, the frequent occurrence of *Fusarium* sp. in roots of soybean has made us a little worrisome; therefore, further research evaluation is needed. In general isolating potential bacterial or fungal endophytes has an edge over the existing soil rhizospheric PGPMs due to having their specialized lifestyle, niche, and survival, and (once they are) inside the plant, there is a possibility that they will survive pretty longer even at high temperature and confer tolerance to plants. Nevertheless, the need of recovering potential candidates bearing genes (dispersed throughout the chromosome) conferring tolerance through higher accumulation of ACC deaminase, IAA, amino acids (e.g., proline) and carbohydrates (e.g., trehalose) etc., will stay. The potential candidates need to be thoroughly characterized biochemically under in vitro (based on stress physiological parameters), eventually to evaluate and release as potential endophytes for field application under drought stress conditions. Besides above, the mechanisms elicited by endophyte such as triggering osmotic response, phytohormone regulation, biocontrol activity, and induction of novel genes which play a vital role are also required to be deciphered for ensuring plant survival under drought stress. Taking the present leads accessible, concentrated future research is desired in terms of identification and characterization of the right kind of microbes, and addressing the issue of delivery systems, quality, and field evaluation of potential organisms is a need of an hour required to sustain the productivity of crops.

Acknowledgements Authors are grateful to Director, ICAR–Indian Institute of Soybean Research, Indore, for providing necessary infrastructure, facilities, and funding from ICAR–extramural network subproject to MPS and fellowship to DC; senior author is gratefully acknowledged.

References

Agre P (2006) The aquaporin water channels. Proc Am Thorac Soc 3(1):5–13

Allen MF (1982) Influence of vesicular-arbuscular mycorrhizae on water movement through *Bouteloua gracilis* (H.B.K.) Lag ex Steud. New Phytol 91:191–196

Allen MF, Boosalis MG (1983) Effects of two species of VA mycorrhizal fungi on drought tolerance of winter wheat. New Phytol 93:67–76

Allen MF, Smith WK, Moore TS Jr, Christensen M (1981) Comparative water relations and photosynthesis of mycorrhizal and non-mycorrhizal *Bouteloua gracilis*. New Phytol 88:683–693

Anjum SA, Xie X, Wang L, Saleem MF, Man C, Lei W (2011) Morphological, physiological and biochemical responses of plants to drought stress. Afr J Agric Res 6:2026–2032

Annapurna K, Ramadoss D, Bose P, Vithal Kumar L (2013) *In situ* localization of *Paenibacillus polymyxa* HKA-15 in roots and root nodules of soybean (*Glycine max* L.) Plant Soil 373:641–648. https://doi.org/10.1007/s11104-013-1825-7

Arnold AE, Mejía LC, Kyllo D, Rojas EI, Maynard Z, Robbins N, Herre EA (2003) Fungal endophytes limit pathogen damage in a tropical tree. Proc Natl Acad Sci USA 100:5649–15654

Aroca R, Porcel R, Ruiz-Lozano JM (2007) How does arbuscular mycorrhizal symbiosis regulate root hydraulic properties and plasma membrane aquaporins in *Phaseolus vulgaris* under drought, cold or salinity stresses? New Phytol 173:808–816

Aroca R, Ruiz-Lozano JM, Zamarreno AM, Paz JA, Garcia-Mina JM, Pozo MJ, Lopez-Raez JA (2013) Arbuscular mycorrhizal symbiosis influences strigolactone production under salinity and alleviates salt stress in lettuce plants. J Plant Physiol 170:47–55

Arshad M, Shaharoona B, Mahmood T (2008) Inoculation with *Pseudomonas* spp. containing ACC-deaminase partially eliminates the effects of drought stress on growth, yield, and ripening of pea (*Pisum sativum* L.) Pedosphere 18:611–620

Asaf S, Khan MA, Khan AL, Waqas M, Shahzad R, Kim AY, Kang SM, Lee IJ (2017) Bacterial endophytes from arid land plants regulate endogenous hormone content and promote growth in crop plants: an example of *Sphingomonas* sp. and *Serratia marcescens*. J Plant Interact 12:31–38

Bacon CW, White JF (2000) Microbial endophytes. Marcel Dekker, New York, p 199

Bai Y, D'Aoust F, Smith DL, Driscoll BT (2002) Isolation of plant growth promoting *Bacillus* strains from soybean root nodules. Can J Microbiol 48(3):230–238

Bano A, Fatima M (2009) Salt tolerance in *Zea mays* (L.) following inoculation with *Rhizobium* and *Pseudomonas*. Biol Fertil Soils 45:405–413

Bashan Y, Holguin G, de Bashan LE (2004) Azospirillum-plant relationships: physiological, molecular, agricultural, and environmental advances. Can J Microbiol 50:521–577

Bashan Y, de-Bashan LE, Prabhu SR, Hernandez JP (2014) Advances in plant growth-promoting bacterial inoculant technology: formulations and practical perspectives (1998–2013). Plant Soil 378:1–33

Belimov AA, Dodd IC, Hontzeas N, Theobald JC, Safronova VI, Davies WJ (2009) Rhizosphere bacteria containing 1-aminocyclopropane-1-carboxylate deaminase increase yield of plants grown in drying soil via both local and systemic hormone signaling. New Phytol 181:413–423

Berard A, Ben Sassi M, Kaisermann A, Renault P (2015) Soil microbial community responses to heat wave components: drought and high temperature. Climate Res 66:243–264

Bertalan M, Albano R, de Padua V, Rouws L, Rojas C, Hemerly A et al (2009) Complete genome sequence of the sugarcane nitrogen-fixing endophyte *Gluconacetobacter diazotrophicus* Pal5. BMC Genomics 10:450

Bilal S, Khan AL, Shahzad R, Asaf S, Kang SM, Lee IJ (2017) Endophytic *Paecilomyces formosus* LHL10 augments *Glycine max* L. Adaptation to Ni-contamination through affecting endogenous phytohormones and oxidative stress. Front Plant Sci 8:870. https://doi.org/10.3389/fpls.2017.00870

Brandl MT (2006) Fitness of human enteric pathogens on plants and implications for food safety. Annu Rev Phytopathol 44:367–392. https://doi.org/10.1146/annurev.phyto.44.070505.143359

Chaves M, Maroco J, Pereira J (2003) Understanding plant responses to drought from genes to whole plant. Funct Plant Biol 30:239–264

Chen M, Wei H, Cao J, Liu R, Wang Y, Zheng C (2007) Expression of *Bacillus subtilis* proAB genes and reduction of feedback inhibition of proline synthesis increases proline production and confers osmotolerance in transgenic *Arabidopsis*. J Biochem Mol Biol 40:396–403

Chen Y, Mao W, Tao H, Zhu W, Qi X, Chen Y, Li H, Zhao C, Yang Y, Hou Y, Wang C, Li N (2011) Structural characterization and antioxidant properties of an exopolysaccharide produced by the mangrove endophytic fungus *Aspergillus* sp. Y16. Bioresour Technol 102(17):8179–8184

Close TJ (1996) Dehydrins emergence of a biochemical role of a family of plant dehydration proteins. Physiol Plant 97:795–803

Cohen AC, Travaglia CN, Bottini R, Piccoli PN (2009) Participation of abscisic acid and gibberellins produced by endophytic *Azospirillum* in the alleviation of drought effects in maize. Botany 87:455–462. https://doi.org/10.1139/B09-023

Cohen AC, Bottini R, Pontin M, Berli FJ, Moreno D, Boccanlandro H, Travaglia CN, Piccoli PN (2015) *Azospirillum brasilense* ameliorates the response of *Arabidopsis thaliana* to drought mainly via enhancement of ABA levels. Physiol Plant 153:79–90

Contesto C, Milesi S, Mantelin S, Zancarini A, Desbrosses G, Varoquaux F, Bellini C, Kowalczyk M, Touraine B (2010) The auxin-signaling pathway is required for the lateral root response of arabidopsis to the rhizobacterium *Phyllobacterium brassicacearum*. Planta 232:1455–1470

Cooper G (2009) The cell: a molecular approach. ASM Press, Washington, DC, p 544. ISBN: 978-0-87893-300-6

Costa JM, Loper JE (1994) Characterization of siderophore production by the biological-control agent *Enterobacter cloacae*. Mol Plant Microbe Interact 7:440–448

Dalal JM, Kulkarni NS (2012) Isolation and identification of endophytic microorganisms of soybean (*Glycine max* (L.) Merril). Biomed Pharmacol J 5:2

Dalal JM, Kulkarni NS (2014) Population variance and diversity of endophytic fungi in soybean (*Glycine max* (L) Merril). Res Rev J Bot Sci 3(4):33–39

Daszkowska-Golec A (2016) The role of abscisic acid in drought stress: how aba helps plants to cope with drought stress. In: Hossain MA, Wani SH, Bhattachajee S, Burritt DJ, Tran LS (eds) Drought stress tolerance in plants, vol 2. Springer International Publishing, Cham, pp 123–151. https://doi.org/10.1007/978-3-319-32423-4_5

De Souza Leite T, Cnossen-Fassoni A, Pereira OL, Mizubuti ES, de Araújo EF, de Queiroz MV (2013) Novel and highly diverse fungal endophytes in soybean revealed by the consortium of two different techniques. J Microbiol 51(1):56–69

Dodd IC, Zinovkina NY, Safronova VI, Belimov AA (2010) Rhizobacterial mediation of plant hormone status. Ann Appl Biol 157:361–379

Duffy BK, Defago G (1999) Environmental factors modulating antibiotic and siderophore biosynthesis by *Pseudomonas fluorescens* biocontrol strains. Appl Environ Microbiol 65:2429–2438

Egamberdieva D, Wirth S, Behrendt U, Abd-Allah EF, Berg G (2016) Biochar treatment resulted in a combined effect on soybean growth promotion and a shift in plant growth promoting rhizobacteria. Front Microbiol 7:209

Elbeltagy A, Nishioka K, Suzuki H, Sato T, Sato YI, Morisaki H, Mitsui H, Minamisawa K (2000) Isolation and characterization of endophytic bacteria from wild and traditionally cultivated rice varieties. Soil Sci Plant Nutr 46:617–629

Evelin H, Giri B, Kapoor R (2012) Contribution of Glomus intraradices inoculation to nutrient acquisition and mitigation of ionic imbalance in NaCl –stressed *Trigonella foenum* – graecum. Mycorrhiza 22:203–217

Farooq M, Wahid A, Kobayashi N, Fujita D, Basra SMA (2009) Plant drought stress: effects, mechanisms and management. Agron Sustain Dev 29:185–212

Fatima Z, Zia M, Chauhary MF (2006) Effect of *Rhizobium* and phosphorus on growth of soybean (*Glycine max*) and survival of *Rhizobium* and P solubilizing bacteria. Pak J Bul 38(2):459–464

Fernandes EG, Pereira OL, da Silva CC, Bento CB, de Queiroz MV (2015) Diversity of endophytic fungi in *Glycine max*. Microbiol Res 181:84–92. https://doi.org/10.1016/j.micres.2015.05.010
Figueiredo MVB, Burity HA, Martinez CR, Chanway CP (2008) Alleviation of drought stress in the common bean (*Phaseolus vulgaris* L.) by co-inoculation with *Paenibacillus polymyxa* and *Rhizobium tropici*. Appl Soil Ecol 40:182–188
Fouts DE, Tyler HL, DeBoy RT, Daugherty S, Ren Q, Badger JH, Durkin AS, Huot H, Shrivastava S, Kothari S, Dodson RJ, Mohamoud Y, Khouri H, Roesch LF, Krogfelt KA, Struve C, Triplett EW, Methe BA (2008) Complete genome sequence of the N_2 fixing broad host range endophyte *Klebsiella pneumoniae* 342 and virulence predictions verified in mice. PLoS Genet 4:e1000141
Frederick JR, Camp CR, Bauer PJ (2001) Drought-stress effects on branch and main stem seed yield and yield components of determinate soybean. Crop Sci 41:759–763
Gaiero JR, McCall CA, Thompson KA, Day NJ, Best AS, Dunfield KE (2013) Inside the root microbiome: bacterial root endophytes and plant growth promotion. Am J Bot 100:1738–1750
Gasser I, Cardinale M, Muller H, Heller S, Eberl L, Lindenkamp N, Kaddor C, Steinbuchel A, Berg G (2011) Analysis of the endophytic lifestyle and plant growth promotion of *Burkholderia terricola* ZR2-12. Plant Soil 347:125–136
Germida JJ, Siciliano SD, Freitas JR, Seib AM (1998) Diversity of root associated bacteria associated with field grown canola (*Brassica napus* L.) and wheat (*Triticum aestivum* L.) FEMS Microbiol Ecol 26:43–50
Gill SS, Gill R, Trivedi KD, Anjum AN, Sharma KK, Ansari MW, Ansari AA, Johri KA, Prasad R, Pereira E, Varma A, Tuteja N (2016) *Piriformospora indica* potential and significance in plant stress tolerance. Front Microbiol 7:332
Glick BR (1995) The enhancement of plant growth by free living bacteria. Can J Microbiol 41:109–117
Glick BR (2005) Modulation of plant ethylene levels by the bacterial enzyme ACC deaminase. FEMS Microbiol Lett 251:1–7
Glick BR, Penrose DM, Li J (1998) A model for lowering of plant ethylene concentrations by plant growth promoting bacteria. J Theoret Biol 190:62–68
Gong H, Zhu X, Chen K, Wang S, Zhang C (2005) Silicon alleviates oxidative damage of wheat plants in pots under drought. Plant Sci 169:313–321
Grover M, Ali SZ, Sandhya V, Rasul A, Venkateswarlu B (2011) Role of microorganisms in adaptation of agriculture crops to abiotic stresses. World J Microbiol Biotechnol 27:1231–1240
Grumberg BC, Urcelay C, Shroeder MA, Vargas-Gil S, Luna CM (2015) The role of inoculum identity in drought stress mitigation by arbuscular mycorrhizal fungi in soybean. Biol Fertil Soils 51:1–10
Hallmann J, Quadt-Hallmann A, Mahaffee WF, Kloepper JW (1997) Bacterial endophytes in agricultural crops. Can J Microbiol 43:895–914
Hamayun M, Khan SA, Ahmad N, Tang DS, Kang SM, Na CI, Sohn EY, Hwang YH Shin DH, Lee BH, Kim JG, Lee IJ (2009) *Cladosporium sphaerospermum* as a new plant growth promoting endophyte from the roots of *Glycine max* (L.) Merr. World J Microbiol Biotechnol 25:627–632
Hardoim PR, van Overbeek LS, van Elsas JD (2008) Properties of bacterial endophytes and their proposed role in plant growth. Trends Microbiol 16:463–471
Hendry GA (2005) Oxygen free radical process and seed longevity. Seed Sci J 3:141–147
Hung PQ, Annapurna K (2004) Isolation and characterization of endophytic bacteria in soybean (*Glycine* sp.) Omonrice 12:92–101
Impullitti AE, Malvick DK (2013) Fungal endophyte diversity in soybean. J Appl Microbiol 114:1500–1506
Jain M, Mathur G, Koul S, Sarin NB (2001) Ameliorative effects of proline on salt stress induced lipid peroxidation in cell lines of groundnuts (*Arachis hypogaea* L.) Plant Cell Rep 20:463–468
Ji SH, Gururani MA, Chun SC (2014) Isolation and characterization of plant growth promoting endophytic diazotrophic bacteria from Korean rice cultivars. Microbiol Res 169:83–98. https://doi.org/10.1016/j.micres.2013.06.003

Joshi OP, Bhatia VS (2003) Stress management in soybean. In: Singh H, Hegde DM (eds) Souvenir. National seminar on stress management in oilseeds for attaining self-reliance in vegetable oils. Indian Society of Oilseeds Research, Hyderabad, pp 13–25

Kaneko T, Minamisawa K, Isawa T, Nakatsukasa H, Mitsui H, Kawaharada Y, Nakamura Y, Watanabe A, Kawashima K, Ono A, Shimizu Y, Takahashi C, Minami C, Fujishiro T, Kohara M, Katoh M, Nakazaki N, Nakayama S, Yamada M, Tabata S, Sato S (2010) Complete genomic structure of the cultivated rice endophyte *Azospirillum* sp. B510. DNA Res 17:37–50

Kang SM, Radhakrishnan R, Khan AL, Kim MJ, Park JM, Kim BR, Shin DH, Lee IJ (2014) Gibberellin secreting rhizobacterium, *Pseudomonas putida* H-2-3 modulates the hormonal and stress physiology of soybean to improve the plant growth under saline and drought conditions. Plant Physiol Biochem 84:115–124

Kaushal M, Wani SP (2016) Plant-growth-promoting rhizobacteria: drought stress alleviators to ameliorate crop production in drylands. Ann Microbiol 66:35–45

Khan AL, Hussain J, Al-Harrasi A, Al-Rawahi A, Lee IJ (2013) Endophytic fungi: a source of gibberellins and crop resistance to abiotic stress. Crit Rev Biotechnol 35:62–74

Kohler J, Caravaca F, Alguacil MM, Roldán A (2009) Elevated CO_2 increases the effect of an arbuscular mycorrhizal fungus and a plant growth promoting rhizobacterium on structural stability of a semiarid agricultural soil under drought conditions. Soil Biol Biochem 41:1710–1716

Krause A, Ramakumar A, Bartels D, Battistoni F, Bekel T, Boch J, Bohm M, Friedrich F, Hurek T, Krause L, Linke B, McHardy AC, Sarkar A, Schneiker S, Syed AA, Thauer R, Vorholter FJ, Weidner S, Puhler A, Reinhold-Hurek B, Kaiser O, Goesmann A (2006) Complete genome of the mutualistic, N_2 fixing grass endophyte *Azoarcus* sp. strain BH72. Nat Biotechnol 24:1385–1391

Kuklinsky Sobral J, Araujo WL, Mendes R, Geraldi IO, Pizzirani Kleiner AA, Azevedo JL (2004) Isolation and characterization of soybean associated bacteria and their potential for plant growth promotion. Environ Microbiol 6:1244–1251

Kwak MJ, Song JY, Kim SY, Jeong H, Kang SG, Kim BK, Kwon SK, Lee CH, Yu DS, Park SH, Kim JF (2012) Complete genome sequence of the endophytic bacterium *Burkholderia* sp. strain KJ006. J Bacteriol 194:4432–4433

Levy I, Krikun J (1980) Effect of vesicular-arbuscular mycorrhiza in Citrus jambhiri water relations. New Phytol 85:25–32

Li J, Wang E, Chen W, Chen W (2008) Genetic diversity and potential for promotion of plant growth detected in nodule endophytic bacteria of soybean grown in Heilongjiang province of China. Soil Biol Biochem 40:238–246

Li T, Hu YJ, Hao ZP, Li H, Chen BD (2013) Aquaporin genes GintAQPF1 and GintAQPF2 from glomus intraradices contribute to plant drought tolerance. Plant Signal Behav 8:e24030. https://doi.org/10.4161/psb.24030

Lobell DB, Schlenker W, Costa-Roberts J (2011) Climate trends and global crop production since 1980. Science 333:616–620

Loon LC, Bakker PAHM, Pieterse CMJ (1998) Systemic resistance induced by rhizosphere bacteria. Annu Rev Phytopathol 36:453–483

Madhaiyan M, Poonguzhali S, Ryu J, Sa T (2006) Regulation of ethylene levels in canola (*Brassica campestris*) by 1-aminocyclopropane- 1-carboxylate deaminase-containing *Methylobacterium fujisawaense*. Planta 224:268–278

Mahapatra S, Banerjee D (2013) Evaluation of in vitro antioxidant potency of exopolysaccharide from endophytic *Fusarium solani* SD5. Int J Biol Macromol 53:62–66

Malfanova N, Kamilova F, Validov S, Shcherbakov A, Chebotar V, Tikhonovich I, Lugtenberg B (2011) Characterization of *Bacillus subtilis* HC8, a novel plant beneficial endophytic strain from giant hogweed. J Microb Biotechnol 4:523–532

Marquez LM, Redman RS, Rodriguez RJ, Roossinl MJ (2007) A virus in a fungus in a plant: three-way symbiosis required for thermal tolerance. Science 315:513–515

Marquez-Santacruz HA, Hernandez-Leon R, Orozco-Mosqueda MC, Velazquez-Sepulveda I, Santoyo G (2010) Diversity of bacterial endophytes in roots of Mexican husk tomato plants (*Physalis ixocarpa*) and their detection in the rhizosphere. Genet Mol Res 9:2372–2380

Mathimaran N, Sharma MP, Mohan Raju B, Bagyaraj DJ (2017) Mycosphere Essay 17 Arbuscular mycorrhizal symbiosis and drought tolerance in crop plants. Mycosphere 8(3):361–376

Meneses CHSG, Rouws LFM, Simoes-Araújo V, idal MS, Baldani JI (2011) Exopolysaccharide production is required for biofilm formation and plant colonization by the nitrogen-fixing endophyte *Gluconacetobacter diazotrophicus*. Mol Plant Microbe Interact 24:1448–1458

Miller WA, Roy KW (1982) Mycoflora of soybean leaves, pods, and seeds. Can J Bot 60:2716–2723

Miller G, Susuki N, Ciftci-Yilmaz S, Mittler R (2010) Reactive oxygen species homeostasis and signalling during drought and salinity stresses. Plant Cell Environ 33:453–467

Nair A, Abraham TK, Jaya DS (2008) Studies on the changes in lipid peroxidation and antioxidants in drought stress induced Cowpea (*Vigna unguiculata* L.) varieties. J Environ Biol 29:689–691

Neilands JB (1995) Siderophores: structure and function of microbial iron transport compounds. J Biol Chem 270:26723–26726

Nhu VTP, Diep CN (2017) Isolation and characterization of endophytic bacteria in soybean (*Glycine max* L. (Merrill)) cultivated on alluvial soil of Can Tho city, Vietnam. Int J Innov Eng Technol 8:202–221. https://doi.org/10.21172/ijiet.83.028

Onofre-Lemus J, Hernandez-Lucas I, Girard L, Caballero-Mellado J (2009) ACC (1-aminocyclopropane-1-carboxylate) deaminase activity, a widespread trait in *Burkholderia* species, and its growth-promoting effect on tomato plants. Appl Environ Microbiol 75:6581–6590

Pimentel IC, Glienke-Blanco C, Gabardo J, Stuart MR, Azevedo JL (2006) Identification and colonization of endophytic fungi from soybean (*Glycine max* (L.) Merril) under different environmental conditions. Braz Arch Biol Technol 49:705–711

Porcel R, Zamarreno AM, Garcia Mina JM, Aroca R (2014) Involvement of plant endogenous ABA in *Bacillus megaterium* PGPR activity in tomato plants. BMC Plant Biol 14:36

Potts M (1999) Mechanisms of desiccation tolerance in cyanobacteria. Eur J Phycol 34:319–328

Rao NSS (1982) Phosphate solubilization by soil microorganisms. In: Rao NSS (ed) Advances in agricultural microbiology. Oxford and IBH Publishing, New Delhi, pp 295–305

Rapparini F, Penuelas J (2014) Mycorrhizal fungi to alleviate drought stress on plant growth. In: Miransari M (ed) Use of microbes for the alleviation of soil stresses, vol 1. Springer, New York, pp 21–42

Reid CPP (1979) Mycorrhizae and water stress. In: Reidacher A, Gagnaire-Michard G (eds) Root physiology and symbiosis. IUFRO Proc, Nancy, pp 392–408

Romero FM, Marina M, Pieckenstain FL (2014) The communities of tomato (*Solanum lycopersicum* L.) leaf endophytic bacteria, analyzed by 16S-ribosomal RNA gene pyrosequencing. FEMS Microbiol Lett 351:187–194

Russo ML, Pelizza SA, Cabello MN, Stenglein SA, Vianna MF, Scorsetti AC (2016) Endophytic fungi from selected varieties of soybean (*Glycine max* L. Merr.) and corn (*Zea mays* L.) grown in an agricultural area of Argentina. Rev Argent Microbiol 48:154–160

Ruth B, Khalvati M, Schmidhalter U (2011) Quantification of mycorrhizal water uptake via high resolution on line water content sensors. Plant and Soil 342:459–468

Safir GR, Boyer JS, Gerdemann JW (1971) Mycorrhizal enhancement of water transports in soybean. Science 172:581–583

Sahay N, Varma A (1999) *Piriformospora indica*, a new biological hardening tool for micropropagated plants. FEMS Lett 181:297–302

Sandhya V, Ali SKZ, Grover M, Reddy G, Venkateswarlu B (2009) Alleviation of drought stress effects in sunflower seedlings by the exopolysaccharides producing *Pseudomonas putida* strain GAP-P45. Biol Fertil Soils 46:17–26

Schulz B, Boyle C, Draeger S, Römmert A-K, Krohn K (2002) Endophytic fungi: a source of novel biologically active secondary metabolites. Mycol Res 106:996–1004

Sediyama T, Teixeira RC, Barros HB (2009) Origem, evoluçao e importancia economica. In: Sediyama T (ed) Tecnologias de produçao e usos da soja. Mecenas, Londrina, pp 1–5

Senthilkumar M, Swarnalakshmi K, Govindasamy V, Lee YK, Annapurna K (2009) Biocontrol potential of soybean bacterial endophytes against charcoal rot fungus, *Rhizoctonia bataticola*. Curr Microbiol 58:288–293

Shahollari B, Varma A, Oelmuller R (2005) Expression of a receptor kinase in *Arabidopsis* roots is stimulated by the basidiomycete *Piriformospora indica* and the protein accumulates in Triton X-100 insoluble plasma membrane microdomains. J Plant Physiol 162:945–958

Sharma S, Kulkarni J, Jha B (2016) Halotolerant rhizobacteria promote growth and enhance salinity tolerance in peanut. Front Microbiol 7:1600. https://doi.org/10.3389/fmicb.2016.01600

Shi Y, Yang H, Zhang T, Sun J, Lou K (2014) Illumina-based analysis of endophytic bacterial diversity and space-time dynamics in sugar beet on the north slope of Tianshan mountain. Appl Microbiol Biotechnol 98:6375–6385

Sieverding E (1981) Influence of soil water regimes in VA mycorrhiza: I. Effect on plant growth, water utilization and development of mycorrhiza. Z Acker Pflanzenbau 150:400–411

Simpson D, Daft MJ (1990) Increasing between water stress and different mycorrhizal colonization inocula on plant growth and mycorrhizal development in maize and sorghum. Plant Soil 121:179–186

Sionet N, Kramer PJ (1977) Effect of water stress during different stages of growth of soybean. Agron J 69:274–278

Smirnoff N, Cumbs QJ (1989) Hydroxyl radical scavenging activity of compatible solutes. Phytochemistry 28:1057–1060

Snellgrove RC, Splittstoesser WE, Stribley DP, Tinker PB (1982) The distribution of carbon and the demand of the fungal symbionts in leek plants with vesicular arbuscular mycorrhizas. New Phytol 92:75–87

Stoltzfus JR, So R, Malarvithi PP, Ladha JK, de Bruijn FJ (1997) Isolation of endophytic bacteria from rice and assessment of their potential for supplying rice with biologically fixed nitrogen. Plant Soil 194:25–36

Suman A, Verma P, Yadav AN (2016) Endophytic microbes in crops: diversity and beneficial impact for sustainable agriculture. In: Singh DP, Singh HB, Prabha R (eds) Microbial inoculants in sustainable agricultural productivity. Springer, New York, pp 117–143

Sun Y, Cheng Z, Glick BR (2009) The presence of a1-aminocyclopropane-1-carboxylate (ACC) deaminase deletion mutation alters the physiology of the endophytic plant growth promoting bacterium *Burkholderia phytofirmans* PsJN. FEMS Microbiol Lett 296:131–136

Taghavi S, Garafola C, Monchy S, Newman L, Hoffman A, Weyens N, Barac T, Vangronsveld J, vander Lelie D (2009) Genome survey and characterization of endophytic bacteria exhibiting a beneficial effect on growth and development of poplar trees. Appl Environ Microbiol 75:748–757

Taiz L, Zeiger E (2010) Plant physiology, 5th edn. Sinecure Associates, Sunderland

Tenguria RK, Firodiya A (2013) Diversity of endophytic fungi in leaves of *Glycine max* (L.) merr. from central region of Madhya Pradesh. World J Pharm Pharm Sci 2(6):5928–5934

Torres MJ, Perez Brandan C, Petroselli G, Erra-Balsells R, Audisio MC (2016) Antagonistic effects of *Bacillus subtilis* subsp. subtilis and *B. amyloliquefaciens* against *Macrophomina phaseolina*: SEM study of fungal changes and UV-MALDI-TOF MS analysis of their bioactive compounds. Microbiol Res 182:31–39

Varma A, Bakshi M, Binggan L, Hartmann A, Oelmueller R (2012) *Piriformospora indica*: a novel plant growth-promoting mycorrhizal fungus. Agric Res 1:117–131

Vassileva M, Vassilev N, Azcon R (1998) Rock phosphate solubilization by *Aspergillus niger* on olive cake-based medium and its further application in a soil plant system. World J Microbiol Biotechnol 14:281–284

Wahid A, Close T (2007) Expression of dehydrins under heat stress and their relationship with water relations of sugarcane leaves. Biol Plant 51:104–109

Wahid A, Rasul E (2005) Photosynthesis in leaf, stem, flower and fruit. In: Pessarakli M (ed) Hand book of photosynthesis, vol 2E. CRC Press, Florida, pp 479–497

Wakelin SA, Warren RA, Harvey PR, Ryder MH (2004) Phosphate solubilization by *Penicillium* sp. closely associated with wheat roots. Biol Fertil Soils 40:36–43

Wang P, Yin L, Liang D, Li C, Ma F, Yue Z (2012) Delayed senescence of apple leaves by exogenous melatonin treatment: toward regulating the ascorbate–glutathione cycle. J Pineal Res 53:11–20

Weilharter A, Mitter B, Shin MV, Chain PS, Nowak J, Sessitsch A (2011) Complete genome sequence of the plant growth promoting endophyte *Burkholderia phytofirmans* strain PsJN. J Bacteriol 193:3383–3384

Wisniewski-Dye F, Borziak K, Khalsa-Moyers G, Alexandre G, Sukharnikov LO, Wuichet K, Hurst GB, McDonald WH, Robertson JS, Barbe V, Calteau A, Rouy Z, Mangenot S, Prigent-Combaret C, Normand P, Boyer M, Siguier P, Dessaux Y, Elmerich C, Condemine G, Krishnen G, Kennedy I, Paterson AH, Gonzalez V, Mavingui P, Zhulin IB (2011) *Azospirillum* genomes reveal transition of bacteria from aquatic to terrestrial environments. PLoS Genet 7 (12):e1002430

Yan Y, Yang J, Dou Y, Chen M, Ping S, Peng J, Lu W, Zhang W, Yao Z, Li H, Liu W, He S, Geng L, Zhang X, Yang F, Yu H, Zhan Y, Li D, Lin Z, Wang Y, Elmerich C, Lin M, Jin Q (2008) Nitrogen fixation island and rhizosphere competence traits in the genome of root associated *Pseudomonas stutzeri* A1501. Proc Natl Acad Sci 105:7564–7569

Yancey PH, Clark ME, Hand SC, Bowlus RD, Somero GN (1982) Living with water stress: evolution of osmolyte system. Science 217(4566):1214–1222

Yang JW, Yu SH, Ryu CM (2009) Priming of defense-related genes confers root colonizing bacilli-elicited induced systemic resistance in pepper. Plant Pathol J 25:303–440

Zachow C, Fatehi J, Cardinale M, Tilcher R, Berg G (2010) Strain-specific colonization pattern of Rhizoctonia antagonists in the root system of sugar beet. FEMS Microbiol Ecol 74:24–35

Zhao FL, Lai X (2017) Antagonistic endophytic bacteria associted with nodules of soybean (*Glycine max* L.) and plant growth-promoting properties. Braz J Microbiol. https://doi.org/10.1016/j.bjm.2017.06.007

Zhang YZ, Chen WF, Li M, Sui XH, Liu H-C, Zhang XX, Chen WX (2012) Bacillus endoradicis sp. nov., an endophytic bacterium isolated from soybean root. Int J Syst Evol Microbiol 62:359–336

Zhu JK (2002) Salt and drought stress signal transduction in plants. Annu Rev Plant Biol 53:247–273

Zhu X, Song F, Liu S (2011) Arbuscular mycorrhiza impacts on drought stress of maize plants by lipid peroxidation, proline content and activity of antioxidant system. J Food Agric Environ 9:583–587

Zinniel DK, Lambrecht P, Harris NB, Feng Z, Kuczmarski D, Higley P, Ishimaru CA, Arunakumari A, Barletta RG, Vidaver AK (2002) Isolation and characterization of endophytic colonizing bacteria from agronomic crops and prairie plants. Appl Environ Microbiol 68:2198–2208

Chapter 12
Perspectives of Rhizobacteria with ACC Deaminase Activity in Plant Growth Under Abiotic Stress

Richa Raghuwanshi and Jay Kishor Prasad

12.1 Introduction

Plant growth in its entire lifetime is challenged by many biotic and abiotic stresses. The abiotic stresses may be extremes of temperature, salt, high light, flooding, drought, presence of toxic metals and organic contaminants, radiation, and wounding, and the biotic stresses may include insect predation and attack by various pathogens like viruses, bacteria, and fungi (Abeles et al. 1992). Most of the adverse effects of stress on plant metabolism occurs in the form of osmotic stress or salt toxicity, ROS production, ethylene production, and nutrient imbalance, which overall affect the plant physiology and inhibit seedling growth, vigor, flowering, and fruit setting (Sairam and Tyagi 2004).

Soil which adheres the plant and provides water and nutrients is rich in microbial diversity. The soil microbial diversity includes the bacteria, actinomycetes, fungi, algae, and protozoa. A fertile soil per gram contains 9×10^7 bacteria, 4×10^6 actinomycetes, 2×10^5 fungi, 3×10^4 algae, 5×10^3 protozoa, and 3×10^1 nematodes (Alexander 1991). The rhizospheric bacterial count is 10–1000 times higher than the count in bulk soil as the root exudates contain carbohydrates (sugars and oligosaccharides), organic acids, vitamins, nucleotides, flavonoids, enzymes, hormones, and volatile compounds that diffuse into the rhizosphere and support microbial growth and activity. Plant growth-promoting rhizobacteria (PGPR) are a group of free-living saprophytic bacteria living in plant rhizosphere that aggressively colonize the root system and promote plant growth and act as biocontrol agents against plant diseases (Kloepper and Beauchamp 1992). Plant growth-promoting bacteria promote plant growth and development through many direct and indirect

R. Raghuwanshi (✉)
Department of Botany, Mahila Mahavidyalaya, Banaras Hindu University, Varanasi, India

J. K. Prasad
Department of Botany, Institute of Science, Banaras Hindu University, Varanasi, India

B. Giri et al. (eds.), *Root Biology*, Soil Biology 52,
https://doi.org/10.1007/978-3-319-75910-4_12

mechanisms. The indirect mechanisms include their ability to act as biocontrol agent on phytopathogens, and the direct include fixation of nitrogen, enhanced nutrient uptake through iron sequestration and phosphate solubilization, and production of hormones like indole-3-acetic acid (IAA) and cytokinin (Glick et al. 1999). Plant growth-promoting rhizobacteria elicit the so-called induced systemic tolerance (IST) in plants under different abiotic stresses by altering the plant metabolism. Production of IAA is a common growth-promoting trait observed in up to 80% of the soil bacteria and bacterial endophytes (Patten and Glick 1996). Besides the above benefits, PGPR also benefit the plants by lowering plant ethylene levels through the activity of 1-aminocyclopropane-1-carboxylate (ACC) deaminase enzyme present in them (Glick et al. 1998). The ethylene level gets elevated during stresses which may be abiotic like temperature, salt, high light, flooding, drought, toxic metals, organic contaminants, and radiation or biotic like attack of viruses, bacteria, and fungi (Abeles et al. 1992). Ethylene was originally regarded as a "stress hormone" due to its accelerated synthesis by plant in response to stress signals (Kende 1993; Johnson and Ecker 1998). Ethylene hormone induces physiological changes in plant growth like overcoming dormancy, differentiation, formation of adventitious roots, abscission of leaf and fruit, induction of flowering and femaleness in dioecious plants, senescence, and fruit ripening (Arshad and Frankenberger 2002; Owino et al. 2006). However, high level of ethylene leads to senescence and abnormal root growth. As ethylene production in plant roots gets accelerated under biotic and abiotic stress factors which have an inhibitory effect on root growth which in turn leads to abnormal plant growth, it becomes vital to regulate the ethylene production in the rhizosphere of plant to achieve normal growth and development. Bacterial strains with ACC deaminase activity are capable of overcoming the ethylene-induced negative responses in plants to a great extent. Bacterial ACC deaminase activity is a widespread character of the rhizospheric bacteria most commonly observed in bacteria residing in stressful conditions (Timmusk et al. 2011). ACC deaminase activity of bacteria endows plants with the capability to withstand the stress better and therefore survive in harsh environmental conditions. Inoculation with PGPR containing ACC deaminase activity has come up as an alternative sustainable approach in improving plant growth and development under stress conditions by reducing stress-induced ethylene production.

12.2 Mechanism of Action

Ethylene is an endogenously produced gaseous plant growth hormone by plants. Plants undergoing any stressed situation show an increased production of ethylene, and for this reason it is also known as a stress hormone. On the onset of stress, an initial small peak of ethylene of low magnitude for a few hours is observed, and then a second much larger peak of high magnitude for 1–3 days is observed (Stearns and Glick 2003; Pierik et al. 2006; Van Loon et al. 2006). The second peak initiates protective response in plants, like transcription of pathogenesis-related genes and acquired resistance (Ciardi et al. 2000; Van Loon and Glick 2004). The second ethylene peak is so large that processes such as senescence, chlorosis, and abscission are initiated, the overall effect of which is generally inhibitory to plant survival. Plant can be released from the inhibitory levels of ethylene by degrading its precursor like *S*-adenosylmethionine (derived from L-methionine) or ACC, which effectively reduces the ethylene levels. Among these, the enzymes ACC deaminase, *S*-adenosylmethionine (SAM) hydrolase, and SAM decarboxylase are much being worked on.

Normally the root exudates are estimated to contain 5–30% of the photosynthetically fixed carbon. The plant root exudates contain tryptophan which is taken by the rhizospheric bacteria and used as a precursor in the synthesis of indole-3-acetic acid (IAA), some of which is taken up by the plant. This bacterial IAA together with endogenously synthesized plant IAA stimulates cell proliferation and cell elongation. It also induces the transcription of ACC synthase which leads to the formation of 1-aminocyclopropane-1-carboxylic acid (ACC) which is the intermediate precursor of ethylene in higher plants (Yang and Hoffman 1984). Small amount of this ACC is exuded from the seeds or roots (Penrose et al. 2001; Grichko and Glick 2001a) which may be taken up by the rhizospheric bacteria. The enzyme present in rhizobacteria ACC deaminase hydrolyzes the ethylene precursor ACC into ammonia and α-ketobutyrate (Glick et al. 1994, 1998; Mayak et al. 1999; Shaharoona et al. 2006). This activity of the rhizobacteria decreases the amount of ACC, and in turn the ethylene level, in the spermosphere and rhizoplane and thereby eliminates its potential inhibitory effects on plants (Glick et al. 1998). Thus the rhizobacterium acts as a sink for the plant-synthesized ACC (Fig. 12.1). Plants inoculated with PGPR possessing ACC deaminase activity are relatively more tolerant to environmental stress (Naveed et al. 2014). When plants get exposed to stress conditions, the gaseous hormone ethylene endogenously regulates plant homeostasis, resulting in reduced root and shoot growth.

The increased ethylene levels in plant cause a feedback inhibition of the IAA signal transduction pathway and thereby limit the ACC synthase transcription (Burg and Burg 1966; Glick et al. 2007) by IAA. Bacteria capable of synthesizing IAA and possessing ACC deaminase activity are more beneficial in plant growth promotion as they do not allow the elevated levels of ethylene being formed in the plants. This prevents feedback inhibition of IAA signal transduction by ethylene, and the plant growth continues in response to the auxin.

12.3 Biochemistry of ACC Deaminase

ACC deaminase belongs to the tryptophan synthase beta superfamily of pyridoxal phosphate-binding proteins (Glick et al. 2007). ACC deaminase is a multimeric enzyme that is cytoplasmically localized. ACC deaminase subunit mass is approximately 35–42 kD, while its native size is estimated to be approximately 100–112 kD (Sheehy et al. 1991; Jacobson et al. 1994; Hontzeas et al. 2004a). The affinity of this

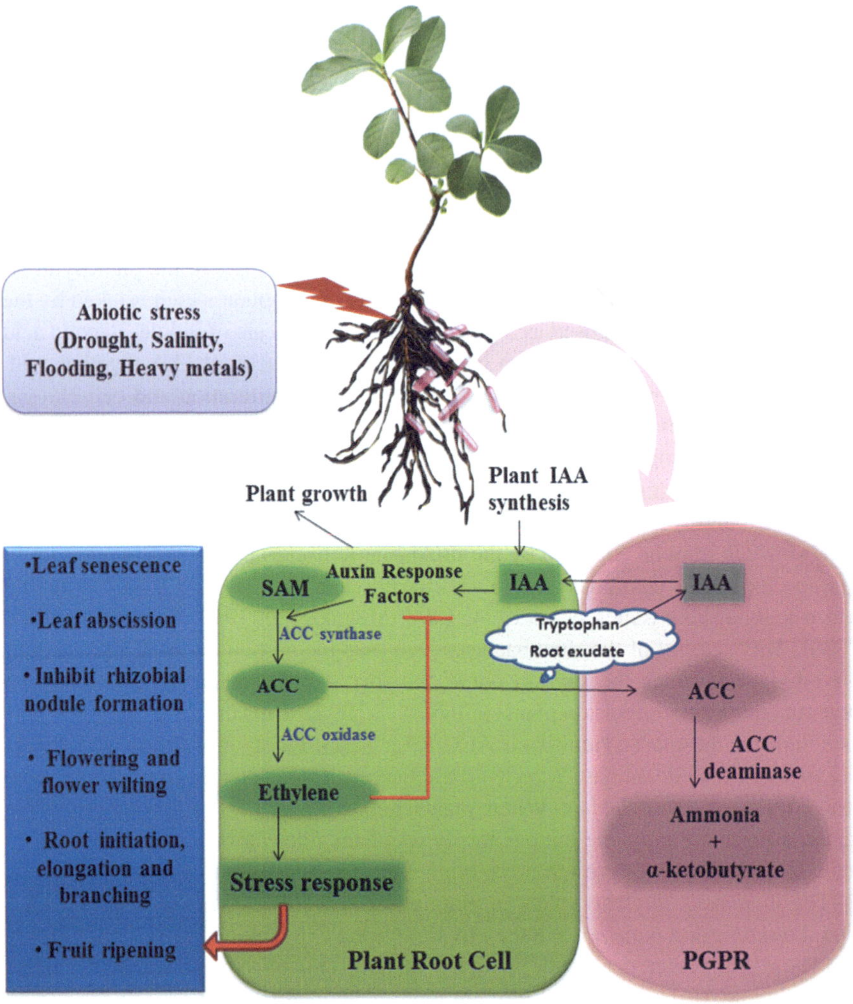

Fig 12.1 A schematic model of interactions between plant gene expression and ACC deaminase-producing bacteria during stress

enzyme for the substrate is not particularly high (Km = 1.5–6.0 mM). The coenzyme pyridoxal phosphate is a cofactor of ACC deaminase (Honma 1985), and even ACC synthase, the enzyme that catalyzes the formation of ACC, requires pyridoxal phosphate for its enzyme activity. ACC deaminase enzyme is present in bacteria at a very low amount, and at the same time, ACC oxidase has a much higher affinity for ACC compared to ACC deaminase (Glick et al. 1998). The ethylene levels in bacteria depend upon the ratio of the two, i.e., ACC oxidase and ACC deaminase (Glick et al. 1998). However, ACC deaminase synthesis is induced by ACC, at levels as low as 100 nM (Jacobson et al. 1994), with full induction requiring up to 10 h. The amino acids L-Ala, DL-Ala, and DL-Val can also induce enzyme activity to a small extent, and g-aminoisobutyric acid can induce activity to almost the same level as ACC (Honma 1983). Maximal enzyme activity typically occurs at 30 °C and pH 8.5. The affinity for the substrate ACC and the competitive inhibitors L-Ala and L-Ser is also highest at pH 8.5 (Hontzeas et al. 2006). *acdS* genes that have been traced in some stramenopiles, bacteria, and various fungi (*Ascomycota* and *Basidiomycota*) are believed to have a common ancestor (Nascimento et al. 2014). The genes are commonly transmitted vertically in various microorganisms, and occasional horizontal gene transfer is also observed including inter-kingdom transfer events. ACC deaminase genes (including both the structural gene acdS and the regulatory gene acdR) have been found in many different rhizobacteria (rhizospheric, endophytic, and rhizobia), including *Azospirillum* spp., *Rhizobium* spp., *Agrobacterium* spp., *Achromobacter* spp., *Burkholderia* spp., *Ralstonia* spp., *Pseudomonas* spp., and *Enterobacter* spp. (Blaha et al. 2006). More importantly, even if some strains of a particular genus and species have an acdS gene, not all strains do.

12.4 Role of ACC Deaminase Bacteria in Ameliorating Various Stress Responses in Plants

A common invariable observation in plants exposed to stress is an increased ethylene level, which leads to damage. Ethylene production upregulates in response to the presence of metals, organics, salt, temperature extremes, drought, ultraviolet light, damage by insects, nematode, and phytopathogens (Abeles et al. 1992). The second ethylene peak observed in plants exposed to stress is more detrimental to plant growth as this initiates processes such as senescence, chlorosis, and leaf abscission. Any treatment whether it is chemical or biological that lowers the magnitude of the second peak of ethylene reduces the damage caused to the plant as a stress consequence. Microorganisms exhibit wide range (>100-fold) in ACC deaminase activity, and it has been observed that organisms that express high ACC deaminase are beneficial as they are nonspecific toward their host (Glick 2005). This group encompasses most of the rhizospheric and phyllospheric microbes along with the endophytes. Such microbes reduce the ethylene by acting as a sink for ACC produced as a consequence of stress. Addition of an ACC deaminase producing PGPR and its negative mutant strain in canola roots (Hontzeas et al. 2004b) showed down regulation of genes involved in ethylene induced plant stress responses and up-regulation of genes involved in plant growth. The results supported that plant growth-promoting bacteria expressing ACC deaminase are able to overcome the stress response in plants. Different ACC deaminase-producing bacteria have been demonstrated for their efficacy in protecting plants against yield loss induced by various abiotic stresses as listed in Table 12.1.

Table 12.1 Reports on PGPR showing alleviation of abiotic stress impacts on plants mediated by ACC deaminase activity

S. no.	ACC deaminase-producing bacteria	Abiotic stress	Host plant	References
1	*Achromobacter piechaudii* ARV8	Drought, salt	*Lycopersicon esculentum*	Mayak et al. (2004a, b)
2	*Achromobacter xylosoxidans* (SF2), *Bacillus pumilus* (SF3 and SF4)	Drought	*Helianthus annuus*	Castillo et al. (2013)
3	*Azospirillum brasilense*	Drought	*Phaseolus vulgaris*	German et al. (2000)
4	*Azospirillum brasilense* Sp245	Drought	*Triticum aestivum*	Creus et al. (2004)
5	*Azospirillum lipoferum AZ1, A. lipoferum AZ9, A. lipoferum AZ45*	Drought	*Triticum aestivum*	Arzanesh et al. (2011)
6	*Bacillus cereus strain AR156, B. subtilis strain SM21, Serratia* sp. *Strain XY21*	Drought	*Cucumis sativus*	Wang et al. (2012)
7	*Bacillus subtilis*	Drought	*Trigonella foenum-graecum*	Barnawal et al. (2013)
8	*Bacillus subtilis* B26	Drought	*Phleum pratense*	Bourque et al. (2016)
9	*Bacillus licheniformis* K11	Drought	*Capsicum annuum*	Lim and Kim (2013)
10	*Burkholderia phytofirmans* PsJN, *Enterobacter* sp. FD17	Drought	*Zea mays*	Naveed et al. (2014)
11	*Paenibacillus polymyxa*	Drought	*Arabidopsis*	Timmusk and Wagner (1999)
12	*Proteus penneri strain (Pp1), Pseudomonas aeruginosa strain (Pa2), Alcaligenes faecalis strain (AF3)*	Drought	*Zea mays*	Naseem and Bano (2014)
13	*Pseudomonas aeruginosa* GGRJ21	Drought	*Vigna radiata*	Sarma and Saikia (2014)
14	*P. fluorescens* ACC-5	Drought	*Pisum sativum*	Zahir et al. (2008)
15	*Pseudomonas* sp.	Drought	*Pisum sativum*	Arshad et al. (2008)
16	*Pseudomonas syringae, Pseudomonas fluorescens*	Drought	*Zea mays*	Zafar-ul-Hye et al. (2014)
17	*Variovorax paradoxus* 5C-2	Drought	*Pisum sativum*	Jiang et al. (2012)
18	*Arthrobacter protophormiae*	Salt	*Pisum sativum*	Barnawal et al. (2014)
19	*Brachybacterium paraconglomeratum* SMR20	Salt	*Chlorophytum*	Barnawal et al. (2016)
20	*Enterobacter hormaechei*	Salt	*Lycopersicon esculentum*	Egamberdieva et al. (2014)
21	*P. fluorescens* YsS6	Salt	*Lycopersicon esculentum*	Ali et al. (2014)

(continued)

Table 12.1 (continued)

S. no.	ACC deaminase-producing bacteria	Abiotic stress	Host plant	References
22	*Pseudomonas putida*	Salt	*Vigna radiata*	Mayak et al. (1999)
23	*Pseudomonas* sp. ST3	Salt	*Vigna unguiculata*	Trung et al. (2016)
24	*P. putida* UW4	Salt	*Lycopersicon esculentum*	Yan et al. (2014)
25	*Pseudomonas putida* UW4	Salt	*Brassica napus*	Cheng et al. (2007)
26	*Pseudomonas fluorescens*	Salt	*Arachis hypogea*	Saravanakumar and Samiyappan (2007)
27	*Bacillus licheniformis HSW-16*	Salt	*Triticum aestivum*	Singh and Jha (2016)
28	*Pseudomonas putida* UW4	Flood	*Lycopersicon esculentum*	Grichko and Glick (2001a)
29	*P. fluorescens* REN_1	Flood	*Oryza sativa*	Etesami et al. (2014)
30	*Alcaligenes* sp., *Bacillus pumilus*	Heavy metals	*Brassica napus*	Belimov et al. (2001)
31	*E. cloacae* CAL2	Arsenate	*Brassica napus*	Nie et al. (2002)
32	*Enterobacter intermedius MH8b*	Zn toxicity	*Sinapis alba*	Płociniczak et al. (2013)
33	*Kluyvera ascorbata* SUD165	Nickel	*Brassica napus*	Burd et al. (1998)
34	*P. putida UW4, P. putida HS-2*	Nickel	*Brassica napus*	Farwell et al. (2007)
35	*K. ascorbata* SUD165/26	Lead	*Lycopersicon esculentum*	Burd et al. (2000)
36	*Sinorhizobium* sp. Pb002	Lead	*Brassica juncea*	Di Gregorio et al. (2006)
37	*Burkholderia* sp. J62	Lead	*Lycopersicon esculentum*	Jiang et al. (2008)
38	*Pseudomonas koreensis* AGB-1	Cd, AS, Cu, Pb and Zn toxicity	*Miscanthus sinensis*	Babu et al. (2015)
39	*Variovorax paradoxus*	Cadmium	*Brassica juncea*	Belimov et al. (2005)
40	*Burkholderia phytofirmans* PsJN	Low temperature	*Vitis vinifera*	Ait Bakra et al. (2006)
41	*P. putida*	Low temperature	*Lycopersicon esculentum*	Cheng et al. (2007)
42	*Enterobacter aerogenes* NBRIK24	Fly-ash soil	*Brassica juncea*	Kumar et al. (2008)

12.4.1 Salt Stress

Salinity, which was a natural feature of ecosystems in arid and semiarid regions, has now become a major constrain due to the anthropogenic activities, primarily due to irrigation of agricultural fields (Abrol et al. 1988). Of the total global cultivable area, 20% is under salinity stress, and this is continuously increasing as a direct consequence of irrigation (Flowers 2004). Around 800 million hectares of land is estimated to be affected by salinity throughout the world (FAO 2008). Salinity stress creates an oxidative burst in cells resulting in an increased accumulation of reactive oxygen species (ROS) which affects the plasma membrane, cell metabolism, and homeostasis. Salt stress imbalances the ethylene production and causes its overproduction which accelerates leaf and petal abscission and organ senescence, leading to premature death (Cheng et al. 2007; Mayak et al. 2004a, b; Zahir et al. 2009).

Reducing the ethylene level, one can alleviate some of the effects of stresses on plants (Glick 2004). Plant losses approximately 40% of photosynthates, through root exudates (Lynch and Whipps 1991), and it has been estimated that during stress much of the released carbon is in the form of ACC, which is a precursor of ethylene, and is exuded from plant roots (Bayliss et al. 1997). Thus, PGPR, with ACC deaminase activity, can be used to convert ACC to ammonia and α-ketobutyrate that are used up by the plant as a nitrogen source simultaneously reducing the negative effects of salinity stress (Cheng et al. 2007; Mayak et al. 2004a, b; Zahir et al. 2009). However, the efficiency of PGPR depends on environmental factors such as the climate, weather conditions, soil characteristics, and interaction with other indigenous microbial flora in the soil (Giongo et al. 2008; Sinha and Raghuwanshi 2015). Salt-tolerant ACC deaminase-producing bacteria can survive well in a saline environment, and their beneficial properties help plants to overcome stress effects (Mayak et al. 2004a, b). Halotolerant bacteria are a group of microorganisms able to grow in media containing a wide range of NaCl up to 1–33% or in the absence of NaCl (Larsen 1986). A significant decrease in the level of ethylene was observed in tomato plants exposed to high salt concentration on inoculation with *Achromobacter piechaudii* ARV8, an isolate obtained from the rhizosphere of *Lycium shawii* plant wildly growing in the Arava region of Israel (Mayak et al. 2004b). The inoculated tomato seedlings showed an increased fresh and dry weight, but this however did not reduce the content of sodium in the plant. Plants inoculated with *Achromobacter piechaudii* ARV8 had four times higher biomass compared to controls, as there was a significant reduction of the ethylene level (Mayak et al. 2004b). Similar effect of the bacterial strain, i.e., lowering the ethylene level, was observed in peppers and tomatoes growing under drought stress (Mayak et al. 2004a). Studies done on maize plant growing in saline–sodic soil when treated with fertilizer along with ACC deaminase-producing *Pseudomonas* strains showed 198% augmented plant dry weight (Zafar-ul-Hye et al. 2014). Studies done on wild-type bacterial endophytes showed protection against salt stress in plants by limiting the buildup of salt and thereby improving plant survival. Inoculating ACC deaminase bacteria do not alter the sodium level in plants, but the uptake of phosphorous and potassium gets slightly increased, which supports plant growth under salt stress. Similar reports by Saravanakumar and Samiyappan (2007) revealed *Pseudomonas fluorescens* strain TDK1 possessing ACC deaminase activity not only enhanced the resistance toward salinity in groundnut plants but also increased yield. Compared to mutant-inoculated or non-inoculated plants, the plants inoculated with ACC deaminase-producing strain show augmented level of chlorophyll content. High chlorophyll content has been linked with stress tolerance in many plants (Vurukonda et al. 2016). Encouraging results were also obtained by the tripartite interaction of *Arthrobacter protophormiae*, *Rhizobium leguminosarum*, and *Glomus mosseae* which increased plant weight by 53%, reduced proline content and lipid peroxidation, and increased pigment content under 200 mM salt condition (Barnawal et al. 2013). Inoculating plants with wild-type ACC deaminase-producing strain tend to prevent salt buildup in plant tissues; however, few contradictory results have also been observed where more salt was deposited per gram of dry biomass in the plants inoculated with the ACC deaminase-producing strains. Study done to evaluate the growth of canola in the presence of wild-type ACC deaminase-containing plant growth-promoting rhizospheric bacterium *P. putida* UW4 showed the accumulation of much higher concentrations of sodium in the shoots compared to the plants treated with ACC deaminase mutants (Cheng et al. 2007).

Pea crop has been reported to suffer approximately 50% yield loss at 100 mM NaCl (Subbarao and Johansen 1994). Ethylene formed in response to salt stress inhibits the development of rhizobial infection threads in *Pisum sativum* cv. Sparkle (Lee and LaRue 1992). Plants undergoing symbiotic association with microbes, like *Rhizobium*, and mycorrhizal fungi also show a slight increase in ethylene levels during the establishment period. As the nitrogen fixation is a high energy-demanding process, there are fair chances of ethylene production by the plant, which may lead to nodule senescence (Murset et al. 2012). During this period the ACC deaminase-producing bacteria residing in the rhizosphere help in establishment of symbiosis by locally lowering the ethylene levels.

Thus effects of soil salinity on crop productivity can be alleviated by bacterial inoculations having ACC deaminase activity. Microbe-assisted plant stress management has emerged as an important strategy, and their role in improving growth and productivity has been well established (Venkateswarlu et al. 2008; Yang et al. 2009).

12.4.2 Waterlogging Stress

Soil flooding or waterlogging causes major changes in the normal functioning of plant roots (Jackson and Drew 1984) as the gas diffusion rates get reduced in flooded soil (Jackson 1985), and at the same time, respiration by microorganisms and plant roots leads to a rapid buildup anaerobic conditions in the soil. Anaerobic conditions of the soil lead to toxicity primarily due to Fe^{2+}, Mn^{+} and sulfide and due to accumulated gases like carbon dioxide, methane, ethane, and ammonia (Ernst

1990). This also affects some of the vital processes like ion uptake in root (Jackson and Drew 1984). These stressful conditions trigger the synthesis of enzyme ACC synthase as well as other stress proteins in the plant which elevate the level of ACC in its roots (Li et al. 2012). This newly synthesized ACC cannot be converted to ethylene in the roots, as ethylene synthesis requires oxygen, so the ACC is transported to the shoots where under aerobic environment the ACC gets converted into ethylene (Bradford and Yang 1980; Else and Jackson 1988) causing epinasty (wilting), leaf chlorosis, necrosis, and stunted growth. Accumulation of ethylene in plants may also lead to adoptive response like shoot elongation (Voesenek and Blom 1989) and formation of aerenchyma (Armstrong et al. 1994), and a few studies also report the formation of adventitious roots (Drew 1992). Waterlogging induces several physiological alterations like reduced photosynthetic rate, stomatal closure, plant growth inhibition, and low yield. A sustainable solution to this problem comes from the ACC deaminase-producing plant growth-promoting bacteria (Barnawal et al. 2012; Grichko and Glick 2001a; Li et al. 2013) which mitigate the stress by lowering the ethylene level in plants and making them more capable to withstand flood (Saleem et al. 2007). Studies have shown that flooded plants inhabiting ACC deaminase-producing microbes are able to overcome the flood response partially. Even plants are genetically engineered to express this enzyme in root-specific manner resulting in less accumulation of ethylene in the roots and thereby minimizing the adverse effects of flooding (Grichko and Glick 2001a, b).

12.4.3 Metal and Organic Pollutants

Industrial revolution has accelerated the toxic metal accumulation rate in the biosphere and has come up as a serious current environmental problem. Metal in soil beyond a limit becomes toxic to plant growth as they interfere with normal growth and development. Soils dumped with heavy metals also cause a severe stress induction in plants that leads to the synthesis of stress ethylene up to an inhibitory level. Plants interact with these heavy metals present in the environment through phytostabilization, phytoextraction, and phytovolatilization (Pilon-Smits 2005). The easy and preferred way to get rid of the metals is phytoextraction. Cleaning heavy metal pollutants through plants, i.e., phytoremediation (Salt et al. 1995), is an eco-friendly and cost-effective approach compared to the traditional soil remediation approaches of metal removal through chemical and physical extraction. However, the limitation lies that not all plants are capable to naturally tolerate and accumulate heavy metals. Many plants effective in phytoremediation are small sized and slow in growth, which limit their practical use (Khan et al. 2000). An effective plant to remediate the soil must be tolerant to one or more pollutants, highly competitive, and fast growing and produce a high biomass. Healthy and robust plants are preferred as they have better ability to phytoremediate metal contaminants. The commonly used plants in heavy metal accumulation belong to the Brassicaceae family (Kumar et al. 1995). The literature is well documented with the role of metal-resistant ACC deaminase-producing bacteria in improving plant growth by decreasing the stress effects due to ethylene. The first report on role of ACC deaminase-containing bacterium in phytoremediation of nickel-contaminated soil indicated that toxicity of nickel toward canola plants was reduced in the presence of the bacteria (Burd et al. 1998). Bacteria increased the uptake of Cd in *Brassica napus* (Sheng and Xia 2006) and Ni in *Alyssum murale* (Abou-Shanab et al. 2006). Rhizobial microfloras are known to affect heavy metals mobility and availability to the plant through release of chelating agents, acidification, and redox changes (Abou-Shanab et al. 2003; Smith and Read 1997). The root-associated ACC deaminase-producing bacteria not only reduce the ethylene levels but also provide multifaceted benefits to the plant (Glick 1995; Glick et al. 1999). Bacteria produce indole-3-acetic acid, siderophores, and solubilize phosphate, which stimulate plant growth (Glick 1995; Chabot et al. 1996a, b; Rajkumar et al. 2006). Heavy metal-contaminated soil often become iron depleted, and this effect of heavy metals can be overcome by inoculating ACC deaminase and siderophore-producing bacteria (Burd et al. 1998, 2000; Reed and Glick 2005). Thus, besides reducing the ethylene level in plants, microbes also enhance the mobility and availability of minerals to the plants (Abou-Shanab et al. 2003; Idris et al. 2004) which improve plant growth.

Treatment of plants with ACC deaminase-producing plant growth-promoting bacteria not only relieves the plant with the growth inhibition effects of ethylene but also allows the plant to grow normal and restore the nutrient cycling (Huang et al. 2004, 2005; Reed and Glick 2005; Greenberg et al. 2006). Therefore, bacterial strains utilized in plant growth promotion under metal stress should be screened for their abilities to resist the targeted toxic metal, synthesize IAA to promote root growth (Patten and Glick 2002), secrete siderophore (Burd et al. 2000) that helps plants to acquire iron from the metal-contaminated soil, and possess ACC deaminase activity which can prevent the building up of inhibitory levels of ethylene in the plants (Glick et al. 1998).

Organic pollutants in the soil, if present above a permissible limit, inhibit plant growth and drag the plant toward senescence by accelerated ethylene production (Abeles et al. 1992). Phytoremediation has come up as a technology to clean soil contaminated with organic oil spills, polycyclic aromatic hydrocarbons (PAHs), and polycyclic biphenyls (PCB), and is being practiced at commercial scale. PGPR possessing ACC deaminase activity has multifold benefits in phytoremediation of organic-, metal-, and salt-contaminated soils. Reduction in stress ethylene partially alleviates the damage caused by the target contaminant (Mayak et al. 2004a, b). Therefore the growth of plants exposed to organic contaminants in the soil should be facilitated by the presence of ACC deaminase-containing plant growth-promoting bacteria. In fact, this strategy of bacterially assisted phytoremediation appears to be particularly effective for removal and/or degradation of organic contaminants from impacted soils. *Helianthus annuus* L. seedlings inoculated with *Achromobacter xylosoxidans* (SF2) and *Bacillus pumilus* (SF3 and SF4) bacterial strains increased the production of auxins, salicylic acid, abscisic acid, jasmonic acid, as well as the plant dry matter. High salicylic acid concentration in stressed seedlings played key role in abiotic stress tolerance (Castillo et al. 2013). Plant growth-promoting bacteria improve plant competitiveness and responses in a stressed ecosystem (Egamberdiyeva and Hoflich 2004).

12.4.4 Drought

Drought stress can adversely affect plant growth and yield and is one of the most fatal reasons for economic losses in agriculture and forestry. Its affects plant water relations at the cellular and whole plant levels, altering the plant physiology and leading to specific and nonspecific phenotype (Pereyra et al. 2009; Arzanesh et al. 2011). In order to combat drought, plants adopt altered gaseous exchange and water relation strategies (Sinha and Raghuwanshi 2016). Bacteria adhering to plant roots containing ACC deaminase enzyme hydrolyze ACC and use it as the source of carbon and nitrogen (Glick 2014), and the process continues until a dynamic equilibrium between the roots and rhizosphere bacteria is maintained and the modulated root system starts normal functioning under low water condition. It has been well documented that inoculation of plants with certain PGPR at seedling stage improves biomass production through their effects on root system, which enhance plant growth and yield (Prasad et al. 2017).

Different ACC deaminase-producing bacteria have been demonstrated for their efficacy in protecting plants against yield loss induced by drought stress (Table 12.1). Studies done by Mayak et al. (2004a, b) on ACC deaminase PGPR *Achromobacter piechaudii* ARV8 showed that during water stress although the bacterium did not influence the water content of plants, it improved the recovery of plants when watered. Exposure of *Bacillus subtilis*-inoculated plants to 8 weeks drought stress led to significant increase in shoot and root biomass by 26.6 and 63.8%, and the photosynthesis and stomatal conductance too got enhanced by 55.2% and 214.9%, respectively (Bourque et al. 2016). *Azospirillum brasilense* sp. 245 uninoculated seeds of *Triticum aestivum* when sown under drought conditions had a yield loss of 26.5% and got reduced to 14.1% on inoculation with *Azospirillum brasilense* sp. 245. Grain Mg and K diminished in nonirrigated, non-inoculated plots. Grains harvested from *Azospirillum*-inoculated plants had significantly higher Mg, K, and Ca than non-inoculated plants (Creus et al. 2004). Cucumber plants treated with a consortium of three plant growth-promoting rhizobacterial strains (*Bacillus cereus* AR156, *Bacillus subtilis* SM21, and *Serratia* sp. XY21) induced tolerance to drought stress. The treatment decreased the leaf monodehydroascorbate (MDA) content and increased the leaf proline content and the root recovery intension by 3.45-fold and 50%, respectively. It also maintained the leaf chlorophyll content in cucumber plants under drought stress (Wang et al. 2012). *Azospirillum lipoferum* strain B3 having phosphate-solubilizing and ACC deaminase activities, when inoculated in wheat under drought, produced the highest amounts of N and auxin and increased wheat yield up to 109% (Arzanesh et al. 2011). Many studies have proven the positive effects of ACC deaminase bacterial activity on plant biomass, leaf area, and transpiration ratio of plants under drought (Saleem et al. 2007). Inoculation with

ACC deaminase bacterial activity has restored nodulation in pea plants under drought which is comparable with the well-irrigated plants (Arshad et al. 2008).

12.5 Conclusion and Future Prospects

It is well proven that bacteria with ACC deaminase activity are potent in improving plant growth and productivity under varied abiotic stress. Knowing the potentially serious environmental health damage caused by the excessive use of chemicals and pesticides in the agricultural sector, we need a major paradigm shift in agricultural practices. As the cost of engineering and developing transgenic plants that are able to defend well the variety of pathogens and other abiotic stresses, it is rather economical to isolate and screen an efficient plant growth-promoting bacterium able to combat the adverse conditions. The major challenge in the large-scale application of these bacteria is their survival under varied geographical and harsh environmental conditions, but a potential solution to the problem can be the exploitation of a potent endophytic plant growth-promoting bacteria. Unrevealing the fundamental mechanism of action of these bacteria will facilitate the wider application of this technology and overcome the bottleneck.

References

Abeles FB, Morgan PW, Saltveit ME Jr (1992) Ethylene in plant biology, 2nd edn. Academic, New York

Abou-Shanab RA, Angle JS, Delorme TA, Chaney RL, van Berkum P, Moawad H, Ghanem K, Ghozlan HA (2003) Rhizobacterial effects on nickel extraction from soil and uptake by *Alyssum murale*. New Phytol 158:219–224

Abou-Shanab RA, Angle JS, Chaney RL (2006) Bacterial inoculants affecting nickel uptake by *Alyssum murale* from low, moderate and high Ni soils. Soil Biol Biochem 38:2882–2889

Abrol IP, Yadav JSP, Massoud FI (1988) Salt affected soils and their management. Food and Agriculture Organization (FAO), UN, Soils Bulletin, Rome, p 39

Ait Bakra E, Nowak J, Clément C (2006) Enhancement of chilling resistance of inoculated grapevine plantlets with a plant growth promoting rhizobacterium *Burkholderia phytofirmans* strain PsJN. Appl Environ Microbiol 72:7246–7252

Alexander M (1991) Introduction to soil microbiology, 2nd edn. Krieger, Malabar

Ali S, Charles TC, Glick BR (2014) Amelioration of high salinity stress damage by plant growth-promoting bacterial endophytes that contain ACC deaminase. Plant Physiol Biochem 80: 160–167

Armstrong W, Bundle R, Jackson MB (1994) Mechanisms of flood tolerance in plants. Ada Bot Neerl 43:307–358

Arshad M, Shaharoona B, Mahmood T (2008) Inoculation with *Pseudomonas* spp. containing ACC-deaminase partially eliminates the effects of drought stress on growth, yield, and ripening of pea (*Pisum sativum* L.) Pedosphere 18:611–620

Arzanesh MH, Alikhani HA, Khavazi K, Rahimian HA, Miransari M (2011) Wheat (*Triticum aestivum* L.) growth enhancement by *Azospirillum* sp. under drought stress. World J Microbiol Biotechnol 27:197–205

Babu AG, Shea PJ, Sudhakar D, Jung I-B, Oh BT (2015) Potential use of *Pseudomonas koreensis* AGB-1 in association with *Miscanthus sinensis* to remediate heavy metal(loid)-contaminated mining site soil. J Environ Manage 151:160–166

Barnawal D, Bharti N, Maji D, Chanotiya CS, Kaira A (2012) 1-Aminocyclopropane-1-carboxylic acid (ACC) deaminase-containing rhizobacteria protect *Ocimum sanctum* plants during waterlogging stress *via* reduced ethylene generation. Plant Physiol Biochem 58:227–235

Barnawal D, Maji D, Bharti N, Chanotiya CS, Kalra A (2013) ACC deaminase-containing *Bacillus subtilis* reduces stress ethylene-induced damage and improves mycorrhizal colonization and rhizobial nodulation in *Trigonella foenum-graecum* under drought stress. J Plant Growth Regul 32:809–822

Barnawal D, Bharti N, Maji D, Chanotiya CS, Kalra A (2014) ACC deaminase-containing *Arthrobacter protophormiae* induces NaCl stress tolerance through reduced ACC oxidase activity and ethylene production resulting in improved nodulation and mycorrhization in *Pisum sativum*. J Plant Physiol 171:884–894

Barnawal D, Bharti N, Tripathi A, Panday SS, Chanotiya CS, Kalra A (2016) ACC-deaminase-producing endophyte *Brachybacterium paraconglomeratum* strain SMR20 ameliorates *Chlorophytum* salinity stress via altering phytohormone generation. J Plant Growth Regul 35:553–564

Bayliss C, Bent E, Culham DE, MacLellan S, Clarke AJ, Brown GL, Wood JM (1997) Bacterial genetic loci implicated in the *Pseudomonas putida* GR12-2R3-canola mutualism: identification of an exudate-inducible sugar transporter. Can J Microbiol 43:809–818

Belimov AA, Safronova VI, Sergeyeva TA, Egorova TN, Matveyeva VA, Tsyganov VE (2001) Characterization of plant growth promoting rhizobacteria isolated from polluted soils and containing 1- aminocyclopropane-1-carboxylate deaminase. Can J Microbiol 47:242–252

Belimov AA, Hontzeas N, Safronova VI, Demchinskaya SV, Piluzza G, Bullitta S (2005) Cadmium-tolerant plant growth-promoting bacteria associated with roots of Indian mustard (*Brassica juncea* L Czern.) Soil Biol Biochem 37:241–250

Blaha D, Prigent-Combaret C, Mirza MS, Moënne-Loccoz Y (2006) Phylogeny of the 1-aminocyclopropane-1-carboxylic acid deaminase-encoding gene acdS in phytobeneficial and pathogenic Proteobacteria and relation with strain biogeography. FEMS Microbiol Ecol 56:455–470

Bourque FG, Bertrand A, Claessens A, Aliferis KA, Jabaji S (2016) Alleviation of drought stress and metabolic changes in Timothy (*Phleum pratense* L.) colonized with *Bacillus subtilis* B26. Front Plant Sci 7:584

Bradford KJ, Yang SF (1980) Stress-induced ethylene production in the ethylene-requiring tomato mutant diageotropica. Plant Physiol 65:327–330

Burd GI, Dixon DG, Glick BR (1998) A plant growth-promoting bacterium that decreases nickel toxicity in seedlings. Appl Environ Microbiol 64:3663–3668

Burd GI, Dixon DG, Glick BR (2000) Plant growth-promoting bacteria that decrease heavy metal toxicity in plants. Can J Microbiol 46:237–245

Burg SP, Burg EA (1966) The interaction between auxin and ethylene and its role in plant growth. Am J Bot 55:262–269

Castillo P, Escalante M, Gallardo M, Alemano S, Abdala G (2013) Effects of bacterial single inoculation and co-inoculation on growth and phytohormone production of sunflower seedlings under water stress. Acta Physiol Plant 35:2299–2309

Chabot R, Antoun H, Cescas MP (1996a) Growth promotion of maize and lettuce by phosphate-solubilizing *Rhizobium leguminosarum* biovar *phaseoli*. Plant Soil 184:311–321

Chabot R, Antoun H, Kloepper JW, Beauchamp CJ (1996b) Root colonization of maize and lettuce by bioluminescent phosphate solubilizing *Rhizobium leguminosarum* biovar phaseoli. Appl Environ Microbiol 62:2767–2772

Cheng Z, Park E, Glick BR (2007) 1-Aminocyclopropane-1-carboxylate deaminase from *Pseudomonas putida* UW4 facilitates the growth of canola in the presence of salt. Can J Microbiol 53:912–918

Ciardi JA, Tieman DM, Lund ST, Jones JB, Stall RE, Klee HJ (2000) Response to *Xanthomonas campestris* pv. *vesicatoria* in tomato involves regulation of ethylene receptor gene expression. Plant Physiol 123:81–92

Creus CM, Sueldo RJ, Barassi CA (2004) Water relations and yield in *Azospirillum*-inoculated wheat exposed to drought in the field. Can J Bot 82:273–281

Di Gregorio S, Barbafieri M, Lampis S, Sanangelantoni AM, Tassi E, Vallini G (2006) Combined application of Triton X-100 and *Sinorhizobium* sp. Pb002 inoculum for the improvement of lead phytoextraction by *Brassica juncea* in EDTA amended soil. Chemosphere 63:293–299

Drew MC (1992) Soil aeration and plant root metabolism. Soil Sci 154:259–268

Egamberdieva D, Botir H, Abeer H, Abd-Allah EF (2014) Characterization of salt tolerant *Enterobacter hormaechei* strain associated with tomato root grown in arid saline soil. J Pure Appl Microbiol 8:4231–4239

Egamberdiyeva D, Hoflich G (2004) Effect of plant growth-promoting bacteria on growth and nutrient uptake of cotton and pea on a semi-arid region of Uzbekistan. J Arid Environ 56:293–301

Else MA, Jackson MB (1988) Transport of 1-aminocyclopropane-1-carboxylic acid (ACC) in the transpiration stream of tomato (*Lycopersicon esculentum*) in relation to foliar ethylene production and petiole epinasty. Aust J Plant Physiol 25:453–458

Ernst WHO (1990) Ecophysiology of plants in waterlogged and flooded environments. Aquat Bot 38:73–90

Etesami H, Mirsyed Hosseini H, Alikhani HA, Mohammadi L (2014) Bacterial biosynthesis of 1-aminocyclopropane-1-carboxylate (ACC) deaminase and indole-3-acetic acid (IAA) as endophytic preferential selection traits by rice plant seedlings. J Plant Growth Regul:1–17

FAO (2008) FAO land and plant nutrition management service. http://www.fao.org/ag/agl/agll/spush

Farwell AJ, Vesely S, Nero V, Rodriguez H, McCormack K, Shah S (2007) Tolerance of transgenic canola plants (*Brassica napus*) amended with plant growth-promoting bacteria to flooding stress at a metal contaminated field site. Environ Pollut 147:540–545

Flowers T (2004) Improving crop salt tolerance. J Exp Bot 55:307–319

Frankenberger WT (2002) Environmental chemistry of arsenic. Marcel Dekker, New York, p 410

German MA, Burdman S, Okon Y, Kigel J (2000) Effects of *Azospirillum brasilense* on root morphology of common bean (*Phaseolus vulgaris* L.) under different water regimes. Biol Fertil Soils 32:259–264

Giongo A, Ambrosini A, Vargas LK, Freire JRJ, Bodanese-Zanettini MH, Passaglia LMP (2008) Evaluation of genetic diversity of *Bradyrhizobia* strains nodulating soybean [*Glycine max* (L.) Merrill] isolated from South Brazilian fields. Appl Soil Ecol 38:261–269

Glick BR (1995) The enhancement of plant growth by free-living bacteria. Can J Microbiol 41:109–117

Glick BR (2004) Bacterial ACC deaminase and the alleviation of plant stress. Adv Appl Microbiol 56:291–312

Glick BR (2005) Modulation of plant ethylene levels by the bacterial enzyme ACC deaminase. FEMS Microbiol Lett 251:1–7

Glick BR (2014) Bacteria with ACC deaminase can promote plant growth and help to feed the world. Microbiol Res 169:30–39

Glick BR, Jacobson CB, Schwarze MMK Pasternak JJ (1994) 1-aminocyclopropane-1-carboxylic acid deaminase mutants of the plant growth promoting rhizobacteria *Pseudomonas putida* GR12-2 do not stimulate canola root elongation. Can J Microbiol 40:911–915

Glick BR, Penrose DM, Li J (1998) A model for the lowering of plant ethylene concentrations by plant growth promoting bacteria. J Theor Biol 190:63–68

Glick BR, Patten CL, Holguin G, Penrose DM (1999) Biochemical and genetic mechanisms used by plant growth-promoting bacteria. Imperial College Press, London

Glick BR, Cheng Z, Czarny J, Duan J (2007) Promotion of plant growth ACC deaminase-containing soil bacteria. Eur J Plant Pathol 119:329–339

Greenberg BM, Huang XD, Gurska Y, Gerhardt KE, Wang W, Lampi MA, Zhang C, Khalid A, Isherwood D, Chang P, Wang H, Dixon DG, Glick BR (2006) Successful field tests of a multi-process phytoremediation system for decontamination of persistent petroleum and organic contaminants. In: Proceedings of the 29th Arctic and Marine oil spill program technical seminar, pp 389–400

Grichko VP, Glick BR (2001a) Amelioration of flooding stress by ACC deaminase-containing plant growth-promoting bacteria. Plant Physiol Biochem 39:11–17

Grichko VP, Glick BR (2001b) Flooding tolerance of transgenic tomato plants expressing the bacterial enzyme ACC deaminase controlled by the 35S, rolD or PRB-1b promoter. Plant Physiol Biochem 39:19–25

Honma M (1983) Enzymatic determination of 1-aminocyclopropane-1-carboxylate deaminase. Agric Biol Chem 47:617–618

Honma M (1985) Chemically reactive sulfhydryl groups of 1-aminocyclopropane-1-carboxylate deaminase. Agric Biol Chem 49:567–571

Hontzeas N, Zoidakis J, Glick BR, Abu-Omar MM (2004a) Expression and characterization of 1-aminocyclopropane-1-carboxylate deaminase from the rhizobacterium *Pseudomonas putida* UW4: a key enzyme in bacterial plant growth promotion. Biochim Biophys Acta 1703:11–19

Hontzeas N, Saleh SS, Glick BR (2004b) Changes in gene expression in canola roots induced by ACC deaminase-containing plant growth-promoting bacteria. Mol Plant Microbe Interact 17: 865–871

Hontzeas N, Hontzeas CE, Glick BR (2006) Reaction mechanisms of the bacterial enzyme 1-aminocyclopropane-1-carboxylate deaminase. Biotechnol Adv 24:420–426

Huang XD, El-Alawi Y, Penrose DM, Glick BR, Greenberg BM (2004) Responses of plants to creosote during phytoremediation and their significance for remediation processes. Environ Pollut 130:453–463

Huang XD, El-Alawai Y, Gurska J, Glick BR, Greenberg BM (2005) A multi-process phytoremediation system for decontamination of persistent total petroleum hydrocarbons (TPHs) from soils. Microchem J 81:139–147

Idris EES, Bochow H, Ross H, Borriss R (2004) Use of *Bacillus subtilis* as biocontrol agent. Phytohormone like action of culture filtrates prepared from plant growth-promoting *Bacillus amyloliquefaciens* FZB24, FZB42, FZB45 and *Bacillus subtilis* FZB37. J Plant Dis Prot 111: 583–597

Jackson MB (1985) Ethylene and responses of plants to soil waterlogging and submergence. Annu Rev Plant Physiol 36:145–174

Jackson MB, Drew MC (1984) Effects of flooding on growth and metabolism of herbaceous plants. In: Kozlowski TT (ed) Flooding and plant growth. Academic, New York, pp 47–128

Jacobson CB, Pasternak JJ, Glick BR (1994) Partial purification and characterization of ACC deaminase from the plant growth-promoting rhizobacterium *Pseudomonas putida* GR12-2. Can J Microbiol 40:1019–1025

Jiang CY, Sheng XF, Qian M, Wang QY (2008) Isolation and characterization of a heavy metal-resistant *Burkholderia* sp. from heavy metal-contaminated paddy field soil and its potential in promoting plant growth and heavy metal accumulation in metal polluted soil. Chemosphere 72:157–164

Jiang F, Chen L, Belimov AA, Shaposhnikov AI, Gong F, Meng X, Hartung W, Jeschke DW, Davies WJ, Dodd IC (2012) Multiple impacts of the plant growth-promoting rhizobacterium *Variovorax paradoxus* 5C-2 on nutrient and ABA relations of *Pisum sativum*. J Exp Bot 63:6421–6430

Johnson PR, Ecker JR (1998) The ethylene gas signal transduction pathway: a molecular perspective. Annu Rev Genet 32:227–254

Kende H (1993) Ethylene biosynthesis. Annu Rev Plant Physiol Plant Mol Biol 44:283–307

Khan AG, Kuek C, Chaudhry TM, Khoo CS, Hayes WJ (2000) Role of plants, mycorrhizae and phytochelators in heavy metal contaminated land remediation. Chemosphere 41:197–207

Kloepper JW, Beauchamp CJ (1992) A review of issues related to measuring colonization of plant roots by bacteria. Can J Microbiol 38:1219–1232

Kumar PBAN, Dushenkov V, Motto H, Raskin I (1995) Phytoextraction-the use of plants to remove heavy metals from soils. Environ Sci Technol 29:1232–1238

Kumar KV, Singh N, Behl HM, Srivastava S (2008) Influence of plant growth promoting bacteria and its mutant on heavy metal toxicity in *Brassica juncea* grown in fly ash amended soil. Chemosphere 72:678–683

Larsen H (1986) Halophilic and halotolerant microorganisms an overview and historical perspective. FEMS Microbiol Rev 39:3–7

Lee KH, LaRue TA (1992) Exogenous ethylene inhibits nodulation of *Pisum sativum* L. cv Sparkle. Plant Physiol 100:1759–1763

Li K, Pang CH, Ding F, Sui N, Feng ZT, Wang BS (2012) Overexpression of Suaeda salsa stroma ascorbate peroxidase in *Arabidopsis* chloroplasts enhances salt tolerance of plants. S Afr J Bot 78:235–245

Li J, McConkey BJ, Cheng Z, Guo S, Glick BR (2013) Identification of plant growth-promoting rhizobacteria-responsive proteins in cucumber roots under hypoxic stress using a proteomic approach. J Proteomics 84:119–131

Lim JH, Kim SD (2013) Induction of drought stress resistance by multi-functional PGPR *Bacillus licheniformis* K11 in pepper. Plant Pathol J 29:201–208

Lynch JM, Whipps JM (1991) Substrate flow in the rhizosphere. In: Keister DL, Cregan B (eds) The rhizosphere and plant growth. Kluwer, Dordrecht, pp 15–24

Mayak S, Tirosh T, Glick BR (1999) Effect of wild-type and mutant plant growth-promoting rhizobacteria on the rooting of mung bean cuttings. J Plant Growth Regul 18:49–53

Mayak S, Tirosh T, Glick BR (2004a) Plant growth-promoting bacteria that confer resistance to water stress in tomato and pepper. Plant Sci 166:525–530

Mayak S, Tirosh T, Glick BR (2004b) Plant growth-promoting bacteria that confer resistance in tomato to salt stress. Plant Physiol Biochem 42:565–572

Murset V, Hennecke H, Pessi G (2012) Disparate role of rhizobial ACC deaminase in root-nodule symbioses. Symbiosis 57:43–50

Nascimento FX, Rossi MJ, Soares CRFS, McConkey BJ, Glick BR (2014) New insights into 1-aminocyclopropane-1-carboxylate (ACC) deaminase phylogeny, evolution and ecological significance. PLoS One 9:99168

Naseem H, Bano A (2014) Role of plant growth-promoting rhizobacteria and their exopolysaccharide in drought tolerance in maize. J Plant Interact 9:689–701

Naveed M, Mitter B, Reichenauer TG, Wieczorek K, Sessitsch A (2014) Increased drought stress resilience of maize through endophytic colonization by *Burkholderia phytofirmans* PsJN and *Enterobacter* sp. FD17. Environ Exp Bot 97:30–39

Nie L, Shah S, Rashid A, Burd GI, Dixon DG, Glick BR (2002) Phytoremediation of arsenate contaminated soil by transgenic canola and the plant growth-promoting bacterium *Enterobacter cloacae* CAL2. Plant Physiol Biochem 40:355–361

Owino WO, Manabe Y, Mathooko FM, Kubo Y, Inaba A (2006) Regulatory mechanisms of ethylene biosynthesis in response to various stimuli during maturation and ripening in fig fruit (*Ficus carica* L.) Plant Physiol Biochem 44:335–342

Patten CL, Glick BR (1996) Bacterial biosynthesis of indole-3-acetic acid. Can J Microbiol 42: 207–220

Patten CL, Glick BR (2002) The role of bacterial indole acetic acid in the development of the host plant root system. Appl Environ Microbiol 68:3795–3801

Penrose DM, Moffatt BA, Glick BR (2001) Determination of 1-aminocyclopropane-1-carboxylic acid (ACC) to assess the effects of ACC deaminase-containing bacteria on roots of canola seedlings. Can J Microbiol 47:77–80

Pereyra MA, Ballesteros FM, Creus CM, Sueldo RJ, Barassi CA (2009) Seedlings growth promotion by *Azospirillum brasilense* under normal and drought conditions remains unaltered in Tebuconazole-treated wheat seeds. Eur J Soil Biol 45:20–27

Pierik R, Tholen D, Poorter H, Visser EJW, Voesenek LACJ (2006) The Janus face of ethylene: growth inhibition and stimulation. Trends Plant Sci 11:176–183
Pilon-Smits E (2005) Phytoremediation. Annu Rev Plant Biol 56:15–39
Płociniczak T, Sinkkonen A, Romantschuk M, Piotrowska-Seget Z (2013) Characterization of *Enterobacter intermedius* MH8b and its use for the enhancement of heavy metals uptake by *Sinapis alba* L. Appl Soil Ecol 63:1–7
Prasad JK, Gupta SK, Raghuwanshi R (2017) Screening multifunctional plant growth promoting rhizobacteria strains for enhancing seed germination in wheat (*Triticum aestivum* L.). Int J Agric Res 12:64–72
Rajkumar M, Nagendran R, Kui JL, Wang HL, Sung ZK (2006) Influence of plant growth promoting bacteria and Cr (VI) on the growth of Indian mustard. Chemosphere 62:741–748
Reed MLE, Glick BR (2005) Growth of canola (*Brassica napus*) in the presence of plant growth-promoting bacteria and either copper or polycyclic aromatic hydrocarbons. Can J Microbiol 51:1061–1069
Sairam RK, Tyagi A (2004) Physiology and molecular biology of salinity stress tolerance in plants. Curr Sci 86:407–421
Saleem M, Arshad M, Hussain S, Bhatti AS (2007) Perspective of plant growth promoting rhizobacteria (PGPR) containing ACC deaminase in stress agriculture. J Ind Microbiol Biotechnol 34:635–648
Salt DE, Blaylock M, Kumar PBAN, Dushenkov V, Ensley BD, Chet I, Raskin I (1995) Phytoremediation: a novel strategy for the removal of toxic metals from the environment using plants. Biotechnology 13:468–474
Saravanakumar D, Samiyappan R (2007) ACC deaminase from *Pseudomonas fluorescens* mediated saline resistance in groundnut (*Arachis hypogea*) plants. J Appl Microbiol 102(5):1283–1292
Sarma RK, Saikia R (2014) Alleviation of drought stress in mung bean by strain *Pseudomonas aeruginosa* GGRJ21. Plant and Soil 377:111–126
Shaharoona B, Arshad M, Zahir ZA (2006) Effect of plant growth promoting rhizobacteria containing ACC-deaminase on maize (*Zea mays* L.) growth under axenic conditions and on nodulation in mung bean (*Vigna radiata* L.) Lett Appl Microbiol 42:155–159
Sheehy RE, Honma M, Yamada M, Sasaki T, Martineau B, Hiatt WR (1991) Isolation, sequence, and expression in *Escherichia coli* of the *Pseudomonas* sp. strain ACP gene encoding 1-aminocyclopropane-1-carboxylate deaminase. J Bacteriol 173:5260–5265
Sheng XF, Xia JJ (2006) Improvement of rape (*Brassica napus*) plant growth and cadmium uptake by cadmium-resistant bacteria. Chemosphere 64:1036–1042
Singh RP, Jha PN (2016) A halotolerant bacterium *Bacillus licheniformis* HSW-16 augments induced systemic tolerance to salt stress in wheat plant (*Triticum aestivum*). Front Plant Sci 7:1890
Sinha S, Raghuwanshi R (2015) Influence of water availability on microbial populations of rhizospheric soils of *E. Prostrata* (L.) L and *A. paniculata* Burm Wall Nees plant under different biofertilizer treatment. In: Khan MMAA, John KG, Dilfuza E, Abid M, Maheshwari RK, Naqvi TS (eds) New dimensions in microbiology. Lenin Media, Delhi, pp 43–52
Sinha S, Raghuwanshi R (2016) Synergistic effect of arbuscular mycorrhizal helper bacteria on physiological mechanism to tolerate drought in *Eclipta prostrate* (L.). Pure. Appl Microbiol 10:1117–1129
Smith SE, Read DJ (1997) Mycorrhizal symbiosis. Academic, San Diego. ISBN: 0-12-652840-3
Stearns J, Glick BR (2003) Transgenic plants with altered ethylene biosynthesis or perception. Biotechnol Adv 21:193–210
Subbarao GV, Johansen C (1994) Strategies and scope for improving salinity tolerance in crop plants. In: Pessarakli M (ed) Handbook of plant and crop stress. Marcel Dekker, New York, pp 1069–1087
Timmusk S, Wagner EGH (1999) The plant-growth-promoting rhizobacterium *Paenibacillus polymyxa* induces changes in *Arabidopsis thaliana* gene expression: a possible connection between biotic and abiotic stress responses. Mol Plant Microbe Interact 12:951–959

Timmusk S, Paalme V, Pavlicek T, Bergquist J, Vangala A, Danilas T, Nevo E (2011) Bacterial distribution in the rhizosphere of wild barley under contrasting microclimates. PLoS One 6:17968

Trung NT, Hieu HV, Thuan NH (2016) Screening of strong 1-aminocyclopropane-1-carboxylate deaminase producing bacteria for improving the salinity tolerance of cowpea. Appl Micro Open Access 2:1000111

Van Loon LC, Glick BR (2004) Increased plant fitness by rhizobacteria. In: Sandermann H (ed) Molecular ecotoxicology of plants. Springer, Berlin, pp 177–205

Van Loon LC, Geraats BPJ, Linthorst HJM (2006) Ethylene as a modulator of disease resistance in plants. Trends Plant Sci 11:184–191

Venkateswarlu B, Desai S, Prasad YG (2008) Agriculturally important microorganisms for stressed ecosystems: challenges in technology development and application. In: Khachatourians GG, Arora DK, Rajendran TP, Srivastava AK (eds) Agriculturally important microorganisms. Academic World, Bhopal, pp 225–246

Voesenek LACJ, Blom CWPM (1989) Growth responses of *Rumex* species in relation to submergence and ethylene. Plant Cell Environ 12:433–439

Vurukonda SSKP, Vardharajula S, Shrivastava M, Skz A (2016) Multifunctional *Pseudomonas putida* strain FBKV2 from arid rhizosphere soil and its growth promotional effects on maize under drought stress. Rhizosphere 1:4–13

Wang C, Yang W, Wang C, Gu C, Niu D, Liu H, Wang Y, Guo J (2012) Induction of drought tolerance in cucumber plants by a consortium of three plant growth-promoting rhizobacterium strains. PLoS One 7:52565

Yan J, Smith MD, Glick BR, Liang Y (2014) Effects of ACC deaminase containing rhizobacteria on plant growth and expression of Toc GTPases in tomato (*Solanum lycopersicum*) under salt stress. Botany 92:775–781

Yang SF, Hoffman NE (1984) Ethylene biosynthesis and its regulation in higher plants. Annu Rev Plant Physiol 35:155–189

Yang J, Kloepper J, Ryu CN (2009) Rhizosphere bacteria help plants tolerate abiotic stress. Trends Plant Sci 14:1–4

Zafar-ul-Hye M, Farooq HM, Zahir ZA (2014) Combined application of ACC-deaminase biotechnology and fertilizers improves maize productivity subjected to drought stress in salt affected soils. Int J Agric Biol 16:591–596

Zahir ZA, Munir A, Asghar HN, Shaharoona B, Arshad M (2008) Effectiveness of rhizobacteria containing ACC deaminase for growth promotion of peas (*Pisum sativum*) under drought conditions. J Microbiol Biotechnol 18:958–963

Zahir AZ, Ghani U, Naveed M, Nadeem SM, Asghar HN (2009) Comparative effectiveness of *Pseudomonas* and *Serratia* sp. containing ACC-deaminase for improving growth and yield of wheat (*Triticum aestivum* L.) under salt-stressed conditions. Arch Microbiol 191:415–424

Chapter 13
Root–Microbe Interactions: Understanding and Exploitation of Microbiome

Amita Sharma and Rajnish Kumar Verma

13.1 Introduction

Rhizosphere represents the zone of soil around the roots which is biologically active and greatly influenced by the presence of roots, root exudates, respiration, and biogeochemical reactions (Pinton et al. 2001; Narula et al. 2009). This zone contains great diversity of microbes which perform equally diverse ecological functions (Dighton 2014; Havlicek and Mitchell 2014; Gera Hol et al. 2014; Krumins 2014; Pérès 2014; Termorshuizen 2014; Pavlović et al. 2014; Orgiazzi et al. 2016). Rhizosphere possesses abundant species of bacteria, fungi, protozoa, nematodes, herbivorous insects, and their predators. All of them act as important members of the food webs operating below the ground and contribute significantly to various ecological functions such as decomposition, mineralization, plant health, and productivity. The rhizosphere has three distinct zones, namely, ectorhizosphere, rhizoplane, and endorhizosphere. The physical and chemical properties of the soil are altered by the interactions between the roots, microbes, and soil particles which in turn regulate the diversity of microbes in the soil (Lavelle and Spain 2005; Bonkowski et al. 2009; Johnson and Rasmann 2015; Johnson et al. 2016; Vos and Kazan 2016).

The interactions between soil organisms and roots are special in nature and can be categorized into three groups: (1) neutral interactions, wherein both of the partners are neither benefitted nor harmed; (2) negative/pathogenic interactions, wherein one of the partners is harmed; and (3) positive interactions, wherein either both partners are benefitted due to association (symbiosis) or only one partner gets benefit with no

A. Sharma
Department of Agriculture, Shaheed Udham Singh College of Research and Technology, Tangori, Punjab, India

R. K. Verma (✉)
Department of Botany, Dolphin PG College of Science and Agriculture, Chunni Kalan, Punjab, India

B. Giri et al. (eds.), *Root Biology*, Soil Biology 52,
https://doi.org/10.1007/978-3-319-75910-4_13

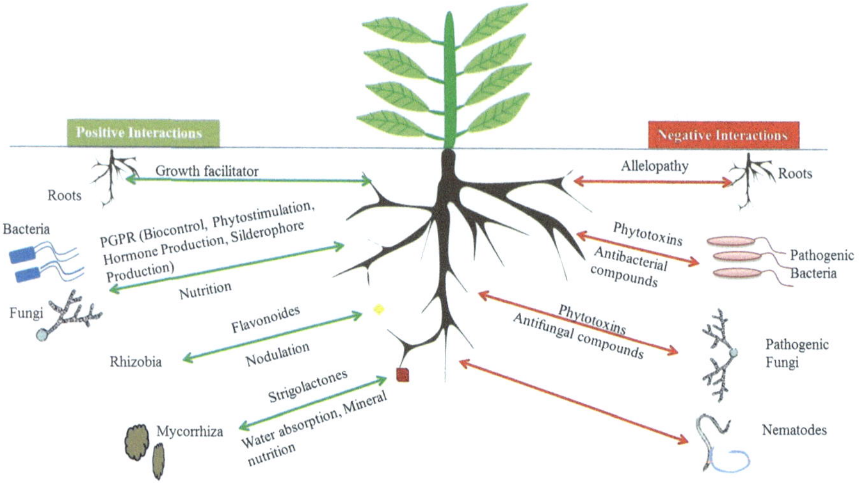

Fig. 13.1 Representation of positive (left side) and negative (right side) interactions between roots and organisms mediated by root exudates

harm to the other partner (associative). In symbiotic positive interaction, there is close association of plant with plant growth-promoting rhizobacteria (PGPR), epiphytes, and mycorrhizal fungi. Plants provide root exudates to the soil microbiota which act as food and signaling molecules. In return, soil microbes play a crucial role in a myriad of beneficial processes like decomposition, nutrient cycling, enhancement in nutrient availability and absorption, suppression of diseases, and stress tolerance to various abiotic and biotic factors (Morrissey et al. 2004; Mendes et al. 2011; Badri et al. 2013a, b; Zolla et al. 2013).

13.2 What Are Root Exudates or Rhizodeposition?

The roots of the plants release a wide range of chemicals into the soil called root exudates including amino acids, secondary metabolites, soluble sugars, dead cell lysates, mucilage, etc. (Fig. 13.1). Rhizodeposition is the term given to describe the total carbon (C) transferred from roots of plant into soil in the form of exudates (Grayston et al. 1996). The type of root exudates determines the plant–microbe and plant–plant interactions in the rhizosphere. Root exudates can be classified into two categories: low and high molecular weight compounds. The former class include amino acids, sugars, organic acids, and secondary metabolites, whereas the latter include polysaccharides and proteins. The composition of root exudates depends upon various factors like species, variety, developmental stage of plant, and various environmental factors like pH, temperature, soil type, moisture content, and the presence of microbes. All these factors play a crucial role in determining specifically the strength and type of microbes in the rhizosphere (Badri and Vivanco 2009; Chaparro et al. 2013).

13.2.1 Carbohydrates and Amino Acids

Root exudates like strigolactone5-deoxystrigol, sugars, and carbohydrates play a crucial role in symbiotic association between non-leguminous plants and mycorrhizal fungi (Fang and St. Leger 2010; Kiers et al. 2011). Similarly, the roots and seed coats of leguminous plants secrete an amino acid canavanine which shows antimicrobial activity against most of the rhizosphere bacteria but not against rhizobia thus selecting these beneficial microbes for association (Cai et al. 2009). Studies have suggested that root exudates play an important role mediating interaction of plant roots with PGPRs. The predominant members of root exudates that attract PGPRs are carbohydrates and amino acids.

13.2.2 Proteins

The studies on how proteins determine the plant–microbes interaction are very limited. Few reports have suggested secreted proteins play a key role in the recognition process of non-pathogenic and pathogenic microbes (Wen et al. 2007; De-la-Pena and Vivanco 2010). Lectins are the proteins which are most studied as the proteins playing crucial role in defense responses of plants and recognition process in the symbiotic associations (De Hoff et al. 2009). Other proteins involved in rhizospheric interactions include chitinases, glucanases, myrosinases, peroxidases, etc. (De-la-Pena et al. 2010). Recently, a new protein called arabinogalactan protein (AGP) has been identified, playing a crucial role determining the interactions between roots of plants and microbes present in the atmosphere. The AGPs have the ability to attract beneficial microorganisms, and they repel plant pathogenic organisms (Xie et al. 2012; Nguema-Ona et al. 2013).

13.2.3 Secondary Metabolites

Plants secrete a wide range of secondary metabolites like terpenoids, phenolics, flavonoids, glucosinolates, nitrogen-containing compounds, alkaloids, etc. into the soil, which mediate plant–microbes interactions. These compounds benefit the plants in several ways like they help in defense mechanism against pathogens and act as chemoattractants to draw beneficial bacteria toward the plants. The roots of maize release benzoxazinoids which attract beneficial bacteria like rhizobacteria (Neal et al. 2012). In a similar way, flavonoids released by leguminous plants act as chemoattractants which draw rhizobacteria toward the roots by altering the expression of nod genes that play crucial role in nodule formation (Abdel-Lateif et al. 2012). Studies have shown that *Arabidopsis thaliana* upon infection with foliar pathogens secreted malic acid in the rhizosphere that recruited *Bacillus subtilis* (PGPR) to the root surface. The PGPR *B. subtilis* in turn activated Salicylic acid and Abscisic acid signaling pathway in *Arabidopsis* to restrict the pathogen entry (Rudrappa et al. 2008; Kumar et al. 2012).

13.3 Root–Root Interactions

The plant species release a wide range of chemical compounds mediating interactions with other plant species (Ma et al. 2014). The phenomenon wherein the chemicals produced by plants influence the growth, development, reproduction, and productivity is known as allelopathy. These chemical compounds are called allelochemicals, and these interactions may be positive or negative. The exploitation of these interactions would enhance the ecosystem productivity and diversity. In the agricultural system, the positive effects of the root exudates include weed control, disease control, crops management, enhancement in yields, etc. (Reich et al. 2004; Gao et al. 2014; Cheng and Cheng 2015). In a recent study, the intercropping of maize (*Zea mays* L.) and faba beans (*Vicia faba* L.) resulted in enhanced yields as well as increased nodulation in faba beans (Li et al. 2016). The negative interactions include toxin production, soil sickness, etc. which reduce the growth of neighboring species. To achieve sustainable development in agriculture sector, the exploitation of the cultivation systems based on the inhibitory and stimulating effects of the allelopathic plants is fascinating area of research. The utilization of the allelochemicals as herbicide, insecticides, growth regulators, and antimicrobial compounds is gaining attention to regulate the growth and development of plants (Cheng and Cheng 2015).

13.4 Root–Organism Interactions

A diverse array of chemicals are secreted by plant roots thereby allowing interactions with a wide range of soil organisms. Soil organisms include insects, nematodes, microorganisms like bacteria and mycorrhizal fungi, and invertebrates which show unique interactions with plant roots in response to specific exudates. The dynamic interactions between soil organisms, plant roots, and various other biotic/abiotic factors determine soil structure, plant root traits, plant health, distribution of various species, etc. (Johnson et al. 2016; Vos and Kazan 2016; Tsunoda et al. 2017).

13.4.1 Microorganisms

The belowground microbial species may provide benefit or cause harm to the plants. Microorganisms which confer a benefit to the plants in one way or another ultimately enhancing the plants' productivity and performance are considered beneficial. They have a role in (1) promoting uptake of nutrients and water, (2) providing resistance against biotic and abiotic factors, and (3) influencing plant hormonal balance. Well-known beneficial microbes include plant growth-promoting rhizobacteria (PGPR), rhizobia, and mycorrhizal fungi. They provide resistance against diseases by improving the immune system and enhance root growth by improving the nutrient uptake (Hayat et al. 2010). Numerous fungal and bacterial diseases have been studied so far which decrease the productivity of plants in a direct or indirect way. The root pathogenic microbes exert a great impact on the economy of agricultural systems. Plants produce antimicrobial compounds to combat these pathogens by suppressing their growth and colonization in the roots. The root exudates are released in a tight regulated manner as growth of both beneficial and pathogenic microbes is influenced based on root chemicals (Baetz and Martinoia 2014; Baetz 2016).

13.4.2 Other Soil Organisms

Nematodes represent a diverse class of organisms with great abundance in the rhizosphere. There are two groups of nematodes which interact with plant roots: entomopathogenic nematodes and parasitic nematodes. The former group indirectly affects the plants by attacking the plant-feeding insects. The latter group directly influences the plants and interacts with the fine roots.

The diversity and abundance of numerous invertebrates like mites, earthworms, etc. are shaped by the root exudates. These invertebrate decomposers are found in abundance both in the leaf litter layer and in the rhizosphere (Chomel et al. 2016). The composition of the root exudates determines the degree of decomposition of the leaf litter as the plant chemicals in the exudates prevent the colonization of the decomposer invertebrates.

The soil organisms like herbivorous insects cause a significant damage to the plants. They mostly attack the less defended roots, or they have the ability to neutralize the plant defensive compounds. The insects show different feeding habits like sucking, chewing, mining, etc. thereby causing damage to root transport system. They exert a strong negative impact on the plants productivity growth and performance. The herbivorous insects cause the greatest damage among all as they likely affect the root base just under the soil surface and destroy the core of the vascular system (Tsunoda et al. 2014a, b, c).

13.4.3 Impact of Root Exudates on Soil Organisms

The studies related to one-to-one interaction (plant–microbe) provide a good insight to understand rhizosphere interactions, but in nature, plants are exposed to all types of microorganisms: pathogenic as well as beneficial. Therefore, studies at the microbe level could provide a true picture of root–microbe interactions taking place in the rhizosphere, instead of studies at the species level. Plants select beneficial bacteria and repel the pathogenic microorganisms, thereby altering the diversity and composition of microbiota of the rhizosphere in a manner specific to the plant species. Thus, plants have ability to determine and shape the composition of microbial communities in the soil rhizosphere by releasing specific root exudates (Houlden et al. 2008; Bakker et al. 2012; Chaparro et al. 2012). Studies have suggested that a significant change in the rhizosphere microbial communities has been observed upon induction of the systemic acquired resistance (SAR) response of the plant via activation of jasmonate signaling pathway. The alteration in the microbial community composition involved the enrichment of bacteria associated with defense responses like *Bacillus*, *Paenibacillus*, *Lysinibacillus*, etc. (Carvalhais et al. 2013).

Numerous studies have suggested that phenolic compounds play a determining role to shape the microbial communities in rhizosphere. The phenolic compounds are utilized by a wide range of soil microbes as substrate or signaling molecules in comparison to other groups like sugars, amino acids, etc. A great abundance of unique taxonomic groups was observed in the presence of phenolic compounds released in the rhizosphere (Fang et al. 2013; Michalet et al. 2013).

The root exudates have a direct impact on the plant performance thereby suppressing the growth of root attackers like pathogenic fungi, bacteria, insects, and nematodes (Cannesan et al. 2012). Moreover, the roots play a crucial role to attract beneficial microorganisms, like rhizobacteria and mycorrhizal fungi. The plant root exudates also improve the plant performance in indirect manner by attracting the predators or enemies of the root-attacking organisms. The mechanism is still unclear how root exudates regulate this process (Baetz 2016). The physical and chemical properties of the soil are altered by root exudates thus influencing the diversity and abundance of numerous organisms (Lambers et al. 2013; Read et al. 2016).

Microbial population in the rhizosphere can also influence the root exudation by particular plant species both qualitatively and quantitatively. The colonization of arbuscular mycorrhizal fungi resulted in increased secretion of phenolic compounds, nitrogen-containing compounds, and hormones like gibberellins and decreased secretion of sugars, phosphorus, and potassium (Jones et al. 2004; Matilla et al. 2010). Furthermore, plants secrete different compounds as root exudates like phytoalexins, defense proteins, oxalic acid, etc. in response to a pathogen attack (Steinkellner et al. 2007).

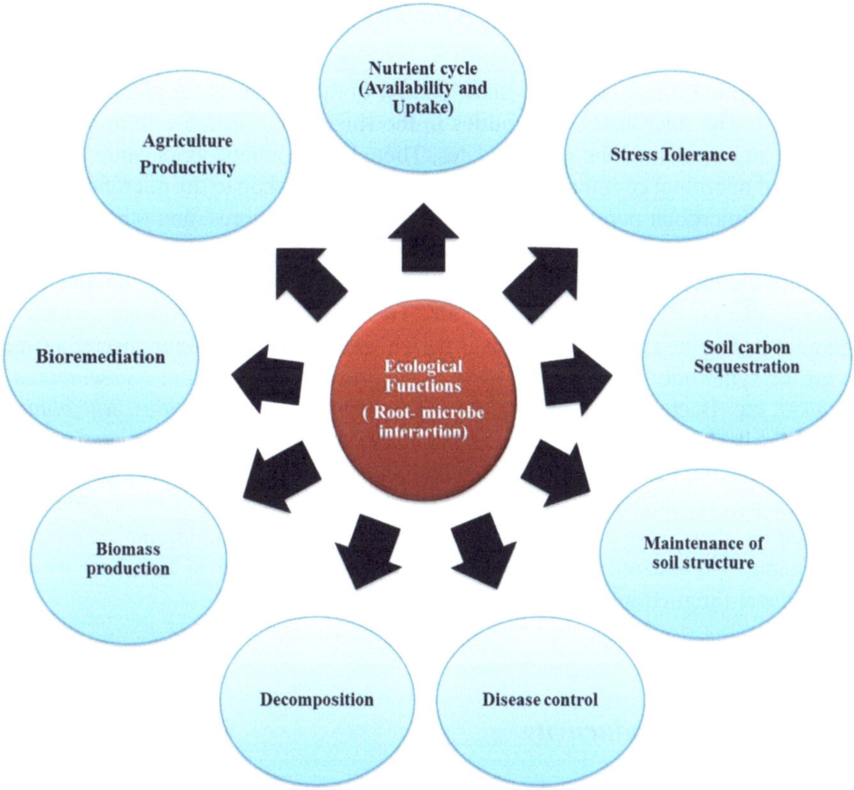

Fig. 13.2 An overview of ecological functions mediated by root–microbe interactions

13.5 Ecological Functions: Roots–Microbes Interaction in the Rhizosphere

The plant-associated microbial communities in collective are known as plant microbiome, and plants rely to some extent on microbial population for specific functions. The root–microbe interactions encourages various ecological functions including beneficial and protective associations, availability of vital nutrients, improved efficiency of nutrient uptake, soil carbon regulation, stress tolerance, bio-control, and maintenance of soil structure (Fig. 13.2) (Cordero and Datta 2016; Lareen et al. 2016).

13.5.1 Nutrient Cycling and Nutrient Uptake (Biofertilizers)

Root exudation is the process in which 30–40% of the fixed C of the plant is released in the soil. The microbial communities in the rhizosphere increase in number and diversity in response to the root exudates. There is a symbiotic association between plants and microbial communities. The plants provide carbon to the microbes, and in return, the microbial population provides nitrogen, phosphorus, and other minerals by decomposition of the organic matter present in the soil. The microbes also influence the morphology and physiology of roots thereby enhancing the uptake of minerals like N, P, and K (Cocking 2003; Trolove et al. 2003). Nitrogen which is often limited in the agricultural fields is replenished by nitrogen-fixing bacteria (free living or symbiotic) like *Burkholderia*, *Azotobacter*, *Azospirillum*, *Psuedomonas*, *Frankia*, etc. Bacteria such as *Psuedomonas*, *Actinomycetes*, *Bacillus*, *Rhizobium*, etc. have been reported to increase the availability of phosphorus in the soil. Mycorrhizal fungi also increase the availability of phosphorus as orthophosphates by releasing organic acids and enzymes (acid phosphatases). Chemoattractants released by plants play a crucial role in determining the association between plants and fungi. Strigolactones released in the root exudates help to find host by the mycorrhizal fungi (Bouwmeester et al. 2007).

13.5.2 Plant Productivity

The growing human population, climate change, and its impact on agricultural practices are presenting a grim situation, which need to be addressed to ensure global food security. Extensive use of agrochemicals results in creating an imbalance in the rhizospheric microbial population. Plant pathogens also have a negative impact on the agricultural productivity as they contaminate the groundwater and reduce the content of organic matter in the soil. The microbial population in the rhizosphere benefits the plants by increasing crop yields promoting growth and development. The research investigating the relation between the manipulation of soil microbial communities and impact on plant growth is accelerating providing insights on applications of these beneficial microbes in agriculture. In addition to the synergistic action of the microbes in the rhizosphere, the soil properties also determine the development of root–microbe interactions. The beneficial microbial population develops in the rhizosphere suppressing the disease-causing pathogens thereby providing a stable environment for root–microbe interactions. The focus of the growing research bodies is to throw light on the importance of these root–microbe interactions in relation to development and productivity of plants (Sen 2003; Pereg and McMillan 2015; Finkel et al. 2017).

13.5.3 Soil Carbon Sequestration

The microbial population act as carbon sinks by regulating the carbon flow in the soil rhizosphere. The primary production by leaves is the determining factor in the soil carbon sequestration, but it is also influenced by the diversity and abundance of the soil microbial population. The microbial community is involved in the regulation of soil carbon storage by the processes like immobilization of carbon and mineralization. The action of microbes and plant roots regulating the sequestration of carbon in the soil is a promising and emerging approach to tackle the problem of global climate change. The regulation of carbon storage in the soil in turn influences the nutrient cycling thus affecting the plants productivity and ecosystem balance (Lal 2003; Velmourougane et al. 2017).

13.5.4 Improvement and Maintenance of Soil structure

The microbes in the rhizosphere possess the ability to maintain the soil structure by altering its physical and chemical properties. The microorganisms produce glomalin (a glycoprotein), which aids in stabilization of the structure of soil aggregates. The root microbes are also involved in the accumulation of nitrogen and phosphorus, organic matter, and decontamination of heavy metals. The biotic interactions taking place in the rhizosphere has a great influence on the host plants as well as the competing plants, thus affecting the aboveground diversity and abundance of plant species (Barea et al. 2002; Bever 2003).

13.5.5 Stress Tolerance

The role of rhizosphere microbes in stress resistance against soil organisms has been highlighted by numerous reports. The PGPRs and PGPF (Plant growth promoting fungi) confer tolerance against stresses by direct suppression of pathogen's growth or by competing for nutrients with the pathogen. Plants immune system is activated by release of different elicitors during herbivore attack (Pieterse et al. 2014; Pineda et al. 2010, 2015; Schmelz 2015; Rasmann et al. 2017). Microbiome is involved in the activation of the immune system of the plant to acquire systemic resistance [or systemic acquired resistance (SAR)] against a broad range of plant pathogens. The hormones ethylene, salicylic acid, and jasmonic acid produced in the plants in response to microbial interactions are crucial mediators in the induced systemic resistance (ISR) (Van der Ent et al. 2009). The PGPR *Stenotrophomonas rhizophila* is reported to provide resistance against both abiotic as well as biotic stresses (Alavi et al. 2013).

13.5.6 Biocontrol

The rhizospheric microbes that have the potential to suppress or limit the growth and survival of pathogens are gaining immense attention as they represent an environment-friendly and cost-effective alternative to chemical pesticides. Numerous studies have reported the mechanisms involved in plant pathogen suppression like production of hydrolytic enzymes (lipases, proteases, chitinases), hydrogen cyanide, toxins, antibiotics, antimicrobial compounds, etc. Some biocontrol agents activate the plant defense system thereby increasing the production of defense-related genes. A number of microbes have been in use commercially as biocontrol agents like Kodiak (*Bacillus subtilis* GB03), Soilgard (*Trichoderma virens*), Afla-guard (*Aspergillus flavus* NRRL 21882), etc. (Naznin et al. 2014; Lareen et al. 2016).

13.5.7 Bioremediation

The remediation of polluted sites has become a growing area of research to clean the environment and improve the quality of life. There are different methods for the remediation of the polluted sites including physical, chemical, and biological methods. The high costs of physical and chemical methods have forced the scientific community to focus and develop biological methods as a better alternative. Numerous bacterial species have been reported to possess the potential to be utilized for bioremediation processes, viz., *Paenibacillus* sp., *Pseudomonas* sp., *Haemophilus* sp., *Mycobacterium* sp., etc. (Bisht et al. 2015). A newer field rhizoremediation is emerging these days which involves the use of both plants and associated rhizospheric microbes for the remediation process. It is a creative approach exploiting the root–microbe interactions for the clearance of pollutants in the contaminated sites. Rhizoremediation is also known as rhizosphere remediation or rhizodegradation. It results in the increased degradation of the tough pollutants by plant roots-associated microbes under the influence of that plant species. In some cases, root exudates possess the ability to induce the expression of microbial enzymes involved in the degradation of recalcitrant pollutants (Olson et al. 2003). It has been reported that the grasses and their associated rhizospheric bacteria are potential candidates to be utilized for remediation of soils contaminated with diesel (Mezzari et al. 2011). Root–microbe interactions play a crucial role in the decontamination of the pollutants. The plants absorb the pollutants and stabilize them in waxes; the microbes associated with the plants improve the pollutant absorption capacity and promote the growth of the plants (Weyens et al. 2009a, b, 2015).

13.5.8 Biomass Production

Recently the research has focused on the exploitation of the plant–microbe interactions to enhance biomass production for industrial purposes. The growth of the plants is promoted by rhizospheric microbes by direct and indirect mechanisms. The direct effects include solubilization of the minerals and facilitation of their uptake. Studies have suggested the role of endophytes in promoting plant growth. Phytohormones and other plant growth substances are produced by microbes that stimulate plant development. The endophytes compete with the pathogens for space and nutrients thereby suppressing the growth of the pathogens. Some microbes produce antimicrobial compounds to suppress the growth of plant pathogens. Thus, suppression of the pathogen attack and development of stress tolerance indirectly promote the healthy growth of the plant. All these factors contribute to the enhanced biomass production which could be utilized at commercial scale to boost up numerous industrial processes. The plant-associated microbes improve plant establishment and promote sustainable production of energy crops on arable and marginal lands (Weyens et al. 2009a, 2015).

13.6 Novel Approaches to Study Root–Microbes Interactions

Most of the studies related to microbiome in the rhizosphere are limited to cultivation processes. Among these processes, BIOLOG and spread plate count methods are the most widely used methods to determine the microbial diversity in the rhizosphere (Singh et al. 2004). But the major challenge is still the inability to culture 99% of the microbial population. Rapid advances in genomics and molecular biology elucidated full genomic sequences of microbes providing insights into root–microbe communications. To correctly recognize the structural aspects of belowground biotic interactions mediated by root chemicals, appropriate sampling scales and methods are needed. The recent techniques like PCR, DNA hybridization, fluorescent in situ hybridization (FISH), fluorescence microscopy, microarray, mass spectrometry, phospholipid fatty acid analysis (PLFA), and re-association studies are well established techniques to analyze the uncultured microbiome of the rhizosphere. The DNA microarray technique has been developed to identify the microbial population and their ecological functions. This approach helps to assess the growth rate of microbes and substrate utilization by the microbial communities. The FISH analysis helps to identify uncultured bacteria phylogenetically utilizing fluorescent probes (Alavi et al. 2013). The combination of FISH and microautoradiography is a powerful tool for detection and quantification of active population which utilizes a specific substrate. Mass spectrometric methods have been developed for the investigation of the communications in the rhizosphere. This technique provides quantification of the compounds present in the root exudates which provide better understanding of the rhizospheric interactions. Recent advancement in the microscopic techniques has facilitated the studies related to understanding root–microbe interactions. Various fluorescent markers and reporter gene systems are available in market to be studied by fluorescence and confocal microscopy (Cardinale 2014). The novel methodologies revealing the rhizosphere ecology will provide an opportunity to optimize plant health, growth, and development in context to beneficial microbial population in the rhizosphere. A wide range of chemical compounds have been elucidated in different rhizospheric environments by mass spectrometric methods utilizing different techniques for separation like capillary electrophoresis, liquid chromatography, gas chromatography, etc. (Rugova et al. 2017). Despite extensive research underway, still there is still no single method present which can be used to elucidate all the root–microbe interactions. To answer specific questions, a combination of molecular biology technique and genomics provides a better understanding of the complexities of rhizosphere interactions at molecular, physiological, and ecological levels.

13.7 Conclusions and Future Perspectives

The climate change and growing human population are posing a risk to the global food security highlighting the need to come up with novel processes increasing plant productivity. Microbial population in the rhizosphere serves numerous ecological functions that determine nutrient cycling, maintenance of soil structure, carbon regulation, disease control, soil productivity, and biodegradation of pollutants. The basic mechanisms of root–microbe interactions are revealed by several studies, but still the understanding of the complexity of plant–microbe communications is not complete. A deeper understanding of the rhizosphere microbial communities and their interactions with plants would allow better exploitation of this underutilized resource. Recently, novel approaches have come up throwing light on molecular basis of communication that can provide a platform to improvise various agricultural practices, industrial processes, etc. The exploitation of root–microbe interactions to achieve better results in the ecological processes like decontamination, soil carbon sequestration, and disease control is a fascinating field for future perspective.

References

Abdel-Lateif K, Bogusz D, Hocher V (2012) The role of flavonoids in the establishment of plant roots endosymbioses with arbuscular mycorrhiza fungi, rhizobia and Frankia bacteria. Plant Signal Behav 7:636–641. https://doi.org/10.4161/psb.20039

Alavi P, Starcher MR, Zachow C, Müller H, Berg G (2013) Root-microbe systems: the effect and mode of interaction of stress protecting agent (SPA) *Stenotrophomonas rhizophila* DSM14405T. Front Plant Sci 4:141

Badri DV, Vivanco JM (2009) Regulation and function of root exudates. Plant Cell Environ 32:666–681. https://doi.org/10.1111/j.1365-3040.2009.01926.x

Badri DV, Chaparro JM, Zhang R, Shen Q, Vivanco JM (2013a) Application of natural blends of phytochemicals derived from the root exudates of *Arabidopsis* to the soil reveal that phenolic-related compounds predominantly modulate the soil microbiome. J Biol Chem 288:4502–4512. https://doi.org/10.1074/jbc.M112.433300

Badri DV, Zolla G, Bakker MG, Manter DK, Vivanco JM (2013b) Potential impact of soil microbiomes on the leaf metabolome and on herbivore feeding behavior. New Phytol 198:264–273. https://doi.org/10.1111/nph.12124

Baetz U (2016) Root exudates as integral part of belowground plant defence. In: CNF V, Kazan K (eds) Belowground defence strategies in plants. Springer, Berlin, pp 45–67. https://doi.org/10.1007/978-3-319-42319-7_3

Baetz U, Martinoia E (2014) Root exudates: the hidden part of plant defense. Trends Plant Sci 19:90–98. https://doi.org/10.1016/j.tplants.2013.11.006

Bakker M, Manter D, Sheflin A, Weir T, Vivanco J (2012) Harnessing the rhizosphere microbiome through plant breeding and agricultural management. Plant Soil 360:1–13. https://doi.org/10.1007/s11104-012-1361-x

Barea JM, Azcón R, Azcón-Aguilar C (2002) Mycorrhizosphere interactions to improve plant fitness and soil quality. Anton Leeuw 81:343–335

Bever JD (2003) Soil community feedback and the coexistence of competitors: conceptual framework and empirical tests. New Phytol 157:465–473

Bisht S, Pandey P, Bhargava B, Sharma S, Kumar V, Sharma KD (2015) Bioremediation of polyaromatic hydrocarbons (PAHs) using rhizosphere technology. Braz J Microbiol 46:7–21

Bonkowski M, Villenage C, Griffiths B (2009) Rhizosphere fauna: the functional and structural diversity of intimate interactions of soil fauna with plant roots. Plant Soil 321:213–233

Bouwmeester HJ, Roux C, López Ráez JA, Bécard G (2007) Rhizosphere communication of plants, parasitic plants and AM fungi. Trends Plant Sci 12:224–230. https://doi.org/10.1016/j.tplants.2007.03.009

Cai T, Cai W, Zhang J, Zheng H, Tsou AM, Xiao L, Zhong Z, Zhu J (2009) Host legume-exuded antimetabolites optimize the symbiotic rhizosphere. Mol Microbiol 73:507–517. https://doi.org/10.1111/j.1365-2958.2009.06790.x

Cannesan MA, Durand C, Burel C, Gangneux C, Lerouge P, Ishii T, Laval K, Follet-Gueye ML, Driouich A, Vicré-Gibouin M (2012) Effect of arabinogalactan proteins from the root caps of pea and *Brassica napus* on *Aphanomyces euteiches* zoospore chemotaxis and germination. Plant Physiol 159:1658–1670. https://doi.org/10.1104/pp.112.198507

Cardinale M (2014) Scanning a microhabitat: plant-microbe interactions revealed by confocal laser microscopy. Front Microbiol 5:94. https://doi.org/10.3389/fmicb.2014.00094

Carvalhais LC, Dennis PG, Fan B, Fedoseyenko D, Kierul K, Becker A, Wiren N, Borriss R, Baxter I (2013) Linking plant nutritional status to plant-microbe interactions. PLoS One 8(7):e68555

Chaparro J, Sheflin A, Manter D, Vivanco J (2012) Manipulating the soil microbiome to increase soil health and plant fertility. Biol Fertil Soils 48:489–499. https://doi.org/10.1007/s00374-012-0691-4

Chaparro JM, Badri DV, Bakker MG, Sugiyama A, Manter DK, Vivanco JM (2013) Root exudation of phytochemicals in Arabidopsis follows specific patterns that are developmentally programmed and correlate with soil microbial functions. PLoS One 8:e55731. https://doi.org/10.1371/journal.pone

Cheng F, Cheng Z (2015) Research progress on the use of plant allelopathy in agriculture and the physiological and ecological mechanisms of allelopathy. Front Plant Sci 6:1020. https://doi.org/10.3389/fpls.2015.01020

Chomel M, Guittonny-Larchevêque M, Fernandez C, Gallet C, Des Rochers A, Paré D, Jackson BG, Baldy V (2016) Plant secondary metabolites: a key driver of litter decomposition and soil nutrient cycling. J Ecol 104:1527–1541. https://doi.org/10.1111/1365-2745.12644

Cocking EC (2003) Endophytic colonisation of plant roots by nitrogen-fixing bacteria. Plant Soil 252:169–175

Cordero OX, Datta MS (2016) Microbial interactions and community assembly at microscales. Curr Opin Microbiol 31:227–234

De Hoff P, Brill L, Hirsch A (2009) Plant lectins: the ties that bind in root symbiosis and plant defense. Mol Genet Genomics 282:1–15. https://doi.org/10.1007/s00438-009-0460-8

De-la-Pena C, Vivanco JM (2010) Root-microbe interactions: the importance of protein secretion. Curr Proteonomics 7:265–274

De-la-Pena C, Badri DV, Lei Z, Watson BS, Brandão MM, Silva-Filho MC, Sumner LW, Vivanco JM (2010) Root secretion of defense-related proteins is development-dependent and correlated with flowering time. J Biol Chem 285:30654–30666

Dighton J (2014) Introduction: soils and their promotion of plant growth. In: Dighton J, Krumins JA (eds) Interactions in soil: promoting plant growth, biodiversity, community and ecosystems. Springer, Berlin, pp 1–26. https://doi.org/10.1007/978-94-017-8890-8_1

Fang W, St. Leger RJ (2010) Mrt, a gene unique to fungi, encodes an oligosaccharide transporter and facilitates rhizosphere competency in *Metarhizium robertsii*. Plant Physiol 154:1549–1557. https://doi.org/10.1104/pp.110.163014

Fang C, Zhuang Y, Xu T, Li Y, Li Y, Lin W (2013) Changes in rice allelopathy and rhizosphere microflora by inhibiting rice phenylalanine ammonia-lyase gene expression. J Chem Ecol 39:204–212. https://doi.org/10.1007/s10886-013-0249-4

Finkel OM, Castrillo G, Paredes SH, González IS, Dang JL (2017) Understanding and exploiting plant beneficial microbes. Curr Opin Plant Biol 38:155–163

Gao X, Wu M, Xu R, Wang X, Pan R, Kim H-J, Liao H (2014) Root interactions in a Maize/Soybean intercropping system control soybean soil-borne disease, red crown rot. PLoS One 9 (5):e95031. https://doi.org/10.1371/journal.pone.0095031

Gera Hol WH, de Boer W, Medina A (2014) Beneficial interactions in the rhizosphere. In: Dighton J, Krumins JA (eds) Interactions in soil: promoting plant growth, biodiversity, community and ecosystems 1. Springer, Berlin, pp 59–80. https://doi.org/10.1007/978-94-017-8890-8_3

Grayston SJ, Vaughan D, Jones D (1996) Rhizosphere carbon flow in trees, in comparison with annual plants: the importance of root exudation and its impact on microbial activity and nutrient availability. Appl Soil Ecol 5:29–56

Havlicek E, Mitchell EAD (2014) Soils supporting biodiversity. In: Dighton J, Krumins JA (eds) Interactions in soil: promoting plant growth, biodiversity, community and ecosystems. Springer, Berlin, pp 27–58. https://doi.org/10.1007/978-94-017-8890-8_2

Hayat R, Ali S, Amara U, Khalid R, Ahmed I (2010) Soil beneficial bacteria and their role in plant growth promotion: a review. Ann Microbiol 60:579–598. https://doi.org/10.1007/s13213-010-0117-1

Houlden A, Timms-Wilson TM, Day MJ, Bailey MJ (2008) Influence of plant developmental stage on microbial community structure and activity in the rhizosphere of three field crops. FEMS Microbiol Ecol 65(2):193–201. https://doi.org/10.1111/j.1574-6941.2008.00535.x

Johnson SN, Rasmann S (2015) Root-feeding insects and their interactions with organisms in the rhizosphere. Annu Rev Entomol 60:517–535

Johnson SN, Erb M, Hartley SE (2016) Roots under attack: contrasting plant responses to below- and aboveground insect herbivory. New Phytol 210:413–418

Jones DL, Hodge A, Kuzyakov Y (2004) Plant and mycorrhizal regulation of rhizodeposition. New Phytol 163:459–480. https://doi.org/10.1111/j.1469-8137.2004.01130.x

Kiers ET, Duhamel M, Beesetty Y, Mensah JA, Franken O, Verbruggen E, Fellbaum CR, Kowalchuk GA, Hart MM, Bago A, Palmer TM, West SA, Vandenkoornhuyse P, Jansa J, Bücking H (2011) Reciprocal rewards stabilize cooperation in the mycorrhizal symbiosis. Science 333:880–882. https://doi.org/10.1126/science.1208473

Krumins JA (2014) The positive effects of trophic interactions in soil. In: Dighton J, Krumins JA (eds) Interactions in soil: promoting plant growth, biodiversity, community and ecosystems 1. Springer, Berlin, pp 81–94. https://doi.org/10.1007/978-94-017-8890-8_4

Kumar AS, Lakshmanan V, Caplan JL, Powell D, Czymmek KJ, Levia DF, Bais HP (2012) Rhizobacteria *Bacillus subtilisrestricts* foliar pathogen entry through stomata. Plant J 72:694–706. https://doi.org/10.1111/j.1365-313X.2012.05116.x

Lal R (2003) Global potential of soil carbon sequestration to mitigate the greenhouse effect. Crit Rev Plant Sci 22:151–184

Lambers H, Clements JC, Nelson MN (2013) How a phosphorus-acquisition strategy based on carboxylate exudation powers the success and agronomic potential of lupines (*Lupinus*, Fabaceae). Am J Bot 100:263–288. https://doi.org/10.3732/ajb.1200474

Lareen A, Burton F, Schäfer P (2016) Plant root-microbe communication in shaping root microbiomes. Plant Mol Biol 90:575–587

Lavelle P, Spain AV (2005) Soil ecology. Springer, Dordrecht, pp 1–654

Li B, Li Y-Y, Wu H-M, Zhang F-F, Li C-J, Li X-X, Lambers H, Li L (2016) Root exudates drive interspecific facilitation by enhancing nodulation and N2 fixation. PNAS 113:6496–6501

Ma Y, Zhang M, Li Y, Shui J, Zhou Y (2014) Allelopathy of rice (*Oryza sativa* L.) root exudates and its relations with *Orobanche Cumana* Wallr. and *Orobanche minor* Sm. germination. J Plant Interact 9:722–730. https://doi.org/10.1080/17429145.2014.912358.

Matilla MA, Ramos JL, Bakker PAHM, Doornbos R, Badri DV, Vivanco JM, Ramos-González MI (2010) *Pseudomonas putida* KT2440 causes induced systemic resistance and changes in Arabidopsis root exudation. Environ Microbiol Rep 2:381–388. https://doi.org/10.1111/j.17582229.2009.00091.x

Mendes R, Kruijt M, de Bruijn I, Dekkers E, van der Voort M, Schneider JH, Piceno YM, DeSantis TZ, Andersen GL, Bakker PA, Raaijmakers JM (2011) Deciphering the rhizosphere microbiome for disease suppressive bacteria. Science 332:1097–1100. https://doi.org/10.1126/science.1203980

Mezzari MP, Zimermann DMH, Corseuil HX, Nogueira AV (2011) Potential of grasses and rhizosphere bacteria for bioremediation of diesel-contaminated soils. Rev Bras Ciênc Solo 35:2227–2236

Michalet S, Rohr J, Warshan D, Bardon C, Roggy J-C, Domenach A-M, Czarnes S, Pommier T, Combourieu B, Guillaumaud N, Bellvert F, Comte G, Poly F (2013) Phytochemical analysis of mature tree root exudates in situ and their role in shaping soil microbial communities in relation to tree N-acquisition strategy. Plant Physiol Biochem 72:169–177. https://doi.org/10.1016/j.plaphy.2013.05.003

Morrissey JP, Dow JM, Mark GL, O'Gara F (2004) Are microbes at the root of a solution to world food production? EMBO Rep 5:922–926. https://doi.org/10.1038/sj.embor.7400263

Narula N, Kothe E, Behl RK (2009) Role of root exudates in plant-microbe interactions. J Appl Bot Food Qual 82:122–130

Naznin HA, Kiyohara D, Kimura M, Miyazawa M, Shimizu M, Hyakumachi M, Yang C-H (2014) Systemic resistance induced by volatile organic compounds emitted by plant growth-promoting fungi in *Arabidopsis thaliana*. PLoS One 9(1):e86882. https://doi.org/10.1371/journal.pone.0086882

Neal AL, Ahmad S, Gordon-Weeks R, Ton J (2012) Benzoxazinoids in root exudates of maize attracts *Pseudomonas putida* to the rhizosphere. PLoS One 7:e35498. https://doi.org/10.1371/journal.pone.0035498

Nguema-Ona E, Vicré-Gibouin M, Cannesan M-A, Driouich A (2013) Arabinogalactan proteins in root–microbe interactions. Trends Plant Sci 18(8):440–449. https://doi.org/10.1016/j.tplants.2013.03.006

Olson PE, Reardon KF, Pilon-Smits EAH (2003) Ecology of rhizosphere bioremediation. In: McCutcheon SC, Schnoor JL (eds) Phytoremediation: transformation and control of contaminants. Wiley, Hoboken, pp 317–357. https://doi.org/10.1002/047127304X.ch10

Orgiazzi A, Bardgett RD, Barrios E, Behan-Pelletier V, Briones MJI, Chotte J-L, De Deyn GB, Eggleton P, Fierer N, Fraser T, Hedlund K, Jeffery S, Johnson NC, Jones A, Kandeler E, Kaneko N, Lavelle P, Lemanceau P, Miko L, Montanarella L, Moreira FMS, Ramirez KS, Scheu S, Singh BK, Six J, van der Putten WH, Wall DH (2016) Global soil biodiversity Atlas. European Commission, Publications Office of the European Union, Luxembourg, 176 pp

Pavlović P, Muscolo A, Sidari M, Mitrović M (2014) Non-trophic interactions: allelopathy. In: Dighton J, Krumins JA (eds) Interactions in soil: promoting plant growth, biodiversity, community and ecosystems 1. Springer, Berlin, pp 139–162. https://doi.org/10.1007/978-94-017-8890-8_7

Pereg L, McMillan M (2015) Scoping the potential uses of beneficial microorganisms for increasing productivity in cotton cropping systems. Soil Biol Biochem 80:349–358

Pérès G (2014) Soils suppressing biodiversity. In: Dighton J, Krumins JA (eds) Interactions in soil: promoting plant growth, biodiversity, community and ecosystems 1. Springer, Berlin, pp 95–118. https://doi.org/10.1007/978-94-017-8890-8_5

Pieterse CMJ, Zamioudis C, Berendsen RL, Weller DM, Van Wees SC, Bakker PA (2014) Induced systemic resistance by beneficial microbes. Annu Rev Phytopathol 52:347–375

Pineda A, Zheng SJ, Van Loon JJA, Pieterse CMJ, Dicke M (2010) Helping plants to deal with insects: the role of beneficial soil-borne microbes. Trends Plant Sci 15:507–514

Pineda A, Soler R, Pozo MJ, Rasmann S, Turlings TCJ (2015) Editorial: Above-belowground interactions involving plants, microbes and insects. Front Plant Sci 6:318. https://doi.org/10.3389/fpls.2015.00318

Pinton R, Varanini Z, Nannipieri P (2001) The rhizosphere. Biochemistry and organic substances at the soil-plant interface. Marcel Dekker, New York, pp 1–472

Rasmann S, Bennett A, Biere A, Karley A, Guerrieri E (2017) Root symbionts: powerful drivers of plant above- and belowground indirect defences. Insect Sci 24:1–14. https://doi.org/10.1111/1744-7917.12464

Read DS, Matzke M, Gweon HS, Newbold LK, Heggelund L, Ortiz MD, Lahive E, Spurgeon D, Svendsen C (2016) Soil pH effects on the interactions between dissolved zinc, non-nano- and nano-ZnO with soil bacterial communities. Environ Sci Pollut Res 23:4120–4128. https://doi.org/10.1007/s11356-015-4538-z

Reich PB, Tilman D, Naeem S, Ellsworth DS, Knops J, Craine J, Wedin D, Trost J (2004) Species and functional group diversity independently influence biomass accumulation and its response to CO_2 and N. Proc Natl Acad Sci USA 101:10101–10106. https://doi.org/10.1073/pnas.0306602101

Rudrappa T, Czymmek KJ, Paré PW, Bais HP (2008) Root-secreted malic acid recruits beneficial soil bacteria. Plant Physiol 148(3):1547–1556. https://doi.org/10.1104/pp.108.127613

Rugova A, Puschenreiter M, Koellensperger G, Hanna S (2017) Elucidating rhizosphere processes by mass spectrometry – a review. Anal Chim Acta 956:1–13

Schmelz EA (2015) Impacts of insect oral secretions on defoliation-induced plant defense. Curr Opin Insect Sci 9:7–15

Sen R (2003) The root–microbe–soil interface: new tool for sustainable plant production. New Phytol 157:391–398

Singh BK, Millard P, Whiteley AS, Murrell JC (2004) Unravelling rhizosphere-microbial interactions: opportunities and limitations. Trends Microbiol 12:386–393. https://doi.org/10.1016/j.tim.2004.06.008

Steinkellner S, Lendzemo V, Langer I, Schweiger P, Khaosaad T, Toussaint J-P, Vierheilig H (2007) Flavonoids and strigolactones in root exudates as signals in symbiotic and pathogenic plant–fungus interactions. Molecules 12:1290–1306. https://doi.org/10.3390/12071290

Termorshuizen AJ (2014) Root pathogens. In: Dighton J, Krumins JA (eds) Interactions in soil: promoting plant growth, biodiversity, community and ecosystems 1. Springer, Berlin, pp 119–138. https://doi.org/10.1007/978-94-017-8890-8_6

Trolove SN, Hedley MJ, Kirk GJD, Bolan NS, Loganathan B (2003) Progress in selected areas of rhizosphere on P acquisition. Aust J Soil Res 41:471–499

Tsunoda T, Kachi N, Suzuki J-I (2014a) Availability and temporal heterogeneity of water supply affect the vertical distribution and mortality of a belowground herbivore and consequently plant growth. PLoS One 9:e100437. https://doi.org/10.1371/journal.pone.0100437

Tsunoda T, Kachi N, Suzuki J-I (2014b) Effects of belowground vertical distribution of a herbivore on plant biomass and survival in *Lolium perenne*. Ecol Res 29:351–355. https://doi.org/10.1007/s11284-014-1133-6

Tsunoda T, Kachi N, Suzuki J-I (2014c) Effects of belowground herbivory on the survival and biomass of *Lolium perenne* and *Plantago lanceolata* plants at various growth stages. Botany 92:737–741. https://doi.org/10.1139/cjb-2014-0045

Tsunoda T, Kachi N, Suzuki J-I (2017) Belowground herbivory decreases shoot water content and biomass of *Lolium perenne* seedlings under nutrient-poor conditions. Botany 95:29–36. https://doi.org/10.1139/cjb-2016-0076

Van der Ent S, Van Wees SCM, Pieterse CMJ (2009) Jasmonate signaling in plant interactions with resistance-inducing beneficial microbes. Phytochemistry 70:1581–1588. https://doi.org/10.1016/j.phytochem.2009.06.009

Velmourougane K, Prasanna R, Saxena AK (2017) Agriculturally important microbial biofilms: present status and future prospects. J Basic Microbiol 57:548–573. https://doi.org/10.1002/jobm.201700046

Vos CNF, Kazan K (2016) Belowground defence strategies in plants. Springer, Berlin

Wen F, Van Etten HD, Tsaprailis G, Hawes MC (2007) Extracellular proteins in pea root tip and border cell exudates. Plant Physiol 143:773–783. https://doi.org/10.1104/pp.106.091637

Weyens N, van der Lelie D, Taghavi S, Newman L, Vangronsveld J (2009a) Exploiting plant–microbe partnerships for improving biomass production and remediation. Trends Biotechnol 27:591–598

Weyens N, van der Lelie D, Taghavi S, Vangronsveld J (2009b) Phytoremediation: Plant-endophyte partnerships take the challenge. Curr Opin Biotechnol 20:248–254

Weyens N, Thijs S, Popek R, Witters N, Przybysz A, Espenshade J, Gawronska H, Vangronsveld J, Gawronski SW (2015) The role of plant–microbe interactions and their exploitation for phytoremediation of air pollutants. Int J Mol Sci 16:25576–25604. https://doi.org/10.3390/ijms161025576

Xie F, Williams A, Edwards A, Downie JA (2012) A plant arabinogalactan like glycoprotein promotes a novel type of polar surface attachment by *Rhizobium leguminosarum*. Mol Plant Microbe Interact 25:250–258. https://doi.org/10.1094/MPMI-08-11-0211

Zolla G, Badri DV, Bakker MG, Manter DK, Vivanco JM (2013) Soil microbiomes vary in their ability to confer drought tolerance to *Arabidopsis*. Appl Soil Ecol 68:1–9. https://doi.org/10.1016/j.apsoil.2013.03.007

Chapter 14
Unfolding the Role of Rhizomicrobiome Toward Sustainable Agriculture

Sanjana Kaul, Suruchi Gupta, Tanwi Sharma, and Manoj K. Dhar

14.1 Introduction

Rhizomicrobiome is an important component of the plant ecosystem which influences the plant health in natural and stressed conditions. It is composed of the diverse microbial communities whose presence is influenced by the root exudates. The rhizosphere and its microbial community constitute the rhizomicrobiome. The term rhizosphere was coined by plant physiologist Lorenz Hiltner. Rhizosphere has been described as the area around plant root inhabited by unique microbes under the influence of root exudates (Hartmann et al. 2008). "Rhizospheric effect" acts as the driving force for composing the diverse and rich rhizospheric microbial population. It is the phenomenon by which active microbial populations are attracted toward the organic matter released by growing plant roots (Morgan and Whipps 2001). Bacteria, fungi, archaea, protozoa, algae, nematodes, viruses, oomycetes, and microarthropods are the important components of the rhizomicrobiome. However, the bacterial domain is the dominant one followed by fungi, actinomycetes, and other groups. Rhizosphere can be differentiated into three regions, viz., endorhizosphere, rhizoplane, and ectorhizosphere. The endorhizosphere includes the area of cortex and endodermis occupied by microbes. The rhizoplane is the medial zone which includes the root epidermis and mucilage. Ectorhizosphere is the outermost zone and extends from the rhizoplane out into the bulk soil.

Indiscriminate use of fertilizers, particularly the nitrogenous and phosphorus compounds, has led to substantial pollution of soil, air, and water in modern cultivation process. Excessive use of these chemicals put forth deleterious effects on the microorganisms present in the soil affects the fertility of soil and also depletes

S. Kaul (✉) · S. Gupta · T. Sharma · M. K. Dhar
School of Biotechnology, University of Jammu, Jammu, Jammu and Kashmir, India

B. Giri et al. (eds.), *Root Biology*, Soil Biology 52,
https://doi.org/10.1007/978-3-319-75910-4_14

the environment (Youssef and Eissa 2014). In addition, production of chemical fertilizers not only depletes nonrenewable resources but also is costly and poses human and environmental hazards (Gupta et al. 2015). To alleviate this problem, we have to depend on the approach of sustainable agriculture. Rhizosphere research is a leading way toward achieving this goal. The major influence that rhizomicrobiome has on plants is an important tool to guard the health of plants in eco-friendly manner.

Rhizospheric microbes are found to be involved in plethora of beneficial interactions with that of the host plant, viz., plant growth promotion, biocontrol, siderophore production, nutrient acquisition, phytohormone production, and stress resistance or tolerance toward various biotic and abiotic factors, and the list continues. In addition to it, they are also found to be involved in soil health management (Welbaum et al. 2004). Various reports are available where rhizosphere-inhabiting microbes have been successfully used as plant inoculants. It includes bacterial genera *Bacillus*, *Pseudomonas*, *Azospirillum*, *Serratia*, *Streptomyces*, and *Stenotrophomonas* (Jacobsen et al. 2004; Ryan et al. 2009). Rhizosphere-inhabiting fungal species belonging to the genera *Ampelomyces*, *Coniothyrium*, and *Trichoderma* has also been reported to possess plant growth-promoting potential (Harman et al. 2004). Such microbes benefit the host plant by different complex mechanisms like biofertilization, phytostimulation, and biocontrol. A consortium of five native root-associated bacterial isolates was found to bestow the tobacco host plant with resistance against fungal wilt disease (Santhanam et al. 2015). Biofertilization involves increasing the nutrient availability to the host plant; phytostimulation involves plant growth promotion by production of phytohormones, whereas biocontrol involves disease control by direct production of bioactive metabolites and lytic enzymes or indirectly by inducing plant defense response. Root-associated microbes are reported to increase the resilience of host plants (Santhanam et al. 2015). The indelible and profound role of rhizomicrobiome on plant growth promotion speaks highly about the importance of keen studies in this field around the world.

14.2 Composition of Rhizosphere Microbial Community

The rhizosphere microflora is diverse and dynamic, thereby comprising of bacteria, fungi, nematodes, protozoa, algae, and microarthropods (Raaijmakers et al. 2009). But the population shows dominance of species belonging to proteobacteria and actinobacteria (Adesemoye and kloepper 2009). Rhizospheric bacteria that possess plant growth promotion potential are referred to as plant growth-promoting rhizobacteria (PGPR) (Kloepper and Schroth 1978). Diversity of rhizospheric microflora varies with different plant species. *Comamonadaceae*, *Flavobacteriaceae*, and *Rhizobiaceae* bacterial families were observed to dominate the root microbiota of *Hordeum vulgare* (Bulgarelli et al. 2015), whereas that of *Thymus zygis* was found to be dominated by *Bradyrhizobiaceae*, *Nocardioidaceae*, and *Geodermatophilaceae* (Pascual et al. 2016). Rhizospheric microbial communities not only vary radially including endorhizosphere, rhizoplane, and ectorhizosphere but also in specific root locations along the root axis. Molecular fingerprints have revealed the presence of distinct microbial communities in different root zones like that of emerging roots and their tips and elongating roots and site of emergence of lateral roots and older roots (Yang and Crowley 2000). Rhizospheric microbial communities are also found to be affected by nutritional status, age, stress, or health of the host plant (Yang and Crowley 2000).

14.3 Factors Affecting Composition of Root Microbiota

Rhizosphere actually represents a dynamic environment; therefore the composition of its inhabitants is affected by numerous factors, viz., type of soil, plant species, age of plant, root exudates, endophytes, and many more. Complex microbe–microbe as well as host–microbe interactions is also responsible for microbial community differentiation at root–soil interface (Bulgarelli et al. 2015). Root exudates are the primary factors affecting the composition of rhizospheric microbial population. The composition of root exudates is different in different plants as it greatly depends on the nutritional status of the plant. Qualitative and quantitative composition of root exudates is also affected by various factors like pH, soil type, oxygen, nutrient and light availability, as well as presence of microbes. Therefore, these exudates alter the chemistry of rhizospheric soil, thereby attracting the selected microbes to harbor the niche. Rhizosphere community is also influenced by the iron nutritional status of the host plant. Denaturing gradient gel analysis of rhizosphere has revealed that distinct microbial communities are present in different root locations (Yang and Crowley 2000). Plant iron nutritional status contributes about 20–40% of the total variation in rhizosphere community structure (Yang and Crowley 2000).

14.4 Benefits of Rhizospheric Microbial Communities to Plants

The rhizosphere, that is, the narrow zone surrounding and influenced by plant roots, is a hot spot for numerous organisms and is considered as one of the most complex ecosystems on earth (Mendes et al. 2013). Rhizomicrobiome has been well studied for its beneficial effect on plant growth and health (Fig. 14.1). The microbial community includes the nitrogen-fixing bacteria, mycorrhizal fungi, plant growth-promoting rhizobacteria (PGPR), biocontrol microorganisms, mycoparasitic fungi, and protozoa. Among the microbial communities of rhizosphere, bacteria constitute 90–95% of the rhizospheric population. It is because of their high growth rate and ability to use different carbon and nitrogen sources that rhizobacteria are found in abundance (Noumavo et al. 2016; Timmusk et al. 2017). The interaction

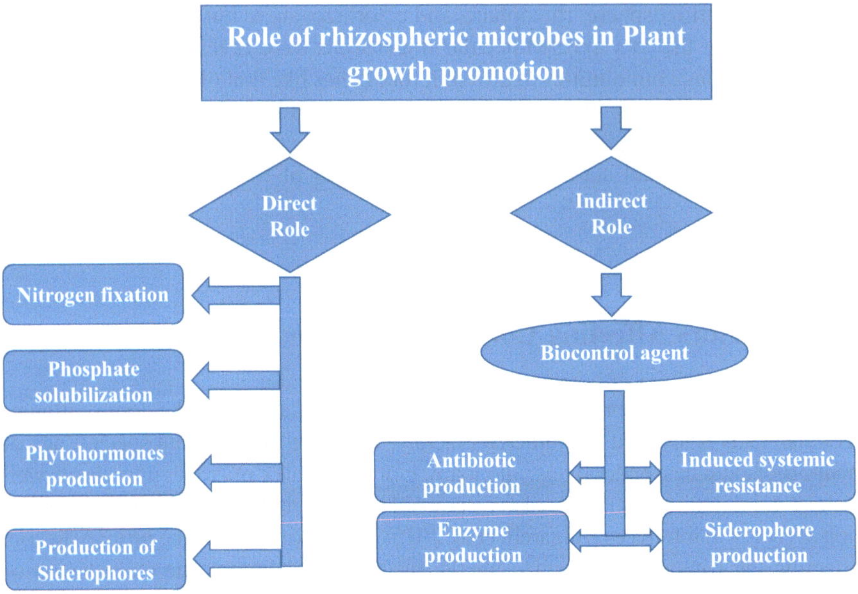

Fig 14.1 Benefits of rhizosphere microbial community in plant growth promotion

between plant and microbes in the rhizosphere is the basis of plant health and soil fertility (Hayat et al. 2010). Rhizomicrobiome mediates plant growth promotion by nutrient mobilization, mineralization, soil organic matter decomposition, nitrogen fixation, and phosphate and potassium mineralization (Satyaprakash et al. 2017). This special ecological niche supports a group of metabolically versatile microorganisms (Chauhan et al. 2017).

Among all the microbes that constitute the rhizosphere, bacteria outnumber the rest of the microbial species. Soil bacteria that colonize rhizosphere of plants and are directly or indirectly involved in plant growth promotion and development are defined as plant growth-promoting bacteria (PGPB) or plant growth-promoting rhizobacteria (PGPR) (Satyaprakash et al. 2017). Some of the PGPRs can be called as biofertilizers that enhance plant growth via providing required nutrition under certain conditions, while some PGPRs are termed as biocontrol agents or biopesticides that help in suppressing or controlling the plant diseases (Timmusk et al. 2017). PGPRs comprise different genera, like *Azospirillum*, *Nitrobacter*, *Bacillus*, *Pseudomonas*, *Bradyrhizobium*, *Acinetobacter*, *Klebsiella*, *Mesorhizobium*, *Rhizobium*, *Burkholderia*, *Micrococcus*, *Azotobacter*, *Erwinia*, etc., that are proficient to colonize the root surface (Benedetto et al. 2017; Moustaine et al. 2017). Generally, PGPRs function in three different ways: synthesizing particular compounds for the plants, facilitating the uptake of certain nutrients from the soil, and lessening or preventing the plants from diseases. Plant growth promotion and development can be facilitated both directly and indirectly (Hayat et al. 2010).

14.4.1 Direct Role of Rhizosphere Microbes in Plant Growth Promotion

Direct plant growth promotion by PGPRs includes production of plant hormones and solubilization of mineral phosphates and other nutrients. PGPRs retain more soil organic N, and other nutrients in the plant–soil system, thus reducing the need for fertilizer N and P and enhancing release of the nutrients.

14.4.1.1 Phosphate Solubilization

Phosphorus (P) functions as the second largest nutrient after nitrogen and is essential for plant growth and development. It takes part in some major metabolic processes, such as signal transduction, macromolecular biosynthesis, energy transfer, photosynthesis, respiration, and macromolecular biosynthesis (Li et al. 2017). This leads to strong demand for the application of P fertilizer. However, frequent application of P fertilizer is not only expensive but also eco-unfriendly. It may lead to algal blooms by causing the eutrophication of lakes and the loss of soil fertility (Naml et al. 2017). In order to meet the high phosphorus input for better crop yields in eco-friendly way, phosphorus-solubilizing bacteria (PSB) and plant growth-promoting rhizobacteria (PGPR) are being emphasized to utilize phosphorus fixed in the soil.

Even though organic and inorganic forms of phosphorus are plentifully available in soil, yet plants are not able to utilize it as 95–99% phosphate present is in the insoluble, immobilized, and precipitated form (Ahemad and Kibret 2014). To overcome this problem, different strategies are employed by plant growth-promoting rhizobacteria present in the soil that solubilizes phosphorus, which is otherwise unavailable to the plant and thus helps the plant absorb phosphorus (Gupta et al. 2015). A number of rhizospheric microbial species like fungi, bacteria, actinomycetes, algae, etc. have the capacity to solubilize phosphorus. However, phosphate-solubilizing bacteria constitutes 1–50% among whole microbial population of rhizosphere, whereas phosphate-solubilizing fungi constitute only 0.1–0.5% (Satyaprakash et al. 2017). Different bacterial genera and species have been reported to have phosphate solubilization ability. These include *Arthrobacter*, *Bacillus*, *Beijerinckia*, *Burkholderia*, *Enterobacter*, *Erwinia*, *Pseudomonas*, *Rhizobium*, *Rhodococcus*, *Serratia*, etc. (Gupta et al. 2015). Some of the recent studies have also shown the potential of phosphate-solubilizing bacteria in increasing the solubility of P (Kumar et al. 2011; Muleta et al. 2013; Haile et al. 2016; Chauhan et al. 2017). Phosphate-solubilizing potential of fungi has been reported by Elias et al. (2016). The fungal isolates were obtained from rhizosphere soil samples of haricot bean, faba bean, cabbage, tomato, and sugarcane. The isolated fungi belonged to genera *Aspergillus*, *Penicillium*, and *Fusarium*. Among them *Aspergillus* and *Penicillium* solubilized maximum amount of phosphorus 728.77 μg/mL and 514.44 μg/mL, respectively (Elias et al. 2016). In an another study, *Penicillium* sp. was among the isolated fungi obtained from the rhizosphere of healthy crop plants of Mysore that possessed phosphate-solubilizing ability (Mahadevamurthy et al. 2016). Li et al. (2017) have reported the potential of six bacterial strains including *Paenibacillus* sp. B1 strain, *Pseudomonas* sp. strains (B10, B14, SX1, and SX2), and *Sphingobium* sp. SX14 strain isolated from the maize rhizosphere. All the strains could solubilize inorganic P ($Ca_3(PO_4)_2$, $FePO_4$, and $AlPO_4$), and only B1 and B10 could solubilize organic P (lecithin). Shakeela et al. (2017) have studied phosphate-solubilizing potential of plant growth-promoting rhizobacteria (PGPR) isolated from *Picrorhiza kurroa*. *Bacillus subtilis* showed maximum P-solubilization among 40 isolates. For more exclusive and extensive examples on phosphate solubilization by rhizosphere microbes, a review article by Satyaprakash et al. (2017) and Ahemad and Kibret (2014) can be referred.

14.4.1.2 Nitrogen Fixation

Nitrogen is one of the essential nutrients for plant growth. Atmospheric nitrogen is not available to plants for direct use. Nitrogen-based fertilizers are being widely used to fulfil the nitrogen need of crops (Santi et al. 2013). But in the present age when we talk about sustainable development and sustainable agriculture, the use of chemical fertilizers is not apt (Gupta et al. 2015). Biological nitrogen fixation (BNF) represents an environment-friendly alternative to chemical fertilizers. Crop inoculation by biological nitrogen-fixing plant growth-promoting rhizomicrobes provides an integrated approach to maintain the nitrogen level in agricultural soil. Biological nitrogen fixation (BNF) changes the atmospheric nitrogen into plant utilizable forms using a complex enzyme system known as nitrogenase (Gaby and Buckley 2012). Rhizobacterial genera generally involved in nitrogen fixation include *Azoarcus*, *Azotobacter*, *Acetobacter*, *Azospirillum*, *Burkholderia*, *Diazotrophicus*, *Enterobacter*, *Gluconacetobacter*, *Pseudomonas*, and *Cyanobacteria* (Bhattacharyya and Jha 2012).

14.4.1.3 Phytohormones Production

Plant growth and development are regulated by a wide range of rhizospheric microorganisms that have the ability to produce phytohormones such as auxins, cytokinins, gibberellins, and ethylene. The phytohormones synthesized by the rhizobacteria can affect cell proliferation by overproduction of lateral roots and root hair and an increase in the nutrient and water uptake, thereby promoting plant growth. Indoleacetic acid (IAA) is the most common natural auxin found in plants among other plant growth regulators (Gupta et al. 2015). Synthesis of phytohormone auxin by the microbes has been known for long time. The microorganisms isolated from rhizosphere region of various crops have an ability to produce indoleacetic acid as secondary metabolite due to rich supply of substrates. It is assumed that over 80%

of the bacteria isolated from the rhizosphere are capable of synthesizing IAA (Spaepen and Vanderleyden 2011). IAA acts in conjunction with endogenous IAA in plants to stimulate cell proliferation and enhances the host's uptake of minerals and nutrients from the soil. It helps in the production of longer roots with increased number of root hairs and root laterals which are involved in nutrient uptake. IAA stimulates cell elongation by modifying certain conditions like increase in osmotic contents of the cell, increase in permeability of water into cell, decrease in wall pressure, and an increase in cell wall synthesis, inhibits or delays abscission of leaves, and induces flowering and fruiting (Mohite 2013). The main precursor for the synthesis of IAA in bacteria is tryptophan, an amino acid commonly found in root exudates. The biosynthesis of indoleacetic acid by plant growth-promoting rhizobacteria involves formation via indole-3-pyruvic acid and indole-3-acetic aldehyde, which is the most common mechanism in bacteria like *Pseudomonas*, *Rhizobium*, *Bradyrhizobium*, *Agrobacterium*, *Enterobacter*, and *Klebsiella*. Different studies have been carried out that demonstrate the potential of rhizospheric microbes to produce significant amount of IAA (Mohite 2013; Bal et al. 2013; Felestrino et al. 2017; Raut et al. 2017; Syamala and Sivaji 2017a). Likewise, in addition to auxins, cytokinins and gibberellins have been reported to be produced by several plant growth-promoting rhizobacteria *Azotobacter* sp., *Rhizobium* sp., *Pantoea agglomerans*, *Rhodospirillum rubrum*, *Pseudomonas fluorescens*, *Bacillus subtilis*, etc. (Gupta et al. 2015).

Ethylene is an essential plant growth hormone produced endogenously by almost all plants. At low concentrations, ethylene can induce multifarious physiological changes in plants like stimulation of seed germination, initiation of root growth, fruit ripening, and activation of other phytohormone synthesis (Ahemad and Kibret 2014; Gupta et al. 2015). However, the moderate or high levels of ethylene produced under stress can halt certain processes such as root elongation or nitrogen fixation in legumes and cause premature senescence that affects the overall plant growth. The plants synthesize 1-aminocyclopropane-1-carboxylate (ACC), which is the precursor for ethylene, in response to exposure to various types of environmental stress, such as cold, drought, and infections with pathogens and due to heavy metals (Vejan et al. 2016). ACC can be converted into α-ketoglutarate and ammonia by the enzyme, 1-aminocyclopropane-1-carboxylate (ACC) deaminase. Plant growth-promoting rhizobacteria thus play a very significant role in plant growth promotion by synthesizing this enzyme. Consequently, through this mechanism the PGPR-producing ACC deaminase regulates the ethylene level in the plant and prevents the growth inhibition caused by high levels of ethylene (Noumavo et al. 2016). This enzyme is expressed in several rhizobacteria (Jha et al. 2012). Barnawal et al. (2017) have reported ACC deaminase-containing plant growth-promoting rhizobacteria (PGPR) that enhance the tolerance of *Papaver somniferum* plant against downy mildew and could also be useful in reducing ethylene-induced damage in the event of abiotic stress. Similarly, ACC deaminase activity of rhizobacteria *Enterobacter cancerogenus* isolated from rhizosphere of *Jatropha* has been documented by Jha et al. (2012).

14.4.1.4 Siderophore Production

The fourth most abundant element on earth is iron which is regarded as an essential micronutrient for plants. The predominant form of iron in nature is ferric ion or Fe^{+3} which is only sparingly soluble. This is the reason that plants are able to assimilate iron in extremely low amounts. Microorganisms play a very significant role in mediating the iron uptake by plants. Specialized mechanisms have been evolved by the microbes for the assimilation of iron that includes the production of siderophores (Ahemad and Kibret 2014; Ahmed and Holmstrom 2014). Siderophores are low molecular weight iron-chelating agents that solubilize iron (III). These contain side chains and functional groups that can provide high affinity set of ligands to coordinate ferric ions (Beneduzi et al. 2012). Depending on the characteristic functional group, siderophores are divided into three main families, i.e., hydroxamates, catecholates, and carboxylates. Siderophores have been implicated for both direct and indirect enhancement of plant growth by plant growth-promoting rhizobacteria. The direct benefits of bacterial siderophores are by enhancement of iron uptake by plants and affect the growth of plants. It has been demonstrated by using radiolabeled ferric siderophores as a sole source of iron that plants are able to take up the labeled iron with the help of a large number of plant growth-promoting rhizobacteria like *Aeromonas*, *Azadirachta*, *Azotobacter*, *Bacillus*, *Burkholderia*, *Pseudomonas*, *Rhizobium*, *Serratia*, and *Streptomyces* sp. The PGPR also enhance chlorophyll level in the inoculated plants as compared to the uninoculated ones (Sujatha and Ammani 2013; Gupta et al. 2015).

14.4.2 Indirect Role of Rhizosphere Microbes in Plant Growth Promotion as Biocontrol Agents

Rhizospheric bacteria help in plant growth promotion in an indirect way by acting as a biocontrol agent. Rhizospheric microbes that reduce the ability or severity of plant diseases are regarded as biocontrol agents (Beneduzi et al. 2012). Biocontrol is the phenomenon in which organisms are used as natural inhibitors of pests/phytopathogens and thereby help the plant physiology and promote its health (Singh et al. 2017). Phytopathogenic microorganisms are a major cause of reduction in the productivity of crops and other plants. Chemical pesticides and fungicides are often used to control the spread of diseases caused by pathogenic organisms. However, excess use of these pesticides and fungicides are chronic threat to sustainable agriculture and ecosystem stability worldwide. They disrupt environment, affect soil fertility, and consequently show harmful effects on human health, along with contaminating groundwater. Plant growth-promoting rhizobacteria is a promising and environmentally friendly approach to obtain sustainable plant growth indirectly (Ahemad and Kibret 2014).

14.4.2.1 Biocontrol Mechanism/Mode of Action of Rhizospheric Communities

Plant growth-promoting rhizobacteria associated with various agricultural crops have shown to promote plant growth by suppressing deleterious pathogenic microflora. *Bacillus* and pseudomonads have emerged as the biggest and potentially most promising groups among PGPRs involved in biocontrol of diseases (Dorjey et al. 2017). Several studies on *Bacillus* and pseudomonads reported their effectiveness as a potential antagonist against pathogens implicated to different plant diseases. Pathogen biocontrol implicates diverse features of rhizobacteria. In general, competition for nutrients, niche exclusion, induced systemic resistance, and production of antifungal metabolites including antibiotics, bacteriocins, and lytic enzymes are the chief modes of biocontrol activity in PGPR (Gupta et al. 2015; Salomon et al. 2017).

14.4.2.1.1 Antibiotic Production

Production of antibiotics is one of the most well-studied and common biocontrol mechanisms employed by plant growth-promoting rhizobacteria against phytopathogens. Antibiotic encompasses a heterogeneous group of low molecular weight organic compounds that are deleterious to the growth or activities of microorganism. A variety of antibiotics have been isolated from fungal and bacterial strains that belong to different classes, including polyketides like 2,4-diacetylphloroglucinol (DAPG), pyoluteorin, mupirocin, heterocyclic nitrogenous compounds including phenazine derivatives: pyocyanin, phenazine-1-carboxylic acid; phenazine-1-carboxylate, phenazine-1-carboxamide and hydroxyphenazines, phenylpyrrole (pyrrolnitrin), cyclic lipopeptides, lipopeptides, etc. Apart from the production of antibiotic, some rhizobacteria are also capable of producing antagonistic volatile compound including hydrogen cyanide (HCN), aldehydes, alcohols, ketones, etc. (Ahemad and Kibret 2014; Gupta et al. 2015; Ulloa-ogaz et al. 2015; Singh et al. 2017).

Each of these antibiotics has a different mode of action, some attack the cellular membranes and inhibit the synthesis of pathogen cell wall, and others have inhibitory effects on the ribosome or other cellular constituents (Ulloa-ogaz et al. 2015; Singh et al. 2017). A cascade of endogenous signals such as sensor kinases, *N*-acetyl homoserine lactones, and sigma factors regulate synthesis of antibiotics (Fernando et al. 2005). In addition to direct anti-pathogenic action, antibiotics also serve as determinants in triggering induced systemic resistance (ISR) in plant system and contribute to disease suppression by conferring a competitive advantage to biocontrol agent (Fernando et al. 2005; Singh et al. 2017).

14.4.2.1.2 Induced Systemic Response

Induced systemic resistance is a physiological state of enhanced defensive ability developed by plant against subsequent biotic challenges stimulated in response to specific environmental stimuli (Gupta et al. 2015). Rhizobacteria can also provide systemic resistance against a range of plant pathogens including bacterial, fungal, and viral pathogens and in some instances even against the damage caused by insects and nematodes. Rhizobacterial ISR resembles pathogen-induced systemic acquired resistance (SAR). Common feature in both is that induced resistance makes uninfected plant part more resistant to plant pathogens. Ethylene and jasmonate are the main hormones involved in ISR that stimulate the host plant's defense responses against a variety of plant pathogens (Beneduzi et al. 2012).

ISR involves jasmonate and ethylene signaling within the plant, and these hormones stimulate the host plant's defense responses against a variety of plant pathogens (Beneduzi et al. 2012). Many individual bacterial components such as lipopolysaccharides (LPS), siderophores, 4-diacetylphloroglucinol, homoserine lactones, acetoin, 2,3-butanediol, etc. can induce ISR (Ahemad and Kibret 2014; Gupta et al. 2015). Failure to elicit ISR in certain hosts may be due to absence of production of inducing components in rhizosphere or inability of a particular plant species to perceive such compounds (Beneduzi et al. 2012).

14.4.2.1.3 Enzyme Production

In addition to antibiosis and induced systemic resistance, hydrolytic enzymes produced by rhizobacteria play a very significant role in plant growth promotion by protecting the plants from biotic stress by suppression of pathogenic fungi. Certain enzymes synthesized by rhizobacterial strains include chitinases, dehydrogenase, β-glucanase, lipases, phosphatases, proteases, etc. (Gupta et al. 2015). Fungal pathogens are more susceptible to the attack of enzymes produced by rhizobacteria. Chitin and β-glucan are the chief components of fungal cell wall; therefore, chitinase and β-glucanase producing rhizobacteria in particular could inhibit the growth of fungal pathogens (Vejan et al. 2016). Different studies have reported the isolation of lytic enzymes producing rhizobacteria that aid the plants against disease-causing pathogens. For instance, *Pseudomonas fluorescens* isolated from the soil of tomato rhizosphere has been found to inhibit growth of *Fusarium udum* causing *Fusarium* wilt. The strain is known to produce chitinase and β-glucanase enzymes in addition to IAA and siderophores (Kumar et al. 2010). Similarly, in another study the role of chitinase and β-glucanase has been highlighted in inhibiting phytopathogens *Rhizoctonia solani* and *Phytophthora capsici* (Arora et al. 2008).

14.4.2.1.4 Siderophore Production

Siderophores have been considered as an eco-friendly alternative to pesticides and play an important role in biological control mechanism. It confers competitive advantage to PGPR over other microbes that share the same ecological niche (Beneduzi et al. 2012). By sequestering $Fe3^{+}$ in the area around the root, siderophore-producing plant growth-promoting rhizobacteria can prevent the proliferation of pathogenic microorganisms and thus promote plant growth in an indirect way. Perusal of the literature has indicated that rhizospheric species of *Pseudomonas* have been known for more than three decades for protecting the plants against pathogens by the production of siderophores. Siderophores produced by the *Pseudomonas* rhizobacteria are known for their high affinity to Fe ion as compared to other rhizobacteria. The best known examples of siderophore-producing pseudomonads as biocontrol agents are *Pseudomonas fluorescens*, *P. putida*, *P. syringae*, etc. (Shanmugaiah et al. 2015). Pyoverdine and pyochelin are the important siderophores reported from *Pseudomonas* sp. by which they control the growth of pathogens. Ahmed and Holmstrom (2014) in their review have quoted the role of pyoverdine by *Pseudomonas* sp. in controlling the pathogen *Fusarium oxysporum*, a causal agent of wilt disease of potato (Schippers et al. 1987). Similarly, *Pseudomonas putida* has been reported to inhibit *Fusarium oxysporum* by producing pseudobacin siderophore (Kloepper et al. 1980). In addition to pseudomonads, other siderophore-producing rhizobacteria have also been reported to be used as biocontrol agents. *Bacillus subtilis* is among the beneficial rhizobacteria that can protect plant against pathogenic attack and also helps in promoting plant growth. Patil et al. (2014) have reported siderophore-producing *B. subtilis* that possessed antiphytopathogenic activity against wilt and dry rot causing fungi in chickpea. Likewise, *Bacillus subtilis* that has biocontrol effect on *Fusarium* wilt of pepper has been documented by Yu et al. (2011).

Recent studies on rhizobacteria suggest that these can be considered as treasured biocontrol agents that play important and indispensable role in promoting plant growth by suppressing the disease incidence and induce resistance in plants against pathogens (Table 14.1).

14.5 Techniques to Study Rhizosphere Microbial Communities

Rhizospheric microbes are found to be involved in positive, negative, as well as neutral relationships with the host plant. In order to unravel the plant–microbe interactions at root level, it is important to study the system as a whole. For understanding the ecology and role of such microbes, deep knowledge of the plant–microbe interactions involved are needed to be addressed. Plant growth-promoting rhizomicrobes can be successfully used as biofertilizers since they are endowed with the properties like biofertilization, phytostimulation,

Table 14.1 Rhizobacteria as a source of biocontrol agents against plant pathogens (2012–2017)

Pathogen/causal organism	Disease caused by pathogen	Biocontrol agent	Source of biocontrol agent	Mechanism of biocontrol agent/effect on pathogen	References
Rhizoctonia solani, *Fusarium solani*, *Fusarium oxysporum*, *Macrophomina phaseolina*	Root rot of Strawberry	*Trichoderma album*, *T. harzianum*, *T. viride*, *T. hamatum*	Rhizosphere of strawberry	In vitro antagonistic activity	Ahmed and El-Fiki (2017)
Colletotrichum acutatum, *Verticillium dahliae*, *Phytophthora cinnamomi*, *Phytophthora cactorum*, *Botryotinia fuckheliana*, *Fusarium oxysporum*	Not mentioned	*Aeromonas* sp.	Rhizosphere of rice	Production of hydrolytic enzymes and plant growth promotion	Aarab et al. (2017)
Fusarium oxysporum	Fusarium wilt	*Pseudomonas fluorescens*, *Pseudomonas putida*	Not specified	Induced systemic resistance	Boukerma et al. (2017)
Ralstonia solanacearum	Bacterial wilt	*Streptomyces toxytricini*, *Stenotrophomonas maltophilia*, *Bacillus pseudomycoides*, *Brevibacillus brevis*	Not specified	Antibiosis	Hassan et al. (2017)
Fusarium oxysporum	Castor wilt	*Bacillus* sp.	Rhizosphere of Castor	Antifungal secondary metabolites	Janga et al. (2017)
Pythium myriotylum, *Rhizoctonia solani*, *Phytophthora infestans*, *Sclerotium rolfsii*, *Fusarium oxysporum*	Not specified	*Serratia* sp.	Rhizosphere of *Bacopa monnieri*	Prodigiosin pigment with antifungal activity	Jimtha et al. (2017)
Fusarium oxysporum	Fusarium wilt	*Burkholderia* sp.	Root nodules of fenugreek (*Trigonella foenum-graecum*)	Chitinase, β-1,3-glucanase, ACC deaminase activity, and plant growth promotion	Kumar et al. (2017)
Macrophomina phaseolina, *Fusarium oxysporum*, *Alternaria alternata*, *Colletotrichum* sp.	Not mentioned	*Bacillus* sp.	Rhizosphere of rice	Siderophore production and lytic enzymes	Mandal et al. (2017)

Alternaria alternata, Helminthosporium oryzae, Penicillium digitatum, Fusarium oxysporum	Not mentioned	*Bacillus subtilis, Cellulosimicrobium cellulans*	Rhizosphere of Tomato	Antifungal activity and plant growth promotion	Sarbadhikary and Mandal (2017)
Erwinia chrysanthemi	Soft rot of aloe vera	*Pseudomonas fluorescens, Bacillus subtilis*	Rhizosphere of aloe vera	Siderophore production	Syamala and Sivaji (2017b)
Meloidogyne javanica, Ditylenchus spp. (nematodes)	Not mentioned	*Pseudomonas* sp., *Bacillus* sp.	Rhizosphere of garlic	Production of chitinase, other lytic enzymes, and cell wall-degrading substances	Turatto et al. (2017)
Sclerotinia sclerotiorium	Stem rot of carnations	*Bacillus* sp.	Rhizosphere of different plants	Antibiotic production and volatile metabolites	Vinodkumar et al. (2017)
Rhizoctonia solani	Root rot of tomato	*Bacillus thuringiensis, Bacillus subtilis, Enterobacter cloacae*	Rhizosphere of tomato	Antagonism by diffusible and volatile metabolites	Abdeljalil et al. (2016)
Fusarium oxysporum	Fusarium wilt of chickpea	*Bacillus aneurinilyticus, Pseudomonas luteola, Pseudomonas fluorescens, Bacillus firmus*	Rhizosphere of Chickpea	Antifungal volatile and extracellular compounds	Abed et al. (2016)
Ascochyta rabiei	Ascochyta blight	*Pseudomonas fluorescens, P. putida, Burkholderia multivorans, Mesorhizobium ciceri*	Rhizosphere of chickpea	Production of hydrolytic enzymes like chitinases, proteases	Azizpour and Rouhrazi (2016)
Fusarium oxysporum, Fusarium solani, Phoma herbarum	Root rot of *Panax notoginseng*	*Bacillus siamensis, Bacillus atrophaeus*	Rhizospheric soil	Not specified	Fan et al. (2016)
Xanthomonas axonopodis pv. *phaseoli*	Common bacterial blight	Not identified	Rhizosphere of bean	Not specified	Giorgio et al. (2016)
Fusarium oxysporum	Wilt in tomato	*Bacillus amyloliquefaciens*	Rhizospheric soil	Antifungal secondary metabolite production	Gowtham et al. (2016)
Colletotrichum gloeosporioides, Fusarium oxysporum	Disease of coffee	*Pseudomonas putida*	Rhizosphere of *Coffee arabica*	Synergistic effects of secondary metabolites, lytic enzymes, and siderophore	Kejela et al. (2016)

(continued)

Table 14.1 (continued)

Pathogen/causal organism	Disease caused by pathogen	Biocontrol agent	Source of biocontrol agent	Mechanism of biocontrol agent/effect on pathogen	References
Colletotrichum gloeosporioides, *C. acutatum*, *Botryosphaeria dothidea*	Anthracnose, white rot	*Paenibacillus polymyxa*, *Bacillus subtilis*	Not mentioned	Production of chitinase, amylase, and protease enzymes	Kim et al. (2016)
Xanthomonas campestris	Black rot of cabbage	Rhizobacteria AP136, AP188, AP209, AP213, AP217, AP218, AP219, AP282, AP295, AP305 (*not identified*)	Not mentioned	Induced systemic resistance	Liu et al. (2016)
Fusarium oxysporum f. sp. *melonis*	Vascular wilt	*Pseudomonas fluorescens*, *Burkholderia* sp., *Bacillus* sp., *Streptomyces* sp.	Rhizosphere of melon	Production of antibiotic, siderophores, and exogenous cell liquid	Mahdikhani and Davoodi (2016)
Fusarium solani	Fusarium wilt of tomato	*Bacillus cereus*, *Bacillus pumilus*	Rhizospheric soil	Production of siderophores and antifungal compounds	Mangalanayaki and Durga (2016)
Heterobasidion annosum, *Armillaria mellea*	Not mentioned	*Pseudomonas fluorescens*, *Bacillus simplex*	Rhizosphere of *Pinus radiata*	Not mentioned	Mesanza et al. (2016)
Sclerotinia sclerotiorum	Stem rot of tomato	*B. subtilis*, *B. thuringiensis*, *B. amyloliquefaciens*, *Enterobacter cloacae*	Rhizospheric of tomato	Inhibitory effects of diffusible volatile metabolites and plant growth promotion	Ouhaibi-Ben Abdeljalil et al. (2016)
Pyricularia sp.	Rice blast	Members of the genus *Bacillus*	Rhizosphere of rice	Production of hydrolytic enzymes protease, glucanase, and cellulase and siderophore	Rais et al. (2016)
Sclerotium rolfsii	Stem rot of groundnut	*Bacillus thuringiensis*	Rhizosphere of healthy plants	Volatile metabolites and siderophore	Rakh and Dalvi (2016)
Ralstonia solanacearum	Bacterial wilt	*Bacillus* sp., *Pseudomonas* spp., and *Serratia* spp.	Rhizosphere of potato	Production of siderophore and chitinase and plant growth promotion	Tahir et al. (2016)

Dematophora necatrix	Apple plant pathogen (disease not mentioned)	*Pseudomonas putida*	Rhizosphere of apple	Not mentioned	Verma et al. (2016)
Botrytis cinerea	Gray mold disease	*Bacillus* sp.	Soil samples collected in garlic fields	Induced expression of pathogenesis-related proteins and plant growth promotion	Xu et al. (2016)
Phytophthora palmivora	Black pod rot	*Pseudomonas chlororaphis*	Rhizosphere of *Theobroma cacao*	Not specified	Acebo-Guerrero et al. (2015)
Pythium myriotylum	Soft rot in ginger	*Bacillus amyloliquefaciens*, *Serratia marcescens*	Rhizosphere of different varieties of ginger	Production of secondary metabolites and competing for colonization sites, nutrients	Dinesh et al. (2015)
Xanthomonas gardneri	Tomato bacterial spot	*Streptomyces setonii*, *Bacillus cereus*, *Serratia marcescens*	Rhizospheric soil	Induced systemic resistance	Ferraz et al. (2015)
Fusarium oxysporum, *Macrophomina phaseolina*, *Fusarium solani*	Root-infecting pathogens	*Paenibacillus illinoisensis*, *Bacillus subtilis*, *Pseudomonas psychrotolerans*	Rhizosphere of groundnut	Production of antifungal metabolites and plant growth promotion	Inam-ul-Haq et al. (2015)
Phytophthora capsici	Crown rot in cucumber	*Pseudomonas stutzeri*, *Bacillus subtilis*, *Stenotrophomonas maltophilia*, *Bacillus amyloliquefaciens*	Rhizosphere of cucumber	Antibiosis, competitive colonization, and plant growth promotion	Islam et al. (2016)
Rhizoctonia solani	Rice sheath blight	*Bacillus cereus*, *Enterobacter* sp., *Aeromonas hydrophila*, *Enterobacter* sp.	Rhizosphere of rice	Production of siderophore	Naureen et al. (2015)
Ralstonia solanacearum	Bacterial wilt of Eucalyptus	*Bacillus thuringiensis*, *Bacillus cereus*	Rhizosphere of Tomato and Eucalyptus	Production of volatile compounds	Santiago et al. (2015)

(continued)

Table 14.1 (continued)

Pathogen/causal organism	Disease caused by pathogen	Biocontrol agent	Source of biocontrol agent	Mechanism of biocontrol agent/effect on pathogen	References
Fusarium oxysporum, *Verticillium* sp.	Root-infecting fungi	*Pseudomonas* sp., *Bacillus* sp.	Rhizosphere of chickpea	Inhibition of mycelial growth, production of siderophores, and plant growth promotion	Shahzaman et al. (2015)
Rhizoctonia oryzae-sativae	Aggregate sheath spot	*Pseudomonas aeruginosa*	Rhizosphere of cereal crops	Antagonism and plant growth promotion	Gad et al. (2014)
Aspergillus niger	Collar rot	*Pseudomonas* sp.	Rhizosphere of groundnut	Antifungal secondary metabolites	Lukkani and Reddy (2014)
Rhizoctonia solani	Not mentioned	Not identified	Rhizosphere of sugarcane	Not specified	Patel et al. (2014)
Rosellinia necatrix	White root rot of avocado	*Trichoderma atroviride*, *Pseudomonas chlororaphis*, *Pseudomonas pseudoalcaligenes*	Not mentioned	Not mentioned	Ruano-Rosa et al. (2014)
Stemphylium lycopersici	Gray leaf spot	*Brevibacterium iodinum*	Rhizosphere of *Elymus tsukushiensis*	*Systemic acquired resistance and pathogenesis-related (PR)* proteins	Son et al. (2014)
Ralstonia solanacearum	Bacterial wilt	*Pseudomonas aeruginosa*, *Pseudomonas putida*	Rhizospheric soil	*Not mentioned*	Maji and Chakrabartty (2014)
Rhizoctonia solani	Scurf and stem canker diseases of potato	*Bacillus* sp., *Pseudomonas* sp.	Rhizospheric soil	Siderophore and antibiotics	Kumar et al. (2013)
Ralstonia solanacearum	Bacterial wilt	*Bacillus cereus*, *Pseudomonas putida*	Rhizosphere of potato	Antibiosis	Kurabachew and Wydra (2013)

Phytophthora capsici	Phytophthora blight and anthracnose of pepper	*Pseudomonas otitidis*, *P. putida*, *Tsukamurella tyrosinosolvens*, *Novosphingobium capsulatum*	Not mentioned	Not specified	Sang et al. (2013)
Fusarium oxysporum, *Rhizoctonia solani*, *Sclerotinia sclerotiorum*	Not mentioned	*Bacillus subtilis*	Rhizosphere of tomato	Antibiosis	Walia et al. (2013)
Colletotrichum acutatum	Anthracnose of pepper	*Bacillus*, *Paenibacillus*	Rhizospheric soil	Production of volatile compounds and antibiosis	Lamsal et al. (2012)
Macrophomina phaseolina, *Rhizoctonia bataticola*, *Rhizoctonia solani*, *F. oxysporum*, *C. gloeosporioides*, *F. solani*, *Sclerotium rolfsii*, *Cercospora capsici*, *Alternaria sesame*, *Xanthomonas axonopodis pv. punicae*	Not mentioned	*Pseudomonas fluorescens*	Rhizosphere of various crop plants	Antagonism by dual culture method	Manjunatha et al. (2012)
Erwinia carotovora	Potato soft rot	*Bacillus* sp.	Rhizospheric of various crop plants	Antagonistic activity	Rahman et al. (2012)
Fusarium oxysporum, *Aspergillus* sp.	Not mentioned	*Pseudomonas fluorescens*	Rhizosphere of wheat	Not specified	Showkat et al. (2012)

rhizoremediation, and phytopathogen biocontrol, but their actual application in the field is very limited (Babalola 2010). This is due to the fact that they show very low efficiency at field level as compared to their actual efficiency. Therefore, keen studies on all aspects of the ecology of plant- and rhizosphere-inhabited microbes as a system are the need of the hour. Various advanced tools and techniques can prove to be helpful to understand such interactions, and thus positive interactions can be exploited for sustainable agriculture. Omic as well as metaomic techniques can be successfully used to study the plant rhizosphere system as a whole. There is lack of information regarding the diversity, competence, colonization, distribution, communication, etc. of the microbe in rhizosphere. Advancement in omic technologies has accelerated the studies regarding host–microbe interactions. Genomic and metagenomic techniques can be used to unravel the actual composition of rhizosphere-associated microbes. Pascual et al. (2016) have successfully used genomic as well as metagenomic technique to explore the rhizosphere bacterial community of the medicinal plant, *Thymus zygis*. Transcriptome, metatranscriptome, proteome, and metaproteome techniques can also be successfully used to identify the core functional communities of the rhizosphere system (Kothari et al. 2016). Additionally, metabolomics and secretomics techniques can further add to our knowledge about the complex interactions involved (Savka et al. 2013). Classical techniques like DGGE, T-RFLP, ARDRA, DNA cloning, and Sanger sequencing are still being potentially used in rhizosphere studies, whereas modern omic tools like next-generation sequencing (NGS), FISH, SIP, microarray, etc. can also be successfully used for quick and comprehensive study of the rhizomicrobiomes (Hao and Xiao 2017). The modern tools and techniques can lead us to the successful construction of biased phytospheres (Savka et al. 2013) and thus help us in better understanding the basis of interactions between host plant and rhizospheric microbes. This can help designing the strategies for improved crop yield and hence sustainable agriculture.

14.6 Conclusions

Rhizomicrobiome represents a complex and dynamic system involving continuous interactions between plant–microbe as well as microbe–microbe. The composition and functional role of rhizomicrobes is governed by various factors as discussed above. Modern tools and techniques are continuously improving our knowledge about this complex niche. Integrated omic tools along with their metaomic partners and modeling techniques can prove to be helpful in solving the complex riddle of rhizosphere system as a whole. By using such knowledge, efficient strategies are needed to be designed for rhizosphere management. Successful understanding and thus feasible exploitation of rhizomicrobes as biofertilizers at ground level over chemical fertilizers will surely prove to be a step forward toward sustainable development in general and sustainable agriculture in particular.

Acknowledgments The authors are thankful to the Department of Biotechnology, Govt. of India, for facilities for conducting work on microbial endobiome of medicinal plants. SK acknowledges DBT, Govt. of India, for funding of the research project on bioprospecting fungal endophytes of medicinal plants (BT/PR9538/NDB/39/425/2013). SG is thankful to the Department of Biotechnology, Govt. of India, for fellowship, DBT JRF, and SRF (DBT-JRF/2011-12/252). Coordinator Bioinformatics Centre, DBT-BIF, School of Biotechnology, is also acknowledged for providing facilities.

References

Aarab S, Ollero FJ, Megias M, Laglaoui A, Bakkali M, Arakrak A (2017) Simultaneous P-solubilizing and biocontrol activity of rhizobacteria isolated from rice rhizosphere soil. In: Kumar V, Kumar M, Sharma S, Prasad R (eds) Probiotics in agroecosystem. Springer, Singapore, pp 207–215

Abdeljalil NOB, Vallance J, Gerbore J, Bruez E, Martins G, Rey P, Daami RM (2016) Biocontrol of *Rhizoctonia* root rot in tomato and enhancement of plant growth using rhizobacteria naturally associated to tomato. J Plant Pathol Microbiol 7:356. https://doi.org/10.4172/2157-7471.1000356

Abed H, Rouag N, Mouatassem D, Rouabhi A (2016) Screening for *Pseudomonas* and *Bacillus* antagonistic rhizobacteria strains for the biocontrol of *Fusarium* wilt of chickpea. Eur J Soil Sci 5(3):166–182

Acebo-Guerrero Y, Hernandez-Rodríguez A, Vandeputte O, Miguelez-Sierra Y, Heydrich-Pérez M, Ye L, Cornelis P, Bertin P, El Jaziri M (2015) Characterization of *Pseudomonas chlororaphis* from *Theobroma cacao* L. rhizosphere with antagonistic activity against *Phytophthora palmivora* (Butler). J Appl Microbiol 119(4):1112–1126

Adesemoye A, Kloepper J (2009) Plant–microbe interactions in enhanced fertilizer-use efficiency. Appl Microbiol Biotechnol 85:1–12

Ahemad M, Kibret M (2014) Mechanisms and applications of plant growth promoting rhizobacteria: current perspective. J King Saud Univ Sci 26(1):1–20

Ahmed MF, El-Fiki IA (2017) Effect of biological control of root rot diseases of Strawberry using *Trichoderma* spp. Sciences 7(3):482–492

Ahmed E, Holmstrom SJ (2014) Siderophores in environmental research: roles and applications. Microb Biotechnol 7(3):196–208

Ali M, Shahid AA, Haider MS (2016) Isolation and in-vitro screening of potential antagonistic rhizobacteria against *Pythium debaryanum*. Pak J Phytopathol 28(2):231–240

Apastambh AR, Tanveer K, Baig MM (2016) Isolation and characterization of Plant growth promoting rhizobacteria from Banana rhizosphere. Int J Curr Microbiol App Sci 5(2):59–65

Arora NK, Khare E, Oh JH, Kang SC, Maheshwari DK (2008) Diverse mechanisms adopted by fluorescent *Pseudomonas* PGC2 during the inhibition of *Rhizoctonia solani* and *Phytophthora capsici*. World J Microbiol Biotechnol 24(4):581–585

Azizpour N, Rouhrazi K (2016) Isolation and characterization of rhizosphere bacteria for the biocontrol of the *Asochyta Rabiei* in Iran. Adv Plants Agric Res 3(4):00104

Babalola OO (2010) Beneficial bacteria of agricultural importance. Biotechnol Lett 32(11):1559–1570

Bal HB, Das S, Dangar TK, Adhya TK (2013) ACC deaminase and IAA producing growth promoting bacteria from the rhizosphere soil of tropical rice plants. J Basic Microbiol 53(12):972–984

Barnawal D, Pandey SS, Bharti N, Pandey A, Ray T, Singh S, Kalra A (2017) ACC deaminase-containing plant growth-promoting rhizobacteria protect *Papaver somniferum* from downy mildew. J Appl Microbiol 122(5):1286–1298

Beneduzi A, Ambrosini A, Passaglia LM (2012) Plant growth-promoting rhizobacteria (PGPR): their potential as antagonists and biocontrol agents. Genet Mol Biol 35(4):1044–1051

Bhattacharyya PN, Jha DK (2012) Plant growth-promoting rhizobacteria (PGPR): emergence in agriculture. World J Microbiol Biotechnol 28(4):1327–1350

Boukerma L, Benchabane M, Charif A, Khelifi L (2017) Activity of plant growth promoting rhizobacteria (PGPRs) in the biocontrol of tomato *Fusarium* wilt. Plant Protect Sci 53(2):74–84

Bulgarelli D, Ruben G, Munch PC, Weiman A, Droge J, Pan Y, McHardy AC, Schulze-Lefert P (2015) Structure and function of the bacterial root microbiota in wild and domesticated barley. Cell Host Microbe 17(3):392–403

Chauhan A, Guleria S, Balgir PP, Walia A, Mahajan R, Mehta P, Shirkot CK (2017) Tricalcium phosphate solubilization and nitrogen fixation by newly isolated *Aneurinibacillus aneurinilyticus* CKMV1 from rhizosphere of *Valeriana jatamansi* and its growth promotional effect. Braz J Microbiol 48(2):294–304

Di Benedetto NA, Corbo MR, Campaniello D, Cataldi MP, Bevilacqua A, Sinigaglia M, Flagella Z (2017) The role of plant growth promoting bacteria in improving nitrogen use efficiency for sustainable crop production: a focus on wheat. AIMS Microbiol 3(3):413–434

Dinesh R, Anandaraj M, Kumar A, Bini YK, Subila KP, Aravind R (2015) Isolation, characterization, and evaluation of multi-trait plant growth promoting rhizobacteria for their growth promoting and disease suppressing effects on ginger. Microbiol Res 173:34–43

Dorjey S, Dolkar D, Sharma R (2017) Plant growth promoting rhizobacteria *Pseudomonas*: a review. Int J Curr Microbiol App Sci 6(7):1335–1344

Elias F, Woyessa D, Muleta D (2016) Phosphate solubilization potential of rhizosphere fungi isolated from plants in Jimma Zone, Southwest Ethiopia. Int J Microbiol. https://doi.org/10.1155/2016/5472601

Fan ZY, Miao CP, Qiao XG, Zheng YK, Chen HH, Chen YW, Xu LH, Zhao LX, Guan HL (2016) Diversity, distribution, and antagonistic activities of rhizobacteria of *Panax notoginseng*. J Ginseng Res 40(2):97–104

Felestrino EB, Santiago IF, da Silva FL, Rosa LH, Ribeiro SP, Moreira LM (2017) Plant growth promoting bacteria associated with *Langsdorffia hypogaea*-rhizosphere-host biological interface: a neglected model of bacterial prospection. Front Microbiol. https://doi.org/10.3389/fmicb.2017.00172

Fernando WD, Ramarathnam R, Krishnamoorthy AS, Savchuk SC (2005) Identification and use of potential bacterial organic antifungal volatiles in biocontrol. Soil Biol Biochem 37(5):955–964

Ferraz HG, Resende RS, Moreira PC, Silveira PR, Milagres EA, Oliveira JR, Rodrigues FA (2015) Antagonistic rhizobacteria and jasmonic acid induce resistance against tomato bacterial spot. Bragantia 74(4):417–427

Gaby JC, Buckley DH (2012) A comprehensive evaluation of PCR primers to amplify the nifH gene of nitrogenase. Plos One 7(7):e42149

Gad MA, Deka M, Ibrahim NA, Mahmoud SS, Kharwar RN, Bora TC (2014) Biocontrol of phytopathogenic fungi of rice crop using plant growth-promoting rhizobacteria. In: Kharwar RN, Upadhyay RS, Dubey NK, Raghuwanshi R (eds) Microbial diversity and biotechnology in food security. Springer, New Delhi, pp 225–234

Giorgio A, Cantore PL, Shanmugaiah V, Lamorte D, Iacobellis NS (2016) Rhizobacteria isolated from common bean in southern Italy as potential biocontrol agents against common bacterial blight. Eur J Plant Pathol 144(2):297–309

Gowtham HG, Hariprasad P, Nayak SC, Niranjana SR (2016) Application of rhizobacteria antagonistic to *Fusarium oxysporum* f. sp. lycopersici for the management of *Fusarium* wilt in tomato. Rhizosphere 2:72–74

Gupta G, Parihar SS, Ahirwar NK, Snehi SK, Singh V (2015) Plant growth promoting rhizobacteria (PGPR): current and future prospects for development of sustainable agriculture. J Microb Biochem Technol 7(2):96–102

Haile D, Mekbib F, Assefa F (2016) Isolation of phosphate solubilizing bacteria from white Lupin (*Lupinus albus* L.) rhizosphere soils collected from Gojam, Ethiopia. J Fertil Pestic 7:172. https://doi.org/10.4172/2471-2728.1000172

Harman GE, Howell CR, Viterbo A, Chet I, Lorito M (2004) Trichoderma species-opportunistic, avirulent plant symbionts. Nat Rev Microbiol 2(11):43–56

Hao D, Xiao P (2017) Rhizosphere microbiota and microbiome of medicinal plants: from molecular biology to omics approaches. Chin Herb Med 9:199–217

Hartmann A, Rothballer M, Schmid M, Hiltner L (2008) A pioneer in rhizosphere microbial ecology and soil bacteriology research. Plant Soil 312:7–14

Hassan EA, Balabel NM, Ahmed AE, Eid NA (2017) Relationship between *Ralstonia solanacearum* and bioagents recovered from different habitats. Int J Sci Eng Res 8(1):91–104

Hayat R, Ali S, Amara U, Khalid R, Ahmed I (2010) Soil beneficial bacteria and their role in plant growth promotion: a review. Ann Microbiol 60(4):579–598

Inam-Ul-Haq M, Tahir M, Hayat R, Khalid R, Ashfaq M (2015) Bioefficacy of rhizobacterial against root infecting fungal pathogens of Chickpea (*Cicer arietinum* L.) J Plant Pathol Microbiol S3:011. https://doi.org/10.4172/2157-7471.S3-011

Islam S, Akanda AM, Prova A, Islam MT, Hossain MM (2016) Isolation and identification of plant growth promoting rhizobacteria from cucumber rhizosphere and their effect on plant growth promotion and disease suppression. Front Microbiol 6:1–10. https://doi.org/10.3389/fmicb.2015.01360

Jacobsen BJ, Zidack NK, Larson BJ (2004) The role of *Bacillus*- based biological control agents in integrated pest management systems. Phytopathology 94:1272–1275

Janga MR, Raoof MA, Ulaganathan K (2017) Effective biocontrol of *Fusarium* wilt in castor (*Ricinius communis* L.) with *Bacillus* sp. in pot experiments. Rhizosphere 3:50–52

Jha CK, Annapurna K, Saraf M (2012) Isolation of rhizobacteria from *Jatropha curcas* and characterization of produced ACC deaminase. J Basic Microbiol 52(3):285–295

Jimtha CJ, Jishma P, Sreelekha S, Chithra S, Radhakrishnan EK (2017) Antifungal properties of prodigiosin producing rhizospheric *Serratia* sp. Rhizosphere 30(3):105–108

Kejela T, Thakkar VR, Thakor P (2016) *Bacillus* species (BT42) isolated from *Coffea arabica* L. rhizosphere antagonizes *Colletotrichum gloeosporioides* and *Fusarium oxysporum* and also exhibits multiple plant growth promoting activity. BMC Microbiol 16(1):277

Kim YS, Balaraju K, Jeon Y (2016) Effects of rhizobacteria *Paenibacillus polymyxa* APEC136 and *Bacillus subtilis* APEC170 on biocontrol of postharvest pathogens of apple fruits. J Zhejiang Univ Sci 17(12):931–940

Kloepper JW, Schroth MN (1978) Plant growth promoting rhizobacteria on radish. In: Station de pathologie végétale et phyto-bacteriologie (eds) Proceedings of the 4th conference plant pathogenic bacteria, Angers, INRA, pp 879–882

Kloepper JW, Leong J, Teintze M, Schiroth MN (1980) Enhanced plant growth by siderophores produced by plant growth promoting rhizobacteria. Nature 286:885–886

Kothari R, Singh RP, Kothari V (2016) Application of next generation sequencing technologies in revealing plant-microbe interactions. Next Gen Seq Appl 3:e108. https://doi.org/10.4172/2469-9853.1000e108

Kumar H, Bajpai VK, Dubey RC, Maheshwari DK, Kang SC (2010) Wilt disease management and enhancement of growth and yield of *Cajanus cajan* (L) var. Manak by bacterial combinations amended with chemical fertilizer. Crop Prot 29(6):591–598

Kumar A, Prakash A, Johri BN (2011) *Bacillus* as PGPR in crop ecosystem. In: Maheshwari DK (ed) Bacteria in agrobiology: crop ecosystems. Springer, Berlin, pp 37–59

Kumar SS, Rao MRK, Kumar RD, Panwar S, Prasad CS (2013) Biocontrol by plant growth promoting rhizobacteria against black scurf and stem canker disease of potato caused by *Rhizoctonia solani*. Arch Phytopathol Plant Prot 46(4):487–502

Kumar H, Dubey RC, Maheshwari DK (2017) Seed-coating fenugreek with *Burkholderia* rhizobacteria enhances yield in field trials and can combat *Fusarium* wilt. Rhizosphere 30 (3):92–99

Kurabachew H, Wydra K (2013) Characterization of plant growth promoting rhizobacteria and their potential as bioprotectant against tomato bacterial wilt caused by *Ralstonia solanacearum*. Biol Control 67(1):75–83

Lamsal K, Kim SW, Kim YS, Lee YS (2012) Application of rhizobacteria for plant growth promotion effect and biocontrol of anthracnose caused by *Colletotrichum acutatum* on pepper. Mycobiology 40(4):244–251

Li Y, Liu X, Hao T, Chen S (2017) Colonization and maize growth promotion induced by phosphate solubilizing bacterial isolates. Int J Mol Sci 18(7):1–16

Liu K, Garrett C, Fadamiro H, Kloepper JW (2016) Induction of systemic resistance in Chinese cabbage against black rot by plant growth-promoting rhizobacteria. Biol Control 99:8–13

Loper JE, Gross H (2007) Genomic analysis of antifungal metabolite production by *Pseudomonas fluorescens* Pf-5. Eur J Plant Pathol 119:265–278

Lukkani NJ, Reddy ES (2014) Evaluation of plant growth promoting attributes and biocontrol potential of native fluorescent *Pseudomonas* spp. against *Aspergillus niger* causing collar rot of ground nut. IJPAES 4(4):256–262

Lupwayi NZ, Rice WA, Clayton GW (1998) Soil microbial diversity and community structure under wheat as influenced by tillage and crop rotation. Soil Biol Biochem 30:1733–1741

Mahadevamurthy M, Sidappa M, Thriveni MC, Mythrashree SR, Amruthesh KN (2016) Isolation of phosphate solubilizing fungi from rhizosphere soil and its effect on seed growth parameters of different crop plants. J Appl Biol Biotechnol 4(6):22–26

Mahaffee WF, Kloepper JW (1997) Temporal changes in the bacterial communities of soil, rhizosphere, and endorhiza associated with fieldgrown cucumber (*Cucumis sativus* L.) Microbiol Ecol 34:210–223

Mahdikhani M, Davoodi A (2016) Evaluation of biocontrol potential of rhizosphere antagonist bacterial strains on *Fusarium* wilt and plant growth in muskmelon plants. Am Eur J Sustain Agric 10(6):15–23

Maji S, Chakrabartty PK (2014) Biocontrol of bacterial wilt of tomato caused by *Ralstonia solanacearum* by isolates of plant growth promoting rhizobacteria. Aust J Crop Sci 8 (2):208–214

Mandal S, Dutta P, Majumdar S (2017) Plant growth promoting and antagonistic activity of *Bacillus* strains isolated from rice rhizosphere. Int J Pharm Biomed Sci 8(1):408–415

Mangalanayaki R, Durga R (2016) Antagonistic effect of *Bacillus* species in biocontrol of plant pathogen *Fusarium*. World J Pharm Pharm Sci 5(6):956–966

Manjunatha H, Naik MK, Patil MB, Lokesha R, Vasudevan SN (2012) Isolation and characterization of native fluorescent pseudomonads and antagonistic activity against major plant pathogens. Karnataka J Agric Sci 25(3):346–349

Mendes R, Garbeva P, Raaijmakers JM (2013) The rhizosphere microbiome: significance of plant beneficial, plant pathogenic, and human pathogenic microorganisms. FEMS Microbiol Rev 37 (5):634–663

Mesanza N, Iturritxa E, Patten CL (2016) Native rhizobacteria as biocontrol agents of *Heterobasidion annosum* s.s and *Armillaria mellea* infection of *Pinus radiata*. Biol Control 101:8–16

Mohite B (2013) Isolation and characterization of indole acetic acid (IAA) producing bacteria from rhizospheric soil and its effect on plant growth. J Soil Sci Plant Nutr 13(3):638–649

Morgan JAW, Whipps JM (2001) Methodological approaches to the study of rhizosphere carbon flow and microbial population dynamics. In: Pinton A, Varanini Z, Nannipieri P (eds) The rhizosphere: biochemistry and organic substances at the soil-plant interface. Marcel Dekker, New York, pp 373–409

Moustaine M, Elkahkahi R, Benbouazza A, Benkirane R, Achbani EH (2017) Effect of plant growth promoting rhizobacterial (PGPR) inoculation on growth in tomato (*Solanum lycopersicum* L.) and characterization for direct PGP abilities in Morocco. Int J Environ Agric Biotechnol 2:590–596

Muleta D, Assefa F, Borjesson E, Granhall U (2013) Phosphate-solubilising rhizobacteria associated with *Coffea arabica* L. in natural coffee forests of southwestern Ethiopia. J Saudi Soc Agric Sci 12(1):73–84

Naml A, Mahmood A, Sevilir B, Ozkır E (2017) Effect of phosphorus solubilizing bacteria on some soil properties, wheat yield and nutrient contents. Eur J Soil Sci 6(3):249–258

Naureen Z, Hafeez FY, Hussain J, Al Harrasi A, Bouqellah N, Roberts MR (2015) Suppression of incidence of *Rhizoctonia solani* in rice by siderophore producing rhizobacterial strains based on competition for iron. ESJ 11(3):186–207

Noumavo PA, Agbodjato NA, Baba-Moussa F, Adjanohoun A, Baba-Moussa L (2016) Plant growth promoting rhizobacteria: beneficial effects for healthy and sustainable agriculture. Afr J Biotechnol 15(27):1452–1463

Ouhaibi-Ben Abdeljalil N, Vallance J, Gerbore J, Rey P, Daami-Remadi M (2016) Bio-suppression of *Sclerotinia* stem rot of tomato and biostimulation of plant growth using tomato-associated rhizobacteria. J Plant Pathol Microbiol 7:331. https://doi.org/10.4172/2157-7471.1000331

Pascual J, Blanco S, Garcia-Lopez M, García-Salamanca A, Bursakov SA, Genilloud O (2016) Assessing bacterial diversity in the rhizosphere of *Thymus zygis* growing in the Sierra Nevada national park (Spain) through culture dependent and independent approaches. Plos One 11(1): e0146558. https://doi.org/10.1371/journal.pone.0146558

Patel M, Sheth U, Hajela P, Joshi B, Animasaun DA (2014) Isolation and diversity analysis of Rhizobacteria from sugarcane and its biocontrol potential against *Rhizoctonia solani* a common plant pathogen. Int J Curr Microbiol Appl Sci 3(12):69–76

Patil S, Bheemaraddi CM, Shivannavar TC, Gaddad MS (2014) Biocontrol activity of siderophore producing *Bacillus subtilis* CTS-G24 against wilt and dry root rot causing fungi in chickpea. J Agric Vet Sci 7(9):63–68

Raaijmakers JM, Paulitz TC, Steinberg C, Alabouvette C, Moenne-Loccoz Y (2009) The rhizosphere: a playground and battlefield for soilborne pathogens and beneficial microorganisms. Plant Soil 321(1–2):341–361

Rahman MM, Ali ME, Khan AA, Akanda AM, Uddin MK, Hashim U, Abd Hamid SB (2012) Isolation, characterization, and identification of biological control agent for potato soft rot in Bangladesh. Sci World J. https://doi.org/10.1100/2012/723293

Rais A, Shakeel M, Hafeez FY, Hassan MN (2016) Plant growth promoting rhizobacteria suppress blast disease caused by *Pyricularia oryzae* and increase grain yield of rice. BioControl 61 (6):769–780

Rakh RR, Dalvi SM (2016) Antagonism of *Bacillus thuringiensis* NCIM2130 against *Sclerotium rolfsii* Sacc. a stem rot pathogen of groundnut. Int J Curr Microbiol App Sci 5(8):501–513

Raut V, Shaikh I, Naphade B, Prashar K, Adhapure N (2017) Plant growth promotion using microbial IAA producers in conjunction with *Azolla*: a novel approach. Chem Biol Technol Agric 4(1):1–11

Reed SC, Cleveland CC, Townsend AR (2011) Functional ecology of free-living nitrogen fixation: a contemporary perspective. Annu Rev Ecol Evol Syst 42:489–512

Ruano-Rosa D, Cazorla FM, Bonilla N, Martin-Perez R, De Vicente A, Lopez-Herrera CJ (2014) Biological control of avocado white root rot with combined applications of *Trichoderma* spp. and rhizobacteria. Eur J Plant Pathol 138(4):751–762

Ryan PR, Dessaux Y, Thomashow LS, Weller DM (2009) Rhizosphere engineering and management for sustainable agriculture. Plant Soil 321(1–2):363–383

Salomon MV, Pinter IF, Piccoli P, Bottini R (2017) Use of plant growth-promoting rhizobacteria as biocontrol agents: induced systemic resistance against biotic stress in plants. In: Kalia V (ed) Microbial applications, vol 2. Springer, New Delhi, pp 133–152

Sang MK, Shrestha A, Kim DY, Park K, Pak CH, Kim KD (2013) Biocontrol of *Phytophthora* blight and anthracnose in pepper by sequentially selected antagonistic rhizobacteria against *Phytophthora capsici*. Plant Pathol J 29(2):154–167

Santhanam R, Luu VT, Weinhold A, Goldberg J, Youngjoo O, Baldwin IT (2015) Native root-associated bacteria rescue a plant from a sudden-wilt disease that emerged during continuous cropping. Proc Natl Acad Sci 112(36):E5013–E5020

Santi C, Bogusz D, Franche C (2013) Biological nitrogen fixation in non-legume plants. Ann Bot 111(5):743–767

Santiago TR, Grabowski C, Rossato M, Romeiro RS, Mizubuti ES (2015) Biological control of eucalyptus bacterial wilt with rhizobacteria. Biol Control 80:14–22

Sarbadhikary SB, Mandal NC (2017) Field application of two plant growth promoting rhizobacteria with potent antifungal properties. Rhizosphere 3:170–175

Satyaprakash M, Nikitha T, Reddi EUB, Sadhana B, Vani SS (2017) Phosphorous and phosphate solubilising bacteria and their role in plant nutrition. Int J Curr Microbiol App Sci 6 (4):2133–2144

Savka MA, Dessaux Y, McSpadden Gardener BB, Mondy S, Kohler PRA, de Bruijn FJ, Rossbach S (2013) The "biased rhizosphere" concept and advances in the omics era to study bacterial competitiveness and persistence in the phytosphere. In: de Bruijn FJ (ed) Molecular microbial ecology of the rhizosphere, vol 1 and 2. Wiley, Hoboken. https://doi.org/10.1002/9781118297674.ch110

Schippers B, Bakker AW, Bakker PAH (1987) Interactions of deleterious and beneficial rhizosphere microorganisms and the effect of cropping practices. Annu Rev Phytopathol 25:339–358

Schrey SD, Tarkka MT (2008) Friends and foes: streptomycetes as modulators of plant disease and symbiosis. Antonie Van Leeuwenhoek 94:11–19

Shahzaman SH, Inam-ul-Haq M, Mukhtar TA, Naeem M (2015) Isolation, identification of antagonistic rhizobacterial strains obtained from chickpea (*Cicer arietinum* L.) field and their in-vitro evaluation against fungal root pathogens. Pak J Bot 47(4):1553–1558

Shakeela S, Padder SA, Bhat ZA (2017) Isolation and characterization of plant growth promoting rhizobacteria associated with medicinal plant *Picrorhiza Kurroa*. J Pharmacogn Phytochem 6 (3):157

Shanmugaiah V, Nithya K, Harikrishnan H, Jayaprakashvel M, Balasubramanian N (2015) Bio-control mechanisms of siderophores against bacterial plant pathogens. Sustainable approaches to controlling plant pathogenic bacteria. CRC Press, Boca Raton, pp 167–190

Showkat S, Murtaza I, Laila O, Ali A (2012) Biological control of *Fusarium oxysporum* and *Aspergillus* sp. by *Pseudomonas fluorescens* isolated from wheat rhizosphere soil of Kashmir. J Pharm Biol Sci 1(4):24–32

Singh H, Jaiswal V, Singh S, Tiwari SP, Singh B, Katiyar D (2017) Antagonistic compounds producing plant growth promoting rhizobacteria: a tool for management of plant disease. J Adv Microbiol 3(4):1–12

Son JS, Sumayo M, Hwang YJ, Kim BS, Ghim SY (2014) Screening of plant growth-promoting rhizobacteria as elicitor of systemic resistance against gray leaf spot disease in pepper. Appl Soil Ecol 73:1–8

Spaepen S, Vanderleyden J (2011) Auxin and plant-microbe interactions. Cold Spring Harb Perspect Biol 3(4):a001438

Sujatha N, Ammani K (2013) Siderophore production by the isolates of fluorescent Pseudomonads. IJCRR 5(20):1–7

Syamala M, Sivaji M (2017a) Functional characterization of various plant growth promoting activity of *Pseudomonas fluorescens* and *Bacillus subtilis* from *Aloe vera* rhizosphere. J Pharmacogn Phytochem 6(3):120–122

Syamala M, Sivaji M (2017b) Isolation and functional characterization of plant growth promoting rhizobacteria against soft rot in Aloe vera (L). J Entomol Zool Stud 5(5):187–191

Tahir MI, Inam-ul-Haq M, Ashfaq M, Abbasi NA, Butt H, Ghazal H (2016) Screening of effective antagonists from potato rhizosphere against bacterial wilt pathogen. Int J Biosci 8(2):228–240

Timmusk S, Behers L, Muthoni J, Muraya A, Aronsson AC (2017) Perspectives and challenges of microbial application for crop improvement. Front Plant Sci 8:1–10

Turatto MF, Dourado FD, Zilli JE, Botelho GR (2017) Control potential of *Meloidogyne javanica* and *Ditylenchus* spp. using fluorescent *Pseudomonas* and *Bacillus* spp. Braz J Microbiol. https://doi.org/10.1016/j.bjm.2017.03.015

Ulloa-Ogaz AL, Munoz-Castellanos LN, Nevarez-Moorillon GV (2015) Biocontrol of phytopathogens: antibiotic production as mechanism of control. In: The battle against microbial pathogens: basic science, technological advances and educational programes. Formatex Research Center, Spain, pp 305–309

Vejan P, Abdullah R, Khadiran T, Ismail S, Nasrulhaq Boyce A (2016) Role of plant growth promoting rhizobacteria in agricultural sustainability-a review. Molecules 21(5):573

Verma PP, Thakur S, Kaur M (2016) Antagonism of *Pseudomonas putida* against *Dematophora necatrix* a major apple plant pathogen and its potential use as a biostimulent. J Pure Appl Microbiol 10(4):2717–2727

Vessey JK (2003) Plant growth promoting rhizobacteria as biofertilizers. Plant Soil 255(2):571–586

Vinodkumar S, Nakkeeran S, Renukadevi P, Malathi VG (2017) Biocontrol potentials of antimicrobial peptide producing *Bacillus* species: multifaceted antagonists for the management of stem rot of carnation caused by *Sclerotinia sclerotiorum*. Front Microbiol 8:446. https://doi.org/10.3389/fmicb.2017.00446

Walia A, Mehta P, Chauhan A, Shirkot CK (2013) Antagonistic activity of plant growth promoting rhizobacteria isolated from tomato rhizosphere against soil borne fungal plant pathogens. Int J Agric Environ Biotechnol 6(4):571–580

Welbaum GE, Sturz AV, Dong Z, Nowak J (2004) Managing soil microorganisms to improve productivity of agro-ecosystems. Crit Rev Plant Sci 23(2):175–193

Xiang N, Lawrence KS, Kloepper JW, Donald PA, McInroy JA (2017) Biological control of *Heterodera glycines* by spore-forming plant growth-promoting rhizobacteria (PGPR) on soybean. PLoS One 12(7):e0181201. https://doi.org/10.1371/journal.pone.0181201

Xu SJ, Park DH, Kim JY, Kim BS (2016) Biological control of Gray mold and growth promotion of Tomato using *Bacillus* spp. isolated from soil. Trop Plant Pathol 41(3):169–176

Yang CH, Crowley DE (2000) Rhizosphere microbial community structure in relation to root location and plant iron nutritional status. Appl Environ Microbiol 66(1):345–351

Youssef MMA, Eissa MFM (2014) Biofertilizers and their role in management of plant parasitic nematodes: a review. J Biotechnol Pharm Res 5(1):1–6

Yu X, Ai C, Xin L, Zhou G (2011) The siderophore-producing bacterium, *Bacillus subtilis* CAS15, has a biocontrol effect on *Fusarium* wilt and promotes the growth of pepper. Eur J Soil Biol 47 (2):138–145

Zahran HH (2001) Rhizobia from wild legumes: diversity, taxonomy, ecology, nitrogen fixation and biotechnology. J Biotechnol 91(2):143–153

Chapter 15
Morphological and Physiological Aspects of Symbiotic Plant–Microbe Interactions and Their Significance

Surinder Kaur and Gurpreet Kaur

15.1 Introduction

Symbiosis, i.e., "the permanent association between two or more specifically distinct organisms, at least during a part of the life cycle," is an important aspect of life on earth. Evolutionarily, plants require positive associations with specialized microbes to sustain their normal growth and development in certain ecological niches. Plants form two such important interactions with microorganisms, viz., mycorrhiza, and with N_2-fixing *Rhizobium*. In the former, the roots of plants are infected by certain fungi that help them acquire phosphate from the soil (Smith and Read 1997; Smith et al. 2003). Fungus in return obtains organic nutrition in the form of carbohydrates and other growth substances from plants and also ideal ecological niche, essential for their growth and development. Two major types of mycorrhizal fungi that are important for plant growth and development are ectomycorrhiza (ECM) and endomycorrhiza (AMF). Through their roles in uptake of nutrients, arbuscular mycorrhizae (AM) fungi were perhaps important in the colonization of land by plants (Heckman et al. 2001). In AMF, arbuscules (branched, microscopic haustorial structures) are the major sites of nutrient exchange and contain mycorrhiza-specific plant phosphate transporters at their edge (Harrison et al. 2002; Manchanda and Garg 2007). In the latter, agricultural symbiosis, occurs between rhizobial bacteria and roots of legumes, such as soybeans and pea, where the plant provides its beneficial partner with carbohydrates, together with other nutrients, and in return receives valuable fixed nitrogen in the form of ammonia and amino acids from it (Udvardi and Day 1997). These associations can increase agricultural production and improve soil fertility and therefore have great prospective as a supplementary, renewable, and ecofriendly source of plant nutrients (Marx 2004). Both associations are extreme in terms of host specificity (Manchanda and Garg 2007). Nodulation almost exclusively

S. Kaur (✉) · G. Kaur
Department of Botany, SGTB Khalsa College, University of Delhi, Delhi, India

B. Giri et al. (eds.), *Root Biology*, Soil Biology 52,
https://doi.org/10.1007/978-3-319-75910-4_15

occurs in legumes, whereas AMF show little host specificity, and more than 80% of terrestrial plants form symbiotic association with fungi of the phylum Glomeromycota (Schüßler et al. 2001; Garg et al. 2006). Both these associations, i.e., plant–fungus and legume–*Rhizobium*, act synergistically.

Mycorrhizal fungi improve nitrogen and phosphorus uptake; the higher levels of phosphorus in the plants facilitate the action of nitrogenase enzyme leading to fixation of more nitrogen, which further promotes the development of mycorrhizae in the plant. Certain strains of *Rhizobium* are able to tolerate abiotic stress such as low water content, salt stress, acid and alkaline pH, presence of fertilizers, and heavy metals and effectively fix nitrogen through symbiotic association (Zahran 1999). *Piriformospora indica* is a unique endophytic fungus which colonizes roots of many plants species and establishes symbiotic association with these. The fungus confers tolerance to biotic and abiotic stresses and promotes plant growth and development especially in nutrient-deficient soils (Johnson et al. 2014). In this chapter, an effort has been made to summarize and highlight morphological and structural features of these plant–microbe symbionts along with their ecological significance.

15.2 *Rhizobium*–Legume Symbiosis

Mutualism is a type of symbiotic relation between associating species for mutual benefit to both partners. One of the families in plants, Fabaceae, has an evolved symbiotic relation with a class of proteobacteria which are nitrogen-fixing soil bacteria capable of forming root or stem nodules in leguminous plants and exceptionally with a nonlegume, *Parasponia* (Akkermans et al. 1978; Trinick 1979; Jordan 1982). Most of species involved in symbiosis with legumes belong to class *Alphaproteobacteria*, which includes genera *Azorhizobium*, *Allorhizobium*, *Mesorhizobium*, *Rhizobium*, *Bradyrhizobium*, and *Sinorhizobium* (Dreyfus and Dommergues 1981; Dreyfus et al. 1988; Jordan 1982; Chen et al. 1988; Jarvis et al. 1997; De Lajudie et al. 1998), while some belong to class Betaproteobacteria as *Burkholderia* and *Cupriavidus* (Gyaneshwar et al. 2011).

After establishment of symbiosis, the bacteria fix atmospheric nitrogen and provide it to the host plant. This provides an advantage to the host plant which allows it to grow without the addition of nitrogen fertilizers. Thus, biological nitrogen fixation is a profitable proposition for plants as there is huge increase in agronomic yield since it alone contributes to 65% of the worldwide nitrogen invested in agriculture (Brelles-Marino and Ane 2008).

Bradyrhizobium sp. forms symbiotic association with *Parasponia andersonii* (Ulmaceae), which is the only non-legume plant having symbiotic nitrogen fixation along with root nodule formation (Akkermans et al. 1978; Trinick 1979; Jordan 1982). There are some non-legumes having endosymbiotic association with *Rhizobium*, but they do not form nodules in the host plant, for example, in rice, *Rhizobium oryzicola*, *R. pseudoryzae*, and *R. rhizoryzae*; *R. oryzae* (Peng et al. 2008; Zhang et al. 2011, 2015) and *R. populi* in *Populus euphratica* (Rozahon et al. 2014) have symbiotic association without nodulation.

The genus *Rhizobium* was described for the first time by Frank in 1889. The bacteria involved in legume symbiosis are categorized as diazotrophs because they catalyze conversion of atmospheric free nitrogen to ammonia, a form readily absorbed by plants (Lipsanen and Lindström 1988). It is a Gram-negative, free-living, saprotrophic soil bacterium which encodes an enzyme nitrogenase responsible for the nitrogen-fixing ability. Various steps are involved in invasion of leguminous roots. These include host bacterial recognition, infection process, nodule formation, colonization in nodules of the legumes, bacteroid formation, and nitrogen fixation (Sharifi 1983; Dreyfus et al. 1988).

15.2.1 Host Bacterial Specificity

There is high level of specificity in interaction between host plant and bacterial species, and only the compatible bacterial species are able to induce nodulation in plant. This specific interaction has been reported in various symbiont pairs, as in *Cicer arietinum*, only *Rhizobium ciceri*, and, in *Parasponia*, only *Bradyrhizobium japonicum* are able to form nodules in roots (Trinick 1979; Tillard and Drevon 1988).

Such symbiotic interaction is initiated as host plants roots secrete (iso)flavonoids (Redmond et al. 1986; Le Strange et al. 1990) and lectins (Bohlool and Schmidt 1974; Diaz et al. 1986, 1989) in the rhizosphere which are recognized by compatible bacterial species. Using immunoassays, it was found that there is a strong relation between infection by *Rhizobium leguminosarum* and presence of lectins on the surface of roots of *Pisum sativum*. Lectin was found concentrated more at tip of growing root hairs and on epidermal cells of young root hairs where it was present as dense small patches and is not distributed uniformly (Diaz et al. 1986).

On recognition of compatible species, there is induction of bacterial genes called nodulation factors (Nod factors) which lead to formation of nodules in host roots. Thus, there is an exchange of chemical signals for recognition of both mutualistic partners (Fig. 15.1). Also, in this symbiotic association, host plant provides carbon source to the bacteroids to meet its energy need for bacteroid differentiation and nitrogen fixation, which is malate in most species (Salminen and Streeter 1992), while glutamate (Bergersen and Turner 1988) and sucrose (Gordon et al. 1999) have also been reported. These bacteroids need an environment with low oxygen levels for optimal expression of enzymes of the nitrogenase complex and thus nitrogen fixation. The anaerobic conditions within the nodule are maintained by a plant-based oxygen-binding protein, the leghemoglobin (Downie and Oldroyd 2004; Ott et al. 2005).

Luteolin, a flavonoid isolated from *Medicago sativa*, dihydroxyflavone from *Trifolium repens*, and isoflavones from *Glycine max* (Kosslak et al. 1987) were the first few Nod factor inducers which were characterized (Redmond et al. 1986; Peters et al. 1986). Later other compounds such as jasmonates (Rosas et al. 1998), xanthones (Mabood et al. 2006), and lectins (Bohlool and Schmidt 1974) have also been shown to stimulate nod gene expression. Phenolics like vanillin from a nonlegume crop,

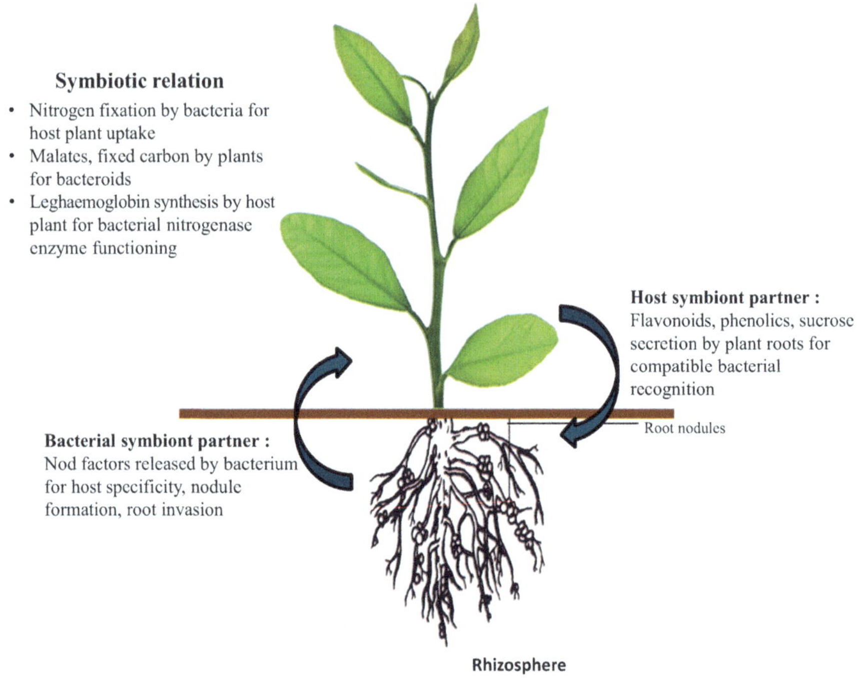

Fig 15.1 Bacterial host specificity in legume–*Rhizobium* interaction showing exchange of signals from both the partners. Flavonoids and lectins are released by plant roots in root exudate and bacteria release nod factors in the rhizosphere

wheat, were shown to induce transcription of nod genes in *Rhizobium* sp. (Le Strange et al. 1990). Promoter of nodA gene of *Rhizobium leguminosarum* has been found to be directly upregulated by flavonoids and flavones released by plants (Zaat et al. 1987). Aldonic acids, monosaccharides with a carboxyl group, were also observed to be inducers of nod genes in *Rhizobium lupini*, *Mesorhizobium loti*, and *Sinorhizobium meliloti* (Gagnon and Ibrahim 1998). Induction of bacterial gene expression by plant exudates provides an insight into plant–microbe interactions during establishment of symbiosis.

Nod factors which initiate the root invasion by bacteria are chemically lipochito-oligosaccharides (Truchet et al. 1991; Denarie et al. 1996). *Rhizobium meliloti* establishes its host specificity by a sulfated and acylated glucosamine oligosaccharide signal (Lerouge et al. 1990). Lipopolysaccharide (LPS) which is essentially present in Gram-negative bacteria may also additionally contribute at various stages of host–bacteria recognition (Noel et al. 1986; Carlson et al. 1987). Expression of Nod genes is an indispensable factor in nodulation of specific host plants (Kondorosi et al. 1984) as mutations in these genes alter the host specificity and thus alter the host range (Rhijn and Vanderleyden 1995).

15.2.2 Establishment of Endosymbiosis

Once nodulation factors (NFs) are recognized by host, there is induction of high levels of calcium as well as influx of calcium ions from extracellular environment (Ehrhardt et al. 1996; Cárdenas et al. 1999) and higher pH (elevated H^+ levels) in the cells of root hairs, which brings about changes in root hair cytoskeleton (Sieberer et al. 2005). Ca^{2+} spikes were observed in all legume–*Rhizobium* symbiosis, but in a non-nodulating legume, *Cercis*, no such Ca^{2+} spikes were observed in response to Nod factors of its symbiont partner, *Sinorhizobium fredii*. *Parasponia andersonii*, a nonlegume shows calcium oscillations in response to *S. fredii*, but its close relative, *Trema tomentosa*, a non-nodulating species does not have such spikes (Granqvist et al. 2015). Additionally, in most cases, there is change in pH of host cells which makes it slightly alkaline. In alfaalfa root cells, there is rapid change in pH by 0.2–0.3 units in response to *Rhizobium meliloti* nod factors (Felle et al. 1996). This is immediately followed by initiation of root hair curling that further traps rhizobia and initiates infection process (Downie et al. 1985). The bacteria enter root hair via infection threads which are inward growth of the root hair cell membrane. The hydrolytic enzymes encoded by *Rhizobium* seem to carry out penetration into the root hair by degradation of its cell wall (Newcomb 1979). Here the bacteria, being endosymbiotic, enter plant cells by endocytosis from the infection threads (Callaham and Torrey 1981; Dazzo et al. 1984). The bacteria which get differentiated into its nitrogen-fixing form within root nodule cells of host are termed as bacteroids (Bergersen 1974), and the host–bacteroid association together is known as symbiosome (Roth et al. 1988; Oke and Long 1999). Hydrogen peroxide seems to control the process of bacterial differentiation into nitrogen-fixing, symbiotic form (Puppo et al. 2013). Symbiosome has a plant-derived membrane around each bacteroid in the infection thread forming an organelle in cytosol (Roth et al. 1988). The membrane originates from three sources: the host infection-thread membrane, ER of plant cell, and de novo synthesis by Golgi and ER (Roth and Stacey 1989). Upon infection, a single bacteroid is enclosed in a single-membrane envelope; later more than one bacteroid found within many symbiosome compartments and the membrane which surrounds bacteria undergoing senescence begin disintegration and dissolution (Tu 1977).

Simultaneously, there are increased cell divisions in root cortex cells (Dudley et al. 1987; Oldroyd and Downie 2008) and establishment of a nodule primordium which undergo continuous mitotic activity to form specialized structures called root nodules which act as localized sites for symbiotic nitrogen fixation in most legumes (Libbenga and Harkes 1973). Each cell of nodule is filled with thousands of symbiosomes (Roth et al. 1988). During symbiosome formation, it has been observed in pea and soybean root nodules that host cells undergo numerous changes in ultrastructure like extensive arrangements of rough endoplasmic reticulum along with vacuole formation (Kijne and Pluvque 1979). The form of bacteroids varies considerably ranging from long-rod, short-rod, long-club, short-club, spherical, ellipsoid, pear-shaped, and L- and Y-shaped, which eventually assume pear or spherical shapes.

15.2.3 Root Morphogenesis for Nodule Formation

In most of the legumes, infection initiates from entry of infection thread which carries bacteria via extension of the root hair into the root cortex cells via breaks in root hair epidermis and root hair (Oldroyd and Downie 2008). The root cells susceptible to invasion are immediately beneath root tip where root hairs are still growing. In response to bacterial invasion, root hairs deform and curl which is mediated by changes in cytoskeleton in root cells (Cárdenas et al. 1998). This curling of root hair further traps bacterial cells in the rhizosphere (Callaham and Torrey 1981). The infection thread is thus an invagination of root cell's plasma membrane itself. As infection thread grows toward the pericycle and cortex, root cortical cells dedifferentiate and enter active cell division to establish nodule primordium (Oldroyd and Downie 2008). The important role of the presence of this specialized structure is because of the fact that nitrogenase enzyme is extremely sensitive to aerobic conditions and nodule maintains the compartment with low oxygen concentrations.

Nodules can be divided into two types, determinate and indeterminate (Fig. 15.2), depending on the transient and persistent nature of host cell proliferation (Oke and Long 1999). Determinate nodules have no meristem and contain homogenous population of symbiotic cells, for example, in *Phaseolus vulgaris* and *Lotus japonicus* roots, determinate nodules are formed, wherein the meristematic activity ceases early during nodule development, and the nodule assumes a spherical shape. On the other hand, active cell division is maintained in indeterminate nodules, for example, nodules of *Medicago sativa*, *M. truncatula*, *Pisum sativum*, and *Vicia sativa* (Sprent 1980; Ferguson 2013). A nodule meristem is present in the apical region which by constant production of new cells and continuous growth form elongated nodule. The type of nodule to be formed is specified by the host plant (Sprent 1980). Cells in outer cortex of root and not pericycle assume meristematic activity and subsequently form nodule-like structure (Newcomb 1979). Indeterminate nodules of *Pisum sativum*

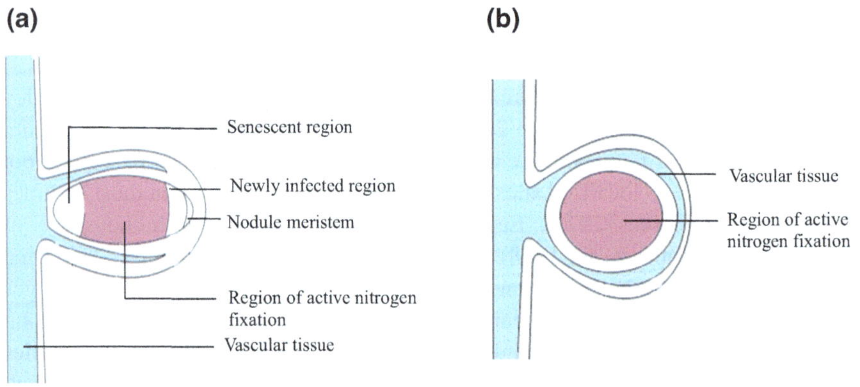

Fig 15.2 Indeterminate (**a**) and determinate (**b**) root nodules showing regions of active nitrogen fixation (marked in pink) with high expression of leghemoglobin protein

possess an active meristem that continues to divide and add new cells, few of which may be subsequently infected. Bacterial cells divide and grow in large numbers as infection thread progresses, and they lose their dividing ability as they colonize the inner cortex and get differentiated as bacteroids (Libbenga and Harkes 1973). Determinate nodule is differentiated into two zones: central infection zone, containing both infected and uninfected cells, surrounding layers of uninfected cells of outer cortex (Udvardi and Poole 2013). In the outer cortex cells, there is expansion of infection thread, while the cells of inner cortex carry out rapid cell division. Metabolites from plant body are transported to the nodule through the vascular tissue reaching the cortex (Rae et al. 1992).

15.2.4 Nitrogen Fixation in Nodules

Genes required for nodulation (Nod) and nitrogen fixation (Nif) are present in a large-sized plasmid (>100 Kb) called symbiosis (*Sym*) plasmid (Nuti et al. 1979; Prakash et al. 1981; Banfalvi et al. 1981; Rosenberg et al. 1981). Rhizobial enzyme nitrogenase, which converts atmospheric dinitrogen to ammonia, is a complex, made up of six protein subunits (two each of NifH, NifD, and NifK), and contains iron–sulfur clusters and two iron–molybdenum cofactors (Fe_7MoS_9N) called FeMo cofactors, which is the site of nitrogen reduction (Downie 2014). The iron–protein component is the smaller part of the enzyme, which gets reduced and transfers electrons to the molybdenum–iron part of the protein which is the larger component and is the catalytic site where dinitrogen binds and actually gets reduced (Dixon and Kahn 2004). These nitrogenase metallocenters are very sensitive to the presence of oxygen and must be active only in an environment with a low level of oxygen. Legume nodules resolve this requirement of low free O_2 levels by expressing high levels of leghemoglobins. These are oxygen-chelating proteins which maintain very low intracellular levels of O_2, while the bacteroids continue to carry out oxidative phosphorylation using a cytochrome oxidase (with a high affinity for O_2). Thus, net effect allows high levels of ATP synthesis required by bacteria for nitrogen fixation yet no oxidative damage to nitrogenase. Additionally, oxygen diffusion into the nodule is also regulated by an oxygen diffusion barrier at nodule periphery (Dixon and Kahn 2004).

Ammonia that is produced by the bacteroid diffuses out to plant cytosol due to concentration gradient of ammonia (Udvardi and Day 1990). The fixed nitrogen is transported to other parts of plants in various forms as ureides, allantoin, allantoic acid, amides, and amino acids, specially glutamine and asparagine (Sprent 1980).

15.2.5 Molecular Interactions Involved During Endosymbiosis and Nodulation

There are specific interactions at root surface of host with *Rhizobium* species determined by the presence of polysaccharides (lipochitooligosaccharides) on surface of the bacteria and flavonoids in the exudates of plant roots (Truchet et al. 1991; van Brussel et al. 1992). Genetic loci of Nod genes on Sym plasmid have been cloned and well-characterized in *Rhizobium meliloti* (Debelle et al. 1986) and *R. phaseoli* (Downie et al. 1985). Nod genes are grouped in two categories, the common and host-specific nod genes (Kondorosi et al. 1984). The common nodABC genes containing operon are conserved across all *Rhizobium* sp. (Torok et al. 1984; Martinez et al. 1993), and these can be interchanged between different rhizobial species, while it is not possible for host-specific nod genes, which determine host specificity for the bacterial strain (Kondorosi et al. 1984).

Bacteria need essentially six genes, nodABCDEF (Marvel et al. 1987), that are arranged as three different operons, nodABC, nodD, and nodFE (Rossen et al. 1985; Debelle et al. 1986). These are not conserved in structure and function among rhizobia, but they are still necessary for the nodule formation (Kondorosi et al. 1984). Additionally, nodH and nodQ are required which are involved in determination of host range in *R. leguminosarum* bv. *viciae* and recognize plant-specific signals present in exudates (Faucher et al. 1989).

The NodD gene is regulatory nod gene as it codes for a protein which binds to conserved region upstream of the nod operons and activates their expression. Its product acts as a transcriptional activator of other nod genes in the presence of plant signals as flavonoids and lectins (Rossen et al. 1985; Rostas et al. 1986). NodD has regulatory function across all species of *Rhizobium*, *Bradyrhizobium*, and *Azorhizobium* strains. The number of nodD gene copies varies among different species (Rhijn et al. 1993). NodD protein product belongs to LysR family of transcriptional activators which has common features as they are activated by chemicals and carry a helix-turn-helix motif for specific DNA-binding ability. The protein acts as cytoplasmic membrane-localized receptor where it interacts with plant-released inducers and transduces the signal to nod operon ABC, thus transcriptionally activating the operon (Fisher et al. 1988). Subtractive hybridization studies have been carried out to identify genes upregulated during root nodule development (Gamas et al. 1996). NodABC operon controls curling of host root hair, rhizobial penetration, formation of infection thread, and nodule development, while nodFE unit regulates invasion of infection thread within root hair (Downie et al. 1985; Debelle et al. 1986).

Other nod genes involved were also later discovered. The protein coded by nodM was found to be involved in transcriptional regulation of operons nodA, B, C, I, J, F, and E (Spaink et al. 1987). This Nod factor further activates transcription of many downstream genes in host plant cell NSP1, NSP2, CYCLOPS/IPD3, ERN1, and NIN which are involved in organogenesis and meristem formation (Laloum et al. 2014). Thus, there is change in levels of phytohormones as cytokinins, ABA, gibberellins, and auxins within nodules during this phase (Phillips and Torrey 1970; Phillips 1971; Williams and Mallorca 1982; Badenoch-Jones et al. 1984; Dangar and Basu 1987). A study using *Rhizobium japonicum* cultures and pea epicotyl showed that that there are high levels of GA in nodule compared to the rest of root tissues which is produced by bacteroid itself and not by the host cell (Williams and Mallorca 1982). Similarly, cytokinin was isolated from a nodulating *Rhizobium japonicum* strain which induced cell divisions in a soybean callus tissue grown in cytokinin lacking medium (Phillips and Torrey 1970).

There is also upregulation of genes involved in production of leghemoglobin protein (Garg and Jain 2013). In *Lotus japonicus*, using cDNA arrays approximately 860 genes were found to be highly upregulated in nodules than in roots. One-third of these expressed genes were involved in metabolism and transport, and over 100 in signaling, or in transcriptional or posttranscriptional regulation of gene expression. Genes required for pathways such as glycolysis, CO_2 fixation, amino acid biosynthesis, and purine, haem, and redox metabolism were also found to be upregulated in nodules (Colebatch et al. 2004).

An induction of 18–20 nodule-specific proteins in host cells called nodulins was observed in root nodules of *Glycine max* in response to *Rhizobium japonicum* association. These nodulins constituted 7–11% of the total protein synthesized in cytoplasm of the host cell which was not present in the uninfected roots, bacteroids, and free-living *Rhizobium*. Thus, there is induction of specific nodulins which are essential for development of symbiosis in the legume root nodules (Legocki and Verma 1980). These genes are structurally and functionally conserved in *Rhizobium*, *Azorhizobium*, and *Bradyrhizobium* spp. (Dobert et al. 1994).

15.2.6 Autoregulation of Nodulation in Legumes

The host plant itself determines the number of nodules that are formed through active signaling pathway called autoregulation of nodulation (AON). It is a negative self-regulating mechanism to control number of nodules when not required. It involves transfer of a signal from older nodule primordia to cortical cells of newly forming nodule. The regulation of root nodules involves nitrate inhibition and is shoot controlled (Francisco and Akao 1993). In mutants lacking autoregulation, hyper-nodulation is observed wherein the roots form excess nodules.

There is differentiation of few regions in root cells into regulatory regions which involves root to shoot and shoot to root signaling (Reid et al. 2011). At molecular level, the regulation seems to be through short peptides (related to the nonsymbiotic CLAVATA3 peptide of *Arabidopsis*) which are synthesized in these regulatory cell clusters. Autoregulation of nodulation is systemic regulatory pathway, while nodulation initiation is local mechanism. Both pathways involve by a peptide CLAVATA/ESR-related (CLE) protein which represses excessive nodulation via negative feedback loops. These peptides are 12–13 amino acids long and have leucine-rich repeat (LRR) receptor kinase through which it acts (Hastwell et al. 2015).

The CLE peptides mediate nodule inhibition responses to control the number of nodules according to plant requirements (Reid et al. 2011).

15.3 Ecological Aspects of Stem and Root Nodulation

15.3.1 Stem Nodulation: Ecological Significance

Although roots are the sites for nodulation in most plants, there are reports of stem nodulation as in *Sesbania rostrata* (Dreyfus and Dommergues 1981). *Azorhizobium caulinodans* is the specific symbiont partner which infects these adventitious roots and has ability to form both stem and root nodules on *Sesbania*. The nodules develop from dormant root primordia abundantly present in stem which get activated in the presence of certain transcripts as H4-1Sr (Goormachtig et al. 1997). Using 16S rRNA sequence, it was found that stem-nodulating strains are phylogenetically very diverse compared to various root-nodulating *Rhizobium* sp., and they form a separate subbranch on the phylogenetic tree (Dreyfus et al. 1988). These N_2-fixing nodules are formed on subepidermal primordia of the adventive roots on the stems (Alazard 1985; Ladha et al. 1992). Twenty-one species of the genus *Aeschynomene* (Alazard 1985; Becker et al. 1988) and three species of the genus *Sesbania* (Dreyfus and Dommergues 1981) are some of the species reported to show stem nodulation. *Sesbania* shows stem nodulation in response to flooding and waterlogged conditions (Trinchant and Rigaud 1989). Stem nodulation, thus, seems to be an evolutionary adaptation to waterlogged conditions and has immense ecological significance. There are some unique features of plants having stem nodulation: rapid growth of plants, more number of sites from which infection can take place, photosynthetic nature of stem nodule, and thus ability to carry out dual functions, photosynthesis and nitrogen fixation (Ladha et al. 1992).

15.3.2 Ecological Significance of Root Nodulation

A few strains of *Rhizobium* have been shown to tolerate various stress conditions as salt and drought stress and carry out nitrogen fixation for host plant, thus improving productivity (Zahran 1999). An acid-tolerant *Rhizobium leguminosarum* biovar *trifolii* strain was shown to have less membrane permeability and high proton expulsion ability compared to acid-sensitive strain (Chen et al. 1993). Thus, there is an immense scope of its application in reclaiming degraded lands.

A study on biotic stress revealed that in *Phaseolus lunatus*, the nitrogen fixed during rhizobial symbiosis was being sanctioned for production of nitrogen-containing cyanogenic defense compounds against an insect herbivore, Mexican bean beetle (*Epilachna varivestis* Muls); thus symbiosis was involved in improving plant growth in dual ways, plant growth and better resistance to biotic stress (Thamer et al. 2011).

15.3.3 Evolutionary Significance of Rhizobium*–Plant Symbiosis: Coevolution of Symbionts*

The symbiotic relation between members of Fabaceae and group "rhizobia," which is a polyphyletic group (Jarvis et al. 1997), is an ideal model to understand the evolutionary basis of symbiosis (De Mita 2007). It is hypothesized that coevolution has taken place through mutual interaction of host plants and *Rhizobium* sp. in the soil making it one of the most efficient symbiotic associations even though both partners can survive in the absence of each other. Evolution of nod genes especially those involved in nodulation might be present in a single bacterial lineage, and there would have been horizontal gene transfer among the different rhizobial species (Hirsch et al. 1995).

Plant-rhizobial association was originally necrotrophic, however evolutionary mechanisms seem to be in favour of biotrophy. Since there is establishment of interdependence between *Rhizobium* and legumes over evolutionary time scale, there has been modification of both leghaemoglobin of host and nitrogen fixing mechanisms of *Rhizobium* to enhance nitrogen fixation (Sharifi 1983).

According to one school of thought, rhizobia repress the host immune system for its entry and establishment inside plant cells as symbiont. For further successful establishment, the host has developed mechanisms of nutrient supply to the symbiotic partner. The secretion of various chemicals by plant cells further promotes invasion by bacteria. For selection of partner, signals produced by the symbiont are used by the host and vice versa (Noe and Hammerstein 1994). There are evidences that interaction with *Rhizobium* has a beneficial effect on legume growth (Kaschuk et al. 2010). In one such study, 15 plants from 12 genotypes of *Medicago truncatula* were grown with a mixture of three *Rhizobium* strains. Most *M. truncatula* plants developed a larger number of nodules with more beneficial *Rhizobium* strains and hence give evidence for partner choice. Also, there was increased frequency of beneficial *Rhizobium* strains in the rhizosphere suggesting how partner's selection also affects bacterial fitness in competitive environment (Heath and Tiffin 2009).

There are many mutant strains of *Rhizobium* which are noncompatible with host but still elicit nodule formation, although such nodules have been found to be ineffective in nitrogen fixation capability and are devoid of leghemoglobin. These nodules possess distinct structure compared to normal nodules and undergo all initial recognition processes as bacterial colonization around roots, root hair curling, and root cortex cell divisions. However, these lack infection thread formation and production of leghemoglobin and ureide (Vandenbosch et al. 1985).

Although host symbionts provide important services to bacterial partner, it seems to penalize rhizobia that fail to fix N_2 inside their root nodules. It was shown by series of experiments that noncooperation (analogous to cheating) leads to reduction in reproductive rate of bacteria by 50% and there was reduced supply of sanctions against these bacteria (Kiers et al. 2003). In this mutualistic interaction, there may be conflict of interest resources if certain bacterial individuals "cheat" by not fixing nitrogen while continuing to accept carbon source from host partner. There are many evidences that support the possibility of "defaulter" rhizobia strains which provide little or no evident benefit to host in the population. So, effective strains have to overcome competition over these strains which are selected by the host. This is done by host withholding resources from noneffective partners (unproductive symbionts) which is determined by host's incentives, thus screening away nonprofitable members and selecting right mutualists (Archetti et al. 2011).

15.4 Mycorrhizal Associations

Mycorrhiza (pl. mycorrhizae, mycorrhizas; Gr. mykes, fungus; rhiza, roots) is a symbiotic, nonpathogenic, or weakly pathogenic association of a fungus and plant roots or underground organs. The association serves mutual benefits: carbohydrates synthesized by plants are used by the fungus and in return receive nutrients especially nitrogen and phosphorus from the fungi (Smith and Read 2008). Fungus uses these carbohydrates for its growth and in the synthesis of molecules like glomalin, the release of which in the soil environment improves the soil structure and increases organic matter content (Kaur et al. 2014). It has been estimated that about 80% of the N and P contents of plant have been acquired from mycorrhiza indicating their importance in plant nutrition (van der Heijden et al. 2008). Besides providing nutritional benefits to their hosts, mycorrhizal fungi also contribute toward ecological significance such as decreased susceptibility to biotic diseases, tolerance to heavy metals, drought tolerance, etc. (Luo et al. 2009; Beniwal et al. 2010). These dual organisms were first described in detail by Frank (1885) who observed it on the roots of several trees of temperate forests. In 1887, Frank divided these mycorrhizae into two types: (a) ectotrophic and (b) endotrophic. In the former, the fungus forms a sheath on the surface of the root and hyphae grow in the soil and between the outer cortical cells of the root. In the latter, the fungal hyphae enter the cortical cells of the root and colonize the roots of a host plant intracellularly enveloped by the plasmalemma of the host (Eshel and Beeckman 2013).

Mycorrhizae are extremely common in the context of phylogeny and ecology (Kistner and Parniske 2002; Bonfante and Genre 2010). These symbiotic relationships are estimated to occur in about 92% plant families and at least 80% of the vascular plants (Wang and Qui 2006). These include angiosperms, gymnosperms, pteridophytes, and some bryophytes (especially liverworts) (Read et al. 2000a, b). However, plants from the families Brassicaceae, Chenopodiaceae, and Proteaceae rarely, if ever, have mycorrhizal association (Vierheilig et al. 2003).

Based on fossil records and phylogenetic analyses, it has been suggested that more than 450 million years ago certain early Devonian period plants have established a close association with filamentous, i.e., mycorrhizal, fungi (Srivastava et al. 1996; Schüßler et al. 2009; Lee et al. 2012; Torres-Cortes et al. 2015). It is believed that translocation of plants from water to land has been possible because of this association only (Simon et al. 1993; Read et al. 2000a, b; Lum and Hirsch 2003; Bücking et al. 2012).

Mycorrhizal fungi have been classified on the basis of the extent of plant root penetration, production of external mantle or sheath, and inter- and intracellular structures formed by them inside the plant root (Morton and Benny 1990; Walker 1992; Smith and Read 1997; Manoharachary et al. 2000). Seven different types of mycorrhizal fungi have been recognized (Harley and Smith 1983), such as endomycorrhiza, ectomycorrhiza, ectendomycorrhiza, arbutoid mycorrhiza, monotropoid mycorrhiza, ericoid mycorrhiza, and orchid mycorrhiza (Table 15.1). Among them, two major types of mycorrhizal fungi that are important for plant growth and development are ectomycorrhiza (EM/ECM) and endomycorrhiza or arbuscular mycorrhiza (AM). These two types are different as the ectomycorrhizal fungi typically form a thick sheath, or mantle, of mycelium around roots, and some of the mycelium penetrates between cortical cells. The hyphae instead of penetrating individual cells within the root cortex get surrounded

Table 15.1 The major types of mycorrhizae and their ecological significance

Mycorrhizal plants	Typical host plant	Fungi involved	Major significance
Ectomycorrhizae	Forest trees, mainly in temperate and boreal regions	*Amanita*, *Boletus*, *Lactarius*, *Pisolithus*, and *Rhizopogon* (Agaricomycetes) and *Tuber* (Ascomycetes)	Nitrogen uptake from soil
Arbuscular mycorrhizae	Many crop plants	Glomeromycota	Phosphorus uptake from soil
Ectendomycorrhizae (Srivastava et al. 1996; Smith and Read 1997)	Mainly pines, spruce, and larch	*Wilcoxina* (Ascomycetes)	Mineral nutrient uptake from soil
Orchidaceous mycorrhizae (Singh and Varma 2000)	Orchids (Orchidaceae)	*Armillaria*, *Mycena*, *Rhizoctonia*, (Agaricomycetes), *Russula* (Russulales), *Thelephora* (Thelephorales)	Fungi supply the plant with sugars or in some orchids, plants obtain sugars from ectomycorrhizal fungi attached to trees
Ericoid Mycorrhizae (Read 1996; Selosse et al. 2007)	Heathland plants of family Ericaceae, e.g., *Calluna*, *Erica*, *Vaccinium*, etc.	*Hymenoscyphus* (Helotiales) and *Pseudogymnoascus*	Nitrogen and phosphorus uptake from soil
Monotropoid mycorrhizae (Manoharachary et al. 2002)	Non-photosynthetic plants of family Monotropaceae, e.g., *Monotropa*	*Boletus edulis* (Agaricomycetes)	Plants obtain carbohydrate from ectomycorrhizal fungi, itself attached to trees
Arbutoid mycorrhizae (Molina and Trappe 1982)	*Arbutus*, *Arctostaphylos*, *Pyrola*	Basidiomycota, similar to ectomycorrhizal fungi	Mineral nutrient uptake from soil

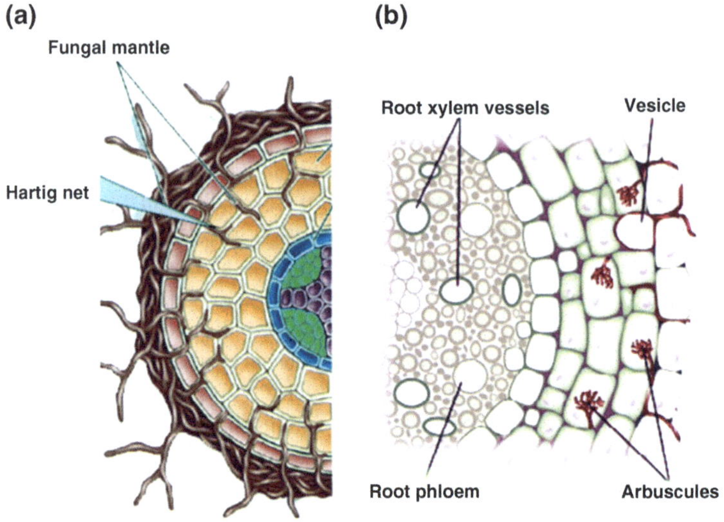

Fig 15.3 (**a**) Ectomycorrhiza (ECM) and (**b**) arbuscular mycorrhizal fungi (AMF); note that mycelium remains outside the root system and hyphae penetrate into root tissue through intracellular spaces; in AMF hyphae are intracellular and form vesicle and arbuscules

by a network of hyphae called the Hartig net for mutual exchange (Fig. 15.3a). The hyphae ensheath root tips forming a typical mantle characterized by differences in color, thickness, presence, and lengths of emanating hyphae (Sethi and Walia 2018). The hyphae of AM/endomycorrhizal fungi grow within the root itself and extend outward from the root into surrounding soil. The hyphae enter the root through either the epidermis or root hairs and invaginate the cell membrane after penetrating through cell wall. The hyphae grow between and into root cortical cells, where they form dichotomously branched treelike structure, arbuscules, and form oval structures, vesicles (Fig. 15.3b). AM fungi develop mainly small spores in the soil, whereas ectomycorrhizal fungi develop aboveground fruit bodies in the surrounding area of trees (Redecker and Schüßler 2014).

While AM are formed with roots of the majority of higher plant species including grasses, herbs, and trees, EM have only been detected in woody species (Brundrett 2009). Most tree species of the boreal and temperate climatic zone have association with EM fungi, whereas AM associations are widespread in the tropical ecosystem. However, there are exceptions. On a global scale, more than 7500 different EM species have been identified, and estimates suggest that the total numbers will be in a range from 25,000 to 30,000 (Rinaldi et al. 2008). With the advent of new methods, knowledge on the composition and diversity of mycorrhizal species in the different ecosystems is currently rapidly increasing suggesting that EM diversity is similar in temperate, tropical, and boreal forests (Tedersoo and Nara 2010; Lang et al. 2011).

15.5 Ectomycorrhiza

Ectomycorrhizae (ECM) (also known as ectotrophic mycorrhizae or sheathing mycorrhizae) are found mainly on woody plants, including many species of coniferous and broad-leaved trees in temperate and boreal regions. They are estimated to form symbiotic association with about 6000 plant species (Brundrett 2009), viz., on trees such as *Pinus* (pines, Pinaceae), *Picea* (spruce, Pinaceae), *Quercus* (oak, Fagaceae), *Fagus* (beech, Fagaceae), *Betula* (birch, Betulaceae), and *Eucalyptus* (eucalyptus, Myrtaceae). The fungi involved in ectomycorrhizal associations are principally members of Agaricomycetes (e.g., *Amanita*, *Boletus*, *Russula*, *Hebeloma*, *Lactarius*, *Pisolithus*, etc.) and few Ascomycota, such as truffles (*Tuber*) (Boroujeni and Hemmatinezhad 2015).

In this association, the fungal hyphae form a sheath-like covering or mantle on the surface of root and enhance the surface area of the roots, with hyphae forming the branching network in the intercellular surface between the epidermis and cortex, known as Hartig net named after Robert Hartig, who is considered the father of forest biology (Blasius et al. 1986). The most apparent feature of ECM is that although the major part of the fungus is in close contact with the root cells, there is no penetration of the host cells (Dighton 2009). It means that the bulk of the fungal mass is located outside the plant roots, hence the name ectotrophic (outside-feeding) or ectomycorrhizae for these mycorrhizae. In this association, the fungus helps in the nutrient and water uptake for the host and, in return, obtains carbohydrate from the host plant. The fungal mycelium, outside the root, forms an extensive network within the soil and leaf litter. The fungal hyphae often grow into the root system of adjacent plant and create a new mycorrhizal association, thereby, linking the different plants through a common mycorrhizal network. This mycorrhizal network helps in the movement of nutrients between different plants and thereby promotes ecosystem succession (Wilson et al. 2006).

The hyphae grow outward from the mantle replacing the root hairs for the absorption of minerals from the soil. Because of the absence of root hairs and the entire root being enclosed by the sheath, there is no direct contact between younger roots and the soil. Thus, all mineral nutrients and water absorbed by the roots have to pass through the fungal sheath. This uptake is facilitated by the extensive network of individual hyphae or aggregated mycelial cords or rhizomorphs, which radiate from the surface of the root sheath into the soil and transport nutrients back to the mycorrhizal sheath.

The fungus colonizes the root in a series of complex events. In the first precontact phase or early phase, the fungus is attracted toward the root. On contact with the root hair, hyphae start growing along with it until they find the surface of the main root (Jacobs et al. 1989). On the main root, morphogenetic changes get initiated leading to the development of mantle. The mantle forms an interphase between the root and the soil. It is a sheath-like covering on the outer surface of the root, leading to root bifurcation and clustering. Sometimes, they are referred as pseudoparenchymatous, as they resemble the parenchymatous tissue of the plant (Dighton 2009). The active mycorrhizal zone appears several millimeters behind the root tip. In older roots it persists even after the association becomes inactive and functions as storage structure and propagule (Massicotte et al. 1987). Once the inner layer of mantle is fully established, the hyphae start penetrating the epidermal cells. Ectomycorrhizal fungi break down the wall polymers locally by secreting certain hydrolytic enzymes like cellulase and pectinase. Soon the fungus colonizes the root cell, and the symbiotic association between the two gets established. Some ectomycorrhizal fungi are also known to produce certain proteins, the ectomycorrhizins, which favor the development of fungal hyphae inside the plant cell. For the development of Hartig net, the hyphae from the inner layer of mantle penetrate mechanically and enter the middle lamella present between the epidermal and cortical cells. Sometime, the fungus releases certain hydrolytic enzymes which solubilize the middle lamella and activate the process.

Once colonization of a certain root part is accomplished and the mantle and Hartig net are well developed, the exchange of nutrients begins between both partners in the Hartig net, leading to the functional mycorrhiza, called late phase (Martin 2007). The tip of the hyphae, forming the Hartig net, accumulates in the mitochondria and endoplasmic reticulum in large number, suggesting that the transfer of nutrients is localized to this part. The hyphae arising from the mantle are long and grow several centimeters in the soil (Anderson and Cairney 2007). Besides providing nutritional benefits, the mantle layer may provide tolerance against heavy metals and high salt concentration and may exclude nematodes from the root. Once colonization is accomplished or equilibrium is established in terms of colonization, the fungus does not penetrate any further. The biotrophic (equilibrium) phase, where efficient nutrient exchange occurs, may be maintained for a certain period of time, depending on the species involved (Smith and Read 2008), until the mycorrhiza undergoes senescence (Felten et al. 2012). Outside the mantle, the mycelium may extend into the soil by a few centimeters or much farther if a particular fungus has the capacity to form mycelial cords or rhizomorphs. This extensive network of rhizomorph and mycelial cords ramifies through the soil from the mycorrhizal sheath. This network can link many different plants within a habitat—even plants of different species, because of the general lack of host specificity of these fungi (Arnebrant et al. 1993; He et al. 2006).

Through this networking, the ectomycorrhizal fungi play several important roles in ecosystem functioning, such as:

- Ectomycorrhizae may connect many of the plants in a community, i.e., the young tree seedlings can be linked to a "mother" tree by a common mycorrhizal network, so that the nutrients can be tapped by seedlings growing nearby in shade from the one growing in sun, thus allowing the seedling to survive under conditions where otherwise they might be superseded by mature plants (Monika et al. 2015).
- An estimated 70–90% of ectomycorrhizal rootlets die and are replaced each year. If these rootlets were not connected, they would decompose, and at least some of the nutrients would be leached from the soil. The mycelial connections could help

to retain mineral nutrients by withdrawing them from the degenerating mycorrhizae to others that are still healthy.
- As the mycelial cords and network of hyphae extend several meters beyond the root zone, it can exploit large volume of soil for nutrients and help in improving the water status of the soil. This role can be largely important in soils with poor water retention, such as former mining sites where trees are planted for land reclamation or under conditions of drought.
- Ectomycorrhizae are efficient in cycling of nutrients, i.e., capturing substances such as phosphates, nitrogen compounds, and cations that are released by exudation from the leaf canopy and leaf fall and returning them to the root system.
- Where phosphates and water are transported to the plant by the fungus, the fungus benefits from these associations by obtaining carbohydrates from the plant. Trees invest a considerable amount of photosynthate to support the fungal biomass—conservatively estimated at 10% or more of the annual photosynthetic production of a tree. Fruiting bodies of Basidiomycetes absorb these carbohydrates maximally. Some achlorophyllous plants may also fulfil their carbon needs by plugging their roots into this mycorrhizal network. This strategy also observed in orchid mycorrhizae is known as mycoheterotrophy.
- Ectomycorrhizal fungi have been found to be particularly prevalent in litter layer (humus) in comparison with lower soil horizons. In these layers nutrients vary widely due to leaf fall, precipitation, activities of animals, etc.; thus they may be particularly useful to the tree in such environments. Since these mycorrhizae can grow well in nutrient-poor or dry soils, it is possible that they have the capacity to produce hydrolytic enzymes and are capable of solubilizing phosphorus and nitrogen from complex sources.
- Many ectomycorrhizal fungi can grow on simple organic media in culture (France and Reid 1984), but they have poor or no ability to degrade cellulose and lignin, as seen in wood decomposer fungi of the Ascomycota and Basidiomycota. So, the ectomycorrhizal fungi seem to be ecologically adapted to grow as symbionts.

15.5.1 Benefits of Ectomycorrhizae

15.5.1.1 Plant Nutrition

Root proliferation in plants is relatively slow. ECM enhances the plant growth, through improvement in nutrients uptake. ECM absorb and stores plant nutrients like nitrogen, phosphorus, potassium and calcium etc. in their mantle thereby help in establishment of high yielding forests, land recuperation and establishment of exotic plant species. Influence of ECM on root exudates is important for the nutrient exchange and mineralization process in the soil (Moore et al. 2011). The mycorrhizal association increases the surface area of the plant roots, due to which its access to soil resources is increased. Extraradical mycelium also plays a direct role in nutrient mobilization by secreting enzymes, absorption, and transport of elite nutrients. They secrete significant quantities of chitinase, phosphatase, and protease, which help to dissolve hard-to-capture nutrients such as organic phosphorus, nitrogen, and iron. Most of the ECM secrete acid phosphatase in the mycosphere, which hydrolyze the organic phosphate into orthophosphate (Bolan 1991).

15.5.1.2 Resistance to Abiotic Stress

ECM association helps plants to overcome different kinds of stress such as soil salinity, alkalinity, acidity and drought conditions (Boroujeni and Hemmatinezhad 2015). The ECM plants exhibit better growth than the non-mycorrhizal ones especially in the arid and semiarid regions where low moisture and high temperature are very critical for survival and growth of the plants (Sandeep et al. 2015). Studies have shown that the ectomycorrhizal fungi provide resistant to drought stress in the seedlings of spruce plant. The ECM association enhances the absorbing area of root surface and improves the soil root contact. Besides, the fungal hyphae can penetrate the soil pore more easily than the root hair, thereby, increasing the access to deep water system. ECM symbiosis with poplar species has been found to improved salt tolerance in salt stress soils. The fungus, by changing the concentration of nutrients and phytohormones and ratios of fatty acids, alters the leaf physiology leading to prevent chlorosis and leaf shedding. ECM fungus *Scleroderma bermudense* was able to improve salt stress in *Coccoloba uvifera* (sea grape) seedlings (Bandou et al. 2006).

15.5.1.3 Phytoremediation

Phytoremediation, a process that includes all kinds of biological, physical, and chemical activities carried out by plants, is used for the remediation of soil contaminated with pollutants such as heavy metals, salts, organic compounds, etc. (Coninx et al. 2017). Fast-growing plants with deep root system and high transpiration rate are generally ideal for this purpose. They produce certain root exudates which enhance the growth of microbial community and create the favorable environment by altering pH and osmotic potential that stimulate the degradation of toxic compounds. Majority of such trees live in symbiotic association with ectomycorrhizal fungi. For example, *Populus* forms symbiotic association with more than 60 different ectomycorrhizal fungal species. Ectomycorrhizal symbiosis between *Populus* and *Paxillus involutus* has shown improved plant growth and increase in phytostabilization potential (Bojarczuk et al. 2015; Szuba et al. 2017). ECM increase the plant tolerance to heavy metals by forming a sheath-like covering that acts as a filter, thereby restricting their movement to root cortex.

Ectomycorrhizal fungi and their extraradical mycelium increase the surface area for the establishment of microbial communities. Their synergistic, competitive, and antagonistic effect shifts the soil microflora toward those communities capable of degrading the soil pollutants. Studies on gray poplar (*P.* × *canescens*), natural hybrid between *P. alba* (white poplar) and *P. tremula* (European aspen), showed that it accumulates high amounts of Cd and Zn when grown in metal-polluted soil (Durand et al. 2011; He et al. 2013) than stressed non-mycorrhizal roots. Inoculation of young poplar plants with mycorrhizal fungi may be used for bioremediation of soils contaminated with toxic metals (Bojarczuk et al. 2015). ECM symbionts can alter mechanism of heavy metal detoxification by influencing physiological and molecular processes of the poplar (Luo et al. 2014).

15.5.1.4 Resistance to Biotic Stress

Ectomycorrhizal fungi are known to inhibit the growth of numerous soilborne pathogens on root surface of host plants by producing certain compounds or antifungal and antibiotic substances (Moore et al. 2011). Compact arrangement of hyphae in fungal mantle protects the root against nematodes and soilborne pathogens and stimulates the plant growth by reducing the severity of diseases. Studies have shown that ECM fungi such as *Lactarius deliciosus* and *Boletus* sp. antagonize plant pathogen *Rhizoctonia solani*. *Lactarius camphoratus* and *Cortinarius* sp. have been found to produce antifungal antibiotics known as "chloromycorrhiza" and "mycorrhizin A" against the phytopathogens like *Rhizoctonia solani*, *Pythium debaryanum*, and *Fusarium oxysporum* (Sandeep et al. 2015).

15.5.1.5 Growth Hormones

Plants with mycorrhizal associations exhibit higher content of growth regulators like cytokinins and auxins as compared to the non-mycorrhizal ones. It indicates that benefits to the higher plants provided by the fungal symbiosis are not just limited to inorganic and organic nutrients from the soil. The fungal symbiont provides the host plant with growth hormones, auxin, cytokinins, and gibberellins, and also growth-regulating B vitamins, thereby producing above normal level of these potent substances, which in turn influence the growth and development of host plant (Sandeep et al. 2015).

15.6 Endomycorrhizal Fungi (AMF)

Before 1974 the term vesicular–arbuscular mycorrhizae was used commonly for the mycorrhizal associations where vesicles and arbuscules were observed in the roots. Later, it was recognized that some fungi form mycorrhizae with arbuscules only and lack the ability to form vesicles in roots of the plants (e.g., *Gigaspora* and *Scutellospora*). It was therefore proposed that the more general term, "arbuscular mycorrhizae (AM)" should be used. Arbuscular mycorrhiza (AM) represents a symbiosis between terrestrial plant roots and fungi of phylum Glomeromycota. The symbiosis derives its name from the Latin word arbusculum (little tree) for typical tuft-like structures formed by fine dichotomously branched fungal hyphae and the Greek word for fungus roots. These are the most common type of mycorrhizae with worldwide distribution. Up to around 80% of investigated species, mainly land plants, among them vascular plants, are known to form AM (Brundrett 2009; Bonfante and Genre 2010). The remainder either are non-mycorrhizal or form one of the other types of mycorrhiza, i.e., ectomycorrhiza, orchid mycorrhiza, or ericoid mycorrhiza. However, many gymnosperms, pteridophytes, and some bryophytes also form AM or AM-like associations. Thus, AM are found ubiquitously in soils wherever their hosts are available (Gerdeman 1968).

On the basis of the molecular studies, AMF have been assigned to a monophyletic group, the Glomeromycota, which is one of the smallest fungal groups. It presently includes 5 orders, 14 families, 29 genera, and approximately 230 described species (Morton and Redecker 2001; Spain et al. 2006; Palenzuela et al. 2008; Oehl et al. 2008, 2011a, b; Schüßler and Walker 2010). Almost all members of this group such as *Acaulospora*, *Archaeospora*, *Gigaspora*, *Entrophospora*, *Glomus*, *Paraglomus*, and *Scutellospora* form mycorrhizae with crop plants and have not been grown in axenic culture, away from their host plants (Rosendahl 2008). Therefore, they are considered to be wholly dependent on plants for their carbon and energy sources.

Roots containing AM fungi show no outward signs of infection. Instead, they look like normal roots, and the extent of colonization by fungal hyphae can only be judged by special techniques. One special method for this is to use differential interference contrast microscope. But the more common method is to treat roots with strong alkali, to destroy the plant protoplasm, and then to stain the roots with a fungal dye such as trypan blue. Fungal hyphae can be observed clearly in such preparations colonizing the roots extensively by growing among the root cortical cells, often producing large, swollen vesicles, and by penetrating individual root cortical cells to form treelike branching structures termed arbuscules (Vierheilig et al. 2005). The arbuscules are sites of nutrient exchange between the fungus and the host (Miyasaka et al. 2003). After germination, the germ tubes show limited growth in vitro, and extended growth can only be observed if the fungus is in association with living root tissues. Therefore, it can be inferred that such mycorrhizal fungi are obligate mutualistic symbionts. AM associations are very ancient, and structures resembling extant arbuscules have been discovered in the fossilized rhizome tissues of early vascular plants.

The AM fungi play an important role in overcoming biotic stresses faced by plants. These fungi increase growth of plant by enhancing uptake of soil nutrient, increasing plant tolerance to drought, and protecting plants against pathogens, nematodes, and insects (Whipps 2004; Pozo and Azcón-Aguilar 2007; Gianinazzi et al. 2010; Jung et al. 2012; Jeffries and Barea 2012).

15.6.1 Systematic Position of AMF

Considerable changes have taken place in the classification of AMF in the last few decades. Earlier in the 1990s, they were classified in Zygomycota exclusively on the basis of phenotypic characteristics like spore morphology and the structure and development of spore wall (Gerdemann and Trappe 1974; Walker 1992). Recent studies based on nuclear-encoded rRNA gene markers suggested that these fungal symbionts form a monophyletic group of true fungi, the phylum Glomeromycota. The phylum Glomeromycota was earlier divided into four orders, namely, Diversisporales, Archaesporales, Glomerales, and Paraglomerales (Schüßler et al. 2001). One of the most widely accepted classification system for AMF was provided by Oehl et al. (2011a, b), who proposed the establishment of a fifth order, namely, Gigasporales (Fig. 15.4). They based their classification on morphological and genetic features (Pagano et al. 2016). They have described four families, Diversisporaceae, Acaulosporaceae, Sacculosporaceae, and Pacisporaceae, within the order Diversisporales (Błaszkowski 2012; Medina et al. 2014), and five families, Scutellosporaceae, Gigasporaceae, Intraomatosporaceae, Dentiscutataceae, and Racocetraceae, within the order Gigasporales (Silva et al. 2012; Pontes et al. 2013; Marinho et al. 2014). Glomerales describes two families, the Enthrophosporaceae and Glomeraceae (Sieverding et al. 2014; Błaszkowski et al. 2015). The order

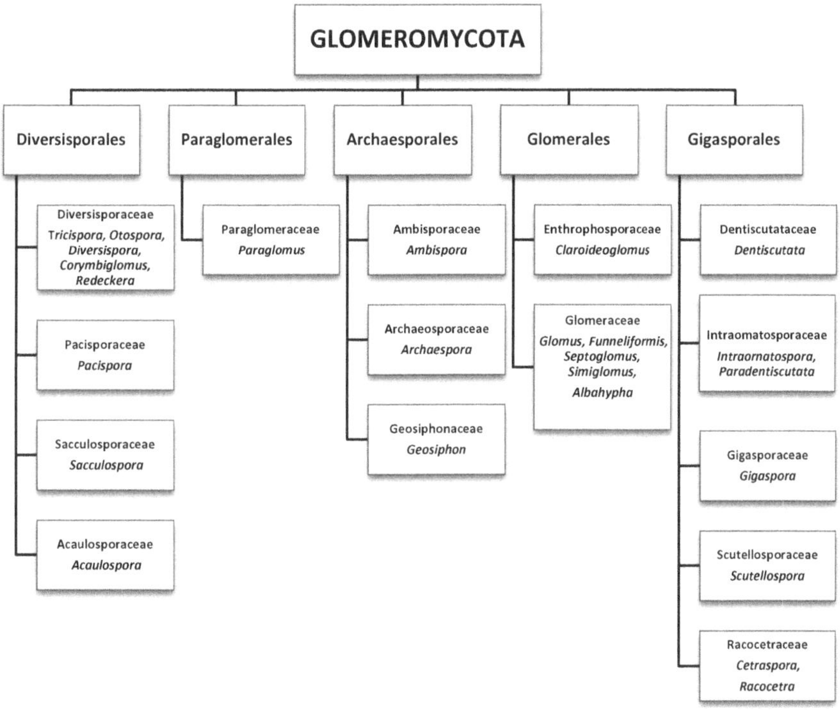

Fig 15.4 Classification of phylum Glomeromycota (Oehl et al. 2011a, b; Goto et al. 2012)

Archaeosporales describes three families, the Ambisporaceae, Geosiphonaceae, and Archaeosporaceae (Oehl et al. 2015). The fifth order Paraglomerales includes only one family, Paraglomeraceae.

15.6.2 Host Ranges and Communities of Arbuscular Mycorrhizae

The AM fungi have astonishingly wide host ranges, i.e., they are nonspecific in their host range and can colonize both herbaceous and woody plants. They are found everywhere where hosts to this symbiosis occur. Non-AM plants may have other kinds of mycorrhiza. Some families such as the Brassicaceae, Chenopodiaceae, and Cyperaceae although regarded as non-mycorrhizal have members that show colonization with these fungi (Smith and Read 2008). Different species of plants react differently to different mycorrhizal fungi. Plants in natural communities have been observed to be colonized by different strains of AM fungi. At least in artificial inoculations, an AM fungus obtained from one type of plant was found to colonize the roots of many unrelated plants. Moreover, the diversity of AM fungi can influence the plant biodiversity in natural ecosystems. With an increase in AM fungal diversity, there is an increase in plant biodiversity and productivity. Majority of fungi can grow at water potentials beyond the tolerance level of plants, so the arbuscular mycorrhizal fungi can be significant in semiarid environments, such as the deserts (Kamalvanshi et al. 2012).

15.6.3 Morphology and Reproduction

Because of their very peculiar evolutionary history, underground lifestyle, and genetic makeup, AM fungi are gifted with extraordinary biological characters (Bonfante and Genre 2008; Parniske 2008; Barea and Azcón-Aguilar 2012). The mycelium formed within the root tissues is coarse, aseptate, and coenocytic, contains hundreds of nuclei, and produces very large multinucleate spores with resistant thick walls containing chitin (Smith and Read 2008). Outside the root, the fungus produces large, distinctive spores which are easily visible to the naked eye. These can vary in diameter from 10 to 1000 μm, in color from hyaline to black and in surface texture from smooth to highly ornamented. These spores germinate and infect the roots from an appressorium-like infection structure on the root surface. Spore germination may be improved by plant-produced factors such as strigolactones. These compounds have been found to induce spore germination or hyphal branching near a potential host thereby increasing the chance of colonization (Akiyama et al. 2005; Besserer et al. 2006). From the initial entry points, the fungus grows extensively between the cells of the root cortex, often producing large, balloon-shaped, thick-walled vesicles, also known as intraradical vesicles, within the root. These vesicles are multinucleate and store large amount of lipids. Their primary function is believed to be storage; however, vesicles can also function as reproductive propagules for the fungus. In some plants, the fungal mycelium penetrates cortical cells of the root and forms extensive intracellular coils. More commonly, the hyphae penetrating host cells branch repeatedly to form dichotomous treelike structures, termed arbuscules. The individual arbuscules have a short active life only for less than 15 days and then degenerate, being replaced by new arbuscules, in other parts of the root. The fine tips of arbuscules are digested by the host cell so that only irregular clumps of fungal matrix remain. The plant and the fungal plasma membranes are separated by an apoplastic compartment known as periarbuscular space. The arbuscules thus can be considered a type of haustorium through which exchange of nutrient between the fungus and the root cells takes place (Bonfante-Fasolo and Grippiolo 1982). Consistent with this, the invaded cells remain alive because the plasmalemma invaginates to surround all the individual branches of the arbuscule and the membrane itself is never penetrated. From the infected root, thick-walled, coarse, mycelium comes out, which grows up to several centimeters into the surrounding soil and produces large (about 400 μm diameter), globose, thick-walled spores, sometimes known as chlamydospores. These spores contain lipid droplets, glycogen, proteins, and trehalose. They may be produced singly or in clusters and are often naked, but in some genera (e.g., *Glomus*) they may be covered by a weft of hyphae to form a sporocarp.

In the absence of any morphological proof for sexual reproduction in Glomeromycota, spores are assumed to be formed asexually (Redecker and Schüßler 2014). Although close examination of nuclear passage at the time of spore formation did not show sexual processes, combined studies including microscopic examination and molecular genetics have provided proof for genetic recombination (Sanders and Croll 2010) in the model AMF *Rhizopogon irregularis*. Being obligatorily biotrophic microbes (Schüßler and Walker 2011), AM fungi are unculturable and are not able to complete their life cycle without colonizing a host plant. This has hindered the study of the biology and the biotechnological applications of these fungi (Bago and Cano 2005; Rosendahl 2008).

15.6.4 Significance of Arbuscular Mycorrhizae

15.6.4.1 Absorption of Nutrients

Many studies have shown that the main role of AM fungi is to provide plants with water and mineral nutrients, especially phosphorus from the soil; in return the plant provides the fungus with sugars. Arbuscular mycorrhizal fungi exhibit higher metabolic rate and diffused distribution in the upper soil layers and play a significant role in the uptake and accumulation of minerals from soil and their translocation to the host plant (Bücking et al. 2012). In fact they serve as a very efficient extension of host root system. They absorb both macronutrients (P, N, K, and Ca) and micronutrients (Zn, Cu, and Mn) from soil and translocate them to the host. The mineral nutrients like P, Zn, and Cu, which do not readily diffuse through the soil due to their poor diffusion, become unavailable to host plant. The extraradical hyphae of AM fungi tend to proliferate several centimeters beyond the nutrient depletion zone, increasing the absorbing surface of host root. AM fungal hyphae extend into the soil penetrating the zone of nutrient depletion and enhance the effective uptake of immobile nutrients, therefore managing the deficiency of such elements in host plant. Phosphorus (P) is a vital soil nutrient for normal functioning of plant. The deficiency of P in soil may stop normal growth and development of plant. In terms of plant nutrition, phosphorus is second only to nitrogen as the major mineral nutrient that plants require, and yet phosphorus is a highly immobile element (Turk et al. 2006). When added to soil in the form of soluble phosphate fertilizers, the phosphate ion readily combines with calcium and other divalent cations, to form insoluble inorganic phosphates, or it combines with organic matter to produce insoluble organic phosphates. The natural rate of release of phosphate is thus extremely slow and is often a limiting factor for plant growth. The AM fungi produce extensive hyphal networks in soil, providing a large surface area for absorption of phosphorus. These fungi also release acid phosphatases to cleave phosphate from organic matter (Harrison and van Buuren 1995). They absorb phosphate in excess of requirements and store it in the form of polyphosphates, which can then be released to the plant when required. As a consequence of the enhanced nutrient uptake, the association also increases water uptake, reduces fertilizer input, reduces heavy metal and salt toxicity, etc.

15.6.4.2 Protection Against Pathogens

Plants colonized with AM fungi exhibit increased tolerance against certain root-borne diseases (Borowicz 2001). Several reports indicate positive effects of AM fungi on root-borne fungal diseases like wilt and root rot and stem- and leaf-borne diseases. The severity of nematode infection found to be reduced in the plants colonized with AM fungi (Veresoglou and Rillig 2012; Schouteden et al. 2015). The effects of mycorrhizal fungi on root pathogenic bacteria *Pseudomonas syringae* and tomato showed significant reduction in damage when plants are exposed to mycorrhizal colonization.

15.6.4.3 Salinity Stress

The significance of soil salinity for agricultural yield is huge. Due to the effects of high concentrations of salts on availability, uptake, transport, or physiological inactivation of a given nutrient, plants are deprived of essential mineral nutrients under saline conditions. This in turn causes the decrease in growth rate and net assimilation capacity (Hasanuzzaman et al. 2009, 2013) and also affects photosynthesis and other variables leading to huge losses in productivity (Mathur et al. 2007; Raziuddin et al. 2011). AM fungi alleviate adverse effects of salinity stress and improve salt tolerance of host plants by enhancing selective uptake of nutrients and prevention of nutritional disorder, accumulation of osmoregulators, and enhanced activities of antioxidant enzymes and molecules. AM hypha readily extends the fungal colony, and upon perception of signals from the nutrient ions, it produces branched absorbing structures or spores, which absorb and translocate these nutrients to the host (Hameed et al. 2014). Therefore, under saline conditions, mycorrhizal plants can potentially access nutrients from a larger area than the non-mycorrhizal controls, offering huge benefit to host plants by improving the uptake of essential nutrients.

15.6.4.4 Drought Stress

Arbuscular mycorrhizal fungi have the ability to improve plant biomass and nutrient uptake under drought conditions (Al-Karaki and Zak 2004; Gholamhoseini et al. 2013; Kapoor et al. 2013, Augé et al. 2015). The improved drought tolerance in mycorrhizal plants may be due to enhanced P nutrition, changes in root hydraulic conductance, soil–water relations and increased soil aggregate stability, greater soil available water, improved stomatal conductance and plant–water potential components (Auge 2001). Mycorrhiza-mediated changes in plant–water relations under drought stress conditions may involve complex interactions among multiple mechanisms (Wu and Xia 2003, 2004). However, the primary impact of this symbiosis changes in stomatal conductance and transpiration. Under drought stress, mycorrhizal and non-mycorrhizal plants differentially regulate the expression of several stress-related genes such as aquaporin in root tissue. Aquaporins are membrane intrinsic proteins which facilitate water uptake in plant root, following an osmotic gradient.

15.6.4.5 Heavy Metal

Heavy metals have been found naturally in a variety of habitats including agroecosystems, where these elements constitute a potential hazard for soil and plants. However, some of the heavy metals are serving as essential plant micronutrients such as copper, zinc, iron, and manganese which are required for normal functioning and improved plant growth (Karimi et al. 2011). On the other hand, some of the heavy metals such as mercury, lead, and cadmium have no biological functions but have been reported in plant tissues (Khodaverdiloo and Samadi 2011). Excessive levels of these elements in soil generally affect normal functioning and growth of plant, which, therefore, has been considered a serious matter of concern (Pandolfini et al. 1997; Keller et al. 2002; Voegelin et al. 2003; Kabata-Pendias and Mukherjee 2007). High concentrations of heavy metals in plant tissues influence the structures of enzymes, consequently affecting the structure of proteins and cell membrane, and also the permeability and functions of plants membranes (Giller et al. 1998). Besides, higher accumulation of heavy metals induces oxidative stress, which on the other hand affects plant growth and development.

15.7 Plant Growth-Promoting Root Endophyte: *Piriformospora indica*

The plant–mycorrhiza symbiosis is an integral and essential part of plant growth and development. These symbionts improve the growth of crops in nutrient-deficient soils with lower inputs of chemical fertilizers (Gianinazzi et al. 1995; Varma et al. 1999). In case where native mycorrhizal fungal inoculum is low, inoculation of fungi is the best method. Since AMF are obligate symbionts, they do not grow like any other fungi apart from their hosts. Because of the absence of a reliable pure culture, commercial production is the greatest hindrance in use and application of mycorrhizal biotechnology (Sudha et al. 1999). An adaptable and cultivable plant growth-promoting endophytic fungus, *Piriformospora*, was discovered from the rhizosphere of the woody shrubs *Prosopis juliflora* and *Ziziphus nummularia* in the sandy desert soil of Rajasthan, India (Verma et al. 1998). The fungus lacks host specificity and has been found to colonize monocot as well as dicot roots. It colonizes with a wide range of hosts including bryophytes, pteridophytes, and gymnosperms (Varma et al. 2001; Fakhro et al. 2009; Oelmüller et al. 2009; Qiang et al. 2011). The fungus promotes uptake of nutrients and enhances the tolerance of host plants against biotic (pathogens) and abiotic stresses (salinity, drought, etc.). Inoculation of plants with the fungus improves plant growth and yield (Varma et al. 2012). The fungus, named for its characteristic spore morphology, provides a model organism for the study of plant–microbe interactions and a new means for improving plant production. Molecular phylogenetics have shown that the fungus belongs to Basidiomycota (order Sebacinales, class Agaricomycetes) (Weiβ et al. 2004).

P. indica forms both inter- and intracellular hyphae in the root cortex. Within the cortical cells, the fungus often forms thick hyphal coils or branched structures and spore or vesicle-like structures within or between the cortical cells. Hyphae multiply within the host cortical tissues like AMF and never cross through the endodermis. Fungal spores are multinucleated, and hyphae are dimorphic and have dolipore septum with a continuous and straight parenthesome on either side (Sudha et al. 1999). It improves the overall growth of different grasses, trees, and herbaceous species, and the biggest advantage is that it can be cultivated on a number of complex and synthetic media.

P. indica produces pear-shaped, autofluorescent chlamydospores at the apex of hyphae; these are mostly flat, submerged into the substratum, and white and almost hyaline. The hyphae are thin walled, diameters ranging from 0.7 to 3.5 μm. The mycelia are often twisted and overlap each other, and connections are often observed. The cytoplasm of the chlamydospores is heavily packed with granular materials and usually contains 8–25 nuclei (Varma et al. 2012). Interestingly, the host range of

P. indica is very much similar to that of AMF. *P. indica* colonizes the roots of plants as diverse as *Oryza sativa*, *Zea mays* (Poaceae), *Solanum melongena*, *Nicotiana tabacum* (Solanaceae), *Glycine max*, *Cicer arietinum* (Fabaceae), *Petroselinum crispum* (Apiaceae), *Artemisia annua* (Asteraceae) and *Bacopa monnieri* (Plantaginaceae). Like arbuscular mycorrhizal fungi, *P. indica* has a great potential application in the pursuit of physiological and agronomical useful characters for crop improvement. In contrast to AMF, *P. indica* can be cultivated axenically on synthetic medium. It is therefore often used as biofertilizer and bioprotector in many crop species (Ansari et al. 2014). Its use in agriculture to improve water absorption, mineral uptake, photosynthesis, plant growth and development, and crop fitness has been well studied. The fungus colonizes the orchids resulting in better growth and increased rate of survival of seeds (Singh et al. 2000). The influence of *P. indica* on the growth of *Arabidopsis thaliana* plants under normal and salt stress conditions has been investigated, and it was found that *P. indica* colonization promotes plant growth and development by increasing biomass, lateral root density, and chlorophyll content under both conditions (Abdelaziz et al. 2017).

15.7.1 Significance of *Piriformospora indica*

15.7.1.1 Abiotic and Biotic Stress Tolerance

Piriformospora indica has been shown to enhance plant tolerance to a number of abiotic stresses like salinity, heavy metal toxicity, and low temperature (Baltruschat et al. 2008; Sun et al. 2010; Unnikumar et al. 2013; Ansari et al. 2013). *P. indica* colonization-mediated high-salinity tolerance was reported in *Triticum aestivum* (Zarea et al. 2012), that of drought stress tolerance in *Arabidopsis* seedlings (Sherameti et al. 2008), *Hordeum vulgare* (Waller et al. 2005), and strawberry (Husaini et al. 2012).

Interaction of *P. indica* with a diverse group of microorganisms such as *Pseudomonas fluorescens* (rhizobacteria), *Chlamydomonas reinhardtii*, *Aspergillus niger*, and *Rhizopus stolonifer* has been reported to improve plants against environmental stresses (Pham et al. 2004; Porras-Alfaro and Bayman 2011). Roots of *Hordeum vulgare* (barley) invaded with *P. indica* were found to be resistant against *Fusarium culmorum* infections and in shoots against *Blumeria graminis* (Waller et al. 2005; Deshmukh and Kogel 2007) and reduce the severity of disease caused by *V. dahliae* (Fakhro et al. 2010). The growth of pathogenic fungi such as *A. sydowii* and *R. stolonifer* has been reported to be entirely hindered by *P. indica*.

15.7.1.2 Acquisition of Phosphorous in Plants

Being one of the most essential mineral nutrients, phosphorus plays diverse regulatory, structural, and energy transfer roles (Balemi and Negisho 2012). Plants cannot directly utilize P present in the soil as it is mostly in the form of scarcely soluble complexes affecting therefore crop production (Balemi and Negisho 2012). Plants acquire P from the soil either directly by its own transporters or indirectly through mycorrhizal associations (Yadav et al. 2010). *P. indica* produces considerable amounts of acid phosphatases which facilitate the host plant to get sufficient amount of insoluble or complex forms of phosphate stored in the soil (Singh et al. 2000).

15.7.1.3 Improve Crop Plant Yield

In addition to plant growth and tolerance to biotic and abiotic stresses, the fungus stimulates the excess biomass production, early flowering, and seed production imparting biological hardening to tissue culture-raised plants (Yadav et al. 2010; Das et al. 2012). It improves crop yield by increasing the vegetative tissue yield, number of flowers (Rai et al. 2001; Dolatabadi et al. 2011), or seed weight (Peskan-Berghofer et al. 2004; Barazani et al. 2005). An increase in yield of barley due to a higher number of ears (Waller et al. 2005) and that of number of pods per plant and number of seeds per pod in green gram was found to be higher (Ray and Valsalakumar 2010).

15.8 Conclusion and Future Perspective

Symbiotic relations hold a lot of promise to agricultural industry as there is huge difference in terms of crop yield as the plant growth is proportional to levels of nitrogen and nutrients availability. Symbiosis or mutualism is a good strategy employed by plants for better adaptation and survival in competitive environment. Thus, in this chapter we attempted to highlight the importance of mutualistic plant–microbe interaction, viz., *Rhizobium*, mycorrhizae, and an endophyte *Piriformospora indica*. These endosymbionts are natural biofertilizers that are economical and safer source of plant nutrition compared to chemical fertilizers. The plants show improved growth and development and become resistant to biotic and abiotic stresses. These microbes increase agriculture production and improve soil fertility. Symbiotic association between species provides useful clue to understand evolutionary significance of these species. The evolutionary significance of cheating *Rhizobium* strains and induction of nonproductive nodules yet seem to be a mystery, which needs to be unraveled. Future researches need to be directed to understand various models on this symbiotic interaction and its long-term maintenance during the course of evolution. Thus, use of biotechnological advancements especially in areas of transcriptomics and proteomics can unravel the gene expression patterns in both partners during symbiotic interaction and thus would provide a better insight into the mechanism.

References

Abdelaziz ME, Kim D, Ali S, Fedoroff NV, Al-Babili S (2017) The endophytic fungus *Piriformospora indica* enhances *Arabidopsis thaliana* growth and modulates Na+/K+ homeostasis under salt stress conditions. Plant Sci 263:107–115

Akiyama K, Matsuzaki K, Hayashi H (2005) Plant sesqueterpenes induce hyphal branching in arbuscular mycorrhizal fungi. Nature 435:824–827

Akkermans ADL, Abdulkadir S, Trinick MJ (1978) N2-fixing root nodules in Ulmaceae: *Parasponia* or (and) *Trema* spp. Plant Soil 49:711–715

Alazard D (1985) Stem and root nodulation in *Aeschynomene* spp. Appl Environ Microbiol 50: 732–734

Al-Karaki MG, Zak B (2004) Field response of wheat to arbuscular mycorrhizal fungi and drought stress. Mycorrhiza 14:263–269

Anderson IC, Cairney JWG (2007) Ectomycorrhizal fungi: exploring the mycelial frontier. FEMS Microbiol Rev 31:388–406

Ansari MW, Bains G, Shukla A, Pant RC, Tuteja N (2013) Low temperature stress ethylene and not *Fusarium* might be responsible for mango malformation. Plant Physiol Biochem 69:34–38

Ansari MW, Gill SS, Tuteja N (2014) *Piriformospora indica* a powerful tool for crop improvement. Proc Indian Natl Sci Acad 80:317–324

Archetti M, Scheuring I, Hoffman M, Frederickson ME, Pierce NE, Yu DW (2011) Economic game theory for mutualism and cooperation. Ecol Lett 14:1300–1312

Arnebrant K, Hans EK, Finlay RD, Söderström B (1993) Nitrogen translocation between *Alnus glutinosa* (L.) Gaertn. seedlings inoculated with *Frankia* sp. and *Pinus contorta* Doug, ex Loud seedlings connected by a common ectomycorrhizal mycelium. New Phytol 124(2):231–242

Auge RM (2001) Water relations, drought and vesicular-arbuscular mycorrhizal symbiosis. Mycorrhiza 11:3–42

Augé RM, Toler HD, Saxton AM (2015) Arbuscular mycorrhizal symbiosis alters stomatal conductance of host plants more under drought than under amply watered conditions: a meta-analysis. Mycorrhiza 25:13–24

Badenoch-Jones J, Summons RE, Rolfe BG, Letham DS (1984) Phytohormones, *Rhizobium* mutants, and nodulation in legumes IV Auxin metabolites in pea root nodules. J Plant Growth Regul 3:23–39

Bago B, Cano C (2005) Breaking myths on arbuscular mycorrhizas in vitro biology. In: Declerck S, Strullu DG, Fortin A (eds) Soil biology, vol 4. Springer, Berlin, pp 111–138

Balemi T, Negisho K (2012) Management of soil phosphorus and plant adaptation mechanisms to phosphorus stress for sustainable crop production: a review. J Soil Sci Plant Nutr 12:547–562

Baltruschat H, Fodor J, Harrach BD, Niemczyk E, Barna B, Gullner G, Janeczko A, Kogel KH, Schäfer P, Schwarczinger I, Zuccaro A, Skoczowski A (2008) Salt tolerance of barley induced by the root endophyte *Piriformospora indica* is associated with a strong increase in antioxidants. New Phytol 180:501–510

Bandou E, Lebailly F, Muller F, Dulormne M, Toribio A, Chabrol J, Courtecuisse R, Plenchette C, Prin Y, Duponnois R, Thiao M, Sylla S, Dreyfus B, Bâ AM (2006) The ectomycorrhizal fungus *Scleroderma bermudense* alleviates salt stress in seagrape (*Coccoloba uvifera* L.) seedlings. Mycorrhiza 16(8):559–565

Banfalvi Z, Sakanyan V, Koncz C, Kiss A, Dusha I, Kondorosi A (1981) Location of nodulation and nitrogen fixation genes on a high molecular weight plasmid of *R. meliloti*. Mol Gen Genet 184:318–325

Barazani O, Benderoth M, Groten K, Kuhlemeier C, Baldwin IT (2005) *Piriformospora indica* and *Sebacina vermifera* increase growth performance at the expense of herbivore resistance in *Nicotiana attenuata*. Oecologia 146:234–243

Barea JM, Azcón-Aguilar C (2012) Evolution, biology and ecological effects of arbuscular mycorrhizas. In: Camisao AF, Pedroso CC (eds) Symbiosis: evolution, biology and ecological effects. Nova, Hauppauge, pp 1–34

Becker M, Ladha JK, Ottow JCG (1988) Stem nodulating legumes as green manure for lowland rice. Philipp J Crop Sci 13:121–127

Beniwal RS, Langenfield-Heyser R, Polle A (2010) Ectomycorrhiza and hydrogel protect hybrid poplar from water deficit and unravel plastic responses of xylem anatomy. Environ Exp Bot 69:189–197

Bergersen FJ (1974) Formation and function of bacteroids. In: Quispel A (ed) The biology of nitrogen fixation. North-Holland, Amsterdam, pp 473–498

Bergersen FJ, Turner GL (1988) Glutamate as a carbon source for N_2-fixing bacteroids prepared from soybean nodules. J Gen Microbiol 134:2441–2448

Besserer A, Puech-Pages V, Kiefer P, Gomez-Rolden V, Jauneau A, Roy S, Porttais J-C, Roux C, Becard G, Sejalon-Delmas N (2006) Strigolactones stimulate arbuscular mycorrhizal fungi by activating mitochondria. PLoS Biol 4:e226. https://doi.org/10.1372/jpournal.pbio.0040226

Blasius D, Feil W, Kottke I, Oberwinkler F (1986) Hartig net structure and formation in fully ensheathed ectomycorrhizas. Nord J Bot 6:837–842

Błaszkowski J (2012) Glomeromycota. W. Szafer Institute of Botany, Polish Academy of Sciences, Kraków, p 303

Błaszkowski J, Chwat G, Goralska A, Ryska P, Kovacs GM (2015) Two new genera: *Dominikia* and *Kamienska* and *D. disticha* sp nov in Glomeromycota. Nova Hedwigia 100:225–238

Bohlool BB, Schmidt EL (1974) Lectins: a possible basis for specificity in the *Rhizobium*–legume symbiosis. Science 185:269–271

Bojarczuk K, Karliński L, Hazubska-Przybył T, Kieliszewska-Rokicka B (2015) Influence of mycorrhizal inoculation on growth of micropropagated *Populus* × *canescens* lines in metal-contaminated soils. New For 46(2):195–215

Bolan NS (1991) A critical review of the role of mycorrhizal fungi in the uptake of phosphorus by plants. Plant Soil 134:189–207

Bonfante P, Genre A (2008) Plants and arbuscular mycorrhizal fungi: an evolutionary-developmental perspective. Trends Plant Sci 13:492–498

Bonfante P, Genre A (2010) Mechanisms underlying beneficial plant–fungus interactions in mycorrhizal symbiosis. Nat Commun 1:48

Bonfante-Fasolo P, Grippiolo R (1982) Ultrastructural and cytochemical changes in the wall of a vesicular-arbuscular mycorrhizal fungus during symbiosis. Can J Bot 60:2303–2312

Boroujeni DS, Hemmatinezhad B (2015) Review of application and importance of ectomycorrhiza fungi and their role in the stability of ecosystems. Biosci Biotechnol Res Asia 12(1):153–158

Borowicz VA (2001) Do arbuscular mycorrhizal fungi alter plant–pathogen relations? Ecology 82: 3057–3068

Brelles-Marino GQ, Ane JM (2008) Nod factors and the molecular dialogue in the rhizobia–legume interaction. In: Couto GN (ed) Nitrogen fixation research progress. Nova Science, New York, pp 173–227

Brundrett MC (2009) Mycorrhizal associations and other means of nutrition of vascular plants: understanding the global diversity of host plants by resolving conflicting information and developing reliable means of diagnosis. Plant Soil 320:37–77

Bücking H, Liepold E, Ambilwade P (2012) The role of the mycorrhizal symbiosis in nutrient uptake of plants and the regulatory mechanisms underlying these transport processes. In: Dhal NK, Sahu SC (eds) Plant science. Intech, Rijeka. ISBN: 978-953-51-0905-1

Callaham DA, Torrey JG (1981) The structural basis for infection of root hairs of *Trifolium repens* by *Rhizobium*. Can J Bot 59(9):1647–1664

Cárdenas L, Vidali L, Dominguez J, Perez H, Sánchez F, Hepler PK, Quinto C (1998) Rearrangement of actin microfilaments in plant root hairs responding to *Rhizobium etli* nodulation signals. Plant Physiol 116:871–877

Cárdenas L, Feijo JA, Kunkel JG, Sánchez F, Holdaway-Clarke T, Hepler PK, Quinto C (1999) *Rhizobium* nod factors induce increases in intracllular free calcium and extracellular calcium influxes in bean root hairs. Plant J 19:347–352

Carlson RS, Kalembasa S, Tunoroski D, Packori P, Noel KD (1987) Characterization of the lipopolysaccharide from a *Rhizobium phaseoli* mutant that is defective in infection thread development. J Bacteriol 169:4923–4928

Chen H, Richardson AE, Rolfe BG (1993) Studies on the physiological and genetic basis of acid tolerance in *Rhizobium leguminosarum* biovar *trifolii*. Appl Environ Microbiol 59:1798–1804
Chen WX, Yan GH, Li JL (1988) Numerical taxonomic study of fast-growing soybean rhizobia and a proposal that *Rhizobium fredii* be assigned to *Sinorhizobium* gen nov. Int J Syst Bacteriol 38(4):392–397
Colebatch G, Desbrosses G, Ott T, Krusell L, Montanari O, Kloska S, Kopka J, Udvardi MK (2004) Global changes in transcription orchestrate metabolic differentiation during symbiotic nitrogen fixation in *Lotus japonicus*. Plant J 39:487–512
Coninx L, Martinova V, Rineau F (2017) Mycorrhiza-assisted phytoremediation. In: Cuypers A, Vangronsveld J (eds) Phytoremediation, vol 83. Elsevier, Amsterdam, pp 127–188
Dangar TK, Basu PS (1987) Studies on plant growth substances, IAA metabolism and nitrogenase activity in root nodules of *Phaseolus aureus* Roxb.var. mungo. Biol Plant 29:350–354
Das A, Kamal S, Shakil Najam A, Sherameti I, Oelmuller R, Dua M, Tuteja N, Johri AK, Varma A (2012) The root endophyte fungus *Piriformospora indica* leads to early flowering, higher biomass and altered secondary metabolites of the medicinal plant, *Coleus forskohlii*. Plant Signal Behav 7:1–10
Dazzo FB, Truchet GL, Sherwood JE, Hrabak EM, Abe M, Pankratz SH (1984) Specific phases of root hair attachment in the *Rhizobium trifolii*–clover symbiosis. Appl Environ Microbiol 48: 1140–1150
De Mita (2007) Investigation of the demographic and selective forces shaping the nucleotide diversity of genes involved in nod factor signaling in *Medicago truncatula*. Genetics 177:2123–2133
de Lajudie P, Laurent-Fulele E, Willems A, Torck U, Coopman R, Collins MD, Kersters K, Dreyfus B, Gillis M (1998) *Allorhizobium undicola* gen nov, sp nov, nitrogen-fixing bacteria that efficiently nodulate *Neptunia natans* in Senegal. Int J Syst Bacteriol 48:1277–1290
Debelle F, Rosenberg C, Vasse J, Maillet F, Martinez E, Dénarie J, Truchet G (1986) Assignment of symbiotic developmental phenotypes to common and specific nodulation (*nod*) genetic loci of *Rhizobium meliloti*. J Bacteriol 168:1075–1086
Denarie J, Debelle F, Prome JC (1996) *Rhizobium* lipo-chitooligosaccharide nodulation factors: signaling molecules mediating recognition and morphogenesis. Annu Rev Biochem 65:503–535
Deshmukh SD, Kogel KH (2007) *Piriformospora indica* protects barley from root rot caused by *Fusarium graminearum*. J Plant Dis Protect 114:263–268
Diaz CL, Van Spronsen PC, Bakhuizen R, Logman GJJ, Lugtenberg EJJ, Kijne JW (1986) Correlation between infection by *Rhizobium leguminosarum* and lectin on the surface of *Pisum sativum* roots. Planta 168:350–359
Diaz CL, Melchers LS, Hooykaas PJJ, Lugtenberg BJJ, Kijne JW (1989) Root lectin as a determinant of host-plant specificity in *Rhizobium*-legume symbiosis. Nature 338:579–581
Dighton J (2009) Mycorrhizae. In: Encyclopedia of microbiology, 3rd edn. Elsevier, SanDiego
Dixon R, Kahn D (2004) Genetic regulation of biological nitogen fixation. Nat Rev Microbiol 2: 621–631
Dobert RC, Breil BT, Triplett EW (1994) DNA sequence of the common nodulation genes of *Bradyrhizobium elkanii* and their phylogenetic relationships to those of other nodulating bacteria. Mol Plant-Microbe Interact 7:564–572
Dolatabadi HK, Goltapeh EM, Jaimand K, Rohani N, Varma A (2011) Effects of *Piriformospora indica* and *Sebacina vermifera* on growth and yield of essential oil in fennel (*Foeniculum vulgare*) under greenhouse conditions. J Basic Microbiol 51:33–39
Downie JA (2014) Legume nodulation. Curr Biol 24:184–190
Downie JA, Oldroyd GE (2004) Calcium, kinases and nodulation signalling in legumes. Nat Rev Mol Cell Biol 5:566–576
Downie JA, Rossen L, Knight CD, Robertson JG, Wells B, Johnston AW (1985) *Rhizobium leguminosarum* genes involved in early stages of nodulation. J Cell Sci Suppl 2:347–354
Dreyfus BL, Dommergues YR (1981) Nitrogen-fixing nodules induced by *Rhizobium* on the stem of the tropical legume *Sesbania rostrata*. FEMS Microbiol Lett 10:313–317
Dreyfus BL, Garcia J, Gillis M (1988) Characterization of *Azorhizobium caulinodans* gen nov sp nov, a stemnodulating nitrogen-fixing bacterium isolated from *Sesbania rostrata*. Int J Syst Bacteriol 38:89–98

Dudley ME, Jacobs TW, Long SR (1987) Microscopic studies of cell divisions induced in alfalfa roots by *Rhizobium meliloti*. Planta 171:289–301
Durand TC, Baillif P, Albe'Ric P, Carpin S, Label P, Hausman JF, Morabito D (2011) Cadmium and zinc are differentially distributed in *Populus tremula* × *P. alba* exposed to metal excess. Plant Biosyst 145:397–405
Ehrhardt DW, Wais R, Long SR (1996) Calcium spiking in plant root hairs responding to *Rhizobium* nodulation signals. Cell 85:673–681
Eshel A, Beeckman T (2013) Plant roots: the hidden half, 4th edn. CRC Press, Boca Raton
Fakhro A, Andrade-Linares DR, von Bargen S, Bandte M, Büttner C, Grosch R, Schwarz D, Franken P (2009) Impact of *Piriformospora indica* on tomato growth and on interaction with fungal and viral pathogens. Mycorrhiza 20:191–200
Fakhro A, Andrade-Linares DR, von Bargen S, Bandte M, Buttner C, Grosch R, Schwarz D, Franken P (2010) Impact of *Piriformospora indica* on tomato growth and on interaction with fungal and viral pathogens. Mycorrhiza 20:191–200
Faucher C, Gamut S, Dénarie J, Truchet G (1989) The *nodH* and *nodQ* host range genes of *Rhizobium meliloti* behave as avirulence genes in *R. leguminosarum* bv *viciae* and determine changes in the production of plant-specific extracellular signals. Mol Plant Microbe Interact 2:291–300
Felle HH, Kondorosi E, Kondorosi A, Schultze M (1996) Rapid alkalization of root hairs in response to rhizobial lipochitooligosaccharide signals. Plant J 10:295–301
Felten J, Martin F, Legué V (2012) Signalling in ectomycorrhizal symbiosis. In: Perotto S, Baluška F (eds) Signaling and communication in plant symbiosis. Springer, Berlin, pp 123–142
Ferguson BJ (2013) Rhizobia and legume nodulation genes. In: Maloy S, Hughes K (eds) Brenner's encyclopedia of genetics. Academic, New York, pp 236–239
Fisher RF, Egelhoff TT, Mulligan JT, Long SR (1988) Specific binding of protein from *Rhizobium meliloti* cell-free extracts containing NodD to DNA sequences upstream of inducible nodulation genes. Genes Dev 2:282–293
France RC, Reid CPP (1984) Pure culture growth of ectomycorrhizal fungi on inorganic nitrogen sources. Microb Ecol 10:187–195
Francisco PB, Akao S (1993) Autoregulation and nitrate inhibition of nodule formation in soybean cv Enrei and its nodulation mutants. J Exp Bot 44:547–553
Frank AB (1885) U ¨ ber die auf Würzelsymbiose beruhende Ehrna ¨hrung gewisser Ba ¨um durch unterirdische Pilze. Ber Deut Bot Ges (in German) 3:128–145
Gagnon H, Ibrahim RK (1998) Aldonic acids: a novel family of nod gene inducers of *Mesorhizobium loti*, *Rhizobium lupini* and *Sinorhizobium meliloti*. Mol Plant-Microbe Interact 11: 988–998
Gamas P, Niebel FDC, Lescure N, Cullimore JV (1996) Use of a subtractive hybridization approach to identify new *Medicago truncatula* genes induced during root nodule development. Mol Plant Microbe Interact 9:233–242
Garg R, Jain M (2013) Transcriptome analyses in legumes: a resource for functional genomics. Plant Genome 6. https://doi.org/10.3835/plantgenome2013.04.0011
Garg N, Geetanjali, Kaur A (2006) Arbuscular mycorrhiza: nutritional aspects. Arch Agron Soil Sci 52:593606
Gerdeman JW (1968) Vesicular- arbuscular mycorrhiza and plant growth. Annu Rev Phytopathol 6: 297–418
Gerdemann JW, Trappe JM (1974) The Endogonaceae in the Pacific northwest. Mycol Mem 5:1–76
Gholamhoseini M, Ghalavand A, Dolatabadian A, Jamshidi E, Khodaei Joghan A (2013) Effects of arbuscular mycorrhizal inoculation on growth, yield, nutrient uptake and irrigation water productivity of sunflowers grown under drought stress. Agric Water Manag 117:106–114
Gianinazzi S, Trouvelot A, Lovato P, Van Tuinen D, Franken P, Gianinazzi-Pearson V (1995) Arbuscular mycorrhizal fungi in plant production of temperate agroecosystems. Crit Rev Biotechnol 15:305–311
Gianinazzi S, Gollotte A, Binet M-N, van Tuinen D, Redecker D, Wipf D (2010) Agroecology: the key role of arbuscular mycorrhizas in ecosystem services. Mycorrhiza 20:519–530

Giller KE, Witter E, McGrath SP (1998) Toxicity of heavy metals to microorganisms and microbial processes in agricultural soils. Soil Biol Biochem 30(10-11):1389–1414

Goormachtig S, Alves-Ferreira M, Van Montagu M, Engler G, Holsters M (1997) Expression of cell cycle genes during *Sesbania rostrata* stem nodule development. Mol Plant Microbe Interact 10:316–325

Gordon AJ, Minchin FR, James CL, Komina O (1999) Sucrose synthase in legume nodules is essential for nitrogen fixation. Plant Physiol 120:867–878

Goto BT, Silva GA, Assis DMA, Silva DKA, Souza RG, Ferreira ACA, Jobim K, Mello CMA, Vieira HEE, Maia LC, Oehl F (2012) Intraornatosporaceae (Gigasporales), a new family with two new genera and two new species. Mycotaxon 119:117–132

Granqvist E, Sun J, Op den Camp R, Pujic P, Hill L, Normand P, Oldroyd GED (2015) Bacterial-induced calcium oscillations are common to nitrogen-fixing associations of nodulating legumes and non-legumes. New Phytol 207(3):551–558

Gyaneshwar P, Hirsch AM, Moulin L, Chen WM, Elliott GN (2011) Legume-nodulating beta-proteobacteria: diversity, host range, and future prospects. Mol Plant Microbe Interact 24: 1276–1288

Hameed A, Egamberdieva D, Abd Allah EF, Hashem A, Kumar A, Ahmad P (2014) Salinity stress and arbuscular mycorrhizal symbiosis in plants. In: Miransari M (ed) Use of microbes for the alleviation of soil stresses, vol 1. Springer Science, New York, pp 139–159

Harley JL, Smith SE (1983) Mycorrhizal symbiosis. Academic, London

Harrison MJ, van Buuren ML (1995) A phosphate transporter from the mycorrhizal fungus *Glomus versiforme*. Nature 378:626–629

Harrison MJ, Dewbre GR, Liu J (2002) A phosphate transporter from Medicago truncatula involved in the acquisition of phosphate released by arbuscular mycorrhizal fungi. Plant Cell 14: 2413–2429

Hasanuzzaman M, Fujita M, Islam MN, Ahamed KU, Nahar K (2009) Performance of four irrigated rice varieties under different levels of salinity stress. Int J Integr Biol 6:85–90

Hasanuzzaman M, Gill SS, Fujita M (2013) Physiological role of nitric oxide in plants grown under adverse environmental conditions. In: Tuteja N, Gill SS (eds) Plant acclimation to environmental stress. Springer Science, New York, pp 269–322

Hastwell AH, Gresshoff PM, Ferguson BJ (2015) The structure and activity of nodulation-suppressing CLE peptide hormones of legumes. Funct Plant Biol 42:229–238

He X, Bledsoe CS, Zasoski RJ, Southworth D (2006) Rapid nitrogen transfer from ectomycorrhizal pines to adjacent ectomycorrhizal and arbuscular mycorrhizal plants in a California oak woodland. New Phytol 170(1):143–115

He JL, Li H, Luo J, Ma CF, Li SJ, Qu L, Gai Y, Jiang XN, Janz D, Polle A, Tyree M, Luo ZB (2013) A transcriptomic network underlies microstructural and physiological responses to cadmium in *Populus* x *canescens*. Plant Physiol 162:424–439

Heath KD, Tiffin P (2009) Stabilizing mechanisms in a legume-*Rhizobium* mutualism. Evolution 63:652–662

Heckman DS, Geiser DM, Eidell BR, Stauffer RL, Kardos NL, Hedges SB (2001) Molecular evidence for the early colonization of land by fungi and plants. Science 293:1129–1133

Hirsch AM, McKhann HI, Reddy A, Liao JY, Fang YW, Marshall CR (1995) Assessing horizontal transfer of Nifhdk genes in eubacteria – nucleotide sequence of Nifk from Frankia strain Hfpcci3. Mol Biol Evol 12:16–27

Husaini AM, Abdin MZ, Khan S, Xu YW, Aquil S, Anis M (2012) Modifying strawberry for better adaptability to adverse impact of climate change. Curr Sci 102:1660–1673

Jacobs PF, Peterson RL, Massicotte HB (1989) Altered fungal morphogenesis during early stages of ectomycorrhiza formation in *Eucalyptus pilularis*. Scanning Microsc 3:249–255

Jarvis BDW, Van Berkum P, Chen WX, Nour SM, Fernandez MP, Cleyet-marel JC, Gillis M (1997) Transfer of *Rhizobium huakuii*, *Rhizobium ciceri*, *Rhizobium mediterraneum*, and *Rhizobium tiashanense* to *Mesorhizobium* gen. Nov. Int J Syst Bacteriol 47(3):895–898

Jeffries P, Barea JM (2012) Arbuscular mycorrhiza – a key component of sustainable plant-soil ecosystems. In: Hock B (ed) Fungal associations, The Mycota, vol IX, 2nd edn. Springer, Berlin, pp 51–75

Johnson JM, Alex T, Oelmüller R (2014) *Piriformospora indica*: the versatile and multifunctional root endophytic fungus for enhanced yield and tolerance to biotic and abiotic stress in crop plants. J Trop Agric 52(2):103–122

Jordan DC (1982) Transfer of *Rhizobium* japonicum to *Bradyrhizobium* gen nov, a genus of slow-growing, root nodule bacteria from leguminous plants. Int J Syst Bacteriol 32(1):136–139

Jung SC, Martinez-Medina A, Lopez-Raez JA, Pozo MJ (2012) Mycorrhiza-induced resistance and priming of plant defenses. J Chem Ecol 38:651–664

Kabata-Pendias A, Mukherjee AB (2007) Trace element from soil to huma. Springer, Heidelberg, 550 p

Kamalvanshi M, Kumar A, Jha A, Dhyani SK (2012) Occurrence of arbuscular mycorrhizal fungi in rhizosphere of *Jatropha curcas* L. in arid and semi arid regions of India. Indian J Microbiol 52(3):492–494

Kapoor R, Evelin H, Giri B (2013) Arbuscular mycorrhiza approaches for abiotic stress tolerance in crop plants for sustainable agriculture. In: Tuteja N, Gill SS (eds) Plant acclimation to environmental stress. Springer, New York, pp 359–401

Karimi A, Khodaverdilo H, Sepiheri M, Sadaghiani MR (2011) Arbuscular mycorrhizal fungi and heavy metal contaminated soil. Afr J Microbiol Res 5(13):1571–1576

Kaschuk G, Leffelaar PA, Giller KE, Alberton O, Hungria M, Kuyper TW (2010) Responses of legumes to rhizobia and arbuscular mycorrhizal fungi: a meta-analysis of potential photosynthate limitation of symbioses. Soil Biol Biochem 42:125–127

Kaur R, Singh A, Kang JS (2014) Influence of different types mycorrhizal fungi on crop productivity. Curr Agric Res 2(1):51–54

Keller C, McGrath SP, Dunham SJ (2002) Trace metal leaching through a soil–grassland system after sewage sludge application. J Environ Qual 31:1550–1560

Khodaverdiloo H, Samadi A (2011) Batch equilibrium study on sorption, desorption, and immobilization of cadmium in some semiarid-zone soils as affected by soil properties. Soil Res 49(5): 444–454

Kiers ET, Rousseau RA, West SA, Denison RF (2003) Host sanctions and the legume-rhizobium mutualism. Nature 425:78–81

Kijne JW, Pluvque K (1979) Ultrastructural study of the endomembrane system in infected cells of pea and soybean root nodules. Physiol Plant Pathol 14:339–345

Kistner C, Parniske M (2002) Evolution of signal transduction in intracellular symbiosis. Trends Plant Sci 7:511–518

Kondorosi E, Banfalvi Z, Kondorosi A (1984) Physical and genetic analysis of a symbiotic region of *Rhizobium meliloti*: identification of nodulation genes. Mol Gen Genet 193:445–452

Kosslak RM, Bookland R, Barkei J, Paaren HE, Appelbaum ER (1987) Induction of *Bradyrhizobium japonicum* common nod genes by isoflavones isolated from *Glycine max*. Proc Natl Acad Sci 84:7428–7432

Ladha JK, Pareek RP, Becker M (1992) Stem-nodulating legume-*Rhizobium* symbiosis and its agronomic use in lowland rice. Adv Soil Sci 20:147–192

Laloum T, Baudin M, Frances L, Lepage A, Billault-Penneteau B, Cerri MR (2014) Two CCAAT box-binding transcription factors redundantly regulate early steps of the legume-rhizobia endosymbiosis. Plant J 79:757–768

Lang C, Seven J, Polle A (2011) Host preferences and differential contributions of deciduous tree species shape mycorrhizal species richness in mixed Central European forest. Mycorrhiza 21: 297–308

Le Strange KK, Bender GL, Djordjevic MA, Rolfe BG, Redmond JW (1990) The *Rhizobium* strain NGR234 nodD1 gene product responds to activation by simple phenolic compounds vanillin and isovanillin present in wheat seedling extracts. Mol Plant Microbe Interact 3:214–220

Lee BR, Muneer S, Avice JC, Jin Jung W, Kim TH (2012) Mycorrhizal colonisation and P-supplement effects on N uptake and N assimilation in perennial ryegrass under well-watered and drought-stressed conditions. Mycorrhiza 22:525–534

Legocki RP, Verma DP (1980) Identification of "nodule-specific" host proteins (nodulins) involved in the development of *Rhizobium*-legume symbiosis. Cell 20(1):153–163

Lerouge P, Roche P, Faucher C, Maillet F, Truchet G, Promé JC, Dénarié J (1990) Symbiotic host-specificity of *Rhizobium meliloti* is determined by a sulphated and acylated glucosamine oligosaccharide signal. Nature 344:781–784

Libbenga KR, Harkes PAA (1973) Initial proliferation of cortical cells in the formation of root nodules in *Pisum sativum* L. Planta 114:17–28

Lipsanen P, Lindström K (1988) Infection and root nodule structure in the *Rhizobium galegae* sp nov-*Galega* sp symbiosis. Symbiosis 6:81–96

Lum MR, Hirsch AM (2003) Roots and their symbiotic microbes: strategies to obtain nitrogen and phosphorus in a nutrientlimiting environment. J Plant Growth Regul 21:368–382

Luo ZB, Janz D, Jiang X, Göbel C, Wildhagen H, Tan Y, Rennenberg H, Feussner I, Polle A (2009) Upgrading root physiology for stress tolerance by ectomycorrhizas: Insights from metabolite and transcriptional profiling into reprogramming for stress anticipation. Plant Physiol 151(4): 1902–1917

Luo ZB, Wu C, Zhang C, Li H, Lipka U, Polle A (2014) The role of ectomycorrhizas in heavy metal stress tolerance of host plants. Environ Exp Bot 108:47–62

Mabood F, Souleimanov A, Khan W, Smith DL (2006) Jasmonates induce Nod factor production by *Bradyrhizobium japonicum*. Plant Physiol Biochem 44:759–765

Manchanda G, Garg N (2007) Endomycorrhizal and rhizobial symbiosis: how much do they share? J Plant Interact 2(2):7988

Manoharachary C, Laxami AN, Kunwar AK (2000) Microbial ecology of polluted soil: some aspects. In: Mukerji KG, Chamola BP, Sharma AK (eds) Glimpses in botany. APH, New Delhi, pp 314–325

Manoharachary C, Kunwar IK, Mukerji KG (2002) Some aspects of monotropoid mycorrhizas. In: Mukerji KG et al (eds) Techniques in mycorrhizal studies. Kluwer Academic, Drodrecht, pp 435–441

Marinho F, Silva GA, Ferreira ACA, Veras JSN, Sousa NMF, Goto BT, Maia LC, Oehl F (2014) *Bulbospora minima*, new genus and new species in the Glomeromycetes from semi-arid Northeast Brazil. Sydowia 66:313–323

Martin KJ (2007) Introduction to molecular analysis of ectomycorrhizal communities. Soil Sci Soc Am J 71:601–610

Martinez E, Puopot R, Prome JC, Pardo MA, Segovia L, Truchet G, Denarie J (1993) Chemical signaling of *Rhizobium* nodulating bean. In: Palacios R, Mora J, Newton WE (eds) New horizons in nitrogen fixation. Kluwer Academic, Dordrecht, pp 171–176

Marvel DJ, Torrey JG, Ausubel FM (1987) *Rhizobium* symbiotic genes required for nodulation of legume and nonlegume hosts. Proc Natl Acad Sci 84(5):1319–1323

Marx J (2004) The roots of plant-microbe collaborations. Science 304:234–236

Massicotte HB, Ackerley CA, Peterson RL (1987) The root-fungus interface as a indicator of symbiont interaction in ectomycorrhizae. Can J Forest Res 17:846–854

Mathur N, Singh J, Bohra S, Vyas A (2007) Arbuscular mycorrhizal status of medicinal halophytes in saline areas of Indian Thar Desert. Int J Soil Sci 2:119–127

Medina J, Cornejo P, Borie F, Meyer S, Palenzuela J, Vieira HEE, Ferreira ACA, Silva GA, Sánchez-Castro I, Oehl F (2014) *Corymbiglomus pacificum*, a new glomeromycete from a saline lakeshore in Chile. Mycotaxon 127:173–183

Miyasaka SC, Habte M, Friday JB, Johnson EV (2003) Manual on arbuscular mycorrhizal fungus production and inoculation techniques. Soil Crop Manage 5:4

Molina R, Trappe JM (1982) Lack of mycorrhizal specificity by the ericaceous hosts *Arbutus menziesii* and *Arctostaphylos uva-ursi*. New Phytol 90:495–509

Monika A, Gorzelak AKA, Pickles BJ, Suzanne WS (2015) Inter-plant communication through mycorrhizal networks mediates complex adaptive behaviour in plant communities. AoB Plants 2:7

Moore D, Robson GD, Anthony TAPJ (2011) 21st Century guidebook to fungi. Cambridge University Press, Cambridge

Morton JB, Benny GL (1990) Revised classification of arbuscular mycorrhizal fungi (Zygomycetes): a new order, Glomales, two new suborders, Glomineae and Gigasporineae, and two new families, Acaulosporaceae and Gigasporaceae, with an emendation of Glomaceae. Mycotaxon 37:471–491

Morton JB, Redecker D (2001) Two new families of Glomales, Archaeosporaceae and Paraglomaceae, with two new genera *Archaeospora* and *Paraglomus*, based on concordant molecular and morphological characters. Mycologia 93:181–195

Newcomb W (1979) Control of morphogenesis and differentiation of pea root nodules. In: Newton WE, Orme-Johnson WH (eds) Nitrogen fixation symbiotic associations and cyanobacteria, vol 2. University Park Press, Baltimore, pp 87–102

Noe R, Hammerstein P (1994) Biological markets–supply-and-demand determine the effect of partner choice in cooperation, mutualism and mating. Behav Ecol Sociobiol 35:1–11

Noel KD, Vanden Bosch KA, Kulpaca B (1986) Mutations in *Rhizobium phaseoli* that lead to arrested development of infection threads. J Bacteriol 168:1392–1401

Nuti MP, Lepidi AA, Prakash RK, Schilperoort RA, Cannon FC (1979) Evidence for nitrogen fixation (nif) genes on indigenous *Rhizobium* plasmids. Nature 282:533–535

Oehl F, Souza F, Sieverding E (2008) Revision of *Scutellospora* and description of five new genera and three new families in the arbuscular mycorrhiza forming Glomeromycetes. Mycotaxon 106: 311–360

Oehl F, Silva GA, Goto BT, Maia LC, Sieverding E (2011a) Glomeromycota: two new classes and a new order. Mycotaxon 116:365–379

Oehl F, Sieverding E, Palenzuela J, Ineichen K, Alves da Silva G (2011b) Advances in Glomeromycota taxonomy and classification. IMA Fungus 2(2):191–199

Oehl F, Sánchez-Castro I, Palenzuela J, Silva GA (2015) *Palaeospora spainii*, a new arbuscular mycorrhizal fungus from Swiss agricultural soils. Nova Hedwigia 101:89–102

Oelmüller R, Sherameti I, Tripathi S, Varma A (2009) *Piriformospora indica*, a cultivable root endophyte with multiple biotechnological applications. Symbiosis 49:1–17

Oke V, Long SR (1999) Bacteroid formation in the *Rhizobium*-legume-symbiosis. Curr Opin Microbiol 2:641–646

Oldroyd GE, Downie JA (2008) Coordinating nodule morphogenesis with rhizobial infection in legumes. Annu Rev Plant Biol 59:519–546

Ott T, van Dongen JT, Günther C, Krusell L, Desbrosses G, Vigeolas H, Bock V, Czechowski T, Geigenberger P, Udvardi MK (2005) Symbiotic leghemoglobins are crucial for nitrogen fixation in legume root nodules but not for general plant growth and development. Curr Biol 15:531–535

Pagano MC, Oehl F, Silva, GA, Maia LC, Silva DK, Cabello MN (2016) Advances in arbuscular mycorrhizal taxonomy. In: Pagano M (ed) Recent advances on mycorrhizal fungi. Fungal biology. Springer, Cham

Palenzuela J, Ferrol N, Boller T, Azcón-Aguilar C, Oehl F (2008) *Otospora bareai*, a new fungal species in the Glomeromycetes from a dolomitic shrub-land in the National Park of Sierra de Baza (Granada, Spain). Mycologia 100:282–291

Pandolfini T, Gremigni P, Gabbrielli R (1997) Biomonitoring of soil health by plants. In: Pankhurst CE, Doube BM, Gupta VVSR (eds) Biological indicators of soil health. CAB International, New York, pp 325–347

Parniske M (2008) Arbuscular mycorrhiza: the mother of plant root endosymbioses. Nat Rev Microbiol 6:763–775

Peng G, Yuan Q, Li H, Zhang W, Tan Z (2008) *Rhizobium oryzae* sp nov, isolated from the wild rice *Oryza alta*. Int J Syst Evol Microbiol 58:2158–2163

Peskan-Berghofer T, Shahollari B, Giong PH, Hehl S, Markert C, Blanke V, Kost G, Varma A, Oelmuller R (2004) Association of *Piriformospora indica* with *Arabidopsis thaliana* roots represents a novel system to study beneficial plant–microbe interactions and involves early plant protein modifications in the endoplasmic reticulum and at the plasma membrane. Physiol Plant 122:465–477

Peters NK, Frost JW, Long SR (1986) A plant flavone, luteolin, induces expression of *Rhizobium meliloti* nodulation genes. Science 233:977–980

Pham GH, Kumari R, Singh A, Sachdev M, Prasad R, Kaldorf M, Buscot F, Oelmüller R, Tatjana P, Weiss M, Hampp R, Varma A (2004) Axenic cultures of *Piriformospora indica*. In: Varma A, Abbott L, Werner D, Hampp R (eds) Plant surface microbiology. Springer, Heidelberg, pp 593–616

Phillips DA (1971) Abscisic acid inhibition of root nodule initiation in *Pisum sativum*. Planta 100:181–190

Phillips DA, Torrey JG (1970) Cytokinin production by *Rhizobium japonicum*. Physiol Plant 23: 1057–1063

Pontes JS, Sánchez-Castro I, Palenzuela J, Maia LC, Silva GA, Oehl F (2013) *Scutellospora alterata*, a new gigasporalean species from the semi-arid Caatinga biome in Northeastern Brazil. Mycotaxon 125:169–181

Porras-Alfaro A, Bayman P (2011) Hidden fungi, emergent properties: endophytes and microbiomes. Annu Rev Phytopathol 49:291–315

Pozo MJ, Azcón-Aguilar C (2007) Unravelling mycorrhiza-induced resistance. Curr Opin Plant Biol 10:393–398

Prakash RK, Schilperoort RA, Nuti MP (1981) Large plasmids of fast-growing rhizobia: homology studies and location of structural nitrogen fixation (nif) genes. J Bacteriol 145:1129

Puppo A, Pauly N, Boscari A, Mandon K, Brouquisse R (2013) Hydrogen peroxide and nitric oxide: key regulators of the legume *Rhizobium* and mycorrhizal symbioses. Antioxid Redox Signal 18:2202–2219

Qiang X, Weiss M, Kogel KH, Schäfer P (2011) *Piriformospora indica* – a mutualistic basidiomycete with an exceptionally large plant host range. Mol Plant Pathol 13:508–518

Rae AL, Bonfante-Fasolo P, Brewin NJ (1992) Structure and growth of infection threads in the legume symbiosis with *Rhizobium leguminosarum*. Plant J 2:385–395

Rai M, Acharya D, Singh A, Varma A (2001) Positive growth responses of the medicinal plants *Spilanthes calva* and *Withania somnifera* to inoculation by *Piriformospora indica* in a field trial. Mycorrhiza 11:123–128

Ray JG, Valsalakumar N (2010) Arbuscular mycorrhizal fungi and *Piriformospora indica* individually and in combination with Rhizobium on green gram. J Plant Nutr 33:285–298

Raziuddin F, Hassan G, Akmal M, Shah SS, Mohammad F, Shafi M, Bakht J, Zhou W (2011) Effects of cadmium and salinity on growth and photosynthesis parameters of Brassica species. Pak J Bot 43(1):333–340

Read DJ (1996) The structure and function of the ericoid mycorrhizal root. Ann Bot 77:365–374

Read DJ, Duckett JG, Francis R, Ligrone R, Russel A (2000a) Symbiotic fungal associations in "lower" land plants. Philos Trans R Soc Biol Sci 355:815–831

Read DJ, Duckett JG, Francis R, Ligrone R, Russell A (2000b) Symbiotic fungal associations in 'lower' land plants. Philos Trans R Soc Lond B Biol Sci 355:815–830

Redecker D, Schüßler A (2014) Glomeromycota. In: Mclaughlin DJ, Spatafora JW (eds) The Mycota: a comprehensive treatise on fungi as experimental systems for basic and applied research, systematics and evolution Part A. Springer, New York

Redmond JW, Batley M, Djordjevic MA, Innes RW, Kuempel PL, Rolfe BG (1986) Flavones induce expression of nodulation genes in *Rhizobium*. Nature 323:632–635

Reid DE, Ferguson BJ, Hayashi S, Lin YH, Gresshoff PM (2011) Molecular mechanisms controlling legume autoregulation of nodulation. Ann Bot 108:789–795

Rhijn V, Vanderleyden J (1995) The *Rhizobium*-plant symbiosis. Microbiol Rev 59:124–142

Rhijn V, Feys PJS, Verreth BC, Vanderleyden J (1993) Multiple copies of nodD in *Rhizobium tropici* CIAT899 and BR816. J Bacteriol 175:438–447

Rinaldi AC, Comandini O, Kuyper TW (2008) Ectomycorrhizal fungal diversity: separating the wheat from the chaff. Fungal Divers 33:1–45

Rosas JC, Castro JA, Robleto EA, Handelsman J (1998) A method for screening *Phaseolus vulgaris* L germplasm for preferential nodulation with a selected *Rhizobium etli* strain. Plant Soil 203: 71–78

Rosenberg C, Boistard P, Denarie J, Casse-Delbart F (1981) Genes controlling early and late functions in symbiosis are located on a megaplasmid in *Rhizobium meliloti*. Mol Gen Genet 184:326–333

Rosendahl S (2008) Communities, populations and individuals of arbuscular mycorrhizal fungi. New Phytol 178:253–266

Rossen L, Shearman CA, Johnston AWB, Downie JA (1985) The *nodD* gene of *Rhizobium leguminosarum* is autoregulatory and in the presence of plant exudate induces the *nodA,B,C* genes. EMBO J 4:3369–3373

Rostas K, Kondorosi E, Horvath B, Simoncsits A, Kondorosi A (1986) Conservation of extended promoter regions of nodulation genes in *Rhizobium*. Proc Natl Acad Sci 83:1757–1761

Roth LE, Stacey G (1989) Bacterium release into host cells of nitrogen-fixing soybean nodules: the symbiosome membrane comes from three sources. Eur J Cell Biol 49(1):13–23

Roth LE, Jeon K, Stacey G (1988) Homology in endosymbiotic systems. In: Palcios R, Verma DPS, St Paul MN (eds) The term "symbiosome". Molecular genetics of plant microbe interactions. ADS Press, St Paul, pp 220–225

Rozahon M, Ismayil N, Hamood B, Erkin R, Abdurahman M, Mamtimin H, Abdukerim M, La R, Rahman E (2014) *Rhizobiun populi* sp. nov., an endophytic bacterium isolated from *Populus euphratica*. Int J Syst Evol Microbiol 64:3215–3221

Salminen SO, Streeter JG (1992) Labeling of carbon pools in *Bradyrhizobium japonicum* and *Rhizobium leguminosarum* bv *viciae* bacteroids following incubation of intact nodules with $^{14}CO_2$. Plant Physiol 100:597–604

Sandeep C, Mohanb V, Viswanatha S (2015) Significance of ectomycorrhizae in forest ecosystems of India. Int J Plant Anim Environ Sci 5:23–31

Sanders IR, Croll D (2010) Arbuscular mycorrhiza: a challenge to understand the genetics of the fungal partner. Ann Rev Genet 44:271–292

Schouteden N, De Waele D, Panis B, Vos CM (2015) Arbuscular mycorrhizal fungi for the biocontrol of plant-parasitic nematodes: a review of the mechanisms involved. Front Microbiol 6:1280

Schüßler A, Walker C (2010) The Glomeromycota: a species list with new families and new genera. Published in libraries at The Royal Botanic Garden Kew, Botanische Staatssammlung Munich, and Oregon State University

Schüßler A, Walker C (2011) Evolution of the plant symbiotic fungal phylum, Glomeromycota. In: Poggeler S, Wostemeyer J (eds) Evolution of fungi and fungal-like organisms. Springer, Berlin, pp 163–185

Schüßler A, Schwarzott D, Walker C (2001) A new fungal phylum, the Glomeromycota: phylogeny and evolution. Mycol Res 105:1413–1421

Schüßler A, Krüger M, Walker C (2009) Phylogeny, evolution and origin of the 'plant-symbiotic' phylum Glomeromycota. In: Wöstemeyer J, Martin W (eds) The Mycota XIV – evolution of fungi and fungal-like organisms. Springer, Berlin

Selosse MA, Setaro S, Glatard F, Richard F, Urecelay C, Weiss M (2007) Sebacinales are common mycorrhizal associates of Ericaceae. New Phytol 174:864–878

Sethi IK, Walia SK (2018) Text book of Fungi and their allies, 2nd edn. Scientific International, New Delhi

Sharifi E (1983) Parasitic origins of nitrogen-fixing *Rhizobium*-legume symbioses: a review of the evidence. Biosystems 16:269–289

Sherameti I, Tripathi S, Varma A, Oelmuller R (2008) The root-colonizing endophyte *Pirifomospora indica* confers drought tolerance in *Arabidopsis* by stimulating the expression of drought stress-related genes in leaves. Mol Plant Microbe Interact 21:799–807

Sieberer BJ, Timmers AC, Emons AM (2005) Nod factors alter the microtubule cytoskeleton in *Medicago truncatula* root hairs to allow root hair reorientation. Mol Plant Microbe Interact 18: 1195–1204

Sieverding E, Silva GA, Berndt R, Oehl F (2014) *Rhizoglomus*, a new genus in the Glomeraceae. Mycotaxon 129:373–386

Silva GA, Maia LC, Oehl F (2012) Phylogenetic systematics of the Gigasporales. Mycotaxon 122: 207–222

Simon L, Bousquet J, Levesque RC, Lalonde M (1993) Origin and diversification of endomycorrhizal fungi and coincidence with vascular land plants. Nature 363:67–69

Singh A, Varma A (2000) Orchidaceous mycorrhizal fungi. In: Mukerji KG (ed) Mycorrhizal biology. Kluwer Academic/Plenum, New York, pp 265–288

Singh A, Sharma J, Rexer KH, Varma A (2000) Plant productivity determinants beyond minerals, water and light. *Piriformospora indica*: a revolutionary plant growth promoting fungus. Curr Sci 79:101–106

Smith SE, Read DJ (1997) Mycorrhizal symbiosis, 2nd edn. Academic, San Diego

Smith SE, Read DJ (2008) Mycorrhizal symbiosis, 3rd edn. Academic, London

Smith SE, Smith AF, Jakobsen I (2003) Mycorrhizal fungi can dominate phosphate supply to plants irrespective of growth responses. Plant Physiol 133:16–20

Spain JL, Sieverding E, Oehl F (2006) *Appendicispora*, a new genus in the arbuscular mycorrhizal-forming Glomeromycetes, with a discussion of the genus *Archaeospora*. Mycotaxon 97: 163–182

Spaink HP, Okker RJH, Wijffelman CA, Pees E, Lugtenberg BJJ (1987) Promoters in nodulation region of the *Rhizobium leguminosarum* Sym plasmid pRL1Jl. Plant Mol Biol 9:29–37

Sprent (1980) Root nodule anatomy, type of export product and evolutionary origin of some Leguminosae. Plant Cell Environ 3:35–43

Srivastava D, Kapoor R, Srivastava SK, Mukerji KG (1996) Vesicular arbuscular mycorrhiza-an overview. In: Mukerji KG (ed) Concepts in mycorrhizal research, vol 813. Kluwer Academic, Dordrecht, pp 1–39

Sudha KH, Narula A, Kumar S, Srivastava PS, Varma A (1999) Mycorrhiza aided biological hardening of in vitro raised plantlets. In: Tewari JP, Lakhanpal TN, Singh J, Gupta R, Chamola BP (eds) Advances in microbial biotechnology. APH, Delhi, pp 469–485

Sun CA, Johnson J, Cai DG, Sherameti I, Oelmuller R, Lou BG (2010) *Piriformospora indica* confers drought tolerance in Chinese cabbage leaves by stimulating antioxidant enzymes, the expression of drought-related genes and the plastid-localized CAS protein. J Plant Physiol 167:1009–1017

Szuba A, Karliński L, Krzesłowska M, Hazubska-Przybył T (2017) Inoculation with a Pb-tolerant strain of *Paxillus involutus* improves growth and Pb tolerance of *Populus* × *canescens* under in vitro conditions. Plant Soil 412(1–2):253–266

Tedersoo L, Nara K (2010) General latitudinal gradient of biodiversity is reversed in ectomycorrhizal fungi. New Phytol 185:351–354

Thamer S, Schädler M, Bonte D, Ballhorn DJ (2011) Dual benefit from a belowground symbiosis: nitrogen fixing rhizobia promote growth and defense against a specialist herbivore in a cyanogenic plant. Plant Soil 34:1209–1219

Tillard P, Drevon JJ (1988) Nodulation and nitrogenase activity of chickpea cultivar INRA199 inoculated with different strains of *Rhizobium ciceri*. Agronomie 8:387–392

Torok I, Kondorosi E, Stepkowski T, Pósfai J, Kondorosi A (1984) Nucleotide sequence of *Rhizobium meliloti* nodulation genes. Nucleic Acids Res 12(24):9509–9524

Torres-Cortes G, Ghignone S, Bonfante P, Schüßler A (2015) Mosaic genome of endobacteria in arbuscular mycorrhizal fungi: transkingdom gene transfer in an ancient mycoplasma-fungus association. Proc Natl Acad Sci U S A 112(25):7785–7790

Trinchant JC, Rigaud J (1989) Alternative energy-yielding substrates for bacteroids isolated from stem and root nodules of *Sesbania rostrata* submitted to O_2 restricted conditions. Plant Sci 59:141–149

Trinick MJ (1979) Structure of nitrogen-fixing nodules formed by *Rhizobium* on roots of *Parasponia andersonii* Planch. Can J Microbiol 25:565–578

Truchet G, Roche P, Lerouge P, Vasse J, Camut S, de Billy F, Promé JC, Dénarié J (1991) Sulphated lipooligosaccharide signals of *Rhizobium meliloti* elicit root nodule organogenesis in alfafa. Nature 351:670–673

Tu JC (1977) Structural organization of the rhizobial root nodule of alfalfa. Can J Bot 55:35–43

Turk MA, Asaf TA, Hameed KM, Al-Tawaha AM (2006) Significance of mycorrhizae. World J Agric Sci 2(1):16–20

Udvardi MK, Day DA (1990) Ammonia ('4C-methylamine) transport across the bacteroid and peribacteroid membranes of soybean root nodules. Plant Physiol 94:71–76

Udvardi MK, Day DA (1997) Metabolite transport across symbiotic membranes of legume nodules. Annu Rev Plant Physiol Plant Mol Biol 48:493–523

Udvardi M, Poole PS (2013) Transport and metabolism in legume-rhizobia symbioses. Annu Rev Plant Biol 64:781–805

Unnikumar KR, Sowjanya SK, Varma A (2013) *Piriformospora indica*: a versatile root endophytic symbiont. Symbiosis 60:107–113

Van Brussel AAN, Bakhulzen R, Van Spronsen PC, Spaink HP, Tak T, Lugtenberg BJJ, Keijne JW (1992) Induction of pre-infection thread structures in the host plant by lipo-oligosaccharides of *Rhizobium*. Science 257:70–72

van der Heijden MGA, Bardgett RD, van Straalen NM (2008) The unseen majority: soil microbes as drivers of plant diversity and productivity in terrestrial ecosystems. Ecol Lett 11:296–310

Vandenbosch KA, Noel KD, Kaneko Y, Newcomb EH (1985) Nodule initiation elicited by noninfective mutants of *Rhizobium phaseoli*. J Bacteriol 162(3):950–959

Varma A, Verma S, Sudha SN, Bütehorn B, Franken P (1999) *Piriformospora indica*, a cultivable plant-growth-promoting root endophyte. Appl Environ Microbiol 65(6):2741–2744

Varma A, Singh A, Sudha M, Sahay NS, Sharma J, Roy A, Kumari M, Rana D, Thakran S, Deka D, Bharti K, Hurek T, Blechert O, Rexer KH, Kost G, Hahn A, Maier W, Walter M, Strack D, Kranner I (2001) *Piriformospora indica*: a cultivable mycorrhiza-like endosymbiotic fungus. In: Hock B (ed) The Mycota IX. Springer, Berlin, pp 125–150

Varma A, Bakshi M, Lou B, Hartmann B, Oelmueller R (2012) *Piriformospora indica*: a novel plant growth-promoting mycorrhizal fungus. Agric Res 1:117. https://doi.org/10.1007/s40003-012-0019-5

Veresoglou SD, Rillig MC (2012) Suppression of fungal and nematode plant pathogens through arbuscular mycorrhizal fungi. Biol Lett 8:214–216

Verma S, Varma A, Rexer K-H, Hassel A, Kost G, Sarbhoy A, Bisen P, Bütehorn B, Franken P (1998) *Piriformospora indica*, gen. et sp. nov., a new root-colonizing fungus. Mycologia 90: 898–905

Vierheilig H, Lerat S, Piché Y (2003) Systemic inhibition of arbuscular mycorrhiza development by root exudates of cucumber plants colonized by *Glomus mosseae*. Mycorrhiza 13:167–170

Vierheilig H, Schweigerb P, Brundrettc M (2005) An overview of methods for the detection and observation of arbuscular mycorrhizal fungi in roots. Physiol Plant 125:393–404

Voegelin A, Barmettler K, Kretzschmar R (2003) Heavy metal release from contaminated soils: comparison of column leaching and batch extraction results. J Environ Qual 32:865–875

Walker C (1992) Systematics and taxonomy of the arbuscular endomycorrhizal fungi (Glomales)-a possible way forward. Agronomie 12:887–897

Waller F, Achatz B, Baltruschat H, Fodor J, Becker K, Fischer M, Heier T, Huckelhoven R, Neumann C, Wettstein D, Franken P, Kogel KH (2005) The endophytic fungus *Piriformospora indica* reprograms barley to salt-stress tolerance, disease resistance, and higher yield. Proc Natl Acad Sci USA 102:13386–13391

Wang W, Qui YL (2006) Phylogenetic distribution and evolution of mycorrhizas in land plants. Mycorrhiza 16(5):299–363

Weiβ M, Selosse MA, Rexer KH, Urban A, Oberwinkler F (2004) Sebacinales: a hitherto overlooked cosm of heterobasidiomycetes with a broad mycorrhizal potential. Mycol Res 108:1002–1010

Whipps JM (2004) Prospects and limitations for mycorrhizas in biocontrol of root pathogens. Can J Bot 82:1198–1227

Williams PM, Mallorca MS (1982) Abscisic acid and gibberellin-like substances in roots and root nodules of *Glycine max*. Plant Soil 65:19–26

Wilson GWT, Hartnett DC, Rice WC (2006) Mycorrhizal-mediated phosphorus transfer between tallgrass prairie plants *Sorghastrum nutans* and *Artemisia ludoviciana*. Funct Ecol 20:427–435

Wu QS, Xia RX (2003) Research and application on vesicular-arbuscular mycorrhiza of fruit trees. Plant Physiol Commun 39:536–540

Wu QS, Xia SR (2004) The relation between vesicular -arbuscular mycorrhizae and water metabolism in plants. Chin Agric Sci Bull 1:188–192

Yadav V, Kumar M, Deep DK, Kumar H, Sharma R, Tripathi T, Tuteja N, Saxena AK, Johri AK (2010) A phosphate transporter from the root endophytic fungus *Piriformospora indica* plays a role in phosphate transport to the host plant. J Biol Chem 285:26532–26544

Zaat SAJ, Wijffelman CA, Spaink HP, van Brussel AAN, Okker RJH, Lugtenberg BJJ (1987) Induction of the nodA promoter of *Rhizobium leguminosarum* sym plasmid pRL1JI by plant flavanones and flavones. J Bacteriol 169:198–204

Zahran HH (1999) *Rhizobium*-Legume symbiosis and nitrogen fixation under severe conditions and in an arid climate. Microbiol Mol Biol Rev 3(4):968–989

Zarea MJ, Hajinia S, Karimi N, Goltapeh EM, Rejali F, Varma A (2012) Effect of *Piriformospora indica* and *Azospirillum* strains from saline or non-saline soil on mitigation of the effects of NaCl. Soil Biol Biochem 45:139–146

Zhang XX, Sun L, Ma XT, Sui XH, Jiang RB (2011) *Rhizobium pseudoryzae* sp. nov., isolated from the rhizosphere of rice. Int J Syst Evol Microbiol 61:2425–2429

Zhang YJ, Zheng WT, Everall I, Young JP, Zhang XX (2015) *Rhizobium anhuiense* sp. nov., isolated from effective nodules of *Vicia faba* and *Pisum sativum*. Int J Syst Evol Microbiol 65: 2960–2967

Chapter 16
Impact of Climate Change on Root–Pathogen Interactions

Parinita Singh, Touseef Hussain, Seema Patel, and Nadeem Akhtar

16.1 Introduction

Global food insecurity in the face of ever-increasing human population is one of the most pressing socioeconomic issues of current times. Food insecurity is marked by the dramatic imbalance between the high demand of food and low agricultural harvest. Environmental degradation, climate change, and plant pathogens are the three major threats to food productivity, which can cause global food crisis (Wheeler and von Braun 2013). The unpredictable climate patterns such as elevated temperature, irregularity of precipitation, and increasing concentration of greenhouse gases in the atmosphere observed in current times are an outcome of intensive agricultural practices, rapid industrialization, and deforestation (Lawrence and Vandecar 2015).

Soil is inhabited by a diversity of microorganisms, which form the ubiquitous biotic components of the former. The massive usage of fertilizers and pesticides is manipulating soil pH and biodiversity (Pratibha and Shah 2012). Roots are very important part of the plants due to their proximity to the soil nutrient pool. The root systems not only anchor the plant, and absorb water and nutrition, but they interact with the soil organisms as well. The rhizosphere is populated with tens of thousands of microbial species, jostling for access into the roots (Berendsen et al. 2012). In fact, the microbial community, often attributed as the second genome of the plant, is of

P. Singh
Department of Biotechnology, Chandigarh University, Mohali, Punjab, India

T. Hussain
Department of Botany, Aligarh Muslim University, Aligarh, Uttar Pradesh, India

S. Patel
Bioinformatics and Medical Informatics Research Centre, San Diego State University, San Diego, CA, USA

N. Akhtar (✉)
Department of Animal Biosciences, University of Guelph, Guelph, ON, Canada

B. Giri et al. (eds.), *Root Biology*, Soil Biology 52,
https://doi.org/10.1007/978-3-319-75910-4_16

significant importance for plant health (Berendsen et al. 2012). These microorganisms are either benign or pernicious to plants, deciding their agronomic efficiency. Plant roots exude metabolites which act as signals modulating bacterial gene expression, thus controlling microbe–plant interactions (Mark et al. 2005). Plant hormones coordinate among themselves to vanquish pathogens. Ethylene promotes endogenous reactive oxygen species (ROS) formation and lesion propagation; jasmonic acid limits the lesion spreading; abscisic acid regulates stomata; and salicylic acid is required for the programmed cell death.

Some of the soil microbes establish symbiotic or commensal relation with the host plant via the roots. Mycorrhiza has significant contribution to plant health, which occurs via salicylic acid- and jasmonate-dependent signaling pathways (Pozo and Azcón-Aguilar 2007). This cross talk with fungal partners enables the host plant to have a robust defense arsenal (Pozo and Azcón-Aguilar 2007). The bacterium *Rhizobium* and leguminous plants forge a symbiotic relation. Root nodules caused by these bacteria can fix atmospheric nitrogen (Oke and Long 1999). The roots are vulnerable as well, as some microbial interactions are pathogenic. Phytopathogens can affect crop plants in a myriad of adverse ways. Soilborne fungal pathogens which include *Rhizoctonia*, *Fusarium*, *Pythium*, *Verticillium*, *Phytophthora*, *Sclerotinia*, *Rosellinia*, etc. exist in the form of spores or dormant propagules, until a favorable condition prevails. These pathogens invade the roots of susceptible host plants either to colonize or infect them. The pathogenic fungi affect the plants in multiple ways, which includes mycotoxin elaboration. While *Fusarium* sp. is associated with wilting, root rot, and head blight (Matny 2015), *Rhizoctonia* sp. is responsible for root rot and damping-off (Karima El-Gamal 2012). Among a number of food crops, banana (Mostert et al. 2017) has been severely affected by fungal pathogens. So, plant disease management strategies are of paramount importance to ensure optimal food production.

16.2 The Concept of Food Security

Food security is defined as "a situation that exists when all people, at all times, have physical, social and economic access to sufficient, safe and nutritious food that meets their dietary needs and food preferences for an active and healthy life" (Porter et al. 2014). It is a multifaceted phenomenon that broadly covers food availability, food access, food utilization, and stability, considered as "the four pillars of food security."

Almost 12% of the landscape on Earth is covered by cropland which is being constantly degraded by anthropogenic practices. According to Food and Agriculture Organization (FAO), till 2050, the annual cereal production must hike to 3 billion tons from the present figure of 2.1 billion tons, to feed the population of 9.1 billion which is 34% higher than today's population, which explains the importance of global food security and the alarming need for crop protection and improvement.

A staple food crop is a plant that is cultivated to make up a major proportion of a population's diet. Considering the per capita calorie intake, the major food crops around the globe are wheat, rice, maize, and soybean, collectively known as the "the big four," which constitute the maximum proportion of staple food to the greater part of the community (Fischer et al. 2014). Other than these cereals, roots and tubers, pulses, sugar crops, nuts, oil crops, etc. form the 18 major crop types, according to global distribution of crops.

16.2.1 Food Security Crisis in Developing Economies Like India

The growing economies, like India, with its ever-expanding population to feed, suffers from a food security crisis, despite having a wide range of favorable geographic domains to support a miscellany of vegetations. India has the second largest arable land area in the world, with almost 54.6% population engaged in this sector, even though agriculture accounts for roughly 17% of the nation's gross domestic product (GDP), as stated by the Ministry of Agriculture and Farmers Welfare in Annual Report 2016–2017. Shrinking size of land holdings, lack of quality infrastructure, unavailability of quality farm inputs at reasonable prices, and insufficiency of innovative market policies, are among the prime factors responsible for such low agricultural produce.

The key cereal crops in the country are rice, wheat, and maize. India stands second in the production of rice, wheat, and vegetables in the world, after China, despite having approximately 34% more arable land than the latter. Sugarcane, potatoes, and groundnuts also account for huge agricultural produce in India, according to the average annual production of the country. As evident from the relatively smaller share in global economy in terms of agricultural exports, whose net value is only 9% (9.3 billion dollars) of the total produce, the share of India in global agricultural produce is not satisfactory.

16.2.2 Climate Change as a Threat to Food Security

Human activities involving poor agricultural practices, urbanization, deforestation, industrialization, and environmental pollution, ultimately lead to global warming, the key driver of climate change (Vermeulen et al. 2012). The alteration in climate patterns has a significant inverse effect on food security. It gradually elevates the frequency and intensity of natural disasters which directly or indirectly affects the food supply and demand, more prominently in local food systems. Persistent droughts, frequent storms and floods, salinity, soil erosion and land degradation, ecological imbalance (Ahuja et al. 2010), all occurring due to climate change, destroying crops, demolish the infrastructure and livelihood, aggravating poverty and hunger. The modified soil attributes due to the changed climatic factors affect microbial communities at the rhizosphere. Plant pathogens become more hostile, as the abiotic factors as temperature, drought, and pH fluctuation are stressors to their metabolic systems.

16.3 Host, Pathogen, and Environment: The Disease Triangle

Pathogenicity is the ability of an organism to interfere with one or more functions of the host plant. The emergence and the severity of a disease can be controlled or modified by changing the external factors like a favorable environment. The environment, host, and an infectious agent form a triangle. Anthropogenic activities exacerbate the effect of environmental stressors on the host–pathogen dynamics, constituting the "disease quadrangle."

16.3.1 Effects of Environment on Host–Pathogen Interaction

Disease regimes of crop pathosystems are highly prone to altered climatic conditions which can aggravate the disease symptoms (Elad and Pertot 2014). Co-evolution between the plants and their adversaries is well-known. Climate change has a tendency to re-modify the coevolutionary relationship between the host and its pathogens, which is an essential factor for determining host specificity, particularly in the case of biotrophic fungi. The Intergovernmental Panel on Climate Change (IPCC) 2007 enunciates that the level of CO_2 has increased by 30% in the atmosphere and average temperature of the Earth has elevated by 0.3–0.6 °C till now (Chakraborty et al. 2000). Changes in environmental variables such as CO_2 and ozone levels, temperature, and water stress directly affect certain attributes of the disease triangle, such as the colonization of pathogen in the host, the extent of disease manifestation, modification of the host defense system, and photosynthetic efficiency in the host (Eastburn et al. 2011). Hence, it is a well-established fact that the physiological interactions between the other two entities of the disease triangle are greatly affected by the surrounding environment.

16.3.2 Major Crop Diseases and Harvest Loss

Agricultural crops suffer immense damage from both biotic and abiotic factors (Oerke 2006). Rice crop endures the highest percentage of crop loss worldwide, *i.e.*, 37.4%, among the three widely grown crops in the world, followed by maize

(31.2%) and wheat (28.2%), mainly due to the postharvest losses. Annually, almost 68% of the total global harvest is lost due to pests and pathogens. Developing economies, like India, and African countries are more prone to postharvest losses due to the lack of infrastructure in supply chain, hence, more damage occurs during storage and transportation. Erratic monsoon and rain are factors in crop spoilage as well. Hence, crop protection needs more emphasis than crop improvement to deal with the food security crisis. Microbial pathogens constitute 16% of the overall crop loss (including pre- and postharvest losses) in which 70–80% destruction is caused by fungal and oomycetes pathogens (Bebber and Gurr 2015). The principal cause of these postharvest losses (quantitative and qualitative both) is various forms of fungal diseases ranging from rust and smuts to downy mildew and late blight.

16.4 Roots as the Most Favored Target for Fungal Pathogens

As mentioned before, roots are immensely important for the functioning of all the aerial parts of a plant as they are involved in anchorage as well as water and nutrient uptake from the soil. Rhizosphere is the region of soil where parasitic and mutualistic microbes reside. Among these organisms, the major ones are fungi, nematodes, bacteria, protozoa, algae, and microarthopods (Prashar et al. 2014). Fungi and oomycetes are the most obstinate of most soilborne pathogens. Unlike bacterial and viral pathogens, which are generally non-sporulating and require an opening or wound to enter the host, the fungi can produce spores and sustain in soil for very long periods until favorable microenvironment is attained to germinate. Fungi are eukaryotic, multicellular, filamentous heterotrophic organisms with chitinous cell wall, which absorbs nutrients from the surrounding substrates by a network of hyphae known as mycelium, whereas oomycetes are phylogenetically more related to brown algae but possess the morphology of fungi with a cellulosic cell wall. All the fungal pathogens are broadly categorized into two groups: necrotrophs and biotrophs. The majority of soilborne pathogenic fungi is necrotrophic in nature with a wide host range due to the lack of host specificity, as they do not require a living host cell to survive and kill them by releasing toxins and enzymes. Biotrophic fungi, on the other hand, have co-evolved with the host over time, and require a living host for survival. Due to this specificity, these fungi have a very narrow host range. The uncompromising specificity of biotrophs to a particular host is evident with the fact that the population of pathogenic fungi declines drastically after the harvesting season of the susceptible host crop terminates. Resistance to necrotrophic fungi in plants is not regulated by one specific gene which makes it more difficult to devise methods of their biological control at a genetic level. Root exudates play a very crucial role in the pathogenicity of these soilborne fungi (Broeckling et al. 2008). Due to the fluctuation in soil temperature owing to the changing environment, survival of these pathogens occurs via the formation of resistant propagules which find shelter inside the roots or in soil as chlamydospores, conidia, or sclerotia in the form of dormant entities. Root exudates, released near the dormant propagule, act as a source of stimulus and trigger the fungi to reach the plant by chemotaxis during favorable conditions (Broeckling et al. 2008). Upon germination, the zoospore attaches to the root surface and enters the epidermal cells of root tips or root hair by degrading cell wall with the help of enzymes. Pathogenic diseases of roots are asymptomatic at first, since roots are subterranean and the traits of root diseases are generally confused with the symptoms originated due to abiotic stress, especially, the lack of nutrients. Necrotrophs are capable of completely destroying root hair which tremendously decreases the water and nutrient absorption capacity of crop plants, leading to nutrient deficiency, stunted growth, drought stress, and reduced pathogenic tolerance (Mengiste 2012).

16.4.1 Major Root Pathogens of Significant Crops

Plants in both cultivated and wild populations carry inherent disease resistance to pathogens (by elaborating chitinases, etc.) (Bishop et al. 2000); even then, the examples of devastating plant disease impacts are innumerable [e.g., Irish Potato Famine (caused by *Phytophthora infestans* during 1845–1847)]. These pathogens belong to heterogeneous groups, among which fungal pathogens are the most prominent ones. The major root pathogens of commercial and cereal crops across the globe are the various species of *Fusarium*, *Rhizoctonia*, and *Pythium*. These soil-inhabiting fungi are primarily necrotrophic in nature and lead to root and stem rot of plants along with numerous other diseases which include wilting, damping-off, tissue necrosis, and seedling blight, just to name a few. The top ten fungal pathogens have been identified as *Magnaporthe oryzae*, *Botrytis cinerea*, *Puccinia* spp., *Fusarium graminearum*, *Fusarium oxysporum*, *Blumeria graminis*, *Mycosphaerella graminicola*, *Colletotrichum* spp., *Ustilago maydis*, and *Melampsora lini* (Dean et al. 2012). Table 16.1 presents a list of fungal pathogens, groups they belong to, diseases they cause, target host ranges, and symptoms the hosts show.

Presence of symptoms due to these pathogens can be localized or very extensive, depending upon the incidence of disease. For instance, root rot often leads to stunted growth, wilting, and chlorosis of parts above the ground or may remain asymptomatic at times. Hence, it is difficult to ascertain the nature of pathogen by examining the discernible symptoms. Another obstruction in the management or control of pathogen arises due to the synergistic effect posed by them, while microbes exist as complex communities in nature; the multispecies interaction is very likely to occur which reforms the dynamics of host–pathogen interaction and makes it even more strenuous to devise measures of control.

Table 16.1 Major fungal root pathogens and diseases they cause in significant crops

Sr. no.	Pathogen	Pathogenic group	Diseases	Target host range	Impact/ symptoms
1.	*Fusarium* sp.	Ascomycetes (fungi)	Head blight (*F. graminearum*), Fusarium wilt (*F. oxysporum*), foot and root rot (*F. culmorum*)	Wheat, maize, legumes, banana, tomato, etc.	Discoloration of internodes and stem bases and blotches on leaves. Contamination of grain from mycotoxins
2.	*Rhizoctonia* sp.	Basidiomycetes (fungi)	Sheath spot (*R. oryzae*), root rot, leaf blight, and black scurf (*R. solani*)	Rice, potato, corn, tobacco soybean, radish, clover, etc.	Reduced nodulation of roots with appearance of lesions, yellowing, and wilting of leaves
3.	*Pythium* sp.	Oomycetes (heterokont/ fungi)	Seedling blight (*P. ultimum*) and root rot (*P. ultimum* and *P. graminicola*)	Maize, rice, sugarcane, beans, turf grass, manila grass, etc.	Feeder root necrosis, dwarfing of plant leading to damping-off

16.4.2 *Effects of* Fusarium *sp. over Crops and Humans*

Fusarium is a genus of soilborne fungus which is filamentous in nature and categorized under hyphomycetes. It constitutes of predominant and emerging plant pathogens, mycotoxin producers, opportunistic human pathogens, and nonpathogenic strains as well. The most widespread species, *F. oxysporum*, is divided into 120 formae speciales on the basis of the target host. The evolution of these formae speciales over time along with the host is evident from the fact that 5% of the total genome comprises of transposable elements from other host species.

The general mechanism of infection involves attachment and subsequent penetration in host roots, the colonization of vascular system, especially xylem, and the production of phytotoxins. The plant–fungal signal transduction cascade controls the interaction with host. Extracellular cell wall-degrading enzymes are released by *F. oxysporum* to facilitate penetration into the root cortex, and further invasion into the vascular system.

Fungal pathogens have devised counter mechanisms to tackle phytoanticipins (antimicrobial compounds synthesized during normal plant development) and phytoalexins (compounds synthesized de novo during pathogenic attack) through enzyme detoxification and non-degradative tolerance, which makes them resistant to host defense mechanism. This leads to the infection of *Fusarium* in host, characterized by vascular wilt. As the prevalence continues, symptoms start to arise, which include the emergence of brown lesions on roots, stem bases, and internodes, leading to tissue necrosis near the lesions. Chlorosis in discontinued patches, stunted growth, leaf blotches, and blackened primary and secondary roots are some other symptoms of *Fusarium* infection (Wagacha et al. 2012). Drought and warm climate favor the disease incidence and proliferation of *Fusarium* via the formation of sexual fruiting structures, perithecia, and double-walled asexual chlamydospores which remain dormant in host crop debris for very long periods of time and act as an inoculum for next crop cycle, depending upon the prevalence of infection (Urban et al. 2003).

Fusarium sp. comes under the category of field fungi which produces mycotoxins while infecting the crop. These mycotoxins are trichothecenes, deoxynivalenol, toxin T-2, zearalenone, fumonisin B1, etc. These mycotoxins contaminate food and feed; ingestion of which leads to several acute and chronic effects. Some health issues include indigestion and gastrointestinal problems due to the increased permeability of mucosal epithelial membrane in gut. Chronic exposure to these toxins alters the rate of cytokinin production in immune-compromised patients (deoxynivalenol increases TGF-β and IFN-γ expression, whereas fumonisins decrease IL-8 expression in gut), further reducing the disease resistance ability. Hence, *Fusarium* is also considered as an emerging human pathogen, whose consequences are partially understood till now.

16.4.3 **Rhizoctonia** *as a Pathogen in Crops*

Rhizoctonia is a necrotrophic soilborne fungus which is omnipresent in all soil types. The organism belongs to basidiomycota group, which are non-sporulating in nature but form hyphae and sclerotia or hyphal propagules. More than 100 subspecies of *Rhizoctonia solani* are known to be facultative plant pathogens and are divided into 14 anastomosis groups (AG), named as, AG1 to AG13 and AGB1, based upon their host range and hyphal incompatibility (Anderson 1982; Foley et al. 2013). Anastomosis groups are interspecific groups (IGs) formed by the fusion between branches of same or different fungal hyphae. *R. solani* is capable of infecting the host in juvenile stages, such as seed development and seed germination. These groups are known to inflict an extensive range of monocots and dicots. AG1 IA alone infects 27 families of monocots and dicots and is considered as the most virulent group of *R. solani* (Ghosh et al. 2014). The major disease caused by this group is sheath blight of rice, which is responsible for 50% of yield loss across the globe if favorable conditions are sustained. AG2 and AG4 trigger damping-off stem and root rot in *Brassicaceae* family, leading to stunted growth, watery lesions on roots and hypocotyls, cortex tissue maceration, and root tissue necrosis. AG8 is associated with root rot and bare patch disease of grasses, i.e., *Poaceae* family, whereas AG3 is responsible for infection in potato and other tubers. The virulence of these groups varies according to the nature of the host.

The infection process of *Rhizoctonia* involves four stages, broadly characterized into adhesion, penetration, colonization, and host reaction. The virulence of *Rhizoctonia solani*, along with the emergence of disease symptoms, depends upon the extent of production of host-specific RS toxin which is a carbohydrate with glucose, mannose, *N*-acetylgalactosamine, and *N*-acetylglucosamine residues (Vidhyasekaran et al. 1997). On a molecular basis, certain effector proteins are introduced into the host to create a host–parasite relationship. These effector proteins, GT family 2 (glycosyltransferase), CtaG/cox11 (cytochrome C oxidase assembly protein), and I9 (peptidase inhibitor), are not host-derived, and they initiate immunity in the host (Stergiopoulos and de Wit 2009). To overcome the situation, protein kinase HOG1 (high osmolarity glycerol 1) gene encodes for protein responsible for regulation of fungal adaptive mechanism in case of immune stress by controlling cell morphogenesis through G1 cyclin expression. Hog1 is a central signaling mediator in osmoregulation and responds to external stresses and stimuli (Brewster and Gustin 2014). Initiation begins with the growth of longitudinal hyphae along the seed, parallel to the sutures of epidermal cells. From these, hyphae, a thick mass of swollen cells appears, forming the infection cushions in secondary branches of hyphae. Emergence of lesions takes place just beneath these infection cushions which gradually reaches the vascular bundle of the stem in the phloem. If the infection is severe enough, xylem is also affected, and confinement of tissues with the bunch of hyphae takes place which eventually leads to tissue necrosis due to the digestion of hypocotyls and radical by enzymes. After damping-off of sprouts, growth of mycelia starts over the dead tissues, and formation of sclerotia takes place which remain dormant for longer periods, further infecting the surrounding plants.

16.5 Climate Change and Its Effect on Root Pathogens

Climate is the long-term weather patterns prevailing in a region, measured in terms of average precipitation, mean temperature, sunshine hours, and extreme weather frequency over the time. Prolonged variation in Earth's mean temperature as well as local changes in weather patterns is termed as "climate change." Climate change is responsible for various morphological and physiological changes in host and pathogens, which leads to the alteration of classical disease triangle and heightens the risk of disease prevalence.

Global warming is a major cause of climate change on this planet. Long summer season and mild winter, due to increasing temperature, is responsible for increased pathogenic instances. The impact is such that more aggressive and resistant strains of plant pathogens are being observed in nature with a greater host range than before, which tend to drift from agricultural ecosystems to natural flora, where the impacts and behavior of these pathogens are unknown. In this way, plant pathogens appear to be the biological indicators of climate change (Garrett et al. 2009). Several factors, viz., chemical fertilizers, atmospheric gases, variation in the temperature, and moisture and water stress, influence the climatic conditions and thus affect the incidence and severity of plant diseases.

16.5.1 Fertilizers and Severity of Diseases in Plants

Chemical fertilizers and pesticides are used to supplement nutrient-deficient soils or to protect crops from pathogenic attacks. However, this practice leads to several issues as well. The heavy usage of agricultural inputs leads to the release of nitrogenous emission into the air and water, leading to pollution and eutrophication. Other than that, these fertilizers tend to promote pathogens.

Increased nutrient availability interferes with the dormancy of fungal pathogens and facilitates their propagation. Ammonia-rich fertilizers are a readily available source of macro- and micronutrients, which reduce the soil pH and initiate sporulation in *Fusarium*. On the other hand, a high level of calcium in the soil inhibits the growth of the same fungi in plants. In a previous study, it has been found that calcium increases ethylene production and reduced abscisic acid production from plants, which are responsible for the suppression of fungal pathogens (Zielińska and Michniewicz 2001). Hence, every disease system is to be speculated in a different manner while making inferences about the effects of fertilizers on plant diseases.

Persistent use of chemical herbicides is also related to the infection caused by fungal pathogens. Along with the development of herbicide-resistant varieties, which lead to crop losses in itself, these weeds are constantly inflicted with fungal pathogens where from the spores are then transferred to healthy crops. Also, the excessive usage of chemical pesticides eventually leads to the development of pesticide-resistant strains of pathogens due to continuous evolution and adaptation (Boyd et al. 2013). Decreased sensitivity and effectiveness of these chemicals in the soil has intensified the issue of emerging plant pathogens with an extended host range, and their management and control requires immediate attention.

16.5.2 Effects of Atmospheric Gases over Host–Pathogen Interaction and Disease Incidence

Anthropogenic activities are a major source of gaseous emissions in the atmosphere, aggravated by the pervasive fossil fuel combustion, fertilizer application industrial activities, and aircraft exhaust emission (Noyes et al. 2009; Pinder et al. 2012). Carbon dioxide (CO_2), oxides of nitrogen, and ozone (O_3) are the three crucial gases responsible for "thickening of Earth's blanket" which causes significant increase in global average temperature (Florides and Christodoulides 2009). The concentration of O_3 in atmosphere is increasing at an annual rate of 2.5%, and it is expected to reach more than 60 ppb by 2050 (Reference). Changes in the level of gases alter the host–pathogen interactions. CO_2 and O_3 behave antagonistically and alter the growth, water use efficiency, and photosynthetic capacity of plants. The former increases the efficiency of host, while the latter shows a negative impact on it. Microarray studies on soybean have shown that accumulated CO_2 and O_3 in the troposphere led to a downregulation of defense and hormone signaling, which was absent in ambient atmosphere. The study reported that elevated CO_2 may decrease the resistance of soybean to herbivory by suppressing the ability to mount an effective defense (Casteel et al. 2008). Host responses also determine the severity of impact by these gases. Elevated CO_2 and O_3 levels strongly obstruct the stomatal opening by production of thicker wax layers and decrease the chances of infections at leaf surfaces (Vanhatalo et al. 2001). Studies performed on *Fusarium oxysporum*, using controlled phytotron chambers followed by validation of substrate fungal profiles, revealed that elevated CO_2 levels alter the host transcriptome regulation mechanism, allowing them to reshape their root architecture and root exudates according to the modified soil profile (Chitarra et al. 2015). Thus, increased CO_2 levels along with a favorable temperature inhibit the general metabolism of host, hereby increasing the severity of disease by the fungal pathogen. Various other studies assert that increased levels of CO_2 exhibit no significant effects on the growth of fungal pathogens, until accompanied by an ambient. However, it is evident that CO_2 greatly improves the photosynthetic efficiency of host that means more accumulation of biomass and sugar content, which corresponds to more nutritional benefits for sugar-dependent, necrotrophic fungi and, hence, better probability of disease incidence. *R. solani*-caused sheath blight of rice, which is an important soilborne disease, was found to be more detrimental at a CO_2 concentration of 280 ppm, higher than the optimum value (Kobayashi et al. 2006). Ozone tends to limit the biochemical photosynthetic machinery of host plants, and hence the enhancement in ozone levels decreases the chances of disease incidence or pathogenesis, but it leads to premature leaf senescence and appearance of lesions (Gielen et al. 2007). Ozone causes an oxidative burst in hosts and affects the interactions between the phytohormones such as salicylic acid, jasmonic acid, ethylene, and abscisic acid (Kangasjärvi et al. 2005). As a defense mechanism, host transcribes the gene for ROS scavengers (Kangasjärvi et al. 2005); hence, the effect of free oxygen neutralizes, and the probability of pathogenesis increases. Therefore, the combined effects of both the gases are found to be in the favor of pathogens.

16.5.3 Effects of Varying Temperature over Plant Pathogenesis

The sensitivity of host–pathogen interaction is highly dependent on ambient temperature, which is different for every host–pathogen system. In the last 50 years, significant increment in the temperature has been observed at a rate of 0.13 °C per decade. Increasing temperature due to global warming and climate change is associated with the change in precipitation patterns, reduced crop duration, and polar migration of pathogens due to shorter winters and longer summers.

Elevated temperature has multiple effects on plant physiology. It influences the synthesis of secondary metabolites which impose pharmacological impacts in plants, enhance the production of lignin content in fodder crops, and also influence the disease resistance of host plant. High temperature is known to favor spore formation and rapid evolutionary of pathogens. Asexual spores produced by various fungi are highly sensitive to temperature and moisture levels. Colder sub-tropic regions are more prone to disease severity in case of elevated temperature than warm and humid sub-tropics, which has been confirmed by studies conducted on *Rhizoctonia solani* sheath blight-affected rice crops in Japan.

Fusarium pseudograminearum is responsible for crown rot disease in barley and wheat along with production of deoxynivalenol mycotoxin. The production of deoxynivalenol and pathogen infection was found to be reduced with increasing temperature, which was highest at 15 °C. Since the effects of elevated temperature on disease index depend upon various factors, including host defense mechanism, outcomes can vary. But, it has been observed that increased temperature and CO_2 levels favor the propagation of pathogens, reducing crop productivity. Among other parameters, the photosynthetic rate, transpiration rate, and total antioxidant capacity are affected by the elevated temperature and CO_2 (Chang et al. 2016). Even these two factors are altering the interactions between plants and insects, which is threatening food security and ecosystems (DeLucia et al. 2012).

16.5.4 Effects of Moisture and Water Stress on Disease Severity in Host

Climate change and increased temperature levels directly govern the recurring changes in precipitation patterns, leading to intensified moisture content in atmosphere in some regions and natural extremities, like droughts and water stress in others (Trenberth 2011). Moisture is the supreme environmental factor which governs the frequency and severity of soilborne fungal diseases and other root pathogens in crops. Fungal spore germination and infection are greatly enhanced by elevated relative moisture in soil and free surface water (Guzman-Plazola et al. 2003). An assessment done over AG8 anastomosis group of *R. solani* which causes root rot disease in wheat revealed that high relative humidity (75% water holding capacity) can cause high disease occurrence. High moisture content alters the microbial activity in soil and also stimulates development of larger canopies in crops, encouraging foliar infection by pathogens, as in the case of vegetable root rot, sheath blight, and powdery mildew.

In contrast to this, water deficit stress hampers both the metabolism of plants (Bray 2001) and pathogens. It also greatly reduces the disease incidence by root pathogens as low moisture levels inhibit their sustenance and minimize the chances of contact between pathogens and roots. The efficacy of plant water uptake is also affected due to drought conditions. High moisture is also beneficial for the multiplication of symbiotic mycorrhizal fungi which regulates the stomatal conductance and transpiration rate in plants during water stress. Extremely low soil humidity due to salt stress hinders the mycorrhizal association, leaving the plants vulnerable to drought conditions (Evelin et al. 2009).

16.6 Biological Control of Root Pathogens as an Alternate Strategy to Disease Management

Plant disease management is aimed at lowering the economic losses. The two principles involved in disease management are prevention and curative action, employed before and after the disease. Modern disease management works upon the concept of exclusion and resilience. Exclusion is the preventive measures taken to control the introduction of pathogens in the agroecosystem, assuming that the dispersal range of most pathogens is relatively less. Other auxiliary methods for disease management involve the eradication of pathogens, protection of plants by manual intervention, chemicals, introduction of disease-resistant varieties, and biological control as a part of integrated disease management (Khoury and Makkouk 2010).

Microbial biocontrol agents are benign alternatives to chemical pesticides, as they target the pests, without jeopardizing human health and the environment. Entomopathogenic fungi have an important position among all the biocontrol agents because of their route of pathogenicity, broad host range, and the ability to control both sap-sucking pests such as mosquitoes and aphids as well as pests with chewing mouthparts (Khan et al. 2012). The entomopathogenic fungi act as biopesticides by secreting lipases, proteases, chitinases, β-galactosidase, β-glutaminase, catalase, etc. which degrade the insect exoskeletons, breach the cuticle to enter the insect hemocoel, and disrupt signaling system (Mondal et al. 2016). Improvements are needed in this "green technology" to fulfil the requirements for high market share.

Biological control of pathogens aims at the use of microbes to terminate the plant diseases. The use of mycorrhiza and *Bacillus* sp. as fungal biocontrol agents has been underrated until now. *Bacillus* species such as *B. subtilis*, *B. licheniformis*, *B. thuringiensis*, and *B. amyloliquefaciens* are capable of producing antifungal antibiotics which are helpful in controlling root pathogens and postharvest loss (Cawoy et al. 2015). The examples of these antifungal agents are iturins, mycosubtilins, bacillomycins, fungistatins, surfactins, subsporins, etc. (Hsieh et al. 2008). *Pseudomonas* sp. is a highly effective biocontrol agent for *Fusarium* sp. Strain 158 of *Pseudomonas* interferes with the translation of Tri5 gene encoding for trichodiene synthase in trichothecene biosynthetic pathway. Trichothecenes are sesquiterpenoid mycotoxins (of which deoxynivalenol and nivalenol are derivatives) produced by *Fusarium* sp. (Malmierca et al. 2013). *Pseudomonas fluorescens* is also known to control sheath blight in rice caused by *R. solani*, by means of its siderophores, volatile metabolic compounds, and antibiotics (Velazhahan et al. 1999). Eight isolates from *P. fluorescens* are capable of inhibiting the germination of sclerotia and their subsequent lysis (Thrane et al. 2001). Another strain of

Pseudomonas, CMR12a, inhibits the hyphal growth in anastomosis groups (AG-2 and AG-4)-mediated root rot disease, by releasing phenazine and cyclic lipopeptide derivatives (D'aes et al. 2011). *Streptomyces* sp. Di-944 is another antagonist against *R. solani* to control damping-off disease in tomato (Sabaratnam and Traquair 2002).

As many as 7 species of ectomycorrhiza and 13 species of *Glomerales* are involved in the control of rhizosphere pathogens, involving *F. oxysporum*, *F. moniliforme*, *R. solani*, and *Phytophthora* sp. Reduced disease incidence in case of wilt and root rot caused by *F. oxysporum* was observed in in vitro studies due to the antagonistic biocontrol effects of *Gigaspora margarita*, *Glomus fasciculatum*, *G. intraradices*, *Clitocybe claviceps*, *Laccaria bicolor*, and *Paxillus involutus*. Similar results have been obtained in case of root rot, weak root rot, pod rot, and stem rot caused by different groups of *Rhizoctonia solani*. The potential mycorrhizal biocontrol agents for *R. solani* are *Glomus coronatum*, *G. clarum*, *G. mosseae*, *Laccaria laccata*, and *Pisolithus tinctorius*. Biocontrol effect was observed on a number of monocot and dicot plants (Kobra et al. 2009). Apart from these, other microbial antagonist has also been employed to eliminate the notorious soilborne fungal pathogens, *Fusarium* and *Rhizoctonia*. Terpenoid-based metabolic intermediates of *Trichoderma virens*, namely, desoxyhemigossypol (dHG) and hemigossypol (HG), types of sesquiterpenes, exhibit great inhibitory response against *R. solani* in the roots of cotton (*Gossypium hirsutum*) plants (Howell et al. 2000). *Verticillium dahlia* Kleb. which affects pepper yield can be tackled with mycorrhiza. The fungal pathogen inhibitory effect of mycorrhiza is straightforward but hinges on environmental factors as well. Mycorrhiza can be more beneficial for plant growth and physiology under dry conditions than in soil moisture condition (Garmendia et al. 2005).

Root exudates, such as free amino acids and phenolic acids (salicylic acid, *p*-hydroxybenzoic acid, phthalic acid, ferulic acid, etc.), released in rhizosphere are important for the colonization of pathogens in roots. The exudates facilitate the mobilization of poorly soluble nutrients in the rhizosphere (Carvalhais et al. 2011). HPLC analysis of root exudates isolated from watermelon vine illustrates positive response against *F. oxysporum* in terms of spore germination. This fungus cases Fusarium wilt of watermelon. It indicates the role of root exudates in disease development in root which leads to yield losses thereafter. Adopting intercropping systems is also beneficial in the control of root pathogens (Zarea et al. 2011). Water melon and aerobic rice intercropping system has significant inhibitory effects on Fusarium wilt. Interestingly, nonpathogenic strains of *F. oxysporum* (CS-20 and CS-24) and *F. solani* (CS-1) were also found to sustain systemic disease resistance in tomato from Fusarium wilt and cause disease suppression under different environmental conditions (Olivain et al. 1995). It has been found that increasing plant community diversity reduced the co-infection of plant by two or more pathogen groups (Rottstock et al. 2014).

The insights discussed in this chapter can be contributive to the elimination of plant pathogens and yield enhancement without destroying the climate further. Figure 16.1 presents a schematic diagram of host–pathogen–environment–anthropogenic activity nexus.

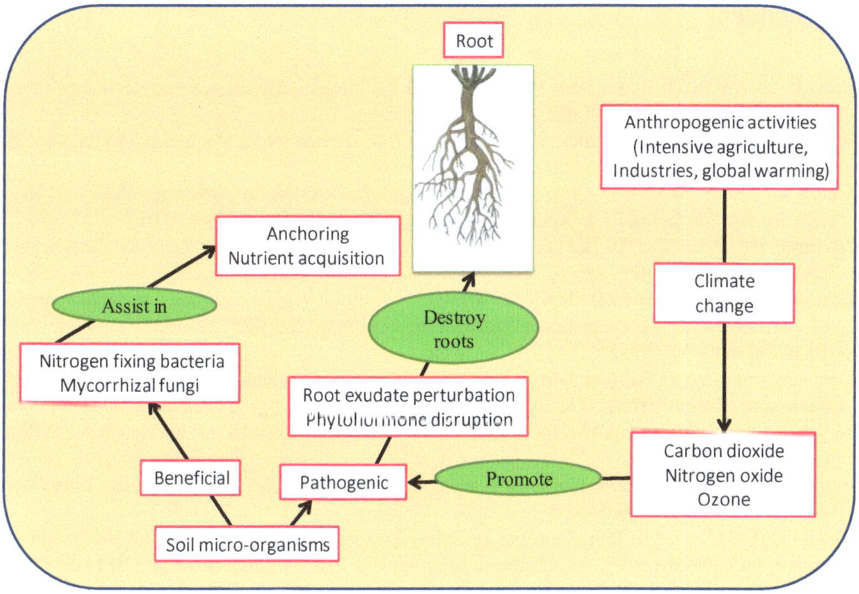

Fig. 16.1 The interactions between host–pathogen–environment–anthropogenic activities

16.7 Conclusion

Root pathogens are one of the most alarming issues associated with food insecurity. The diseases caused by these pathogens are exacerbated by the changing climate, a resultant of anthropogenic activities. Fungal pathogens, which contribute to almost 70% of the yield losses, interact with their host plants by signaling mechanisms carried out with the help of root exudates in the rhizosphere. Changes in microenvironment alter the host–pathogen signaling pathways, which depend upon a number of factors functioning concertedly. Effects of temperature, water stress, and atmospheric gases are interrelated to each other, and one aggravates the impact of other over disease severity which is the reason why in vitro analysis and field studies generate different results. The paucity of research on the combined effects of these factors makes it difficult to outline a strategy for disease control and management. The commercial use of biocontrol agents as an effective alternative to chemical inputs is another hindrance in disease management, as their efficacy is also impeded by the rapidly changing environment. The use of modern analytical techniques along with additional field testing data, acquired by extensive research, is required to hypothesize the precise effects of these climate changes over root–pathogen interaction.

Acknowledgment We are grateful for the suggestions we received from Er. Gourav Dhiman, Assistant Professor, Chandigarh University, Punjab, in completion of this chapter.

References

Ahuja I, de Vos RCH, Bones AM, Hall RD (2010) Plant molecular stress responses face climate change. Trends Plant Sci 15:664–674

Anderson NA (1982) The genetics and pathology of *Rhizoctonia solani*. Annu Rev Phytopathol 20 (1):329–347

Bebber DP, Gurr SJ (2015) Crop-destroying fungal and oomycete pathogens challenge food security. Fungal Genet Biol 74:62–64. https://doi.org/10.1016/j.fgb.2014.10.012

Berendsen RL, Pieterse CMJ, Bakker PAHM (2012) The rhizosphere microbiome and plant health. Trends Plant Sci 17:478–486

Bishop JG, Dean AM, Mitchell-Olds T (2000) Rapid evolution in plant chitinases: molecular targets of selection in plant-pathogen coevolution. Proc Natl Acad Sci 97:5322–5327. https://doi.org/10.1073/pnas.97.10.5322

Boyd LA, Ridout C, O'Sullivan DM et al (2013) Plant-pathogen interactions: disease resistance in modern agriculture. Trends Genet 29:233–240

Bray EA (2001) Plant response to water-deficit stress. Encyclopedia of life sciences. Wiley, Chichester, pp 1–7

Brewster JL, Gustin MC (2014) Hog1: 20 years of discovery and impact. Sci Signal 7:re7. https://doi.org/10.1126/scisignal.2005458

Broeckling CD, Broz AK, Bergelson J et al (2008) Root exudates regulate soil fungal community composition and diversity. Appl Environ Microbiol 74:738–744. https://doi.org/10.1128/AEM.02188-07

Carvalhais LC, Dennis PG, Fedoseyenko D et al (2011) Root exudation of sugars, amino acids, and organic acids by maize as affected by nitrogen, phosphorus, potassium, and iron deficiency. J Plant Nutr Soil Sci 174:3–11. https://doi.org/10.1002/jpln.201000085

Casteel CL, O'Neill BF, Zavala JA et al (2008) Transcriptional profiling reveals elevated CO_2 and elevated O_3 alter resistance of soybean (*Glycine max*) to Japanese beetles (*Popillia japonica*). Plant Cell Environ 31:419–434. https://doi.org/10.1111/j.1365-3040.2008.01782.x

Cawoy H, Debois D, Franzil L et al (2015) Lipopeptides as main ingredients for inhibition of fungal phytopathogens by *Bacillus subtilis*/amyloliquefaciens. J Microbial Biotechnol 8:281–295. https://doi.org/10.1111/1751-7915.12238

Chakraborty S, Tiedemann A, Teng P (2000) Climate change: potential impact on plant diseases. Environ Pollut 108:317–326. https://doi.org/10.1016/S0269-7491(99)00210-9

Chitarra W, Siciliano I, Ferrocino I, et al (2015) Effect of elevated atmospheric CO_2 and temperature on the disease severity of rocket plants caused by fusarium wilt under phytotron conditions. PLoS One. https://doi.org/10.1371/journal.pone.0140769

D'aes J, Hua GKH, De Maeyer K et al (2011) Biological control of *Rhizoctonia* root rot on bean by phenazine- and cyclic lipopeptide-producing Pseudomonas CMR12a. Phytopathology 101:996–1004. https://doi.org/10.1094/PHYTO-11-10-0315

Dean R, Van Kan JAL, Pretorius ZA et al (2012) The top 10 fungal pathogens in molecular plant pathology. Mol Plant Pathol 13:804–804

DeLucia EH, Nabity PD, Zavala JA, Berenbaum MR (2012) Climate change: resetting plant-insect interactions. Plant Physiol 160:1677–1685. https://doi.org/10.1104/pp.112.204750

Chang JD, Mantri N, Sun B et al (2016) Effects of elevated CO_2 and temperature on *Gynostemma pentaphyllum* physiology and bioactive compounds. J Plant Physiol 196–197:41–52. https://doi.org/10.1016/j.jplph.2016.02.020

Eastburn DM, McElrone AJ, Bilgin DD (2011) Influence of atmospheric and climatic change on plant-pathogen interactions. Plant Pathol 60:54–69

Elad Y, Pertot I (2014) Climate change impacts on plant pathogens and plant diseases. J Crop Improv 28:99–139. https://doi.org/10.1080/15427528.2014.865412

Evelin H, Kapoor R, Giri B (2009) Arbuscular mycorrhizal fungi in alleviation of salt stress: a review. Ann Bot 104:1263–1280

Fischer T, Byerlee D, Edmeades G (2014) Crop yields and global food security. Aust Cent Int Agric Res 660. https://doi.org/10.1093/erae/jbv034

Florides GA, Christodoulides P (2009) Global warming and carbon dioxide through sciences. Environ Int 35:390–401

Foley RC, Gleason CA, Anderson JP, Hamann T, Singh KB (2013) Genetic and genomic analysis of *Rhizoctonia solani* interactions with Arabidopsis; evidence of resistance mediated through NADPH oxidases. PLoS One 8(2):e56814

Garmendia I, Goicoechea N, Aguirreolea J (2005) Moderate drought influences the effect of arbuscular mycorrhizal fungi as biocontrol agents against Verticillium-induced wilt in pepper. Mycorrhiza 15:345–356. https://doi.org/10.1007/s00572-004-0336-z

Garrett KA, Nita M, De Wolf ED et al (2009) Plant pathogens as indicators of climate change. Clim Change 425–437. https://doi.org/10.1016/B978-0-444-53301-2.00025-7

Ghosh S, Gupta SK, Jha G (2014) Identification and functional analysis of AG1-IA specific genes of *Rhizoctonia solani*. Curr Genet 60:327–341. https://doi.org/10.1007/s00294-014-0438-x

Gielen B, Löw M, Deckmyn G et al (2007) Chronic ozone exposure affects leaf senescence of adult beech trees: a chlorophyll fluorescence approach. J Exp Bot 58:785–795. https://doi.org/10.1093/jxb/erl222

Guzman-Plazola RA, Davis RM, Marois JJ (2003) Effects of relative humidity and high temperature on spore germination and development of tomato powdery mildew (*Leveillula taurica*). Crop Prot 22:1157–1168. https://doi.org/10.1016/S0261-2194(03)00157-1

Karima HEH, El-Gamal NG (2012) In vitro study on *Fusarium solani* and *Rhizoctonia solani* isolates causing the damping off and root rot diseases in tomatoes. Nat Sci 10(11):16–25

Howell CR, Hanson LE, Stipanovic RD, Puckhaber LS (2000) Induction of Terpenoid synthesis in cotton roots and control of *Rhizoctonia solani* by seed treatment with *Trichoderma virens*. Phytopathology 90:248–252. https://doi.org/10.1094/PHYTO.2000.90.3.248

Hsieh FC, Lin TC, Meng M, Kao SS (2008) Comparing methods for identifying Bacillus strains capable of producing the antifungal lipopeptide iturin A. Curr Microbiol 56:1–5. https://doi.org/10.1007/s00284-007-9003-x

Kangasjärvi J, Jaspers P, Kollist H (2005) Signalling and cell death in ozone-exposed plants. Plant Cell Environ 28:1021–1036

Khan S, Guo L, Maimaiti Y et al (2012) Entomopathogenic fungi as microbial biocontrol agent. Mol Plant Breed. https://doi.org/10.5376/mpb.2012.03.0007

Khoury WE, Makkouk K (2010) Integrated plant disease management in developing countries. J Plant Pathol 92:35–42

Kobayashi T, Ishiguro K, Nakajima T et al (2006) Effects of elevated atmospheric co(2) concentration on the infection of rice blast and sheath blight. Phytopathology 96:425–431. https://doi.org/10.1094/PHYTO-96-0425

Kobra N, Jalil K, Youbert G (2009) Effects of three *Glomus* species as biocontrol agents against *Verticillium*-induced wilt in cotton. J Plant Prot Res 49:185–189. https://doi.org/10.2478/v10045-009-0027-z

Lawrence D, Vandecar K (2015) Effects of tropical deforestation on climate and agriculture. Nat Clim Chang 5:27–36

Malmierca MG, Cardoza RE, Alexander NJ et al (2013) Relevance of trichothecenes in fungal physiology: disruption of tri5 in *Trichoderma arundinaceum*. Fungal Genet Biol 53:22–33. https://doi.org/10.1016/j.fgb.2013.02.001

Mark GL, Dow JM, Kiely PD et al (2005) Transcriptome profiling of bacterial responses to root exudates identifies genes involved in microbe-plant interactions. Proc Natl Acad Sci 102:17454–17459. https://doi.org/10.1073/pnas.0506407102

Mengiste T (2012) Plant immunity to necrotrophs. Annu Rev Phytopathol 50:267–294. https://doi.org/10.1146/annurev-phyto-081211-172955

Mondal S, Baksi S, Koris A, Vatai G (2016) Journey of enzymes in entomopathogenic fungi. Pac Sci Rev A Nat Sci Eng 18:85–99. https://doi.org/10.1016/j.psra.2016.10.001

Mostert D, Molina AB, Daniells J et al (2017) The distribution and host range of the banana Fusarium wilt fungus, *Fusarium oxysporum* F. Sp. Cubense, in Asia. PLoS One. https://doi.org/10.1371/journal.pone.0181630

Matny ON (2015) Fusarium head blight and crown rot on wheat & barley: losses and health risks. Adv Plants Agric Res 2:2–7. https://doi.org/10.15406/apar.2015.02.00039

Noyes PD, McElwee MK, Miller HD et al (2009) The toxicology of climate change: environmental contaminants in a warming world. Environ Int 35:971–986

Oerke E-C (2006) Crop losses to pests. J Agric Sci 144:31. https://doi.org/10.1017/S0021859605005708

Oke V, Long SR (1999) Bacterial genes induced within the nodule during the Rhizobium-legume symbiosis. Mol Microbiol 32:837–849. https://doi.org/10.1046/j.1365-2958.1999.01402.x

Olivain C, Steinberg C, Alabouvette C (1995) Evidence of induced resistance in tomato inoculated by non-pathogenic strains of *Fusarium oxysporum*. In: Manka M (ed) Environmental biotic factors in integrated plant disease control. The Polish Phytopathological Society, Poznan, pp 427–430

Pinder RW, Davidson EA, Goodale CL et al (2012) Climate change impacts of US reactive nitrogen. Proc Natl Acad Sci USA 109:7671–7675. https://doi.org/10.1073/pnas.1114243109

Porter JR, Xie L, Challinor AJ et al (2014) Food security and food production systems. In: Climate change 2014: impacts, adaptation, and vulnerability. Part A: Global and sectoral aspects. Contribution of working group II to the fifth assessment report of the intergovernmental panel on climate change, pp 485–533

Pozo MJ, Azcón-Aguilar C (2007) Unraveling mycorrhiza-induced resistance. Curr Opin Plant Biol 10:393–398

Prashar P, Kapoor N, Sachdeva S (2014) Rhizosphere: its structure, bacterial diversity and significance. Rev Environ Sci Biotechnol 13:63–77

Pratibha P, Shah S (2012) Impact of fertilizers and pesticides on soil microflora in agriculture. In: Lichtfouse E (ed) Sustainable agriculture reviews. Springer, New York, pp 331–361

Rottstock T, Joshi J, Kummer V, Fischer M (2014) Higher plant diversity promotes higher diversity of fungal pathogens, while it decreases pathogen infection per plant. Ecology 95:1907–1917. https://doi.org/10.1890/13-2317.1

Sabaratnam S, Traquair JA (2002) Formulation of a streptomyces biocontrol agent for the suppression of Rhizoctonia Damping-off in tomato transplants. Biol Control 23:245–253. https://doi.org/10.1006/bcon.2001.1014

Stergiopoulos I, de Wit PJGM (2009) Fungal effector proteins. Annu Rev Phytopathol 47:233–263. https://doi.org/10.1146/annurev.phyto.112408.132637

Thrane C, Nielsen MN, Sørensen J, Olsson S (2001) Pseudomonas fluorescens DR54 reduces sclerotia formation, biomass development, and disease incidence of *Rhizoctonia solani* causing damping-off in sugar beet. Microb Ecol 42:438–445. https://doi.org/10.1007/s00248-001-0007-y

Trenberth KE (2011) Changes in precipitation with climate change. Climate Res 47:123–138. https://doi.org/10.3354/cr00953

Urban M, Mott E, Farley T, Hammond-Kosack K (2003) The *Fusarium graminearum* MAP 1 gene is essential for pathogenicity and development of perithecia. Mol Plant Pathol 4:347–359. https://doi.org/10.1046/j.1364-3703.2003.00183.x

Vanhatalo M, Huttunen S, Bäck J (2001) Effects of elevated [CO_2] and O_3 on stomatal and surface wax characteristics in leaves of pubescent birch grown under field conditions. Trees Struct Funct 15:304–313. https://doi.org/10.1007/s004680100105

Velazhahan R, Samiyappan R, Vidhyasekaran P (1999) Relationship between antagonistic activities of *Pseudomonas fluorescens* isolates against *Rhizoctonia solani* and their production of lytic enzymes. Z Pflanzenkr Pflanzenschutz 106:244–250

Vermeulen SJ, Campbell BM, Ingram JSI (2012) Climate change and food systems. Annu Rev Environ Resour 37(1):195–222

Vidhyasekaran P, Ponmalar TR, Samiyappan R et al (1997) Host-specific toxin production by *Rhizoctonia solani*, the rice sheath blight pathogen. Phytopathology 87:1258–1263. https://doi.org/10.1094/PHYTO.1997.87.12.1258

Wagacha JM, Oerke EC, Dehne HW, Steiner U (2012) Colonization of wheat seedling leaves by Fusarium species as observed in growth chambers: a role as inoculum for head blight infection? Fungal Ecol 5:581–590. https://doi.org/10.1016/j.funeco.2012.02.002

Wheeler T, von Braun J (2013) Climate change impacts on global food security. Science 341 (80):508–513. https://doi.org/10.1126/science.1239402

Zarea MJ, Karimi N, Goltapeh EM, Ghalavand A (2011) Effect of cropping systems and arbuscular mycorrhizal fungi on soil microbial activity and root nodule nitrogenase. J Saudi Soc Agric Sci 10:109–120. https://doi.org/10.1016/j.jssas.2011.04.003

Zielińska M, Michniewicz M (2001) The effect of calcium on the production of ethylene and abscisic acid by fungus *Fusarium culmorum* and by wheat seedlings infected with that pathogen. Acta Physiol Plant 23:79–85. https://doi.org/10.1007/s11738-001-0026-9

Chapter 17
Arbuscular Mycorrhizal Fungi and Their Responses to Nutrient Enrichment

Haishui Yang, Michelle Schroeder-Moreno, Bhoopander Giri, and Shuijin Hu

17.1 Introduction

Mycorrhizal symbiosis between plant roots and soil fungi are widely spread in terrestrial ecosystems, from forests, grasslands, and croplands to even deserts (Brundrett 2009; Smith and Read 2008). In this mutualistic symbiosis, the fungi trade nutrients, e.g., N and P, for carbon from photosynthesis of plants (Smith and Read 2008). Annually, plants might allocate ~4–20% photosynthates to their associated mycorrhizal fungi (Eissenstat et al. 1993). In the past decades, four major types of mycorrhizal associations were described based on their structure and functions, including arbuscular mycorrhiza (AM), ectomycorrhiza (EM), orchid mycorrhiza, and ericoid mycorrhiza (van der Heijden et al. 2015). Both AM and EM fungi live inside the root cortex and their hyphae function and live in the soil. However, the hyphae of AM fungi can penetrate the epidermal and cortical cell walls and form dichotomously branched arbuscules in such types of root cells (possibly the exchanging site between C and nutrients), while the mycelium of EM fungi do not penetrate into their host's root cells but form Hartig net between epidermal and cortical root cells (Smith and Read 2008). Orchid mycorrhizas are only formed

---

H. Yang
College of Agriculture, Nanjing Agricultural University, Nanjing, Jiangsu, China

M. Schroeder-Moreno
Department of Crop and Soil Sciences, North Carolina State University, Raleigh, NC, USA

B. Giri
Department of Botany, Swami Shraddhanand College, University of Delhi, Delhi, India

S. Hu (✉)
College of Resources and Environmental Sciences, Nanjing Agricultural University, Nanjing, Jiangsu, China

Department of Entomology and Plant Pathology, North Carolina State University, Raleigh, NC, USA

B. Giri et al. (eds.), *Root Biology*, Soil Biology 52,
https://doi.org/10.1007/978-3-319-75910-4_17

between roots in the plant family Orchidaceae and various basidiomycete fungi, while ericoid mycorrhizas are the symbiotic relationship between members of the plant family Ericaceae and several lineages of soil fungi (McCormick et al. 2012; Selosse et al. 2007). It is estimated that ~74% of land plants form arbuscular mycorrhizas with Glomeromycotan fungi, ~2% of land plants form ectomycorrhizas, and ~9% of land plants form orchid mycorrhizal associations, but only ~1% plants form ericoid mycorrhizas (Brundrett 2009).

The mycelia of arbuscular mycorrhizal fungi grow from plant roots to explore and exploit soil volume for nutrients, which are subsequently delivered to their host plants (van der Heijden et al. 2015). Mycorrhizal fungi are estimated to be responsible for a large contribution to nitrogen (N) and phosphorus (P) uptake of plant requirements in the natural ecosystems (Hobbie and Hobbie 2006; van der Heijden et al. 2006a, b), as up to 90% of P is delivered from AM hyphae/pathway to plants (Jakobsen et al. 1992; van der Heijden et al. 1998b). AM-mediated N uptake, however, is still unresolved. Several studies reported that AM fungi had no effect on plant N adsorption (van der Heijden et al. 2006a, b; Reynolds et al. 2005), while other studies found that AM fungi could facilitate N acquisition of plants under some conditions (Thirkell et al. 2016; Yang et al. 2017). Importantly, AM fungi can only adsorb and transfer inorganic forms of nutrients, e.g., ammonium, nitrates, and phosphates (van der Heijden et al. 2015). AM fungi cannot directly acquire nutrients from organic matter but must go through priming of other microbes for organic matter decomposition (Cheng et al. 2012; Hodge et al. 2001). However, EM excrete a wide range of extracellular enzymes to degrade organic matter and thus directly forage N (Tunlid et al. 2016). EM fungi may be responsible for up to ~80% of all N acquired by plants in boreal and temperate forests (Hobbie and Hobbie 2006; MacFall et al. 1992).

In this chapter, we only focus on AM fungi based on the following considerations that (1) we have already known that EM fungi are highly sensitive to nutrients (Ekblad et al. 2016; Pritchard et al. 2014; Ulm et al. 2017); (2) EM fungi are saprotrophic and can mobilize nutrients through decomposing organic matter (Koide et al. 2011); and (3) there are many known but varying results about AM fungi responses to nutrients.

17.2 Pathways of Mycorrhizal Nutrient Acquisition, Conversion, and Transfer

Mycorrhizal-mediated nutrient uptake has been gradually deciphered with the development of molecular tools. AM fungi-mediated facilitation of P uptake may be achieved through the following pathways. First, a large fraction of P is fixed in organic forms or metal complexes in soils (Javot et al. 2007), and AM colonization could induce the secretion of organic acids, e.g., oxalate or citric acids from roots (Koide and Kabir 2000; Zhang et al. 2016). AM fungi may also be able to prime

P-solubilizing bacteria to release phosphatase (Zhang et al. 2014a). Soil organic P is hydrolyzed into phosphate through the abovementioned processes. The inorganic phosphate (Pi) is subsequently transferred into AM hyphae through P transporters (e.g., GvPT and GiPT), which are anchored in fungal mycelia membrane and then polymerized into poly-P by a series of polymerases (Harrison and Vanbuuren 1995; Maldonado-Mendoza et al. 2001). Poly-P is subsequently transported to the symbiotic interface, arbuscules depolarized into monomer Pi by alkaline phosphatase (Zhu et al. 2007), and finally translocated into plant root cells through phosphorus transporters (e.g., StPT5, LePT3, OsPT11, and MtPT4) (Harrison et al. 2002; Nagy et al. 2005; Paszkowski et al. 2002; Xu et al. 2007).

Mycorrhizal-mediated N adsorption occurs through several potential mechanisms as follows. AM fungi cannot directly decompose organic matter for N but can recruit other decomposing soil microbes through releasing small organic molecules (Cheng et al. 2012). The released inorganic N (e.g. NH_4^+ or NO_3^-) or minor amounts of organic N (e.g., amino acid) will be subsequently sensed and transferred into mycorrhizal hyphae through high-affinity transporters for NO_3^-, NH_4^+, or amino acids (e.g., *GintAMT1* or *GiNT* for AM fungi and *HcGAP1* for EM fungi) (Lopez-Pedrosa et al. 2006; Muller et al. 2007; Tian et al. 2010). When nitrate is transported into AM hyphae, it will be transformed into ammonia by nitrate reductase and nitrite reductase (Guescini et al. 2007; Kaldorf et al. 1998). Ammonia from nitrate or directly adsorbing from soils will be transformed into arginine by GS/GOGAT enzymes in the external hyphae (Tian et al. 2010). Arginine is subsequently translocated to the internal hyphae and then broken down as ornithine and ammonia in arbuscules (Govindarajulu et al. 2005). Ammonia is finally transported into root cells through ammonia channels (e.g., AMT) (Fochi et al. 2017; Guether et al. 2009; Koegel et al. 2013).

17.3 Why Are We Interested in Impacts of Nutrient Enrichment on Mycorrhizas?

One main function for mycorrhizal symbiosis is facilitating nutrient uptake (e.g., P and N) for host plants under nutrient deficiency conditions. However, researchers are also interested in the impacts of nutrient enrichment on mycorrhizas. Increased chemical fertilizer applications are one of the core results of the Green Revolution and has contributed to increase yields of crops worldwide (Tilman et al. 2002). Since 1960, global application of N fertilizers has increased sevenfold, and P fertilizer applications increased 3.5-fold in 1995 (Tilman et al. 2002). Applications of both N and P fertilizers have been predicted to increase another threefold by 2050 (Tilman et al. 2001). However, further increased fertilizer applications are not likely to be as effective at increasing yields, and use efficiency of fertilizers declines at higher levels of addition (Tilman et al. 2002). The use efficiency of crops is estimated at only 30–50% for N fertilizers and ~45% for P fertilizers in modern intensive agricultural systems (Smil 1999, 2003). Therefore, excessive applications of chemical fertilizers may lead to N and P enrichment in agriculture, thereby negatively impacting mycorrhizas and their functioning.

Extensive fertilization is unsustainable for agricultural production from both economic and environmental perspectives. As mentioned above, the nutrient use efficiency is very low (<50%) for current crops around the world (Smil 1999, 2003). Greater than 50% of chemical fertilizers applied are not utilized by crops (Ju et al. 2009), which can accumulate in soils or lost in the environment. Such high losses of nutrients can cause severe environmental consequences, particularly for water contamination (e.g., eutrophication in lakes or bays and excessive nitrate concentration in drinking water), and cause great economic burdens for farmers (Monteagudo et al. 2012; Withers et al. 2014). Therefore, improvements in nutrient use efficiency can make crop production more sustainable, especially through tapping the potential of AM mycorrhizal symbiosis (Gosling et al. 2006).

Other human activities, such as utilization of fossil fuels for industrial production and transportation and burning of crop residues, have also greatly increased biologically reactive N entering into the atmosphere (Galloway et al. 2008; Huang et al. 2002; Zhang et al. 2014b). In particular, these activities increase the atmospheric levels of reactive nitrogen by a rate of ~25 Tg N year^{-1} from 1995 to 2005, which are subsequently deposited into plants and soils (Cofala et al. 2007; Galloway et al. 2008). High N deposition can result in drastic changes in ecosystems structure and functioning, altering plant community composition and reducing plant diversity (Bobbink et al. 2010; Schlesinger 2009).

17.4 Experimental Results of Nutrient Enrichment on Mycorrhizas

According to the predictions of the functional equilibrium model, carbon allocation to AM structures will be reduced when soils are sufficiently fertilized, because mycorrhizal delivery of soil resources is no longer a value to host plants (Johnson et al. 2003). In the past decades, a series of studies have been conducted to test how nutrient enrichment affects mycorrhiza in different ecosystems. Overall, nutrient enrichment is shown to be negative for mycorrhiza formation as has been synthesized in the case of N and P fertilization, leading to significant decreases in mycorrhizal abundance (Treseder 2004).

17.4.1 Forest and Shrubland Ecosystems

N and P additions can suppress arbuscular mycorrhizas in forest and shrub systems. Camenzind et al. (2014) reported that additions of N and P fertilizers significantly reduced AM fungal root colonization in a *Graffenrieda emarginata*-dominated tropical montane forest. AM fungal root colonization, hyphal biomass, storage, and lipid storage structures also declined in response to N addition with 30 kg ha^{-1} year^{-1} in a northern hardwood forest (van Diepen et al. 2007). Interestingly, AM fungal abundance was increased by 1.56-fold with P addition of 10 kg ha^{-1} year^{-1} but decreased by 27.45% due to N addition of 50 kg ha^{-1} year^{-1} at the elevation of 2000 m in a tropical montane forest, which might be caused by N/P co-limitation in this site (Camenzind et al. 2016). Nitrogen amendment (60 kg N ha^{-1} year^{-1}) was found to significantly reduce AM fungal spore density and root colonization in a coastal sage scrub ecosystem (Egerton-Warburton and Allen 2000).

17.4.2 Grassland Ecosystems

Nutrient enrichment effects on mycorrhiza have been extensively studied in grassland ecosystems, especially for N. For example, Antoninka et al. (2011) reported that 7-year N additions (40 kg ha^{-1} year^{-1}) reduced AM fungal spore abundance but did not affect the spore volume and hyphal density in a grassland ecosystem. In a semiarid grassland ecosystem, it was further found that N amendments (100 kg ha^{-1} year^{-1}) significantly reduced dominant AM fungal species (*Glomus intraradices* and *G. fasciculatum*) (Porras-Alfaro et al. 2007). Saito et al. (2011) found that N additions (20 mg kg^{-1} soil) increased AM fungal sporulation for highly mycorrhizal-responsive plant species but inhibited it for less mycorrhizal-responsive plant species. In a semiarid steppe ecosystem, Kim et al. (2015) found that AM extraradical hyphal density was significantly decreased by N additions (100 kg ha^{-1} year^{-1}). However, mycorrhizal inhibition by nitrogen enrichment might also be dependent on phosphorus availability. For example, Johnson et al. (2003) found that N enrichment (100–170 kg ha^{-1} year^{-1}) strongly decreased AM fungal structures (vesicles and coils) under the lowest soil N:P conditions but increased AM fungal structures under the highest soil N:P conditions from a wide range of grasslands. Johnson et al. (2003) found extraradical mycorrhizal structures (hyphae and spores) to be more responsive to N (100–170 kg ha^{-1} year^{-1}) than P enrichment (10–200 kg ha^{-1} year^{-1}). In a temperate steppe ecosystem, 6-year P fertilization (50 kg ha^{-1} year^{-1}) but not N (100 kg ha^{-1} year^{-1}) additions decreased AM fungal root colonization and extraradical hyphal density in *Artemisia frigida*-, *Stipa krylovii*-, and *Cleistogenes squarrosa*-dominated plant community (Chen et al. 2014). Nitrogen enrichment was also found to decrease mycorrhizal colonization under extraordinarily P-rich soils (120 g kg^{-1}) (Blanke et al. 2005). Thus, soil stoichiometry between N and P might be an important determinant for mycorrhizal response to nutrient enrichment.

17.4.3 Agroecosystems

A large quantity of studies showed that increasing N or P availability could reduce crop responses to AM fungi. For example, Azcón et al. (2003) reported that high N and P availability in the soil reduced macro- and micronutrients in mycorrhizal lettuce (with a rate of 9 mM vs. 1 mM for N and 0.5 mM vs. 0.1 mM for P). Al-Karaki and Clark (1999) reported that high P level (1.6 m mol kg^{-1}) significantly decreased seed dry weight and seed P content by AM fungi when compared to no P addition. At high N or P conditions, crops might reduce C allocation to their associated AM fungal symbionts and thus decreasing mycorrhizal abundance (Nagy et al. 2009). For example, an application of 90 kg N ha^{-1} $year^{-1}$ as ammonium nitrate reduced AM fungal spore density in the rhizosphere of *Zea mays* and *Medicago sativa* in a Mediterranean agroecosystem (Avio et al. 2013). Wang et al. (2017) found that increasing P inputs to a level of 75–100 kg ha^{-1} $year^{-1}$ significantly reduced AM colonization of *Z. mays* in an experimental field (a plot size of 35 m^2). Kahiluoto et al. (2001) reported that 20 years of P fertilization (45 kg ha^{-1} $year^{-1}$) significantly reduced mycorrhizal root colonization and spore density of *Linum usitatissimum* L., *Trifolium pratense* L., and *Hordeum vulgare* L., respectively. However, it is very likely that more complicated factors might impact AM response to nutrient enrichment, such as the background soil nutrient availability (N or P limited when added), the fertilizer type (organic or mineral), the species diversity of AM fungi inoculated, as well as the host crop varieties. All of these might change the outcome of fertilizer enrichment on AM fungi. For example, Gryndler et al. (2006) found that mineral fertilization showed reversed effects on AM fungal external biomass when compared with manure application. Some other studies also reported that modern wheat cultivar was shown to be less dependent and responsive to AM fungi than their ancestors (Hetrick et al. 1993; Zhu et al. 2001).

17.5 Potential Mechanisms that Underlie Mycorrhizal Responses to Nutrient Enrichment

Mycorrhizal responses to nutrient enrichment conforms to the predictions of functional equilibrium model that high nutrient availability might reduce carbon allocation from plants to AM fungi, which thus declines the fungal abundance (van Diepen et al. 2007). However, the underlying mechanisms are still not well-known, particularly about how plants precisely control carbon allocation to different AM fungal species. Here, we will discuss some conceptual models and the underlying molecular mechanisms.

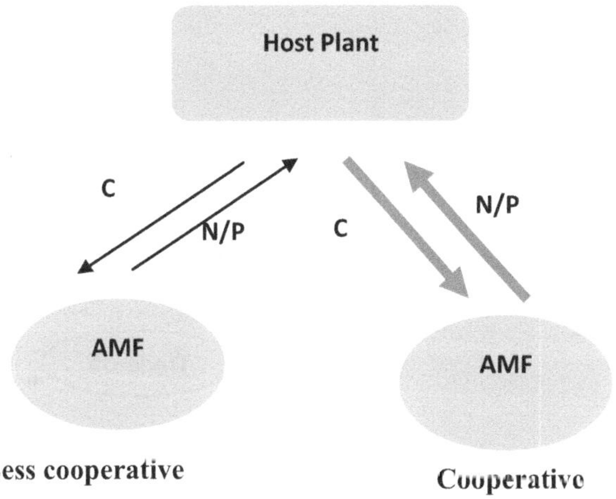

Fig. 17.1 Conceptual model of reciprocal reward strategy for mycorrhizal symbiosis (modified according to Kiers et al. 2011). The black arrows represent the carbon/nutrient flow between host and less beneficial AMF, while the gray arrows stand for the more beneficial interactions between AMF and their hosts

17.5.1 Plant Control of C Allocation to AMF: Conceptual Models

A long-standing controversy is that how plants may control their AM fungal partners as well as how to maintain fair trade for resources (e.g., N and P) between both partners (Kiers and van der Heijden 2006). In a split root experiment, Bever et al. (2009) demonstrated that plant could prefer to allocate carbon to more beneficial AM symbionts. However, they did not provide evidence to explain how plants achieved such goals. Based on results from more precisely controlled experiments (including isotope-labeling and triple split-plate system), Kiers et al. (2011) proposed a reciprocal reward model for stabilizing cooperation between mycorrhizal partners (Fig.17.1) in which plants can detect, discriminate, and reward the best fungal partners with more carbohydrates, and in turn their fungal partners enforce mutual cooperation by increasing nutrient transfer only to those roots providing more carbohydrates.

Mycorrhizal symbiosis is primarily asymmetric, namely, most plants can complete their life cycle without mycorrhizal fungi but the fungi cannot (Werner and Kiers 2015). So, there might be an active pathway for which plants need to evaluate how to allocate carbon to AM fungi (Nagy et al. 2009). Fellbaum et al. (2012) further provided experimental evidence supporting such predictions that carbon flux was an activator of mycorrhizal pathway for nitrogen. Mycorrhizal-mediated nitrogen transfer would be triggered only when carbon was delivered by plants across the mycorrhizal interface. This finding suggests that plants may actively control carbon allocation or at least play a role in the initial priming for carbon consumption. Host plants may be capable of evaluating nutrient requirements and decide to whether activate the mycorrhizal pathway. When nutrients are limited and the mycorrhizal pathway is required, more photosynthesis-fixed carbon will be allocated to mycorrhizal interface than to root cells, thus stimulating mycorrhizal nutrient uptake

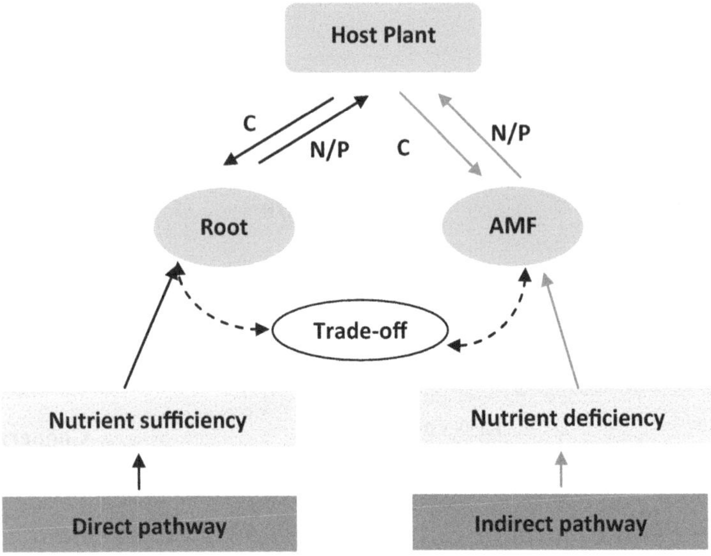

Fig. 17.2 Conceptual model of nutrient uptake pathways

through exploring more extensive soil volumes. When nutrients are sufficient, carbon allocation to mycorrhizas might be reduced and thus reducing AM abundance.

The root system and its associated mycorrhizal fungi are two key components in nutrient adsorption for plants (Smith and Read 2008). However, both root turnover and mycorrhizal proliferation are carbon-consuming processes, with an estimated ~30–50% of photosynthesized carbon allocated to root systems in forests and even more in grasses (Gill and Jackson 2000), and mycorrhizal carbon may account for ~20% of total net primary production (van der Heijden et al. 2008). Thus, how plants balance root development and mycorrhizal proliferation is a central question in root biology. Koide (2000) proposed a functional complementarity model in which AM fungi possess functions that complement those of roots. This model described that root hairs and fine roots were capable of adsorbing nutrients from soils close to the root surface (*direct* nutrient uptake pathway), while mycorrhizal fungi could transport nutrients from beyond the reach of root systems (*in direct* nutrient uptake pathway). Based on this model, we further constructed a conceptual model and argue that host plants could trade off carbon allocation between fine root development and mycorrhizal formation, but this trade-off depends on nutrient availability (Fig.17.2; also see Unger et al. 2016).

Under nutrient-sufficient conditions, nutrients are available for fine roots, and nutrient requirements for plant growth can be easily satisfied with root nutrient uptake. Thus C allocation is only needed to sustain fine root turnover. However, when availability of nutrients close to fine roots is low, such as with P, plants have to take up two possible strategies: one is to elongate fine roots, and the other is to activate the mycorrhizal pathway. Here, a trade-off might occur for plants. If the plant invests C in fine roots but not receiving as much nutrients back as investing the same C in mycorrhizas, then the selection might be for C allocation to mycorrhizas.

17.6 Molecular Evidence of Mycorrhizal Responses to Nutrient Enrichments

Nitrogen and phosphorus are indispensable macroelements for plant growth. For mycorrhizal plants, a crucial process is that how both partners sense the availability of nutrients surrounding root hairs or fungal hyphae in the soil. Generally, two types of proteins anchored in cell membrane are functioned as nutrient sensors and transporters in plant root or fungal mycelium (Amtmann et al. 2006; Scheible and Rojas-Triana 2015). However, some proteins exhibited a dual role of sensing and transporting for nutrients, which is named as transceptor (transporter–receptor) (Scheible and Rojas-Triana 2015).

For host plants, Pi transporters have been identified in different plant species, all of which belong to PHT gene family, e.g., *OsPht*1;1~*OsPht*1;13 and *OsPht*2;1 in *Oryza sativa* L., *TaPht*1;1~*TaPht*1;11 in *Triticum aestivum* L., and *GmPht*1;1~*GmPht*1;14 in *Glycine max* (Linn.) Merr. (Goff et al. 2002; Qin et al. 2012; Teng et al. 2017). However, Pi sensors/transceptors have not yet been reported or identified in plants, despite evidence provided for Pi sensing in root tips (Scheible and Rojas-Triana 2015; Svistoonoff et al. 2007). For plant-sensing ammonium (NH_4^+), an ammonium transceptor*AM1*;1 was identified in *Arabidopsis thaliana* (Lanquar et al. 2009). The activation of *AM1*;1 requires effective interactions between a trimmer of subunits. Conformational change accompanies ammonium transport with *AM1*;1 transceptor. The allosteric regulation is mediated by a cytosolic C-terminal trans-activation domain, which carries a conserved Thr (T460) in a critical position. Phosphorylation of T460 can lead to inactivation of the trimmeric complex, but this process is dependent on NH_4^+ concentrations. Higher NH_4^+ concentrations can trigger phosphorylation of T460 in *AM1*,1, which functions to prevent ammonium from accumulating at toxic levels in root cells. NO_3 sensors have also been found in plant root cells, e.g., *NTR*1,1 (Munos et al. 2004). *NTR*1,1 is a transporting transceptor, which is not only a NO_3^- sensing receptor but also can transport NO_3^- and facilitate the uptake of auxin (Krouk et al. 2010).

For mycorrhizal fungal partners, some high-affinity phosphate transporters have been identified, e.g., GiPT, GvPT, and GmosPT (Benedetto et al. 2005; Harrison and Vanbuuren 1995; Maldonado-Mendoza et al. 2001). However, how AM fungi sense P and whether these phosphate transporters act as sensors remain unclear in the past decades. Recently, researchers found that such transporters resemble Pho84 from yeast, which might have a dual role of sensing and transporting for phosphate (Tisserant et al. 2012). Xie et al. (2016) confirmed such predictions and found that GigmPT functions as a transceptor in *Gigaspora margarita*, which can activate the phosphate-signaling pathway and protein kinase A (PKA) signaling cascade. GigmPT showed similar DNA sequences and protein structure with Pho84 sensor in yeast (Popova et al. 2010), and the potential mechanisms might be as follows: GigmPT is induced under Pi-deficient conditions; however, when Pi becomes available, Pi transport will cause conformational changes of this protein and activates PKA, which, in turn, might phosphorylate GigmPT. Phosphorylated GigmPT will be ubiquitinated and finally degraded. Mycorrhizal fungi also facilitate N uptake, and some N transporters have been identified, e.g., GintAMT1, GiNT, and HcGAP1 (Lopez-Pedrosa et al. 2006; Muller et al. 2007; Tian et al. 2010). However, how AM fungi sense N signal is still not well characterized. In yeast, Mep2 protein has been found to function as an ammonium sensor under ammonium-limiting conditions (Lorenz and Heitman 1998). Javelle et al. (2003) argued that high-affinity ammonium transporters from AM fungi could act in a similar manner as yeast to sense N availability in the soil environment.

17.7 Responses of the Mycorrhizal Fungal Community to Nutrient Enrichment

The mycorrhizal community structure (e.g., diversity and composition) may determine a series of cascading functions of plant systems. For example, AM fungal diversity might mitigate plant–plant competition, promote plant diversity, reduce ecosystem variability, and increase productivity (van der Heijden et al. 1998b; Wagg et al. 2011). Thus, such importance of mycorrhizal functioning has attracted researchers to decipher how nutrient enrichment affects AM fungal community structure in the past decades. To date, many studies have described AM community responses to various nutrient inputs across many ecosystems and attempted to decipher the underlying mechanisms as described below.

17.7.1 Forests and Shrublands

In a northern hardwood forest, a 12-year continuous N additions ($NaNO_3$ with a rate of 30 kg ha^{-1} $year^{-1}$) significantly altered the AMF community composition, but the AMF diversity was unaffected (van Diepen et al. 2011). In a coastal sage scrub ecosystem, N enrichment (60 kg ha^{-1} $year^{-1}$) was shown to shift AMF composition through a displacement of *Gigasporaceae* by *Glomeraceae*, which thus led to a reduction in species richness and diversity (Egerton-Warburton and Allen 2000). In a tropical montane forest, AMF species richness was significantly reduced by N (50 kg ha^{-1} $year^{-1}$) and P (10 kg ha^{-1} $year^{-1}$) additions as well as their combinations, mainly through replacing rare AM fungal species (Camenzind et al. 2014). More importantly, this study further elucidated that Diversiporales richness was mainly reduced by N amendment while *Glomerales* was more sensitive to P additions after 2 years.

17.7.2 Grasslands

Positive, neutral, and negative effects have all been reported for responses of AMF community to N and P enrichment in grassland ecosystems, but the outcome or direction of the response depends on the beginning nutrient availability/limitation and the specific nutrient. For the positive response, Kim et al. (2015) reported that N additions (100 kg ha^{-1} $year^{-1}$) significantly increased AMF diversity but not species richness, as well as altered species composition through shifts in plant community in a semiarid steppe ecosystem. Xiang et al. (2016) additionally reported positive responses of AMF species richness and phylogenetic diversity to N and P additions in the nutrient-limited Qinghai–Tibet Plateau alpine grasslands. Porras-Alfaro et al. (2007) also found that N amendments decreased dominant AMF species, but may have reduced suppression to subdominant or rare species, and thus increased total diversity in semiarid grassland. For negative effects, Chen et al. (2014) found that N fertilization (100 kg ha^{-1} $year^{-1}$) in a temperate steppe ecosystem significantly reduced AM fungal species richness, but did not affect the diversity, while P fertilization (50 kg ha^{-1} $year^{-1}$) had no effect on either of two parameters, leading to alterations of AMF community composition by N, but not by P fertilization. From a cross-site grassland experiment, Egerton-Warburton et al. (2007) found that N fertilization (>100 kg ha^{-1} $year^{-1}$) reduced AM fungal species richness and diversity namely because the abundance of Glomeraceae was higher in P-rich soils while there was more Gigasporaceae abundant species in P-poor soils. However, Antoninka et al. (2011) found that long-term N fertilization (40 kg ha^{-1} $year^{-1}$) did not affect AMF spore species richness in a grassland ecosystem. Such large variation might determine by the background N or P availability; thus the baseline value of nutrient is necessary for explaining the response of AM fungal community to nutrient enrichment.

17.7.3 Agroecosystems

A 55-year experiment in Skåne, Sweden, with continuous N×P full-factorial treatments showed that N fertilization altered AM fungal species composition and reduced diversity, but P fertilization showed no effect in a crop rotation of spring barley (*Hordeum vulgare* L.)–white mustard (*Sinapis alba* L.) or spring oilseed rape (*Brassica napus* L.)–winter wheat (*Triticum aestivum* L.)–sugar beet (*Beta vulgaris* L.) (Williams et al. 2017). N fertilization increased the abundance of *Funneliformis* sp.1 and *Rhizophagus irregularis* sp.2 but decreased the abundance of *Claroideoglomus* sp.3, *C. glomus* sp.5 and *Funneliformis mosseae* (Williams et al. 2017). A 90-year experiment with continuous N and P fertilizer treatments since 1914 at Hokkaido University (Sapporo, Japan) showed that both P (100 kg ha^{-1} $year^{-1}$) and N (100 kg ha^{-1} $year^{-1}$) fertilization altered AM fungal composition and reduced diversity, but P fertilization led to a larger reduction in the AM fungal diversity than N fertilization (Cheng et al. 2013). A 21-year experiment in a wheat (*Triticum aestivum* L)–maize (*Zea mays* L.) rotation system showed that long-term balanced NP fertilization decreased AM fungal species richness and diversity, as well as altered community composition by increasing *Glomeraceae* species but reducing the number of the dominant *Gigasporaceae* species (Lin et al. 2012). Thus, continuous fertilization might provide a selective pressure to AM fungal species, and the N- or P-tolerating ones will be maintained, but others might be discarded in the cropping systems.

17.8 Responses of Mycorrhizal Functioning to Nutrient Enrichment

Nutrient enrichment exerts profound effects on the intra- and extraradical abundance, diversity, and species composition of AM fungi. Such changes in mycorrhizas might first induce plant responses at individual levels, e.g., stress tolerance, and then cause cascading ecosystem responses, such as changes to plant productivity and diversity, soil aggregation, and carbon storage.

Nutrient enrichment may decrease the stress tolerance effects on individual hosts mediated by AM fungi. For some abiotic stresses, such as drought, salt, or heavy metal, mycorrhizal colonization might activate genes responsible for such tolerance, i.e., aquaporins, proline, or metallothionein (Reddy et al. 2016; Ruiz-Lozano et al. 2006). For some biotic stresses, mycorrhizal colonization might induce systematic resistance in host plants, such as releasing volatiles to repel insect herbivores (Bennett and Bever 2009), activating chitinase expression to inhibit nematode growth (Li et al. 2006), or upregulating pathogenesis-related genes (Campos-Soriano et al. 2012). In addition, competitive inhibition for carbon resource or colonization site is another mechanism for mycorrhizal-mediated resistance to biotic stresses (Borowicz 2001). However, as shown in some cases, increased nutrient inputs can reduce mycorrhizal root colonization, which may weaken the induction and activation of systematic resistance or tolerance toward stresses under nutrient-rich conditions.

Shifts in the mycorrhizal fungal community in response to nutrient enrichment can also lead to changes in plant community. Plant species has been shown to differ in their responsiveness to AM fungi, and different fungal species have distinct effects on plant growth variables (van der Heijden et al. 1998a, 2006a, b). Thus, shifts in mycorrhizal species composition might potentially drive changes in plant community composition through modifying plant competition. For example, distinct competitive outcome was shown between mycorrhizal and non-mycorrhizal plants when inoculated with *Gigaspora margarita* and *Glomus intraradices* (Facelli et al. 2010). Mycorrhizal fungi markedly increased competitiveness of a pioneer tree (*Rhus chinensis*) on a late-pioneer (*Celtis sinensis*) or mid-successional tree (*Cinnamomum camphora*), but the competitive strength was dependent on fungal identity (Shi et al. 2016). Mycorrhizal fungal identity was shown to have a large impact on competitive interactions between a grass and a legume by favoring the latter (Wagg et al. 2011). These experiments implied that nutrient enrichment-induced community shifts in mycorrhizal fungi might contribute to plant–plant interactions and subsequently regulate plant community structure and functions.

Reduction in mycorrhizas as a result of nutrient enrichment may negatively affect soil structure and reduce soil carbon storage. Mycorrhizal fungal hyphae can excrete a long carbon chain glycoprotein, glomalin, which can promote adhesion of microaggregates (Rillig et al. 2015). At the same time, AM fungal extraradical hyphae can also enmesh microaggregates into macroaggregates (Peng et al. 2013). Additionally, glomalin has been shown to be difficult to degrade in soils, and thus glomalin-C would be sequestrated in the soils for a long time (Wilson et al. 2009). Mycorrhizal-mediated soil aggregation might also provide physical protection for organic carbon from microbial decomposition (Rillig 2004).

17.9 What Controls Mycorrhizal Community Composition in Roots and Soils?

Changes in AM fungal composition can be explained through responses to nutrient enrichment of both partners. Here, we proposed two models: a plant-centric model and a fungal-centric model.

17.9.1 Plant-Centric Model

This model predicts that N or P enrichment first induces shifts in plant community composition and subsequently drives changes in their associated mycorrhizal fungal community. Many studies have shown that nutrient enrichment, in particular N, can significantly alter plant community composition and reduced plant diversity (Bobbink et al. 2010; Clark and Tilman 2008; Stevens et al. 2004). However, host preference has also been shown for mycorrhizal fungi to some extent. For example, distinctive mycorrhizal species composition was found to be among co-existing trees, forbs, or grasses, respectively (Husband et al. 2002; Vandenkoornhuyse et al. 2002, 2003). Plant composition was also shown to greatly affect mycorrhizal fungal diversity and led to distinct community composition (Johnson et al. 2004). Thus, alteration in plant composition induced by nutrient enrichment might drive changes in mycorrhizal fungal community because of the asymmetric nature of mycorrhiza, which we define as *driving effect.*

17.9.2 Fungal-Centric Model

This model can be derived by two hypotheses. One is that nutrient enrichment promotes AM fungal competition for plant photosynthate (here we defined it as *competitive effect*); the other is that different mycorrhizal fungal species have different sensitivities and uptake capacity for different nutrients (here we define it as *selective effect*). For the competitive effect, plants might reduce carbon allocation to their selective AM fungi because nutrient enrichment decreases the value of mycorrhiza for nutrient uptake (Johnson et al. 2003). Consequently, if host carbon becomes a scarce resource for AM fungi, competitive exclusion might reduce or eliminate rare species but maintain dominant ones (Knegt et al. 2016). Such competitive exclusion for carbon could lead to reduced AM fungal species richness and diversity over time (Liu et al. 2015). For the selective effect, experimental evidence has shown that Diversiporales is sensitive to N enrichment but tolerant to P, while *Glomerales* is sensitive to P but not to N (Lin et al. 2012; Williams et al. 2017). Thus, nutrient enrichment might select some N- or P-loving mycorrhizal fungal species and exclude some sensitive taxa, the idea which is especially supported by farmlands, e.g., dominance and persistence of some specific AM fungal species in high N or P availability (Cheng et al. 2013; Lin et al. 2012).

17.10 Future Directions

While a number of studies have examined mycorrhizal responses to nutrient enrichment in various settings, significant knowledge gaps still persist. This limits our capacity to predict the impact of nutrient inputs on mycorrhizas and potentially to manage nutrient inputs to maximize mycorrhizal benefits to plants. First, it is still unclear whether there are some unifying mechanisms/patterns that underlie arbuscular mycorrhizal responses to nutrient enrichment, including C and nutrient trade-offs, thresholds, and molecular controlling mechanisms. Secondly, although increasing experimental evidence has shown that nutrient enrichment alters AM fungal community composition (Kim et al. 2015; van Diepen et al. 2011; Williams et al. 2017), much less is known about the identity of fungal species under impact. There is limited evidence showing that *Glomerales* taxa are sensitive to P but not to N additions, while *Diversisporales* seemed to be the opposite. Yet, many questions remain: Does this represent a general pattern? What about other AM taxa responses? Third, AM fungal diversity showed positive, neutral, or negative responses to nutrient enrichment. We do not know what factors lead to such variations and how initial soil conditions contribute to this variability. Fourth, many previous studies so far have focused on additions of single nutrient with one high rate. This raises questions to whether there is a nutrient threshold for mycorrhizal responses. If yes, does the threshold vary with the fungal species or taxa? Experiments that examine mycorrhizal responses across a gradient nutrient input are needed in the future, especially in field conditions and under different N and P limitations. Moreover, how AM fungal response to one nutrient depends on the availability of other nutrients warrants further study. These knowledge gaps highlight the need for factorial experiments. Lastly, how other environmental factors, such as water availability and temperature, may modulate arbuscular mycorrhizal responses to nutrient enrichment requires examination. This knowledge is essential for us to understand mycorrhizal functions in a globe with multifaceted changes.

References

Al-Karaki GN, Clark R (1999) Mycorrhizal influence on protein and lipid of durum wheat grown at different soil phosphorus levels. Mycorrhiza 9:97–101

Amtmann A, Hammond JP, Armengaud P, White PJ (2006) Nutrient sensing and signalling in plants: potassium and phosphorus. In: Callow JA (ed) Advances in botanical research. Incorporating advances in plant pathology, vol 43. Academic, London, pp 209–257

Antoninka A, Reich PB, Johnson NC (2011) Seven years of carbon dioxide enrichment, nitrogen fertilization and plant diversity influence arbuscular mycorrhizal fungi in a grassland ecosystem. New Phytol 192:200–214

Avio L, Castaldini M, Fabiani A, Bedini S, Sbrana C, Turrini A, Giovannetti M (2013) Impact of nitrogen fertilization and soil tillage on arbuscular mycorrhizal fungal communities in a Mediterranean agroecosystem. Soil Biol Biochem 67:285–294

Azcón R, Ambrosano E, Charest C (2003) Nutrient acquisition in mycorrhizal lettuce plants under different phosphorus and nitrogen concentration. Plant Sci 165:1137–1145

Benedetto A, Magurno F, Bonfante P, Lanfranco L (2005) Expression profiles of a phosphate transporter gene (*Gmos*PT) from the endomycorrhizal fungus *Glomus mosseae*. Mycorrhiza 15:620–627

Bennett AE, Bever JD (2009) Trade-offs between arbuscular mycorrhizal fungal competitive ability and host growth promotion in *Plantago lanceolata*. Oecologia 160:807–816

Bever JD, Richardson SC, Lawrence BM, Holmes J, Watson M (2009) Preferential allocation to beneficial symbiont with spatial structure maintains mycorrhizal mutualism. Ecol Lett 12:13–21

Blanke V, Renker C, Wagner M, Fullner K, Held M, Kuhn AJ, Buscot F (2005) Nitrogen supply affects arbuscular mycorrhizal colonization of *Artemisia vulgaris* in a phosphate-polluted field site. New Phytol 166:981–992

Bobbink R, Hicks K, Galloway J, Spranger T, Alkemade R, Ashmore M, Bustamante M, Cinderby S, Davidson E, Dentener F, Emmett B, Erisman JW, Fenn M, Gilliam F, Nordin A, Pardo L, De Vries W (2010) Global assessment of nitrogen deposition effects on terrestrial plant diversity: a synthesis. Ecol Appl 20:30–59

Borowicz VA (2001) Do arbuscular mycorrhizal fungi alter plant-pathogen relations? Ecology 82:3057–3068

Brundrett MC (2009) Mycorrhizal associations and other means of nutrition of vascular plants: understanding the global diversity of host plants by resolving conflicting information and developing reliable means of diagnosis. Plant Soil 320:37–77

Camenzind T, Hempel S, Homeier J, Horn S, Velescu A, Wilcke W, Rillig MC (2014) Nitrogen and phosphorus additions impact arbuscular mycorrhizal abundance and molecular diversity in a tropical montane forest. Glob Chang Biol 20:3646–3659

Camenzind T, Homeier J, Dietrich K, Hempel S, Hertel D, Krohn A, Leuschner C, Oelmann Y, Olsson PA, Suarez JP, Rillig MC (2016) Opposing effects of nitrogen versus phosphorus additions on mycorrhizal fungal abundance along an elevational gradient in tropical montane forests. Soil Biol Biochem 94:37–47

Campos-Soriano L, Garcia-Martinez J, Segundo BS (2012) The arbuscular mycorrhizal symbiosis promotes the systemic induction of regulatory defence-related genes in rice leaves and confers resistance to pathogen infection. Mol Plant Pathol 13:579–592

Chen YL, Zhang X, Ye JS, Han HY, Wan SQ, Chen BD (2014) Six-year fertilization modifies the biodiversity of arbuscular mycorrhizal fungi in a temperate steppe in Inner Mongolia. Soil Biol Biochem 69:371–381

Cheng L, Booker F, Tu C, Burkey K, Zhou L, Shew H, Rufty T, Hu S (2012) Arbuscular mycorrhizal fungi increase organic carbon decomposition under elevated CO_2. Science 337:1084

Cheng Y, Ishimoto K, Kuriyama Y, Osaki M, Ezawa T (2013) Ninety-year-, but not single, application of phosphorus fertilizer has a major impact on arbuscular mycorrhizal fungal communities. Plant Soil 365:397–407

Clark CM, Tilman D (2008) Loss of plant species after chronic low-level nitrogen deposition to prairie grasslands. Nature 451:712–715

Cofala J, Amann M, Klimont Z, Kupiainen K, Höglund-Isaksson L (2007) Scenarios of global anthropogenic emissions of air pollutants and methane until 2030. Atmos Environ 41:8486–8499

Egerton-Warburton LM, Allen EB (2000) Shifts in arbuscular mycorrhizal communities along an anthropogenic nitrogen deposition gradient. Ecol Appl 10:484–496

Egerton-Warburton LM, Johnson NC, Allen EB (2007) Mycorrhizal community dynamics following nitrogen fertilization: a cross-site test in five grasslands. Ecol Monogr 77:527–544

Eissenstat DM, Graham JH, Syvertsen JP, Drouillard DL (1993) Carbon economy of sour orange in relation to mycorrhizal colonization and phosphorus status. Ann Bot 71:1–10

Ekblad A, Mikusinska A, Ågren GI, Menichetti L, Wallander H, Vilgalys R, Bahr A, Eriksson U (2016) Production and turnover of ectomycorrhizal extramatrical mycelial biomass and necromass under elevated CO_2 and nitrogen fertilization. New Phytol 211:874–885

Facelli E, Smith SE, Facelli JM, Christophersen HM, Smith FA (2010) Underground friends or enemies: model plants help to unravel direct and indirect effects of arbuscular mycorrhizal fungi on plant competition. New Phytol 185:1050–1061

Fellbaum CR, Gachomo EW, Beesetty Y, Choudhari S, Strahan GD, Pfeffer PE, Kiers ET, Buecking H (2012) Carbon availability triggers fungal nitrogen uptake and transport in arbuscular mycorrhizal symbiosis. Proc Natl Acad Sci USA 109:2666–2671

Fochi V, Chitarra W, Kohler A, Voyron S, Singan VR, Lindquist EA, Barry KW, Girlanda M, Grigoriev I, Martin F, Balestrini R, Perotto S (2017) Fungal and plant gene expression in the *Tulasnella calospora-Serapias vomeracea* symbiosis provides clues about nitrogen pathways in orchid mycorrhizas. New Phytol 213:365–379

Galloway JN, Townsend AR, Erisman JW, Bekunda M, Cai ZC, Freney JR, Martinelli LA, Seitzinger SP, Sutton MA (2008) Transformation of the nitrogen cycle: recent trends, questions, and potential solutions. Science 320:889–892

Gill RA, Jackson RB (2000) Global patterns of root turnover for terrestrial ecosystems. New Phytol 147:13–31

Goff SA, Ricke D, Lan TH, Presting G, Wang RL, Dunn M et al (2002) A draft sequence of the rice genome (*Oryza sativa* L. ssp japonica). Science 296:92–100

Gosling P, Hodge A, Goodlass G, Bending GD (2006) Arbuscular mycorrhizal fungi and organic farming. Agr Ecosyst Environ 113:17–35

Govindarajulu M, Pfeffer P, Jin H, Abubaker J, Douds D, Allen J, Bücking H, Lammers P, Shachar-Hill Y (2005) Nitrogen transfer in the arbuscular mycorrhizal symbiosis. Nature 435:819–823

Gryndler M, Larsen J, Hršelová H, Řezáčová V, Gryndlerová H, Kubát J (2006) Organic and mineral fertilization, respectively, increase and decrease the development of external mycelium of arbuscular mycorrhizal fungi in a long-term field experiment. Mycorrhiza 16:159–166

Guescini M, Zeppa S, Pierleoni R, Sisti D, Stocchi L, Stocchi V (2007) The expression profile of the Tuber borchii nitrite reductase suggests its positive contribution to host plant nitrogen nutrition. Curr Genet 51:31–41

Guether M, Neuhauser B, Balestrini R, Dynowski M, Ludewig U, Bonfante P (2009) A mycorrhizal-specific ammonium transporter from *Lotus japonicus* acquires nitrogen released by arbuscular mycorrhizal fungi. Plant Physiol 150:73

Harrison MJ, Vanbuuren ML (1995) A phosphate transporter from the mycorrhizal fungus *Glomus versiforme*. Nature 378:626–629

Harrison MJ, Dewbre GR, Liu J (2002) A phosphate transporter from *Medicago truncatula* involved in the acquisition of phosphate released by arbuscular mycorrhizal fungi. Plant Cell 14:2413

Hetrick B, Wilson G, Cox T (1993) Mycorrhizal dependence of modern wheat cultivars and ancestors: a synthesis. Can J Bot 71:512–518

Hobbie JE, Hobbie EA (2006) N-15 in symbiotic fungi and plants estimates nitrogen and carbon flux rates in Arctic tundra. Ecology 87:816

Hodge A, Campbell CD, Fitter AH (2001) An arbuscular mycorrhizal fungus accelerates decomposition and acquires nitrogen directly from organic material. Nature 413:297–299

Huang RJ, Zhang YL, Bozzetti C, Ho KF, Cao JJ, Han YM et al (2002) Molecular diversity of arbuscular mycorrhizal fungi and patterns of host association over time and space in a tropical forest. Mol Ecol 11:2669–2678

Husband R, Herre EA, Turner SL, Gallery R, Young JPW (2002) Molecular diversity of arbuscular mycorrhizal fungi and patterns of host association over time and space in a tropical forest. Mol Ecol 11:2669–2678

Jakobsen I, Abbott LK, Robson AD (1992) External hyphae of vesicular-arbuscular mycorrhizal fungi associated with *Trifolium subterraneum* L. I: spread of hyphae and phosphorus inflow into roots. New Phytol 120:371–380

Javelle A, Andre B, Marini AM, Chalot M (2003) High-affinity ammonium transporters and nitrogen sensing in mycorrhizas. Trends Microbiol 11:53–55

Javot H, Penmetsa RV, Terzaghi N, Cook DR, Harrison MJ (2007) A Medicago truncatula phosphate transporter indispensable for the arbuscular mycorrhizal symbiosis. Proc Natl Acad Sci USA 104:1720–1725

Johnson NC, Rowland DL, Corkidi L, Egerton-Warburton LM, Allen EB (2003) Nitrogen enrichment alters mycorrhizal allocation at five mesic to semiarid grasslands. Ecology 84:1895–1908

Johnson D, Vandenkoornhuyse PJ, Leake JR, Gilbert L, Booth RE, Grime JP, Young JPW, Read DJ (2004) Plant communities affect arbuscular mycorrhizal fungal diversity and community composition in grassland microcosms. New Phytol 161:503–515

Ju XT, Xing GX, Chen XP, Zhang SL, Zhang LJ, Liu XJ, Cui ZL, Yin B, Christie P, Zhu ZL (2009) Reducing environmental risk by improving N management in intensive Chinese agricultural systems. Proc Natl Acad Sci 106:3041–3046

Kahiluoto H, Ketoja E, Vestberg M, Saarela I (2001) Promotion of AM utilization through reduced P fertilization 2. Field studies. Plant Soil 231:65–79

Kaldorf M, Schmelzer E, Bothe H (1998) Expression of maize and fungal nitrate reductase genes in arbuscular mycorrhiza. Mol Plant Microbe Interact 11:439

Kiers ET, van der Heijden MGA (2006) Mutualistic stability in the arbuscular mycorrhizal symbiosis: exploring hypotheses of evolutionary cooperation. Ecology 87:1627–1636

Kiers ET, Duhamel M, Beesetty Y, Mensah JA, Franken O, Verbruggen E, Fellbaum CR, Kowalchuk GA, Hart MM, Bago A, Palmer TM, West SA, Vandenkoornhuyse P, Jansa J,

Bucking H (2011) Reciprocal rewards stabilize cooperation in the mycorrhizal symbiosis. Science 333:880–882
Kim YC, Gao C, Zheng Y, He XH, Yang W, Chen L, Wan SQ, Guo LD (2015) Arbuscular mycorrhizal fungal community response to warming and nitrogen addition in a semiarid steppe ecosystem. Mycorrhiza 25:267–276
Knegt B, Jansa J, Franken O, Engelmoer DJ, Werner GD, Bücking H, Kiers ET (2016) Host plant quality mediates competition between arbuscular mycorrhizal fungi. Fungal Ecol 20:233–240
Koegel S, Lahmidi NA, Arnould C, Chatagnier O, Walder F, Ineichen K, Boller T, Wipf D, Wiemken A, Courty P-E (2013) The family of ammonium transporters (AMT) in *Sorghum bicolor*: two AMT members are induced locally, but not systemically in roots colonized by arbuscular mycorrhizal fungi. New Phytol 198:853–865
Koide RT (2000) Functional complementarity in the arbuscular mycorrhizal symbiosis. New Phytol 147:223–235
Koide RT, Kabir Z (2000) Extraradical hyphae of the mycorrhizal fungus *Glomus intraradices* can hydrolyse organic phosphate. New Phytol 148:511–517
Koide RT, Fernandez CW, Peoples MS (2011) Can ectomycorrhizal colonization of Pinus resinosa roots affect their decomposition? New Phytol 191:508–514
Krouk G, Lacombe B, Bielach A, Perrine-Walker F, Malinska K, Mounier E, Hoyerova K, Tillard P, Leon S, Ljung K, Zazimalova E, Benkova E, Nacry P, Gojon A (2010) Nitrate-regulated auxin transport by NRT1.1 defines a mechanism for nutrient sensing in plants. Dev Cell 18:927–937
Lanquar V, Loque D, Hormann F, Yuan LX, Bohner A, Engelsberger WR, Lalonde S, Schulze WX, von Wiren N, Frommer WB (2009) Feedback inhibition of ammonium uptake by a phospho-dependent allosteric mechanism in *Arabidopsis*. Plant Cell 21:3610–3622
Li HY, Yang GD, Shu HR, Yang YT, Ye BX, Nishida I, Zheng CC (2006) Colonization by the arbuscular mycorrhizal fungus *Glomus versiforme* induces a defense response against the root-knot nematode *Meloidogyne incognita* in the grapevine (*Vitis amurensis* Rupr.), which includes transcriptional activation of the class III chitinase gene VCH3. Plant Cell Physiol 47:154–163
Lin XG, Feng YZ, Zhang HY, Chen RR, Wang JH, Zhang JB, Chu HY (2012) Long-term balanced fertilization decreases arbuscular mycorrhizal fungal diversity in an arable soil in North China revealed by 454 pyrosequencing. Environ Sci Technol 46:5764–5771
Liu YJ, Mao L, Li JY, Shi GX, Jiang SJ, Ma XJ, An LZ, Du GZ, Feng HY (2015) Resource availability differentially drives community assemblages of plants and their root-associated arbuscular mycorrhizal fungi. Plant Soil 386:341–355
Lopez-Pedrosa A, Gonzalez-Guerrero M, Valderas A, Azcon-Aguilar C, Ferrol N (2006) GintAMT1 encodes a functional high-affinity ammonium transporter that is expressed in the extraradical mycelium of *Glomus intraradices*. Fungal Genet Biol 43:102–110
Lorenz MC, Heitman J (1998) The MEP2 ammonium permease regulates pseudohyphal differentiation in *Saccharomyces cerevisiae*. EMBO J 17:1236–1247
MacFall JS, Slack SA, Wehrli S (1992) Phosphorus distribution in red pine roots and the ectomycorrhizal fungus Hebeloma arenosa. Plant Physiol 100:713–717
Maldonado-Mendoza I, Dewbre G, Harrison M (2001) A phosphate transporter gene from the extra-radical mycelium of an arbuscular mycorrhizal fungus *Glomus intraradices* is regulated in response to phosphate in the environment. Mol Plant Microbe Interact 14:1140–1148
McCormick MK, Taylor DL, Juhaszova K, Burnett RK, Whigham DF, O'Neill JP (2012) Limitations on orchid recruitment: not a simple picture. Mol Ecol 21:1511–1523
Monteagudo L, Luis Moreno J, Picazo F (2012) River eutrophication: irrigated vs. non-irrigated agriculture through different spatial scales. Water Res 46:2759–2771
Muller T, Avolio M, Olivi M, Benjdia M, Rikirsch E, Kasaras A, Fitz M, Chalot M, Wipf D (2007) Nitrogen transport in the ectomycorrhiza association: the *Hebeloma cylindrosporum-Pinus pinaster* model. Phytochemistry 68:41–51

Munos S, Cazettes C, Fizames C, Gaymard F, Tillard P, Lepetit M, Lejay L, Gojon A (2004) Transcript profiling in the chl1-5 mutant of *Arabidopsis* reveals a role of the nitrate transporter NRT1.1 in the regulation of another nitrate transporter, NRT2.1. Plant Cell 16:2433–2447
Nagy R, Karandashov V, Chague V, Kalinkevich K, Tamasloukht M, Xu G, Jakobsen I, Levy A, Amrhein N, Bucher M (2005) The characterization of novel mycorrhiza-specific phosphate transporters from *Lycopersicon esculentum* and *Solanum tuberosum* uncovers functional redundancy in symbiotic phosphate transport in solanaceous species. Plant J 42:236
Nagy R, Drissner D, Amrhein N, Jakobsen I, Bucher M (2009) Mycorrhizal phosphate uptake pathway in tomato is phosphorus-repressible and transcriptionally regulated. New Phytol 181:950–959
Paszkowski U, Kroken S, Roux C, Briggs SP (2002) Rice phosphate transporters include an evolutionarily divergent gene specifically activated in arbuscular mycorrhizal symbiosis. Proc Natl Acad Sci USA 99:13324
Peng SL, Guo T, Liu GC (2013) The effects of arbuscular mycorrhizal hyphal networks on soil aggregations of purple soil in southwest China. Soil Biol Biochem 57:411–417
Popova Y, Thayumanavan P, Lonati E, Agrochao M, Thevelein JM (2010) Transport and signaling through the phosphate-binding site of the yeast Pho84 phosphate transceptor. Proc Natl Acad Sci USA 107:2890–2895
Porras-Alfaro A, Herrera J, Natvig DO, Sinsabaugh RL (2007) Effect of long-term nitrogen fertilization on mycorrhizal fungi associated with a dominant grass in a semiarid grassland. Plant Soil 296:65–75
Pritchard SG, Taylor BN, Cooper ER, Beidler KV, Strand AE, McCormack ML, Zhang S (2014) Long-term dynamics of mycorrhizal root tips in a loblolly pine forest grown with free-air CO_2 enrichment and soil N fertilization for 6 years. Glob Chang Biol 20:1313–1326
Qin L, Zhao J, Tian J, Chen LY, Sun ZA, Guo YX, Lu X, Gu MA, Xu GH, Liao H (2012) The high-affinity phosphate transporter GmPT5 regulates phosphate transport to nodules and nodulation in soybean. Plant Physiol 159:1634–1643
Reddy MS, Kour M, Aggarwal S, Ahuja S, Marmeisse R, Fraissinet-Tachet L (2016) Metal induction of a *Pisolithus albus* metallothionein and its potential involvement in heavy metal tolerance during mycorrhizal symbiosis. Environ Microbiol 18:2446–2454
Reynolds HL, Hartley AE, Vogelsang KM, Bever JD, Schultz PA (2005) Arbuscular mycorrhizal fungi do not enhance nitrogen acquisition and growth of old-field perennials under low nitrogen supply in glasshouse culture. New Phytol 167:869
Rillig MC (2004) Arbuscular mycorrhizae and terrestrial ecosystem processes. Ecol Lett 7:740–754
Rillig MC, Aguilar-Trigueros CA, Bergmann J, Verbruggen E, Veresoglou SD, Lehmann A (2015) Plant root and mycorrhizal fungal traits for understanding soil aggregation. New Phytol 205:1385–1388
Ruiz-Lozano JM, Porcel R, Aroca R (2006) Does the enhanced tolerance of arbuscular mycorrhizal plants to water deficit involve modulation of drought-induced plant genes? New Phytol 171:693–698
Saito M, Oba H, Kojima T (2011) Effect of nitrogen on the sporulation of arbuscular mycorrhizal fungi colonizing several gramineous plant species. Soil Sci Plant Nutr 57:29–34
Scheible W-R, Rojas-Triana M (2015) Sensing, signalling, and control of phosphate starvation in plants: molecular players and applications. In: Plaxton WC, Lambers H (eds) Phosphorus metabolism in plants, vol 48. Wiley, Hoboken, pp 25–63
Schlesinger WH (2009) On the fate of anthropogenic nitrogen. Proc Natl Acad Sci USA 106:203–208
Selosse MA, Setaro S, Glatard F, Richard F, Urcelay C, Weiss M (2007) Sebacinales are common mycorrhizal associates of Ericaceae. New Phytol 174:864–878
Shi NN, Gao C, Zheng Y, Guo LD (2016) Arbuscular mycorrhizal fungus identity and diversity influence subtropical tree competition. Fungal Ecol 20:115–123
Smil V (1999) Nitrogen in crop production: an account of global flows. Global Biogeochem Cycles 13:647–662

Smil V (2003) Phosphorus in the environment: natural flows and human interferences. Annu Rev Energy Environ 25:53–88

Smith SE, Read DJ (2008) Mycorrhizal symbiosis, 3rd edn. Academic, San Diego

Stevens CJ, Dise NB, Mountford JO, Gowing DJ (2004) Impact of nitrogen deposition on the species richness of grasslands. Science 303:1876–1879

Svistoonoff S, Creff A, Reymond M, Sigoillot-Claude C, Ricaud L, Blanchet A, Nussaume L, Desnos T (2007) Root tip contact with low-phosphate media reprograms plant root architecture. Nat Genet 39:792–796

Teng W, Zhao YY, Zhao XQ, He X, Ma WY, Deng Y, Chen XP, Tong YP (2017) Genome-wide identification, characterization, and expression analysis of PHT1 phosphate transporters in wheat. Front Plant Sci 8:543

Thirkell TJ, Cameron DD, Hodge A (2016) Resolving the 'nitrogen paradox' of arbuscular mycorrhizas: fertilization with organic matter brings considerable benefits for plant nutrition and growth. Plant Cell Environ 39:1683

Tian C, Kasiborski B, Koul R, Lammers PJ, Bücking H, Shachar-Hill Y (2010) Regulation of the nitrogen transfer pathway in the arbuscular mycorrhizal symbiosis: gene characterization and the coordination of expression with nitrogen flux. Plant Physiol 153:1175–1187

Tilman D, Fargione J, Wolff B, D'Antonio C, Dobson A, Howarth R, Schindler D, Schlesinger WH, Simberloff D, Swackhamer D (2001) Forecasting agriculturally driven global environmental change. Science 292:281

Tilman D, Cassman KG, Matson PA, Naylor R, Polasky S (2002) Agricultural sustainability and intensive production practices. Nature 418:671–677

Tisserant E, Kohler A, Dozolme-Seddas P, Balestrini R, Benabdellah K, Colard A et al (2012) The transcriptome of the arbuscular mycorrhizal fungus *Glomus intraradices* (DAOM 197198) reveals functional tradeoffs in an obligate symbiont. New Phytol 193:755–769

Treseder KK (2004) A meta-analysis of mycorrhizal responses to nitrogen, phosphorus, and atmospheric CO_2 in field studies. New Phytol 164:347–355

Tunlid A, Floudas D, Koide RT, Martin F (2016) Molecular mycorrhizal symbiosis: 15. Soil organic matter decomposition mechanisms in ectomycorrhizal fungi. Wiley, Hoboken

Ulm F, Gouveia C, Dias T, Cruz C (2017) N fertilization in a Mediterranean ecosystem alters N and P turnover in soil, roots and the ectomycorrhizal community. Soil Biol Biochem 113:60–70

Unger S, Friede M, Hundacker J, Volkmar K, Beyschlag W (2016) Allocation trade-off between root and mycorrhizal surface defines nitrogen and phosphorus relations in 13 grassland species. Plant Soil 407:279–292

van der Heijden MGA, Boller T, Wiemken A, Sanders IR (1998a) Different arbuscular mycorrhizal fungal species are potential determinants of plant community structure. Ecology 79:2082–2091

van der Heijden MGA, Klironomos JN, Ursic M, Moutoglis P, Streitwolf-Engel R, Boller T, Wiemken A, Sanders IR (1998b) Mycorrhizal fungal diversity determines plant biodiversity, ecosystem variability and productivity. Nature 396:69–72

van der Heijden MGA, Streitwolf-Engel R, Riedl R, Siegrist S, Neudecker A, Ineichen K, Boller T, Wiemken A, Sanders IR (2006a) The mycorrhizal contribution to plant productivity, plant nutrition and soil structure in experimental grassland. New Phytol 172:739–752

van der Heijden MGA, Streitwolf-Engel R, Riedl R, Siegrist S, Neudecker A, Ineichen K, Boller T, Wiemken A, Sanders IR (2006b) The mycorrhizal contribution to plant productivity, plant nutrition and soil structure in experimental grassland. New Phytol 172:739–752

van der Heijden MGA, Bardgett RD, Van Straalen NM (2008) The unseen majority: soil microbes as drivers of plant diversity and productivity in terrestrial ecosystems. Ecol Lett 11:296–310

van der Heijden MGA, Martin FM, Selosse M-A, Sanders IR (2015) Mycorrhizal ecology and evolution: the past, the present, and the future. New Phytol 205:1406–1423

van Diepen LTA, Lilleskov EA, Pregitzer KS, Miller RM (2007) Decline of arbuscular mycorrhizal fungi in northern hardwood forests exposed to chronic nitrogen additions. New Phytol 176:175–183

van Diepen LTA, Lilleskov EA, Pregitzer KS (2011) Simulated nitrogen deposition affects community structure of arbuscular mycorrhizal fungi in northern hardwood forests. Mol Ecol 20:799–811
Vandenkoornhuyse P, Husband R, Daniell TJ, Watson IJ, Duck JM, Fitter AH, Young JPW (2002) Arbuscular mycorrhizal community composition associated with two plant species in a grassland ecosystem. Mol Ecol 11:1555–1564
Vandenkoornhuyse P, Ridgway KP, Watson IJ, Fitter AH, Young JPW (2003) Co-existing grass species have distinctive arbuscular mycorrhizal communities. Mol Ecol 12:3085–3095
Wagg C, Jansa J, Stadler M, Schmid B, van der Heijden MGA (2011) Mycorrhizal fungal identity and diversity relaxes plant-plant competition. Ecology 92:1303–1313
Wang C, White PJ, Li C (2017) Colonization and community structure of arbuscular mycorrhizal fungi in maize roots at different depths in the soil profile respond differently to phosphorus inputs on a long-term experimental site. Mycorrhiza 27:369–381
Werner GDA, Kiers ET (2015) Partner selection in the mycorrhizal mutualism. New Phytol 205:1437–1442
Williams A, Manoharan L, Rosenstock NP, Olsson PA, Hedlund K (2017) Long-term agricultural fertilization alters arbuscular mycorrhizal fungal community composition and barley (*Hordeum vulgare*) mycorrhizal carbon and phosphorus exchange. New Phytol 213:874–885
Wilson GWT, Rice CW, Rillig MC, Springer A, Hartnett DC (2009) Soil aggregation and carbon sequestration are tightly correlated with the abundance of arbuscular mycorrhizal fungi: results from long-term field experiments. Ecol Lett 12:452–461
Withers PJA, Neal C, Jarvie HP, Doody DG (2014) Agriculture and eutrophication: where do we go from here? Sustainability 6:5853–5875
Xiang X, Gibbons SM, He J-S, Wang C, He D, Li Q, Ni Y, Chu H (2016) Rapid response of arbuscular mycorrhizal fungal communities to short-term fertilization in an alpine grassland on the Qinghai-Tibet Plateau. Peerj 4:e2226
Xie XA, Lin H, Peng XW, Xu CR, Sun ZF, Jiang KX, Huang A, Wu XH, Tang NW, Salvioli A, Bonfante P, Zhao B (2016) Arbuscular mycorrhizal symbiosis requires a phosphate transceptor in the *Gigaspora margarita* fungal symbiont. Mol Plant 9:1583–1608
Xu GH, Chague V, Melamed-Bessudo C, Kapulnik Y, Jain A, Raghothama KG, Levy AA, Silber A (2007) Functional characterization of LePT4: a phosphate transporter in tomato with mycorrhiza-enhanced expression. J Exp Bot 58:2491
Yang HS, Zhang Q, Koide RT, Hoeksema JD, Tang J, Bian X, Hu SJ, Chen X (2017) Taxonomic resolution is a determinant of biodiversity effects in arbuscular mycorrhizal fungal communities. J Ecol 105:219–228
Zhang L, Fan J, Ding X, He X, Zhang F, Feng G (2014a) Hyphosphere interactions between an arbuscular mycorrhizal fungus and a phosphate solubilizing bacterium promote phytate mineralization in soil. Soil Biol Biochem 74:177–183
Zhang Y, Zang GQ, Tang ZH, Chen XH, Yu YS (2014b) Burning straw, air pollution, and respiratory infections in China. Am J Infect Control 42:815–815
Zhang L, Xu M, Liu Y, Zhang F, Hodge A, Feng G (2016) Carbon and phosphorus exchange may enable cooperation between an arbuscular mycorrhizal fungus and a phosphate-solubilizing bacterium. New Phytol 210:1022–1032
Zhu Y, Smith S, Barritt A, Smith F (2001) Phosphorus (P) efficiencies and mycorrhizal responsiveness of old and modern wheat cultivars. Plant Soil 237:249–255
Zhu HH, Yao Q, Sun XT, Hu YL (2007) Colonization, ALP activity and plant growth promotion of native and exotic arbuscular mycorrhizal fungi at low pH. Soil Biol Biochem 39:942–950

Chapter 18
Relationship Between Arbuscular Mycorrhizas and Plant Growth: Improvement or Depression?

Li-Hui Lü, Ying-Ning Zou, and Qiang-Sheng Wu

18.1 Introduction

Arbuscular mycorrhizal fungi (AMF) are a type of soil fungi belonging to phylum *Glomeromycota* and can associate with the roots of most plants, in which fungal hyphae penetrate the root cortical cells to form inner structures known as arbuscules, hyphae, and/or vesicles (Fig. 18.1a) and outer structures known as extraradical hyphae and entry points (Fig. 18.1b). In the symbiosis system, host plants provide necessary carbon source and energy for mycorrhizal fungi. In return, mycorrhizas form well-developed extraradical hyphae to enlarge the absorption range of plant roots, resulting in the nutrient enhancement of host plants (Zhang et al. 2012b). Hence, arbuscular mycorrhizas play an important role in nutrient acquisition of the host plant (Mohammad et al. 2004; Li et al. 2011). AMF not only senses the signals secreted by the host plant roots but also guides the hyphae into the "presymbiotic growth stage." AMF also secretes certain factors that are identified by the root system. These mycorrhizal factors can be recognized by plants. Earlier studies had shown that mycorrhizas improved the utilization of nutrients for plants in soils (Wu and Tan 2005; Wu and Zou 2009; Zhang et al. 2012a, b).

Symbiosis between legumes and *Rhizobium* can be described as a typical model. *Rhizobium* absorbs nitrogen from the air and some of which are used for their own consumption; the rest is provided for hosts. The vast majority of legumes can be used to become nitrogenous nutrients, while leguminous plants are also not stingy and produce nonnitrogenous carbohydrate nutrients for the use of rhizobia openly, which is typical of mutualistic association (Simms and Taylor 2002). Arbuscular mycorrhizas as mutualistic symbiosis can help host plants to absorb nutrient and water from the soil to the plant partner, resist the effects of adverse environments, and

L.-H. Lü · Y.-N. Zou · Q.-S. Wu (✉)
College of Horticulture and Gardening, Yangtze University, Jingzhou, Hubei, China

Institute of Root Biology, Yangtze University, Jingzhou, Hubei, China

B. Giri et al. (eds.), *Root Biology*, Soil Biology 52,
https://doi.org/10.1007/978-3-319-75910-4_18

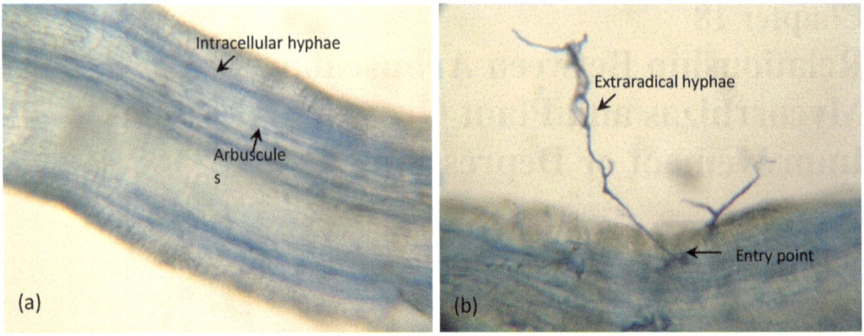

Fig. 18.1 Root colonization [(**a**) intracellular hyphae and arbuscules; (**b**) extraradical hyphae and entry point] of trifoliate orange by an arbuscular mycorrhizal fungus, *Funneliformis mosseae*

improve soil environment, thereby promoting plant growth and development (Deshmukh et al. 2006; Purin and Rillig 2007). In addition, AMF also stimulates endogenous phytohormone synthesis of host plants to increase total biomass and improve stress resistance. Excreta of AMF has a positive effect on soil aggregates that effect on soil stability and soil physical–chemical traits (Bever et al. 2001; Wu et al. 2016b). Based on individual level of plants, studies indicated that the maintenance of the AMF–plant symbiosis system depends on the two-way reciprocal mechanism and that the plant can identify the AMF that is beneficial to it and give the photosynthetic product priority to such AMF. On the contrary, AMF tends to deliver soil nutrients to plants that provide the most photosynthetic products, and this two-way reciprocal mechanism ensures the stability of the symbiotic system (Kiers et al. 2011).

Besides improved plant growth, mycorrhizal inoculation also has the inhibited effects on plant growth. For example, mycorrhizal presence significantly inhibited lemon growth under high phosphorus (5 mmol/L KH_2PO_4) supply (Peng et al. 1993). At high CO_2 concentrations, inoculation with *Glomus intraradices* heavily inhibited plant growth of *Citrus aurantium* (Jifon et al. 2002). It seems that mycorrhiza-modulated responses of host plant growth are a complex issue. In fact, the AMF responses of plant growth may be related to the changes in mutualism, commensalism, and parasitism. In this chapter, we simply discussed the promoted and inhibited effects of AMF on host plant growth and outlined the relevant mechanisms.

18.2 Improvement of Plant Growth by Mycorrhiza

As stated above, mycorrhizal workers often find the promoted effects of AMF inoculation on plant growth (Fig. 18.2). Here, we simply outlined the relevant mechanisms.

Fig. 18.2 The effect of inoculation with different AMF on growth of trifoliate orange seedlings

18.2.1 Increasing Nutrient Acquisition

It is well known that AMF-promoted growth of host plants is associated with AMF-increased nutrient acquisition. AMF is regarded as an "organ" of the nutrient absorption of plants. It has been estimated that about 75–90% P and 5–80% N in plants are contributed by AMF (Li et al. 1991; van Der Heijden et al. 2008). AMF also increases the absorption of K, Fe, Zn, Cu, and Mo from the soil to the fungal partner. Marschner and Dell (1994) estimated that mycorrhizas contributed 80% of P, 25% of N, 10% of K, 25% of Zn, and 60% of Cu, respectively. In the study of N, it was found that N species affected the absorption efficiency of AMF. The results of Tanaka and Yano (2005) indicated the nitrogen uptake of plants from $(NH_4)_2SO_4$ was ten times as much as $NaNO_3$ after inoculating with *Glomus aggregatum*. Besides, extraradical hyphae of AMF also take part in the metabolic process of N. Meanwhile, AMF can make plants assimilate more available P, leading to improved P nutrition. The study of Graham and Timmer (1985) showed that citrus mycorrhizal seedlings could take in more P nutrition from insoluble P fertilizer after applying soluble and insoluble P fertilizer into citrus mycorrhizal seedlings severally. Mycorrhiza-promoted P acquisition is related to plant status. For example, irrespective of inoculation with *Glomus mosseae* to citrus seedlings, there was no significant difference between P level in shoot and root of mycorrhizal seedlings versus non-mycorrhizal seedlings due to narrow phloem (Koch and Johnson 1984).

Furthermore, AMF can form enormous extraradical hyphae networks, extending to 11.7 cm from the root and thus expanding the nutrient uptake range (Effendy and Wijayani 2008). In addition, the transmission rate of soil nutrients in the mycelia network is much higher than that of the nutrient directly in the soil (Zhang et al. 2010). The hyphae bridges are established between different host plants, which increase the substance (carbohydrates, mineral nutrients, water) exchange among plants, thereby, reducing nutrient loss and accelerating nutrient cyclic utilization (Hart and Reader 2005). As a result, such enhancement of nutrient acquisition by mycorrhization is closely associated with growth improvement of host plants.

18.2.2 Enhancing Water Absorption from Soils

Mycorrhizas expanded the absorption area of the root system of the host plant, and the number and length of mycorrhizal hyphae were more than the host plant roots. Developed hyphae in the roots can form loose mycelium, whilst parts of the hyphae also invaded the root, forming a huge composite network (Simard et al. 2012; Wu et al. 2009). Studies in the past showed that the mycorrhizal hyphae and hyphal network had the functioning on absorbing and transferring water in various ecosystems. Extraradical hyphae of AMF enter into the tiny soil pores that plant roots do not pass to take in capillary water, in favor of AMF relief and lower plants wilting coefficient (Bolgiano et al. 1983). Meanwhile, Marulanda et al. (2003) argued that the efficiency of enhanced water uptake by different AMF species against lettuce (*Lactuca sativa*) is closely related to the biomass of extraradical hyphae. Levy and Krikun (1980) showed that mycorrhizas could enhance the transpiration and stomatal conductance of host plants, making the water transport more smooth and fast during water stress and normal water after inoculating *G. fasciculatus* on citrus. This suggests that mycorrhizas may provide a special water channel that reduces the transport distance and resistance of water. Allen (2006) through the analysis has found that extraradical hyphae were directly involved in water transportation; the diaphragm resistance to water flow resulted in limiting the ability of hyphae to transport water. And thus, it is estimated to be 25 cm/h, 131 nL/h, or 100 nL/h from different inference conclusion. Therefore, mycorrhizal hyphae have positive effects on water absorption of host plants irrespective of any water conditions.

AMF inoculation has an indirect influence on water metabolism of host plants under normal water condition, which have been confirmed on various plants, such as onion (Nelsen and Safir 1982), apple (Liu 1989), sunflower (Morte et al. 2000), and citrus (Wu et al. 2007). In low phosphorus soils, AMF can improve the water status of hosts by changing the physiological condition, but the similar effect does not appear in high phosphorus soils. Nelsen and Safir (1982) first reported that onion had higher leaf water potential and transpiration rate inoculated with *G. etunicatum*

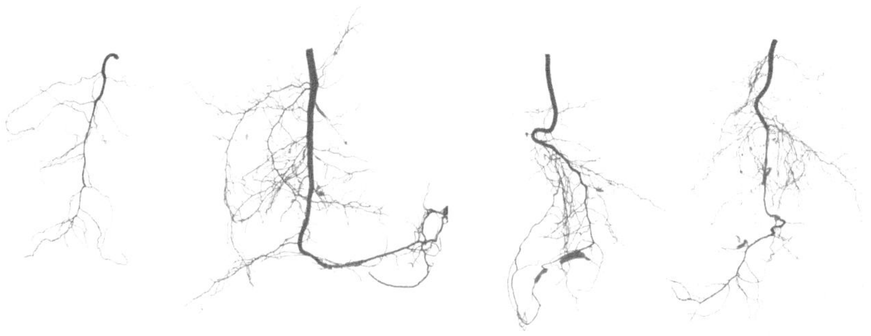

Fig. 18.3 The effect of inoculation with different AMF on root morphology of trifoliate orange seedlings

under low phosphorus soil conditions. Subsequently, Morte et al. (2000) also obtained similar results that inoculation with the AMF on sunflower significantly improved stomatal conductance of host plants and reduced stomatal resistance and natural saturation deficiencies, thereby enhancing water transport and promoting plant growth. Wu et al. (2007) also confirmed that the effects of AMF on the red tangerine seedlings under normal water and *G. mosseae* and *G. geosporum* could significantly increase transpiration rate.

18.2.3 Greater Root Morphology

AMF-promoted plant vegetative growth is related with increased root growth, which is conducive to use and store deep water in the soil for plants, thereby maintaining good water (Subrammanian and Charest 1999). In trifoliate orange seedlings, inoculation with *Glomus mosseae*, *Paraglomus occultum*, and *Glomus versiforme* significantly increased root total length, total projected area, surface area, and volume but decreased root diameter (Wu et al. 2011) (Fig. 18.3). In white clover, *Rhizoglomus intraradices*, *Diversispora versiformis*, and *Paraglomus occultum* significantly induced greater root total length, projected area, and volume (Lü and Wu 2017). The AMF effects were heavily dependent on AMF species used. Yao et al. (2009) found more fine roots and less coarse roots in AM plants. In addition, mycorrhizal colonization also increased root branches of *Vitis vinifera* (Schellenbaum et al. 1991). Greater root morphology of AM plants can ensure host plants to explore more water and nutrients, thereby, keeping a kind of greater plant growth behavior. AMF-improved root morphology is closely related with mycorrhiza-induced IAA production and mycorrhiza-regulated polyamine metabolism (Wu et al. 2012; Liu and Wu 2017)

The root–shoot ratio of inoculated AMF plants is bigger in comparison with non-AMF plants, which has an advantage in nutrient acquisition. And, AMF can change root architecture of host plants to resist the tolerance of drought (White 1992). Greater root morphology under drought stress is the critical role in enhancing water absorption of mycorrhizal plants, relative to non-mycorrhizal plants (Zou et al. 2017).

18.2.4 Regulating Phytohormone Levels

Endogenous hormones are the vital importance for plant growth. Auxin, cytokinin (CTK), gibberellic acid (GA), ethylene (ET), and abscisic acid (ABA) can regulate the growth and development of plants, control plant morphology and physiological metabolism, and also stimulate mutual recognition and mycorrhizal formation between mycorrhizal fungi and plants (Yu et al. 2009). The effects of *Gigaspora rosea*, *Glomus mosseae*, and *Glomus versghrme* on endogenous hormones in maize and cotton plants were studied by Liu et al. (1999) under the pot conditions of greenhouse. They found that AMF could significantly increase the contents of zeatin, auxin, and GA and decrease content of ABA under well-watered and drought conditions. A significantly higher putrescine (Put) and spermidine (Spd) level was found in *Citrus tangerina* seedlings inoculated with *F. mosseae* (Wu et al. 2012). As a result, with the increase of endogenous hormone levels of mycorrhizal plants, plant biomass and growth vigor were also significantly increased. Dugassa et al. (1996) found that AMF increased the contents of auxin, GA, ethylene, CTK, and ABA in stems and leaves. Moreover, Barea and Azcón-Aguilar (1982) had proven that hyphae of AMF could produce auxin, CTK, and GA. These phytohormones may be the initiating factor of plant growth and development and stress resistance gene expression, which can regulate gene expression and protein synthesis (Yu et al. 2009). It demonstrates that mycorrhizal symbiosis can induce and modulate the phytohormone production to stimulate growth and development of host plants, as well AMF.

In addition, AMF-improved plant growth can be regulated by exogenous phytohormones. In trifoliate orange seedlings inoculated with *Glomus versiforme*, exogenous Put, Spd, and spermine (Spm) were applied into rhizosphere (Wu et al. 2010b). The results showed that Put application, but not Spd and Spm, heavily stimulated root mycorrhizal colonization and numbers of entry points, arbuscules, and vesicles, which further magnified AMF-improved plant growth and root morphology. In another study conducted by Liu et al. (2016), trifoliate orange seedlings were grown in a two-chambered root box separated by 37 μm mesh, where trifoliate orange plants were planted in root+hyphae chamber, and indole butyric acid (IBA), ABA, and JA (each at 0.1 μM concentration) were applied into hyphae chamber. The study showed that exogenous phytohormones, especially IBA, magnified the mycorrhiza-stimulated growth responses. In a word, exogenous phytohormones can stimulate greater mycorrhizal growth of host plants, thereby, further magnifying the AMF-improved growth responses.

18.2.5 Regulating Soil Physicochemical Properties

AMF can improve soil structure and contribute to maintain soil fertility, which indirectly affects plant growth (Jeffries et al. 2003; Wu et al. 2014). Glomalin, secreted by AMF, can glue small soil particles into a diameter of >0.25 mm macroaggregates, further forming the large polymers (Lovelock et al. 2004). The formation of soil aggregate is good at improving soil physical condition and increasing soil stability (Bever et al. 2001; Chaudhary et al. 2009). Long-term fields monitoring experiments showed that the soil hyphal density was positively correlated with soil aggregates and carbon–nitrogen fixation (Wilson et al. 2009; Peng et al. 2012). A potted experiment showed that AMF inoculation decreased the loss of P and NH_4^+ in the soil by 6.0% and 7.5%, respectively (Van Der Heijden 2010), which is the critical factor in maintaining greater soil nutrient status, beneficial to growth of host plants. What's more, glomalin can glue and chelate soil toxic substances to ameliorate soil toxic environments and serve as a carbon source to increase plant biomass (Rillig et al. 2002). As a result, mycorrhizal soils generally possess better soil structure and permeability and thus provide a lot of oxygen for respiration and further enhance carbon accumulation in soil. Therefore, this is no doubt that AMF has a significant effect on improving soil structure, thereby, promoting growth of host plants.

18.2.6 Greater Plant Growth Derived from Osmotic Regulation and Early Warning Under Abiotic and Biotic Stresses

Osmotic solute changes such as soluble sugar, amino acids, and glycine betaines directly affect the absorption ratio to mineral nutrients in plants (Duke et al. 1986; Feng et al. 2002; Sharifi et al. 2007). The growth of AMF consumes host carbohydrates, leading to less accumulation of low-molecular organic substances in root cells. As a result, the intracellular osmotic potential is increased, resulting in the enhanced ability of plants to fight against osmotic stress, further stimulating plant growth (Ruiz-Lozano and Azcón 1995). Mycorrhizal plants can absorb more P, Cu, and Mg and reduce the absorption of Na and Cl, thereby alleviating toxic effects on plant growth (Evelin et al. 2009; Wu et al. 2010a). Sannazzaro et al. (2006) also confirmed that inoculation with *Glomus intraradices* significantly increased the ratio of K^+/Na^+ under salt stress in leguminous (*Lotus glaber*). In addition, the ability of different AMF has different resistance to salt stress. *Glomus intraradices* isolated from plants with higher salt tolerance was more effective to alleviate salt damage in the salt-tolerant plants than in non-salt-tolerant plants, which may be due to the long-term adaptability (Estrada et al. 2013). In addition, mycorrhizal plants could induce the roots releasing more H^+ into mycorrhizosphere, as observed higher root H^+ efflux rates in *F. mosseae*-colonized trifoliate orange versus in non-AMF plants under soil salinity (Wu et al. 2013). The acidic rhizosphere caused by mycorrhiza is important to secondary active transporter of organic and inorganic nutrients, turgor regulation, and in the regulation of cell wall plasticity, as suggested in "acid-growth theory" (Wu and Zou 2013).

AMF can activate defense reactions of host plants and increase defensive enzyme activities to protect host plants escaping pathogens like viruses and bacteria, which is beneficial to plant growth under pathogens conditions. Earlier studies had shown that higher chitinase activities in the roots of mycorrhizal plants limited the growth and development of root pathogens (Gianinazzi-Pearson et al. 1996; Dumas-Gaudot et al. 1996; Elsharkawy et al. 2012). AMF reduces the harm of nematodes by altering plant root exudates, and roots secrete abundant substances to regulate growth of AM symbiosis (Buwalda et al. 1984). Inoculation with AMF can increase the production of secondary metabolites such as jasmonic acid and the metabolism of carbon and nitrogen, thereby enhancing plant resistance to fungal diseases (Li et al. 2013). In addition, mycorrhizal hyphal networks can communicate the signals of intruders like *Acyrthosiphon pisum* among different plants for early warning (Babikova et al. 2013).

18.3 No or Depressed Effects of Mycorrhiza on Plant Growth

Besides growth promotion under mycorrhization, we also occasionally found no positive effect or inhibited effect on plant growth (Fig. 18.4). Possibly, the negative effect of mycorrhiza on plant growth is not reported by researchers. In general, the negative effect of mycorrhizas occurs in high phosphorus conditions, especially

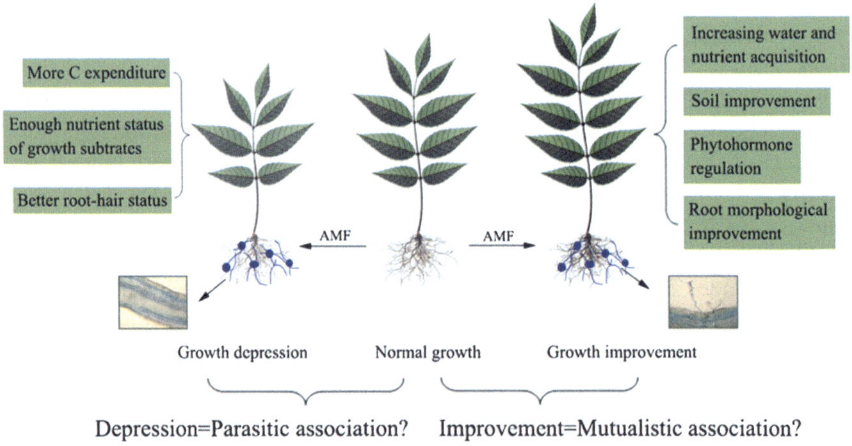

Fig. 18.4 The schematic diagram of mycorrhizal effects on growth of host plants

when photosynthate or light level is limited, such as young plants (Reynolds et al. 2005). Earlier study by Bethlenfalvay et al. (1983) showed that, in soybean grown in 0, 25, 50, 100, or 200 mg hydroxyapatite [HAP, $Ca_{10}(PO_4)_6(OH)_2$] per pot, *Glomus fasciculatum*-colonized plants showed 20, 25, and 38% growth retardation under 0, 100, and 200 mg HAP, relative to non-colonized controls. At 50 mg HAP, growth of mycorrhizal plant was significantly enhanced. The study of Peng et al. (1993) showed that lemon growth under mycorrhization was inhibited under high phosphorus supply. The growth of *C. aurantium* was inhibited by 18% with *Glomus intraradices* inoculation at high CO_2 concentrations, while at normal CO_2 concentration, the growth of *C. aurantium* was promoted by 15% under mycorrhization (Jifon et al. 2002). Studies of *Citrus tangerina* seedlings inoculated with AMF showed that *Glomus versiforme* inhibited the plant height, stem diameter, dry weight of shoots and roots, and dry weight of plants.

It is well known that 3–20% of plant photosynthates can be expended for mycorrhizal symbiosis. Hence, the improvement and inhibition of plant growth under mycorrhization is due to the competition between host plant and AM fungus for the carbon source (Buwalda and Goh 1982). Mycorrhizal plants need to consume more C sources and accumulate fatty acids in the roots, thereby increasing the root or rhizosphere respiration and reducing the contents of soluble starch in the roots. In the early stage of AMF infection, plant defense systems are activated, and fungal symbiotic consumes part of the energy. Hence, AMF and plant roots have a C competition in the process of pre-symbiosis. As reported by Buwalda and Goh (1982), total oxidizable C, soluble sugar content, and C/N ratio were lower in *Gigaspora margarita*-colonized perennial ryegrass plants, indicating a competition of mycorrhizas with the host for photosynthetically derived C, finally causing growth depression. We conclude that (1) if the expenditure of AM fungus in carbohydrates does not affect the request of host plant in carbohydrates, the establishment of AM symbiosis is mutualistic association, which can stimulate growth performance of host plants and (2) if the C expenditure of AM fungus affects the normal C request of host plants, AM presence may be a parasitic association.

In addition to the C competition between host plants and AM fungus, nutrient status of growth substrates heavily regulates the shift of mutualistic association and parasitic association. As reported by Peng et al. (1993), growth of lemon under mycorrhization was inhibited under high P supply. Similarly, at high CO_2 concentrations, growth of *C. aurantium* was depressed by inoculation with *Glomus intraradices* (Jifon et al. 2002). Possibly, the host plant roots can absorb enough P from growth substrates and do not request AMF functioning on nutrient acquisition from soils. We guess that (1) if growth substrates have adequate soil fertility to fulfil the request of roots, AM functioning will be weak in nutrient acquisition, resulting in no or depressed effects on growth of host plants and (2) if growth substrates have deficient soil fertility and do not fulfil the request of roots, AM functioning will be strengthened in nutrient acquisition, resulting in growth improvement of host plants.

Mycorrhizal hyphae and root hairs of host plants collectively absorb soil nutrients at the root surface (Wu et al. 2016a). In a general rule, the plant species with abundant root hairs are less dependent on mycorrhizal symbiosis for nutrient acquisition (Itoh and Barber 1983). As proposed by Baylis (1975), root hair length and abundance may indicate the degree of mycorrhizal dependence and mycorrhizal responses. Plants with few root hairs are strongly mycorrhiza-dependent, while those with a huge number of root hairs are less dependent on mycorrhizal symbioses (Novero et al. 2008). Typically non-mycorrhizal plants, including rushes, sedges, and grasses, have highly developed root hairs (Hetrick et al. 1988). It concludes that plants with high mycorrhizal dependence and less root hairs have strongly positive responses to mycorrhization, and plants with low mycorrhizal dependence and abundant root hairs have weak responses to mycorrhization.

18.4 Conclusion and Future Prospects

In general, AMF has improved effects on the growth of host plants, which is related to the promotion of water and nutrition absorption, improvement of endogenous hormone levels, enhancement of stressed tolerance, improvement of soil physico-chemical properties, and root morphological modification (Fig. 18.4). Occasionally, AMF represents no or depressed effects on plant growth, which is associated with C expenditure, nutrient status of growth substrates, and root hair status. Essentially, AMF effects on plant growth that is involved in mutualistic or parasitic association (Fig. 18.4). Future prospects in several fields are needed to keep a watchful eye:

1. The relevant mechanisms of mycorrhiza-improved plant growth at the molecular level need to be studied.
2. The critical value of AMF-promoted/inhibited plant growth about the soil fertility level should be clear and definite.
3. Transition between mutualistic and parasitic association will be paid attention to mycorrhizal works.
4. Expect the emergence of negative reports in the literature, and also keep a watchful eye on the underlying mechanisms of AMF-induced negative effects on plant growth.

Acknowledgments This study was supported by the Plan in Scientific and Technological Innovation Team of Outstanding Young Scientist, Hubei Provincial Department of Education (T201604).

References

Allen MF (2006) Water dynamics of mycorrhizas in arid soils. In: Gadd GM (ed) Fungi biogeochemical cycle. Cambridge University Press, Cambridge, pp 74–97

Babikova Z, Gilbert L, Bruce TJ, Birkett M, Caulfield JC, Woodcock C, Pickett JA, Johnson D (2013) Underground signals carried through common mycelial networks warn neighbouring plants of aphid attack. Ecol Lett 16:835–843

Barea JM, Azcón-Aguilar C (1982) Production of plant growth-regulating substances by the vesicular-arbuscular mycorrhizal fungus *Glomus mosseae*. Appl Environ Microbiol 43:810–813

Baylis GTS (1975) The magnolioid mycorrhiza and mycotrophy in root systems derived from it. In: Sanders FE, Mosse B, Tinker PB (eds) Endomycorrhizas. Academic, London, pp 373–389

Bethlenfalvay GJ, Bayne HG, Pacovsky RS (1983) Parasitic and mutualistic associations between a mycorrhizal fungus and soybean: The effect of phosphorus on host plant-endophyte interactions. Physiol Plant 57:543–548

Bever JD, Schultz PA, Pringle A, Morton JB (2001) Arbuscular mycorrhizal fungi: more diverse than meets the eye, and the ecological tale of why: the high diversity of ecologically distinct species of arbuscular mycorrhizal fungi within a single community has broad implications for plant ecology. AIBS Bull 51:923–931

Bolgiano NC, Safir GR, Warncke DD (1983) Mycorrhizal infection and growth of onion in the field in relation to phosphorus and water availability. J Am Soc Hortic Sci 108:819–825

Buwalda JG, Goh KM (1982) Host-fungus competition for carbon as a cause of growth depressions in vesicular-arbuscular mycorrhizal ryegrass. Soil Biol Biochem 14:103–106

Buwalda JG, Stribley DP, Tinker PB (1984) The development of endomycorrhizal root systems V. The detailed pattern of development of infection and the control of Infection level by host in young leek plants. New Phytol 96:411–427

Chaudhary VB, Bowker MA, O'Dell TE, Grace JB, Redman AE, Rillig MC, Johnson NC (2009) Untangling the biological contributions to soil stability in semiarid shrublands. Ecol Appl 19:110–122

Deshmukh S, Hückelhoven R, Schäfer P, Imani J, Sharma M, Weiss M, Waller F, Kogel KH (2006) The root endophytic fungus *Piriformospora indica* requires host cell death for proliferation during mutualistic symbiosis with barley. Proc Natl Acad Sci USA 103:18450–18457

Dugassa GD, Von Alten H, Schonbeck F (1996) Effect of arbuscular mycorrhiza (AM) on health of *Linum usitatissimum* L. infected by fungal pathogens. Plant Soil 185:173–182

Duke ER, Johnson CR, Koch KE (1986) Accumulation of phosphorus, dry matter and betaine during NaCl stress of split-root citrus seedlings colonized with vesicular-arbuscular mycorrhizal fungi on zero, one or two halves. New Phytol 104:583–590

Dumas-Gaudot E, Slezack S, Dassi B, Pozo MJ, Gianinazzi-Pearson V, Gianinazzi S (1996) Plant hydrolytic enzymes (chitinases and β-1, 3-glucanases) in root reactions to pathogenic and symbiotic microorganisms. Plant Soil 185:211–221

Effendy M, Wijayani BW (2008) Study of the external hyphae of AMF in understanding the function to contribution of p sorption by plants using the thin section method. J Tanah Tropika 13:241–252

Elsharkawy MM, Shimizu M, Takahashi H, Hyakumachi M (2012) The plant growth-promoting fungus *Fusarium equiseti* and the arbuscular mycorrhizal fungus *Glomus mosseae* induce systemic resistance against cucumber mosaic virus in cucumber plants. Plant Soil 361:397–409

Estrada B, Barea JM, Aroca R, Ruiz-Lozano JM (2013) A native *Glomus intraradices* strain from a Mediterranean saline area exhibits salt tolerance and enhanced symbiotic efficiency with maize plants under salt stress conditions. Plant Soil 366:333–349

Evelin H, Kapoor R, Giri B (2009) Arbuscular mycorrhizal fungi in alleviation of salt stress: a review. Ann Bot 104:1263–1280

Feng G, Zhang F, Li X, Tian C, Tang C, Rengel Z (2002) Improved tolerance of maize plants to salt stress by arbuscular mycorrhiza is related to higher accumulation of soluble sugars in roots. Mycorrhiza 12:185–190

Gianinazzi-Pearson V, Dumas-Gaudot E, Gollotte A, Alaoui AT, Gianinazzi S (1996) Cellular and molecular defence-related root responses to invasion by arbuscular mycorrhizal fungi. New Phytol 133:45–57

Graham JH, Timmer LW (1985) Rock phosphate as a source of phosphorus for vesicular-arbuscular mycorrhizal development and growth of citrus in a soilless medium. J Am Soc Hortic Sci 110:489–492

Hart MM, Reader RJ (2005) The role of the external mycelium in early colonization for three arbuscular mycorrhizal fungal species with different colonization strategies. Pedobiologia 49:269–279

Hetrick BAD, Kitt DG, Wilson GT (1988) Mycorrhizal dependence and growth habit of warm-season and cool-season tallgrass prairie plants. Can J Bot 66:1376–1380
Itoh S, Barber SA (1983) Phosphorus uptake by six plant species as related to root hairs. Agron J 75:457–461
Jeffries P, Gianinazzi S, Perotto S, Turnau K, Barea JM (2003) The contribution of arbuscular mycorrhizal fungi in sustainable maintenance of plant health and soil fertility. Biol Fertil Soils 37:1–16
Jifon JL, Graham JH, Drouillard DL, Syvertsen JP (2002) Growth depression of mycorrhizal Citrus seedlings grown at high phosphorus supply is mitigated by elevated CO_2. New Phytol 153:133–142
Kiers ET, Duhamel M, Beesetty Y, Mensah JA, Franken O, Verbruggen E, Fellbaum CR, Kowalchuk GA, Hart MM, Bago A, Palmer TM, West SA, Vandenkoornhuyse P, Jansa J, Bücking H (2011) Reciprocal rewards stabilize cooperation in the mycorrhizal symbiosis. Science 333:880–882
Koch KE, Johnson CR (1984) Photosynthate partitioning in split-root citrus seedlings with mycorrhizal root systems. Plant Physiol 75:26–30
Levy Y, Krikun J (1980) Effect of vesicular-arbuscular mycorrhiza on *Citrus jambhiri* water relations. New Phytol 85:25–31
Li XL, George E, Marschner H (1991) Extension of the phosphorus depletion zone in VA–mycorrhizal white clover in a calcareous soil. Plant Soil 136:41–48
Li H, Ye ZH, Chan WF, Chen XW, Wu FY, Wu SC, Wong MH (2011) Can arbuscular mycorrhizal fungi improve grain yield, as uptake and tolerance of rice grown under aerobic conditions? Environ Pollut 159:2537–2545
Li YJ, Liu ZL, Hou HY, Lei H, Zhu XC, Li XH, He XY, Tian CJ (2013) Arbuscular mycorrhizal fungi-enhanced resistance against *Phytophthora sojae* infection on soybean leaves is mediated by a network involving hydrogen peroxide, jasmonic acid, and the metabolism of carbon and nitrogen. Acta Physiol Plant 35:3465–3475
Liu RJ (1989) Effects of vesicular-arbuscular mycorrhizas and phosphorus on water status and growth of apple. J Plant Nutr 12:997–1017
Liu CY, Wu QS (2017) Responses of plant growth, root morphology, chlorophyll and indoleacetic acid to phosphorus stress in trifoliate orange. Biotechnology 16:40–44
Liu RJ, Li M, Meng XX, Liu X, Li XL (1999) Effects of AM fungi on endogenous hormones in corn and cotton plants. Mycosystema 19:91–96
Liu CY, Srivastava AK, Zhang DJ, Zou YN, Wu QS (2016) Exogenous phytohormones and mycorrhizas modulate root hair configuration of trifoliate orange. Not Bot Horti Agrobo 44:548–556
Lovelock CE, Wright SF, Clark DA, Ruess RW (2004) Soil stocks of glomalin produced by arbuscular mycorrhizal fungi across a tropical rain forest landscape. J Ecol 92:278–287
Lü LH, Wu QS (2017) Mycorrhizas promote plant growth, root morphology and chlorophyll production in white clover. Biotechnology 16:34–39
Marschner H, Dell B (1994) Nutrient uptake in mycorrhizal symbiosis. Plant Soil 159:89–102
Marulanda A, Azcon R, Ruiz-Lozano AM (2003) Contribution of six arbuscular mycorrhizal fungi isolates to water uptake by *Lactuca sativa* L. plants under drought stress. Physiol Plant 119:526–533
Mohammad A, Mitra B, Khan AG (2004) Effects of sheared-root inoculum of *Glomus intraradices* on wheat grown at different phosphorus levels in the field. Agric Ecosyst Environ 103:245–249
Morte A, Lovisolo C, Schubert A (2000) Effects of drought stress on growth and water relation of the mycorrhizal association *Helianthemum almeriense-Terfezia claveryi*. Mycorrhiza 10:115–119
Nelsen CE, Safir GR (1982) The water relations of well-watered, mycorrhizal and non-mycorrhizal onion plants. J Am Soc Hortic Sci 107:71–74
Novero M, Genre A, Szczyglowski K, Bonfante P (2008) Root hair colonization by mycorrhizal fungi. In: Emons AMC, Ketelaar T (eds) Root hairs. Springer, Berlin, pp 315–338

Peng S, Eissenstat DM, Graham JH, Williams K, Hodge NC (1993) Growth depression in mycorrhizal citrus at high-phosphorus supply (analysis of carbon costs). Plant Physiol 101:1063–1071
Peng SL, Shen H, Zhang YT, Guo T (2012) Compare different effect of arbuscular mycorrhizal colonization on soil structure. Acta Ecol Sin 32:863–870
Purin S, Rillig MC (2007) Parasitism of arbuscular mycorrhizal fungi: reviewing the evidence. FEMS Microbiol Lett 279:8–14
Reynolds HL, Hartley AE, Vogelsang KM, Bever JD, Schultz PA (2005) Arbuscular mycorrhizal fungi do not enhance nitrogen acquisition and growth of old-field perennials under low nitrogen supply in glasshouse culture. New Phytol 167:860–880
Rillig MC, Wright SF, Eviner VT (2002) The role of arbuscular mycorrhizal fungi and glomalin in soil aggregation: comparing effects of five plant species. Plant Soil 238:325–333
Ruiz-Lozano JM, Azcón R (1995) Hyphal contribution to water uptake in mycorrhizal plants as affected by the fungal species and water status. Physiol Plant 95:472–478
Sannazzaro AI, Ruiz OA, Albertó EO, Menéndez AB (2006) Alleviation of salt stress in Lotus glaber by *Glomus intraradices*. Plant Soil 285:279–287
Schellenbaum L, Berta G, Ravolanirina F, Tisserant B, Gianinazzi S, Fitter AH (1991) Influence of endomycorrhizal infection on root morphology in a micropropagated woody plant species (*Vitis vinifera* L.). Ann Bot 68:135–141
Sharifi M, Ghorbanli M, Ebrahimzadeh H (2007) Improved growth of salinity-stressed soybean after inoculation with salt pre-treated mycorrhizal fungi. J Plant Physiol 164:1144–1156
Simard SW, Beiler KJ, Bingham MA, Deslippe JR, Philip LJ, Teste FP (2012) Mycorrhizal networks: mechanisms, ecology and modelling. Fungal Biol Rev 26:39–60
Simms EL, Taylor DL (2002) Partner choice in nitrogen-fixation mutualisms of legumes and rhizobia. Integr Comp Biol 42:369–380
Subrammanian KS, Charest C (1999) Acquisition of N by external hyphae of arbuscular mycorrhizal fungus and its impact on physiological responses in maize under drought-stressed and well-watered condition. Mycorrhiza 9:69–75
Tanaka Y, Yano K (2005) Nitrogen delivery to maize via mycorrhizal hyphae depends on the form of N supplied. Plant Cell Environ 28:1247–1254
Van Der Heijden MG (2010) Mycorrhizal fungi reduce nutrient loss from model grassland ecosystems. Ecology 91:1163–1171
Van Der Heijden MG, Bardgett RD, Van Straalen NM (2008) The unseen majority: soil microbes as drivers of plant diversity and productivity in terrestrial ecosystems. Ecol Lett 11:296–310
White RH (1992) Acremonium endophyte effects on tall fescue drought tolerance. Crop Sci 32:1392–1396
Wilson GW, Rice CW, Rillig MC, Springer A, Hartnett DC (2009) Soil aggregation and carbon sequestration are tightly correlated with the abundance of arbuscular mycorrhizal fungi: results from long-term field experiments. Ecol Lett 12:452–461
Wu T, Tan ZY (2005) Vesicular arbuscular mycorrhiza and its function on phosphorus in soil. Hunan Agric Sci 2:41–43. (in Chinese with English abstract)
Wu QS, Zou YN (2009) Mycorrhizal influence on nutrient uptake of citrus exposed to drought stress. Philipp Agric Sci 92:33–38
Wu QS, Zou YN (2013) Mycorrhizal symbiosis alters root H^+ effluxes and root system architecture of trifoliate orange seedlings under salt stress. J Anim Plant Sci 23:143–148
Wu QS, Zou YN, Xia RX, Wang MY (2007) Five *Glomus* species affect water relations of *Citrus tangerine* during drought stress. Bot Stud 48:147–154
Wu QS, Levy Y, Zou YN (2009) Arbuscular mycorrhizae and water relations in citrus. In: Tennant P, Benkeblia N (eds), Citrus II. Tree and forestry science and biotechnology (Special Issue 1). Global Science Press, USA, pp 105–112
Wu QS, Zou YN, He XH (2010a) Contributions of arbuscular mycorrhizal fungi to growth, photosynthesis, root morphology and ionic balance of citrus seedlings under salt stress. Acta Physiol Plant 32:297–304

Wu QS, Zou YN, He XH (2010b) Exogenous putrescine, not spermine or spermidine, enhances root mycorrhizal development and plant growth of trifoliate orange (*Poncirus trifoliata*) seedlings. Int J Agric Biol 12:576–580

Wu QS, Zou YN, He XH, Luo P (2011) Arbuscular mycorrhizal fungi can alter some root characters and physiological status in trifoliate orange (*Poncirus trifoliata* L. Raf.) seedlings. Plant Growth Regul 65:273–278

Wu QS, He XH, Zou YN, Liu CY, Xiao J, Li Y (2012) Arbuscular mycorrhizas alter root system architecture of *Citrus tangerine* through regulating metabolism of endogenous polyamines. Plant Growth Regul 68:27–35

Wu QS, Srivastava AK, Zou YN (2013) AMF–induced tolerance to drought stress in citrus: a review. Sci Hortic 164:77–87

Wu QS, Yuan FY, Fei YJ, Li L, Huang YM (2014) Effects of arbuscular mycorrhizal fungi on aggregate stability, GRSP, and carbohydrates of white clover. Acta Pratac Sin 23:269–275. (in Chinese with English abstract)

Wu QS, Liu CY, Zhang DJ, Zou YN, He XH, Wu QH (2016a) Mycorrhiza alters the profile of root hairs in trifoliate orange. Mycorrhiza 26:237–247

Wu QS, Wang S, Srivastava AK (2016b) Mycorrhizal hyphal disruption induces changes in plant growth, glomalin-related soil protein and soil aggregation of trifoliate orange in a core system. Soil Tillage Res 160:82–91

Yao Q, Wang LR, Zhu HH, Chen JZ (2009) Effect of arbuscular mycorrhizal fungal inoculation on root system architecture of trifoliate orange (*Poncirus trifoliata* L. Raf.) seedlings. Sci Hortic 121:458–461

Yu JX, Li M, Liu RJ (2009) Advances in the study of interactions between mycorrhizal fungi and plant hormones. J Qingdao Agric Univ 26:4–7. (in Chinese with English abstract)

Zhang FS, Shen JB, Zhang JL, Zuo YM, Li L, Chen XP (2010) Rhizosphere processes and management for improving nutrient use efficiency and crop productivity: implications for China. Adv Agron 107:1–32

Zhang X, Wang L, Ma F, Zhang SJ, Xu YN, Li Z, Fu SJ (2012a) Effects of nitrogen and biological fertilizer coupling on rice resource utilization. J Harbin Inst Technol 44:39–42. (in Chinese with English abstract)

Zhang YT, Zhu M, Xian Y, Shen H, Zhao J, Guo T (2012b) Influence of mycorrhizal inoculation on competition between plant species and inorganic phosphate forms. Acta Ecol Sin 32:7091–7101. (in Chinese with English abstract)

Zou YN, Wang P, Liu CY, Ni QD, Zhang DJ, Wu QS (2017) Mycorrhizal trifoliate orange has greater root adaptation of morphology and phytohormones in response to drought stress. Sci Rep 7:41134

Chapter 19
Arbuscular Mycorrhizal Fungi Symbiosis and Conservation of Endangered Tropical Legume Trees

Husna Faad, Faisal Danu Tuheteru, and Asrianti Arif

19.1 Introduction

Indonesia is one of the countries in the world with tropical forests of high value. Biodiversity stretches in Indonesia range from mangrove forests along the coast; lowland tropical forests; mountain forests on the plains of Sumatra, Sulawesi, and Borneo; and subalpine and alpine vegetation in Papua. However, various human activities, such as illegal logging, the practice of shifting cultivation, forest fires, and conversion of forests for mining and large-scale plantations of rubber and oil palm, cause deforestation of tropical forests in Indonesia (Food and Agriculture Organization 2016; Tsujino et al. 2016). In addition, these activities can also impact on the decline of biodiversity such as the threat of tropical tree species (Ghazoul and Sheil 2010).

Reforestation of tropical forests of Indonesia with local and endangered species needs to be tested. The adaptation and success of seedlings in the field are a vital component of reforestation and conservation programs. Therefore, improvement of the quality of forest plant seedlings with beneficial soil microbes such as mycorrhizal fungi in the nursery is an important step in restoration of ecosystems (Urgiles et al. 2009). AMF is reported to increase the growth of various tropical trees at the nursery scale (Tawaraya and Turjaman 2014). AMF application studies in the tropics are generally carried out in controlled conditions (greenhouse, nursery), and still there are few reports about the effects of AMF inoculation on the growth of tropical trees on field conditions (Tawaraya and Turjaman 2014; Husna et al. 2017b). The potential of AMF can be developed for restoration and conservation programs (Solaiman and Mickan 2014; Shah 2014; Wang 2017). According to Jasper (1994) AMF can contribute significantly to the success of revegetation through several

H. Faad (✉) · F. D. Tuheteru (✉) · A. Arif
Faculty of Forestry and Environmental Science, Department of Forestry, Halu Oleo University, Kendari, Southeast Sulawesi, Indonesia

B. Giri et al. (eds.), *Root Biology*, Soil Biology 52,
https://doi.org/10.1007/978-3-319-75910-4_19

mechanisms: (1) to increase growth and survival of plants, through nutrient and water enhancement; (2) to maintain species biodiversity in the ecosystem, by contacting external hyphae from tree to tree; (3) contributing on the efficiency of nutrient recycling and long-term stability; and (4) soil stability.

Zubek et al. (2009) reported that AMF is effective in the maintenance and cultivation of endangered species. In addition, the AMF also significantly accelerate the succession and the success of living species in conservation and rehabilitation programs (Fuchs and Haselwandter 2004, 2008; Panwar and Tarafdar 2006; Zubek et al. 2009; Bothe et al. 2010; Husna et al. 2017b). In Indonesia, the AMF application for the conservation of endangered species has been reported in *Aquilaria malaccensis* and *A. crassna* (Turjaman et al. 2006a), *A. filaria* (Turjaman et al. 2006b), and *Gonystylus bancanus* (Muin 2003). In general, AMF inoculation accelerates threatened species growth and can support species conservation programs in Indonesia.

This paper reviews the results of research and publications related to AMF types that form symbiosis with various plants and various environmental conditions and AMF applications on tree species in Indonesia. Aside from that, AMF inoculation on three endangered legume species, namely, *P. mooniana* (Thw.), *P. indicus* (Willd), and *K. celebica* (Kosterm), has also been revealed in this chapter.

19.2 AMF Diversity in Indonesia

A total of 72 types of AMF have symbiosis association with various types of plants in various environmental conditions in Indonesia. The 72 species were classified into 4 orders, 16 genera, and 8 families (Table 19.1). *Glomeraceae* included a dominant family with 36 species or 53% of the total followed by *Acaulosporaceae* (13 species). Some types of AMF, which have wide dissemination, include *Glomus aggregatum*, *G. fuegianum*, *Funneliformis geosporum*, *Sclerocystis rubiformis*, *Claroideoglomus etunicatum*, *Acaulospora foveata*, *A. tuberculata*, and *A. scrobiculata*. Until now, AMF identification in Indonesia is generally based on spore morphology. The identification of AMF types with molecular DNA approach is still limited with only one research conducted by Arofatullah (2015).

The stated types of AMF are scattered in various habitats such as farmland (Kramadibrata et al. 1995; Haerida and Kramadibrata 2002), fruit land (Silviana et al. 1999; Muliawan et al. 2002; Lucia 2005), forest land (Suciatmih and Kramadibrata 2002; Sabaruddin 2004; Husna et al. 2014), grassland (Widiastuti and Kramadibrata 1992), peatland (Tawaraya et al. 2003; Pangaribuan 2014), and forest conservation (National Park, Nature conserve, Botanical garden) (Kramadibrata 1993, 2012, 2013; Aradea 2004; Husna et al. 2014). AMF is also reported in some forms of land including mountains (Kramadibrata 2013), islands (Kramadibrata 2011, 2016), watershed (Kramadibrata 1993), hills (Arif et al. 1999), and beaches (Delvian 2003). Some researchers report that AMF is symbiotic with plants in degraded or polluted land and forest ecosystems such as those contaminated

Table 19.1 List of arbuscular mycorrhizal fungi found in Indonesia

No.	Type of AMF	References
Archaeosporales		
Ambisporaceae/Ambispora		
1	*A. appendicula*	Husna et al. (2014)
2	*A. leptoticha*	Prasetyo et al. (2010)
3	*A. cf. fecundisporum*	Destifani (2013)
Diversisporales		
Acaulosporaceae/Acaulospora		
4	*A. bireticulata*	Silviana et al. (1999), Fahriny (2013), Destifani (2013)
5	*A. delicata* Walker, Pfeiffer, and Bloss	Widiastuti and Kramadibrata (1992), Kramadibrata et al. (1995), Husna et al. (2014)
6	*A. foveata* Trappe and Janos	Widiastuti and Kramadibrata (1992), Kramadibrata (1993, 2009, 2011, 2012, 2013, 2016), Kramadibrata et al. (1995), Setya et al. (1995), Suciatmih and Kramadibrata (2002), Chairani et al. (2002), Aradea (2004), Lucia (2005), Puspitasari (2005), Kramadibrata et al. (2007), Proborini et al. (2013)
7	*A. longula*	Chairani et al. (2002), Ningsih et al. (2013), Fahriny (2013), Puspitasari (2005)
8	*A. mellea* Spain and Schenck	Widiastuti and Kramadibrata (1992, 1993), Muliawan et al. (2002), Lucia (2005), Arofatullah (2015)
9	*A. morrowiae* Spain and Schenck	Suciatmih and Kramadibrata (2002)
10	*A. rehmii* Sieverd. & S. Toro	Kramadibrata et al. (1995), Haerida dan Kramadibrata (2002), Aradea (2004), Lucia (2005), Kramadibrata (2009), Setya et al. (1995)
11	*A. rugose*	Sabaruddin (2004)
12	*A. scrobiculata* Trappe	Widiastuti dan Kramadibrata (1992, 1993), Setya et al. (1995), Kramadibrata (1993, 2009, 2011, 2012, 2013, 2016), Kramadibrata et al. (1995, 2007), Silviana et al. (1999), Wulandari (2001), Haerida and Kramadibrata (2002), Muliawan et al. (2002), Chairani et al. (2002), Aradea (2004), Lucia (2005), Puspitasari (2005), Prasetyo et al. (2010), Proborini et al. (2013), Ningsih et al. (2013), Fahriny (2013), Destifani (2013), Husna et al. (2014), Kumalawati et al. (2014)
13	*A. spinosa*	Muliawan et al. (2002), Lucia (2005), Arofatullah (2015), Setya et al. (1995)
14	*A. tuberculata*	Kramadibrata (1993, 2009, 2011, 2012, 2013, 2016), Widiastuti and Kramadibrata (1993), Kramadibrata et al. (1995, 2007), Silviana et al. (1999), Arif et al. (1999), Boddington and Dodd (2000), Chairani et al. (2002), Aradea (2004), Lucia (2005), Prastyo (2004), Puspitasari

(continued)

Table 19.1 (continued)

No.	Type of AMF	References
		(2005), Kramadibrata and Gunawan (2006), Wulandari (2001), Ningsih et al. (2013), Proborini et al. (2013), Fahriny (2013), Ginting (2013), Kumalawati et al. (2014)
15	*A. undulata*	Proborini et al. (2013), Puspitasari (2005)
16	*A. walkeri* Kramadibrata and Hedger	Widiastuti and Kramadibrata (1992), Kramadibrata (1993, 2009)
Diversisporaceae/Diversisporales		
17	*D. versiforme*	Muliawan et al. (2002), Chairani et al. (2002), Husna et al. (2014)
Redeckera		
18	*R. canadense*	Husna et al. (2014)
Acaulosporaceae/Entrophospora		
19	*E. infrequens*	Kramadibrata (1993), Proborini et al. (2013)
Gigasporaceae		
Cetraspora		
20	*C. pellucida* (Nicol. and Schenck) Walker and Sanders	Kramadibrata et al. (1995), Lucia (2005), Kramadibrata (2009)
Dentiscutata		
21	*D. heterogama* (Nicol. and Gerd.) Walker and Sanders	Widiastuti and Kramadibrata (1992), Kramadibrata et al. (1995), Boddington and Dodd (2000), Puspitasari (2005), Kramadibrata and Gunawan (2006), Proborini et al. (2013), Kramadibrata (2016)
Gigaspora		
22	*Gi. albida*	Sabaruddin (2004), Lucia (2005), Wijayanti (2006), Proborini et al. (2013)
23	*Gi. cf. decipiens*	Wulandari (2001)
24	*Gi. gigantea* (Nicol. and Gerd.) Gardemann and Trappe	Widiastuti and Kramadibrata (1992), Kramadibrata et al. (1995), Setya et al. (1995), Silviana et al. (1999), Muliawan et al.(2002), Kramadibrata (2009, 2011, 2013, 2016), Puspitasari (2005)
25	*Gi. margarita*	Proborini et al. (2013)
26	*Gi. ramisporophora*	Lucia (2005), Puspitasari (2005)
Racocetra		
27	*R. gregaria* (N. C. Schenck and T. H. Nicolson) Oehl, F. A. Souza and Sieverd	Husna et al. (2014)
28	*R. fulgida* Koske and Walker	Kramadibrata (2009)
Scutellospora		
29	*Sc. auriglobosa* (I. R. Hall) C. Walker and F. E. Sanders*	Puspitasari (2005), Husna et al. (2014)
30	*Sc. biornata*	Kramadibrata (2013)
31	*Sc.* cf. *dipapillosa*	Fahriny (2013)

(continued)

Table 19.1 (continued)

No.	Type of AMF	References
32	*Sc. calospora* (Nicolson and Gerdemann) Walker and Sanders	Setya et al. (1995), Haerida dan Kramadibrata (2002), Wulandari (2001), Chairani et al. (2002), Aradea (2004), Kramadibrata et al. (2007), Prasetyo et al. (2010), Ningsih et al. (2013), Fahriny (2013)
33	*Sc. cerradensis*	Arofatullah (2015)
34	*Sc. erythropus*	Wulandari (2001), Kramadibrata (2013)
35	*Sc. projecturata* Kramadibrata and Walker	Suciatmih and Kramadibrata (2002), Kramadibrata (2011, 2013)
Glomerales/Glomeraceae		
Glomus		
36	*G. aggregatum* Schenck and Smith	Widiastuti and Kramadibrata (1992), Haerida and Kramadibrata (2002), Arif et al. (1999), Kramadibrata (1993, 2009), Suciatmih and Kramadibrata (2002), Chairani et al. (2002), Aradea (2004), Prasetyo et al. (2010), Mbaubedari (2011), Proborini et al. (2013), Ginting (2013), Destifani (2013), Husna et al. (2014)
37	*G. boreale*	Husna et al. (2014), Fahriny (2013)
38	*G. cerebriforme*	Lucia (2005)
39	*G.* cf. *citricolum*	Widiastuti and Kramadibrata (1992)
40	*G. clavisporum* (Trappe) Almeida and Schenck	Widiastuti and Kramadibrata (1992), Kramadibrata (1993), Kramadibrata et al. (1995), Puspitasari (2005)
41	*G. claroideum*	Fahriny (2013)
42	*G. glomerulatum*	Suciatmih and Kramadibrata (2002), Ningsih et al. (2013)
43	*G. macrocarpum* Becker and Gerdemann	Widiastuti and Kramadibrata (1992)
44	*G. fuegianum* (Spegazzini) Trappe and Gerdemann	Widiastuti and Kramadibrata (1992, 1993), Kramadibrata (1993, 2009, 2012, 2013, 2016), Aradea (2004), Sabaruddin (2004), Kramadibrata et al. (2007), Ginting (2013)
45	*G. cf. formosanum* Wu and Chen	Kramidbrata et al. (2007)
46	*G. microaggregatum* Koske, Gemma, and Olexia	Kramadibrata et al. (1995), Kramadibrata (2009, 2012), Puspitasari (2005), Destifani (2013)
47	*G. microcarpum* Tul. and Tul.	Widiastuti and Kramadibrata (1992), Kramadibrata (2012)
48	*G. multicaulis* Gardemann and Bakshi	Kramadibrata (1993, 2009, 2012, 2016)
49	*G. tortuosum*	Puspitasari (2005)
Funneliformis		
50	*F. caledonius*	Sabaruddin (2004), Prasetyo et al. (2010)
51	*F. geosporus*	Kramadibrata (1993, 2011, 2012, 2013, 2016), Sabaruddin (2004), Lucia (2005), Puspitasari (2005), Kramidbrata et al. (2007), Prasetyo et al.

(continued)

Table 19.1 (continued)

No.	Type of AMF	References
		(2010), Ginting (2013), Destifani (2013), Ningsih et al. (2013)
52	*F. halonatus*	Husna et al. (2014)
53	*F. mosseae* (Nicol. and Gerd.) Gerdemann and Trappe	Kramidbrata et al. (2007), Kramadibrata and Gunawan (2006), Prasetyo et al. (2010), Mbaubedari (2011), Proborini et al. (2013)
54	*F. cf. multisubstensum*	Suciatmih and Kramadibrata (2002)
Sclerocystis		
55	*Scl. clavispora* Trappe	Husna et al. (2014), Kramadibrata (2016)
56	*Scl. pachycaulis*	Arif et al. (1999)
57	*Scl. sinuosa* Gerd. and B. K. Bakshi	Sabaruddin (2004), Kramadibrata (2009, 2012, 2016)
58	*Scl rubiformis* Wu and Chen (Almeida and Schenck)	Widiastuti and Kramadibrata (1992, 1993), Kramadibrata (1993, 2009, 2011, 2012, 2013, 2016), Setya et al. (1995), Silviana et al. (1999), Lucia (2005), Chairani et al. (2002), Arif et al. (1999), Sabaruddin (2004), Puspitasari (2005), Proborini et al. (2013), Kumalawati et al. (2014)
59	*Scl. taiwanensis*	Kramadibrata (2016)
Rhizophagus		
60	*R. clarus*	Arofatullah (2015)
61	*R. diaphanus* (C. Cano and Y. Dalpe)	Kramadibrata (2009), Husna et al. (2014)
62	*R. fasciculatum*	Kramadibrata et al. (1995), Setya et al. (1995), Silviana et al. (1999), Muliawan et al. (2002), Chairani et al. (2002), Kramadibrata (2009), Prasetyo et al. (2010), Mbaubedari (2011), Husna et al. (2014)
63	*R. intraradices*	Prasetyo et al. (2010), Proborini et al. (2013)
64	*R. invermaium*	Kramadibrata (2009)
65	*R. manihotis*	Silviana et al. (1999), Boddington and Dodd (2000), Muliawan et al. (2002), Sabaruddin (2004)
66	*R. sinuosus* (Gerdemann and Bakshi) Almeida and Schenck	Widiastuti and Kramadibrata (1993), Kramadibrata (1993, 2009)
Septoglomus		
67	*S. constrictum*	Arif et al. (1999), Kramadibrata and Gunawan (2006), Husna et al. (2014)
68	*S. deserticola*	Puspitasari (2005), Prasetyo et al. (2010)
Claroideglomeraceae/Claroideoglomus		
69	*C. etunicatum* Becker and Gerdemann	Widiastuti and Kramadibrata (1992), Arif et al. (1999), Setya et al. (1995), Silviana et al. (1999), Muliawan et al. (2002), Haerida and Kramadibrata (2002), Suciatmih and Kramadibrata (2002), Chairani et al. (2002), Aradea (2004), Sabaruddin (2004), Puspitasari (2005), Lucia (2005), Kramidbrata et al. (2007),

(continued)

Table 19.1 (continued)

No.	Type of AMF	References
		Prasetyo et al. (2010), Kramadibrata (2011, 2013), Proborini et al. (2013), Ningsih et al. (2013), Husna et al. (2014)
Paraglomerales/Paraglomeraceae		
Paraglomus		
70	*P. albidum* Walker and Rhodes	Aradea (2004), Chairani et al. (2002), Kramadibrata (2009), Fahriny (2013), Ningsih et al. (2013)
71	*P. occultum* (C. Walker) J. B. Morton and D. Redecker	Lucia (2005), Puspitasari (2005), Kramadibrata (2009, 2012)
72	*P. lacteum*	Silviana et al. (1999), Chairani et al. (2002)

with saline (Delvian 2003; Puspitasari 2005), post-coal mining (Margarettha 2011), nickel (Setiadi and Setiawan 2011; Husna et al. 2014; 2015a, b), tin (Novera 2008; Raharja 2015), gold tailings (Djuuna et al. 2010; Suharno et al. 2016), and petroleum contamination (Ervayenri 2005; Faiza et al. 2013).

Fruit species were reported in symbiosis with the AMF among *Durio zibethinus* (Smith et al. 1998), *Garcinia mangostana* (Silviana et al. 1999), *Anacardium occidentale* (Proborini et al. 2013), *Theobroma cacao* (Widiastuti and Kramadibrata 1992), *Nephelium lappaceum* (Muliawan et al. 2002), *Diospyros blanco* (Ningsih et al. 2013), *Sandoricum koetjape* (Ginting 2013), *Citrus* sp. (Suamba et al. 2014), *Arenga pinnata* (Miska et al. 2016), and *Musa* sp. (Rainiyati 2007); grasses such as *Imperata cylindrica* (Widiastuti and Kramadibrata 1992; Dewi et al. 2014) and *Pennisetum purpureum* (Astuti 2000); Indonesian tropical tree species such as *Gonystylus bancanus* (Muin 2003) and *Tectona grandis* (Husna et al. 2006); and food crop such as soybean (Kramadibrata et al. 1995), corn (Widiastuti and Kramadibrata 1992, Haerida and Kramadibrata 2002; Ishaq et al. 2017), peanuts (Marizal and Syariyah 2016), rice (Sunandar 2016), taro (Wulandari 2001), *Dioscorea esculenta* (Nuryana 2016), and cassava *(Manihot esculenta* Crantz) (Aryaji 2017).

19.3 Inoculation of Indonesia Tree Species with AMF

The arbuscular mycorrhizal fungi prospect was developed as biological fertilizer in Indonesia. Based on various research results, AMF application shows that AMF inoculation improves the quality of seedlings and increases the growth of tree species seedlings in various conditions of media and land in the nursery/greenhouse scale or field. In Indonesia, the study of AMF applications is still limited to the scale of greenhouses or nurseries. In a greenhouse, inoculation with *R. clarus* and *Gi. decipiens* increased shoot N and P of two non-forest product species *Dyera polyphylla* and *Aquilaria filaria* (Turjaman et al. 2006a). Similar results were reported by Turjaman et al. (2006b) that colonization by five AM fungi (*Entrophospora* sp., *Gigaspora decipiens*, *Glomus clarum*, *Glomus* sp. ZEA, and *Glomus* sp. ACA) increased plant height, diameter, and shoot and root dry weights. Nitrogen and phosphorus content of the seedlings *Aquilaria malaccensis* and *A. crassna* were also high. Survival rates were higher in the AM-colonized seedlings at 180 days after transplantation than those in the control seedlings. Wulandari et al. (2014) reported that inoculation with *Rhizophagus clarus* and *Gigaspora decipiens* increased shoot height and dry weight; shoot N and P content of *Mallotus paniculatus* and *Albizia saman* 6 months under greenhouse condition. *Glomus clarum* inoculation effectively improved nutrient content (N, P, K) and growth of *Alstonia scholaris* plant at 150 days after inoculation under greenhouse conditions (Turjaman et al. 2007). Tuheteru et al. (2011a) observed that inoculation with indigenous AMF (*Glomus* sp. 1-2 and *Acaulospora* sp. 1-2) increased the height, diameter, total dry weight, and nodulation of *Albizia saponaria* after 3 months grown on ultisol media.

AMF also improves plant growth on soil contaminated with heavy metal and waterlogging conditions (Tuheteru et al. 2011b). AMF inoculation (*Glomus etunicatum*, *G. manihotis*, *Acaulospora tuberculata*, and *Gigaspora rosea*) increases the height, diameter, total dry weight, and nodulation in *Albizia saponaria* while grown on post-nickel mining land. Tuheteru et al. (2017) observed that inoculation with *Acaulospora tuberculate* and *Glomus* sp. significantly increased root and shoot dry weight. Indeed, *Glomus* sp. reduced root's uptake of Fe and Ni by 13% and 3%, respectively, in the post-nickel mining land media. Under waterlogging condition, *Nauclea orientalis* from dry land needed AMF (mycofer IPB) to increase biomass and N accumulation at the age of 3 months (Tuheteru et al. 2015).

On the field scale, inoculation with AM fungi (*Glomus clarum* and *G. aggregatum*) improves early growth (shoot height, stem diameter, leaf number, and shoot and root dry weights) of *Ploiarium alternifolium* and *Calophyllum hosei* in a tropical peat swamp forest (Turjaman et al. 2008). Graham et al. (2013) reported that inoculation with *R. clarus* and *Gi. decipiens* increased levels of P and N in *D. polyphylla* grown on peat swamp forest in Central Kalimantan. The results of Wulandari et al. (2016) under nursery conditions suggests that AMF inoculation (*Rhizophagus clarus, Gigaspora decipiens*, and *Scutellospora* sp.) increased shoot P content and dry weight; stem diameter, shoot N content, shoot P content, shoot dry weight, and survival rate were higher in inoculated seedlings than in control experiments after 7 months of transplanting under field conditions in opencast coal mine at East Kalimantan. Irianto and Santoso (2005) reported that inoculations with *G. aggregatum* increased height and diameter growth of Teak (*Tectona grandis* L.) up to 61% and 47%, respectively, compared to the control after 3 months in the field at Cikampek, West Java.

19.4 Role for Species Conservation Activities in Legume

Many species of trees in the tropics are of economic value. One of the families that has important and useful timber species in Indonesia is Fabaceae (Heyne 1988). Some types of legumes of commercial timber produced in Southeast Asia (Soerianegara and Lemmens 1994; Sosef et al. 1998) and classified as endangered species (IUCN 1994) were *Pericopsis mooniana*, *Pterocarpus indicus*, and *Kalappia celebica*. The descriptions of the third types are presented in Table 19.2.

Table 19.2 Description of three legumes

Description	*Pericopsis mooniana*	*Pterocarpus indicus*	*Kalappia celebica*
Trees	Medium to large trees up to 40 m tall	Medium to large trees up to 40 m tall	Medium to large trees up to 40 m tall
Distribution	Southeast Asia	Southeast Asia	Endemic Sulawesi (Indonesia)
Ecology	pH: 3.7–6.2, a year of rainfall 750–2000 mm, elevation 200–350 m above sea level	Different type of soil, except for the heavy clay	Lowland forest (coastal elevation-500 m asl), acid pH (4–5)
Wood uses	Substitute for luxury and teakwood for cabinets, furniture, veneer, turnery, high quality joinery, flooring	Construction materials weight, floor, veneer, plywood, furniture, musical instruments, ship building, and wooden furniture	Used as beautiful wood for frames, cabinets and furniture building construction, furniture making, frame, ceiling, bridge, flooring, panel
Silviculture	Generative and vegetative	Generative and vegetative	Silvicultural techniques are still limited until now and can be reproduced by cuttings shoots
Status of scarcity	Prone (VU A1c, d)	Prone (VU A1d)	Prone (VU D1 + 2c)
References	Soerianegara and Lemmens (1994), IUCN (1994), UNEP-WCMC (2007)	Soerianegara and Lemmens (1994), Martawijaya et al. (2005), IUCN (1994), UNEP-WCMC (2007)	Whitten et al. (1987), Keßler et al. (2002), Sosef et al. (1998), Arif et al. (2015), IUCN (1994), UNEP-WCMC (2007)

19.5 AMF Diversity with Local Endemic and Endangered Legume Species

In general, endangered plant species are reported to be symbiotic with AMF. Wang and Qiu (2006) reported that 139 species are endangered from 2469 species which establish symbiotic associations with AMF. The study of AMF diversity in the rhizosphere of three endangered legume species in Indonesia has been done.

19.5.1 Pericopsis mooniana *Thw.*

AMF diversity study has been conducted on six hardwood tree-growing habitats in Southeast Sulawesi province. The growing habitat includes among them the following: (1) Lamedai Nature Reserve, Kolaka; (2) Tanggetada natural forest, Kolaka; (3) post-nickel mining land PT Vale Indonesia Tbk, Kolaka; (4) plantation forest of Desa Bali Jaya, Kolaka; (5) campus of Haluoleo University, Kendari; and (6) city forest in Southeast Sulawesi, Kendari. Husna et al. (2014, 2015a, b) reported that the roots of hardwood trees grown on the six locations in Southeast Sulawesi are colonized by AMF. The percentage of AMF colonization did not differ in all habitats with a percentage of >50%. AMF colonization is characterized by the discovery of AMF structure at the root. AMF structures found include internal hyphae, external hyphae, hyphae of niches and vesicles, and arbuscules. A total of 15 different types of AMF were found there and grouped into 5 families and 9 genera: *Glomeraceae* (*Glomus/Rhizophagus*, *Sclerocystis*, *Septoglomus*), *Claroideoglomeraceae* (*Claroideoglomus*), *Gigasporaceae* (*Scutellospora*, *Racocetra*), *Acaulosporaceae* (*Acaulospora*), and *Ambisporaceae* (*Ambispora*). These 15 types are *Glomus aggregatum*, *G. boreale*, *G. canadense*, *G. halonatum*, *G. versiforme*, *Rhizophagus diaphanous*, *R. fasciculatus*, *Sclerocystis clavispora*, *Septoglomus constrictum*, *Claroideoglomus etunicatum*, *Scutellospora auriglobosa*, *Racocetra gregaria*, *Acaulospora delicata*, *A. scrobiculata*, and *Ambispora appendicula*. *Glomeraceae* dominates the AMF with nine types and has a frequency and a relatively high density at all sites. Four new types of local AMF were first discovered in Indonesia, which are *G. canadense*, *G. halonatum*, *R. gregaria*, and *A. appendicula* (Fig. 19.1). The study of AMF diversity in the rhizosphere of *P. mooniana* is still limited in Southeast Sulawesi.

19.5.2 Pterocarpus indicus

Preliminary study of the diversity of the AMF by the authors showed that the AMF were found in two genera, namely, *Glomus* (seven species) and *Acaulospora* (one species), in the rhizosphere of *P. indicus* in the village of Lakapera, Central Buton Regency, Southeast Sulawesi. AMF symbiosis with *Pterocarpus* sp. has also been reported in the village of Banjarsari Enggano Island (Bali) by Kramadibrata (2016).

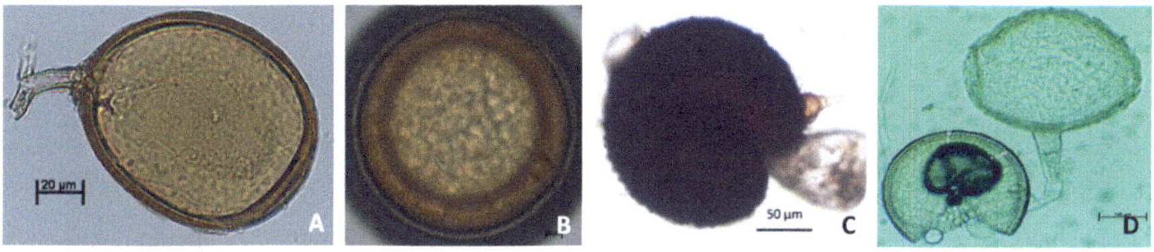

Fig. 19.1 Four types of new local AMF was first discovered in Indonesia, which are *G. canadense* (A), *G. halonatum* (B), *R. gregaria* (C) and *A. appendicula* (D) (Husna et al. 2014)

Kramadibrata (2016) found *Acaulospora scrobiculata*, *A. tuberculata*, *Scl. clavispora*, *Glomus fuegianum*, and *Glomus multicaule* in the rhizosphere of *Pterocarpus* spp. in Indonesia, angsana symbiosis with the AMF has also been reported in Thailand. Namanusart (2003) found AMF spores in the rhizosphere of angsana such as *Acaulospora*, *Gigaspora*, *Glomus*, and *Scutellospora*.

19.5.3 Kalappia celebica

AMF diversity was also found in the rhizosphere of kalapi in community forests in Kolaka district, Tanggetada sub-district, Southeast Sulawesi. The results of the research conducted by Arif et al. (2016) found three fungi genera *Glomus*, *Gigaspora*, and *Acaulospora*. *Glomus* is the dominant genera with seven species of AMF. AMF colonization on kalapi root ranges from 10 to 30%. The structure of AMF found in rooting kalapi is internal hyphae, external hyphae, and vesicles (Arif et al. 2015). To date, the study of AMF symbiosis with kalapi is still limited to the diversity aspect. The study of AMF application on kalapi seedlings has never been carried out.

19.6 Roles of AMF in Leguminous Plant

19.6.1 Pericopsis mooniana *Thw.*

Studies on inoculation effect on the growth of *P. mooniana* by AMF have been done in Indonesia on greenhouse/nursery or fieldscale. In general, improved growth and nutrient status increased in *P. mooniana* seedlings inoculated with AMF in various growing media such as ultisol, inceptisol, and serpentine (post-nickel mining) soil (Table 19.3). The results of both height and diameter measurements of *P. mooniana* inoculated with AMF at 36 months on post-nickel mining land showed that plants inoculated with AMF inoculum of CA Lamedai (non-serpentine) and PT Vale Indonesia (serpentine) increased the increments of height and diameter by 297 and 249% and 125 and 123%, respectively, compared to controls (Fig. 19.2).

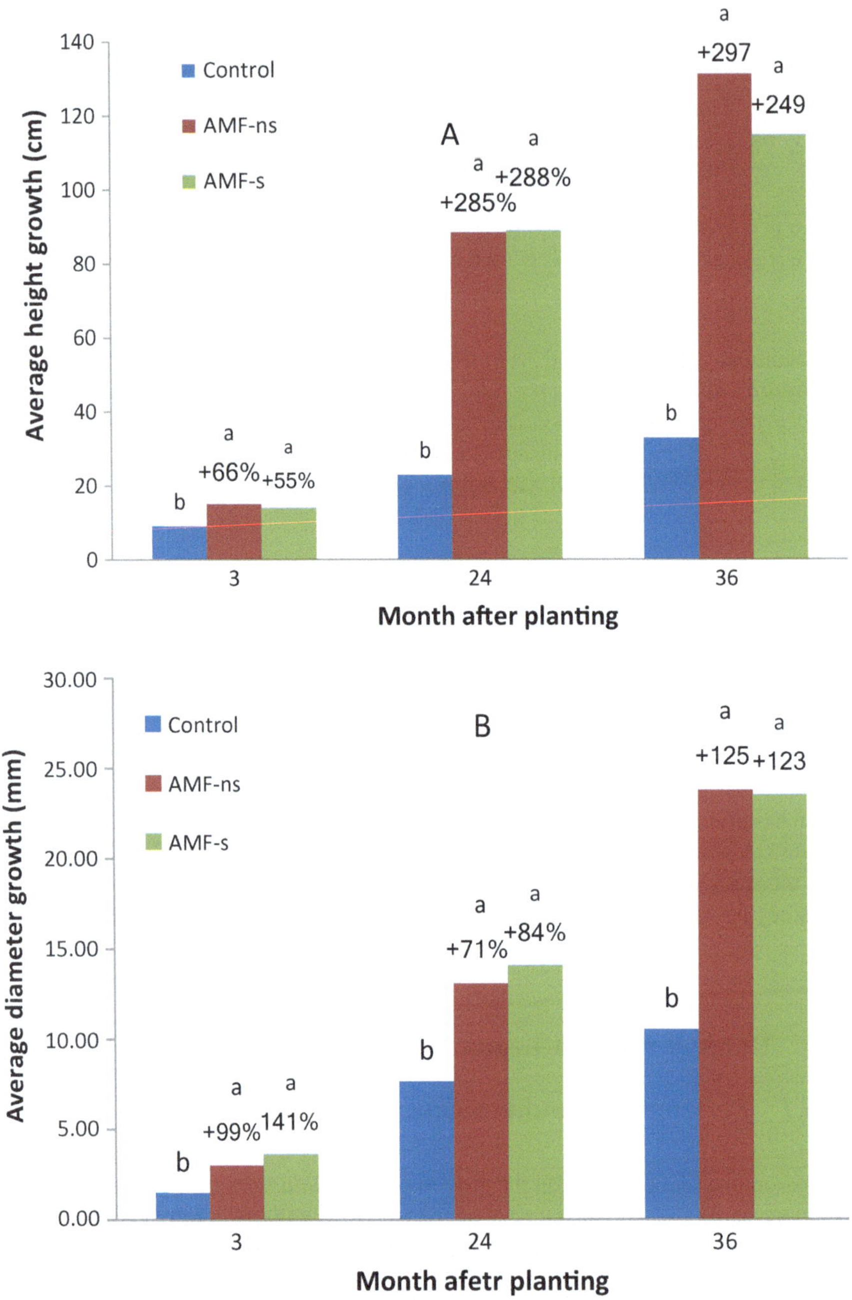

Fig. 19.2 Growth responses in height (**a**) and diameter (**b**) of inoculated *P. mooniana* (Thw.) at 3, 24, and 36 months after planting on post-nickel mining site at PT Vale Indonesia (Tbk.), Pomalaa (Kolaka), Southeast Sulawesi (modified by Husna et al. 2017b)

Table 19.3 Research review on AMF symbiosis with *Pericopsis mooniana* Thw

AMF species	Media	Period (m)	Effect	Experiment type	References
Commercial AMF inoculum "mycofer IPB" (*Glomus etunicatum*, *G. manihotis*, *Acaulospora tuberculata*, and *Gigaspora rosea*)	Ultisol	3	Increased growth and dry weight	Greenhouse	Iskandar (2010)
5.0 g of commercial AMF inoculum "mycofer IPB" (mixed *Glomus manihotis*, *G. etunicatum*, *A. tuberculata*, *Gigaspora margarita*) and the addition of 20.0 g sago dregs	Soil media of post-nickel mining site	3	High growth, dry weight, root nodule and K and Ca uptake, and reduced Ni by 32%	Greenhouse	Husna (2010)
AMF native from *P. mooniana* rhizosphere from Lamedai NR (non-serpentine) and PT Vale Indonesia (serpentine)	Post-nickel mining land	3	Increased survival, growth, biomass, and accumulation of N, P, and K	Field	Husna (2015)
Commercial AMF inoculum "mycofer IPB" (*Glomus etunicatum*, *G. manihotis*, *Acaulospora tuberculata*, and *Gigaspora rosea*)	Soil media of post-nickel mining site	3	High survival rate, increased growth, biomass and nutrient accumulation of N, P, and K, and low Ni content	Greenhouse	Husna et al. (2015a)
AMF native from *P. mooniana* rhizosphere	Inceptisol	5	Increased height, stem diameter, number of leaves, root nodules, total dry weight, total chlorophyll, and P, K, Ca, and Mg content	Greenhouse	Husna et al. (2015b)
AMF native from *P. mooniana* rhizosphere	Soil media of post-nickel mining site	5	increasing growth and biomass of plants; absorption of N, P, and K in three parts of the plants of Ca and Mg in leaf tissues; root nodules and reduce Ni content	Greenhouse	Husna et al. (2016b)

(continued)

Table 19.3 (continued)

AMF species	Media	Period (m)	Effect	Experiment type	References
Glomus sp. (HA)	Ultisol	3	Increased plant height, stem diameter, number of leaves, dry weight (roots, shoots, total), nodulation, root length, and leaf length	Greenhouse	Husna et al. (2017a)
AMF from *P. mooniana* rhizosphere from Lamedai NR (non-serpentine) and PT Vale Indonesia (serpentine)	Post-nickel mining site	24	Increasing height and diameter	Field	Husna et al. (2017b)

Table 19.4 Effect of treatment on root colonization, mycorrhizal inoculation effect (MIE), and nodulation of *P. indicus* seedling roots at 90 days

Treatment	Colonization MA (%)		MIE (%)	Nodule	
Control	0 ± 0.00	c	–	0.33 ± 0.33	e
C. etunicatum	11 ± 2.94	b	52 ± 17.69	5.7 ± 3.18	de
Glomus KDI	24 ± 2.96	a	63 ± 4.010	13.0 ± 4.04	cd
Glomus HA	25 ± 3.93	a	63 ± 17.54	23.3 ± 3.18	bc
Acaulospora HA	25 ± 2.54a		69 ± 02.69	24 ± 4.61	bc
Ac HA+Gl HA	19 ± 2.31a		65 ± 10.63	41 ± 6.65	a
Mixture*	17 ± 2.02	ab	56 ± 17.42	29.7 ± 0.88	ab

*Mixed (*C. etunicatum*, *Glomus* KDI, *Glomus* HA)
**The same letter in the same column indicates not significantly different at the 95% test level based on Duncan's multiple-range test

19.6.2 Pterocarpus indicus

Root colonization by local AMF was <50%, and plants without mycorrhizae were found with no AMF structure in rooting and MIE values of 52–63%. Indigenous inoculation with AMF stimulated the number of nodules with an increase of 1627–8687% compared to controls (Table 19.4). AMF inoculation also increased the height, diameter, root length, total number of leaves, and total dry weight (Table 19.5) and N, P, K, Ca, and Mg nutrient accumulation (Table 19.6). The MIE high value indicates that the local AMF inoculation can be utilized for the production of quality seedlings in the nursery and supporting the success of conservation efforts on a scale of angsana field through ex situ conservation program (Fig. 19.3). Results of Husna et al. (2017a) also reported that AMF inoculation

Table 19.5 Effect of treatment on growth of *P. indicus* seedlings after 90 days

Treatment	Height (cm)		Diameter (mm)		Total number of leaves		Total dry weight (g)		Root length (cm)	
Control (A)	11.4 ± 2.71	b	2.05 ± 0.31	c	6.67 ± 0.67	b	0.72 ± 0.10	b	65.5 ± 14.80	b
C. etunicatum(B)	25.3 ± 4.91	a	3.73 ± 0.43	a	9.0 ± 1.53	ab	2.13 ± 0.31	a	204.8 ± 46.43	a
Glomus KDI (C)	20.5 ± 1.65	a	3.37 ± 0.31	ab	9.0 ± 0.58	ab	2.10 ± 0.36	a	163.7 ± 30.97	a
Glomus HA (D)	22.7 ± 0.98	a	3.43 ± 0.72	ab	10.0 ± 1.15	a	2.08 ± 0.68	a	180.5 ± 37.03	a
Acaulospora HA (E)	24.1 ± 1.93	a	3.55 ± 0.25	ab	21.7 ± 2.03	a	2.31 ± 0.16	a	144 ± 2.01	ab
AcHA+Gl HA (F)	24.5 ± 0.64	a	2.71 ± 0.21	bc	23.7 ± 3.71	a	2.25 ± 0.42	a	134 ± 11.59	ab
Mixture (G)	24.7 ± 0.74	a	3.53 ± 0.56	ab	11.3 ± 0.67	a	2.57 ± 0.51	a	166.6 ± 40.02	a

Table 19.6 Effect of treatment of the uptake of N, P, K, Ca, and Mg of Angsana seedlings at 90 days

Treatment	Nutrient uptake (mg/plant)									
	N		P		K		Ca		Mg	
Control	1.53	b	0.14	b	0.52	b	1.11	b	0.36	c
C. etunicatum	4.72	a	0.33	a	1.59	a	2.50	a	0.84	a
Glomus KDI	4.68	a	0.29	a	1.51	a	2.03	a	0.58	b
Glomus HA	7.28	a	0.37	a	2.17	a	2.56	a	0.79	a
Acaulospora HA	5.83	a	0.31	a	1.77	a	2.15	a	0.75	ab
Ac HA+Gl HA	5.77	a	0.36	a	1.78	a	2.07	a	0.59	ab
Mixture	6.32	a	0.39	a	1.88	a	2.16	a	0.67	ab

Fig. 19.3 Performance of Angsana plant at 3 months old (left) and structure of the AMF in the root (right)

mixture (*Acaulospora delicata*, *A. scrobiculata*) enhanced growth (height and diameter), total dry weight, and nodulation in angsana seedlings.

There was an increased growth of endangered inoculated hardwood and angsana presumably through the absorption of nutrients (particularly phosphates) and nitrogen and water (Smith and Read 2008) as well as plant resistance to biotic and abiotic stresses (Akhtar and Siddiqui 2007; Wu et al. 2013; Zhang et al. 2014; Tuheteru and Wu 2017). In addition, local AMF also stimulates root nodule formation compared to control experiments (Tables 19.3 and 19.4). The results are consistent with some previous research on legumes *Sesbania aegyptiaca* and *Sesbania grandiflora* (Giri and Mukerji 2004), *Cassia siamea* (Giri et al. 2005), *Gliricidia sepium* (Giri 2017), *Pisum sativum* cv. Avola (Geneva et al. 2006), *Albizia saponaria* (Tuheteru et al. 2011a, b), *Pericopsis mooniana* (Husna et al. 2015a, b, 2016a, b), and *Cicer arietinum* L. (Garg and Singla 2016). AMF can supply P for the formation of nodules as well as increase the activity of fixed N_2 by *Rhizobium* (Geneva et al. 2006).

19.7 Conclusion and Future Research

The diversity and identification of the AMF in Indonesia remain limited to spore morphology approach; therefore, in the future, DNA molecular approach for the identification of the AMF in Indonesia needs to be done. Research on AMF diversity in all three legume species threatened with extinction is limited in the province of Southeast Sulawesi. Therefore, it needs a comprehensive research on a variety of habitats and environmental conditions throughout Indonesia. Arbuscular mycorrhizal fungi play an important role in supporting the conservation of Indonesian tropical tree species threatened with extinction, which necessitates inoculating threatened species seedlings with AMF in the nursery for the production of quality seedlings. Examining the effectiveness of the AMF to improve growth of kalapi (*Kalappia celebica* Kosterm) needs to be tried because of a gap in the knowledge.

Acknowledgments This research was supported by a competitive grant (No. 228c.1/UN29.20/PP/2016) and a fundamental grant (No. 229.a.1/UN.29.20/PPM/2016) of the Ministry of Research, Technology and Higher Education. The author would like to thank the leadership of the PT Vale Indonesia (PT) Pomalaa Region, Kolaka, Southeast Sulawesi.

References

Akhtar MS, Siddiqui ZA (2007) Effects of *Glomus fasciculatum* and *Rhizobium* sp. on the growth and root-rot disease complex of chickpea. Arch Phytopathol Plant Protect 40:37–43

Aradea NN (2004) Arbuscular mycorrhizal fungi (AMF) on Iwul (*Orania sylvicola*). Floribunda 2 (6):164–170

Arif S, Yusnaini S, Niswati A, Setiawan A, Tuchida K, Katou T, Touji Y, Nonaka M (1999) Population of arbuscular mycorrhizal fungi (AMF) by different land use in Sumatra, Indonesia comparison of AMF spore numbers in primary forest, secondary forest, fields growing coffee and native grass. Microbes Environ 14(1):9–17

Arif A, Tuheteru FD, Husna (2015) The conservation of endemic and endangered species *Kalappia celebica* Kosterm through cuttings propagation and AMF potential assessment. In: Proceedings ICSD, Universitas Mahasaraswati press. ISBN: 978-602-18622-7-8, Bali, pp 53–65

Arif A, Tuheteru FD, Husna, Kandari AM, Mekuo IS, Masnun (2016) Status and culture of arbuscular mycorrhizal fungi isolated from rhizosphere of endemic and endangered species of Kalapi (*Kalappia celebica* Kosterm). Eur J Sustain Dev 5(4):395–402

Arofatullah NA (2015) Isolation, molecular identification, and *in vitro* propagation of arbuscular mycorrhiza from tea plant rhizosphere. Thesis, Gadjah Mada University

Aryaji L (2017) Kolonisasi dan komposisi genetik jamur mikoriza arbuskula pada 8 varietas ketela pohon (*Manihot esculenta* crantz) di tanah vulkanik. Skripsi. Universitas Gadjah Mada

Astuti WD (2000) Biodiversity of arbuscular mycorrhizal fungi in elephant grass (*Pennisetum purpureum*) rhizosphere in Bogor dan Lembang. Skripsi, Institut Pertanian Bogor

Boddington CL, Dodd JC (2000) The effect of agricultural practices on the development of indigenous arbuscular mycorrhizal fungi. I. Field studies in an Indonesian ultisol. Plant Soil 218:137–144

Bothe H, Turnau K, Regvar M (2010) The potential role of arbuscular mycorrhizal fungi in protecting endangered plants and habitats [review]. Mycorrhiza 20:445–457

Chairani, Gunawan AW, Kramadibrata K (2002) Mikoriza durian di Bogor dan sekitarnya. J Mikrobiol Indones 7:44–46

dan Suciatmih, Kramadibrata K (2002) Arbuscular mycorrhizal fungi at different ecosystems of Gunung Halimun National Park. Bertia Biol 6(1):145–149

Delvian (2003) Diversity of arbuscular mycorrhizal fungi in coastal forest and its potential use. Dissertation, Bogor Agricultural University

Destifani YF (2013) Arbuscular mycorrhizal fungi on Star Apple (*Chrysophyllum cainito*) at IPB Darmaga Campus. Undergraduate Thesis, Bogor Agricultural University

Dewi NKS, Wirawan GP, Sritamin M (2014) Identification of arbuscular mycorrhiza through microscopis methode in rhizosfer of several Grass and Cacao (*Theobroma cacao* L.) E-J Agroteknol 3(4):259–268

Djuuna IAF, Puteh HP, Bachri S (2010) Spatial distribution of arbuscular mycorrhiza (am) fungi in the tailing modADA deposition areas. In: 19th World Congress of soil science, soil solutions for a changing world, 1–6 August 2010, Brisbane. Published on DV

Ervayenri (2005) The beneficial of arbuscular mycorrhizal fungi (AMF) and indigenous plant for the revegetation of the petroleum contaminated land. Dissertation, Bogor Agricultural University

Fahriny RA (2013) Arbuscular mycorrhizal diversity of *Areca* in Bogor Botanical Garden. Undergraduate Thesis, Bogor Agricultural University

Faiza R, Rahayu YS, Yuliani (2013) Identifikasi spora jamur mikoriza vesikular arbuskular (MVA) pada tanah tercemar minyak bumi di Bojonegoro. Lentera Bio 2(1):7–11

Food and Agriculture Organization (2016) State of the world's forests 2016: forest and agriculture: land-use challenges and opportunities. FAO, Roma

Fuchs B, Haselwandter K (2004) Red list plants: colonization by arbuscular mycorrhizal fungi and dark septate endophytes. Mycorrhiza 14:277–281

Fuchs B, Haselwandter K (2008) Arbuscular mycorrhiza of endangered plant species: potential impacts on restoration strategies. In: Varma A (ed) Mycorrhiza: state of the art, genetics and molecular biology, eco-function, biotechnology, eco-physiology, structure and systematics, 3rd edn. Springer, New York, pp 565–579

Garg N, Singla P (2016) Stimulation of nitrogen fixation and trehalose biosynthesis by naringenin (Nar) and arbuscular mycorrhiza (AM) in chickpea under salinity stress. Plant Growth Regul 80:5–22

Geneva M, Zehirov G, Djonova E et al (2006) The effect of inoculation of pea plants with mycorrhizal fungi and *Rhizobium* on nitrogen and phosphorus assimilation. Plant Soil Environ 52(10):435–440

Ghazoul J, Sheil D (2010) Tropical rain forest ecology, diversity and conservation. Oxford University Press, New York, p 516

Ginting F (2013) Arbuscular mycorrhizal fungi on Kechapi (*Sandoricum koetjape*) at LIPI Cibinong. Undergraduate Thesis, Bogor Agricultural University

Giri B (2017) Mycorrhizal dependency and growth response of *Gliricidia sepium* (Jacq.)Kunth ex Walp. under saline condition. Plant Sci Today 4(4):154–160

Giri B, Mukerji KG (2004) Mycorrhizal inoculant alleviates salt stress in *Sesbania aegyptiaca* and *Sesbania grandiflora* under field conditions: evidence for reduced sodium and improved magnesium uptake. Mycorrhiza 14:307–312

Giri B, Kapoor R, Mukerji KG (2005) Effect of the arbuscular mycorrhizae *Glomus fasciculatum* and *G. macrocarpum* on the growth and nutrient content of *Cassia siamea* in a semi-arid Indian wasteland soil. New Forests 29:63–73

Graham LL, Turjaman M, Page SE (2013) *Shorea balangeran* and *Dyera polyphylla* (syn. *Dyera lowii*) as tropical peat swamp forest restoration transplant species: effects of mycorrhizae and level of disturbance. Wetl Ecol Manage 21(5):307–321

Haerida I, Kramadibrata K (2002) Identification of arbuscular mycorrhizal fungi on corn rhizosphere in Java. Floribunda 2(2):33–37

Heyne K (1988) Tumbuhan Berguna Indonesia, Jilid II. Badan Litbang Kehutanan, Jakarta

Husna (2010) Pertumbuhan bibit kayu kuku (*Pericopsis mooniana* THW) melalui aplikasi fungi mikoriza arbuskula (FMA) dan ampas sagu pada media tanah bekas tambang nikel. Thesis, Univeristas Halu Oleo

Husna (2015) Potensi Fungi Mikoriza (FMA) Lokal dalam konservasi ex-situ jenis terancam punah Kayu Kuku [*Pericopsis mooniana* (Thw.) Thw]. Dissertation, Bogor Agricultural University

Husna, Mahfudz, Tuheteru FD (2006) Diversitas mikoriza pada pohon plus jati di Sulawesi Tenggara. J Penelitian Hutan Tanaman 3(1):275–284

Husna, Sri Wilarso Budi R, Mansur I, Kramadibrata K (2014) Fungi mikoriza arbuskula pada rizosfer *Pericopsis mooniana (Thw.)* Thw. di Sulawesi Tenggara. Berita Biol 13(3):263–273

Husna, Budi SW, Mansur I, Kusmana C (2015a) Diversity of arbuscular mycorrhizal fungi in the growth habitat of Kayu Kuku (*Pericopsis mooniana Thw.*) in South east Sulawesi. Pak J Biol Sci 18(1):1–10

Husna, Budi SW, Mansur I, Kusmana C (2015b) Respon pertumbuhan bibit kayu kuku (*Pericopsis mooniana* (Thw.) Thw.) terhadap inokulasi fungi mikoriza arbuskula lokal. Pemuliaan Tanaman Hutan 9(3):131–148

Husna, Tuheteru FD, Khalifah N (2016a) Symbiosis arbuscular mycorrhizal fungi with pioneer plants on nickel post mining land. Presented paper on national seminar of Silviculture IV. Faculty of Forestry, Mulawarman University, Balikpapan (Indonesia) 19–20 Juli 2016

Husna, Budi SW, Mansur I, Kusmana C (2016b) Growth and nutrient status of Kayu Kuku (*Pericopsis mooniana* Thw.) with mycorrhiza in soil media of nickel post mining. Pak J Biol Sci 19:158–170

Husna, Tuheteru FD, Wigati E (2017a) Growth response and dependency of endangered nedum tree species (*Pericopsis mooniana*) affected by indigenous arbuscular mycorrhizal fungi inoculation. Nusantara Biosci 9(1):57–61

Husna, Tuheteru FD, Arif A (2017b) Arbuscular mycorrhizal fungi and plant growth on serpentine soils. In: Wu QS (ed) Arbuscular mycorrhizas and stress tolerance of plants. Springer, Singapore, pp 293–303

[IUCN] International Union for Conservation of Nature and Natural Resources (1994) IUCN Red List of Threatened Species. http://www.iucnredlist.org

Irianto RSB, Santoso E (2005) Effect of arbuscular mycorrhiza fungi inoculation on teak (*Tectona grandis* Linn. F) at Cikampek, West Java. J Forest Res 2(2):1–5

Ishaq L, Tae ASJA, Airthut MA, Bako PO (2017) Abundance of arbuscular mycorrhiza associated with corn planted with traditional and more modern farming system in Kupang, East Nusa Tenggara, Indonesia [Short Communication]. Biodiversitas 18(3):887–892

Iskandar F (2010) Peningkatan kualitas bibit kayu kuku *(Pericopsis mooniana* Thwaites*)* yang diberi fungi mikoriza arbuskula dan tepung tulang. Skripsi, Universitas Halu Oleo

Jasper DA (1994) Management of mycorrhizas in revegetation. In: Robson AD, Abbott LK, Malajczuk N (eds) Management of mycorrhizas in agriculture, horticulture and forestry. Kluwer Academic, Dordrecht, pp 211–219

Keßler PJA, Bos MM, Sierra Daza SEC, Kop A, Willemse LPM, Pitopang R, Gradstein SR (2002) Checklist of woody plants of Sulawesi, Indonesia. BLUMEA Supplement 14. National Herbarium Nederland, Universiteit Leiden branch

Kramadibrata K (1993) Jenis-jenis jamur Glomales dari DAS Cisadane. J Mikrobiol 2(2):24–26

Kramadibrata K (2009) The distribution of Glomeromycota in Cacao rhizosphere in Indonesia. Reinwardtia 12(5):347–356

Kramadibrata K (2011) Arbuscular fungi of bamboo in Sumba Island. Berita Biol 10(5):635–639

Kramadibrata K (2012) Arbuscular fungi in Ujung Kulon National Park. Berita Biol 11(2):205–209

Kramadibrata K (2013) Diversity of glomeromycota in baturaden botanic garden and its surrounding area in Slamet mountain. Berita Biol 12(2):217–227

Kramadibrata K (2016) Diversity of arbuscular fungi in Enggano Island. Berita Biol 15(3):257–265

Kramadibrata K, Gunawan AW (2006) Arbuscular mycorrhizal fungi surrounding tropical kudzu and para grass. J Mikrobiol Indones 11(2):67–71

Kramadibrata K, Riyanti EI, Simanungkalit RDM (1995) Arbuscular mycorrhizal fungi from the rhizospheres of soybean crops in Lampung and West Java. Biotropia 8:30–38
Kramidibrata K, Prastyo H, Gunawan AW (2007) Arbuscular fungi of bamboo in Java. Berita Biol 8(6):531–536
Kumalawati Z, Musa Y, Amin N, Laode A, Ridwan I (2014) Exploration of arbuscular mycorrhizal fungi from sugarcane rhizosphere in South Sulawesi. Int J Sci Technol Res 3(1):201–203
Lucia Y (2005) Cendawan mikoriza arbuskula di bawah tegakan tanaman manggis dan peranannya dalam pertumbuhan bibit manggis (*Garcinia mangostana* L.). Thesis, Institut Pertanian Bogor
Margarettha (2011) Exploration and identification of indigenous mycorrhiza of ex-coal mining soil. Berita Biol 10(5):641
Marizal S, Syariyah A (2016) The diversity of arbuscular fungus (AMF) indigenous in peanuts (*Arachis Hypogea* L.) rhizosphere under different elevation. J Trop Soil 21(2):109–114
Martawijaya A, Kartasujana I, Kadir K, Prawira SA (2005) Atlas Kayu Indonesia Jilid I. 3rd.Pusat Penelitian dan Pengambangan Hasil Hutan, Bogor
Mbaubedari KF (2011) The effect of arbuscular mycorrhizal fungi (AMF) indigenous papua and plantling media to the growth of Agarwood (*Gyrinops versteegii* (gilg) Domke) plantling from in-vitro multiplication. Thesis, Bogor Agricultural University
Miska MEE, Junaedi A, Wachjar A, Mansur I (2016) Characterization of arbuscular mycorrhizal fungus from sufar palm (*Arenga pinnata* (Wurmb) Merr.)0 West Java and Banten. J Silvikultur Trop 7(1):18–23
Muin A (2003) Pertumbuhan anakan ramin (*Gonystylus bancanus* (Miq.)Kurz)) dengan inokulasi cendawan mikoriza arbuskula (CMA) pada berbagai intensitas cahaya dan dosis fosfat alam. Dissertation, Institut Pertanian Bogor
Muliawan J, Gunawan AW, Kramadibrata K (2002) Mycorrhiza of rambutan in Bogor area and its surroundings. J Mikrobiol Indones 7(1):24–25
Namanusart MW (2003) Genetic diversity of arbuscular fungi infected. A Thesis Submitted in Partial Fulfillment of the Requirements for the Degree of Master of Science in Biotechnology Suranaree University of Technology Academic Year 2003. ISBN: 974-533-309-3
Ningsih DR, Kramadibrata K, Gunawan AW (2013) Arbuscular mycorrhizal fungi associated with Bisbul (*Diospyros blancoi*). Biotropia 20(2):112–121
Novera Y (2008) Analysis of vegetation, soil characteristics and arbuscular mycorrhizal fungi (AMF) colonization of post tin mining land in Bangka Island. Undergraduate Thesis, Bogor Agricultural University
Nuryana I (2016) The colonization and diversity of arbuscular mycorrhizal fungi (AMF) in lesser yam (*dioscorea esculenta*) plant's root growing at two altitudes. Thesis, Gadjah Mada University
Pangaribuan N (2014) Trapping of indigenous arbuscular mycorrhiza fungi from physic corn and nuts at peatland West Kalimantan. J Agro 1(1):50–60
Panwar J, Tarafdar JC (2006) Distribution of three endangered medicinal plant species and their colonization with arbuscular mycorrhizal fungi. J Arid Environ 65:337–350
Prasetyo B, Krisnayanti BD, Utomo WH, Anderson CWN (2010) Rehabilitation of artisanal mining gold land in West Lombok, Indonesia: 2. arbuscular mycorrhiza status of tailings and surrounding soils. J Agric Sci 2(2):202–209
Prastyo H (2004) Cendawan mikorza arbuskula pada bambu. Skripsi, Institut Pertanian Bogor
Proborini MW, Sudana M, Suarna W, Ristiati NP (2013) Indigenous vesicular arbuscular mycorrhizal (VAM) fungi in cashew nut (*Anacardium occidentale* L.) plantation at north east-Bali island-Indonesia. J Biol Agric Healthc 3(3):114–121
Puspitasari RT (2005) Diversity of arbuscular mycorrhizal fungi (AMF) at coastal forest Ujung Genteng, Sukabumi-West Java. Thesis, Bogor Agricultural University
Raharja NC (2015) Isolation and identification of indigenous arbuscular mycorrhizal fungi (AMF) in rhizosphere of grass post tin-mining land at East Belitung. Undergraduate Thesis, Bogor Agricultural University

Rainiyati (2007) The status and diversity of arbuscular mycorrhizal fungi (AMF) on Banan cv Raja Nangka and the potential of their application to increase the production of tissue culture-regenerated Banan in Merangin Regency Jambi Province. Dissertation, Bogor Agricultural University

Sabaruddin (2004) Diversity of indigenous arbuscular mycorrhizae under three different land uses. Agrikultura 15:80–85

Setiadi Y, Setiawan A (2011) Study of arbuscular mycorrhizal fungi status at rehabilitation post-nickel mining area (Case Study at PT INCO Tbk. Sorowako, South Sulawesi). J Silvikultur Trop 03(01):88–95

Setya AP, Gunawan AW, Kramadibrata K (1995) Arbuscular mycorrhizal fungi isolated from bambooin Bogor Batanic Gardens. Hayati 2(2):85–86

Shah MA (2014) Mycorrhizas: novel dimensions in the changing world. Springer, New Delhi

Silviana, Gunawan AW, Kramadibrata K (1999) Biodiversity of arbuscular mycorrhizal fungi in the rhizospheres of mangosteen. In: Smith FA, Kramadibrata K, Simanungkalit RDM, Sukarno N, Nuhamara ST (eds) Proceedings of international conference on mycorrhizas in sustainable tropical agriculture and forest ecosystems. The Indonesian Institute University (LIPI), Bogor Agricultural University, and The University of Adelaide, Australia. Bogor-Indonesia, 27–-30 October 1997, Bogor

Smith SE, Read DJ (2008) Mycorrhizal symbiosis, 3rd edn. Academic Press, Boston

Smith HF, O'Connor PJ, Smith SE, Smith FA (1998) Vesicular-arbuscular mycorrhizas of durian and other plants of forest gardens in West Kalimantan, Indonesia. In: Schulte A, Ruhiyat D (eds) Soils of tropical forest ecosystems: characteristics, ecology and management. Springer, New York, pp 192–198

Soerianegara I, Lemmens RHMJ (eds) (1994) Plant resources of South-East Asia 5(1) timber trees: major commercial timbers. PROSEA, Bogor, p 610

Solaiman ZM, Mickan B (2014) Use of Mycorrhiza in sustainable agriculture and land restoration. In: Solaiman ZM et al (eds) Mycorrhizal fungi: use in sustainable agriculture and land restoration. Springer, New York, pp 1–16

Sosef MSM, Hong LT, Prawirohatmodjo S (eds) (1998) Timber trees: lesser-known timbers No 5 (3). PROSEA, Leiden, p 861

Suamba IW, Wirawan IGP, Adiartayasa W (2014) Isolation and identification of mycorrhizal fungi (AMF) by microscopis (AMF) in rhizosphere of citrus plant (*Citrus* sp.) in Kerta Village, Payangan District, Gianyar Regency. E-J Agroteknol Trop 3(4):201–208

Suharno, Kasiamdari RS, Soetarto ES, Sancayaningsih RP (2016) Presence of arbuscular mycorrhizal fungi on fern from tailing deposition area of gold mine in Timika, Indonesia. Int J Environ Bioremediat Biodegrad 4(1):1–7

Sunandar A (2016) Identifikasi Cendawan Mikoriza pada Rizosper Tiga Varietas Padi Lokal Kalimantan Barat. J Biol Educ 3(2):82–86

Tawaraya K, Turjaman M (2014) Use of arbuscular mycorrhizal fungi for reforestation of degraded tropical forests. In: Solaiman ZM et al (eds) Mycorrhizal fungi: use in sustainable agriculture and land restoration. Springer, Berlin, pp 357–373

Tawaraya K, Takaya Y, Turjaman M, Tuah SJ, Limin SH, Tamai Y, Cha JY, Wagatsuma T, Osaki M (2003) Arbuscular mycorrhizal colonization of tree species grown in peat swamp forests of Central Kalimantan, Indonesia. Forest Ecol Manage 182:381–386

Tsujino R, Yumoto T, Kitamura S, Djamaluddin I, Darnaedi D (2016) History of forest loss and degradation in Indonesia. Land Use Policy 57:335–347

Tuheteru FD, Wu QS (2017) Arbuscular mycorrhizal fungi and tolerance of waterlogging stress in plants. In: Wu QS (ed) Arbuscular mycorrhizas and stress tolerance of plants. Springer, Singapore, pp 43–66

Tuheteru FD, Kusmana C, Mansur I, Iskandar (2015) Response of Lonkida (*Nauclea orientalis* L.) towards mycorrhizal inoculum in waterlogged condition. Biotropia 22(1):61–71

Tuheteru FD, Husna, Arif A (2011a) Respon pertumbuhan dan ketergantungan *Albizia saponaria* (Lour.) Miq terhadap inokulasi Fungi Mikoriza Arbuskula local Sultra pada media tanah pascatambang nikel. Berita Biol 10(5):605–612

Tuheteru FD, Husna, Alimuddin LD (2011b) Respon pertumbuhan dan ketergantungan *Albizia saponaria* (Lour.) Miq terhadap fungi mikoriza arbuskula lokal Sultra. Biota 16(2):252–261

Tuheteru FD, Arif A, Widiastuti E, Rahmawati N (2017) Heavy metal uptake by indigenous arbuscular mycorrhizas of *Nauclea orientalis* L. and the potential for phytoremediation of serpentine soil. J Ilmu Kehutanan 2:76–84

Turjaman M, Tamai Y, Santoso E, Osaki M, Tawaraya K (2006a) Arbuscular mycorrhizal fungi incresead early growth of two nontimber forest product species *Dyera polyphylla* and *Aquilaria filaria* under greenhouse conditions. Mycorrhiza 16:459–464

Turjaman M, Santosa E, Sumarna Y (2006b) Arbuscular mycorrhizal fungi increased early growth of gaharu wood species *Aquilaria malaccensis* and *A. crasna* under greenhouse conditions. J Forest Res 3(2):139–148

Turjaman M, Santoso E, Tawaraya K (2007) Arbuscular mycorrhizal fungi increased plant growth and nutrient concentrations of milkwood tropical tree species *Alstonia scholaris* under greenhouse conditions. J Forest Res 4(2):61–71

Turjaman M, Tamai Y, Sitepu IR, Santoso E, Osaki M, Tawaraya K (2008) Improvement of early growth of two tropical peat-swampforest tree species *Ploiarium alternifolium* and *Calophyllum hosei* by two arbuscular mycorrhizal fungiunder greenhouse conditions. New Forest 36:1–12

UNEP-WCMC (2007) Strategies for the sustainable use and management of timber tree species subject to international trade: South East Asia. Cambridge

Urgiles N, Loja'n P, Aguirre N, Blaschke H, Günter S, Stimm B, Kottke I (2009) Application of mycorrhizal roots improves growth of tropical tree seedlings in the nursery: a step towards reforestation with native species in the Andes of Ecuador. New Forest 38:229–239

Wang F (2017) Arbuscular Mycorrhizas and Ecosystem Restoration. In: Wu QS (ed) Arbuscular mycorrhizas and stress tolerance of plants, Springer Singapore, pp 245-292

Wang B, Qiu YL (2006) Phylogenetic distribution and evolution of mycorrhizas in land plants. Mycorrhiza 16:299–363

Whitten AJ, Mustafa M, Henderson GS (1987) The ecology of Sulawesi. Gajah Mada University Press, Yogyakarta

Widiastuti H, Kramadibrata K (1992) Vesicular-arbuscular mycorrhizal fungi on selected acid soil of West Java. Menara Perkebunan 60(1):9–19

Widiastuti H, Kramadibrata K (1993) Identification of vesicular-arbuscular mycorrhizal fungi at selected oil palm estates of West Java. Menara Perkebunan 61(1):13–19

Wijayanti DR (2006) Identification arbuscule mycorrhiza fungi from *Aqularia* spp. Undergraduate Thesis, Bogor Agricultural University

Wu QS, Zou YN, Huang YM (2013) The arbuscular mycorrhizal fungus *Diversispora spurca* ameliorates effects of waterlogging on growth, root system architecture and antioxidant enzyme activities of citrus seedlings. Fungal Ecol 6(1):37–43

Wulandari DR (2001). Mikoriza vesikular arbuskula pada Talas. Skripsi, Institut Pertanian Bogor

Wulandari D, Saridi, Cheng W, Tawaraya K (2014) Arbuscular mycorrhizal colonization enhanced earlygrowth of *Mallotus paniculatus* and *Albizia saman* under nursery conditions in East Kalimantan, Indonesia. Int J Forest Res 1–8

Wulandari D, Saridi, Cheng W, Tawaraya K (2016) Arbuscular mycorrhizal fungal inoculation improves Albizia saman and Paraserianthes falcataria growth in post-opencast coal mine field in East Kalimantan, Indonesia. Forest Ecol Manage 376:67–73

Zhang Z, Zhang J, Huang Y (2014) Effects of arbuscular mycorrhizal fungi on the drought tolerance of *Cyclobalanopsis glauca* seedlings under greenhouse conditions. New Forests 45:545–556

Zubek S, Turnau K, Tsimilli-Michael M, Strasser RJ (2009) Response of endangered plant species to inoculation with arbuscular mycorrhizal fungi and soil bacteria. Mycorrhiza 19:113–123

Chapter 20
From Mycorrhizosphere to Rhizosphere Microbiome: The Paradigm Shift

Manju M. Gupta, Ashima Aggarwal, and Asha

20.1 Introduction

Mycorrhiza is the symbiotic association between the plant roots and the soil fungi. It is estimated that 74% of all plant species form arbuscular mycorrhiza (AM), 9% form orchid mycorrhiza, 2% form ectomycorrhizal (EM), and 1% form ericoid mycorrhiza associations (Van der Heijden et al. 2015). These fungus–root associations play a key role in terrestrial ecosystems as they regulate nutrient and carbon cycles. Mycorrhizal fungi provide up to 80% of the plant's N and P to get bread (carbohydrates) and butter (lipids) in return (Rich et al. 2017). The roots, both mycorrhizal and non-mycorrhizal, are the key source for providing various organic compounds in the habitat in the proximity of, on, and inside the root, which affects the composition, aeration properties, and biological activities of soil.

The term "mycorrhizosphere" is derived from "mycorrhiza" and "rhizosphere" (the region around roots). Since plant roots are commonly mycorrhizal, the rhizosphere concept was widened to include the fungal component of the symbiosis into it (Linderman 2008). Thus, the mycorrhizosphere is the zone influenced by both the root and the mycorrhizal fungus and includes the more specific term "hyphosphere," which refers only to the zone surrounding individual fungal hyphae (Johansson et al. 2004). The microbial habitats in the mycorrhizosphere are further divided into rhizosphere (soil–root interface), rhizoplane (root surface), and root endosphere (inside root). The three sub-habitats usually harbor different microorganisms (Fig. 20.1). Microbiota thriving on rhizoplane and within roots is selected by a host genotype-dependent differentiation (Bertin et al. 2003), which, in turn, influence the

M. M. Gupta (✉) · A. Aggarwal
Department of Botany, Sri Aurobindo College, University of Delhi, Delhi, India

Asha
Department of Energy and Environment, TERI University, Delhi, India

B. Giri et al. (eds.), *Root Biology*, Soil Biology 52,
https://doi.org/10.1007/978-3-319-75910-4_20

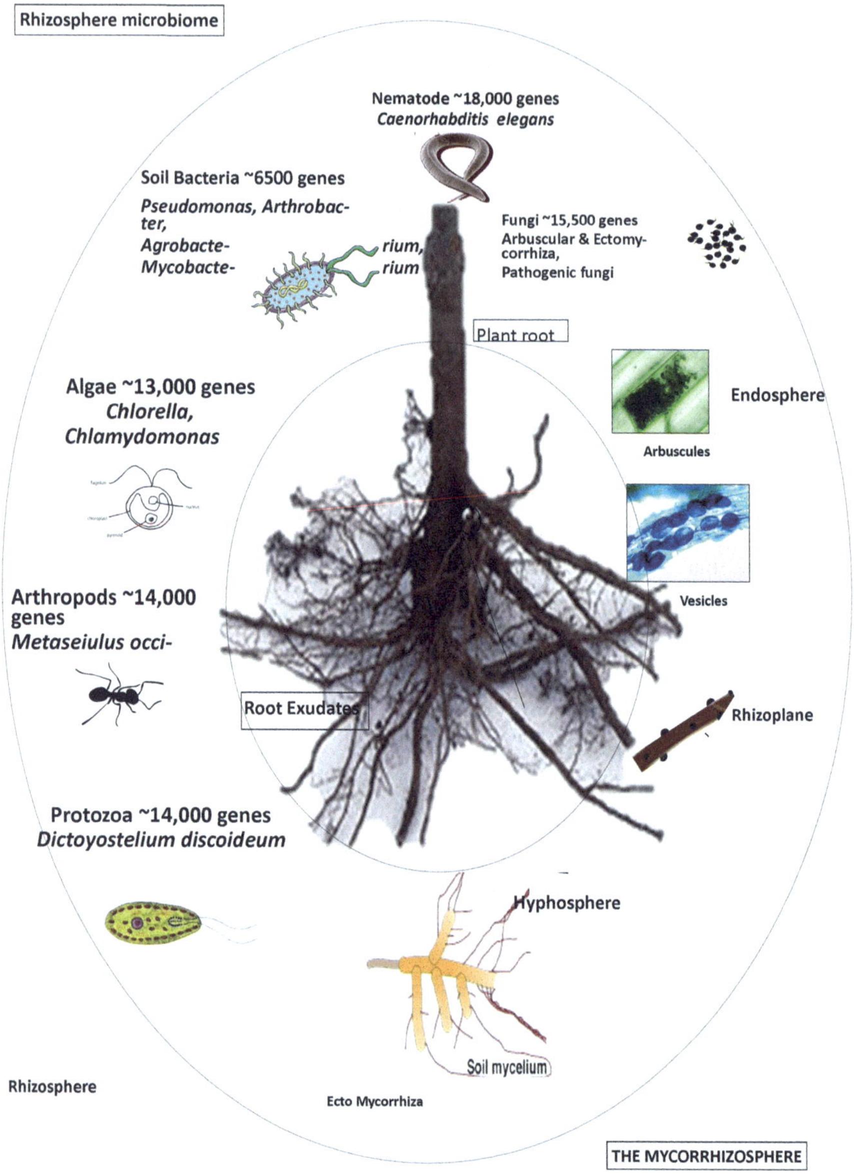

Fig. 20.1 Overview of mycorrhizosphere. The genome data given here is based on information given in Mendes et al. (2013)

plant resistance to pests, support beneficial symbioses, alter the chemical and physical properties of the soil, and inhibit the growth of competing plant species.

The mycorrhizosphere region is characterized by increased microbial activity stimulated by the leakage and exudation of organic substances from the root, called as root exudates (Bansal and Mukerji 1994, 1996; Bansal et al. 2000; Edwards et al. 2015). Plants release 10–20% of their photosynthates as exudates, which alter the physical and chemical properties of soil that in turn provides suitable niches for microbial proliferation (Edwards et al. 2015; Yuan et al. 2016). Root exudates include a wide range of compounds, like carbohydrates, amino acids, organic acids, fatty acids, nucleotides, flavones, vitamins, and enzymes (Bansal and Mukerji 1996). A positive correlation was indicated between mycorrhiza-induced changes in the qualitative and quantitative pattern of root exudation and mycorrhizosphere mycoflora (Bansal and Mukerji 1994).

Rhizosphere microbiome is a relatively new term, which refers to the diverse and dynamic community of microorganisms associated with plant roots that is not much different from (mycor)rhizosphere in its essence. However, it certainly indicates that the microorganisms studied are genomes or virtual taxa, using metagenomic methods. Studies of rhizosphere microbiome present a holistic view of diversity and interaction across the habitat. Consistent with the terminology used for microorganisms colonizing the human body, the collective communities of plant-associated microorganisms are referred as the plant microbiome or as the plants' other genome (Qin et al. 2010). In this context, plants are viewed as "superorganisms" which are partly dependent on their microbiome for specific functions and traits. This includes all plant-associated microbial habitats such as rhizosphere, spermosphere (seed surface), phyllosphere (leaf surface), and the stem microbiome. Recent applications of microbial metagenomics, metatranscriptomics, and metabolomics to plants and their surroundings have confirmed a key role of mycorrhizal fungi, rhizosphere bacteria, and fungi in determining the makeup of rhizosphere microbial community and suggested a world of hitherto undiscovered interactions in the rhizosphere (Dickie et al. 2015). This knowledge is leading to a paradigm-shifting view that plants are to be considered as a meta-organism or holobionts instead of isolated individuals.

Metagenomic analyses have provided a powerful lens for a holistic view of the microbial world in the rhizosphere and improved our understanding of entire rhizosphere functioning and microbial community interactions. Since the taxonomic identification of interacting microflora is not mandatory for biome-level studies, all rhizosphere microflora in soils could be characterized in workable details. This also overcomes the difficulties associated with the study organisms whose culturing is difficult—AM fungi, which cannot be cultured axenically and their taxonomy is difficult (Powell and Bennett 2016). Recent characterization of barcode sequences (Krüger et al. 2012) and development of dedicated environmental sequence databases, such as MaarjAM for AM fungi (Öpik et al. 2010, 2016), have made it possible to study and characterize AM fungal genomes in environmental samples. Powerful amplicon-based deep sequencing techniques provide more detailed and accurate insights into the diversity, structure, and assembly of microbial communities than previous clone library sequencing or PCR–DGGE (denaturing gradient gel electrophoresis) approaches (Guttman et al. 2014). Small subunit ribosomal RNA (16S SSU) gene and nuclear ribosomal internal transcribed spacer (ITS) and large subunit ribosomal RNA (28S LSU) have often been used as barcodes, for amplicon sequencing of bacterial and fungal communities (Qin et al. 2010; Krüger et al. 2012).

It is becoming evident with recent studies that interactions of mycorrhizal microbiomes play an important role in soil nutrient uptake and management of soilborne diseases in sustainable agricultural practices (Berruti et al. 2016). Different levels of interaction in rhizosphere microbes change their nutrition equation, which could be related to plant health (Kiers et al. 2016). It is suggested that these microbiomes are not passive players rather microbes that can alter host development, physiology, and systemic defenses, enable toxin production and disease resistance (Weller et al. 2012), increase host tolerance to stress and drought, modulate niche breadth, and change fitness outcomes in host interactions with competitors, predators, and pathogens (reviewed by Berg et al. 2014). In return, plants deposit their photosynthetically fixed carbon into their direct surroundings (Raaijmakers et al. 2009), thereby feeding the microbial community and influencing their composition and activities.

The present chapter primarily focuses on different aspects of the microbial community associated with mycorrhizal roots including diversity, interaction, and applications in enhancing the crop productivity. The main concepts and recent terminology used in rhizosphere microbiome studies, which are equally applicable to both mycorrhizal and non-mycorrhizal roots, are included.

20.2 Diversity and Interactions Across Mycorrhizal Microbiome

The (mycor)rhizosphere is considered as one of the most complex ecosystems on earth, which harbors numerous microorganisms. The number of (micro)organisms that constitute rhizosphere microbiome is much greater than the number of plant cells (Mendes et al. 2013). In addition, the number of microbial genes in the rhizosphere outnumbers the number of plant genes in a microbiome (Fig. 20.1). Organisms found in the (mycor)rhizosphere include bacteria, fungi, oomycetes, nematodes, protozoa, algae, viruses, archaea, and arthropods (Bansal et al. 2000). The interactions among them can be physical, i.e., for space, or physiological but are directed toward nutrient acquisition. Most members of the rhizosphere microbiome compete for the large amount of nutrients released by the plants as root exudates.

Several criteria have been used to group rhizosphere organisms. Mendes et al. (2013) classified them as "good," "bad," or "ugly" on the basis of their role in the rhizosphere. Microorganisms that have been well studied for their beneficial effects on plant growth and health are classified as "good" component of rhizosphere microbiome. These include the nitrogen-fixing bacteria, mycorrhizal fungi, plant growth-promoting rhizobacteria (PGPR), biocontrol microorganisms, mycoparasitic fungi, and protozoa. Rhizosphere microorganisms that are deleterious to plant growth and health are classified as "bad." These include the pathogenic fungi, oomycetes, bacteria, and nematodes. A third group of microorganisms, those found in the rhizosphere, are the human pathogens. These are classified as "ugly."

Over the past decade, there are an increasing number of reports describing the proliferation of human pathogenic bacteria in the rhizosphere soil (Kumar et al. 2013).

Microbial community present in different sub-habitats of rhizosphere microenvironments are frequently separated into rhizosphere, rhizoplane, and endosphere microflora, each possessing distinct features to which microorganisms have to adapt (Fig. 20.1) (McNear 2013; Van der Heijden and Schlaeppi 2015; Edwards et al. 2015). There is evidence that plant roots select these specific microbes in early growth stages and sustain a relatively stable community irrespective of growth stages (Edwards et al. 2015; Yuan et al. 2016). Vandenkoornhuyse et al. (2015) found that in almost all the cases, the diversity of microbes decreased from rhizosphere to endosphere, suggesting some strong filtering mechanism of habitats. The endosphere of roots has well-adapted microbial communities due to the pressure exerted by the host plant (Hernández et al. 2015). However, a systematic understanding of how overall rhizosphere communities and their members differ from or complement each other, in terms of functioning within the plant, across the plants and between the taxa, is still lacking.

Investigations on diversity and interaction of rhizosphere microbiome hold a great promise in solving the food and grain problem. A survey of research papers, published in the year 2017 (until now), revealed that rhizosphere microbiome has been investigated extensively and holds a potential to increase plant growth and production of important crop plants. Up to 37% of the total studies examined from citations in Google Scholar, JSTOR, and Catalogue, Harvard Library were devoted to application of rhizosphere microbiomes in increasing growth and productivity

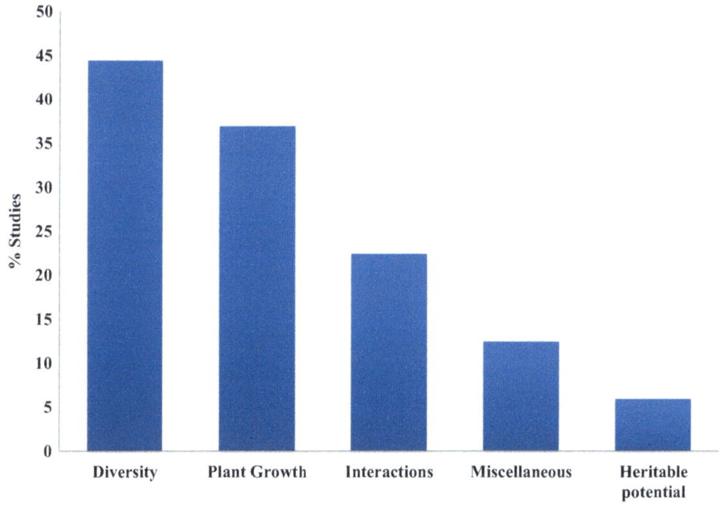

Fig. 20.2 Graphic representation of percentage studies on different aspects of rhizosphere microbiome in the year 2017 (till now) as revealed by Google Scholar, JSTOR, and Catalogue, Harvard Library

(Fig. 20.2). Many more papers have investigated microbial diversity with an indirect aim of increasing crop production.

Different plant species support unique microbiomes. Microbiome structure and function under different natural and agricultural environments have been explored in many plant species, including *Arabidopsis thaliana* (Schlaeppi et al. 2014), barley (*Hordeum vulgare*) (Bulgarelli et al. 2015), soybean (*Glycine max*) (Rascovan et al. 2016), corn (*Zea mays*) (Aira et al. 2010), wheat (*Triticum aestivum*) (Donn et al. 2015), rice (*Oryza sativa*) (Edwards et al. 2015), and cottonwood trees (*Populus trichocarpa*) (Shakya et al. 2013). Efforts are being made to develop a complete catalog of microbial species thriving in the rhizo- and endosphere of some model plants including *Arabidopsis thaliana* and *Populus* spp. (Hacquard and Schadt 2015; Lundberg et al. 2012) and crops such as maize and rice (Edwards et al. 2015; Peiffer et al. 2013) grown in their natural habitats, agricultural soils, or controlled artificial conditions. The rhizosphere microbiome communities associated with different genomic clones of wild-type and transgenic clones were reported to be different in *Populus* sp. (reviewed by Hacquard and Schadt 2015). Studies conducted in *Arabidopsis thaliana* (Lundberg et al. 2012; Wagner et al. 2014), *Zea mays*, and *Populus* demonstrated that within a host species, habitat and soil type, rather than host genetic background, have a larger influence on overall structure of microbiome (Bulgarelli et al. 2012; Lundberg et al. 2012; Peiffer et al. 2013).

Rhizosphere communities differ in different environments. Soil type and plant species are often believed to be the main factors affecting the structure of microbiomes (Agler et al. 2016; Lakshmanan et al. 2014; Lakshmanan 2015). In addition, climatic conditions, biogeography, agricultural practices, and plant domestication have also been suggested to contribute to the variation in the plant microbiome (Coleman-Derr et al. 2016; Pérez-Jaramillo et al. 2016; Sessitsch and Mitter 2015). Rhizosphere and root microbiomes have also been investigated in extreme environments, such as arid and saline soils (Soussi et al. 2015; Coleman-Derr et al. 2016; Fonseca-García et al. 2016; Valverde et al. 2016) and marine plants (Cúcio et al. 2016). These studies helped to clarify how plant in different habitat/niches, host environmental, soil, and geographic factors influence the rhizosphere microbiome community.

In a rhizosphere microbiome, not all of the microbes are needed to fulfil the ecological services to plants. The existence of functional redundancy in microbial communities across diverse environments is common (Dopheide et al. 2015; Alves-de-Souza et al. 2015). Based on relative occurrence of its members, the microbiomes can be classified as core or minimal microbiomes. A core microbiome (CM) is comprised of the members common to two or more microbial assemblages associated with a habitat (Turnbaugh et al. 2007; Hamady and Knight 2009) (Fig. 20.3). There are various ways to define the CM within a habitat using bioinformatic-based approaches. Shade and Handelsman (2012) suggested five parameters, including membership, composition, phylogeny, persistence, and connectivity, to discover the core microbiota based on a Venn diagram analysis. However, taxa occurring with low relative abundances may also be crucial in maintaining the community functions (Shi et al. 2016), thus less abundant taxa should not be overlooked. The concept of

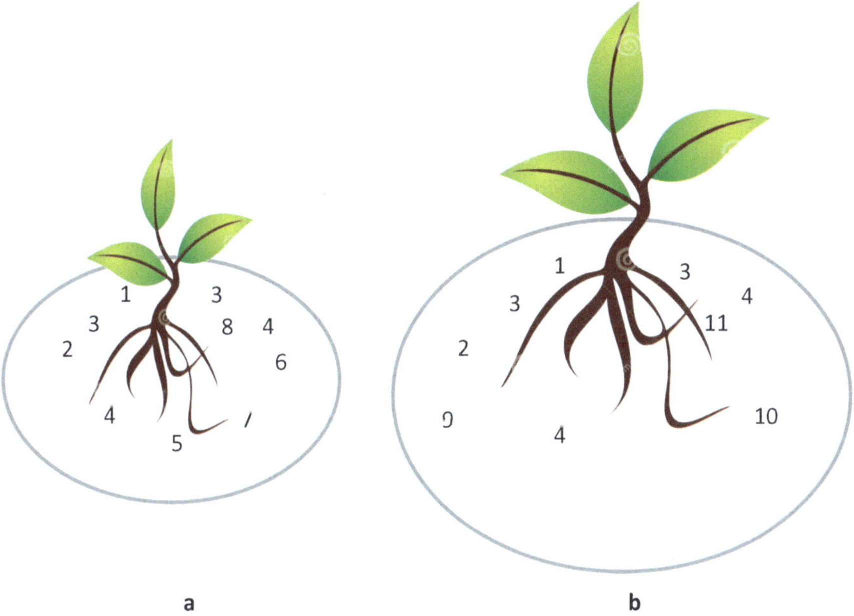

Fig. 20.3 Diagrammatic representation of different microbial communities present in rhizosphere in a given habitat. Figures **a** and **b** represent communities present in natural environment and synthetic community, respectively. Species numbers 1, 2, 3, and 4 are part of core community in the given habitat. Species 1 and 2 represent the minimal community without which plants do not survive; 5–8 are part of accessory microbiome, which can be replaced by other beneficial microbes; and 9–11 are synthetic beneficial microbes

minimal microbiome (MM) implied the smallest but functionally indispensable subset of the total microbiome (Raaijmakers 2015). The MM is composed only of indispensable members that can retain the key features of natural communities and thus are important for community assembly. Accessory microbes are those additional members that are not obligatory for community and could be replaced by other microorganisms. The ultimate goal of identifying such CMs or MMs is to exploit them in reconstruction of synthetic microbial consortium (SMC) with desirable member microbes (Hacquard 2016). SMCs are composed of multiple species with well-defined genetic background and help in accomplishing specific function through interactions among microorganisms.

20.3 Managing Rhizomicrobiome for Better Plant Health: The Rhizosphere Engineering

The rhizosphere microbiome engineering implies a multigenerational, artificial creation or selection of hosts that vary in microbiome content, thereby affecting the host traits (reviewed by Mueller and Sachs 2015; Dessaux et al. 2016; Ahkami et al. 2017). Rhizosphere microbiome diversity and their inheritance had been projected to be equally important as that of plant genome, since the number of genes in plant microbiome is more than the number of genes in a host (Mendes et al. 2013). The plants and the associated microbes are not seen individually as a unit of inheritance and evolution, rather as a holobiont or a superorganism. The approach involves microbial population engineering rather than single strain engineering. The rhizosphere engineering holds great promise for future plant breeding programs and biotechnological application.

It is widely known that plant phenotype is determined by plant genotype and environmental properties. The plant phenotype under the influence of local adaptation to abiotic stress (environment) is also a manifestation of change in subset of microbes associated with it. Microbiome assembly can be very sensitive to host genetic and environmental parameters and can vary even between different plant tissues. The rhizosphere management methods should primarily focus on the hypothesis of increase in yield by altering the dynamics of host genotype–environment–microbe interactions (Busby et al. 2017). Indeed, our ability to manage and manipulate microbiome is limited. There are three main approaches in building a productive microbiome—the first one relies on construction of a high-yielding microbial consortium, and second and third approaches involve manipulating the plant or the superorganism, respectively.

1. Developing microbial consortium—The most direct way to alter the microbiome is through inoculation with several strains or mixed cultures of AM or EM fungi, rhizobia, endophytes, etc. designated as biofertilizers. The concept of SMC is different from co-cultures, mixed cultures, microbial consortia, and other similar concepts in a way that it includes not only living together but also labor division (Fig. 20.3) (Rosier et al. 2006; Großkopf and Soyer 2014; De Roy et al. 2014; Van der Heijden et al. 2016). There are two ways for designing and constructing SMCs (Jiao et al. 2016). The first one is to reengineer naturally occurring microbial consortia, the top-down method. This starts from studies based on multiple omic analysis, macroscopic microbial consortium, and molecular mechanism in a natural field environment. The other one is bottom-up method, which begins with design and construction of artificial microbial consortia, based on engineering principles to obtain microbial consortia with higher efficiency, stability, and controllability. This method is more popular and applied more commonly.

   Products containing one or several species microbial consortia have been commercially available for decades—are in practice of being tried for most of important crops. However, most of these microbial species were isolated under traditional culture conditions, thus did not emulate the soil chemical environment (Verbruggen et al. 2013). Because of this reason, the inoculants often showed promising results under controlled lab and greenhouse conditions but did not consistently produce equivalent under natural field conditions in agricultural soils. Not only do key attributes like pH, nutrient stoichiometry, and texture differ among soils but also the climate regime experienced by microbes in the field spans a broad range. The conditions used to develop the synthetic microbial consortia must overlap with the multidimensional niche of the host plant for them to have a chance to survive, reproduce, and function. Another important issue that should be taken into account is the number of species to be included in SMCs. No general-purpose framework for the reconstruction of SMCs used to promote plant health is yet available (Busby et al. 2017).

Inoculation by recombinant strains of microbes is another strategy to enhance plant performance (reviewed by Quiza et al. 2015). Recombinant strains of mycorrhizal fungi are not yet developed; however, different combinations of recombinant soil bacteria have been extensively studied. For example, soil-polluting compound trichloroethylene (TCE) was removed from soils by using a wheat rhizosphere established by coating seeds with a recombinant, TCE-degrading *Pseudomonas fluorescens* strain that expresses the *tomA*$^{+}$ (toluene *o*-monooxygenase) genes from *Burkholderia cepacia* (Dennis et al. 1998).

2. Engineering plant traits—Plant-based strategies are designed to improve the plant productivity through the selection of a better-adapted microbiome. Approaches for engineering plant traits mainly include host plant genetic modifications and breeding (cultivar selection) (Nogales et al. 2016). Variations are induced by altering the physical and chemical environment in the rhizosphere through plant-affected characters, which change the spectrum of the fitness and interactions among microbes and evolution of new microbes better suited to the rhizosphere environment (Lambers et al. 2009). These changes in microbiome structure and function are usually attributed to differences in root exudate chemistry (Bais et al. 2006; Rasmann and Turlings 2016), in root architecture, and in plant nutrient uptake rates (Bell et al. 2015), which make it possible to engineer these traits into crops through gene-editing tools. Since many genes controlling exudates have been identified, there have been few attempts to engineer the rhizosphere by manipulating the root exudates. One such example is transgenic rice and tomato plants transformed with the *Arabidopsis* vacuolar H^{+}-pyrophosphatase *gene* AVP1, which showed approximately 50% greater citrate and malate efflux than wild types when treated with $AlPO_4$ (Yang et al. 2007). This was interpreted as a means to enhance resistance to Al^{3+} stress and improve the ability to utilize insoluble phosphorus. However, it is important to note that plant engineering to impact rhizosphere could be a very complex process due to degradation or inactivation of the engineered compound in the soil, small rate of exudation to influence the rhizosphere. More studies on the root exudate composition and effect of change of exudate release time and levels on plant development would expedite the application of this approach (Huang et al. 2014).

Selecting a naturally occurring plant species or cultivar with a high capacity to recruit a beneficial microbiome is another approach that has been explored. This approach seems promising as it emulates the interactions that support beneficial microbes in natural systems and which were selected through evolution of the holobiont. Some of the plant traits under breeding selection are already known to be linked to the microbiome. For example, plant phenology, nutrient uptake, and defense have been shown to be influenced by the soil microbiome (Wagner et al. 2014; Panke-Buisse et al. 2015). Plant breeding programs, which include specific microbiome functions, target only a very specific taxa or function of that taxon.

3. The meta-organism route—The meta-organism or superorganism approach is based on the fact that both microbiome and the plants are highly dependent on each other as the microbiome contributes a significant portion of the secondary genome of the host plant. The heritability of the meta-organisms is not solely dependent on the genetics of microbes but the genetics of host plant as well. The study conducted on 27 modern inbred maize rhizosphere revealed that heritability of microbiome also depends on other factors like host plant species and physical and chemical properties of soil (Peiffer et al. 2013). However, this route needs to be explored with more plant species.

20.4 Concluding Remarks

The rhizosphere microbiome includes all the microbial partners of plant root present in the soil. Microbiome research of both mycorrhizal and non-mycorrhizal roots, which targets to increase the crop productivity, should follow five research priorities as summarized by Busby et al. (2017).

1. Develop model host microbiome systems for crop plants and non-crop plants with associated microbial culture collections and reference genomes.
2. Define core microbiomes and metagenomes in model host microbiome systems.
3. Elucidate the rules of synthetic, functionally programmable microbiome assembly.
4. Determine functional mechanisms of plant–microbiome interactions.
5. Characterize and refine plant genotype-by-environment-by-microbiome-by-management interactions.

Considering the wide potential applications of microbiome research, startups focus on the microbiome as an organic solution to increase crop yields. For example, NewLeaf Symbiotics (https://www.crunchbase.com/organization/newleaf-symbiotics#/entity), BioConsortia (http://bioconsortia.com), and Indigo (www.indigoag.com), the start-up companies, have generated several robust pipelines for identifying microbial consortia for improving almost all plant traits. We still look forward for more microbiome-based products to be discovered in the future.

Acknowledgments MMG and AA are grateful to the University Grants Commission for funding the UGC-Major Project MRP-MAJOR-BOTA-2013-21235.

References

Agler MT, Ruhe J, Kroll S, Morhenn C, Kim ST, Weigel D, Kemen EM (2016) Microbial hub taxa link host and abiotic factors to plant microbiome variation. PLoS Biol 14:1–31

Ahkami A, White RA, Handakumbura PP, Jansson C (2017) Rhizosphere engineering: enhancing sustainable plant ecosystem productivity. Rhizosphere 3:233–243

Aira M, GoÂmez-BrandoÂn M, Lazcano C, Bååth E, DomõÂnguez J (2010) Plant genotype strongly modifies the structure and growth of maize rhizosphere microbial communities. Soil Biol Biochem 2:2276–2281

Alves-de-Souza C, Pecqueur D, Le Floc'h F, Mas S, Roques C, Mostajir B (2015) Significance of plankton community structure and nutrient availability for the control of dinoflagellate blooms by parasites: a modeling approach. PLoS One 10:e0127623

Bais HP, Weir TL, Perry LG, Gilory S, Vivanco JM (2006) The role of root exudates in rhizosphere interactions with plants and other organisms. Annu Rev Plant Biol 57:233–266

Bansal M, Mukerji KG (1994) Positive correlation between root exudation and VAM induced changes in rhizosphere mycoflora. Mycorrhiza 5:39–44

Bansal M, Mukerji KG (1996) Root exudates in rhizosphere biology. In: Mukerji KG, Singh VP, Dwivedi S (eds) Concepts in applied microbiology and biotechnology. Aditya, New Delhi, pp 98–120

Bansal M, Naqvi N, Chamola BP, Mukerji KG (2000) Mycorrhizosphere: interaction between rhizosphere microflora and VAM fungi. In: Mukerji KG, Chamola BP, Singh J (eds) Mycorrhizal biology. Kluwer, Dordrecht, pp 143–152

Bell TH, Hurteau BC, Al-Otaibi F, Turmel MC, Yergeau E, Courchesne F, St-Arnaud M (2015) Early rhizosphere microbiome composition is related to the growth and Zn uptake of willows introduced to a former landfill. J Environ Biol 17:3025–3038

Berg G, Rube M, Schloter M, Smalla K (2014) Unraveling the plant microbiome: looking back and future perspectives. Front Microbiol 5:1–7

Berruti A, Lumini E, Balestrini R, Bianciotto V (2016) Arbuscular mycorrhizal fungi as natural biofertilizers: let's benefit from past successes. Front Microbiol 6:1559

Bertin C, Yang X, Weston LA (2003) The role of root exudates and allelochemicals in the rhizosphere. Plant Soil 256:67–83

Bulgarelli D, Rott M, Schlaeppi K, Ver Loren van Themaat E, Ahmadinejad N, Assenza F, Rauf P, Huettel B, Reinhardt R, Schmelzer E, Peplies J, Gloeckner FO, Amann R, Eickhorst T, Schulze-Lefert P (2012) Revealing structure and assembly cues for *Arabidopsis* root-inhabiting bacterial microbiota. Nature 488:91–95

Bulgarelli D, Garrido-Oter R, Münch PC, Weiman A, Dröge J, Pan Y, McHardy AC, Schulze-Lefert P (2015) Structure and function of the bacterial root microbiota in wild and domesticated barley. Cell Host Microbe 17:392–403

Busby PE, Soman C, Wagner MR, Friesen ML, Kremer J, Bennett A, Morsy M, Eisen JA, Leach JE, Dangl JL (2017) Research priorities for harnessing plant microbiomes in sustainable agriculture. PLoS Biol 15:1–14

Coleman-Derr D, Desgarennes D, Fonseca-Garcia C, Gross S, Clingenpeel S, Woyke T, North G, Visel A, Partida-Martinez LP, Tringe SG (2016) Plant compartment and biogeography affect microbiome composition in cultivated and native *Agave* species. New Phytol 209:798–811

Cúcio C, Engelen AH, Costa R, Muyzer G (2016) Rhizosphere Microbiomes of European Seagrasses are selected by the plant, but are not species specific. Front Microbiol 7:440–456

De Roy K, Marzorati M, Van den Abbeele P, Van de Wiele T, Boon N (2014) Synthetic microbial ecosystems: an exciting tool to understand and apply microbial communities. Environ Microbiol 16:1472–1481

Dennis CY, Jennifer AM, Thomas KW (1998) Rhizoremediation of trichloroethylene by a recombinant, root-colonizing *Pseudomonas fluorescens* strain expressing toluene *ortho*-monooxygenase constitutively. Appl Environ Microbiol 64:112–118

Dessaux Y, Grandclement C, Faure D (2016) Engineering the rhizosphere. Trends Plant Sci 21:266–278

Dickie AI, Alexander I, Lennon S, Opik M, Selosse M, Marcel HAG, Martin MF (2015) Evolving insights to understanding mycorrhizas. New Phytol 205:1369–1374

Donn S, Kirkegaard JA, Perera G, Richardson AE, Watt M (2015) Evolution of bacterial communities in the wheat crop rhizosphere. Environ Microbiol 17:610–621

Dopheide A, Lear G, He Z, Zhou J, Lewis GD (2015) Functional gene composition, diversity and redundancy in microbial stream biofilm communities. PLoS One 10:e0123179 (1–21)

Edwards J, Johnson C, Santos-MedellõÂn C, Lurie E, Podishetty NK, Bhatnagar S (2015) Structure, variation and assembly of the root-associated microbiomes of rice. PNAS 112:E91–E20

Fonseca-García C, Coleman-Derr D, Garrido E, Visel A, Tringe SG, Partida-Martínez LP (2016) The cacti microbiome: interplay between habitat-filtering and host-specificity. Front Microbiol 7:150

Großkopf T, Soyer OS (2014) Synthetic microbial community. Curr Opin Microbiol 18:72–77

Guttman DS, McHardy AC, Schulze-Lefert P (2014) Microbial genome-enables insights into plant-microorganism interactions. Nat Rev Genet 15:797–813

Hacquard S (2016) Disentangling the factors shaping microbiota composition across the plant holobiont. New Phytol 209:454–457

Hacquard S, Schadt CW (2015) Towards a holistic understanding of the beneficial interactions across the *Populus* microbiome. New Phytol 205:1424–1430

Hamady M, Knight R (2009) Microbial community profiling for human microbiome projects: tools, techniques and challenges. Genome Res 19:1141–1152

Heijden MG, Martin FM, Selosse MA, Sanders IR (2015) Mycorrhizal ecology and evolution: the past, the present and the future. New Phytol 205:1406–1423

Hernández M, Dumon M, Yuan Q, Conrad R (2015) Different bacterial populations associated with the roots and rhizosphere of rice incorporate plant-derived carbon. Appl Environ Microbiol 81:2244–2253

Huang XF, Chaparro JM, Reardon KF, Zhang RF, Shen QR, Vivanco JM (2014) Rhizosphere interactions root exudates microbes and microbial communities. Botany 92:267–275

Jiao X, Lyu Y, Wu X, Li H, Cheng L, Zhang C, Shen J (2016) Grain production versus resource and environmental costs: towards increasing sustainability of nutrient use in China. J Exp Bot 67:4935–4949

Johansson JF, Paul LR, Finlay RD (2004) Microbial interactions in the mycorrhizosphere and their significance for sustainable agriculture. FEMS Microbiol Ecol 48:1–13

Kiers ET, West SA, Wyatt GA, Gardner A, Bücking H, Werner GD (2016) Misconceptions on the application of biological market theory to the mycorrhizal symbiosis. Nat Plants 2:160–163

Krüger M, Krüger C, Walker C, Stockinger H, Schüßler A (2012) Phylogenetic reference data for systematics and phylotaxonomy of arbuscular mycorrhizal fungi from phylum to species level. New Phytol 193:970–984

Kumar A, Munder A, Aravind R, Eapen SJ, Tümmler B, Raaijmakers JM (2013) Friend or foe: genetic and functional characterization of plant endophytic *Pseudomonas aeruginosa*. Environ Microbiol 15:764–779

Lakshmanan V (2015) Root microbiome assemblage is modulated by plant host factors. Adv Bot Res 75:57–79

Lakshmanan V, Selvaraj G, Bais H (2014) Functional soil microbiome: belowground solutions to an above ground problem. Plant Physiol 166:689–700

Lambers H, Mougel C, Jaillard B, Hinsinger P (2009) Plant-microbe-soil interactions in the rhizosphere: an evolutionary perspective. Plant Soil 321:83–115
Linderman RG (2008) The mycorrhizosphere phenomenon. In: Feldman F, Kapulnik Y, Barr J (eds) Mycorrhiza works. Deutsche Phytomedizinische Gesellschaft, Braunschweig, pp 341–355
Lundberg DS, Lebeis SL, Paredes SH, Yourstone S, Gehring J, Malfatti S, Dangl JL (2012) Defining the core *Arabidopsis thaliana* root microbiome. Nature 488:86–90
McNear DH Jr (2013) The rhizosphere—roots, soil and everything in between. Nat Educ Knowl 4:1
Mendes R, Garbeva P, Raaijmakers JM (2013) The rhizosphere microbiome: significance of plant beneficial, plant pathogenic and human pathogenic microorganisms. FEMS Microbiol Rev 37:634–663
Mueller UG, Sachs JL (2015) Engineering microbiomes to improve plant and animal health. Trends Microbiol 2:606–661
Nogales A, Nobre T, Valadas V, Ragonezi C, Döring M, Polidoros A, Arnholdt-Schmitt B (2016) Can functional hologenomics aid tackling current challenges in plant breeding? Brief Funct Genomics 15:288–297
Öpik M, Vanatoa A, Vanatoa E, Moora M, Davison J, Kalwij JM, Reier Ü, Zobel M (2010) The online database Maarj*AM* reveals global and ecosystemic distribution patterns in arbuscular mycorrhizal fungi (Glomeromycota). New Phytol 188:223–241
Öpik M, Davison J, Moora M, Partel M, Zobel M (2016) Response to comment on "Global assessment of arbuscular mycorrhizal fungus diversity reveals very low endemism". Science 351:826
Panke-Buisse K, Poole AC, Goodrich JK, Ley RE, Kao-Kniffin J (2015) Selection on soil microbiomes reveals reproducible impacts on plant function. ISME J 9:980–989
Peiffer JA, Spor A, Koren O, Jin Z, Tringe SG, Dangl JL, Buckler ES, Ley RE (2013) Diversity and heritability of the maize rhizosphere microbiome under field conditions. PNAS 110:6548–6553
Pérez-Jaramillo JE, Mendes R, Raaijmakers JM (2016) Impact of plant domestication on rhizosphere microbiome assembly and function. Plant Mol Biol 90:635–644
Powell JR, Bennett AE (2016) Unpredictable assembly of arbuscular mycorrhizal fungal communities. Pedobiologia 59:11–15
Qin J, Li R, Raes J, Arumugam M, Burgdorf KS, Manichanh C et al (2010) A human gut microbial gene catalogue established by metagenomic sequencing. Nature 464:59–65
Quiza L, St-Arnaud M, Yergeau E (2015) Harnessing phytomicrobiome signaling for rhizosphere microbiome engineering. Front Plant Sci 6:507–519
Raaijmakers JM (2015) The minimal rhizosphere microbiome. In: Lugtenberg B (ed) Principles of plant-microbe interactions. Springer, Heidelberg, pp 411–417
Raaijmakers JM, Paulitz TC, Steinberg C, Alabouvette C, Moënne-Loccoz Y (2009) The rhizosphere: a playground and battlefield for soilborne pathogens and benefical microorganisms. Plant Soil 321:341–361
Rascovan N, Carbonetto B, Perrig D, Díaz M, Canciani W, Abalo M, Alloati J, González-Anta G, Vazquez MP (2016) Integrated analysis of root microbiomes of soybean and wheat from agricultural fields. Sci Rep 6:28084
Rasmann S, Turlings T (2016) Root signals that mediate mutualistic interactions in the rhizosphere. Curr Opin Plant Biol 32:62–68
Rich MK, Nouri E, Courty PE, Reinhardt D (2017) Diet of arbuscular mycorrhizal fungi—bread & butter? Trends Plant Sci 22:652–660
Rosier CL, Hoye AT, Rillig MC (2006) Glomalin related soil protein: assessment of current detection and quantification tools. Soil Biol Biochem 38:2205–2211
Schlaeppi K, Dombrowski N, Oter RG, van Themaat EVL, Schulze-Lefert P (2014) Quantitative divergence of the bacterial root microbiota in *Arabidopsis thaliana* relatives. PNAS 111:585–592
Sessitsch A, Mitter B (2015) 21st century agriculture: integration of plant microbiomes for improved crop production and food security. Microbial Biotechnol 8:32–33

Shade A, Handelsman J (2012) Beyond the venn diagram: the hunt for a core microbiome. Environ Microbiol 14:4–12

Shakya M, Gottel N, Castro H, Yang ZK, Gunter L, Labbe ÂJ (2013) A multifactor analysis of fungal and bacterial community structure in the root microbiome of mature *Populusdeltoides* trees. PLoS One 8:e76382

Shi S, Nuccio EE, Shi ZJ, He Z, Zhou J, Firestone MK, Johnson N (2016) The interconnected rhizosphere: high network complexity dominates rhizosphere assemblages. Ecol Lett 1–11

Soussi A, Ferjani R, Marasco R, Guesmi A, Cherif H, Rolli E, Mapelli F, Ouzari HI, Daffonchio D, Cherif A (2015) Plant-associated microbiomes in arid lands: diversity, ecology and biotechnological potential. Plant Soil 405:357–370

Turnbaugh PJ, Ley RE, Hamady M, Fraser-Liggett C, Knight R, Gordon JI (2007) The human microbiome project: exploring the microbial part of ourselves in a changing world. Nature 449:804–810

Valverde A, Pieter DM, Oberholster T, Henschel J, LouwM K, Donald C (2016) Specific microbial communities associate with the rhizosphere of *Welwitschia mirabilis*, a living fossil. PLoS One 11:1–11

Vandenkoornhuyse P, Quaiser A, Duhamel M, Le Van A, Dufresne A (2015) The importance of the microbiome of the plant holobiont. New Phytol 206:1196–1206

Van der Heijden MGA, Schlaeppi K (2015) Root surface as a frontier for plant microbiome research. PNAS 112:2299–2300

Van der Heijden MGA, de Bruin S, Luckerhoff L, van Logtestijn RSP, Schlaeppi K (2016) A widespread plant-fungal-bacterial symbiosis promotes plant biodiversity, plant nutrition and seedling recruitment. ISME J 10:389–399

Verbruggen E, Heijden MG, Rillig MC, Kiers ET (2013) Mycorrhizal fungal establishment in agricultural soils: factors determining inoculation success. New Phytol 197:1104–1109

Wagner MR, Lundberg DS, Coleman-Derr D, Tringe SG, Dangl JL, Mitchell-Olds T (2014) Natural soil microbes alter flowering phenology and the intensity of selection on flowering time in a wild *Arabidopsis* relative. Ecol Lett 17:717–726

Weller DM, Mavrodi DV, van Pelt JA, Pieterse CM, van Loon LC, Bakker PA (2012) Induced systemic resistance in *Arabidopsis thaliana* against *Pseudomonas syringae* pv. *tomato* by 2,4-diacetylphloroglucinol-producing *Pseudomonas fluorescens*. Phytopathology 102:403–412

Yang H, Knapp J, Koirala P, Rajagopal D, Peer WA, Silbart LK (2007) Enhanced phosphorus nutrition in monocots and dicots over-expressing a phosphorus-responsive type IH-+-pyrophosphatase. Plant Biotechnol J 5:735–745

Yuan Z, Druzhinina IS, Labbé J, Redman R, Qin Y, Rodriguez R et al (2016) Specialized microbiome of a halophyte and its role in helping non-host plants to withstand salinity. Sci Rep 6:32467

Chapter 21
Growth Response of Different Species and Provenances of Jujube Seedlings to Inoculation with Arbuscular Mycorrhizal Fungi

B. Thioye, A. Kane, S. M. de Faria, D. Fall, D. Sanogo, C. Ndiaye, K. B. Sanon, A. Soule, R. Duponnois, S. N. Sylla, and A. M. Bâ

21.1 Introduction

Domestication of fruit trees could be achieved through a combination of approaches including selection and multiplication of quality planting material, fertilization, irrigation, pruning, and controlled mycorrhization (Bâ et al. 2003). One of the promising approaches that can be used in the domestication of indigenous fruit trees is the arbuscular mycorrhizal (AM) inoculation, since it is established that

B. Thioye
Laboratoire Commun de Microbiologie (LCM) IRD/ISRA/UCAD, Dakar, Sénégal

Laboratoire des Symbioses Tropicales et Méditerranéennes (LSTM), Montpellier, France

Laboratoire de biologie et physiologie végétales, Université des Antilles, Pointe à Pitre, France

A. Kane · C. Ndiaye · S. N. Sylla
Laboratoire Commun de Microbiologie (LCM) IRD/ISRA/UCAD, Dakar, Sénégal

S. M. de Faria
Embrapa Agrobiologia, Seropedica, Rio de Janeiro, Brazil

D. Fall · D. Sanogo
Centre National de Recherches Forestières, Dakar, Sénégal

K. B. Sanon
Institut de l'Environnement et de Recherches Agricoles, Ouagadougou, Burkina Faso

A. Soule
Ecole Normale Supérieure de Nouakchott, Nouakchott, Mauritanie

R. Duponnois
Laboratoire des Symbioses Tropicales et Méditerranéennes (LSTM), Montpellier, France

A. M. Bâ (✉)
Laboratoire des Symbioses Tropicales et Méditerranéennes (LSTM), Montpellier, France

Laboratoire de biologie et physiologie végétales, Université des Antilles, Pointe à Pitre, France
e-mail: amadou.ba@ird.fr

B. Giri et al. (eds.), *Root Biology*, Soil Biology 52,
https://doi.org/10.1007/978-3-319-75910-4_21

many of these fruit trees benefit in terms of growth and mineral nutrition from this symbiotic association (Guissou et al. 1998, 2016; Bâ et al. 2000, 2001; Mathur and Vyas 2000; Sidibé et al. 2012). The AM symbioses are associations between roots of terrestrial plants and members of the fungal phylum *Glomeromycota* (Schüßler et al. 2001). These are the most common and widespread symbiosis involving 80% of land plants and at least 250 morphologically defined arbuscular mycorrhizal fungi (AMF) (Davison et al. 2015). The AMF receive plant-synthesized carbon and increase capacity of plants for nutrient capture through its network of external hyphae (Smith and Read 2008). The AMF are promoted as biofertilizers for sustainable agriculture (Verbruggen et al. 2013; Hart et al. 2015). Nevertheless, they are poorly investigated, particularly with the indigenous fruit trees from the Sahelian and Sudanian zones of West Africa.

Ziziphus mauritiana Lam., commonly named jujube, is one of the indigenous fruit tree species farmers maintain on their farms as a source of food and income in West Africa (Ouédraogo et al. 2006). Jujube is a multipurpose tree providing mainly fruits and fodder. It also planted to reforest degraded soils and as hedgerows to protect crops. Nearly every part of *Ziziphus* plants can be utilized. Due to the high dry weight protein content, leaves are an important source of protein for cattle (Ngwa et al. 2000; Arndt and Kayser 2001). Responses of *Z. mauritiana* to AM inoculation differ with respect to functional compatibility, measured as mycorrhizal formation, root colonization, nutrient absorption, and morphological properties of the root (Guissou et al. 1998; Bâ et al. 2000, 2001; Mathur and Vyas 2000; Sidibé et al. 2012). Guissou et al. (1998) showed that *Rhizophagus irregularis*, isolate IR27 (syn. *Glomus aggregatum* IR27; Bâ et al. 1996), was one of the AMF providing high growth and mineral nutrition benefits for *Z. mauritiana* seedlings. Mycorrhizal dependency (MD) of *Z. mauritiana* as defined by Plenchette et al. (1983) can reach a maximum of 78%. Similar results were found by Guissou et al. (2001) and Sidibé et al. (2012). Bâ et al. (2001, 2003) concluded that the absence of AM inoculation on *Z. mauritiana* seedlings in nursery could lead to higher mortality of outplanted jujube trees in the field. It is also well known that responses of AM inoculation vary greatly from one plant species to another and even between origin and cultivars within a single species (Smith et al. 2009). Guissou et al. (2016) highlighted the importance of considering seed provenance of *Z. mauritiana* when performing preselection of mycotrophic plant candidates prior to large-scale fruit tree propagation in orchards and agroforestry systems.

The complex taxonomy in the genus *Ziziphus* is subjected to debates and could be between 86 and 170 species (Azam-Ali et al. 2006). However, the two major cultivated species in agroforestry systems are *Z. mauritiana*, the Indian jujube, and *Z. jujuba* Mill., the Chinese jujube (Azam-Ali et al. 2006). Many of other species of *Ziziphus* are underutilized crops despite their potential interests in agroforestry systems as fruits, firewood, fodder, medicines, and live hedge (Soule 2011). Except for *Z. mauritiana*, the effectiveness of AMF on growth and mineral nutrition of *Ziziphus* spp. is not known. Arbuscular mycorrhizae may have a particular importance for these multipurpose fruit trees because P is often the limiting nutrient in agroforestry systems. The objective of this work was to investigate the effects of three AMF (*Rhizophagus irregularis* isolate IR27, *Funneliformis mosseae* isolate DAOM227131, and *Rhizophagus irregularis* isolate DAOM197198) on MD and mineral nutrition of seven species of *Ziziphus* (*Z. mauritiana*, *Z. lotus*, *Z. spina-christi*, *Z. mucronata*, *Z. amphibia*, *Z. abyssinica*, and *Z. sphaerocarpa*) and six provenances of *Z. mauritiana* (Senegal, Mali, Mauritania, Burkina Faso, Niger, and India) in greenhouse conditions.

21.2 Materials and Methods

The soil used in the experiment was collected from Sangalkam (14°46′N, 17°13′W), Senegal. It was a sandy soil with 88.8% sand, 5.8% silt, 5.4% clay, 0.6% organic matter, 0.3% total C, 0.02% total N, ratio C/N – 14, 333.5 ppm total K, 41.4 ppm total P, 2.1 ppm P-Bray 1, 1.03 ppm Ca, 0.3 ppm Mg, pH = 6.0 of a soil/water mixture (ratio 1:2, v/v), and pH = 4.6 of a soil/KCl mixture (ratio 1:2, v/v). The soil was passed through a 2 mm sieve, sterilized for 4 h in an autoclave oven system at 180 °C to eliminate native AMF and transferred into plastic bags (1.5 kg soil per plastic bag).

Three isolates of AMF were used: *Rhizophagus irregularis* isolate IR27 (syn. *Glomus aggregatum* IR27) (Bâ et al. 1996), *Funneliformis mosseae* DAOM227131 (TH Nicolson and Gerd.) C. Walker and A. Schüßler (Redecker et al. 2013), and *Rhizophagus irregularis* DAOM197198 (Błaszk., Wubet, Renker and Buscot) C. Walker and A. Schüßler comb. nov. (Stockinger et al. 2009). The AMF were provided by the LCM laboratory (IRD, Dakar, Senegal, certified ISO 9001, version 2000). They were propagated on maize (*Zea mays* L.) for 3 months on sterilized sandy soil in greenhouse conditions. Mycorrhizal inoculation of the soil was achieved by placing 20 g portions of a crude inoculum of AMF consisting of sand, spores, fragments of hyphae, and maize root segments below the seeds during transplanting. The inoculum density of *R. irregularis* IR27, *F. mosseae*, and *R. irregularis* DAOM197198 was calibrated by the most probable number method (Adelman and Morton 1986) as 1635, 1023, and 1347 infective propagules per 20 g of inoculum, respectively. Controls also received 20 g of autoclaved crude inoculum of AMF with a 10 ml water extract of the inoculum by filtration (Whatman No. 1 filter paper) as controls to balance composition of the microbial community between inoculated and non-inoculated plants.

Seeds of seven species of *Ziziphus* (*Z. mauritiana* Lam, *Z. lotus* (L.) Lam, *Z. spina-christi* (L.) Desf., *Z. mucronata* Willd, *Z. amphibia* A. Chev., *Z. abyssinica* A. Rich., and *Z. sphaerocarpa* Tul.) and six provenances of *Z. mauritiana* (Senegal, Mali, Mauritania, Burkina Faso, Niger, and India) (Table 21.1) were surface-sterilized with 1% NaOCl for 15 min, washed several times, and soaked in sterile distilled water for 30 min before being planted in the soil as three per plastic bag (24 cm × 7.5 cm). Plants were grown in the nursery at ISRA/IRD Research Center, Bel Air, Dakar, Senegal (14°44′N, 17°30′W) under natural sunlight (35 °C day, 27 °C night, relative humidity 75%, and 14 h photoperiod).

Table 21.1 Seed provenances and main uses of the six provenances of *Z. mauritiana* and the seven *Ziziphus* spp.

Jujube trees species	Countries of provenances	Seed collection sites	Main uses
Z. lotus	Mauritania	Bouguedra	Food, live hedge
Z. spina-christi	Mauritania	Maghama	Food, timber
Z. mucronata	Mauritania	Maghama	Forage, firewood
Z. amphibia	Mauritania	Bouguedra	Food, medicine
Z. abyssinica	Burkina Faso	Dindéresso	Food, medicine
Z. sphaerocarpa	French West Indies	Bois Jolan	Food
Z. mauritiana	Mauritania	Atar	Food[a], forage
Z. mauritiana	Senegal	Tasset	Food[a], forage
Z. mauritiana	Mali	Segou	Food[a], forage
Z. mauritiana	Burkina Faso	Dori	Food[a], forage
Z. mauritiana	Niger	Sadoré	Food[a], forage
Z. mauritiana	India	Sangalkam	Food[a], forage

[a]Food: the fruit part

After emergence, the seedlings were thinned to one plant per plastic bag. Two greenhouse experiments were set up: (1) a 4×6 factorial design consisting of three AMF and control and six provenances of *Z. mauritiana* and (2) a 4×7 factorial design consisting of three AMF and control and seven *Ziziphus* spp. Each experiment was arranged in a completely randomized design with 15 replicates per treatment combination.

Four months after sowing, plants were harvested to measure height, dry weight of shoots, and roots (48 h at 70 °C). The MD of each plant of *Ziziphus* spp. and provenance was calculated using the formula:

$$\mathrm{MD}\ (\%) = 100 \times \frac{(\mathrm{TDWM} - \mathrm{TDWNM})}{\mathrm{TDWM}}$$

where TDW_M and TDW_{NM} are total dry weight of mycorrhizal and non-mycorrhizal plants, respectively (Plenchette et al. 1983) (Table 21.1).

For mycorrhizal root infection measurement, a part of fresh fine roots was collected from the root system of each seedling. Roots were gently washed under tap water, bleached (KOH, 10%) at 80 °C during 30 min, and stained in 0.05% trypan blue at 80 °C during 35 min following the method of Phillips and Hayman (1970). Percentage of root length colonized by AMF was assessed at 40× magnification using 100 fragments of lateral roots (approximately 1 cm length) on microscopic slides. Mycorrhizal root colonization was evaluated by using the method of Trouvelot et al. (1986). After drying, leaf tissues of each plant were ground, mineralized through heating at 500 °C, and digested in 2 mLHCl (6N) and 10 mL HNO_3. Total K content was determined by the atomic absorbance, and total P content was determined by colorimetry through a spectrophotometer at 660 nm. Analyses were performed in the Agricultural Chemistry Laboratory of Embrapa, Rio de Janeiro, Brazil.

To measure the length of hyphae of each AMF, 2 g of soil samples from each treatment (three replicates per treatment) were blended with 500 mL distilled water, and 30 mL aliquots of this were filtered and stained with trypan blue on gridded membrane filters (1.2 μm) according to the method described by Jakobsen and Rosendahl (1990). The gridded membrane filters were mounted on slides, and hyphae were viewed at 100× by using an optical microscope (Olympus BH4). Many of hyphae occurred on the membranes presented a morphology similar to those produced by members of the Glomeromycota and were attached to spores and auxiliary cells. All intersections between hyphae and a gridded membrane filter were counted in 20 fields of view. We used the method of Newman (1966) to calculate hyphal length (H):

$$H = \frac{\pi N A}{2 L}$$

where N is intersections between the hyphae and the gridded lines, A is the total area of the filter, and L is the total line length of the gridded lines.

Mycorrhizal infection percentages were arcsine transformed to normalize the distribution of data before statistical analysis. The two-way analysis of variance (ANOVA) was performed on all data. Mean values were compared using Tukey's test (honestly significant differences, HSD) at the significance level ($P < 0.05$) with XLSTAT (version 2010, Addinsoft) software. Pearson's correlation coefficient between dependent variables was performed using the same software.

21.3 Results and Discussion

21.3.1 Root Colonization

The two factors (AMF and plants species) had a significant effect ($P < 0.05$) on all parameters studied (Table 21.2). No AM colonization was observed in the non-inoculated controls (Table 21.2). Mycorrhizal infection varied with plants species, and AMF. *R. irregularis* IR27 had colonized better *Z. lotus*, *Z. amphibia*, and *Z. abyssinica* than the other AMF. Mycorrhizal colonization was significantly higher with *R. irregularis* IR27 and *R. irregularis* DAOM197198 in roots of seedlings of *Z. spina-christi*, *Z. mucronata*, *Z. mauritiana*, and *Z. sphaerocarpa* than with *F. mosseae*. Overall, *F. mosseae* showed the lowest AM colonization compared to the other AMF (Table 21.2).

For all the provenances of jujube seedlings, the percentage of mycorrhizal infection was significantly higher ($P < 0.05$) in roots of seedlings inoculated with *R. irregularis* IR27 except for Mali and Mauritania provenances (Table 21.3). No AM infection was observed in the non-inoculated controls (Table 21.3). *R. irregularis* IR27 and *R. irregularis* DAOM197198 had colonized at least 70% and 58% of jujube seedlings roots, respectively. However, *F. mosseae* colonized no

Table 21.2 Effects of inoculation with arbuscular mycorrhizal fungi (AMF) on *Ziziphus* spp. growth parameters (RDW, root dry weight; SDW, shoot dry weight), mycorrhizal dependency (MD), hyphal length, and mycorrhizal infection (MI) after 4 months under greenhouse conditions

Ziziphus spp. with or without AMF	Height (cm)	SDW (g)	RDW (g)	Root/shoot ratios	Total dry biomass (g)	MD (%)	Hyphal length ($cm\ g^{-1}$)	MI (%)
Z. mauritiana								
[1]*R. irregularis*	31.66 ± 3.34 a	0.58 ± 0.03 b–e	1.06 ± 0.31 a–d	1.85 ± 0.72 b–f	1.64 ± 0.31 abc	71.73 ± 1.07 bcd	46.62 ± 0.77ef	78.20 ± 16.29 ab
F. mosseae	26.11 ± 4.63 a–e	0.57 ± 0.06 b–f	0.91 ± 0.19 b–f	1.59 ± 0.29 d–g	1.48 ± 0.24 bcd	69.33 ± 0.66 de	34.10 ± 1.51 i	46.09 ± 14.46 ghi
[2]*R. irregularis*	28.00 ± 4.06 a–d	0.55 ± 0.08 c–g	1.01 ± 0.10 a–e	1.85 ± 0.38 b–f	1.56 ± 0.14 a–d	70.30 ± 1.22 de	63.14 ± 1.94 a	73.78 ± 18.15 bc
Control	13.33 ± 1.13 hi	0.14 ± 0.02 lm	0.31 ± 0.14 ijk	2.24 ± 0.62 a–d	0.45 ± 0.16 h–k	–	–	–
Z. lotus								
[1]*R. irregularis*	12.44 ± 1.50 hi	0.33 ± 0.14 h–l	0.38 ± 0.12 h–k	1.18 ± 0.26 e–j	0.71 ± 0.25 g–j	45.29 ± 1.80 g	41.94 ± 1.79 g	67.05 ± 15.26 cd
F. mosseae	13.55 ± 1.13 hi	0.31 ± 0.13 i–l	0.15 ± 0.04 k	0.50 ± 0.32 j	0.46 ± 0.15 h–k	68.22 ± 1.58 de	40.38 ± 2.70 hi	49.12 ± 9.67 fghi
[2]*R. irregularis*	14.00 ± 1.87 hi	0.43 ± 0.20 e–j	0.30 ± 0.08 ijk	0.67 ± 0.41 h–j	0.73 ± 0.24 g–j	55.74 ± 1.46 f	55.46 ± 1.06bc	53.40 ± 8.96 fg
Control	11.00 ± 1.22 i	0.08 ± 0.01 m	0.13 ± 0.03 k	1.61 ± 0.31 c–g	0.21 ± 0.03 k	–	–	–
Z. spina-christi								
[1]*R. irregularis*	24.11 ± 6.16 b–f	0.57 ± 0.15 b–f	1.06 ± 0.26 a–d	1.85 ± 0.29 b–f	1.64 ± 0.40 abc	69.66 ± 1.06 de	40.38 ± 3.29 hi	85.49 ± 8.52 a
F. mosseae	25.00 ± 4.10 b–f	0.42 ± 0.08 e–j	0.83 ± 0.17 b–g	1.96 ± 0.67 b–e	1.25 ± 0.54 c–f	57.52 ± 1.38 f	42.54 ± 0.49 fg	73.55 ± 2.42 bc
[2]*R. irregularis*	23.66 ± 4.49 c–g	0.44 ± 0.07 e–j	0.68 ± 0.22 e–h	1.57 ± 0.40 d–g	1.12 ± 0.24 d–g	46.14 ± 1.55 g	61.80 ± 1.25 a	84.96 ± 4.87 a
Control	11.55 ± 1.93 i	0.21 ± 0.03 klm	0.63 ± 0.03 f–i	2.98 ± 0.45 a	0.85 ± 0.06 e–h	–	–	–
Z. mucronata								
[1]*R. irregularis*	18.11 ± 2.52 gh	0.17 ± 0.02 klm	0.11 ± 0.03 k	1.62 ± 0.18 c–g	0.29 ± 0.04 jk	68.88 ± 1.43 de	52.78 ± 0.47bc	76.95 ± 4.50 b
F. mosseae	23.00 ± 4.59 d–g	0.56 ± 0.05 c–g	0.28 ± 0.05 jk	0.50 ± 0.07 j	0.84 ± 0.09 e–i	45.40 ± 0.90 g	44.25 ± 1.14efg	51.67 ± 5.47 fgh
[2]*R. irregularis*	20.66 ± 4.55 efg	0.36 ± 0.07 g–k	0.26 ± 0.15 jk	0.85 ± 0.45 g–j	0.63 ± 0.17 h–k	55.02 ± 1.43 f	61.37 ± 1.13 a	74.37 ± 6.02 bc
Control	11.77 ± 1.39 i	0.08 ± 0.01 m	0.11 ± 0.01 k	1.39 ± 0.21 e–i	0.20 ± 0.03 k	–	–	–
Z. sphaerocarpa								
[1]*R. irregularis*	29.66 ± 1.58 ab	0.99 ± 0.19 a	1.00 ± 0.20 a–e	1.01 ± 0.41 f–j	1.99 ± 0.33 a	78.65 ± 1.65 a	51.40 ± 1.54 cd	66.89 ± 10.31 cd
F. mosseae	24.55 ± 3.53 b–f	0.74 ± 0.07 bc	0.78 ± 0.15 c–g	1.05 ± 0.23 f–j	1.53 ± 0.14 a–d	68.81 ± 1.66 de	47.55 ± 0.90 de	64.33 ± 9.90 d
[2]*R. irregularis*	29.22 ± 3.52 abc	0.77 ± 0.15 b	1.02 ± 0.20 a–e	1.32 ± 0.27 e–i	1.79 ± 0.32 ab	74.96 ± 1.99 ab	53.60 ± 2.31bc	62.50 ± 8.60 de
Control	13.88 ± 2.80 hi	0.27 ± 0.03 i–m	0.55 ± 0.20 g–j	2.37 ± 1.10 a–d	0.82 ± 0.45 f–i	–	–	–
Z. amphibia								
[1]*R. irregularis*	29.22 ± 3.10 abc	0.71 ± 0.09 bcd	1.16 ± 0.06 ab	1.63 ± 0.13 c–g	1.87 ± 0.15 ab	74.57 ± 2.65 abc	32.81 ± 2.05 i	79.84 ± 14.45 ab
F. mosseae	24.22 ± 3.64 b–f	0.54 ± 0.14 c–g	0.76 ± 0.15 d–g	1.40 ± 0.30 e–h	1.30 ± 0.23 cde	30.50 ± 0.75 h	40.38 ± 0.59gh	44.02 ± 3.76 hi
[2]*R. irregularis*	28.00 ± 4.18 a–d	0.70 ± 0.10 bcd	0.84 ± 0.03 b–g	1.20 ± 0.14 e–j	1.55 ± 0.12 a–d	67.21 ± 2.16 e	54.28 ± 0.74bc	46.93 ± 11.86 ghi
Control	11.44 ± 1.81 i	0.25 ± 0.02 j–m	0.12 ± 0.03 k	0.48 ± 0.13 j	0.37 ± 0.04 ijk	–	–	–

Z. abyssinica								
[1]*R. irregularis*	20.22 ± 3.52 fg	0.53 ± 0.03 d–h	1.31 ± 0.30 a	2.46 ± 0.80 ab	1.84 ± 0.31 ab	70.55 ± 1.30 cde	56.83 ± 0.65 b	67.46 ± 6.96 cd
F. mosseae	20.33 ± 2.29 efg	0.37 ± 0.10 f–k	1.05 ± 0.36 a–d	2.83 ± 0.40 a	1.42 ± 0.44 bcd	44.14 ± 1.26 g	40.38 ± 1.09 i	43.69 ± 8.63 i
[2]*R. irregularis*	20.33 ± 3.04 efg	0.46 ± 0.08 e–i	1.12 ± 0.21 abc	2.43 ± 0.65 ab	1.58 ± 0.21 a–d	68.15 ± 2.25 de	52.29 ± 1.92 c	56.20 ± 4.94 ef
Factors tested								
AMF	***	***	***	**	***	***	***	***
Ziziphus spp.	***	**	*	*	**	***	***	***
AMF × *Ziziphus* spp.	***	***	**	NS	***	***	***	***

Means in the same column followed by the same letter are not significantly different ($P < 0.05$) according to Tukey's HSD. Significance level was obtained from two-way ANOVA testing the effects of AMF and *Ziziphus* spp. level on growth parameters, mycorrhizal dependency, hyphal length, and mycorrhizal infection. Significant values are indicated as follows: *$P < 0.05$, **$P < 0.01$, and ***$P < 0.001$, *NS* not significant

Data values represent means ± standard errors ($n = 15$)

[1]*R. irregularis* IR27

[2]*R. irregularis* DAOM197198

Table 21.3 Effects of inoculation with arbuscular mycorrhizal fungi (AMF) on *Z. mauritiana* provenance growth parameters (RDW, root dry weight; SDW, shoot dry weight), mycorrhizal dependency (MD), hyphal length, and mycorrhizal infection (MI) after 4 months under greenhouse conditions

Provenance with or without AMF	Height (cm)	SDW (g)	RDW (g)	Root/shoot ratios	Total dry biomass (g)	MD (%)	Hyphal length (cm g^{-1})	MI (%)
Senegal								
[1]*R. irregularis*	32.44 ± 3.85 bcde	0.61 ± 0.04 bcd	1.24 ± 0.11 abc	2.03 ± 0.20 cde	1.85 ± 0.14 abcd	54.49 ± 2.74 cd	39.89 ± 0.60 fgh	76.47 ± 1.05 b
F. mosseae	26.33 ± 4.79 e–h	0.51 ± 0.11 d–h	1.04 ± 0.13 a–f	2.04 ± 0.43 cde	1.54 ± 0.21 b–g	47.40 ± 1.57 ef	38.38 ± 1.48 gh	57.08 ± 1.17 de
[2]*R. irregularis*	27.88 ± 3.77 defg	0.57 ± 0.06 cdef	1.06 ± 0.08 a–f	1.85 ± 0.16 def	1.63 ± 0.14 a–f	50.84 ± 1.25 de	59.18 ± 2.03 b	64.00 ± 2.08 c
Control	16.00v0.70 j	0.20 ± 0.03 jkl	0.50 ± 0.09 g–k	2.50 ± 0.25 bcd	0.70 ± 0.13 jk	–	–	–
Mali								
[1]*R. irregularis*	29.27 ± 4.88 b–f	0.60 ± 0.08 bcde	1.39 ± 0.21 a	2.31 ± 0.79 bcde	1.99 ± 0.23 ab	61.32 ± 1.68 b	42.65 ± 0.74 fg	65.63 ± 1.49 c
F. mosseae	22.88 ± 6.66 f–j	0.38 ± 0.10 f–j	1.21 ± 0.32 abcd	3.18 ± 0.54 ab	1.21 ± 0.38 e–j	36.58 ± 1.72 gh	35.88 ± 1.94 hi	55.50 ± 1.63 ef
[2]*R. irregularis*	26.66 ± 4.52 defg	0.45 ± 0.06 d–i	1.03 ± 0.15 a–f	2.28 ± 0.30 bcde	1.48 ± 0.20 c–g	39.44 ± 1.68 g	52.14 ± 2.41 cd	55.19 ± 1.16 f
Control	17.33 ± 1.00 j	0.16 ± 0.02 l	0.60 ± 0.09 f–k	3.75 ± 0.62 a	0.76 ± 0.10 jk	–	–	–
Mauritania								
[1]*R. irregularis*	28.55 ± 2.64 c–g	0.48 ± 0.03 d–h	1.17 ± 0.30 a–e	2.43 ± 0.69 bcde	1.66 ± 0.30 a–f	50.29 ± 1.23 de	46.40 ± 1.30 ef	76.33 ± 2.51 b
F. mosseae	25.33 ± 3.20 e–i	0.44 ± 0.06 d–i	0.68 ± 0.19 d–j	1.54 ± 0.30 ef	1.13 ± 0.24 f–j	36.78 ± 1.01 gh	32.71 ± 1.17 ij	58.69 ± 1.28 de
[2]*R. irregularis*	22.44 ± 4.06 f–j	0.49 ± 0.08 d–h	0.82 ± 0.10 b–i	1.67 ± 0.37 ef	1.31 ± 0.14 e–i	22.69 ± 2.24 i	60.51 ± 1.97 b	68.38 ± 0.70 c
Control	17.00 ± 0.86 j	0.21 ± 0.01 ijkl	0.66 ± 0.09 e–k	3.14 ± 0.52 ab	0.87 ± 0.09 ijk	–	–	–
Burkina Faso								
[1]*R. irregularis*	33.77 ± 4.07 abcd	0.62 ± 0.18 bc	1.28 ± 0.44 ab	2.06 ± 0.79 cde	1.90 ± 0.56 abc	68.80 ± 1.47 a	48.02 ± 2.80 de	68.69 ± 1.17 c
F. mosseae	28.11 ± 3.96 defg	0.40 ± 0.10 fghi	0.64 ± 0.32 d–j	1.60 ± 0.41 ef	1.04 ± 0.39 hijk	42.52 ± 0.88 f	33.51 ± 1.13 ij	55.88 ± 1.47 ef
[2]*R. irregularis*	26.55 ± 5.02 d–h	0.55 ± 0.08 c–g	1.21 ± 0.15 abcd	2.20 ± 0.22 bcde	1.76 ± 0.22 a–e	57.85 ± 0.98 c	65.42 ± 1.66 a	70.52 ± 1.70 d
Control	19.10 ± 2.10 hij	0.20 ± 0.09 jkl	0.39 ± 0.13 k	1.95 ± 0.94 def	0.60 ± 0.20 k	–	–	–
Niger								
[1]*R. irregularis*	27.78 ± 3.1 defg	0.48 ± 0.09 d–h	1.17 ± 0.30 a–e	2.43 ± 0.52 bcde	1.65 ± 0.37 a–f	49.26 ± 0.99 de	51.18 ± 2.00 cd	55.91 ± 1.96 ef
F. mosseae	26.13 ± 1.83 e–i	0.36 ± 0.07 g–k	0.81 ± 0.20 b–i	2.25 ± 0.39 bcde	1.17 ± 0.24 g–k	39.06 ± 1.13 g	31.12 ± 1.44 jk	44.50 ± 2.97 h
[2]*R. irregularis*	22.69 ± 3.17 f–j	0.46 ± 0.11 dh	0.95 ± 0.16 b–h	2.06 ± 0.26 cde	1.41 ± 0.26 d–h	14.58 ± 1.46 j	56.58 ± 1.27 bc	51.04 ± 0.96 g
Control	18.77 ± 2.77 ij	0.26 ± 0.06 hijk	0.74 ± 0.11 c–j	2.84 ± 0.65 abc	1.00 ± 0.14 hijk	–	–	–
India								
[1]*R. irregularis*	40.33 ± 5.80 a	0.99 ± 0.18 a	1.29 ± 0.30 ab	1.30 ± 0.40 f	2.28 ± 0.43 a	64.27 ± 2.24 ab	44.22 ± 1.66 fg	80.16 ± 2.01 a
F. mosseae	35.33 ± 5.17 ab	0.55 ± 0.14 c–g	1.20 ± 0.08 abcd	2.18 ± 0.22 bcde	1.75 ± 0.14 a–e	46.86 ± 1.84 ef	30.10 ± 2.45 k	65.05 ± 0.91 b
[2]*R. irregularis*	34.77 ± 9.01 abc	0.77 ± 0.21 b	1.03 ± 0.15 a–f	1.33 ± 0.28 f	1.80 ± 0.32 abcd	40.98 ± 1.23 fg	46.73 ± 0.97 ef	69.63 ± 2.20 b
Control	20.55 ± 5.24 ghij	0.40 ± 0.21 fghi	0.66 ± 0.22 ek	1.65 ± 1.13 ef	1.06 ± 0.41 hijk	–	–	–

Factors tested								
AMF	***	***	**	**	***	***	***	***
Provenance	***	*	***	NS	**	***	***	NS
AMF × provenance	***	**	**	NS	**	***	***	**

Means in the same column followed by the same letter are not significantly different ($P < 0.05$) according to Tukey's HSD. Significance level was obtained from two-way ANOVA testing the effects of AMF and provenance level on growth parameters, mycorrhizal dependency, hyphal length, and mycorrhizal infection. Significant values are indicated as follows: *$P < 0.05$, **$P < 0.01$, and ***$P < 0.001$, *NS* not significant

Data values represent means ± standard errors ($n = 15$)

[1]*R. irregularis* IR27

[2]*R. irregularis* DAOM197198

more than 38% of jujube seedlings roots. AMF and provenance factors had significant effect on mycorrhizal infection ($P < 0.05$).

The extent of root colonization found in the inoculated *Z. mauritiana* by *R. irregularis* IR27 was 78.20%. These values are similar to those presented by other authors as being good indicators of the effectiveness of inoculation (Guissou et al. 1998, 2016; Bâ et al. 2000, 2001; Sidibé et al. 2012; Guissou 2009). Moreover, Bâ et al. (2000), Guissou (2009), and Sidibé et al. (2012) found that the absence of the AM inoculation could have a detrimental effect on the growth of *Z. mauritiana* in unsterile nursery soils. Here, we have extended these findings by reporting for the first time, the effectiveness of three AMF (*R. irregularis* IR27, *R. irregularis* DAOM197198, and *F. mosseae*) on seven *Ziziphus* spp. (including *Z. mauritiana*) in greenhouse conditions. All analyzed seedlings of *Ziziphus* spp. were colonized better with *R. irregularis* IR27 (67.10%) and *R. irregularis* DAOM197198 (65.30%) than *F. mosseae* (59.96%).

21.3.2 Hyphal Length

There were some variations in the production of external hyphae by AMF (Table 21.2). The total length of hyphae in the rhizospheric soil samples from inoculated treatments was significantly higher in *Ziziphus* spp. associated with *R. irregularis* DAOM197198 than *R. irregularis* IR27 and *F. mosseae*, except for *Z. abyssinica* (Table 21.2). Irrespective of *Ziziphus* spp., the production of external hyphae increased in the following order: *F. mosseae*, *R. irregularis* IR27, and *R. intraradices*DAOM197198 (Table 21.2).

In our study, the production of hyphal length significantly increased according to fungal species in the following order: *R. irregularis* DAOM197198, *R. irregularis* IR27, and *F. mosseae*. Hence, there was a significant positive correlation ($r = 0.902$, $P < 0.0001$) between development of external hyphae and MD (Table 21.6). The results of the present study are in agreement with those of Yao et al. (2001) who found that the MD of three *Triticum aestivum* cultivars was determined by hyphal development. The mycorrhizal inoculation induced a significant decrease of the root/shoot ratios particularly for the *Z. sphaerocarpa* and *Z. spina-christi* compared with non-mycorrhizal plants. Moreover, there were negative correlations between root/shoot ratios and hyphal length ($r = -0.300$, $P < 0.0001$), mycorrhizal colonization ($r = -0.236$, $P < 0.0001$), and MD ($r = -0.236$, $P < 0.001$) (Table 21.6). These results already occurred with argan tree (Nouaim and Chaussod 1994) can be explained by the higher efficiency of a mycorrhizal root system due to the development of external hyphae. The production of external hyphae could play a key role for uptake and transport of P (Jakobsen et al. 1992). Our findings indicate that total length of hyphae in soil samples from inoculated treatments was significantly higher in the association with *R. irregularis* DAOM197198.

There was a significant positive correlation between MD and hyphal length ($r = 0.822$, $P < 0.0001$) (Table 21.6). Our measurements of hyphal length clearly showed that the effectiveness of one fungus depended on the associated *Ziziphus* spp. The hyphal length differed between AMF in previous studies (Jakobsen et al. 1992; Pearson and Jakobsen 1993) and in the present study. The underground hyphal networks formed by AMF can influence plant growth, nutrient acquisition as well as plant-plant interactions (Smith and Read 2008). In our study, the production of hyphal length significantly decreased according to species in the following order: *R. irregularis* DAOM197198, *R. irregularis* IR27, and *F. mosseae*. *Rhizophagus irregularis* DAOM197198 produced the highest hyphal length and also mobilized more P than the other AMF. Hence, there was a significant positive correlation ($r = 0.908$, $P < 0.0001$) between hyphal length and P concentrations in shoots of *Ziziphus* spp. (Table 21.6). Our results are in accordance with works of Smith et al. (2000) who found that *Medicago truncatula* with *Scutellospora calospora* grew better than the plants growing with *Gigaspora caledonium* that may happen due to the development of external hyphae.

21.3.3 Plant Growth

The growth advantages resulting from inoculation with AM isolates were not the same for each *Ziziphus* spp. in the P-deficient soil (Table 21.2). Non-inoculated *Ziziphus* spp. displayed the lowest growth compared to the inoculated plants. Inoculation with AMF enhanced significantly height of *Ziziphus* spp. as compared with non-inoculated controls, except for *Z. lotus*. There was a positive effect of *R. irregularis* IR27 in the production of biomass of *Ziziphus* spp. when compared to the non-inoculated controls, except for the case of *Z. mucronata*. *Rhizophagus irregularis* DAOM197198 increased biomass production only on *Z. mauritiana*, *Z. amphibia*, and *Z. sphaerocarpa* in comparison to non-inoculated ones. With respect to biomass production, *F. mosseae* showed higher values only on *Z. mauritiana*, *Z. mucronata*, *Z. amphibia*, and *Z. sphaerocarpa* as compared with non-inoculated plants. The lowest effect of AMF inoculation on biomass production of *Ziziphus* spp. was observed with *F. mosseae* except for *Z. mucronata–R. irregularis* IR27 (Table 21.2). *Ziziphus sphaerocarpa* showed the highest SDW values when inoculated with *R. irregularis* IR27 irrespective of plant–fungus combinations. RDW varied with plant species and AMF, and its highest values were observed in *Z. abyssinica–R. irregularis* IR27 combination (Table 21.2). The root/shoot ratios of inoculated and non-inoculated *Ziziphus* spp. were similar, except in the cases of *Z. spina-christi* and *Z. sphaerocarpa*. There was a significant increase in this ratio only in *Z. amphibia* inoculated with *R. irregularis* IR27 and *F. mosseae* (Table 21.2). This could probably be explained by the fact that *R. irregularis* IR27 positively influence RDW and SDW production of *Z. amphibia*. Similar trends were found by Guissou et al. (1998) with *Acaulospora spinosa*, *Glomus manihotis*, and *Glomus aggregatum*.

For the provenances of jujube seedlings, height was significantly higher ($P < 0.05$) in the plants of Senegal, Burkina Faso, and India inoculated with *R.*

irregularis IR27 (Table 21.3). No significant difference was observed between treatments inoculated with *R. irregularis* IR27 and those with *R. irregularis* DAOM197198 of Niger, Mauritania, and Mali provenances seedlings. There was a significant increase in the total dry biomass of jujube seedlings when inoculated with *R. irregularis* IR27 and *R. irregularis* DAOM197198 except the provenance of India where only *R. irregularis* IR27 stimulated significantly the total dry biomass. Arbuscular mycorrhizal fungi had significant effects ($P < 0.05$) on height, collar diameter, and total dry biomass (Table 21.3). We observed a similar effect with the provenance factor except for collar diameter. Of the AMF tested, only *R. irregularis* IR27 and *R. irregularis* DAOM197198 led to increase all parameters measured. Our findings confirm previous studies indicating that *Z. mauritiana* seedlings give the best responses in terms of biomass production particularly with *R. irregularis* IR27 (Guissou et al. 1998; Guissou 2009, 2016; Bâ et al. 2000, 2001; Sidibé et al. 2012).

The MD differed between plant species according to the AMF (Table 21.2). Irrespective of plant–fungus combinations, *Z. sphaerocarpa* showed the highest MD values when inoculated with *R. irregularis* IR27. These MD values ranged from 68.81 to 78.65%. *Ziziphus amphibia* in symbiosis with *F. mosseae* showed the lowest MD values of 30.5%. However, MD values of *Ziziphus* spp. declined according to AMF in the following order: *R. irregularis* IR27, *R. irregularis* DAOM197198, and *F. mosseae* (Table 21.2). There was a significant positive correlation between MD and hyphal length ($r = 0.902$, $P < 0.0001$) and MD and mycorrhizal infection ($r = 0.908$, $P < 0.0001$) (Table 21.6). Consequently, the highest levels of mycorrhizal colonization have fostered the highest values of MD on *Ziziphus* spp. Indeed, MD on *Ziziphus* spp. with *F. mosseae* was 52.45%, whereas the highest MD values with *R. irregularis* IR27 and *R. irregularis* DAOM197198 reached 69.51% and 63.58%, respectively (Table 21.2). However, there was a significant positive correlation between MD and mycorrhizal infection ($r = 0.908$, $P < 0.0001$) (Table 21.6). The MD of the plant species was also different among AMF (Tawaraya 2003). These differences in MD could be due to differences in the development of hyphal length. Mycorrhizal dependency is often attributed to increased P uptake by AMF in P-deficient soils (Tawaraya 2003; Smith and Smith 2012). It is well known that MD values of different plants decreased with the increase in the soil P level (Tawaraya 2003). At a soil P level, the values of MD also vary with host plants and AMF (Bâ et al. 2001; Giri 2017). When the soil P was low, MD of fruit trees differed markedly with AMF (Guissou et al. 1998, 2016; Bâ et al. 2000). *Ziziphus mauritiana* inoculated with *R. irregularis* IR27 showed the highest MD values reaching 74%. In the present study, these features may well have contributed to enhancing the availability of nutrients in the *Ziziphus* spp. and as consequence increase biomass production. In this respect, there was not a significant correlation between MD and K ($r = 0.166$, $P > 0.0001$), suggesting that this nutrient contributed little to the biomass production of *Ziziphus* spp. (Table 21.6).

Mycorrhizal dependency differed between provenances and plants species following the inoculated AMF (Table 21.3). All the provenances of jujube seedlings showed the highest MD values when inoculated with *R. irregularis* IR27 (Table 21.3). The MD of the provenances of jujube plants with *R. irregularis*

IR27 ranged from 63 to 74%, while the MD with *R. irregularis* DAOM197198 and *F. mosseae* was far below around 50%–68% and 34%–56%, respectively (Table 21.3). It was probably due to the significant correlation between MD and hyphal length ($r = 0.617$, $P < 0.0001$) and between MD and mycorrhizal infection ($r = 0.831$, $P < 0.0001$) (Table 21.7).

21.3.4 Nutrient Concentrations in Shoots

The analysis of variance revealed that P and K concentrations varied with *Ziziphus* spp. and AMF (Table 21.4). Three plant–fungus combinations such as *Z. sphaerocarpa–R. irregularis* IR27, *Z. abyssinica–R. irregularis* IR27, and *Z. abyssinica–R. irregularis* DAOM197198 had higher K concentrations in the shoots than the other plant–fungus combinations (Table 21.4). The plant–fungus combinations *Z. abyssinica–R. irregularis* IR27 and *Z. lotus–R. irregularis* DAOM197198 showed the highest P concentrations in shoots compared with the other plant–fungus combinations. In all, *R. irregularis* IR27 and *R. irregularis* DAOM197198 provided more K and P in shoots of *Ziziphus* spp., respectively, than *F. mosseae* (Table 21.4). Shoot P concentrations had significant positive correlations with hyphal length ($r = 0.908$, $P < 0.0001$), mycorrhizal infection ($r = 0.841$, $P < 0.0001$), and MD ($r = 0.822$, $P < 0.0001$), whereas K concentrations did not (Table 21.6).

Acquisition of nutrient ions is considered to be a major factor associated with improved seedling growth of different plants in P-deficient soils (Tawaraya 2003; Mathur and Vyas 1999; Klironomos 2003). In our case, inoculation with AMF increased significantly P and K contents of *Ziziphus* spp. (Table 21.4) and *Z. mauritiana* provenances (Table 21.5). Nutrient uptake in mycorrhizal *Ziziphus* spp. and *Z. mauritiana* provenances was higher as compared with non-mycorrhizal ones. Hence, improved total dry biomass of *Ziziphus* spp. and *Z. mauritiana* provenances could be attributed to improved nutrient uptake by AMF. *Ziziphus* spp. differed in their ability to uptake K and P depending on AMF. *Z. sphaerocarpa* and *Z. abyssinica* in symbiosis with *R. irregularis* IR27 accumulated more K and P than the other plant–fungus combinations. The enhanced nutritional status of *Ziziphus* spp. was marked with *R. irregularis* IR27 and *R. irregularis* DAOM197198 since it showed the highest mycorrhizal colonization with the both AMF, and it accumulated more P and K with *R. irregularis* DAOM197198 and *R. irregularis* IR27, respectively. P absorption probably contributed to this more than the absorption of K (Guissou et al. 1998). Provenances of *Z. mauritiana* also differed in their ability to uptake K and P depending on AMF (Table 21.5). *Rhizophagus irregularis* IR27 was the most efficient AM fungus to absorb K and P whatever the *Z. mauritiana* provenances. Shoot P concentrations had significant positive correlations with hyphal length ($r = 0.655$, $P < 0.0001$), mycorrhizal infection ($r = 0.758$, $P < 0.0001$), and MD ($r = 0.548$, $P < 0.0001$), whereas K concentrations did not (Table 21.7). This result was similar in *Ziziphus* spp. and *Z. mauritiana* provenances

Table 21.4 Effects of inoculation with arbuscular mycorrhizal fungi (AMF) on phosphorus (P) and potassium (K) shoot contents of *Ziziphus* spp. after 4 months under greenhouse conditions

Ziziphus spp. with or without AMF	K (%)	P (%)
Z. mauritiana		
[1]*R. irregularis*	8.11 ± 0.15 klm	2.64 ± 0.04 cd
F. mosseae	7.33 ± 0.32 lmn	1.93 ± 0.04 i
[2]*R. irregularis*	6.52 ± 0.15 no	2.54 ± 0.12 cde
Control	5.48 ± 0.02 o	0.70 ± 0.03 k
Z. lotus		
[1]*R. irregularis*	10.97 ± 0.17 fgh	1.95 ± 0.03 hi
F. mosseae	11.34 ± 0.66 fg	2.32 ± 0.09 efg
[2]*R. irregularis*	12.15 ± 0.43 ef	2.96 ± 0.12 ab
Control	8.48 ± 0.02j–m	0.83 ± 0.03 k
Z. spina-christi		
[1]*R. irregularis*	8.61 ± 0.12 jkl	2.32 ± 0.05 efg
F. mosseae	7.96 ± 0.02 klm	2.12 ± 0.03 f–i
[2]*R. irregularis*	9.73 ± 0.45 hij	2.38 ± 0.04 def
Control	7.14 ± 0.36 mn	0.84 ± 0.04 k
Z. mucronata		
[1]*R. irregularis*	12.71 ± 0.20 de	2.62 ± 0.10 cd
F. mosseae	10.89 ± 0.22 fgh	2.15 ± 0.11 f–i
[2]*R. irregularis*	12.94 ± 0.14 cde	2.23 ± 0.13 fgh
Control	10.48 ± 0.02 gh	0.64 ± 0.04 k
Z. amphibia		
[1]*R. irregularis*	9.73 ± 0.30 hij	2.21 ± 0.10 f–i
F. mosseae	11.15 ± 0.27 fg	2.17 ± 0.09 f–i
[2]*R. irregularis*	8.61 ± 0.54 jkl	2.15 ± 0.15 f–i
Control	8.81 ± 0.56 ijk	0.68 ± 0.04 k
Z. abyssinica		
[1]*R. irregularis*	15.18 ± 0.21 ab	3.18 ± 0.03 a
F. mosseae	13.75 ± 0.83 cd	2.04 ± 0.04 ghi
[2]*R. irregularis*	15.36 ± 0.45 ab	2.29 ± 0.08 efg
Control	10.51 ± 0.40 gh	0.86 ± 0.02 k
Z. sphaerocarpa		
[1]*R. irregularis*	16.28 ± 0.12 a	2.77 ± 0.10 bc
F. mosseae	13.73 ± 0.24 cd	2.80 ± 0.03 bc
[2]*R. irregularis*	14.15 ± 0.37 bc	2.54 ± 0.03 cde
Control	10.12 ± 0.07 ghi	1.30 ± 0.07 j
Factors tested		
AMF	***	***
Ziziphus spp.	**	***
AMF × *Ziziphus* spp.	***	***

Means in the same column followed by the same letter are not significantly different ($P < 0.05$) according to Tukey's HSD. Significance level was obtained from two-way ANOVA testing the effects of AMF and *Ziziphus* spp. level on P and K contents. Significant values are indicated as follows: **$P < 0.01$ and ***$P < 0.001$

Data values represent means ± standard errors ($n = 15$)

[1]*R. irregularis* IR27

[2]*R. irregularis* DAOM197198

Table 21.5 Effects of inoculation with arbuscular mycorrhizal fungi (AMF) on phosphorus (P) and potassium (K) shoot contents of *Z. mauritiana* provenances after 4 months under greenhouse conditions

Provenance de *Z. mauritiana* with or without AMF	K (%)	P (%)
Senegal		
[1]*R. irregularis*	09.43 ± 0.69 cd	3.06 ± 0.05 bc
F. mosseae	06.06 ± 0.21 ij	1.79 ± 0.09 jk
[2]*R. irregularis*	06.44 ± 0.06 hij	2.42 ± 0.05 fg
Control	05.72 ± 0.05 j	1.67 ± 0.01 k
Mali		
[1]*R. irregularis*	08.47 ± 0.15 de	3.22 ± 0.02 b
F. mosseae	08.21 ± 0.14 def	2.21 ± 0.05 hi
[2]*R. irregularis*	07.01 ± 0.20 fghi	2.70 ± 0.03 de
Control	06.11 ± 0.06 ij	1.69 ± 0.02 k
Mauritania		
[1]*R. irregularis*	08.18 ± 0.22 def	2.95 ± 0.08 cd
F. mosseae	06.36 ± 0.27 hij	1.84 ± 0.03 jk
[2]*R. irregularis*	06.73 ± 0.58 ghij	2.52 ± 0.01 ef
Control	05.48 ± 0.02 j	0.95 ± 0.03 m
Burkina Faso		
[1]*R. irregularis*	11.13 ± 0.99 b	3.17 ± 0.17 bc
F. mosseae	08.31 ± 0.31 de	2.15 ± 0.08 i
[2]*R. irregularis*	09.31 ± 0.49 cd	2.23 ± 0.05 hi
Control	07.76 ± 0.06 efg	0.64 ± 0.01 n
Niger		
[1]*R. irregularis*	08.41 ± 0.40 de	2.99 ± 0.09 cd
F. mosseae	07.45 ± 0.16 efgh	2.12 ± 0.02 i
[2]*R. irregularis*	07.22 ± 0.02 e–i	2.21 ± 0.03 hi
Control	06.32 ± 0.25 hij	1.16 ± 0.03 l
India		
[1]*R. irregularis*	13.58 ± 0.39 a	3.97 ± 0.06 a
F. mosseae	10.62 ± 0.18 bc	2.32 ± 0.01 gh
[2]*R. irregularis*	12.84 ± 0.50 a	3.04 ± 0.06 bc
Control	09.95 ± 0.25 bc	1.91 ± 0.07 j
Factors tested		
AMF	**	***
Provenance	***	**
AMF × provenance	**	***

Means in the same column followed by the same letter are not significantly different ($P < 0.05$) according to Tukey's HSD. Significance level was obtained from two-way ANOVA testing the effects of AMF and provenance level on P and K contents. Significant values are indicated as follows: $^{**}P < 0.01$ and $^{***}P < 0.001$

Data values represent means ± standard errors ($n = 15$)

[1]*R. irregularis* IR27

[2]*R. irregularis* DAOM197198

Table 21.6 Correlation coefficients between hyphal length (HL), mycorrhizal infection (MI), mycorrhizal dependency (MD), root/shoot ratio, and nutritional parameters of *Ziziphus* spp. seedlings

	K	P	HL	MI	MD	Root/shoot ratio
K	1					
P	0.365 NS	1				
HL	0.268 NS	0.908*	1			
MI	0.145 NS	0.841*	0.938*	1		
MD	0.166 NS	0.822*	0.902*	0.908*	1	
Root/shoot ratio	−0.051 NS	−0.256 NS	−0.300 NS	−0.236 NS	−0.236 NS	1

NS not significant, $^{*}P < 0.0001$ using Pearson's correlation coefficient

Table 21.7 Correlation coefficients between hyphal length (HL), mycorrhizal infection (MI), mycorrhizal dependency (MD), root/shoot ratio, and nutritional parameters of *Z. mauritiana* provenances

	K	P	HL	MI	MD	Root/shoot ratio
K	1					
P	0.084	1				
HL	0.001	0.655*	1			
MI	0.066	0.758*	0.688*	1		
MD	0.060	0.548*	0.617*	0.831*	1	
Root/shoot ratio	0.216	0.274	0.165	0.203	0.141	1

NS not significant, $^{*}P < 0.0001$ using Pearson's correlation coefficient

(Tables 21.6 and 21.7). Our results supported the findings of Koide and Mosse (2004) who found that there is an evident relationship between the degree to which a root system of a plant is colonized by AMF and the potential for the plant to benefit significantly from the symbiosis. P is critical for plant growth and makes up about 0.2% of dry weight, but it is one of the most difficult nutrients for plants to acquire (Smith et al. 2011).

21.4 Conclusion

The present study clearly showed that AM inoculation promoted *Ziziphus* spp. and *Z. mauritiana* provenances growth by enhancing significantly biomass production and nutrient uptake, particularly P uptake. The differences of MD among the various *Ziziphus* spp. and *Z. mauritiana* provenances tested seem to be due to differences in the development of hyphal length in the soil and in the P uptake by the external hyphae. Moreover, *R. irregularis* IR27 was an efficient fungus for all the *Ziziphus* spp. and *Z. mauritiana* provenances tested. From a practical point of view, this AMF constitutes a promising tool for the production of higher quality nursery stock with potential improved field performance of *Ziziphus* spp. and *Z. mauritiana* provenances in agroforestry systems.

Acknowledgment We dedicate this manuscript to the memory of our dear friend and colleague Dr. Kadidia Bibata Sanon.

References

Adelman MJ, Morton JB (1986) Infectivity of vesicular arbuscular mycorrhizal fungi: influence of host-soil diluent combinations on MPN estimates and percentage colonization. Soil Biol Biochem 18:7–13

Arndt SK, Kayser O (2001) *Ziziphus*, a medicinal plant genus with tradition and future potential. Z Phytother 22:91 (in German)

Azam-Ali S, Bonkoungou E, Bowe C, de Kock C, Godara A, Williams JT (2006) Ber and other jujubes. In: Williams JT, Smith RW, Haq N, Dunsiger Z (eds) Ber and other Jujubes, Southampton. International Centre for Underutilised Crops, University of Southampton, Southampton

Bâ AM, Dalpé Y, Guissou T (1996) Les Glomales d'Acacia holosericea et d'Acacia mangium. Bois For Trop 250:5–18

Bâ AM, Plenchette C, Danthu P, Duponnois R, Guissou T (2000) Functional compatibility of two arbuscular mycorrhizae with thirteen fruit trees in Senegal. Agroforest Syst 50:95–105

Bâ AM, Guissou T, Duponnois R, Plenchette C, Danthu P, Sacko O, Sidibé D, Sylla K, Windou B (2001) Mycorhization contrôlée et fertilisation phosphatée: Application à la domestication du jujubier. Fruits 56:261–269

Bâ AM, Danthu P, Duponnois R, Soleviev P (2003) Domestication of jujube fruit trees (*Zizyphus mauritiana* Lam.). In: Dris R, Niskanen R, Jain SM (eds) Crop management and post harvest handling of horticultural crops. Science Publishers, Enfield, pp 255–279

Davison J, Moora M, Öpik M, Adholeya A, Ainsaar L, Bâ AM, Burla S, Diédhiou AG, Hiiesalu I, Jairus T, Johnson NC, Kane A, Koorem K, Kochar M, Ndiaye C, Pärtel M, Reier Ü, Saks Ü, Singh R, Vasar M, Zobel M (2015) Global assessment of arbuscular mycorrhizal fungus diversity reveals very low endemism. Science 349:970–973

Giri B (2017) Mycorrhizal dependency and growth response of *Gliricidia sepium* (Jacq.) Kunth ex Walp. under saline condition. Plant Sci Today 4(4):154–160

Guissou T (2009) Contribution of arbuscular mycorrhizal fungi to growth and nutrient uptake by jujube and tamarind seedlings in a phosphate (P)-deficient soil. Afr J Microbiol Res 3 (5):297–304

Guissou T, Bâ AM, Ouadba JM, Guinko S, Duponnois R (1998) Responses of *Parkia biglobosa* (Jacq.) Benth., *Tamarindus indica* L. and *Ziziphus mauritiana* Lam. To arbuscular mycorrhizal fungi in a phosphorus-deficient sandy soil. Biol Fertil Soils 26:194–198

Guissou T, Bâ AM, Plenchette C, Guinko S, Duponnois R (2001) Effets des mycorhizes à arbuscules sur la tolérance à un stress hydrique chez quatre arbres fruitiers *Balanites aegyptiaca* (L.) Del., *Parkia biglobosa* (Jacq.) Benth., *Tamarindus indica* L. et *Ziziphus mauritiana* Lam. Sécheresse 12:121–127

Guissou T, Sanon KB, Babana A, Bâ AM (2016) Effect of arbuscular mycorrhizae on growth and mineral nutrition of greenhouse propagated fruit trees from diverse geographic provenances. Biotechnol Agron Soc Environ 20:417–426

Hart M, Ehret DL, Krumbein A, Leung C, Murch S, Turi C, Franken P (2015) Inoculation with arbuscular mycorrhizal fungi improves the nutritional value of tomatoes. Mycorrhiza 25:359–376

Jakobsen I, Rosendahl L (1990) Carbon flow into soil and external hyphae from roots of mycorrhizal cucumber plants. New Phytol 115:77–83. https://doi.org/10.1111/j.1469-8137.1990.tb00924.x

Jakobsen I, Abbott LK, Robson AD (1992) External hyphae of vesicular–arbuscular mycorrhizal fungi associated with *Trifolium subterraneum* L. 1. Spread of hyphae and phosphorus inflow into roots. New Phytol 120:371–380. https://doi.org/10.1111/j.1469-8137.1992.tb01077.x

Klironomos JN (2003) Variation in plant response to native and exotic arbuscular mycorrhizal fungi. Ecology 84:2292–2301. https://doi.org/10.1890/02-0413

Koide RT, Mosse B (2004) A history of research on arbuscular mycorrhiza. Mycorrhiza 14:145–163. https://doi.org/10.1007/s00572-004-0307-4

Mathur N, Vyas A (1999) Improved biomass production, nutrient uptake and establishment of *in vitro* raised *Ziziphus mauritiana* by VA mycorrhiza. J Plant Physiol 155:129–132. https://doi.org/10.1016/S0176-1617(99)80153-9

Mathur N, Vyas A (2000) Influence of arbuscular mycorrhizae on biomass production, nutrient uptake and physiological changes in *Ziziphus mauritiana* Lam. Under water stress. J Arid Environ 45:191–195. https://doi.org/10.1006/jare.2000.0644

Newman EI (1966) A method of estimating the total length of a root in a sample. J Appl Ecol 3:139–145. https://doi.org/10.2307/2401670

Ngwa AT, Pone DK, Mafeni JM (2000) Feed selection and dietary preferences of forage by small ruminants grazing natural pastures in the Sahelian zone of Cameroon. Anim Feed Sci Technol 88:253–266. https://doi.org/10.1016/S0377-8401(00)00215-7

Nouaim R, Chaussod R (1994) Mycorrhizal dependency of micropropagated argan tree (*Argania spinosa*): I. Growth and biomass production. Agr Syst 27:53–65. https://doi.org/10.1007/BF00704834

Ouédraogo SJ, Bayala J, Dembélé C, Kaboré A, Kaya B, Niang A, Somé AN (2006) Establishing jujube trees in sub-Saharan Africa: response of introduced and local cultivars to rock phosphate and water supply in Burkina Faso, West Africa. Agroforest Syst 68:69–80. https://doi.org/10.1007/s10457-006-6843-5

Pearson JN, Jakobsen I (1993) The relative contribution of hyphae and roots to phosphorus uptake by arbuscular mycorrhizal plants measured by dual labelling with 32^{P} and 33^{P}. New Phytol 124:489–494. https://doi.org/10.1111/j.1469-8137.1993.tb03840.x

Phillips JM, Hayman DS (1970) Improved procedures for clearing roots and staining parasitic and vesicular-arbuscular mycorrhizal fungi for rapid assessment of infection. Trans Br Mycol Soc 55:158–161. https://doi.org/10.1016/S0007-1536(70)80110-3

Plenchette C, Fortin JA, Furlan V (1983) Growth responses of several plant species to mycorrhizae in a soil of moderate P-fertility. I: Mycorrhizal dependency under field conditions. Plant Soil 70:199–209. https://doi.org/10.1007/BF02374780

Redecker D, Schüβler A, Stockinger H, Stürmer S, Morton J, Walker C (2013) An evidence-based consensus for the classification of arbuscular mycorrhizal fungi (*Glomeromycota*). Mycorrhiza 23:515–531. https://doi.org/10.1007/s00572-013-0486-y

Schüßler A, Schwarzott D, Walker C (2001) A new fungal phylum, the *Glomeromycota*: phylogeny and evolution. Mycol Res 105:1413–1421. https://doi.org/10.1017/S0953756201005196

Sidibé D, Sanou H, Teklehaimanot Z, Mahamadi D, Kone S (2012) The use of mycorrhizal inoculation in the domestication of *Ziziphus mauritiana* and *Tamarindus indica* in Mali (West Africa). Agroforest Syst 85:519–528

Smith SE, Read DJ (2008) Mycorrhizal symbiosis. Academic, London

Smith SE, Smith FA (2012) Fresh perspectives on the roles of arbuscular mycorrhizal fungi in plant nutrition and growth. Mycologia 104:1–13. https://doi.org/10.3852/11-229

Smith FA, Jakobsen I, Smith SE (2000) Spatial differences in acquisition of soil phosphate between two arbuscular mycorrhizal fungi in symbiosis with *Medicago truncatula*. New Phytol 147:357–366. https://doi.org/10.1046/j.1469-8137.2000.00695.x

Smith SE, Facelli E, Pope S, Smith FA (2009) Plant performance in stressful environments: interpreting new and established knowledge of the roles of arbuscular mycorrhizas. Plant Soil 326:3–20. https://doi.org/10.1007/s11104-009-9981-5

Smith SE, Jakobsen I, Grønlund M, Smith FA (2011) Roles of arbuscular mycorrhizas in plant phosphorus nutrition: interactions between pathways of phosphorus uptake in arbuscular mycorrhizal roots have important implications for understanding and manipulating plant phosphorus acquisition. Plant Physiol 156:1050–1057. https://doi.org/10.1104/pp.111.174581

Soule A (2011) Plantes Ligneuses de Mauritanie: Caractéristiques et Usages (Fondation MON-3)

Stockinger H, Walker C, Schüßler A (2009) *Glomus intraradices* DAOM197198, a model fungus in arbuscular mycorrhiza research, is not *Glomus intraradices*. New Phytol 183:1176–1187

Tawaraya K (2003) Arbuscular mycorrhizal dependency of different plant species and cultivars. Soil Sci Plant Nutr 49:655–668. https://doi.org/10.1080/00380768.2003.10410323

Trouvelot A, Kough JL, Gianinazzi-Pearson V (1986) Mesure du taux de mycorhization VA d'un système radiculaire. Recherche de méthodes d'estimation ayant une signification fonctionnelle. In: Gianinazzi-Pearson V, Gianinazzi S (eds) Physiology and genetic aspects of mycorrhizae. INRA, Paris, pp 217–221

Verbruggen E, Van der Heijden MGA, Rillig MC, Kiers ET (2013) Mycorrhizal fungal establishment in agricultural soils: factors determining inoculation success. New Phytol 197:1104–1109. https://doi.org/10.1111/j.1469-8137.2012.04348.x

Yao Q, Li XL, Feng G, Christie P (2001) Influence of extramatrical hyphae on mycorrhizal dependency of wheat genotypes. Commun Soil Sci Plant Anal 32:3307–3317. https://doi.org/10.1081/CSS-120001122